Student Solutions Manual

Cindy Trimble & Associates

Beginning Algebra

FIFTH EDITION

Elayn Martin-Gay

Upper Saddle River, NJ 07458

Editorial Director, Mathematics: Christine Hoag
Editor-in-Chief: Paul Murphy
Sponsoring Editor: Mary Beckwith
Print Supplement Editor: Georgina Brown
Senior Managing Editor: Linda Mihatov Behrens
Project Manager: Robert Merenoff
Art Director: Heather Scott
Supplement Cover Manager: Paul Gourhan
Supplement Cover Designer: Victoria Colotta
Operations Specialist: Ilene Kahn
Senior Operations Supervisor: Diane Peirano

Pearson Prentice Hall
Pearson Education, Inc.
Upper Saddle River, NJ 07458

Printed in the United States of America

10 9 8 7 6 5 4 3 2 1

ISBN-13: 978-0-13-603108-6 Standalone
ISBN-10: 0-13-603108-0 Standalone
ISBN-13: 978-0-13-603109-3 Component
ISBN-10: 0-13-603109-9 Component

Pearson Education Ltd., *London*
Pearson Education Australia Pty. Ltd., *Sydney*
Pearson Education Singapore, Pte. Ltd.
Pearson Education North Asia Ltd., *Hong Kong*
Pearson Education Canada, Inc., *Toronto*
Pearson Educación de Mexico, S.A. de C.V.
Pearson Education—Japan, *Tokyo*
Pearson Education Malaysia, Pte. Ltd.

Contents

Chapter 1

Exercise Set 1.1

Answers will vary on Exercises 1–21.

Section 1.2

Practice Exercises

1. **a.** $5 < 8$ since 5 is to the left of 8 on the number line.

   b. $6 > 4$ since 6 is to the right of 4 on the number line.

   c. $16 < 82$ since 16 is to the left of 82 on the number line.

2. **a.** $9 \geq 3$ is true, since $9 > 3$ is true.

   b. $3 \geq 8$ is false, since neither $3 > 8$ nor $3 = 8$ is true.

   c. $25 \leq 25$ is true, since $25 = 25$ is true.

   d. $4 \leq 14$ is true, since $4 < 14$ is true.

3. **a.** $3 < 8$

   b. $15 \geq 9$

   c. $6 \neq 7$

4. The integer −52 represents owing the bank 52 dollars.

5. **a.** The natural number is 25.

   b. The whole number is 25.

   c. The integers are 25, −15, −99.

   d. The rational numbers are 25, $\frac{7}{3}$, −15, $\frac{-3}{4}$, −3.7, 8.8, −99.

   e. The irrational number is $\sqrt{5}$.

   f. The real numbers are 25, $\frac{7}{3}$, −15, $\frac{-3}{4}$, $\sqrt{5}$, −3.7, 8.8, −99.

6. **a.** $0 < 3$ since 0 is to the left of 3 on the number line.

   b. $15 > -5$ since 15 is to the right of −5 on the number line.

   c. $3 = \frac{12}{4}$ since $\frac{12}{4}$ simplifies to 3.

7. **a.** $|-8| = 8$ since −8 is 8 units from 0 on a number line.

   b. $|9| = 9$ since 9 is 9 units from 0 on a number line.

   c. $|-2.5| = 2.5$ since −2.5 is 2.5 units from 0 on a number line.

   d. $\left|\frac{5}{11}\right| = \frac{5}{11}$ since $\frac{5}{11}$ is $\frac{5}{11}$ unit from 0 on a number line.

   e. $\left|\sqrt{3}\right| = \sqrt{3}$ since $\sqrt{3}$ is $\sqrt{3}$ units from 0 on a number line.

8. **a.** $|8| = |-8|$ since $8 = 8$.

   b. $|-3| > 0$ since $3 > 0$.

   c. $|-7| < |-11|$ since $7 < 11$.

   d. $|3| > |2|$ since $3 > 2$.

   e. $|0| < |-4|$ since $0 < 4$.

Vocabulary and Readiness Check

1. The <u>whole</u> numbers are {0, 1, 2, 3, 4, ...}.
2. The <u>natural</u> numbers are {1, 2, 3, 4, 5, ...}.
3. The symbols ≠, ≤, and > are called <u>inequality</u> symbols.
4. The <u>integers</u> are {..., −3, −2, −1, 0, 1, 2, 3, ...}.
5. The <u>real</u> numbers are {all numbers that correspond to points on the number line}.
6. The <u>rational</u> numbers are $\left\{\frac{a}{b} \middle| a \text{ and } b \text{ are integers}, b \neq 0\right\}$.

7. The irrational numbers are {nonrational numbers that correspond to points on the number line}.

8. The distance between a number b and 0 on a number line is $|b|$.

Exercise Set 1.2

1. $7 > 3$ since 7 is to the right of 3 on the number line.

3. $6.26 = 6.26$

5. $0 < 7$ since 0 is to the left of 7 on the number line.

7. $-2 < 2$ since -2 is to the left of 2 on the number line.

9. $32 < 212$ since 32 is to the left of 212 on the number line.

11. $2631 > 2456$ since 2631 is to the right of 2456 on the number line.

13. $11 \leq 11$ is true, since $11 = 11$.

15. $10 > 11$ is false, since 10 is to the left of 11 on the number line.

17. $3 + 8 \geq 3(8)$ is false, since 11 is to the left of 24 on the number line.

19. $7 > 0$ is true, since 7 is to the right of 0 on the number line.

21. $30 \leq 45$

23. Eight is less than twelve is written as $8 < 12$.

25. Five is greater than or equal to four is written as $5 \geq 4$.

27. Fifteen is not equal to negative two is written as $15 \neq -2$.

29. 535 represents an altitude of 535 feet. -8 represents 8 feet below sea level.

31. $-21,350$ represents a population decrease of 21,350.

33. 350 represents a deposit of \$350. -126 represents a withdrawal of \$126.

35. The tallest bars represent the greatest number of visitors; 1998, 1999

37. Look for the bars that have heights greater than 280; 1998, 1999, 2000

39. In 2001, there were 279 million visitors. In 2006, there were 273 million visitors.
279 million > 273 million

41. The number 0 belongs to the sets of: whole numbers, integers, rational numbers, and real numbers.

43. The number -2 belongs to the sets of: integers, rational numbers, and real numbers.

45. The number 6 belongs to the sets of: natural numbers, whole numbers, integers, rational numbers, and real numbers.

47. The number $\frac{2}{3}$ belongs to the sets of: rational numbers and real numbers.

49. The number $-\sqrt{5}$ belongs to the sets of: irrational numbers and real numbers.

51. False; rational numbers may be non-integers.

53. True

55. True

57. True

59. False; an irrational number may not be written as a fraction.

61. $-10 > -100$ since -10 is to the right of -100 on the number line.

63. $32 > 5.2$ since 32 is to the right of 5.2 on the number line.

65. $\frac{18}{3} < \frac{24}{3}$ since $6 < 8$.

67. $-51 < -50$ since -51 is to the left of -50 on the number line.

69. $|-5| > -4$ since $5 > -4$.

71. $|-1| = |1|$ since $1 = 1$.

73. $|-2| < |-3|$ since $2 < 3$.

75. $|0| < |-8|$ since $0 < 8$.

77. $-0.04 > -26.7$ since -0.04 is to the right of -26.7 on the number line.

79. The sun is brighter since $-26.7 < -0.04$.

81. The sun is the brightest since -26.7 is to the left of all other numbers listed.

83. $20 \le 25$ has the same meaning as $25 \ge 20$.

85. $6 > 0$ has the same meaning as $0 < 6$.

87. $-12 < -10$ has the same meaning as $-10 > -12$.

89. Answers may vary

Section 1.3

Practice Problems

1. a. $36 = 4 \cdot 9 = 2 \cdot 2 \cdot 3 \cdot 3$

b. $75 = 3 \cdot 25 = 3 \cdot 5 \cdot 5$

2. a. $\frac{63}{72} = \frac{3 \cdot 3 \cdot 7}{2 \cdot 2 \cdot 2 \cdot 3 \cdot 3} = \frac{7}{2 \cdot 2 \cdot 2} = \frac{7}{8}$

b. $\frac{64}{12} = \frac{2 \cdot 2 \cdot 2 \cdot 2 \cdot 2 \cdot 2}{2 \cdot 2 \cdot 3} = \frac{2 \cdot 2 \cdot 2 \cdot 2}{3} = \frac{16}{3}$

c. $\frac{7}{25} = \frac{7}{5 \cdot 5}$

There are no common factors other than 1, so $\frac{7}{25}$ is already in lowest terms.

3. $\frac{3}{8} \cdot \frac{7}{9} = \frac{3 \cdot 7}{8 \cdot 9} = \frac{3 \cdot 7}{2 \cdot 2 \cdot 2 \cdot 3 \cdot 3} = \frac{7}{2 \cdot 2 \cdot 2 \cdot 3} = \frac{7}{24}$

4. a. $\frac{3}{4} \div \frac{4}{9} = \frac{3}{4} \cdot \frac{9}{4} = \frac{3 \cdot 9}{4 \cdot 4} = \frac{27}{16}$

b. $\frac{5}{12} \div 15 = \frac{5}{12} \cdot \frac{1}{15} = \frac{5 \cdot 1}{12 \cdot 15} = \frac{5}{12 \cdot 3 \cdot 5} = \frac{1}{36}$

c. $\frac{7}{6} \div \frac{7}{15} = \frac{7}{6} \cdot \frac{15}{7} = \frac{7 \cdot 15}{6 \cdot 7} = \frac{15}{6} = \frac{3 \cdot 5}{2 \cdot 3} = \frac{5}{2}$

5. a. $\frac{8}{5} - \frac{3}{5} = \frac{8-3}{5} = \frac{5}{5} = 1$

b. $\frac{8}{5} - \frac{2}{5} = \frac{8-2}{5} = \frac{6}{5}$

c. $\frac{3}{5} + \frac{1}{5} = \frac{3+1}{5} = \frac{4}{5}$

d. $\frac{5}{12} + \frac{1}{12} = \frac{5+1}{12} = \frac{6}{12} = \frac{1}{2}$

6. $\frac{2}{3} = \frac{2}{3} \cdot \frac{7}{7} = \frac{2 \cdot 7}{3 \cdot 7} = \frac{14}{21}$

7. a.
$$\begin{aligned}\frac{5}{11} + \frac{1}{7} &= \frac{5 \cdot 7}{11 \cdot 7} + \frac{1 \cdot 11}{7 \cdot 11}\\ &= \frac{35}{77} + \frac{11}{77}\\ &= \frac{35+11}{77}\\ &= \frac{46}{77}\end{aligned}$$

b.
$$\begin{array}{rcrcr} 9\frac{1}{13} & = & 9\frac{2}{26} & = & 8\frac{28}{26} \\ -5\frac{1}{2} & = & -5\frac{13}{26} & = & -5\frac{13}{26} \\ \hline & & & & 3\frac{15}{26} \end{array}$$

c.
$$\begin{aligned}\frac{1}{3} + \frac{29}{30} - \frac{4}{5} &= \frac{10}{30} + \frac{29}{30} - \frac{4 \cdot 6}{5 \cdot 6}\\ &= \frac{10+29}{30} - \frac{24}{30}\\ &= \frac{39-24}{30}\\ &= \frac{15}{30}\\ &= \frac{1}{2}\end{aligned}$$

Vocabulary and Readiness Check

1. A quotient of two numbers, such as $\frac{5}{8}$, is called a fraction.

2. In the fraction $\frac{3}{11}$, the number 3 is called the numerator and the number 11 is called the denominator.

3. To factor a number means to write it as a product.

4. A fraction is said to be simplified when the numerator and the denominator have no common factors other than 1.

5. In $7 \cdot 3 = 21$, the numbers 7 and 3 are called factors and the number 21 is called the product.

6. The fractions $\frac{2}{9}$ and $\frac{9}{2}$ are called reciprocals.

7. Fractions that represent the same quantity are called equivalent fractions.

8. 3 of the 8 equal parts are shaded; $\frac{3}{8}$

9. 1 of the 4 equal parts are shaded; $\frac{1}{4}$

10. 5 of the 7 equal parts are shaded; $\frac{5}{7}$

11. 2 of the 5 equal parts are shaded; $\frac{2}{5}$

Exercise Set 1.3

1. $33 = 3 \cdot 11$

3. $98 = 2 \cdot 49 = 2 \cdot 7 \cdot 7$

5. $20 = 4 \cdot 5 = 2 \cdot 2 \cdot 5$

7. $75 = 3 \cdot 25 = 3 \cdot 5 \cdot 5$

9. $45 = 9 \cdot 5 = 3 \cdot 3 \cdot 5$

11. $\frac{2}{4} = \frac{2}{2 \cdot 2} = \frac{1}{2}$

13. $\frac{10}{15} = \frac{2 \cdot 5}{3 \cdot 5} = \frac{2}{3}$

15. $\frac{3}{7} = \frac{3}{7}$

17. $\frac{18}{30} = \frac{2 \cdot 3 \cdot 3}{2 \cdot 3 \cdot 5} = \frac{3}{5}$

19. $\frac{1}{2} \cdot \frac{3}{4} = \frac{1 \cdot 3}{2 \cdot 4} = \frac{1 \cdot 3}{2 \cdot 2 \cdot 2} = \frac{3}{8}$

21. $\frac{2}{3} \cdot \frac{3}{4} = \frac{2 \cdot 3}{3 \cdot 4} = \frac{2 \cdot 3}{3 \cdot 2 \cdot 2} = \frac{1}{2}$

23. $\frac{1}{2} \div \frac{7}{12} = \frac{1}{2} \cdot \frac{12}{7} = \frac{1 \cdot 12}{2 \cdot 7} = \frac{1 \cdot 2 \cdot 2 \cdot 3}{2 \cdot 7} = \frac{2 \cdot 3}{7} = \frac{6}{7}$

25. $\frac{3}{4} \div \frac{1}{20} = \frac{3}{4} \cdot \frac{20}{1} = \frac{3 \cdot 20}{4 \cdot 1} = \frac{3 \cdot 2 \cdot 2 \cdot 5}{2 \cdot 2} = \frac{3 \cdot 5}{1} = 15$

27. $\frac{7}{10} \cdot \frac{5}{21} = \frac{7 \cdot 5}{10 \cdot 21} = \frac{7 \cdot 5}{2 \cdot 5 \cdot 3 \cdot 7} = \frac{1}{2 \cdot 3} = \frac{1}{6}$

29. $2\frac{7}{9} \cdot \frac{1}{3} = \frac{25}{9} \cdot \frac{1}{3} = \frac{25 \cdot 1}{9 \cdot 3} = \frac{5 \cdot 5 \cdot 1}{3 \cdot 3 \cdot 3} = \frac{25}{27}$

31. $$\begin{aligned} \text{Area} &= \frac{11}{12} \cdot \frac{3}{5} \\ &= \frac{11 \cdot 3}{12 \cdot 5} \\ &= \frac{11 \cdot 3}{2 \cdot 2 \cdot 3 \cdot 5} \\ &= \frac{11}{2 \cdot 2 \cdot 5} \\ &= \frac{11}{20} \text{ sq mi} \end{aligned}$$

33. $\frac{4}{5} - \frac{1}{5} = \frac{4-1}{5} = \frac{3}{5}$

35. $\frac{4}{5} + \frac{1}{5} = \frac{4+1}{5} = \frac{5}{5} = 1$

37. $\frac{17}{21} - \frac{10}{21} = \frac{17-10}{21} = \frac{7}{21} = \frac{7}{3 \cdot 7} = \frac{1}{3}$

39. $\frac{23}{105} + \frac{4}{105} = \frac{23+4}{105} = \frac{27}{105} = \frac{3 \cdot 3 \cdot 3}{3 \cdot 5 \cdot 7} = \frac{3 \cdot 3}{5 \cdot 7} = \frac{9}{35}$

41. $\frac{7}{10} = \frac{7 \cdot 3}{10 \cdot 3} = \frac{21}{30}$

43. $\frac{2}{9} = \frac{2 \cdot 2}{9 \cdot 2} = \frac{4}{18}$

45. $\frac{4}{5} = \frac{4 \cdot 4}{5 \cdot 4} = \frac{16}{20}$

47. $\frac{2}{3}+\frac{3}{7}=\frac{2\cdot 7}{3\cdot 7}+\frac{3\cdot 3}{7\cdot 3}=\frac{14}{21}+\frac{9}{21}=\frac{23}{21}$

49. $2\frac{13}{15}-1\frac{1}{5}=\frac{43}{15}-\frac{6}{5}$
$=\frac{43}{15}-\frac{6\cdot 3}{5\cdot 3}$
$=\frac{43-18}{15}$
$=\frac{25}{15}$
$=1\frac{2}{3}$

51. $\frac{5}{22}-\frac{5}{33}=\frac{5\cdot 3}{22\cdot 3}-\frac{5\cdot 2}{33\cdot 2}$
$=\frac{15}{66}-\frac{10}{66}$
$=\frac{15-10}{66}$
$=\frac{5}{66}$

53. $\frac{12}{5}-1=\frac{12}{5}-\frac{5}{5}=\frac{12-5}{5}=\frac{7}{5}$

55. $1-\frac{3}{10}-\frac{5}{10}=\frac{10}{10}-\frac{3}{10}-\frac{5}{10}$
$=\frac{10-3-5}{10}$
$=\frac{2}{10}$
$=\frac{2}{2\cdot 5}$
$=\frac{1}{5}$

The unknown part is $\frac{1}{5}$.

57. $1-\frac{1}{4}-\frac{3}{8}=\frac{8}{8}-\frac{1\cdot 2}{4\cdot 2}-\frac{3}{8}=\frac{8-2-3}{8}=\frac{3}{8}$

The unknown part is $\frac{3}{8}$.

59. $1-\frac{1}{2}-\frac{1}{6}-\frac{2}{9}=\frac{18}{18}-\frac{1\cdot 9}{2\cdot 9}-\frac{1\cdot 3}{6\cdot 3}-\frac{2\cdot 2}{9\cdot 2}$
$=\frac{18-9-3-4}{18}$
$=\frac{2}{18}$
$=\frac{1}{9}$

The unknown part is $\frac{1}{9}$.

61. $\frac{10}{21}+\frac{5}{21}=\frac{10+5}{21}=\frac{15}{21}=\frac{3\cdot 5}{3\cdot 7}=\frac{5}{7}$

63. $\frac{10}{3}-\frac{5}{21}=\frac{10\cdot 7}{3\cdot 7}-\frac{5}{21}=\frac{70}{21}-\frac{5}{21}=\frac{65}{21}$

65. $\frac{2}{3}\cdot\frac{3}{5}=\frac{2\cdot 3}{3\cdot 5}=\frac{2}{5}$

67. $\frac{3}{4}\div\frac{7}{12}=\frac{3}{4}\cdot\frac{12}{7}=\frac{3\cdot 12}{4\cdot 7}=\frac{3\cdot 3\cdot 4}{4\cdot 7}=\frac{9}{7}$

69. $\frac{5}{12}+\frac{4}{12}=\frac{5+4}{12}=\frac{9}{12}=\frac{3\cdot 3}{3\cdot 4}=\frac{3}{4}$

71. $5+\frac{2}{3}=\frac{15}{3}+\frac{2}{3}=\frac{15+2}{3}=\frac{17}{3}$

73. $\frac{7}{8}\div 3\frac{1}{4}=\frac{7}{8}\div\frac{13}{4}=\frac{7}{8}\cdot\frac{4}{13}=\frac{7\cdot 4}{8\cdot 13}=\frac{7\cdot 4}{2\cdot 4\cdot 13}=\frac{7}{26}$

75. $\frac{7}{18}\div\frac{14}{36}=\frac{7}{18}\cdot\frac{36}{14}=\frac{7\cdot 36}{18\cdot 14}=\frac{7\cdot 2\cdot 18}{18\cdot 2\cdot 7}=1$

77. $\frac{23}{105}-\frac{2}{105}=\frac{23-2}{105}=\frac{21}{105}=\frac{21}{21\cdot 5}=\frac{1}{5}$

79. $1\frac{1}{2}+3\frac{2}{3}=\frac{3}{2}+\frac{11}{3}$
$=\frac{3\cdot 3}{2\cdot 3}+\frac{11\cdot 2}{3\cdot 2}$
$=\frac{9}{6}+\frac{22}{6}$
$=\frac{9+22}{6}$
$=\frac{31}{6}$
$=5\frac{1}{6}$

81. $\frac{2}{3}-\frac{5}{9}+\frac{5}{6}=\frac{2\cdot 2\cdot 3}{3\cdot 2\cdot 3}-\frac{5\cdot 2}{9\cdot 2}+\frac{5\cdot 3}{6\cdot 3}$
$=\frac{12}{18}-\frac{10}{18}+\frac{15}{18}$
$=\frac{12-10+15}{18}$
$=\frac{17}{18}$

83. $5+4\frac{1}{8}+4\frac{1}{8}+15\frac{3}{4}+15\frac{3}{4}+10\frac{1}{2}$
$=\frac{40}{8}+\frac{33}{8}+\frac{33}{8}+\frac{126}{8}+\frac{126}{8}+\frac{84}{8}$
$=\frac{40+33+33+126+126+84}{8}$
$=\frac{442}{8}$
$=55\frac{1}{4}$ feet

85. $5\frac{1}{50}+1\frac{3}{25}=5\frac{1}{50}+1\frac{6}{50}=6\frac{7}{50}$ meters

87. Answers may vary

89. $5\frac{1}{2}-2\frac{1}{8}=\frac{11}{2}-\frac{17}{8}$
$=\frac{11\cdot 4}{2\cdot 4}-\frac{17}{8}$
$=\frac{44}{8}-\frac{17}{8}$
$=\frac{44-17}{8}$
$=\frac{27}{8}$
$=3\frac{3}{8}$ miles

91. $\frac{7}{50}$ are in the physical sciences.

93. $1-\frac{4}{25}-\frac{7}{50}-\frac{7}{50}-\frac{7}{100}-\frac{21}{100}-\frac{3}{100}$
$=\frac{100}{100}-\frac{4\cdot 4}{25\cdot 4}-\frac{7\cdot 2}{50\cdot 2}-\frac{7\cdot 2}{50\cdot 2}-\frac{7}{100}-\frac{21}{100}-\frac{3}{100}$
$=\frac{100-16-14-14-7-21-3}{100}$
$=\frac{25}{100}$
$=\frac{1}{4}$

$\frac{1}{4}$ are in the biological and agricultural sciences.

95. $\frac{960}{3054}=\frac{6\cdot 160}{6\cdot 509}=\frac{160}{509}$ were Old Navy stores.

97. $\text{Area}=\frac{1}{2}bh=\frac{1}{2}\cdot\frac{7}{8}\cdot\frac{4}{9}=\frac{7\cdot 4}{2\cdot 2\cdot 4\cdot 9}=\frac{7}{36}$ sq ft

Section 1.4

Practice Exercises

1. a. $1^3=1\cdot 1\cdot 1=1$

b. $5^2=5\cdot 5=25$

c. $\left(\frac{1}{10}\right)^2=\left(\frac{1}{10}\right)\left(\frac{1}{10}\right)=\frac{1}{100}$

d. $9^1=9$

e. $\left(\frac{2}{5}\right)^3=\left(\frac{2}{5}\right)\left(\frac{2}{5}\right)\left(\frac{2}{5}\right)=\frac{8}{125}$

2. a. $6+3\cdot 9=6+27=33$

b. $4^3\div 8+3=64\div 8+3=8+3=11$

c. $\left(\frac{2}{3}\right)^2\cdot|-8|=\frac{4}{9}\cdot 8=\frac{32}{9}$ or $3\frac{5}{9}$

d. $\frac{9(14-6)}{|-2|}=\frac{9(8)}{2}=\frac{72}{2}=36$

e. $\frac{7}{4}\cdot\frac{1}{4}-\frac{1}{4}=\frac{7}{16}-\frac{4}{16}=\frac{3}{16}$

3. $\frac{6^2-5}{3+|6-5|\cdot 8}=\frac{36-5}{3+|1|\cdot 8}=\frac{31}{3+8}=\frac{31}{11}$

4. $$\begin{aligned}4[25-3(5+3)] &= 4[25-3(8)]\\ &= 4[25-24]\\ &= 4[1]\\ &= 4\end{aligned}$$

5. $\frac{36\div 9+5}{5^2-3}=\frac{4+5}{25-3}=\frac{9}{22}$

6. a. $2x+y=2(2)+5=4+5=9$

b. $\frac{4x}{3y}=\frac{4(2)}{3(5)}=\frac{8}{15}$

c. $\frac{3}{x}+\frac{x}{y}=\frac{3}{2}+\frac{2}{5}=\frac{15}{10}+\frac{4}{10}=\frac{19}{10}$

d. $x^3+y^2=2^3+5^2=8+25=33$

7. $$\begin{aligned}9x-6&=7x\\ 9(4)-6&\stackrel{?}{=}7(4)\\ 36-6&\stackrel{?}{=}28\\ 30&=28 \quad \text{False}\end{aligned}$$
4 is not a solution of $9x-6=7x$.

8. a. Six times a number is $6x$, since $6x$ denotes the product of 6 and x.

b. A number decreased by 8 is $x-8$ because "decreased by" means subtract.

c. The product of a number and 9 is $x\cdot 9$ or $9x$.

d. Two times a number is $2x$, plus 3 is $2x+3$.

e. The sum of 7 and a number x is $7+x$.

9. a. A number x increased by 7 is $x+7$, so $x+7=13$.

b. Two less than a number x is $x-2$, so $x-2=11$.

c. Double a number x is $2x$, added to 9 is $2x+9$, so $2x+9\neq 25$.

d. Five times 11 is 5(11), so $5(11)\geq x$, where x is an unknown number.

Calculator Explorations

1. $5^4=625$

2. $7^4=2401$

3. $9^5=59{,}049$

4. $8^6=262{,}144$

5. $2(20-5)=30$

6. $3(14-7)+21=3(7)+21=21+21=42$

7. $24(862-455)+89=9857$

8. $99+(401+962)=1462$

9. $\frac{4623+129}{36-34}=2376$

10. $\frac{956-452}{89-86}=168$

Vocabulary and Readiness Check

1. In the expression 5^2, the 5 is called the base and the 2 is called the exponent.

2. The symbols ( ), [ ], and { } are examples of grouping symbols.

3. A symbol that is used to represent a number is called a variable.

4. A collection of numbers, variables, operation symbols, and grouping symbols is called an expression.

5. A mathematical statement that two expressions are equal is called an equation.

6. A value for the variable that makes an equation a true statement is called a solution.

7. Deciding what values of a variable make an equation a true statement is called solving the equation.

8. To simplify the expression $1+3\cdot 6$, first multiply.

9. To simplify the expression (1 + 3) · 6, first add.

10. To simplify the expression (20 − 4) · 2, first subtract.

11. To simplify the expression 20 − 4 ÷ 2, first divide.

Exercise Set 1.4

1. $3^5 = 3 \cdot 3 \cdot 3 \cdot 3 \cdot 3 = 243$

3. $3^3 = 3 \cdot 3 \cdot 3 = 27$

5. $1^5 = 1 \cdot 1 \cdot 1 \cdot 1 \cdot 1 = 1$

7. $5^1 = 5$

9. $\left(\frac{1}{5}\right)^3 = \left(\frac{1}{5}\right)\left(\frac{1}{5}\right)\left(\frac{1}{5}\right) = \frac{1 \cdot 1 \cdot 1}{5 \cdot 5 \cdot 5} = \frac{1}{125}$

11. $\left(\frac{2}{3}\right)^4 = \left(\frac{2}{3}\right)\left(\frac{2}{3}\right)\left(\frac{2}{3}\right)\left(\frac{2}{3}\right) = \frac{2 \cdot 2 \cdot 2 \cdot 2}{3 \cdot 3 \cdot 3 \cdot 3} = \frac{16}{81}$

13. $7^2 = 7 \cdot 7 = 49$

15. $4^2 = 4 \cdot 4 = 16$

17. $(1.2)^2 = (1.2) \cdot (1.2) = 1.44$

19. 5 + 6 · 2 = 5 + 12 = 17

21. 4 · 8 − 6 · 2 = 32 − 12 = 20

23. 2(8 − 3) = 2(5) = 10

25. $2 + (5 - 2) + 4^2 = 2 + 3 + 4^2 = 2 + 3 + 16 = 21$

27. $5 \cdot 3^2 = 5 \cdot 9 = 45$

29. $\frac{1}{4} \cdot \frac{2}{3} - \frac{1}{6} = \frac{2}{12} - \frac{1}{6} = \frac{1}{6} - \frac{1}{6} = 0$

31. $\frac{6-4}{9-2} = \frac{2}{7}$

33.
$$\begin{aligned} 2[5 + 2(8 - 3)] &= 2[5 + 2(5)] \\ &= 2[5 + 10] \\ &= 2[15] \\ &= 30 \end{aligned}$$

35. $\frac{19 - 3 \cdot 5}{6 - 4} = \frac{19 - 15}{6 - 4} = \frac{4}{2} = 2$

37. $\frac{|6-2|+3}{8+2 \cdot 5} = \frac{|4|+3}{8+2 \cdot 5} = \frac{4+3}{8+2 \cdot 5} = \frac{4+3}{8+10} = \frac{7}{18}$

39. $\frac{3+3(5+3)}{3^2+1} = \frac{3+3(8)}{3^2+1} = \frac{3+3(8)}{9+1} = \frac{3+24}{9+1} = \frac{27}{10}$

41.
$$\begin{aligned} \frac{6+|8-2|+3^2}{18-3} &= \frac{6+|6|+3^2}{18-3} \\ &= \frac{6+6+3^2}{18-3} \\ &= \frac{6+6+9}{18-3} \\ &= \frac{21}{15} \\ &= \frac{3 \cdot 7}{3 \cdot 5} \\ &= \frac{7}{5} \end{aligned}$$

43. No; since in the absence of grouping symbols we always perform multiplications or divisions before additions or subtractions in any expression.

45. **a.** (6 + 2) · (5 + 3) = 8 · 8 = 64

b. (6 + 2) · 5 + 3 = 8 · 5 + 3 = 40 + 3 = 43

c. 6 + 2 · 5 + 3 = 6 + 10 + 3 = 19

d. 6 + 2 ·(5 + 3) = 6 + 2 · 8 = 6 + 16 = 22

47. Let $y = 3$.
$3y = 3(3) = 9$

49. Let $x = 1$ and $z = 5$.
$\frac{z}{5x} = \frac{5}{5(1)} = \frac{5}{5} = 1$

51. Let $x = 1$.
$3x - 2 = 3(1) - 2 = 3 - 2 = 1$

53. Let $x = 1$ and $y = 3$.
$|2x + 3y| = |2(1) + 3(3)| = |2 + 9| = |11| = 11$

55. Let $y = 3$.
$5y^2 = 5(3)^2 = 5(9) = 45$

57. Let $x = 12$, $y = 8$ and $z = 4$.
$\frac{x}{z} + 3y = \frac{12}{4} + 3(8) = 3 + 24 = 27$

59. Let $x = 12$ and $y = 8$.
$$\begin{aligned} x^2 - 3y + x &= (12)^2 - 3(8) + 12 \\ &= 144 - 24 + 12 \\ &= 132 \end{aligned}$$

61. Let $x = 12$, $y = 8$ and $z = 4$.
$\frac{x^2 + z}{y^2 + 2z} = \frac{(12)^2 + 4}{(8)^2 + 2(4)} = \frac{144 + 4}{64 + 8} = \frac{148}{72} = \frac{37}{18}$

63. Evaluate $16t^2$ for each value of t.
$t = 1$: $16(1)^2 = 16(1) = 16$
$t = 2$: $16(2)^2 = 16(4) = 64$
$t = 3$: $16(3)^2 = 16(9) = 144$
$t = 4$: $16(4)^2 = 16(16) = 256$

Time t (in seconds)	Distance $16t^2$ (in feet)
1	16
2	64
3	144
4	256

65. Let $x = 5$.
$$\begin{aligned} 3x + 30 &= 9x \\ 3(5) + 30 &\stackrel{?}{=} 9(5) \\ 15 + 30 &\stackrel{?}{=} 45 \\ 45 &= 45, \text{ true} \end{aligned}$$
5 is a solution of the equation.

67. Let $x = 0$.
$$\begin{aligned} 2x + 6 &= 5x - 1 \\ 2(0) + 6 &\stackrel{?}{=} 5(0) - 1 \\ 0 + 6 &\stackrel{?}{=} 0 - 1 \\ 6 &= -1, \text{ false} \end{aligned}$$
0 is not a solution of the equation.

69. Let $x = 8$.
$$\begin{aligned} 2x - 5 &= 5 \\ 2(8) - 5 &\stackrel{?}{=} 5 \\ 16 - 5 &\stackrel{?}{=} 5 \\ 9 &= 5, \text{ false} \end{aligned}$$
8 is not a solution of the equation.

71. Let $x = 2$.
$$\begin{aligned} x + 6 &= x + 6 \\ 2 + 6 &\stackrel{?}{=} 2 + 6 \\ 8 &= 8, \text{ true} \end{aligned}$$
2 is a solution of the equation.

73. Let $x = 0$.
$$\begin{aligned} x &= 5x + 15 \\ (0) &\stackrel{?}{=} 5(0) + 15 \\ 0 &\stackrel{?}{=} 0 + 15 \\ 0 &= 15, \text{ false} \end{aligned}$$
0 is not a solution of the equation.

75. More than means to add; $x + 15$

77. Five subtracted from a number is represented as $x - 5$.

79. Three times a number increased by 22 is represented by $3x + 22$.

81. One increased by two equals the quotient of nine and three is $1 + 2 = 9 \div 3$.

83. Three is not equal to four divided by two is represented by $3 \neq 4 \div 2$.

85. The sum of 5 and a number is 20 is represented by $5 + x = 20$.

87. Thirteen minus three times a number is 13 is represented by $13 - 3x = 13$.

89. The quotient of 12 and a number is $\frac{1}{2}$ is represented by $\frac{12}{x} = \frac{1}{2}$.

91. Answers may vary.

93. $(20 - 4) \cdot 4 \div 2 = (16) \cdot 4 \div 2 = 64 \div 2 = 32$

95. Let $l = 8$ and $w = 6$.
$2l + 2w = 2(8) + 2(6) = 16 + 12 = 28$ m

97. Let $l = 120$ and $w = 100$.
$lw = (120)(100) = 12{,}000$ sq ft

99. Let $P = 650$, $T = 3$, and $I = 126.75$.
$\frac{I}{PT} = \frac{126.75}{(650)(3)} = \frac{126.75}{1950} = 0.065 = 6.5\%$

101. Let $m = 84$
$3.00 + 0.12(84) = 3.00 + 10.08 = \$13.08$

Section 1.5

Practice Exercises

1. 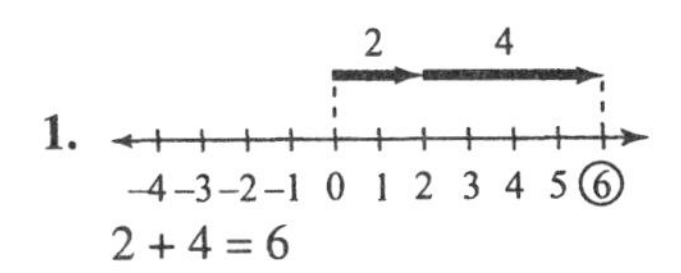

$2 + 4 = 6$

2. –3 –2
–5 –4 –3 –2 –1 0 1 2 3 4 5
$-2 + (-3) = -5$

3. **a.** $-5 + (-8)$
Add the absolute values.
$5 + 8 = 13$
The common sign is negative, so
$-5 + (-8) = -13$.

b. $-31 + (-1)$
Add the absolute values.
$31 + 1 = 32$
The common sign is negative, so
$-31 + (-1) = -32$.

4. 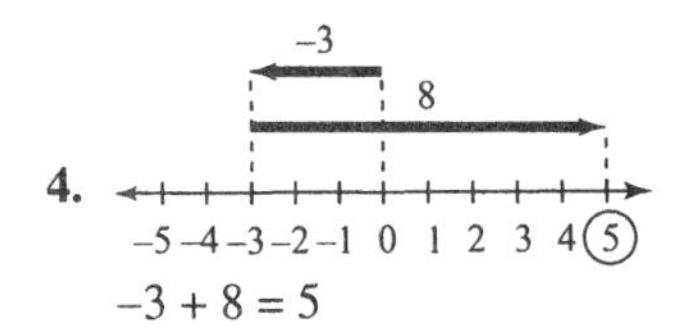

$-3 + 8 = 5$

5. **a.** $15 + (-18)$
Subtract the absolute values.
$18 - 15 = 3$
Use the sign of the number with the largest absolute value.
$15 + (-18) = -3$

b. $-19 + 20 = 20 - 19 = 1$

c. $-0.6 + 0.4 = -(0.6 - 0.4) = -(0.2) = -0.2$

6. **a.** $-\frac{3}{5} + \left(-\frac{2}{5}\right) = -\frac{5}{5} = -1$

b. $3 + (-9) = -6$

c. $2.2 + (-1.7) = 0.5$

d. $-\frac{2}{7} + \frac{3}{10} = -\frac{20}{70} + \frac{21}{70} = \frac{1}{70}$

7. **a.** $8 + (-5) + (-9) = 3 + (-9) = -6$

b. $[-8+5]+\left[-5+|-2|\right] = [-3]+[-5+2]$
$= -3 + [-3]$
$= -6$

8. $-5 + 8 + (-2) = 3 + (-2) = 1$
The overall gain is \$1.

9. **a.** The opposite of $-\frac{5}{9}$ is $\frac{5}{9}$.

b. The opposite of 8 is -8.

c. The opposite of 6.2 is -6.2.

d. The opposite of -3 is 3.

10. **a.** Since $|-15| = 15$, $-|-15| = -15$.

b. $-\left(-\frac{3}{5}\right) = \frac{3}{5}$

c. $-(-5y) = 5y$

d. $-(-8) = 8$

Vocabulary and Readiness Check

1. Two numbers that are the same distance from 0 but lie on opposite sides of 0 are called <u>opposites</u>.

2. The sum of a number and its opposite is always <u>0</u>.

3. If n is a number, then $-(-n) = \underline{n}$.

4. $-80 + (-127) =$ <u>negative number</u>.

5. $-162 + 164 =$ <u>positive number</u>.

6. $-162 + 162 = \underline{0}$.

7. $-1.26 + (-8.3) =$ <u>negative number</u>.

8. $-3.68 + 0.27 =$ <u>negative number</u>.

9. $-\frac{2}{3} + \frac{2}{3} = \underline{0}$.

Exercise Set 1.5

1. $6 + 3 = 9$

3. $-6 + (-8) = -14$

5. $8 + (-7) = 1$

7. $-14 + 2 = -12$

9. $-2 + (-3) = -5$

11. $-9 + (-3) = -12$

13. $-7 + 3 = -4$

15. $10 + (-3) = 7$

17. $5 + (-7) = -2$

19. $-16 + 16 = 0$

21. $27 + (-46) = -19$

23. $-18 + 49 = 31$

25. $-33 + (-14) = -47$

27. $6.3 + (-8.4) = -2.1$

29. $|-8| + (-16) = 8 + (-16) = -8$

31. $117 + (-79) = 38$

33. $-9.6 + (-3.5) = -13.1$

35. $-\frac{3}{8} + \frac{5}{8} = \frac{2}{8} = \frac{1}{4}$

37. $-\frac{7}{16} + \frac{1}{4} = -\frac{7}{16} + \frac{1 \cdot 4}{4 \cdot 4} = -\frac{7}{16} + \frac{4}{16} = -\frac{3}{16}$

39.
$$\begin{aligned} -\frac{7}{10} + \left(-\frac{3}{5}\right) &= -\frac{7}{10} + \left(-\frac{3 \cdot 2}{5 \cdot 2}\right) \\ &= -\frac{7}{10} + \left(-\frac{6}{10}\right) \\ &= -\frac{13}{10} \end{aligned}$$

41. $-15 + 9 + (-2) = -6 + (-2) = -8$

43. $-21 + (-16) + (-22) = -37 + (-22) = -59$

45. $-23 + 16 + (-2) = -7 + (-2) = -9$

47. $|5 + (-10)| = |-5| = 5$

49. $6 + (-4) + 9 = 2 + 9 = 11$

51. $[-17 + (-4)] + [-12 + 15] = [-21] + [3] = -18$

53. $|9 + (-12)| + |-16| = |-3| + 16 = 3 + 16 = 19$

55.
$$\begin{aligned} -1.3 + [0.5 + (-0.3) + 0.4] &= -1.3 + [0.2 + 0.4] \\ &= -1.3 + [0.6] \\ &= -0.7 \end{aligned}$$

57. $-15 + 9 = -6$
The high temperature in Anoka was $-6°$.

59. $-17{,}657 + 1230 = -16{,}427$
You are 16,427 feet below the rim.

61.
$$\begin{aligned} (-155) + (-3895) + (-5200) &= (-4050) + (-5200) \\ &= -9250 \end{aligned}$$
The total net income was –$9250 million.

63. $(-5) + (-2) + (-2) = (-7) + (-2) = -9$
Her score was 9 under par.

65. The opposite of 6 is –6.

67. The opposite of –2 is 2.

69. The opposite of 0 is 0.

71. Since $|-6|$ is 6, the opposite of $|-6|$ is –6.

73. Answers may vary

75. $-|-2| = -2$

77. $-|0| = -0 = 0$

79. $-\left|-\frac{2}{3}\right| = -\frac{2}{3}$

81. Answers may vary

83. Let $x = -4$.
$$\begin{aligned} x + 9 &= 5 \\ (-4) + 9 &\stackrel{?}{=} 5 \\ 5 &= 5, \text{ true} \end{aligned}$$
–4 is a solution of the equation.

85. Let $y = -1$.
$$\begin{aligned} y + (-3) &= -7 \\ (-1) + (-3) &\stackrel{?}{=} -7 \\ -4 &= -7, \text{ false} \end{aligned}$$
–1 is not a solution of the equation.

87. Look for the tallest bar. The temperature is the highest in July.

89. Look for the bar whose length has a positive value closest to 0; October

91. $[(-9.1)+14.4+8.8]\div 3=[5.3+8.8]\div 3$
$=[14.1]\div 3$
$=4.7$

The average was 4.7°F.

93. Since a is a positive number, $-a$ is a negative number.

95. Since a is a positive number, $a+a$ is a positive number.

Section 1.6

Practice Exercises

1. a. $-7-6=-7+(-6)=-13$

b. $-8-(-1)=-8+1=-7$

c. $9-(-3)=9+3=12$

d. $5-7=5+(-7)=-2$

2. a. $8.4-(-2.5)=8.4+2.5=10.9$

b. $-\frac{5}{8}-\left(-\frac{1}{8}\right)=-\frac{5}{8}+\frac{1}{8}=-\frac{4}{8}=-\frac{1}{2}$

c. $-\frac{3}{4}-\frac{1}{5}=-\frac{3}{4}+\left(-\frac{1}{5}\right)$
$=-\frac{15}{20}+\left(-\frac{4}{20}\right)$
$=-\frac{19}{20}$

3. $-2-5=-2+(-5)=-7$

4. a. $-15-2-(-4)+7=-15+(-2)+4+7=-6$

b. $3.5+(-4.1)-(-6.7)=3.5+(-4.1)+6.7$
$=6.1$

5. a. $-4+[(-8-3)-5]=-4+[(-8+(-3))-5]$
$=-4+[(-11)-5]$
$=-4+[-11+(-5)]$
$=-4+[-16]$
$=-20$

b. $|-13|-3^2+[2-(-7)]=13-9+[2+7]$
$=13-9+9$
$=13$

6. a. $\frac{7-x}{2y+x}=\frac{7-(-3)}{2(4)+(-3)}=\frac{7+3}{8+(-3)}=\frac{10}{5}=2$

b. $y^2+x=(4)^2+(-3)=16+(-3)=13$

7. $282-(-75)=282+75=\$357$

8. a. $x=90°-62°=28°$

b. $y=180°-43°=137°$

Vocabulary and Readiness Check

1. 7 minus a number $\underline{7-x}$

2. 7 subtracted from a number $\underline{x-7}$.

3. A number decreased by 7 $\underline{x-7}$

4. 7 less a number $\underline{7-x}$

5. A number less than 7 $\underline{7-x}$

6. A number subtracted from 7 $\underline{7-x}$

Exercise Set 1.6

1. $-6-4=-6+(-4)=-10$

3. $4-9=4+(-9)=-5$

5. $16-(-3)=16+3=19$

7. $\frac{1}{2}-\frac{1}{3}=\frac{1}{2}+\left(-\frac{1}{3}\right)$
$=\frac{1\cdot 3}{2\cdot 3}+\left(-\frac{1\cdot 2}{3\cdot 2}\right)$
$=\frac{3}{6}+\left(-\frac{2}{6}\right)$
$=\frac{1}{6}$

9. $-16-(-18)=-16+18=2$

11. $-6-5=-6+(-5)=-11$

13. $7-(-4)=7+4=11$

15. $-6-(-11)=-6+11=5$

17. $16-(-21)=16+21=37$

19. $9.7-16.1=9.7+(-16.1)=-6.4$

21. $-44-27=-44+(-27)=-71$

23. $-21-(-21)=-21+21=0$

25. $-2.6-(-6.7)=-2.6+6.7=4.1$

27. $-\frac{3}{11}-\left(-\frac{5}{11}\right)=-\frac{3}{11}+\frac{5}{11}=\frac{2}{11}$

29. $$\begin{aligned}-\frac{1}{6}-\frac{3}{4}&=-\frac{1}{6}+\left(-\frac{3}{4}\right)\\&=-\frac{1\cdot 2}{6\cdot 2}+\left(-\frac{3\cdot 3}{4\cdot 3}\right)\\&=-\frac{2}{12}+\left(-\frac{9}{12}\right)\\&=-\frac{11}{12}\end{aligned}$$

31. $8.3-(-0.62)=8.3+0.62=8.92$

33. $8-(-5)=8+5=13$

35. $-6-(-1)=-6+1=-5$

37. $7-8=7+(-8)=-1$

39. $-8-15=-8+(-15)=-23$

41. Answers may vary

43. $$\begin{aligned}-10-(-8)+(-4)-20&=-10+8+(-4)+(-20)\\&=-2+(-4)+(-20)\\&=-6+(-20)\\&=-26\end{aligned}$$

45. $$\begin{aligned}5-9+(-4)-8-8&=5+(-9)+(-4)+(-8)+(-8)\\&=-4+(-4)+(-8)+(-8)\\&=-8+(-8)+(-8)\\&=-16+(-8)\\&=-24\end{aligned}$$

47. $-6-(2-11)=-6-(-9)=-6+9=3$

49. $3^3-8\cdot 9=27-8\cdot 9=27-72=27+(-72)=-45$

51. $2-3(8-6)=2-3(2)=2-6=2+(-6)=-4$

53. $$\begin{aligned}(3-6)+4^2&=[3+(-6)]+4^2\\&=[-3]+4^2\\&=[-3]+16\\&=13\end{aligned}$$

55. $$\begin{aligned}&-2+[(8-11)-(-2-9)]\\&=-2+[(8+(-11))-(-2+(-9))]\\&=-2+[(-3)-(-11)]\\&=-2+[(-3)+11]\\&=-2+[8]\\&=6\end{aligned}$$

57. $$\begin{aligned}|-3|+2^2+[-4-(-6)]&=3+2^2+[-4+6]\\&=3+2^2+[2]\\&=3+4+[2]\\&=7+[2]\\&=9\end{aligned}$$

59. Let $x=-5$ and $y=4$.
$x-y=-5-4=-5+(-4)=-9$

61. Let $x=-5$, $y=4$, and $t=10$.
$$\begin{aligned}|x|+2t-8y&=|-5|+2(10)-8(4)\\&=5+2(10)-8(4)\\&=5+20-32\\&=25-32\\&=25+(-32)\\&=-7\end{aligned}$$

63. Let $x=-5$ and $y=4$.
$\frac{9-x}{y+6}=\frac{9-(-5)}{4+6}=\frac{9+5}{4+6}=\frac{14}{10}=\frac{2\cdot 7}{2\cdot 5}=\frac{7}{5}$

65. Let $x=-5$ and $y=4$.
$y^2-x=4^2-(-5)=16+5=21$

67. Let $x=-5$ and $t=10$.
$$\begin{aligned}\frac{|x-(-10)|}{2t}&=\frac{|-5-(-10)|}{2(10)}\\&=\frac{|-5+10|}{2(10)}\\&=\frac{|5|}{2(10)}\\&=\frac{5}{20}\\&=\frac{5}{4\cdot 5}\\&=\frac{1}{4}\end{aligned}$$

69. The change in temperature is the difference between the last temperature and the first temperature.
$-56 - 44 = -56 + (-44) = -100$
The temperature dropped 100°.

71. Gains: +2
Losses: −5, −20
$2 + (-5) + (-20) = -3 + (-20) = -23$
Total loss of 23 yards

73. $-475 - 94 = -475 + (-94) = -569$
He was born in 569 B.C.

75. Rises: +120
Drops: −250, −178
$120 + (-250) + (-178) = -130 + (-178) = -308$
The overall vertical change was a drop of 308 feet.

77. $19,340 - (-512) = 19,340 + 512 = 19,852$
19,852 feet higher

79. $y = 180 - 50 = 180 + (-50) = 130$
The supplementary angle is 130°.

81. $x = 90 - 60 = 90 + (-60) = 30$
The complementary angle is 30°.

83. Let $x = -4$.

$$x - 9 = 5$$
$$-4 - 9 \stackrel{?}{=} 5$$
$$-4 + (-9) \stackrel{?}{=} 5$$
$$-13 = 5, \text{ false}$$

−4 is not a solution of the equation.

85. Let $x = -2$.

$$-x + 6 = -x - 1$$
$$-(-2) + 6 \stackrel{?}{=} -(-2) - 1$$
$$2 + 6 \stackrel{?}{=} 2 + (-1)$$
$$8 = 1, \text{ false}$$

−2 is not a solution of the equation.

87. Let $x = 2$.

$$-x - 13 = -15$$
$$-2 - 13 \stackrel{?}{=} -15$$
$$-2 + (-13) \stackrel{?}{=} -15$$
$$-15 = -15, \text{ true}$$

2 is a solution of the equation.

89. The change in temperature is the difference between the given month's temperature and the previous month's.
F: $-23.7 - (-19.3) = -23.7 + 19.3 = -4.4°$
Mr: $-21.1 - (-23.7) = -21.1 + 23.7 = 2.6°$
Ap: $-9.1 - (-21.1) = -9.1 + 21.1 = 12°$
Ma: $14.4 - (-9.1) = 14.4 + 9.1 = 23.5°$
Jn: $29.7 - 14.4 = 29.7 + (-14.4) = 15.3°$
Jy: $33.6 - 29.7 = 33.6 + (-29.7) = 3.9°$
Au: $33.3 - 33.6 = 33.3 + (-33.6) = -0.3°$
S: $27.0 - 33.3 = 27.0 + (-33.3) = -6.3°$
O: $8.8 - 27.0 = 8.8 + (-27.0) = -18.2°$
N: $-6.9 - 8.8 = -6.9 + (-8.8) = -15.7°$
D: $-17.2 - (-6.9) = -17.2 + 6.9 = -10.3°$

91. Look for the negative number whose absolute value is the greatest; October

93. True; answers may vary

95. True; answers may vary

97. Negative; $4.362 - 7.0086 = -2.6466$

Integrated Review

1. The opposite of a positive number is a <u>negative</u> number.

2. The sum of two negative numbers is a <u>negative</u> number.

3. The absolute value of a negative number is a <u>positive</u> number.

4. The absolute value of zero is <u>0</u>.

5. The reciprocal of a positive number is a <u>positive</u> number.

6. The sum of a number and its opposite is <u>0</u>.

7. The absolute value of a positive number is a <u>positive</u> number.

8. The opposite of a negative number is a <u>positive</u> number.

	Number	Opposite	Absolute Value
9.	$\frac{1}{7}$	$-\frac{1}{7}$	$\frac{1}{7}$
10.	$-\frac{12}{5}$	$\frac{12}{5}$	$\frac{12}{5}$
11.	3	−3	3
12.	$-\frac{9}{11}$	$\frac{9}{11}$	$\frac{9}{11}$

13. $-19 + (-23) = -42$

14. $7 - (-3) = 7 + 3 = 10$

15. $-15 + 17 = 2$

16. $-8 - 10 = -8 + (-10) = -18$

17. $18 + (-25) = -7$

18. $-2 + (-37) = -39$

19. $-14 - (-12) = -14 + 12 = -2$

20. $5 - 14 = 5 + (-14) = -9$

21. $4.5 - 7.9 = 4.5 + (-7.9) = -3.4$

22. $-8.6 - 1.2 = -8.6 + (-1.2) = -9.8$

23. $-\frac{3}{4}-\frac{1}{7}=-\frac{21}{28}-\frac{4}{28}=-\frac{21}{28}+\left(-\frac{4}{28}\right)=-\frac{25}{28}$

24. $\frac{2}{3}-\frac{7}{8}=\frac{16}{24}-\frac{21}{24}=\frac{16}{24}+\left(-\frac{21}{24}\right)=-\frac{5}{24}$

25. $$\begin{aligned}-9-(-7)+4-6 &= -9+7+4-6\\ &= -9+7+4+(-6)\\ &= -4\end{aligned}$$

26. $$\begin{aligned}11-20+(-3)-12 &= 11+(-20)+(-3)+(-12)\\ &= -9+(-3)+(-12)\\ &= -12+(-12)\\ &= -24\end{aligned}$$

27. $$\begin{aligned}24-6(14-11) &= 24-6[14+(-11)]\\ &= 24-6(3)\\ &= 24-18\\ &= 24+(-18)\\ &= 6\end{aligned}$$

28. $$\begin{aligned}30-5(10-8) &= 30-5[10+(-8)]\\ &= 30-5(2)\\ &= 30-10\\ &= 30+(-10)\\ &= 20\end{aligned}$$

29. $(7-17)+4^2=[7+(-17)]+4^2=(-10)+16=6$

30. $$\begin{aligned}9^2+(10-30) &= 9^2+[10+(-30)]\\ &= 81+(-20)\\ &= 61\end{aligned}$$

31. $$\begin{aligned}|-9|+3^2+(-4-20) &= 9+9+[-4+(-20)]\\ &= 9+9+(-24)\\ &= 18+(-24)\\ &= -6\end{aligned}$$

32. $$\begin{aligned}|-4-5|+5^2+(-50) &= |-4+(-5)|+5^2+(-50)\\ &= |-9|+25+(-50)\\ &= 9+25+(-50)\\ &= 34+(-50)\\ &= -16\end{aligned}$$

33. $$\begin{aligned}-7+[(1-2)+(-2-9)] &= -7+[(-1)+(-11)]\\ &= -7+[-12]\\ &= -19\end{aligned}$$

34. $$\begin{aligned}-6+[(-3+7)+(4-15)] &= -6+[(4)+(-11)]\\ &= -6+(-7)\\ &= -13\end{aligned}$$

35. $1 - 5 = 1 + (-5) = -4$

36. $-3 - (-2) = -3 + 2 = -1$

37. $\frac{1}{4}-\left(-\frac{2}{5}\right)=\frac{1}{4}+\frac{2}{5}=\frac{5}{20}+\frac{8}{20}=\frac{13}{20}$

38. $-\frac{5}{8}-\left(\frac{1}{10}\right)=-\frac{25}{40}-\frac{4}{40}=-\frac{25}{40}+\left(-\frac{4}{40}\right)=-\frac{29}{40}$

39. $$\begin{aligned}&2(19-17)^3-3(-7+9)^2\\ &= 2[19+(-17)]^3-3(-7+9)^2\\ &= 2(2)^3-3(2)^2\\ &= 2(8)-3(4)\\ &= 16-12\\ &= 16+(-12)\\ &= 4\end{aligned}$$

40. $$\begin{aligned}&3(10-9)^2+6(20-19)^3\\ &= 3[10+(-9)]^2+6[20+(-19)]^3\\ &= 3(1)^2+6(1)^3\\ &= 3+6\\ &= 9\end{aligned}$$

41. $x - y = -2 - (-1) = -2 + 1 = -1$

42. $x + y = -2 + (-1) = -3$

43. $y + z = -1 + 9 = 8$

44. $z - y = 9 - (-1) = 9 + 1 = 10$

45. $\frac{|5z-x|}{y-x}=\frac{|5(9)-(-2)|}{-1-(-2)}=\frac{|45+2|}{-1+2}=\frac{|47|}{1}=47$

46. $\frac{|-x-y+z|}{2z}=\frac{|-(-2)-(-1)+9|}{2(9)}$
$=\frac{|2+1+9|}{18}$
$=\frac{|12|}{18}$
$=\frac{12}{18}$
$=\frac{2}{3}$

Section 1.7

Practice Exercises

1. a. $8(-5)=-40$

b. $(-3)(-4)=12$

c. $(-6)(9)=-54$

2. a. $(-1)(-5)(-6)=5(-6)=-30$

b. $(-3)(-2)(4)=6(4)=24$

c. $(-4)(0)(5)=0(5)=0$

d. $(-2)(-3)-(-4)(5)=6-(-20)$
$=6+20$
$=26$

3. a. $(0.23)(-0.2)=-[(0.23)(0.2)]=-0.046$

b. $\left(-\frac{3}{5}\right)\cdot\left(\frac{4}{9}\right)=-\frac{3\cdot 4}{5\cdot 9}=-\frac{12}{45}=-\frac{4}{15}$

c. $\left(-\frac{7}{12}\right)(-24)=\frac{7\cdot 24}{12\cdot 1}=7\cdot 2=14$

4. a. $(-6)^2=(-6)(-6)=36$

b. $-6^2=-(6\cdot 6)=-(36)=-36$

c. $(-4)^3=(-4)(-4)(-4)=16(-4)=-64$

d. $-4^3=-(4\cdot 4\cdot 4)=-[16(4)]=-64$

5. a. The reciprocal of $\frac{8}{3}$ is $\frac{3}{8}$ since $\frac{8}{3}\cdot\frac{3}{8}=1$.

b. The reciprocal of 15 is $\frac{1}{15}$ since $15\cdot\frac{1}{15}=1$.

c. The reciprocal of $-\frac{2}{7}$ is $-\frac{7}{2}$ since $\left(-\frac{2}{7}\right)\left(-\frac{7}{2}\right)=1$.

d. The reciprocal of -5 is $-\frac{1}{5}$ since $(-5)\left(-\frac{1}{5}\right)=1$.

6. a. $\frac{16}{-2}=16\left(-\frac{1}{2}\right)=-8$

b. $24\div(-6)=24\left(-\frac{1}{6}\right)=-4$

c. $\frac{-35}{-7}=\frac{35}{7}=\frac{5\cdot 7}{7}=5$

7. a. $\frac{-18}{-6}=\frac{18}{6}=\frac{3\cdot 6}{6}=3$

b. $\frac{-48}{3}=-\frac{48}{3}=-\frac{3\cdot 16}{3}=-16$

c. $\frac{3}{5}\div\left(-\frac{1}{2}\right)=\frac{3}{5}\cdot(-2)=-\frac{6}{5}$

d. $-\frac{4}{9}\div 8=-\frac{4}{9}\cdot\frac{1}{8}=-\frac{4}{9\cdot 4\cdot 2}=-\frac{1}{9\cdot 2}=-\frac{1}{18}$

8. a. $\frac{0}{-2}=0$

b. $\frac{-4}{0}$ is undefined.

c. $\frac{-5}{6(0)}=\frac{-5}{0}$ is undefined.

9. a. $\frac{(-8)(-11)-4}{-9-(-4)}=\frac{88-4}{-9+4}=\frac{84}{-5}=-\frac{84}{5}$

b. $\frac{3(-2)^3-9}{-6+3}=\frac{3(-8)-9}{-3}$
$=\frac{-24-9}{-3}$
$=\frac{-33}{-3}$
$=11$

10. a. $7y-x=7(-2)-(-5)=-14+5=-9$

b. $x^2-y^3=(-5)^2-(-2)^3$
$=25-(-8)$
$=25+8$
$=33$

c. $\frac{2x}{3y}=\frac{2(-5)}{3(-2)}=\frac{-10}{-6}=\frac{5}{3}$

Calculator Explorations

1. $-38(26-27)=38$

2. $-59(-8)+1726=2198$

3. $134+25(68-91)=-441$

4. $45(32)-8(218)=-304$

5. $\frac{-50(294)}{175-265}=163.\overline{3}$

6. $\frac{-444-444.8}{-181-324}=1.76$

7. $9^5-4550=54,499$

8. $5^8-6259=384,366$

9. $(-125)^2=15,625$

10. $-125^2=-15,625$

Vocabulary and Readiness Check

1. If n is a real number, then $n\cdot 0=\underline{0}$ and $0\cdot n=\underline{0}$.

2. If n is a real number, but not 0, then $\frac{0}{n}=\underline{0}$ and we say $\frac{n}{0}$ is <u>undefined</u>.

3. The product of two negative numbers is a <u>positive</u> number.

4. The quotient of two negative numbers is a <u>positive</u> number.

5. The quotient of a positive number and a negative number is a <u>negative</u> number.

6. The product of a positive number and a negative number is a <u>negative</u> number.

7. The reciprocal of a positive number is a <u>positive</u> number.

8. The opposite of a positive number is a <u>negative</u> number.

Exercise Set 1.7

1. $-6(4)=-24$

3. $2(-1)=-2$

5. $-5(-10)=50$

7. $-3\cdot 4=-12$

9. $-7\cdot 0=0$

11. $2(-9)=-18$

13. $-\frac{1}{2}\left(-\frac{3}{5}\right)=\frac{1\cdot 3}{2\cdot 5}=\frac{3}{10}$

15. $-\frac{3}{4}\left(-\frac{8}{9}\right)=\frac{3\cdot 8}{4\cdot 9}=\frac{24}{36}=\frac{2\cdot 12}{3\cdot 12}=\frac{2}{3}$

17. $5(-1.4)=-7$

19. $-0.2(-0.7)=0.14$

21. $-10(80)=-800$

23. $4(-7)=-28$

25. $(-5)(-5)=25$

27. $\frac{2}{3}\left(-\frac{4}{9}\right)=-\frac{2\cdot 4}{3\cdot 9}=-\frac{8}{27}$

29. $-11(11)=-121$

31. $-\frac{20}{25}\left(\frac{5}{16}\right)=-\frac{20\cdot 5}{25\cdot 16}=-\frac{100}{400}=-\frac{1}{4}$

33. $(-1)(2)(-3)(-5)=-2(-3)(-5)=6(-5)=-30$

35. $(-2)(5)-(-11)(3)=-10-(-33)=-10+33=23$

37. $(-6)(-1)(-2)-(-5)=-12+5=-7$

39. True; example: $(-2)(-2)(-2)=-8$
False; example: $(-2)(-2)(-2)(-2)=16$

41. False

43. $(-2)^4=(-2)(-2)(-2)(-2)$
$=4(-2)(-2)$
$=-8(-2)$
$=16$

45. $-1^5=-(1)(1)(1)(1)(1)=-1$

47. $(-5)^2=(-5)(-5)=25$

49. $-7^2=-(7)(7)=-49$

51. Reciprocal of 9 is $\frac{1}{9}$ since $9\cdot\frac{1}{9}=1$.

53. Reciprocal of $\frac{2}{3}$ is $\frac{3}{2}$ since $\frac{2}{3}\cdot\frac{3}{2}=1$.

55. Reciprocal of -14 is $-\frac{1}{14}$ since $-14\cdot -\frac{1}{14}=1$.

57. Reciprocal of $-\frac{3}{11}$ is $-\frac{11}{3}$ since $-\frac{3}{11}\cdot -\frac{11}{3}=1$.

59. Reciprocal of 0.2 is $\frac{1}{0.2}$ since $0.2\cdot\frac{1}{0.2}=1$.

61. Reciprocal of $\frac{1}{-6.3}$ is -6.3 since

$\frac{1}{-6.3}\cdot -6.3=1$.

63. $\frac{18}{-2}=18\cdot -\frac{1}{2}=-9$

65. $\frac{-16}{-4}=-16\cdot -\frac{1}{4}=4$

67. $\frac{-48}{12}=-48\cdot\frac{1}{12}=-4$

69. $\frac{0}{-4}=0\cdot -\frac{1}{4}=0$

71. $-\frac{15}{3}=-15\cdot\frac{1}{3}=-5$

73. $\frac{5}{0}$ is undefined.

75. $\frac{-12}{-4}=-12\cdot -\frac{1}{4}=3$

77. $\frac{30}{-2}=30\cdot -\frac{1}{2}=-15$

79. $\frac{6}{7}\div -\frac{1}{3}=\frac{6}{7}\cdot\left(-\frac{3}{1}\right)=-\frac{6\cdot 3}{7\cdot 1}=-\frac{18}{7}$

81. $-\frac{5}{9}\div\left(-\frac{3}{4}\right)=-\frac{5}{9}\cdot\left(-\frac{4}{3}\right)=\frac{5\cdot 4}{9\cdot 3}=\frac{20}{27}$

83. $-\frac{4}{9}\div\frac{4}{9}=-\frac{4}{9}\cdot\frac{9}{4}=-1$

85. $\frac{-9(-3)}{-6}=\frac{27}{-6}=-\frac{9}{2}$

87. $\frac{12}{9-12}=\frac{12}{-3}=-4$

89. $\frac{-6^2+4}{-2}=\frac{-36+4}{-2}=\frac{-32}{-2}=16$

91. $\frac{8+(-4)^2}{4-12}=\frac{8+16}{4-12}=\frac{24}{-8}=-3$

93. $\frac{22+(3)(-2)}{-5-2}=\frac{22+(-6)}{-5-2}=\frac{16}{-7}=-\frac{16}{7}$

95. $\frac{-3-5^2}{2(-7)}=\frac{-3-25}{2(-7)}=\frac{-3+(-25)}{-14}=\frac{-28}{-14}=2$

97. $\frac{6-2(-3)}{4-3(-2)}=\frac{6-(-6)}{4-(-6)}=\frac{6+6}{4+6}=\frac{12}{10}=\frac{6}{5}$

99. $\frac{-3-2(-9)}{-15-3(-4)} = \frac{-3-(-18)}{-15-(-12)} = \frac{-3+18}{-15+12} = \frac{15}{-3} = -5$

101. $\frac{|5-9|+|10-15|}{|2(-3)|} = \frac{|-4|+|-5|}{|-6|} = \frac{4+5}{6} = \frac{9}{6} = \frac{3}{2}$

103. Let $x = -5$ and $y = -3$.
$3x+2y = 3(-5)+2(-3) = -15+(-6) = -21$

105. Let $x = -5$ and $y = -3$.

$$\begin{aligned} 2x^2 - y^2 &= 2(-5)^2 - (-3)^2 \\ &= 2(25) - 9 \\ &= 50 + (-9) \\ &= 41 \end{aligned}$$

107. Let $x = -5$ and $y = -3$.
$x^3 + 3y = (-5)^3 + 3(-3) = -125 + (-9) = -134$

109. Let $x = -5$ and $y = -3$.
$\frac{2x-5}{y-2} = \frac{2(-5)-5}{-3-2} = \frac{-10-5}{-3-2} = \frac{-15}{-5} = 3$

111. Let $x = -5$ and $y = -3$.
$\frac{-3-y}{x-4} = \frac{-3-(-3)}{-5-4} = \frac{-3+3}{-5-4} = \frac{0}{-9} = 0$

113. $4(-6203) = -24,812$
The net income will be –$24,812 million.

115. Let $x = 7$.

$$\begin{aligned} -5x &= -35 \\ -5(7) &\stackrel{?}{=} -35 \\ -35 &= -35, \text{ true} \end{aligned}$$

7 is a solution of the equation.

117. Let $x = -20$.

$$\begin{aligned} \frac{x}{10} &= 2 \\ \frac{-20}{10} &\stackrel{?}{=} 2 \\ -2 &= 2, \text{ false} \end{aligned}$$

–20 is not a solution of the equation.

119. Let $x = 5$.

$$\begin{aligned} -3x - 5 &= -20 \\ -3(5) - 5 &\stackrel{?}{=} -20 \\ -15 - 5 &\stackrel{?}{=} -20 \\ -20 &= -20, \text{ true} \end{aligned}$$

5 is a solution of the equation.

121. Answers may vary

123. –1 and 1 are their own reciprocals.

125. Since q is negative, r is negative, and t is positive, then $\frac{q}{r \cdot t}$ is positive.

127. It is not possible to determine whether $q + t$ is positive or negative.

129. Since q is negative, r is negative, and t is positive, then $t(q + r)$ is negative.

131.

$$\begin{aligned} -2 + \frac{-15}{3} &= \frac{-2 \cdot 3}{1 \cdot 3} + \frac{-15}{3} \\ &= \frac{-6+(-15)}{3} \\ &= \frac{-21}{3} \\ &= -7 \end{aligned}$$

133. $2[-5 + (-3)] = 2(-8) = -16$

The Bigger Picture

1. $-0.2(25) - 5$

2. $86 - 100 = -14$

3. $-\frac{1}{7} + \left(-\frac{3}{5}\right) = -\frac{5}{35} - \frac{21}{35} = -\frac{26}{35}$

4. $\frac{-40}{-5} = 8$

5. $(-7)^2 = (-7)(-7) = 49$

6. $-7^2 = -(7 \cdot 7) = -49$

7. $\frac{|-42|}{-|-2|} = \frac{42}{-2} = -21$

8. $\frac{8.6}{0}$ is undefined.

9. $\frac{0}{8.6} = 0$

10. $-25 - (-13) = -25 + 13 = -12$

11. $-8.3 - 8.3 = -16.6$

12. $-\frac{8}{9}\left(-\frac{3}{16}\right)=\frac{3\cdot 8}{9\cdot 16}=\frac{1\cdot 1}{3\cdot 2}=\frac{1}{6}$

13. $2+3(8-11)^3 = 2+3(-3)^3$
$= 2+3(-27)$
$= 2+(-81)$
$= -79$

14. $-2\frac{1}{2}\div\left(-3\frac{1}{4}\right)=-\frac{5}{2}\div\left(-\frac{13}{4}\right)$
$=-\frac{5}{2}\left(-\frac{4}{13}\right)$
$=\frac{4\cdot 5}{2\cdot 13}$
$=\frac{2\cdot 5}{13}$
$=\frac{10}{13}$

15. $20\div 2\cdot 5 = 10\cdot 5 = 50$

16. $-2[(1-5)-(7-17)] = -2[(-4)-(-10)]$
$= -2[-4+10]$
$= -2[6]$
$= -12$

Section 1.8

Practice Exercises

1. a. $x\cdot 8 = \underline{8\cdot x}$

b. $x+17=\underline{17+x}$

2. a. $(2+9)+7=\underline{2+(9+7)}$

b. $-4\cdot(2\cdot 7)=\underline{(-4\cdot 2)\cdot 7}$

3. a. $(5+x)+9=(x+5)+9=x+(5+9)=x+14$

b. $5(-6x)=[5\cdot(-6)]x=-30x$

4. a. $5(x-y)=5(x)-5(y)=5x-5y$

b. $-6(4+2t)=-6(4)+(-6)(2t)=-24-12t$

c. $2(3x-4y-z)=2(3x)+2(-4y)+2(-z)$
$=6x-8y-2z$

d. $(3-y)\cdot(-1)=3(-1)+(-y)(-1)=-3+y$

e. $-(x-7+2s)=(-1)(x-7+2s)$
$=(-1)x+(-1)(-7)+(-1)(2s)$
$=-x+7-2s$

f. $2(7x+4)+6=2(7x)+2(4)+6$
$=14x+8+6$
$=14x+14$

5. a. $5\cdot w+5\cdot 3=5(w+3)$

b. $9w+9z=9\cdot w+9\cdot z=9(w+z)$

6. a. $(7\cdot 3x)\cdot 4=(3x\cdot 7)\cdot 4$; commutative property of multiplication

b. $6+(3+y)=(6+3)+y$; associative property of addition

c. $8+(t+0)=8+t$; identity element for addition

d. $-\frac{3}{4}\cdot\left(-\frac{4}{3}\right)=1$; multiplicative inverse property

e. $(2+x)+5=5+(2+x)$; commutative property of addition

f. $3+(-3)=0$; additive inverse property

g. $(-3b)\cdot 7=(-3\cdot 7)\cdot b$; commutative and associative properties of multiplication

Vocabulary and Readiness Check

1. $x+5=5+x$ is a true statement by the <u>commutative property of addition</u>.

2. $x\cdot 5=5\cdot x$ is a true statement by the <u>commutative property of multiplication</u>.

3. $3(y+6)=3\cdot y+3\cdot 6$ is true by the <u>distributive property</u>.

4. $2\cdot(x\cdot y)=(2\cdot x)\cdot y$ is a true statement by the <u>associative property of multiplication</u>.

5. $x+(7+y)=(x+7)+y$ is a true statement by the <u>associative property of addition</u>.

6. The numbers $-\frac{2}{3}$ and $-\frac{3}{2}$ are called <u>reciprocals or multiplicative inverses</u>.

7. The numbers $-\frac{2}{3}$ and $\frac{2}{3}$ are called <u>opposites or additive inverses</u>.

Exercise Set 1.8

1. $x + 16 = 16 + x$

3. $-4 \cdot y = y \cdot (-4)$

5. $xy = yx$

7. $2x + 13 = 13 + 2x$

9. $(xy) \cdot z = x \cdot (yz)$

11. $2 + (a + b) = (2 + a) + b$

13. $4 \cdot (ab) = 4a \cdot (b)$

15. $(a + b) + c = a + (b + c)$

17. $8 + (9 + b) = (8 + 9) + b = 17 + b$

19. $4(6y) = (4 \cdot 6)y = 24y$

21. $\frac{1}{5}(5y) = \left(\frac{1}{5} \cdot 5\right)y = 1 \cdot y = y$

23. $$\begin{aligned}(13 + a) + 13 &= (a + 13) + 13\\ &= a + (13 + 13)\\ &= a + 26\end{aligned}$$

25. $-9(8x) = (-9 \cdot 8)x = -72x$

27. $\frac{3}{4}\left(\frac{4}{3}s\right) = \left(\frac{3}{4} \cdot \frac{4}{3}\right)s = 1s = s$

29. Answers may vary

31. $4(x + y) = 4x + 4y$

33. $9(x - 6) = 9x - 9 \cdot 6 = 9x - 54$

35. $2(3x + 5) = 2(3x) + 2(5) = 6x + 10$

37. $7(4x - 3) = 7(4x) - 7(3) = 28x - 21$

39. $3(6 + x) = 3(6) + 3x = 18 + 3x$

41. $-2(y - z) = -2y - (-2)z = -2y + 2z$

43. $-7(3y + 5) = -7(3y) + (-7)(5) = -21y - 35$

45. $$\begin{aligned}5(x + 4m + 2) &= 5x + 5(4m) + 5(2)\\ &= 5x + 20m + 10\end{aligned}$$

47. $$\begin{aligned}-4(1 - 2m + n) &= -4(1) - (-4)(2m) + (-4)n\\ &= -4 + 8m - 4n\end{aligned}$$

49. $$\begin{aligned}-(5x + 2) &= -1(5x + 2)\\ &= -1(5x) + (-1)(2)\\ &= -5x - 2\end{aligned}$$

51. $$\begin{aligned}-(r - 3 - 7p) &= -1(r - 3 - 7p)\\ &= -1r - (-1)(3) - (-1)(7p)\\ &= -r + 3 + 7p\end{aligned}$$

53. $$\begin{aligned}\frac{1}{2}(6x + 8) &= \frac{1}{2}(6x) + \frac{1}{2}(8)\\ &= \left(\frac{1}{2} \cdot 6\right)x + \left(\frac{1}{2} \cdot 8\right)\\ &= 3x + 4\end{aligned}$$

55. $$\begin{aligned}-\frac{1}{3}(3x - 9y) &= -\frac{1}{3}(3x) - \left(-\frac{1}{3}\right)(9y)\\ &= \left(-\frac{1}{3} \cdot 3\right)x - \left(-\frac{1}{3} \cdot 9\right)y\\ &= -1 \cdot x + 3 \cdot y\\ &= -x + 3y\end{aligned}$$

57. $$\begin{aligned}3(2r + 5) - 7 &= 3(2r) + 3(5) - 7\\ &= 6r + 15 + (-7)\\ &= 6r + 8\end{aligned}$$

59. $$\begin{aligned}-9(4x + 8) + 2 &= -9(4x) + (-9)(8) + 2\\ &= -36x - 72 + 2\\ &= -36x - 70\end{aligned}$$

61. $$\begin{aligned}-4(4x + 5) - 5 &= -4(4x) + (-4)(5) - 5\\ &= -16x + (-20) + (-5)\\ &= -16x - 25\end{aligned}$$

63. $4 \cdot 1 + 4 \cdot y = 4(1 + y)$

65. $11x + 11y = 11(x + y)$

67. $(-1) \cdot 5 + (-1) \cdot x = -1(5 + x) = -(5 + x)$

69. $30a + 30b = 30(a + b)$

71. $3 \cdot 5 = 5 \cdot 3$; commutative property of multiplication

73. $2 + (x + 5) = (2 + x) + 5$; associative property of addition

75. $9(3 + 7) = 9 \cdot 3 + 9 \cdot 7$; distributive property

77. $(4 \cdot y) \cdot 9 = 4 \cdot (y \cdot 9)$; associative property of multiplication

79. $0 + 6 = 6$; identity element of addition

81. $-4(y + 7) = -4 \cdot y + (-4) \cdot 7$; distributive property

83. $-4 \cdot (8 \cdot 3) = (8 \cdot -4) \cdot 3$; associative and commutative properties of multiplication

85.

Expression	Opposite	Reciprocal
8	-8	$\frac{1}{8}$

87.

Expression	Opposite	Reciprocal
x	$-x$	$\frac{1}{x}$

89.

Expression	Opposite	Reciprocal
$2x$	$-2x$	$\frac{1}{2x}$

91. No

93. Yes

95. Answers may vary

Chapter 1 Vocabulary Check

1. The symbols ≠, <, and > are called inequality symbols.
2. A mathematical statement that two expressions are equal is called an equation.
3. The absolute value of a number is the distance between that number and 0 on the number line.
4. A symbol used to represent a number is called a variable.
5. Two numbers that are the same distance from 0 but lie on opposite sides of 0 are called opposites.
6. The number in a fraction above the fraction bar is called the numerator.
7. A solution of an equation is a value for the variable that makes the equation a true statement.
8. Two numbers whose product is 1 are called reciprocals.
9. In 2^3, the 2 is called the base and the 3 is called the exponent.
10. The number in a fraction below the fraction bar is called the denominator.
11. Parentheses and brackets are examples of grouping symbols.
12. A set is a collection of objects.

Chapter 1 Review

1. $8 < 10$ since 8 is to the left of 10 on the number line.
2. $7 > 2$ since 7 is to the right of 2 on the number line.
3. $-4 > -5$ since -4 is to the right of -5 on the number line.
4. $\frac{12}{2} > -8$ since $6 > -8$.
5. $|-7| < |-8|$ since $7 < 8$.
6. $|-9| > -9$ since $9 > -9$.
7. $-|-1| = -1$ since $-1 = -1$.
8. $|-14| = -(-14)$ since $14 = 14$.
9. $1.2 > 1.02$ since 1.2 is to the right of 1.02 on the number line.
10. $-\frac{3}{2} < -\frac{3}{4}$ since $-\frac{3}{2}$ is to the left of $-\frac{3}{4}$ on the number line.
11. Four is greater than or equal to negative three is written as $4 \geq -3$.
12. Six is not equal to five is written as $6 \neq 5$.
13. 0.03 is less than 0.3 is written as $0.03 < 0.3$.
14. $400 > 155$ or $155 < 400$

15. a. The natural numbers are 1 and 3.

b. The whole numbers are 0, 1, and 3.

c. The integers are –6, 0, 1, and 3.

d. The rational numbers are –6, 0, 1, $1\frac{1}{2}$, 3, and 9.62.

e. The irrational number is π.

f. The real numbers are all numbers in the given set.

16. a. The natural numbers are 2 and 5.

b. The whole numbers are 2 and 5.

c. The integers are –3, 2, and 5.

d. The rational numbers are –3, –1.6, 2, 5, $\frac{11}{2}$, and 15.1.

e. The irrational numbers are $\sqrt{5}$ and 2π.

f. The real numbers are all numbers in the given set.

17. Look for the negative number with the greatest absolute value. The greatest loss was on Friday.

18. Look for the largest positive number. The greatest gain was on Wednesday.

19. $36 = 4 \cdot 9 = 2 \cdot 2 \cdot 3 \cdot 3$

20. $120 = 8 \cdot 15 = 2 \cdot 2 \cdot 2 \cdot 3 \cdot 5$

21. $\frac{8}{15} \cdot \frac{27}{30} = \frac{8 \cdot 27}{15 \cdot 30} = \frac{2 \cdot 4 \cdot 3 \cdot 3 \cdot 3}{3 \cdot 5 \cdot 2 \cdot 3 \cdot 5} = \frac{12}{25}$

22. $\frac{7}{8} \div \frac{21}{32} = \frac{7}{8} \cdot \frac{32}{21} = \frac{7 \cdot 32}{8 \cdot 21} = \frac{7 \cdot 8 \cdot 4}{8 \cdot 3 \cdot 7} = \frac{4}{3}$

23.
$$\begin{aligned} \frac{7}{15} + \frac{5}{6} &= \frac{7 \cdot 2}{15 \cdot 2} + \frac{5 \cdot 5}{6 \cdot 5} \\ &= \frac{14}{30} + \frac{25}{30} \\ &= \frac{14+25}{30} \\ &= \frac{39}{30} \\ &= \frac{3 \cdot 13}{3 \cdot 10} \\ &= \frac{13}{10} \end{aligned}$$

24.
$$\begin{aligned} \frac{3}{4} - \frac{3}{20} &= \frac{3 \cdot 5}{4 \cdot 5} - \frac{3}{20} \\ &= \frac{15}{20} - \frac{3}{20} \\ &= \frac{15-3}{20} \\ &= \frac{12}{20} \\ &= \frac{3 \cdot 4}{5 \cdot 4} \\ &= \frac{3}{5} \end{aligned}$$

25.
$$\begin{aligned} 2\frac{3}{4} + 6\frac{5}{8} &= \frac{11}{4} + \frac{53}{8} \\ &= \frac{11 \cdot 2}{4 \cdot 2} + \frac{53}{8} \\ &= \frac{22}{8} + \frac{53}{8} \\ &= \frac{22+53}{8} \\ &= \frac{75}{8} \\ &= 9\frac{3}{8} \end{aligned}$$

26. $7\frac{1}{6}-2\frac{2}{3}=\frac{43}{6}-\frac{8}{3}$
$=\frac{43}{6}-\frac{8\cdot 2}{3\cdot 2}$
$=\frac{43}{6}-\frac{16}{6}$
$=\frac{43-16}{6}$
$=\frac{27}{6}$
$=\frac{9\cdot 3}{2\cdot 3}$
$=\frac{9}{2}$
$=4\frac{1}{2}$

27. $5\div\frac{1}{3}=5\cdot\frac{3}{1}=15$

28. $2\cdot 8\frac{3}{4}=2\cdot\frac{35}{4}=\frac{2\cdot 35}{2\cdot 2}=\frac{35}{2}=17\frac{1}{2}$

29. $1-\frac{1}{6}-\frac{1}{4}=\frac{12}{12}-\frac{1\cdot 2}{6\cdot 2}-\frac{1\cdot 3}{4\cdot 3}$
$=\frac{12}{12}-\frac{2}{12}-\frac{3}{12}$
$=\frac{12-2-3}{12}$
$=\frac{7}{12}$

The unknown part is $\frac{7}{12}$.

30. $P=2l+2w$
$P=2\left(1\frac{1}{3}\right)+2\left(\frac{7}{8}\right)$
$=\frac{2}{1}\cdot\frac{4}{3}+\frac{2}{1}\cdot\frac{7}{8}$
$=\frac{8}{3}+\frac{14}{8}$
$=\frac{8\cdot 8}{3\cdot 8}+\frac{14\cdot 3}{8\cdot 3}$
$=\frac{64}{24}+\frac{42}{24}$
$=\frac{64+42}{24}$
$=\frac{106}{24}$
$=4\frac{10}{24}$
$=4\frac{5}{12}$ meters
$A=lw$
$A=1\frac{1}{3}\cdot\frac{7}{8}$
$=\frac{4}{3}\cdot\frac{7}{8}$
$=\frac{4\cdot 7}{3\cdot 2\cdot 4}$
$=\frac{7}{6}$
$=1\frac{1}{6}$ sq meters

31. P = the sum of the lengths of the sides
$P=\frac{5}{11}+\frac{8}{11}+\frac{3}{11}+\frac{3}{11}+\frac{2}{11}+\frac{5}{11}=\frac{26}{11}=2\frac{4}{11}$ in.
A = the sum of the two areas, each given by lw
$A=\frac{5}{11}\cdot\frac{5}{11}+\frac{3}{11}\cdot\frac{3}{11}=\frac{25}{121}+\frac{9}{121}=\frac{34}{121}$ sq in.

32. $7\frac{1}{2}-6\frac{1}{8}=\frac{15}{2}-\frac{49}{8}$
$=\frac{15\cdot 4}{2\cdot 4}-\frac{49}{8}$
$=\frac{60}{8}-\frac{49}{8}$
$=\frac{60-49}{8}$
$=\frac{11}{8}$
$=1\frac{3}{8}$ ft

33. $1\frac{1}{8}+1\frac{13}{16}=\frac{9}{8}+\frac{29}{16}$
$=\frac{9\cdot 2}{8\cdot 2}+\frac{29}{16}$
$=\frac{18}{16}+\frac{29}{16}$
$=\frac{18+29}{16}$
$=\frac{47}{16}$
$=2\frac{15}{16}$ lb

34. $1\frac{1}{2}+1\frac{11}{16}+1\frac{3}{4}+1\frac{5}{8}+\frac{11}{16}+1\frac{1}{8}$
$=\frac{3}{2}+\frac{27}{16}+\frac{7}{4}+\frac{13}{8}+\frac{11}{16}+\frac{9}{8}$
$=\frac{3\cdot 8}{2\cdot 8}+\frac{27}{16}+\frac{7\cdot 4}{4\cdot 4}+\frac{13\cdot 2}{8\cdot 2}+\frac{11}{16}+\frac{9\cdot 2}{8\cdot 2}$
$=\frac{24+27+28+26+11+18}{16}$
$=\frac{134}{16}$
$=8\frac{3}{8}$ lb

35. Total weight = weight of girls + weight of boys
$8\frac{3}{8}+2\frac{15}{16}=\frac{67}{8}+\frac{47}{16}$
$=\frac{67\cdot 2}{8\cdot 2}+\frac{47}{16}$
$=\frac{134+47}{16}$
$=\frac{181}{16}$
$=11\frac{5}{16}$ lb

36. Look for the largest number. Jioke weighed the most.

37. Look for the smallest number. Odera weighed the least.

38. $1\frac{13}{16}-\frac{11}{16}=\frac{29}{16}-\frac{11}{16}$
$=\frac{29-11}{16}$
$=\frac{18}{16}$
$=1\frac{2}{16}$
$=1\frac{1}{8}$ lb

39. $5\frac{1}{2}-1\frac{5}{8}=\frac{11}{2}-\frac{13}{8}$
$=\frac{11\cdot 4}{2\cdot 4}-\frac{13}{8}$
$=\frac{44-13}{8}$
$=\frac{31}{8}$
$=3\frac{7}{8}$ lb

40. $4\frac{5}{32}-1\frac{1}{8}=\frac{133}{32}-\frac{9}{8}$
$=\frac{133}{32}-\frac{9\cdot 4}{8\cdot 4}$
$=\frac{133-36}{32}$
$=\frac{97}{32}$
$=3\frac{1}{32}$ lb

41. $2^4=2\cdot 2\cdot 2\cdot 2=16$

42. $5^2=5\cdot 5=25$

43. $\left(\frac{2}{7}\right)^2=\frac{2}{7}\cdot\frac{2}{7}=\frac{4}{49}$

44. $\left(\frac{3}{4}\right)^3=\frac{3}{4}\cdot\frac{3}{4}\cdot\frac{3}{4}=\frac{27}{64}$

45. $6\cdot 3^2+2\cdot 8=6\cdot 9+2\cdot 8=54+16=70$

46. $68-5\cdot 2^3=68-5\cdot 8=68-40=28$

47. $3(1+2\cdot5)+4 = 3(1+10)+4$
$= 3(11)+4$
$= 33+4$
$= 37$

48. $8+3(2\cdot6-1) = 8+3(12-1)$
$= 8+3(11)$
$= 8+33$
$= 41$

49. $\frac{4+|6-2|+8^2}{4+6\cdot4} = \frac{4+|4|+64}{4+24}$
$= \frac{4+4+64}{4+24}$
$= \frac{72}{28}$
$= \frac{4\cdot18}{4\cdot7}$
$= \frac{18}{7}$

50. $5[3(2+5)-5] = 5[3(7)-5]$
$= 5[21-5]$
$= 5[16]$
$= 80$

51. The difference of twenty and twelve is equal to the product of two and four is written as $20 - 12 = 2 \cdot 4$.

52. The quotient of nine and two is greater than negative five is written as $\frac{9}{2} > -5$.

53. Let $x = 6$ and $y = 2$.
$2x + 3y = 2(6) + 3(2) = 12 + 6 = 18$

54. Let $x = 6$, $y = 2$, and $z = 8$.
$x(y+2z) = 6[2+2(8)] = 6[2+16] = 6[18] = 108$

55. Let $x = 6$, $y = 2$, and $z = 8$.
$\frac{x}{y}+\frac{z}{2y} = \frac{6}{2}+\frac{8}{2(2)} = \frac{6}{2}+\frac{8}{4} = 3+2 = 5$

56. Let $x = 6$ and $y = 2$.
$x^2-3y^2 = (6)^2-3(2)^2$
$= 36-3(4)$
$= 36-12$
$= 36+(-12)$
$= 24$

57. Let $a = 37$ and $b = 80$.
$180-a-b = 180-37-80$
$= 180+(-37)+(-80)$
$= 143+(-80)$
$= 63°$

58. Let $x = 3$.
$7x-3 = 18$
$7(3)-3 \stackrel{?}{=} 18$
$21-3 \stackrel{?}{=} 18$
$18 = 18$, true
3 is a solution to the equation.

59. Let $x = 1$.
$3x^2+4 = x-1$
$3(1)^2+4 \stackrel{?}{=} 1-1$
$3+4 \stackrel{?}{=} 0$
$7 = 0$, false
1 is not a solution to the equation.

60. The additive inverse of -9 is 9.

61. The additive inverse of $\frac{2}{3}$ is $-\frac{2}{3}$.

62. The additive inverse of $|-2|$ is -2 since $|-2| = 2$.

63. The additive inverse of $-|-7|$ is 7 since $-|-7| = -7$.

64. $-15 + 4 = -11$

65. $-6 + (-11) = -17$

66. $\frac{1}{16}+\left(-\frac{1}{4}\right) = \frac{1}{16}+\left(-\frac{1\cdot4}{4\cdot4}\right)$
$= \frac{1}{16}+\left(-\frac{4}{16}\right)$
$= -\frac{3}{16}$

67. $-8 + |-3| = -8 + 3 = -5$

68. $-4.6 + (-9.3) = -13.9$

69. $-2.8 + 6.7 = 3.9$

70. $-282 + 728 = 446$ feet

71. $6 - 20 = 6 + (-20) = -14$

72. $-3.1 - 8.4 = -3.1 + (-8.4) = -11.5$

73. $-6-(-11)=-6+11=5$

74. $4-15=4+(-15)=-11$

75. $$\begin{aligned}-21-16+3(8-2)&=-21+(-16)+3[8+(-2)]\\&=-21+(-16)+3[6]\\&=-21+(-16)+18\\&=-37+18\\&=-19\end{aligned}$$

76. $$\begin{aligned}\frac{11-(-9)+6(8-2)}{2+3\cdot4}&=\frac{11+9+6[8+(-2)]}{2+3\cdot4}\\&=\frac{11+9+6[6]}{2+3\cdot4}\\&=\frac{11+9+36}{2+12}\\&=\frac{56}{14}\\&=4\end{aligned}$$

77. Let $x=3$, $y=-6$, and $z=-9$.
$$\begin{aligned}2x^2-y+z&=2(3)^2-(-6)+(-9)\\&=2(9)+6+(-9)\\&=18+6+(-9)\\&=24+(-9)\\&=15\end{aligned}$$

78. Let $x=3$ and $y=-6$.
$$\begin{aligned}\frac{y-x+5x}{2x}&=\frac{y+4x}{2x}\\&=\frac{-6+4(3)}{2(3)}\\&=\frac{-6+12}{6}\\&=\frac{6}{6}\\&=1\end{aligned}$$

79. The multiplicative inverse of -6 is $-\frac{1}{6}$ since $-6\cdot-\frac{1}{6}=1.$

80. The multiplicative inverse of $\frac{3}{5}$ is $\frac{5}{3}$ since $\frac{3}{5}\cdot\frac{5}{3}=1.$

81. $6(-8)=-48$

82. $(-2)(-14)=28$

83. $\frac{-18}{-6}=3$

84. $\frac{42}{-3}=-14$

85. $\frac{4\cdot(-3)+(-8)}{2+(-2)}=\frac{-12+(-8)}{2+(-2)}=\frac{-20}{0}$
The expression is undefined.

86. $\frac{3(-2)^2-5}{-14}=\frac{3(4)-5}{-14}=\frac{12-5}{-14}=\frac{7}{-14}=-\frac{1}{2}$

87. $\frac{-6}{0}$ is undefined.

88. $\frac{0}{-2}=0$

89. $$\begin{aligned}-4^2-(-3+5)\div(-1)\cdot2&=-16-(2)\div(-1)\cdot2\\&=-16+2\cdot2\\&=-16+4\\&=-12\end{aligned}$$

90. $$\begin{aligned}-5^2-(2-20)\div(-3)\cdot3&=-25-(-18)\div(-3)\cdot3\\&=-25-6\cdot3\\&=-25-18\\&=-43\end{aligned}$$

91. Let $x=-5$ and $y=-2$.
$x^2-y^4=(-5)^2-(-2)^4=25-16=9$

92. Let $x=-5$ and $y=-2$.
$x^2-y^3=(-5)^2-(-2)^3=25-(-8)=25+8=33$

93. $\frac{-9+(-7)+1}{3}=\frac{-15}{3}=-5$
Her average score per round was 5 under par.

94. $\frac{-1+0+(-3)+0}{4}=\frac{-4}{4}=-1$
His average score per round was 1 under par.

95. $-6+5=5+(-6)$; commutative property of addition

96. $6\cdot1=6$; multiplicative identity property

97. $3(8 - 5) = 3 \cdot 8 + 3 \cdot (-5)$; distributive property

98. $4 + (-4) = 0$; additive inverse property

99. $2 + (3 + 9) = (2 + 3) + 9$; associative property of addition

100. $2 \cdot 8 = 8 \cdot 2$; commutative property of multiplication

101. $6(8 + 5) = 6 \cdot 8 + 6 \cdot 5$; distributive property

102. $(3 \cdot 8) \cdot 4 = 3 \cdot (8 \cdot 4)$; associative property of multiplication

103. $4 \cdot \frac{1}{4} = 1$; multiplicative inverse property

104. $8 + 0 = 8$; additive identity property

105. $5(y - 2) = 5(y) + 5(-2) = 5y - 10$

106. $-3(z + y) = -3(z) + (-3)(y) = -3z - 3y$

107. $-(7 - x + 4z) = (-1)(7) + (-1)(-x) + (-1)(4z) = -7 + x - 4z$

108. $\frac{1}{2}(6z - 10) = \frac{1}{2}(6z) + \frac{1}{2}(-10) = 3z - 5$

109. $-4(3x + 5) - 7 = -4(3x) + (-4)(5) - 7 = -12x - 20 - 7 = -12x - 27$

110. $-8(2y + 9) - 1 = -8(2y) + (-8)(9) - 1 = -16y - 72 - 1 = -16y - 73$

111. $-|-11| < |11.4|$ since $-|-11| = -11$ and $|11.4| = 11.4$.

112. $-1\frac{1}{2} > -2\frac{1}{2}$ since $-1\frac{1}{2}$ is to the right of $-2\frac{1}{2}$ on the number line.

113. $-7.2 + (-8.1) = -15.3$

114. $14 - 20 = 14 + (-20) = -6$

115. $4(-20) = -80$

116. $\frac{-20}{4} = -5$

117. $-\frac{4}{5}\left(\frac{5}{16}\right) = -\frac{4}{16} = -\frac{1}{4}$

118. $-0.5(-0.3) = 0.15$

119. $8 \div 2 \cdot 4 = 4 \cdot 4 = 16$

120. $(-2)^4 = (-2)(-2)(-2)(-2) = 16$

121. $\frac{-3 - 2(-9)}{-15 - 3(-4)} = \frac{-3 + 18}{-15 + 12} = \frac{15}{-3} = -5$

122. $5 + 2[(7 - 5)^2 + (1 - 3)] = 5 + 2[2^2 + (-2)] = 5 + 2[4 + (-2)] = 5 + 2[2] = 5 + 4 = 9$

123. $-\frac{5}{8} \div \frac{3}{4} = -\frac{5}{8} \cdot \frac{4}{3} = -\frac{20}{24} = -\frac{5}{6}$

124. $\frac{-15 + (-4)^2 + |-9|}{10 - 2 \cdot 5} = \frac{-15 + 16 + 9}{10 - 10} = \frac{1 + 9}{0}$ is undefined.

Chapter 1 Test

1. The absolute value of negative seven is greater than five is written as $|-7| > 5$.

2. The sum of nine and five is greater than or equal to four is written as $(9 + 5) \geq 4$.

3. $-13 + 8 = -5$

4. $-13 - (-2) = -13 + 2 = -11$

5. $12 \div 4 \cdot 3 - 6 \cdot 2 = 3 \cdot 3 - 6 \cdot 2 = 9 - 12 = -3$

6. $(13)(-3) = -39$

7. $(-6)(-2) = 12$

8. $\frac{|-16|}{-8} = \frac{16}{-8} = -2$

9. $\frac{-8}{0}$ is undefined.

10. $\frac{|-6| + 2}{5 - 6} = \frac{6 + 2}{5 + (-6)} = \frac{8}{-1} = -8$

11. $\frac{1}{2}-\frac{5}{6}=\frac{1\cdot 3}{2\cdot 3}-\frac{5}{6}=\frac{3}{6}-\frac{5}{6}=\frac{3-5}{6}=\frac{-2}{6}=-\frac{1}{3}$

12. $-1\frac{1}{8}+5\frac{3}{4}=-\frac{9}{8}+\frac{23}{4}$
$=-\frac{9}{8}+\frac{2\cdot 23}{2\cdot 4}$
$=-\frac{9}{8}+\frac{46}{8}$
$=\frac{-9+46}{8}$
$=\frac{37}{8}$
$=4\frac{5}{8}$

13. $(2-6)\div\frac{-2-6}{-3-1}-\frac{1}{2}=(2-6)\div\frac{-8}{-4}-\frac{1}{2}$
$=-4\div 2-\frac{1}{2}$
$=-2-\frac{1}{2}$
$=-2\frac{1}{2}$

14. $3(-4)^2-80=3(16)-80=48+(-80)=-32$

15. $6[5+2(3-8)-3]=6\{5+2[3+(-8)]+(-3)\}$
$=6\{5+2[-5]+(-3)\}$
$=6\{5+(-10)+(-3)\}$
$=6\{-5+(-3)\}$
$=6\{-8\}$
$=-48$

16. $\frac{-12+3\cdot 8}{4}=\frac{-12+24}{4}=\frac{12}{4}=3$

17. $\frac{(-2)(0)(-3)}{-6}=\frac{0(-3)}{-6}=\frac{0}{-6}=0$

18. $-3>-7$ since -3 is to the right of -7 on the number line.

19. $4>-8$ since 4 is to the right of -8 on the number line.

20. $2<|-3|$ since $2<3$.

21. $|-2|=-1-(-3)$ since $|-2|=2$ and $-1-(-3)=-1+3=2$.

22. $2221 < 10{,}993$ or $10{,}993 > 2221$

23. **a.** The natural numbers are 1 and 7.

b. The whole numbers are 0, 1 and 7.

c. The integers are -5, -1, 0, 1, and 7.

d. The rational numbers are -5, -1, $\frac{1}{4}$, 0, 1, 7, and 11.6.

e. The irrational numbers are $\sqrt{7}$ and 3π.

f. The real numbers are all numbers in the given set.

24. Let $x=6$ and $y=-2$.
$x^2+y^2=(6)^2+(-2)^2=36+4=40$

25. Let $x=6$, $y=-2$ and $z=-3$.
$x+yz=6+(-2)(-3)=6+6=12$

26. Let $x=6$ and $y=-2$.
$2+3x-y=2+3(6)-(-2)$
$=2+18+2$
$=20+2$
$=22$

27. Let $x=6$, $y=-2$ and $z=-3$.
$\frac{y+z-1}{x}=\frac{-2+(-3)-1}{6}=\frac{-5+(-1)}{6}=\frac{-6}{6}=-1$

28. $8+(9+3)=(8+9)+3$; associative property of addition

29. $6\cdot 8=8\cdot 6$; commutative property of multiplication

30. $-6(2+4)=-6\cdot 2+(-6)\cdot 4$; distributive property

31. $\frac{1}{6}(6)=1$; multiplicative inverse property

32. The opposite of -9 is 9.

33. The reciprocal of $-\frac{1}{3}$ is -3.

34. Look for the negative number that has the greatest absolute value. The second down had the greatest loss of yardage.

35. Gains: 5, 29
Losses: −10, −2

$$\begin{aligned}\text{Total gain or loss} &= 5+(-10)+(-2)+29\\ &= (-5)+(-2)+29\\ &= -7+29\\ &= 22 \text{ yards gained}\end{aligned}$$

Yes, they scored a touchdown.

36. Since $-14 + 31 = 17$, the temperature at noon was 17°.

37. $356 + 460 + (-166) = 650$
The net income was $650 million.

38. Change in value per share = −1.50
Change in total value = 280(−1.50) = −420
Total loss of $420

Chapter 2

Section 2.1

Practice Exercises

1. a. The numerical coefficient of t is 1, since t is $1t$.

   b. The numerical coefficient of $-7x$ is -7.

   c. The numerical coefficient of $-\frac{w}{5}$ is $-\frac{1}{5}$, since $-\frac{w}{5}$ means $-\frac{1}{5}\cdot w$.

   d. The numerical coefficient of $43x^4$ is 43.

   e. The numerical coefficient of $-b$ is -1, since $-b$ is $-1b$.

2. a. $-4xy$ and $5yx$ are like terms, since $xy = yx$ by the commutative property.

   b. $5q$ and $-3q^2$ are unlike terms, since the exponents on q are not the same.

   c. $3ab^2$, $-2ab^2$, and $43ab^2$ are like terms, since each variable and its exponent match.

   d. y^5 and $\frac{y^5}{2}$ are like terms, since the exponents on y are the same.

3. a. $4x^2 + 3x^2 = (4+3)x^2 = 7x^2$

   b. $-3y + y = -3y + 1y = (-3 + 1)y = -2y$

   c. $5x - 3x^2 + 8x^2 = 5x + (-3+8)x^2 = 5x + 5x^2$

4. a. $3y + 8y - 7 + 2 = (3+8)y + (-7+2) = 11y - 5$

   b. $$\begin{aligned} 6x - 3 - x - 3 &= 6x - 1x + (-3-3) \\ &= (6-1)x + (-3-3) \\ &= 5x - 6 \end{aligned}$$

   c. $\frac{3}{4}t - t = \frac{3}{4}t - 1t = \left(\frac{3}{4} - 1\right)t = -\frac{1}{4}t$

   d. $$\begin{aligned} 9y + 3.2y + 10 + 3 &= (9 + 3.2)y + (10+3) \\ &= 12.2y + 13 \end{aligned}$$

   e. $5z - 3z^4$
   These two terms cannot be combined because they are unlike terms.

5. a. $3(2x - 7) = 3(2x) + 3(-7) = 6x - 21$

   b. $$\begin{aligned} &-5(3x - 4z - 5) \\ &= -5(3x) + (-5)(-4z) + (-5)(-5) \\ &= -15x + 20z + 25 \end{aligned}$$

   c. $$\begin{aligned} &-(2x - y + z - 2) \\ &= -1(2x - y + z - 2) \\ &= -1(2x) - 1(-y) - 1(z) - 1(-2) \\ &= -2x + y - z + 2 \end{aligned}$$

6. a. $4(9x + 1) + 6 = 36x + 4 + 6 = 36x + 10$

   b. $$\begin{aligned} -7(2x-1) - (6 - 3x) &= -14x + 7 - 6 + 3x \\ &= -11x + 1 \end{aligned}$$

   c. $8 - 5(6x + 5) = 8 - 30x - 25 = -30x - 17$

7. "Subtract $7x - 1$ from $2x + 3$" translates to
   $(2x+3) - (7x-1) = 2x + 3 - 7x + 1 = -5x + 4$

8. a. 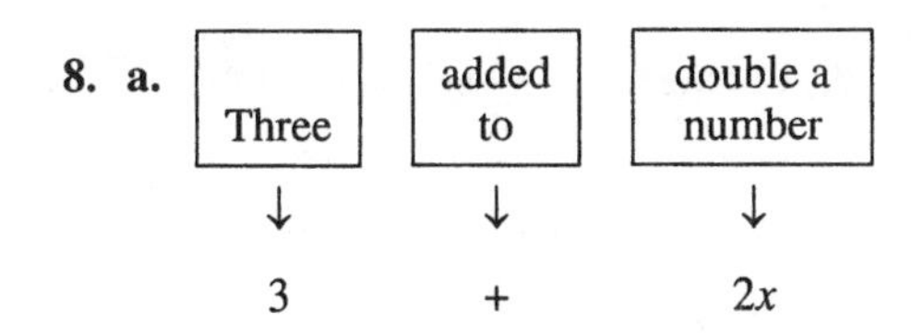

b.

the sum of 5 and a number	subtracted from	six	
↓	↓	↓	
$(5 + x)$	$-$	6	$= 5 + x - 6$

$(5 + x) - 6 = 5 + x - 6 = x - 1$

c.

two	times	the sum of 3 and a number	increased by	4
↓	↓	↓	↓	↓
2	$\cdot$	$(3 + x)$	$+$	4

$2(3 + x) + 4 = 6 + 2x + 4 = 2x + 10$

d.

a number	added to	half the number	added to	5 times the number
↓	↓	↓	↓	↓
x	$+$	$\frac{1}{2}x$	$+$	$5x$

$$x + \frac{1}{2}x + 5x = \frac{13}{2}x$$

Vocabulary and Readiness Check

1. $23y^2 + 10y - 6$ is called an <u>expression</u> while $23y^2$, $10y$, and -6 are each called a <u>term</u>.

2. To simplify $x + 4x$, we <u>combine like terms</u>.

3. The term y has an understood <u>numerical coefficient</u> of 1.

4. The terms $7z$ and $7y$ are <u>unlike</u> terms and the terms $7z$ and $-z$ are <u>like</u> terms.

5. For the term $-\frac{1}{2}xy^2$, the number $-\frac{1}{2}$ is the <u>numerical coefficient</u>.

6. $5(3x - y)$ equals $15x - 5y$ by the <u>distributive</u> property.

7. The numerical coefficient of $-7y$ is -7.

8. The numerical coefficient of $3x$ is 3.

9. The numerical coefficient of x is 1.

10. The numerical coefficient of $-y$ is -1.

11. The numerical coefficient of $-\frac{5y}{3}$ is $-\frac{5}{3}$.

12. The numerical coefficient of $-\frac{2}{3}z$ is $-\frac{2}{3}$.

13. $5y$ and y are like terms.

14. $-2x^2y$ and $6xy$ are unlike terms.

15. $2z$ and $3z^2$ are unlike terms.

16. b^2a and $-\frac{7}{8}ab^2$ are like terms.

Exercise Set 2.1

1. $7y + 8y = (7 + 8)y = 15y$

3. $8w - w + 6w = (8 - 1 + 6)w = 13w$

5. $3b-5-10b-4 = 3b-10b-5-4$
$= (3-10)b-9$
$= -7b-9$

7. $m-4m+2m-6 = (1-4+2)m-6 = -m-6$

9. $5g-3-5-5g = (5g-5g)+(-3-5)$
$= (5-5)g+(-8)$
$= 0g-8$
$= -8$

11. $6.2x-4+x-1.2 = 6.2x+x-4-1.2$
$= (6.2+1)x-5.2$
$= 7.2x-5.2$

13. $6x-5x+x-3+2x = 6x-5x+x+2x-3$
$= (6-5+1+2)x-3$
$= 4x-3$

15. $7x^2+8x^2-10x^2 = (7+8-10)x^2 = 5x^2$

17. $6x+0.5-4.3x-0.4x+3$
$= 6x-4.3x-0.4x+0.5+3$
$= (6-4.3-0.4)x+(0.5+3)$
$= 1.3x+3.5$

19. Answers may vary

21. $5(y - 4) = 5(y) - 5(4) = 5y - 20$

23. $-2(x + 2) = -2(x) + (-2)(2) = -2x - 4$

25. $7(d - 3) + 10 = 7d - 21 + 10 = 7d - 11$

27. $-5(2x-3y+6) = -5(2x)-(-5)(3y)+(-5)(6)$
$= -10x+15y-30$

29. $-(3x - 2y + 1) = -3x + 2y - 1$

31. $5(x+2)-(3x-4) = 5x+10-3x+4$
$= 2x+14$

33. $6x+7$ added to $4x-10$
↓ ↓ ↓
$(6x+7) + (4x-10) = 6x+4x+7-10$
$= 10x-3$

35. $3x-8$ minus $7x+1$
↓ ↓ ↓
$(3x-8) - (7x+1) = 3x-8-7x-1$
$= 3x-7x-8-1$
$= -4x-9$

37. $m-9$ minus $5m-6$
↓ ↓ ↓
$(m-9) - (5m-6) = m-9-5m+6$
$= m-5m-9+6$
$= -4m-3$

39. $2k - k - 6 = (2 - 1)k - 6 = k - 6$

41. $-9x+4x+18-10x = -9x+4x-10x+18$
$= (-9+4-10)x+18$
$= -15x+18$

43. $-4(3y-4)+12y = -4(3y)-(-4)(4)+12y$
$= -12y+16+12y$
$= -12y+12y+16$
$= 16$

45. $3(2x-5)-5(x-4) = 6x-15-5x+20 = x+5$

47. $-2(3x-4)+7x-6 = -6x+8+7x-6 = x+2$

49. $5k - (3k - 10) = 5k - 3k + 10 = 2k + 10$

51. $(3x+4)-(6x-1) = 3x+4-6x+1 = -3x+5$

53. $3.4m-4-3.4m-7 = 3.4m-3.4m-4-7 = -11$

55. $\frac{1}{3}(7y-1)+\frac{1}{6}(4y+7) = \frac{7}{3}y-\frac{1}{3}+\frac{4}{6}y+\frac{7}{6}$
$= \frac{7}{3}y+\frac{2}{3}y-\frac{1}{3}+\frac{7}{6}$
$= \frac{9}{3}y-\frac{2}{6}+\frac{7}{6}$
$= 3y+\frac{5}{6}$

57. $2 + 4(6x - 6) = 2 + 24x - 24 = -22 + 24x$

59. $0.5(m+2)+0.4m = 0.5m+1+0.4m = 0.9m+1$

61. $10 - 3(2x + 3y) = 10 - 6x - 9y$

63. $6(3x-6)-2(x+1)-17x = 18x-36-2x-2-17x$
$= 18x-2x-17x-36-2$
$= -x-38$

65. $\frac{1}{2}(12x-4)-(x+5) = 6x-2-x-5 = 5x-7$

67.

twice a number	decreased by	4
↓	↓	↓
$2x$	$-$	4

69.

seven	added to	double a number
↓	↓	↓
7	+	$2x$

71.

three-fourths of a number	increased by	12
↓	↓	↓
$\frac{3}{4}x$	+	12

73.

5 times a number	added to	−2	added to	7 times a number
↓	↓	↓	↓	↓
$5x$	+	−2	+	$7x$

$5x + (-2) + 7x = 12x - 2$

75.

8	times	the sum of a number and 6
↓	↓	↓
8	·	$(x+6)$

$8(x + 6) = 8x + 48$

77.

double a number	minus	the sum of the number and 10
↓	↓	↓
$2x$	$-$	$(x+10)$

$2x-(x+10)=2x-x-10=x-10$

79.

7	multiplied by	the quotient of a number and 6
↓	↓	↓
7	$\cdot$	$\frac{x}{6}$

$7\left(\frac{x}{6}\right)=\frac{7x}{6}$

81.

2	added to	3 times a number	added to	-9	added to	4 times a number
↓	↓	↓	↓	↓	↓	↓
2	+	$3x$	+	-9	+	$4x$

$2+3x+(-9)+4x=7x-7$

83. $y-x^2=3-(-1)^2=3-1=2$

85. $a-b^2=2-(-5)^2=2-25=-23$

87. $yz-y^2=(-5)(0)-(-5)^2=0-25=-25$

89. $5x+(4x-1)+5x+(4x-1)=5x+4x-1+5x+4x-1$
$=18x-2$
The perimeter is $(18x-2)$ feet.

91. 1 cone + 1 cylinder $\stackrel{?}{=}$ 3 cubes
1 cube + 2 cubes $\stackrel{?}{=}$ 3 cubes
3 cubes = 3 cubes: Balanced

93. 2 cylinders + 1 cube $\stackrel{?}{=}$ 3 cones + 2 cubes
$2\cdot 2$ cubes + 1 cube $\stackrel{?}{=}$ 3 cubes + 2 cubes
4 cubes + 1 cube $\stackrel{?}{=}$ 3 cubes + 2 cubes
5 cubes = 5 cubes: Balanced

95. Answers may vary

97. $12(x+2)+(3x-1)=12x+24+3x-1=15x+23$
The total length is $(15x+23)$ inches.

99. $5b^2c^3+8b^3c^2-7b^3c^2=5b^2c^3+b^3c^2$

101. $3x-(2x^2-6x)+7x^2 = 3x-2x^2+6x+7x^2$
$= 5x^2+9x$

103. $-(2x^2y+3z)+3z-5x^2y$
$= -2x^2y-3z+3z-5x^2y$
$= -7x^2y$

Section 2.2

Practice Exercises

1. $x+3=-5$
$x+3-3=-5-3$
$x=-8$
Check: $x+3=-5$
$-8+3 \stackrel{?}{=} -5$
$-5=-5$
The solution is −8.

2. $y-0.3=-2.1$
$y-0.3+0.3=-2.1+0.3$
$y=-1.8$
Check: $y-0.3=-2.1$
$-1.8-0.3 \stackrel{?}{=} -2.1$
$-2.1=-2.1$
The solution is −1.8.

3. $\frac{2}{5}=x+\frac{3}{10}$
$\frac{2}{5}-\frac{3}{10}=x+\frac{3}{10}-\frac{3}{10}$
$\frac{2}{2}\cdot\frac{2}{5}-\frac{3}{10}=x$
$\frac{4}{10}-\frac{3}{10}=x$
$\frac{1}{10}=x$
Check: $\frac{2}{5}=x+\frac{3}{10}$
$\frac{2}{5} \stackrel{?}{=} \frac{1}{10}+\frac{3}{10}$
$\frac{2}{5}=\frac{2}{5}$
The solution is $\frac{1}{10}$.

4. $4t+7=5t-3$
$4t+7-4t=5t-3-4t$
$7=t-3$
$7+3=t-3+3$
$10=t$
Check: $4t+7=5t-3$
$4(10)+7 \stackrel{?}{=} 5(10)-3$
$40+7 \stackrel{?}{=} 50-3$
$47=47$
The solution is 10.

5. $8x-5x-3+9=x+x+3-7$
$3x+6=2x-4$
$3x+6-2x=2x-4-2x$
$x+6=-4$
$x+6-6=-4-6$
$x=-10$
Check:
$8x-5x-3+9=x+x+3-7$
$8(-10)-5(-10)-3+9 \stackrel{?}{=} -10+(-10)+3-7$
$-80+50-3+9 \stackrel{?}{=} -10+(-10)+3-7$
$-24=-24$
The solution is −10.

6. $4(2a-3)-(7a+4)=2$
$4(2a)+4(-3)-7a-4=2$
$8a-12-7a-4=2$
$a-16=2$
$a-16+16=2+16$
$a=18$
Check by replacing a with 18 in the original equation.

7. $12-x=20$
$12-x-12=20-12$
$-x=8$
$x=-8$
Check: $12-x=20$
$12-(-8) \stackrel{?}{=} 20$
$20=20$
The solution is −8.

8. **a.** The other number is 9 − 2 = 7.

b. The other number is $9-x$.

c. The other piece has length $(9-x)$ feet.

9. The speed of the French TGV is $(s-3.8)$ mph.

Vocabulary and Readiness Check

1. The difference between an equation and an expression is that an equation contains an equal sign, whereas an expression does not.

2. Equivalent equations are equations that have the same solution.

3. A value of the variable that makes the equation a true statement is called a solution of the equation.

4. The process of finding the solution of an equation is called solving the equation for the variable.

5. By the addition property of equality, $x = -2$ and $x + 10 = -2 + 10$ are equivalent equations.

6. The equations $x = \frac{1}{2}$ and $\frac{1}{2} = x$ are equivalent equations. The statement is true.

Exercise Set 2.2

1.
$$\begin{aligned} x+7&=10 \\ x+7-7&=10-7 \\ x&=3 \end{aligned}$$
Check: $x+7=10$
$$3+7 \stackrel{?}{=} 10$$
$$10=10$$
The solution is 3.

3.
$$\begin{aligned} x-2&=-4 \\ x-2+2&=-4+2 \\ x&=-2 \end{aligned}$$
Check: $x-2=-4$
$$-2-2 \stackrel{?}{=} -4$$
$$-4=-4$$
The solution is −2.

5.
$$\begin{aligned} -2&=t-5 \\ -2+5&=t-5+5 \\ 3&=t \end{aligned}$$
Check: $-2=t-5$
$$-2 \stackrel{?}{=} 3-5$$
$$-2=-2$$
The solution is 3.

7.
$$\begin{aligned} r-8.6&=-8.1 \\ r-8.6+8.6&=-8.1+8.6 \\ r&=0.5 \end{aligned}$$
Check: $x-8.6=-8.1$
$$0.5-8.6 \stackrel{?}{=} -8.1$$
$$-8.1=-8.1$$
The solution is 0.5.

9.
$$\begin{aligned} \frac{3}{4}&=\frac{1}{3}+f \\ \frac{3}{4}-\frac{1}{3}&=\frac{1}{3}-\frac{1}{3}+f \\ \frac{9}{12}-\frac{4}{12}&=f \\ \frac{5}{12}&=f \end{aligned}$$
Check: $\frac{3}{4}=\frac{1}{3}+f$
$$\frac{3}{4} \stackrel{?}{=} \frac{1}{3}+\frac{5}{12}$$
$$\frac{3}{4} \stackrel{?}{=} \frac{4}{12}+\frac{5}{12}$$
$$\frac{3}{4} \stackrel{?}{=} \frac{9}{12}$$
$$\frac{3}{4}=\frac{3}{4}$$
The solution is $\frac{5}{12}$.

11.
$$\begin{aligned} 5b-0.7&=6b \\ 5b-5b-0.7&=6b-5b \\ -0.7&=b \end{aligned}$$
Check: $5b-0.7=6b$
$$5(-0.7)-0.7 \stackrel{?}{=} 6(-0.7)$$
$$-3.5-0.7 \stackrel{?}{=} -4.2$$
$$-4.2=-4.2$$
The solution is −0.7.

13.
$$\begin{aligned} 7x-3&=6x \\ 7x-6x-3&=6x-6x \\ x-3&=0 \\ x-3+3&=0+3 \\ x&=3 \end{aligned}$$
Check: $7x-3=6x$
$$7(3)-3 \stackrel{?}{=} 6(3)$$
$$21-3 \stackrel{?}{=} 18$$
$$18=18$$
The solution is 3.

15. Answers may vary

17. $7x + 2x = 8x - 3$
$9x = 8x - 3$
$9x - 8x = 8x - 8x - 3$
$x = -3$
Check: $7x + 2x = 8x - 3$
$7(-3) + 2(-3) \stackrel{?}{=} 8(-3) - 3$
$-21 - 6 \stackrel{?}{=} -24 - 3$
$-27 = -27$
The solution is -3.

19. $\frac{5}{6}x + \frac{1}{6}x = -9$
$\frac{6}{6}x = -9$
$x = -9$
Check: $\frac{5}{6}x + \frac{1}{6}x = -9$
$\frac{5}{6}(-9) + \frac{1}{6}(-9) \stackrel{?}{=} -9$
$-\frac{45}{6} - \frac{9}{6} \stackrel{?}{=} -9$
$-9 = -9$
The solution is -9.

21. $2y + 10 = 5y - 4y$
$2y + 10 = y$
$2y - y + 10 = y - y$
$y + 10 = 0$
$y + 10 - 10 = 0 - 10$
$y = -10$
Check: $2y + 10 = 5y - 4y$
$2(-10) + 10 \stackrel{?}{=} 5(-10) - 4(-10)$
$-20 + 10 \stackrel{?}{=} -50 + 40$
$-10 = -10$
The solution is -10.

23. $-5(n - 2) = 8 - 4n$
$-5n + 10 = 8 - 4n$
$-5n + 5n + 10 = 8 - 4n + 5n$
$10 = 8 + n$
$10 - 8 = 8 - 8 + n$
$2 = n$
Check: $-5(n - 2) = 8 - 4n$
$-5(2 - 2) \stackrel{?}{=} 8 - 4(2)$
$-5(0) \stackrel{?}{=} 8 - 8$
$0 = 0$
The solution is 2.

25. $\frac{3}{7}x + 2 = -\frac{4}{7}x - 5$
$\frac{3}{7}x + 2 + \frac{4}{7}x = -\frac{4}{7}x - 5 + \frac{4}{7}x$
$\frac{7}{7}x + 2 = -5$
$x + 2 - 2 = -5 - 2$
$x = -7$
Check: $\frac{3}{7}x + 2 = -\frac{4}{7}x - 5$
$\frac{3}{7}(-7) + 2 \stackrel{?}{=} -\frac{4}{7}(-7) - 5$
$-3 + 2 \stackrel{?}{=} 4 - 5$
$-1 = -1$
The solution is -7.

27. $5x - 6 = 6x - 5$
$5x - 6 - 5x = 6x - 5 - 5x$
$-6 = x - 5$
$-6 + 5 = x - 5 + 5$
$-1 = x$
Check: $5x - 6 = 6x - 5$
$5(-1) - 6 \stackrel{?}{=} 6(-1) - 5$
$-5 - 6 \stackrel{?}{=} -6 - 5$
$-11 = -11$
The solution is -1.

29. $8y + 2 - 6y = 3 + y - 10$
$2y + 2 = y - 7$
$2y - y + 2 = y - y - 7$
$y + 2 = -7$
$y + 2 - 2 = -7 - 2$
$y = -9$
Check: $8y + 2 - 6y = 3 + y - 10$
$8(-9) + 2 - 6(-9) \stackrel{?}{=} 3 + (-9) - 10$
$-72 + 2 + 54 \stackrel{?}{=} -16$
$-16 = -16$
The solution is -9.

31. $-3(x - 4) = -4x$
$-3x + 12 = -4x$
$-3x + 12 + 3x = -4x + 3x$
$12 = -x$
$x = -12$
Check: $-3(x - 4) = -4x$
$-3(-12 - 4) \stackrel{?}{=} -4(-12)$
$-3(-16) \stackrel{?}{=} 48$
$48 = 48$
The solution is -12.

33.
$$\frac{3}{8}x-\frac{1}{6}=-\frac{5}{8}x-\frac{2}{3}$$
$$\frac{3}{8}x+\frac{5}{8}x-\frac{1}{6}=-\frac{5}{8}x+\frac{5}{8}x-\frac{2}{3}$$
$$\frac{8}{8}x-\frac{1}{6}=-\frac{2}{3}$$
$$x-\frac{1}{6}+\frac{1}{6}=-\frac{2}{3}+\frac{1}{6}$$
$$x=-\frac{4}{6}+\frac{1}{6}$$
$$x=-\frac{3}{6}$$
$$x=-\frac{1}{2}$$

Check: $\frac{3}{8}x-\frac{1}{6}=-\frac{5}{8}x-\frac{2}{3}$
$$\frac{3}{8}\left(-\frac{1}{2}\right)-\frac{1}{6}\stackrel{?}{=}-\frac{5}{8}\left(-\frac{1}{2}\right)-\frac{2}{3}$$
$$-\frac{3}{16}-\frac{1}{6}\stackrel{?}{=}\frac{5}{16}-\frac{2}{3}$$
$$-\frac{9}{48}-\frac{8}{48}\stackrel{?}{=}\frac{15}{48}-\frac{32}{48}$$
$$-\frac{17}{48}=-\frac{17}{48}$$

The solution is $-\frac{1}{2}$.

35.
$$2(x-4)=x+3$$
$$2x-8=x+3$$
$$2x-x-8=x-x+3$$
$$x-8=3$$
$$x-8+8=3+8$$
$$x=11$$

Check: $2(x-4)=x+3$
$$2(11-4)\stackrel{?}{=}11+3$$
$$2(7)\stackrel{?}{=}14$$
$$14=14$$

The solution is 11.

37.
$$3(n-5)-(6-2n)=4n$$
$$3n-15-6+2n=4n$$
$$5n-21=4n$$
$$5n-4n-21=4n-4n$$
$$n-21=0$$
$$n-21+21=0+21$$
$$n=21$$

Check: $3(n-5)-(6-2n)=4n$
$$3(21-5)-(6-2(21))\stackrel{?}{=}4(21)$$
$$3(16)-(6-42)\stackrel{?}{=}84$$
$$48-(-36)\stackrel{?}{=}84$$
$$84=84$$

The solution is 21.

39.
$$-2(x+6)+3(2x-5)=3(x-4)+10$$
$$-2x-12+6x-15=3x-12+10$$
$$4x-27=3x-2$$
$$4x-3x-27=3x-3x-2$$
$$x-27=-2$$
$$x-27+27=-2+27$$
$$x=25$$

Check: $-2(x+6)+3(2x-5)=3(x-4)+10$
$$-2(25+6)+3(2(25)-5)\stackrel{?}{=}3(25-4)+10$$
$$-2(31)+3(50-5)\stackrel{?}{=}3(21)+10$$
$$-62+3(45)\stackrel{?}{=}63+10$$
$$-62+135\stackrel{?}{=}73$$
$$73=73$$

The solution is 25.

41.
$$-11=3+x$$
$$-11-3=3+x-3$$
$$-14=x$$

43.
$$x-\frac{2}{5}=-\frac{3}{20}$$
$$x-\frac{2}{5}+\frac{2}{5}=-\frac{3}{20}+\frac{2}{5}$$
$$x=-\frac{3}{20}+\frac{8}{20}$$
$$x=\frac{5}{20}$$
$$x=\frac{1}{4}$$

45.
$$3x-6=2x+5$$
$$3x-2x-6=2x-2x+5$$
$$x-6=5$$
$$x-6+6=5+6$$
$$x=11$$

47. $13x-9+2x-5=12x-1+2x$
$$15x-14=14x-1$$
$$15x-14-14x=14x-1-14x$$
$$x-14=-1$$
$$x-14+14=-1+14$$
$$x=13$$

49. $7(6+w)=6(2+w)$
$$42+7w=12+6w$$
$$42+7w-6w=12+6w-6w$$
$$42+w=12$$
$$42-42+w=12-42$$
$$w=-30$$

51. $n+4=3.6$
$$n+4-4=3.6-4$$
$$n=-0.4$$

53. $10-(2x-4)=7-3x$
$$10-2x+4=7-3x$$
$$14-2x=7-3x$$
$$14-2x+3x=7-3x+3x$$
$$14+x=7$$
$$14-14+x=7-14$$
$$x=-7$$

55. $\frac{1}{3}=x+\frac{2}{3}$
$$\frac{1}{3}-\frac{2}{3}=x+\frac{2}{3}-\frac{2}{3}$$
$$-\frac{1}{3}=x$$

57. $-6.5-4x-1.6-3x=-6x+9.8$
$$-8.1-7x=-6x+9.8$$
$$-8.1-7x+7x=-6x+7x+9.8$$
$$-8.1=x+9.8$$
$$-8.1-9.8=x+9.8-9.8$$
$$-17.9=x$$

59. $-3\left(x-\frac{1}{4}\right)=-4x$
$$-3x+\frac{3}{4}=-4x$$
$$-3x+4x+\frac{3}{4}=-4x+4x$$
$$x+\frac{3}{4}=0$$
$$x+\frac{3}{4}-\frac{3}{4}=0-\frac{3}{4}$$
$$x=-\frac{3}{4}$$

61. $7(m-2)-6(m+1)=-20$
$$7m-14-6m-6=-20$$
$$m-20=-20$$
$$m-20+20=-20+20$$
$$m=0$$

63. $0.8t+0.2(t-0.4)=1.75$
$$0.8t+0.2t-0.08=1.75$$
$$1t-0.08=1.75$$
$$t-0.08+0.08=1.75+0.08$$
$$t=1.83$$

65. The other number is $20-p$.

67. The length of the other piece is $(10-x)$ feet.

69. The supplement of the angle $x°$ is $(180-x)°$.

71. The length is $\left(m+1\frac{1}{2}\right)$ feet.

73. The area of the Sahara Desert is $7x$ square miles.

75. The reciprocal of $\frac{5}{8}$ is $\frac{8}{5}$ since $\frac{5}{8}\cdot\frac{8}{5}=1$.

77. The reciprocal of 2 is $\frac{1}{2}$ since $2\cdot\frac{1}{2}=1$.

79. The reciprocal of $-\frac{1}{9}$ is -9 since $-\frac{1}{9}\cdot -9=1$.

81. $\frac{3x}{3}=x$

83. $-5\left(-\frac{1}{5}y\right)=y$

85. $\frac{3}{5}\left(\frac{5}{3}x\right) = x$

87. $$\begin{aligned}180-[x+(2x+7)] &= 180-[x+2x+7]\\ &= 180-[3x+7]\\ &= 180-3x-7\\ &= 173-3x\end{aligned}$$
The third angle is $(173 - 3x)°$.

89. Answers may vary

91. $$\begin{aligned}x-4 &= -9\\ x-4+(4) &= -9+(4)\\ x &= -5\end{aligned}$$
The answer is 4.

93. Answers may vary

95. $$\begin{aligned}200+150+400+x &= 1000\\ 750+x &= 1000\\ 750-750+x &= 1000-750\\ x &= 250\end{aligned}$$
The fluid needed by the patient is 250 ml.

97. Answers may vary

99. Check $y = 1.2$: $8.13 + 5.85y = 20.05y - 8.91$
$$\begin{aligned}8.13+5.85(1.2) &\stackrel{?}{=} 20.05(1.2)-8.91\\ 8.13+7.02 &\stackrel{?}{=} 24.06-8.91\\ 15.15 &= 15.15\end{aligned}$$
Solution

101. Check $z = 4.8$:
$$\begin{aligned}7(z-1.7)+9.5 &= 5(z+3.2)-9.2\\ 7(4.8-1.7)+9.5 &\stackrel{?}{=} 5(4.8+3.2)-9.2\\ 7(3.1)+9.55 &\stackrel{?}{=} 5(8.0)-9.2\\ 21.7+9.55 &\stackrel{?}{=} 40.0-9.2\\ 31.2 &\neq 30.8\end{aligned}$$
Not a solution

Section 2.3

Practice Exercises

1. $$\begin{aligned}\frac{4}{5}x &= 16\\ \frac{5}{4}\cdot\frac{4}{5}x &= \frac{5}{4}\cdot 16\\ \left(\frac{5}{4}\cdot\frac{4}{5}\right)x &= \frac{5}{4}\cdot 16\\ 1x &= 20\\ x &= 20\end{aligned}$$

Check: $$\begin{aligned}\frac{4}{5}x &= 16\\ \frac{4}{5}\cdot 20 &\stackrel{?}{=} 16\\ 16 &= 16\end{aligned}$$
The solution is 20.

2. $$\begin{aligned}8x &= -96\\ \frac{8x}{8} &= \frac{-96}{8}\\ x &= -12\end{aligned}$$
Check: $$\begin{aligned}8x &= -96\\ 8(-12) &\stackrel{?}{=} -96\\ -96 &= -96\end{aligned}$$
The solution is -12.

3. $$\begin{aligned}\frac{x}{5} &= 13\\ 5\cdot\frac{x}{5} &= 5\cdot 13\\ x &= 65\end{aligned}$$
Check: $$\begin{aligned}\frac{x}{5} &= 13\\ \frac{65}{5} &\stackrel{?}{=} 13\\ 13 &= 13\end{aligned}$$
The solution is 65.

4. $$\begin{aligned}2.7x &= 4.05\\ \frac{2.7x}{2.7} &= \frac{4.05}{2.7}\\ x &= 1.5\end{aligned}$$
The solution is 1.5.
Check by replacing x with 1.5 in the original equation.

5. $$\begin{aligned}-\frac{5}{3}x &= \frac{4}{7}\\ -\frac{3}{5}\cdot-\frac{5}{3}x &= -\frac{3}{5}\cdot\frac{4}{7}\\ x &= -\frac{12}{35}\end{aligned}$$

Check by replacing x with $-\frac{12}{35}$ in the original equation. The solution is $-\frac{12}{35}$.

6.
$$-y+3=-8$$
$$-y+3-3=-8-3$$
$$-y=-11$$
$$\frac{-y}{-1}=\frac{-11}{-1}$$
$$y=11$$
To check, replace y with 11 in the original equation. The solution is 11.

7. $6b-11b=18+2b-6+9$
$$-5b=21+2b$$
$$-5b-2b=21+2b-2b$$
$$-7b=21$$
$$\frac{-7b}{-7}=\frac{21}{-7}$$
$$b=-3$$
Check by replacing b with -3 in the original equation. The solution is -3.

8.
$$10x-4=7x+14$$
$$10x-4-7x=7x+14-7x$$
$$3x-4=14$$
$$3x-4+4=14+4$$
$$3x=18$$
$$\frac{3x}{3}=\frac{18}{3}$$
$$x=6$$
To check, replace x with 6 in the original equation to see that a true statement results. The solution is 6.

9.
$$4(3x-2)=-1+4$$
$$4(3x)-4(2)=-1+4$$
$$12x-8=3$$
$$12x-8+8=3+8$$
$$12x=11$$
$$\frac{12x}{12}=\frac{11}{12}$$
$$x=\frac{11}{12}$$
To check, replace x with $\frac{11}{12}$ in the original equation to see that a true statement results. The solution is $\frac{11}{12}$.

10. Let x = first integer.
$x + 2$ = second even integer.
$x + 4$ = third even integer.
$x + (x + 2) + (x + 4) = 3x + 6$

Vocabulary and Readiness Check

1. By the multiplication property of equality, $y=\frac{1}{2}$ and $5\cdot y=5\cdot\frac{1}{2}$ are equivalent equations.

2. The equations $\frac{z}{4}=10$ and $4\cdot\frac{z}{4}=10$ are not equivalent equations. The statement is false.

3. The equations $-7x = 30$ and $\frac{-7x}{-7}=\frac{30}{7}$ are not equivalent equations. The statement is false.

4. By the multiplication property of equality, $9x = -63$ and $\frac{9x}{9}=\frac{-63}{9}$ are equivalent equations.

5. $3a=27$
$$a=\frac{27}{3}=9$$

6. $9c=54$
$$c=\frac{54}{9}=6$$

7. $5b=10$
$$b=\frac{10}{5}=2$$

8. $7t=14$
$$t=\frac{14}{7}=2$$

Exercise Set 2.3

1. $-5x=-20$
$$\frac{-5x}{-5}=\frac{-20}{-5}$$
$$x=4$$
Check: $-5x=-20$
$$-5(4)\stackrel{?}{=}-20$$
$$-20=-20$$
The solution is 4.

3. $3x = 0$

$\frac{3x}{3} = \frac{0}{3}$

$x = 0$

Check: $3x = 0$

$3(0) \stackrel{?}{=} 0$

$0 = 0$

The solution is 0.

5. $-x = -12$

$\frac{-x}{-1} = \frac{-12}{-1}$

$x = 12$

Check: $-x = -12$

$-(12) \stackrel{?}{=} -12$

$-12 = -12$

The solution is 12.

7. $\frac{2}{3}x = -8$

$\frac{3}{2}\left(\frac{2}{3}x\right) = \frac{3}{2}(-8)$

$x = -12$

Check: $\frac{2}{3}x = -8$

$\frac{2}{3}(-12) \stackrel{?}{=} -8$

$-8 = -8$

The solution is −12.

9. $\frac{1}{6}d = \frac{1}{2}$

$6\left(\frac{1}{6}d\right) = 6\left(\frac{1}{2}\right)$

$d = 3$

Check: $\frac{1}{6}d = \frac{1}{2}$

$\frac{1}{6}(3) \stackrel{?}{=} \frac{1}{2}$

$\frac{1}{2} = \frac{1}{2}$

The solution is 3.

11. $\frac{a}{2} = 1$

$2\left(\frac{a}{2}\right) = 2(1)$

$a = 2$

Check: $\frac{a}{2} = 1$

$\frac{2}{2} \stackrel{?}{=} 1$

$1 = 1$

The solution is 2.

13. $\frac{k}{-7} = 0$

$-7\left(\frac{k}{-7}\right) = -7(0)$

$k = 0$

Check: $\frac{k}{-7} = 0$

$\frac{0}{-7} \stackrel{?}{=} 0$

$0 = 0$

The solution is 0.

15. $1.7x = 10.71$

$\frac{1.7x}{1.7} = \frac{10.71}{1.7}$

$x = 6.3$

Check: $1.7x = 10.71$

$1.7(6.3) \stackrel{?}{=} 10.71$

$10.71 = 10.71$

The solution is 6.3.

17. $2x - 4 = 16$

$2x - 4 + 4 = 16 + 4$

$2x = 20$

$\frac{2x}{2} = \frac{20}{2}$

$x = 10$

Check: $2x - 4 = 16$

$2(10) - 4 \stackrel{?}{=} 16$

$20 - 4 \stackrel{?}{=} 16$

$16 = 16$

The solution is 10.

19. $-x+2=22$
$-x+2-2=22-2$
$-x=20$
$x=-20$
Check: $-x+2=22$
$-(-20)+2 \stackrel{?}{=} 22$
$20+2 \stackrel{?}{=} 22$
$22=22$
The solution is −20.

21. $6a+3=3$
$6a+3-3=3-3$
$6a=0$
$\frac{6a}{6}=\frac{0}{6}$
$a=0$
Check: $6a+3=3$
$6(0)+3 \stackrel{?}{=} 3$
$0+3 \stackrel{?}{=} 3$
$3=3$
The solution is 0.

23. $\frac{x}{3}-2=-5$
$\frac{x}{3}-2+2=-5+2$
$\frac{x}{3}=-3$
$3\cdot\frac{x}{3}=3\cdot-3$
$x=-9$
Check: $\frac{x}{3}-2=-5$
$\frac{-9}{3}-2 \stackrel{?}{=} -5$
$-3-2 \stackrel{?}{=} -5$
$-5=-5$
The solution is −9.

25. $6z-z=-2+2z-1-6$
$5z=2z-9$
$5z-2z=2z-9-2z$
$3z=-9$
$\frac{3z}{3}=\frac{-9}{3}$
$z=-3$
Check: $6z-z=-2+2z-1-6$
$6(-3)-(-3) \stackrel{?}{=} -2+2(-3)-1-6$
$-18+3 \stackrel{?}{=} -2-6-1-6$
$-15=-15$
The solution is −3.

27. $1=0.4x-0.6x-5$
$1=-0.2x-5$
$1+5=-0.2x-5+5$
$6=-0.2x$
$\frac{6}{-0.2}=\frac{-0.2x}{-0.2}$
$-30=x$
Check: $1=0.4x-0.6x-5$
$1 \stackrel{?}{=} 0.4(-30)-0.6(-30)-5$
$1 \stackrel{?}{=} -12+18-5$
$1=1$
The solution is −30.

29. $\frac{2}{3}y-11=-9$
$\frac{2}{3}y-11+11=-9+11$
$\frac{2}{3}y=2$
$\frac{3}{2}\cdot\frac{2}{3}y=\frac{3}{2}\cdot 2$
$y=3$
Check: $\frac{2}{3}y-11=-9$
$\frac{2}{3}\cdot 3-11 \stackrel{?}{=} -9$
$2-11 \stackrel{?}{=} -9$
$-9=-9$
The solution is 3.

31. $\frac{3}{4}t-\frac{1}{2}=\frac{1}{3}$
$\frac{3}{4}t-\frac{1}{2}+\frac{1}{2}=\frac{1}{3}+\frac{1}{2}$
$\frac{3}{4}t=\frac{5}{6}$
$\frac{4}{3}\cdot\frac{3}{4}t=\frac{4}{3}\cdot\frac{5}{6}$
$t=\frac{10}{9}$

Check: $\frac{3}{4}t - \frac{1}{2} = \frac{1}{3}$

$$\frac{3}{4} \cdot \frac{10}{9} - \frac{1}{2} \stackrel{?}{=} \frac{1}{3}$$
$$\frac{5}{6} - \frac{1}{2} \stackrel{?}{=} \frac{1}{3}$$
$$\frac{2}{6} \stackrel{?}{=} \frac{1}{3}$$
$$\frac{1}{3} = \frac{1}{3}$$

The solution is $\frac{10}{9}$.

33.
$$8x + 20 = 6x + 18$$
$$8x + 20 - 6x = 6x + 18 - 6x$$
$$2x + 20 = 18$$
$$2x + 20 - 20 = 18 - 20$$
$$2x = -2$$
$$\frac{2x}{2} = \frac{-2}{2}$$
$$x = -1$$

35.
$$3(2x + 5) = -18 + 9$$
$$6x + 15 = -9$$
$$6x + 15 - 15 = -9 - 15$$
$$6x = -24$$
$$\frac{6x}{6} = \frac{-24}{6}$$
$$x = -4$$

37.
$$2x - 5 = 20x + 4$$
$$2x - 5 - 2x = 20x + 4 - 2x$$
$$-5 = 18x + 4$$
$$-5 - 4 = 18x + 4 - 4$$
$$-9 = 18x$$
$$\frac{-9}{18} = \frac{18x}{18}$$
$$-\frac{1}{2} = x$$

39.
$$2 + 14 = -4(3x - 4)$$
$$16 = -12x + 16$$
$$16 - 16 = -12x + 16 - 16$$
$$0 = -12x$$
$$\frac{0}{-12} = \frac{-12x}{-12}$$
$$0 = x$$

41.
$$-6y - 3 = -5y - 7$$
$$-6y - 3 + 6y = -5y - 7 + 6y$$
$$-3 = y - 7$$
$$-3 + 7 = y - 7 + 7$$
$$4 = y$$

43.
$$\frac{1}{2}(2x - 1) = -\frac{1}{7} - \frac{3}{7}$$
$$x - \frac{1}{2} = -\frac{4}{7}$$
$$x - \frac{1}{2} + \frac{1}{2} = -\frac{4}{7} + \frac{1}{2}$$
$$x = -\frac{8}{14} + \frac{7}{14}$$
$$x = -\frac{1}{14}$$

45.
$$-10z - 0.5 = -20z + 1.6$$
$$-10z - 0.5 + 20z = -20z + 1.6 + 20z$$
$$10z - 0.5 = 1.6$$
$$10z - 0.5 + 0.5 = 1.6 + 0.5$$
$$10z = 2.1$$
$$\frac{10z}{10} = \frac{2.1}{10}$$
$$z = 0.21$$

47.
$$-4x + 20 = 4x - 20$$
$$-4x + 20 + 4x = 4x - 20 + 4x$$
$$20 = 8x - 20$$
$$20 + 20 = 8x - 20 + 20$$
$$40 = 8x$$
$$\frac{40}{8} = \frac{8x}{8}$$
$$5 = x$$

49.
$$42 = 7x$$
$$\frac{42}{7} = \frac{7x}{7}$$
$$6 = x$$

51.
$$4.4 = -0.8x$$
$$\frac{4.4}{-0.8} = \frac{-0.8x}{-0.8}$$
$$-5.5 = x$$

53.
$$6x + 10 = -20$$
$$6x + 10 - 10 = -20 - 10$$
$$6x = -30$$
$$\frac{6x}{6} = \frac{-30}{6}$$
$$x = -5$$

55.
$$\begin{aligned} 5-0.3k &= 5 \\ 5-5-0.3k &= 5-5 \\ -0.3k &= 0 \\ \frac{-0.3k}{-0.3} &= \frac{0}{-0.3} \\ k &= 0 \end{aligned}$$

57.
$$\begin{aligned} 13x-5 &= 11x-11 \\ 13x-5-11x &= 11x-11-11x \\ 2x-5 &= -11 \\ 2x-5+5 &= -11+5 \\ 2x &= -6 \\ \frac{2x}{2} &= \frac{-6}{2} \\ x &= -3 \end{aligned}$$

59.
$$\begin{aligned} 9(3x+1) &= 4x-5x \\ 27x+9 &= -x \\ 27x+9-27x &= -x-27x \\ 9 &= -28x \\ \frac{9}{-28} &= \frac{-28x}{-28} \\ -\frac{9}{28} &= x \end{aligned}$$

61.
$$\begin{aligned} -\frac{3}{7}p &= -2 \\ -\frac{7}{3}\left(-\frac{3}{7}p\right) &= -\frac{7}{3}(-2) \\ p &= \frac{14}{3} \end{aligned}$$

63.
$$\begin{aligned} -\frac{4}{3}x &= 12 \\ -\frac{3}{4}\left(-\frac{4}{3}x\right) &= -\frac{3}{4}(12) \\ x &= -9 \end{aligned}$$

65.
$$\begin{aligned} -2x+\frac{1}{2} &= \frac{7}{2} \\ -2x+\frac{1}{2}-\frac{1}{2} &= \frac{7}{2}-\frac{1}{2} \\ -2x &= \frac{6}{2} \\ -2x &= 3 \\ \frac{-2x}{-2} &= \frac{3}{-2} \\ x &= -\frac{3}{2} \end{aligned}$$

67.
$$\begin{aligned} 10 &= 2x-1 \\ 10+1 &= 2x-1+1 \\ 11 &= 2x \\ \frac{11}{2} &= \frac{2x}{2} \\ \frac{11}{2} &= x \end{aligned}$$
$\frac{11}{2}$.

69.
$$\begin{aligned} 10-3x-6-9x &= 7 \\ 4-12x &= 7 \\ 4-4-12x &= 7-4 \\ -12x &= 3 \\ \frac{-12x}{-12} &= \frac{3}{-12} \\ x &= -\frac{1}{4} \end{aligned}$$

71.
$$\begin{aligned} z-5z &= 7z-9-z \\ -4z &= 6z-9 \\ -4z-6z &= 6z-6z-9 \\ -10z &= -9 \\ \frac{-10z}{-10} &= \frac{-9}{-10} \\ z &= \frac{9}{10} \end{aligned}$$

73.
$$\begin{aligned} -x-\frac{4}{5} &= x+\frac{1}{2}+\frac{2}{5} \\ -x-\frac{4}{5} &= x+\frac{9}{10} \\ -x-\frac{4}{5}+x &= x+\frac{9}{10}+x \\ -\frac{4}{5} &= 2x+\frac{9}{10} \\ -\frac{8}{10}-\frac{9}{10} &= 2x+\frac{9}{10}-\frac{9}{10} \\ -\frac{17}{10} &= 2x \\ \frac{1}{2}\cdot\left(-\frac{17}{10}\right) &= \frac{1}{2}\cdot 2x \\ -\frac{17}{20} &= x \end{aligned}$$

75.
$$\begin{aligned} -15+37 &= -2(x+5) \\ 22 &= -2x-10 \\ 22+10 &= -2x-10+10 \\ 32 &= -2x \\ \frac{32}{-2} &= \frac{-2x}{-2} \\ -16 &= x \end{aligned}$$

77. Sum = first integer + second integer
Sum = $x + (x + 2) = x + x + 2 = 2x + 2$

79. Sum = first integer + third integer
Sum = $x + (x + 2) = x + x + 2 = 2x + 2$

81. Let x = first room number.
$x + 2$ = second room number
$x + 4$ = third room number
$x + 6$ = fourth room number
$x + 8$ = fifth room number
$x + (x + 2) + (x + 4) + (x + 6) + (x + 8) = 5x + 20$

83. $5x + 2(x - 6) = 5x + 2x - 12 = 7x - 12$

85. $-(x - 1) + x = -x + 1 + x = 1$

87. $(-3)^2 = (-3)(-3) = 9$
$-3^2 = -3 \cdot 3 = -9$
$(-3)^2 > -3^2$

89. $(-2)^3 = (-2)(-2)(-2) = -8$
$-2^3 = -2 \cdot 2 \cdot 2 = -8$
$(-2)^3 = -2^3$

91. $-|-6| = -(6) = -6$
$-|-6| < 6$

93. $6x = \underline{\quad}$
$6(-8) = \underline{\quad}$
$-48 = \underline{\quad}$

95. Answers may vary

97. Answers may vary

99. $9x = 2100$
$$\frac{9x}{9} = \frac{2100}{9}$$
$$x = \frac{700}{3}$$
Each dose should be $\frac{700}{3}$ milligrams.

101. $-3.6x = 10.62$
$$\frac{-3.6x}{-3.6} = \frac{10.62}{-3.6}$$
$$x = -2.95$$

103. $7x - 5.06 = -4.92$
$$7x - 5.06 + 5.06 = -4.92 + 5.06$$
$$7x = 0.14$$
$$\frac{7x}{7} = \frac{0.14}{7}$$
$$x = 0.02$$

Section 2.4

Practice Exercises

1. $2(4a - 9) + 3 = 5a - 6$
$$8a - 18 + 3 = 5a - 6$$
$$8a - 15 = 5a - 6$$
$$8a - 15 - 5a = 5a - 6 - 5a$$
$$3a - 15 = -6$$
$$3a - 15 + 15 = -6 + 15$$
$$3a = 9$$
$$\frac{3a}{3} = \frac{9}{3}$$
$$a = 3$$
Check: $2(4a - 9) + 3 = 5a - 6$
$$2[4(3) - 9] + 3 \stackrel{?}{=} 5(3) - 6$$
$$2(12 - 9) + 3 \stackrel{?}{=} 15 - 6$$
$$2(3) + 3 \stackrel{?}{=} 9$$
$$6 + 3 \stackrel{?}{=} 9$$
$$9 = 1$$
The solution is 3 or the solution set is $\{3\}$.

2. $7(x - 3) = -6x$
$$7x - 21 = -6x$$
$$7x - 21 - 7x = -6x - 7x$$
$$-21 = -13x$$
$$\frac{-21}{-13} = \frac{-13x}{-13}$$
$$\frac{21}{13} = x$$
Check: $7(x - 3) = -6x$
$$7\left(\frac{21}{13} - 3\right) \stackrel{?}{=} -6\left(\frac{21}{13}\right)$$
$$7\left(\frac{21}{13} - \frac{39}{13}\right) \stackrel{?}{=} -\frac{126}{13}$$
$$7\left(-\frac{18}{13}\right) \stackrel{?}{=} -\frac{126}{13}$$
$$-\frac{126}{13} = -\frac{126}{13}$$
The solution is $\frac{21}{13}$.

3.
$$\frac{3}{5}x-2=\frac{2}{3}x-1$$
$$15\left(\frac{3}{5}x-2\right)=15\left(\frac{2}{3}x-1\right)$$
$$15\left(\frac{3}{5}x\right)-15(2)=15\left(\frac{2}{3}x\right)-15(1)$$
$$9x-30=10x-15$$
$$9x-30-9x=10x-15-9x$$
$$-30=x-15$$
$$-30+15=x-15+15$$
$$-15=x$$

Check: $\frac{3}{5}x-2=\frac{2}{3}x-1$
$$\frac{3}{5}\cdot-15-2\stackrel{?}{=}\frac{2}{3}\cdot-15-1$$
$$-9-2\stackrel{?}{=}-10-1$$
$$-11=-11$$

The solution is −15.

4.
$$\frac{4(y+3)}{3}=5y-7$$
$$3\cdot\frac{4(y+3)}{3}=3\cdot(5y-7)$$
$$4(y+3)=3(5y-7)$$
$$4y+12=15y-21$$
$$4y+12-4y=15y-21-4y$$
$$12=11y-21$$
$$12+21=11y-21+21$$
$$33=11y$$
$$\frac{33}{11}=\frac{11y}{11}$$
$$3=y$$

To check, replace y with 3 in the original equation. The solution is 3.

5.
$$0.35x+0.09(x+4)=0.30(12)$$
$$100[0.35x+0.09(x+4)]=100[0.03(12)]$$
$$35x+9(x+4)=3(12)$$
$$35x+9x+36=36$$
$$44x+36=36$$
$$44x+36-36=36-36$$
$$44x=0$$
$$\frac{44x}{44}=\frac{0}{44}$$
$$x=0$$

To check, replace x with 0 in the original equation. The solution is 0.

6.
$$4(x+4)-x=2(x+11)+x$$
$$4x+16-x=2x+22+x$$
$$3x+16=3x+22$$
$$3x+16-3x=3x+22-3x$$
$$16=22$$

There is no solution.

7.
$$12x-18=9(x-2)+3x$$
$$12x-18=9x-18+3x$$
$$12x-18=12x-18$$
$$12x-18+18=12x-18+18$$
$$12x=12x$$
$$12x-12x=12x-12x$$
$$0=0$$

The solution is all real numbers.

Calculator Explorations

1. Solution (−24 = −24)
2. Solution (−4 = −4)
3. Not a solution (19.4 ≠ 10.4)
4. Not a solution (−11.9 ≠ −60.1)
5. Solution (17,061 = 17,061)
6. Solution (−316 = −316)

Vocabulary and Readiness Check

1. $x = -7$ is an equation.
2. $x - 7$ is an expression.
3. $4y - 6 + 9y + 1$ is an expression.
4. $4y - 6 = 9y + 1$ is an equation.
5. $\frac{1}{x}-\frac{x-1}{8}$ is an expression.
6. $\frac{1}{x}-\frac{x-1}{8}=6$ is an equation.
7. $0.1x + 9 = 0.2x$ is an equation.
8. $0.1x^2+9y-0.2x^2$ is an expression.

Exercise Set 2.4

1.
$$\begin{aligned}-4y+10&=-2(3y+1)\\-4y+10&=-6y-2\\-4y+10-10&=-6y-2-10\\-4y&=-6y-12\\-4y+6y&=-6y-12+6y\\2y&=-12\\\frac{2y}{2}&=\frac{-12}{2}\\y&=-6\end{aligned}$$

3.
$$\begin{aligned}15x-8&=10+9x\\15x-8+8&=10+9x+8\\15x&=18+9x\\15x-9x&=18+9x-9x\\6x&=18\\\frac{6x}{6}&=\frac{18}{6}\\x&=3\end{aligned}$$

5.
$$\begin{aligned}-2(3x-4)&=2x\\-6x+8&=2x\\-6x+6x+8&=2x+6x\\8&=8x\\\frac{8}{8}&=\frac{8x}{8}\\1&=x\end{aligned}$$

7.
$$\begin{aligned}5(2x-1)-2(3x)&=1\\10x-5-6x&=1\\4x-5&=1\\4x-5+5&=1+5\\4x&=6\\\frac{4x}{4}&=\frac{6}{4}\\x&=\frac{3}{2}\end{aligned}$$

9.
$$\begin{aligned}-6(x-3)-26&=-8\\-6x+18-26&=-8\\-6x-8&=-8\\-6x-8+8&=-8+8\\-6x&=0\\\frac{-6x}{-6}&=\frac{0}{-6}\\x&=0\end{aligned}$$

11.
$$\begin{aligned}8-2(a+1)&=9+a\\8-2a-2&=9+a\\-2a+6&=9+a\\-2a+6-6&=9+a-6\\-2a&=3+a\\-2a-a&=3+a-a\\-3a&=3\\\frac{-3a}{-3}&=\frac{3}{-3}\\a&=-1\end{aligned}$$

13.
$$\begin{aligned}4x+3&=-3+2x+14\\4x+3&=2x+11\\4x-2x+3&=2x-2x+11\\2x+3&=11\\2x+3-3&=11-3\\2x&=8\\\frac{2x}{2}&=\frac{8}{2}\\x&=4\end{aligned}$$

15.
$$\begin{aligned}-2y-10&=5y+18\\-2y-5y-10&=5y-5y+18\\-7y-10&=18\\-7y-10+10&=18+10\\-7y&=28\\\frac{-7y}{-7}&=\frac{28}{-7}\\y&=-4\end{aligned}$$

17.
$$\begin{aligned}\frac{2}{3}x+\frac{4}{3}&=-\frac{2}{3}\\3\left(\frac{2}{3}x+\frac{4}{3}\right)&=3\left(-\frac{2}{3}\right)\\2x+4&=-2\\2x+4-4&=-2-4\\2x&=-6\\\frac{2x}{2}&=\frac{-6}{2}\\x&=-3\end{aligned}$$

19.
$$\frac{3}{4}x - \frac{1}{2} = 1$$
$$4\left(\frac{3}{4}x - \frac{1}{2}\right) = 4(1)$$
$$3x - 2 = 4$$
$$3x - 2 + 2 = 4 + 2$$
$$3x = 6$$
$$\frac{3x}{3} = \frac{6}{3}$$
$$x = 2$$

21.
$$0.50x + 0.15(70) = 35.5$$
$$100[0.50x + 0.15(70)] = 100(35.5)$$
$$50x + 15(70) = 3550$$
$$50x + 1050 = 3550$$
$$50x + 1050 - 1050 = 3550 - 1050$$
$$50x = 2500$$
$$\frac{50x}{50} = \frac{2500}{50}$$
$$x = 50$$

23.
$$\frac{2(x+1)}{4} = 3x - 2$$
$$4\left[\frac{2(x+1)}{4}\right] = 4(3x - 2)$$
$$2(x+1) = 12x - 8$$
$$2x + 2 = 12x - 8$$
$$2x - 12x + 2 = 12x - 12x - 8$$
$$-10x + 2 = -8$$
$$-10x + 2 - 2 = -8 - 2$$
$$-10x = -10$$
$$\frac{-10x}{-10} = \frac{-10}{-10}$$
$$x = 1$$

25.
$$x + \frac{7}{6} = 2x - \frac{7}{6}$$
$$6\left(x + \frac{7}{6}\right) = 6\left(2x - \frac{7}{6}\right)$$
$$6x + 7 = 12x - 7$$
$$6x - 12x + 7 = 12x - 12x - 7$$
$$-6x + 7 = -7$$
$$-6x + 7 - 7 = -7 - 7$$
$$-6x = -14$$
$$\frac{-6x}{-6} = \frac{-14}{-6}$$
$$x = \frac{7}{3}$$

27.
$$0.12(y-6) + 0.06y = 0.08y - 0.7$$
$$100[0.12(y-6) + 0.06y] = 100[0.08y - 0.7]$$
$$12(y-6) + 6y = 8y - 70$$
$$12y - 72 + 6y = 8y - 70$$
$$18y - 72 = 8y - 70$$
$$18y - 8y - 72 = 8y - 8y - 70$$
$$10y - 72 = -70$$
$$10y - 72 + 72 = -70 + 72$$
$$10y = 2$$
$$\frac{10y}{10} = \frac{2}{10}$$
$$y = \frac{1}{5} = 0.2$$

29.
$$4(3x + 2) = 12x + 8$$
$$12x + 8 = 12x + 8$$
$$12x + 8 - 12x = 12x + 8 - 12x$$
$$8 = 8$$
All real numbers are solutions.

31.
$$\frac{x}{4} + 1 = \frac{x}{4}$$
$$4\left(\frac{x}{4} + 1\right) = 4\left(\frac{x}{4}\right)$$
$$x + 4 = x$$
$$x - x + 4 = x - x$$
$$4 = 0$$
There is no solution.

33.
$$3x - 7 = 3(x + 1)$$
$$3x - 7 = 3x + 3$$
$$3x - 3x - 7 = 3x - 3x + 3$$
$$-7 = 3$$
There is no solution.

35.
$$-2(6x - 5) + 4 = -12x + 14$$
$$-12x + 10 + 4 = -12x + 14$$
$$-12x + 14 = -12x + 14$$
$$-12x + 14 + 12x = -12x + 14 + 12x$$
$$14 = 14$$
All real numbers are solutions.

37.
$$\frac{6(3 - z)}{5} = -z$$
$$5\left[\frac{6(3 - z)}{5}\right] = 5(-z)$$
$$6(3 - z) = 5(-z)$$
$$18 - 6z = -5z$$
$$18 - 6z + 6z = -5z + 6z$$
$$18 = z$$

39.
$$\begin{aligned}
-3(2t-5)+2t &= 5t-4 \\
-6t+15+2t &= 5t-4 \\
-4t+15 &= 5t-4 \\
-4t+15+4 &= 5t-4+4 \\
-4t+19 &= 5t \\
-4t+19+4t &= 5t+4t \\
19 &= 9t \\
\frac{19}{9} &= \frac{9t}{9} \\
\frac{19}{9} &= t
\end{aligned}$$

41.
$$\begin{aligned}
5y+2(y-6) &= 4(y+1)-2 \\
5y+2y-12 &= 4y+4-2 \\
7y-12 &= 4y+2 \\
7y-12+12 &= 4y+2+12 \\
7y &= 4y+14 \\
7y-4y &= 4y+14-4y \\
3y &= 14 \\
\frac{3y}{3} &= \frac{14}{3} \\
y &= \frac{14}{3}
\end{aligned}$$

43.
$$\begin{aligned}
\frac{3(x-5)}{2} &= \frac{2(x+5)}{3} \\
6\left[\frac{3(x-5)}{2}\right] &= 6\left[\frac{2(x+5)}{3}\right] \\
9(x-5) &= 4(x+5) \\
9x-45 &= 4x+20 \\
9x-4x-45 &= 4x-4x+20 \\
5x-45 &= 20 \\
5x-45+45 &= 20+45 \\
5x &= 65 \\
\frac{5x}{5} &= \frac{65}{5} \\
x &= 13
\end{aligned}$$

45.
$$\begin{aligned}
0.7x-2.3 &= 0.5 \\
10(0.7x-2.3) &= 10(0.5) \\
7x-23 &= 5 \\
7x-23+23 &= 5+23 \\
7x &= 28 \\
\frac{7x}{7} &= \frac{28}{7} \\
x &= 4
\end{aligned}$$

47.
$$\begin{aligned}
5x-5 &= 2(x+1)+3x-7 \\
5x-5 &= 2x+2+3x-7 \\
5x-5 &= 5x-5 \\
5x-5x-5 &= 5x-5x-5 \\
-5 &= -5
\end{aligned}$$
All real numbers are solutions.

49.
$$\begin{aligned}
4(2n+1) &= 3(6n+3)+1 \\
8n+4 &= 18n+9+1 \\
8n+4 &= 18n+10 \\
8n+4-4 &= 18n+10-4 \\
8n &= 18n+6 \\
8n-18n &= 18n+6-18n \\
-10n &= 6 \\
\frac{-10n}{-10} &= \frac{6}{-10} \\
n &= -\frac{3}{5}
\end{aligned}$$

51.
$$\begin{aligned}
x+\frac{5}{4} &= \frac{3}{4}x \\
4\left(x+\frac{5}{4}\right) &= 4\left(\frac{3}{4}x\right) \\
4x+5 &= 3x \\
4x+5-4x &= 3x-4x \\
5 &= -x \\
\frac{5}{-1} &= \frac{-x}{-1} \\
-5 &= x
\end{aligned}$$

53.
$$\begin{aligned}
\frac{x}{2}-1 &= \frac{x}{5}+2 \\
10\left(\frac{x}{2}-1\right) &= 10\left(\frac{x}{5}+2\right) \\
5x-10 &= 2x+20 \\
5x-10+10 &= 2x+20+10 \\
5x &= 2x+30 \\
5x-2x &= 2x+30-2x \\
3x &= 30 \\
\frac{3x}{3} &= \frac{30}{3} \\
x &= 10
\end{aligned}$$

55.
$$\begin{aligned}
2(x+3)-5 &= 5x-3(1+x) \\
2x+6-5 &= 5x-3-3x \\
2x+1 &= 2x-3 \\
2x-2x+1 &= 2x-2x-3 \\
1 &= -3
\end{aligned}$$
There is no solution.

57.
$$\begin{aligned}0.06-0.01(x+1)&=-0.02(2-x)\\100[0.06-0.01(x+1)]&=100[-0.02(2-x)]\\6-(x+1)&=-2(2-x)\\6-x-1&=-4+2x\\5-x&=-4+2x\\5-x-2x&=-4+2x-2x\\5-3x&=-4\\5-5-3x&=-4-5\\-3x&=-9\\\frac{-3x}{-3}&=\frac{-9}{-3}\\x&=3\end{aligned}$$

59.
$$\begin{aligned}\frac{9}{2}+\frac{5}{2}y&=2y-4\\2\left(\frac{9}{2}+\frac{5}{2}y\right)&=2(2y-4)\\9+5y&=4y-8\\9+5y-4y&=4y-8-4y\\9+y&=-8\\9+y-9&=-8-9\\y&=-17\end{aligned}$$

61.
$$\begin{aligned}-2y-10&=5y+18\\-2y-10-18&=5y+18-18\\-2y-28&=5y\\-2y-28+2y&=5y+2y\\-28&=7y\\\frac{-28}{7}&=\frac{7y}{7}\\-4&=y\end{aligned}$$

63.
$$\begin{aligned}0.6x-0.1&=0.5x+0.2\\10(0.6x-0.1)&=10(0.5x+0.2)\\6x-1&=5x+2\\6x-5x-1&=5x-5x+2\\x-1&=2\\x-1+1&=2+1\\x&=3\end{aligned}$$

65.
$$\begin{aligned}0.02(6t-3)&=0.12(t-2)+0.18\\100[0.02(6t-3)]&=100[0.12(t-2)+0.18]\\2(6t-3)&=12(t-2)+18\\12t-6&=12t-24+18\\12t-6&=12t-6\\12t-12t-6&=12t-12t-6\\-6&=-6\end{aligned}$$
All real numbers are solutions.

67.

−8	minus	a number
↓	↓	↓
−8	−	x

69.

−3	plus	twice a number
↓	↓	↓
−3	+	$2x$

71.

9	times	a number	plus	20
↓	↓	↓	↓	↓
9	·	$(x$	+	$20) = 9(x+20)$

73.
$$\begin{aligned}x+(2x-3)+(3x-5)&=x+2x-3+3x-5\\&=6x-8\end{aligned}$$
The perimeter is $(6x-8)$ meters.

75. **a.**
$$\begin{aligned}x+3&=x+3\\x+3-x&=x+3-x\\3&=3\\3-3&=3-3\\0&=0\end{aligned}$$
All real numbers are solutions.

b. Answers may vary

c. Answers may vary

77.
$$\begin{aligned}5x+1&=5x+1\\5x+1-5x&=5x+1-5x\\1&=1\end{aligned}$$
All real numbers are solutions. The answer is a.

79.
$$\begin{aligned}2x-6x-10&=-4x+3-10\\-4x-10&=-4x-7\\-4x-10+4x&=-4x-7+4x\\-10&=-7\end{aligned}$$
There is no solution. The answer is b.

81.
$$\begin{aligned}9x-20&=8x-20\\9x-20-8x&=8x-20-8x\\x-20&=-20\\x-20+20&=-20+20\\x&=0\end{aligned}$$
The answer is c.

83. Answers may vary

85. **a.** Since the perimeter is the sum of the lengths of the sides, $x+x+x+2x+2x=28$.

b. $$\begin{aligned} 7x &= 28 \\ \frac{7x}{7} &= \frac{28}{7} \\ x &= 4 \end{aligned}$$

c. $2x = 2(4) = 8$
The lengths are $x = 4$ centimeters and $2x = 8$ centimeters.

87. Answers may vary

89. $$\begin{aligned} 1000(7x-10) &= 50(412+100x) \\ 7000x-10,000 &= 20,600+5000x \\ 7000x-5000x-10,000 &= 20,600+5000x-5000x \\ 2000x-10,000 &= 20,600 \\ 2000x-10,000+10,000 &= 20,600+10,000 \\ 2000x &= 30,600 \\ \frac{2000x}{2000} &= \frac{30,600}{2000} \\ x &= 15.3 \end{aligned}$$

91. $$\begin{aligned} 0.035x+5.112 &= 0.010x+5.107 \\ 1000(0.035x+5.112) &= 1000(0.010x+5.107) \\ 35x+5112 &= 10x+5107 \\ 35x-10x+5112 &= 10x-10x+5107 \\ 25x+5112 &= 5107 \\ 25x+5112-5112 &= 5107-5112 \\ 25x &= -5 \\ \frac{25x}{25} &= \frac{-5}{25} \\ x &= -\frac{1}{5} = -0.2 \end{aligned}$$

93. $$\begin{aligned} x(x-3) &= x^2+5x+7 \\ x^2-3x &= x^2+5x+7 \\ x^2-x^2-3x &= x^2-x^2+5x+7 \\ -3x &= 5x+7 \\ -3x-5x &= 5x-5x+7 \\ -8x &= 7 \\ \frac{-8x}{-8} &= \frac{7}{-8} \\ x &= -\frac{7}{8} \end{aligned}$$

95.
$$\begin{aligned} 2z(z+6) &= 2z^2+12z-8 \\ 2z^2+12z &= 2z^2+12z-8 \\ 2z^2-2z^2+12z &= 2z^2-2z^2+12z-8 \\ 12z &= 12z-8 \\ 12z-12z &= 12z-12z-8 \\ 0 &= -8 \end{aligned}$$
There is no solution.

Integrated Review

1.
$$\begin{aligned} x-10 &= -4 \\ x-10+10 &= -4+10 \\ x &= 6 \end{aligned}$$

2.
$$\begin{aligned} y+14 &= -3 \\ y+14-14 &= -3-14 \\ y &= -17 \end{aligned}$$

3.
$$\begin{aligned} 9y &= 108 \\ \frac{9y}{9} &= \frac{108}{9} \\ y &= 12 \end{aligned}$$

4.
$$\begin{aligned} -3x &= 78 \\ \frac{-3x}{-3} &= \frac{78}{-3} \\ x &= -26 \end{aligned}$$

5.
$$\begin{aligned} -6x+7 &= 25 \\ -6x+7-7 &= 25-7 \\ -6x &= 18 \\ \frac{-6x}{-6} &= \frac{18}{-6} \\ x &= -3 \end{aligned}$$

6.
$$\begin{aligned} 5y-42 &= -47 \\ 5y-42+42 &= -47+42 \\ 5y &= -5 \\ \frac{5y}{5} &= \frac{-5}{5} \\ y &= -1 \end{aligned}$$

7.
$$\begin{aligned} \frac{2}{3}x &= 9 \\ \frac{3}{2}\left(\frac{2}{3}x\right) &= \frac{3}{2}(9) \\ x &= \frac{27}{2} \end{aligned}$$

8.
$$\begin{aligned} \frac{4}{5}z &= 10 \\ \frac{5}{4}\left(\frac{4}{5}z\right) &= \frac{5}{4}(10) \\ z &= \frac{25}{2} \end{aligned}$$

9.
$$\begin{aligned} \frac{r}{-4} &= -2 \\ -4\left(\frac{r}{-4}\right) &= -4(-2) \\ r &= 8 \end{aligned}$$

10.
$$\begin{aligned} \frac{y}{-8} &= 8 \\ -8\left(\frac{y}{-8}\right) &= -8(8) \\ y &= -64 \end{aligned}$$

11.
$$\begin{aligned} 6-2x+8 &= 10 \\ -2x+14 &= 10 \\ -2x+14-14 &= 10-14 \\ -2x &= -4 \\ \frac{-2x}{-2} &= \frac{-4}{-2} \\ x &= 2 \end{aligned}$$

12.
$$\begin{aligned} -5-6y+6 &= 19 \\ -6y+1 &= 19 \\ -6y+1-1 &= 19-1 \\ -6y &= 18 \\ \frac{-6y}{-6} &= \frac{18}{-6} \\ y &= -3 \end{aligned}$$

13.
$$\begin{aligned} 2x-7 &= 2x-27 \\ 2x-2x-7 &= 2x-2x-27 \\ -7 &= -27 \end{aligned}$$
There is no solution.

14.
$$\begin{aligned} 3+8y &= 8y-2 \\ 3+8y-8y &= 8y-8y-2 \\ 3 &= -2 \end{aligned}$$
There is no solution.

15. $$\begin{aligned} -3a+6+5a &= 7a-8a \\ 2a+6 &= -a \\ 2a-2a+6 &= -a-2a \\ 6 &= -3a \\ \frac{6}{-3} &= \frac{-3a}{-3} \\ -2 &= a \end{aligned}$$

16. $$\begin{aligned} 4b-8-b &= 10b-3b \\ 3b-8 &= 7b \\ 3b-3b-8 &= 7b-3b \\ -8 &= 4b \\ \frac{-8}{4} &= \frac{4b}{4} \\ -2 &= b \end{aligned}$$

17. $$\begin{aligned} -\frac{2}{3}x &= \frac{5}{9} \\ -\frac{3}{2}\left(-\frac{2}{3}x\right) &= -\frac{3}{2}\left(\frac{5}{9}\right) \\ x &= -\frac{5}{6} \end{aligned}$$

18. $$\begin{aligned} -\frac{3}{8}y &= -\frac{1}{16} \\ -\frac{8}{3}\left(-\frac{3}{8}y\right) &= -\frac{8}{3}\left(-\frac{1}{16}\right) \\ y &= \frac{1}{6} \end{aligned}$$

19. $$\begin{aligned} 10 &= -6n+16 \\ 10-16 &= -6n+16-16 \\ -6 &= -6n \\ \frac{-6}{-6} &= \frac{-6n}{-6} \\ 1 &= n \end{aligned}$$

20. $$\begin{aligned} -5 &= -2m+7 \\ -5-7 &= -2m+7-7 \\ -12 &= -2m \\ \frac{-12}{-2} &= \frac{-2m}{-2} \\ 6 &= m \end{aligned}$$

21. $$\begin{aligned} 3(5c-1)-2 &= 13c+3 \\ 15c-3-2 &= 13c+3 \\ 15c-5 &= 13c+3 \\ 15c-13c-5 &= 13c-13c+3 \\ 2c-5 &= 3 \\ 2c-5+5 &= 3+5 \\ 2c &= 8 \\ \frac{2c}{2} &= \frac{8}{2} \\ c &= 4 \end{aligned}$$

22. $$\begin{aligned} 4(3t+4)-20 &= 3+5t \\ 12t+16-20 &= 3+5t \\ 12t-4 &= 3+5t \\ 12t-5t-4 &= 3+5t-5t \\ 7t-4 &= 3 \\ 7t-4+4 &= 3+4 \\ 7t &= 7 \\ \frac{7t}{7} &= \frac{7}{7} \\ t &= 1 \end{aligned}$$

23. $$\begin{aligned} \frac{2(z+3)}{3} &= 5-z \\ 3\left[\frac{2(z+3)}{3}\right] &= 3(5-z) \\ 2z+6 &= 15-3z \\ 2z+3z+6 &= 15-3z+3z \\ 5z+6 &= 15 \\ 5z+6-6 &= 15-6 \\ 5z &= 9 \\ \frac{5z}{5} &= \frac{9}{5} \\ z &= \frac{9}{5} \end{aligned}$$

24. $$\begin{aligned} \frac{3(w+2)}{4} &= 2w+3 \\ 4\left[\frac{3(w+2)}{4}\right] &= 4(2w+3) \\ 3w+6 &= 8w+12 \\ 3w-8w+6 &= 8w-8w+12 \\ -5w+6 &= 12 \\ -5w+6-6 &= 12-6 \\ -5w &= 6 \\ \frac{-5w}{-5} &= \frac{6}{-5} \\ w &= -\frac{6}{5} \end{aligned}$$

25.
$$\begin{aligned} -2(2x-5) &= -3x+7-x+3 \\ -4x+10 &= -4x+10 \\ -4x+4x+10 &= -4x+4x+10 \\ 10 &= 10 \end{aligned}$$
All real numbers are solutions.

26.
$$\begin{aligned} -4(5x-2) &= -12x+4-8x+4 \\ -20x+8 &= -20x+8 \\ -20x+20x+8 &= -20x+20x+8 \\ 8 &= 8 \end{aligned}$$
All real numbers are solutions.

27.
$$\begin{aligned} 0.02(6t-3) &= 0.04(t-2)+0.02 \\ 100[0.02(6t-3)] &= 100[0.04(t-2)+0.02] \\ 2(6t-3) &= 4(t-2)+2 \\ 12t-6 &= 4t-8+2 \\ 12t-6 &= 4t-6 \\ 12t-4t-6 &= 4t-4t-6 \\ 8t-6 &= -6 \\ 8t-6+6 &= -6+6 \\ 8t &= 0 \\ \frac{8t}{8} &= \frac{0}{8} \\ t &= 0 \end{aligned}$$

28.
$$\begin{aligned} 0.03(m+7) &= 0.02(5-m)+0.03 \\ 100[0.03(m+7)] &= 100[0.02(5-m)+0.03] \\ 3(m+7) &= 2(5-m)+3 \\ 3m+21 &= 10-2m+3 \\ 3m+21 &= 13-2m \\ 3m+2m+21 &= 13-2m+2m \\ 5m+21 &= 13 \\ 5m+21-21 &= 13-21 \\ 5m &= -8 \\ \frac{5m}{5} &= \frac{-8}{5} \\ m &= -\frac{8}{5} = -1.6 \end{aligned}$$

29.
$$\begin{aligned} -3y &= \frac{4(y-1)}{5} \\ 5(-3y) &= 5\left[\frac{4(y-1)}{5}\right] \\ -15y &= 4y-4 \\ -15y-4y &= 4y-4y-4 \\ -19y &= -4 \\ \frac{-19y}{-19} &= \frac{-4}{-19} \\ y &= \frac{4}{19} \end{aligned}$$

30.
$$\begin{aligned} -4x &= \frac{5(1-x)}{6} \\ 6(-4x) &= 6\left[\frac{5(1-x)}{6}\right] \\ -24x &= 5-5x \\ -24x+5x &= 5-5x+5x \\ -19x &= 5 \\ \frac{-19x}{-19} &= \frac{5}{-19} \\ x &= -\frac{5}{19} \end{aligned}$$

31.
$$\begin{aligned} \frac{5}{3}x-\frac{7}{3} &= x \\ 3\left(\frac{5}{3}x-\frac{7}{3}\right) &= 3(x) \\ 5x-7 &= 3x \\ 5x-5x-7 &= 3x-5x \\ -7 &= -2x \\ \frac{-7}{-2} &= \frac{-2x}{-2} \\ \frac{7}{2} &= x \end{aligned}$$

32.
$$\begin{aligned} \frac{7}{5}n+\frac{3}{5} &= -n \\ 5\left(\frac{7}{5}n+\frac{3}{5}\right) &= 5(-n) \\ 7n+3 &= -5n \\ 7n-7n+3 &= -5n-7n \\ 3 &= -12n \\ \frac{3}{-12} &= \frac{-12n}{-12} \\ -\frac{1}{4} &= n \end{aligned}$$

33.
$$\begin{aligned} \frac{1}{10}(3x-7) &= \frac{3}{10}x+5 \\ 10\left[\frac{1}{10}(3x-7)\right] &= 10\left(\frac{3}{10}x+5\right) \\ 3x-7 &= 3x+50 \\ 3x-7-3x &= 3x+50-3x \\ -7 &= 50 \end{aligned}$$
There is no solution.

34. $\frac{1}{7}(2x-5)=\frac{2}{7}x+1$

$$7\left[\frac{1}{7}(2x-5)\right]=7\left(\frac{2}{7}x+1\right)$$
$$2x-5=2x+7$$
$$2x-5-2x=2x+7-2x$$
$$-5=7$$

There is no solution.

35. $5+2(3x-6)=-4(6x-7)$

$$5+6x-12=-24x+28$$
$$6x-7=-24x+28$$
$$6x-7+24x=-24x+28+24x$$
$$30x-7=28$$
$$30x-7+7=28+7$$
$$30x=35$$
$$\frac{30x}{30}=\frac{35}{30}$$
$$x=\frac{7}{6}$$

36. $3+5(2x-4)=-7(5x+2)$

$$3+10x-20=-35x-14$$
$$10x-17=-35x-14$$
$$10x-17+35x=-35x-14+35x$$
$$45x-17=-14$$
$$45x-17+17=-14+17$$
$$45x=3$$
$$\frac{45x}{45}=\frac{3}{45}$$
$$x=\frac{1}{15}$$

Section 2.5

Practice Exercises

1. Let x = the number.

$$3x-6=2x+3$$
$$3x-6-2x=2x+3-2x$$
$$x-6=3$$
$$x-6+6=3+6$$
$$x=9$$

The number is 9.

2. Let x = the number.

$$3x-4=2(x-1)$$
$$3x-4=2x-2$$
$$3x-4-2x=2x-2-2x$$
$$x-4=-2$$
$$x-4+4=-2+4$$
$$x=2$$

The number is 2.

3. Let x = the length of short piece, then $4x$ = the length of long piece.

$$x+4x=45$$
$$5x=45$$
$$\frac{5x}{5}=\frac{45}{5}$$
$$x=9$$

$4x=4(9)=36$

The short piece is 9 inches and the long piece is 36 inches.

4. Let x = number of Republicans, then $x+6$ = number of Democrats.

$$x+x+6=50$$
$$2x+6=50$$
$$2x+6-6=50-6$$
$$2x=44$$
$$\frac{2x}{2}=\frac{44}{2}$$
$$x=22$$

$x+6=22+6=28$

There were 22 Republican and 28 Democratic Governors.

5. x = degree measure of first angle

$3x$ = degree measure of second angle

$x+55$ = degree measure of third angle

$$x+3x+(x+55)=180$$
$$5x+55=180$$
$$5x+55-55=180-55$$
$$5x=125$$
$$\frac{5x}{5}=\frac{125}{5}$$
$$x=25$$

$3x=3(25)=75$

$x+55=25+55=80$

The measures of the angles are 25°, 75°, and 80°.

6. Let x = the first even integer, then $x + 2$ = the second even integer, and $x + 4$ = the third even integer.

$$x+(x+2)+(x+4)=144$$
$$3x+6=144$$
$$3x+6-6=144-6$$
$$3x=138$$
$$\frac{3x}{3}=\frac{138}{3}$$
$$x=46$$

$x + 2 = 46 + 2 = 48$
$x + 4 = 46 + 4 = 50$
The integers are 46, 48, and 50.

Vocabulary and Readiness Check

1. $2x$; $2x - 31$
2. $3x$; $3x + 17$
3. $x + 5$; $2(x + 5)$
4. $x - 11$; $7(x - 11)$
5. $20 - y$; $\frac{20-y}{3}$ or $(20 - y) \div 3$
6. $-10 + y$; $\frac{-10+y}{9}$ or $(-10 + y) \div 9$

Exercise Set 2.5

1. Let x = the number.

$$2x+7=x+6$$
$$2x+7-x=x+6-x$$
$$x+7=6$$
$$x+6-7=6-7$$
$$x=-1$$

The number is -1.

3. Let x = the number.

$$3x-6=2x+8$$
$$3x-6-2x=2x+8-2x$$
$$x-6=8$$
$$x-6+6=8+6$$
$$x=14$$

The number is 14.

5. Let x = the number.

$$2(x-8)=3(x+3)$$
$$2x-16=3x+9$$
$$2x-2x-16=3x-2x+9$$
$$-16=x+9$$
$$-16-9=x+9-9$$
$$-25=x$$

The number is -25.

7. Let x = the number.

$$4(-2+x)=5x+\frac{1}{2}$$
$$-8+4x=5x+\frac{1}{2}$$
$$-8+4x-4x=5x+\frac{1}{2}-4x$$
$$-8=x+\frac{1}{2}$$
$$-8-\frac{1}{2}=x+\frac{1}{2}-\frac{1}{2}$$
$$-8\frac{1}{2}=x$$
$$-\frac{17}{2}=x$$

The number is $-\frac{17}{2}$.

9. Let x = length of the shorter piece and $2x + 2$ = length of the longer piece.

$$x+2x+2=17$$
$$3x+2=17$$
$$3x+2-2=17-2$$
$$3x=15$$
$$\frac{3x}{3}=\frac{15}{3}$$
$$x=5$$

$2x + 2 = 2(5) + 2 = 12$
The shorter piece is 5 feet and the longer piece is 12 feet.

11. Let x = weight of Armanty meteorite, then $3x$ = weight of Hoba West meteorite.

$$x+3x=88$$
$$4x=88$$
$$\frac{4x}{4}=\frac{88}{4}$$
$$x=22$$

$3x = 3(22) = 66$
The Armanty meteorite weighs 22 tons and the Hoba West meteorite weighs 66 tons.

13. Let x = number of cinema screens in the U.S., then $x + 5806$ = number of cinema screens in China.

$$x + x + 5806 = 78,994$$
$$2x + 5806 = 78,994$$
$$2x + 5806 - 5806 = 78,994 - 5806$$
$$2x = 73,188$$
$$\frac{2x}{2} = \frac{73,188}{2}$$
$$x = 36,594$$

$x + 5806 = 36,594 + 5806 = 42,400$

The U.S. has 36,594 cinema screens, and China has 42,400.

15. Let x = the measure of each of the two equal angles, and $2x + 30$ = the measure of the third.

$$x + x + 2x + 30 = 180$$
$$4x + 30 = 180$$
$$4x + 30 - 30 = 180 - 30$$
$$4x = 150$$
$$\frac{4x}{4} = \frac{150}{4}$$
$$x = 37.5$$

$2x + 30 = 2(37.5) + 30 = 105$

The angles are 37.5°, 37.5°, and 105°.

	First Integer	Next Integers			Indicated Sum
17. Three consecutive integers:	Integer: x	$x + 1$	$x + 2$		Sum of the three consecutive integers simplified: $(x + 1) + (x + 2) = 2x + 3$
19. Three consecutive even integers:	Even integer: x	$x + 2$	$x + 4$		Sum of the first and third even consecutive integers, simplified: $x + (x + 4) = 2x + 4$
21. Four consecutive integers:	Integer: x	$x + 1$	$x + 2$	$x + 3$	Sum of the four consecutive integers, simplified: $x + (x + 1) + (x + 2) + (x + 3) = 4x + 6$
23. Three consecutive odd integers:	Odd integer: x	$x + 2$	$x + 4$		Sum of the second and third consecutive odd integers, simplified: $(x + 2) + (x + 4) = 2x + 6$

25. Let x = the number of the left page and $x + 1$ = the number of the right page.

$$x + x + 1 = 469$$
$$2x + 1 = 469$$
$$2x + 1 - 1 = 469 - 1$$
$$2x = 468$$
$$\frac{2x}{2} = \frac{468}{2}$$
$$x = 234$$

$x + 1 = 234 + 1 = 235$

The page numbers are 234 and 235.

27. Let x = the code for Belgium,
$x + 1$ = the code for France,
$x + 2$ = the code for Spain.

$$x+x+1+x+2=99$$
$$3x+3=99$$
$$3x+3-3=99-3$$
$$3x=96$$
$$\frac{3x}{3}=\frac{96}{3}$$
$$x=32$$

$x + 1 = 32 + 1 = 33$
$x + 2 = 32 + 2 = 34$
The codes are Belgium: 32; France: 33; Spain: 34.

29.
$$x+2x+(1+5x)=25$$
$$8x+1=25$$
$$8x+1-1=25-1$$
$$8x=24$$
$$\frac{8x}{8}=\frac{24}{8}$$
$$x=3$$

$2x = 2(3) = 6$
$1 + 5x = 1 + 5(3) = 16$
The lengths of the pieces are 3 inches, 6 inches, and 16 inches.

31. Let x = the number.

$$10-5x=3x$$
$$10-5x+5x=3x+5x$$
$$10=8x$$
$$\frac{10}{8}=\frac{8x}{8}$$
$$\frac{5}{4}=x$$

The number is $\frac{5}{4}$.

33. Let x = carats in Angola, then
$4x$ = carats in Botswana.

$$x+4x=40,000,000$$
$$5x=40,000,000$$
$$\frac{5x}{5}=\frac{40,000,000}{5}$$
$$x=8,000,000$$

$4x = 4(8,000,000) = 32,000,000$
Botswana produces 32,000,000 carats and Angola produces 8,000,000 carats.

35. Let x = the measure of the smallest angle,
$x + 2$ = the measure of the second, and
$x + 4$ = the measure of the third.

$$x+x+2+x+4=180$$
$$3x+6=180$$
$$3x+6-6=180-6$$
$$3x=174$$
$$\frac{3x}{3}=\frac{174}{3}$$
$$x=58$$

$x + 2 = 58 + 2 = 60$
$x + 4 = 58 + 4 = 62$
The angles are 58°, 60°, and 62°.

37. Let x = first integer (Russia),
$x + 1$ = second integer (Austria),
$x + 2$ = third integer (Canada),
$x + 3$ = fourth integer (United States).

$$x+(x+1)+(x+2)+(x+3)=94$$
$$4x+6=94$$
$$4x+6-6=94-6$$
$$4x=88$$
$$\frac{4x}{4}=\frac{88}{4}$$
$$x=22$$

$x + 1 = 22 + 1 = 23$
$x + 2 = 22 + 2 = 24$
$x + 3 = 22 + 3 = 25$
The number of medals for each country is Russia: 22; Austria: 23; Canada: 24; United States: 25.

39. Let x = the number.

$$3(x+5)=2x-1$$
$$3x+15=2x-1$$
$$3x+15-2x=2x-1-2x$$
$$x+15=-1$$
$$x+15-15=-1-15$$
$$x=-16$$

The number is −16.

41. Let x = votes for Pavich, then
$x + 20,196$ = votes for Weller.

$$x+x+20,196=196,554$$
$$2x+20,196=196,554$$
$$2x+20,196-20,196=196,554-20,196$$
$$2x=176,358$$
$$\frac{2x}{2}=\frac{176,358}{2}$$
$$x=88,179$$

$x + 20,196 = 88,179 + 20,196 = 108,375$
Pavich received 88,179 votes and Weller received 108,375 votes.

43. Let x = smaller angle, then
$3x + 8$ = larger angle.

$$x + (3x + 8) = 180$$
$$4x + 8 = 180$$
$$4x + 8 - 8 = 180 - 8$$
$$4x = 172$$
$$\frac{4x}{4} = \frac{172}{4}$$
$$x = 43$$

$3x + 8 = 3(43) + 8 = 137$

The angles measure 43° and 137°.

45. Let x = the number.

$$\frac{x}{4} + \frac{1}{2} = \frac{3}{4}$$
$$4\left(\frac{x}{4} + \frac{1}{2}\right) = 4\left(\frac{3}{4}\right)$$
$$x + 2 = 3$$
$$x + 2 - 2 = 3 - 2$$

The number is 1.

47. Let x = the measure of each of the two smaller angles, and $2x - 15$ = the measure of each of the larger angles.

$$2x + 2(2x - 15) = 360$$
$$2x + 4x - 30 = 360$$
$$6x - 30 = 360$$
$$6x - 30 + 30 = 360 + 30$$
$$6x = 390$$
$$\frac{6x}{6} = \frac{390}{6}$$
$$x = 65$$

$2x - 15 = 2(65) - 15 = 115$

The smaller angles are each 65° and the larger angles are each 115°.

49. Let x = speed of TGV, then
$x + 3.8$ = speed of Maglev.

$$x + x + 3.8 = 718.2$$
$$2x + 3.8 = 718.2$$
$$2x + 3.8 - 3.8 = 718.2 - 3.8$$
$$2x = 714.4$$
$$\frac{2x}{2} = \frac{714.4}{2}$$
$$x = 357.2$$

$x + 3.8 = 357.2 + 3.8 = 361$

The speed of the TGV is 357.2 mph and the speed of the Maglev is 361 mph.

51. Let x = the number.

$$\frac{1}{3} \cdot x = \frac{5}{6}$$
$$3 \cdot \frac{1}{3} x = 3 \cdot \frac{5}{6}$$
$$x = \frac{5}{2}$$

The number is $\frac{5}{2}$.

53. Let x = number of counties in Montana and $x + 2$ = number in California.

$$x + x + 2 = 114$$
$$2x + 2 = 114$$
$$2x + 2 - 2 = 114 - 2$$
$$2x = 112$$
$$\frac{2x}{2} = \frac{112}{2}$$
$$x = 56$$

$x + 2 = 56 + 2 = 58$

There are 56 counties in Montana and 58 counties in California.

55. Let x = points for Bears, then
$x + 12$ = points for Colts.

$$x + (x + 12) = 46$$
$$2x + 12 = 46$$
$$2x + 12 - 12 = 46 - 12$$
$$2x = 34$$
$$\frac{2x}{2} = \frac{34}{2}$$
$$x = 17$$

$x + 12 = 17 + 12 = 29$

The Bears scored 17 points and the Colts scored 29 points.

57. Let x = smaller angles, then
$x + 76.5$ = third angle.

$$x + x + (x + 76.5) = 180$$
$$3x + 76.5 = 180$$
$$3x + 76.5 - 76.5 = 180 - 76.5$$
$$3x = 103.5$$
$$\frac{3x}{3} = \frac{103.5}{3}$$
$$x = 34.5$$

$x + 76.5 = 34.5 + 76.5 = 111$

The angles measure 34.5°, 34.5°, and 111°.

59. Let x = length of the first piece,
$2x$ = length of the second piece, and
$5x$ = length of the third piece.
$x+2x+5x=40$
$8x=40$
$\frac{8x}{8}=\frac{40}{8}$
$x=5$
$2x = 2(5) = 10$
$5x = 5(5) = 25$
The lengths are 5, 10, and 25 inches.

61. Select the tallest bar. Hawaii spends the most money on tourism.

63. Let x = amount spent by Florida, then
$x + 1.7$ = amount spent by Texas.
$x+x+1.7=60.5$
$2x+1.7=60.5$
$2x+1.7-1.7=60.5-1.7$
$2x=58.8$
$\frac{2x}{2}=\frac{58.8}{2}$
$x=29.4$
$x + 1.7 = 29.4 + 1.7 = 31.1$
Florida spends \$29.4 million and Texas spends \$31.1 million.

65. Answers may vary

67. $2W + 2L = 2(7) + 2(10) = 14 + 20 = 34$

69. $\pi r^2 = \pi\cdot(15)^2 = \pi\cdot 225 = 225\pi$

71. Answers may vary

Section 2.6

Practice Exercises

1. Let $d = 580$ and $r = 5$.
$d=r\cdot t$
$580=5t$
$\frac{580}{5}=\frac{5t}{5}$
$116=t$
It takes 116 seconds or 1 minute 56 seconds.

2. Let $l = 40$ and $P = 98$.
$P=2l+2w$
$98=2\cdot 40+2w$
$98=80+2w$
$98-80=80+2w-80$
$18=2w$
$\frac{18}{2}=\frac{2w}{2}$
$9=w$
The dog run is 9 feet wide.

3. Let $C = 8$.
$F=\frac{9}{5}C+32$
$F=\frac{9}{5}\cdot 8+32$
$F=\frac{72}{5}+\frac{160}{5}$
$F=\frac{232}{5}=46.4$
The equivalent temperature is 46.4°F.

4. Let w = width of sign, then
$5w + 3$ = length of sign.
$P=2l+2w$
$66=2(5w+3)+2w$
$66=10w+6+2w$
$66=12w+6$
$66-6=12w+6-6$
$60=12w$
$\frac{60}{12}=\frac{12w}{12}$
$5=w$
$5w + 3 = 5(5) + 3 = 28$
The sign has length 28 inches and width 5 inches.

5. $I=Prt$
$\frac{I}{Pt}=\frac{Prt}{Pt}$
$\frac{I}{Pt}=r$ or $r=\frac{I}{Pt}$

6. $H=5as+10a$
$H-10a=5as+10a-10a$
$H-10a=5as$
$\frac{H-10a}{5a}=\frac{5as}{5a}$
$\frac{H-10a}{5a}=s$ or $s=\frac{H-10a}{5a}$

7.
$$N = F + d(n-1)$$
$$N - F = F + d(n-1) - F$$
$$N - F = d(n-1)$$
$$\frac{N-F}{n-1} = \frac{d(n-1)}{n-1}$$
$$\frac{N-F}{n-1} = d \text{ or } d = \frac{N-F}{n-1}$$

8.
$$A = \frac{1}{2}a(b+B)$$
$$2 \cdot A = 2 \cdot \frac{1}{2}a(b+B)$$
$$2A = a(b+B)$$
$$2A = ab + aB$$
$$2A - ab = ab + aB - ab$$
$$2A - ab = aB$$
$$\frac{2A-ab}{a} = \frac{aB}{a}$$
$$\frac{2A-ab}{a} = B \text{ or } B = \frac{2A-ab}{a}$$

Exercise Set 2.6

1. Let $A = 45$ and $b = 15$.
$$A = bh$$
$$45 = 15h$$
$$\frac{45}{15} = \frac{15h}{15}$$
$$3 = h$$

3. Let $S = 102$, $l = 7$, and $w = 3$.
$$S = 4lw + 2wh$$
$$102 = 4(7)(3) + 2(3)h$$
$$102 = 84 + 6h$$
$$102 - 84 = 84 - 84 + 6h$$
$$18 = 6h$$
$$\frac{18}{6} = \frac{6h}{6}$$
$$3 = h$$

5. Let $A = 180$, $B = 11$, and $b = 7$.
$$A = \frac{1}{2}h(B+b)$$
$$180 = \frac{1}{2}h(11+7)$$
$$2(180) = 2\left[\frac{1}{2}h(18)\right]$$
$$360 = 18h$$
$$\frac{360}{18} = \frac{18h}{18}$$
$$20 = h$$

7. Let $P = 30$, $a = 8$, and $b = 10$.
$$P = a + b + c$$
$$30 = 8 + 10 + c$$
$$30 = 18 + c$$
$$30 - 18 = 18 - 18 + c$$
$$12 = c$$

9. Let $C = 15.7$, and $\pi \approx 3.14$.
$$C = 2\pi r$$
$$15.7 \approx 2(3.14)r$$
$$15.7 \approx 6.28r$$
$$\frac{15.7}{6.28} \approx \frac{6.28r}{6.28}$$
$$2.5 \approx r$$

11. Let $I = 3750$, $P = 25{,}000$, and $R = 0.05$.
$$I = PRT$$
$$3750 = 25{,}000(0.05)T$$
$$3750 = 1250T$$
$$\frac{3750}{1250} = \frac{1250T}{1250}$$
$$3 = T$$

13. Let $V = 565.2$, $r = 6$, and $\pi \approx 3.14$.
$$V = \frac{1}{3}\pi r^2 h$$
$$565.2 \approx \frac{1}{3}(3.14)(6)^2 h$$
$$565.2 \approx 37.68h$$
$$\frac{565.2}{37.68} \approx \frac{37.68h}{37.68}$$
$$15 \approx h$$

15.
$$f = 5gh$$
$$\frac{f}{5g} = \frac{5gh}{5g}$$
$$\frac{f}{5g} = h$$

17.
$$V = lwh$$
$$\frac{V}{lh} = \frac{lwh}{lh}$$
$$\frac{V}{lh} = w$$

19.
$$3x + y = 7$$
$$3x - 3x + y = 7 - 3x$$
$$y = 7 - 3x$$

21. $A = P + PRT$
$A - P = P - P + PRT$
$A - P = PRT$
$\frac{A-P}{PT} = \frac{PRT}{PT}$
$\frac{A-P}{PT} = R$

23. $V = \frac{1}{3}Ah$
$3V = 3\left(\frac{1}{3}Ah\right)$
$3V = Ah$
$\frac{3V}{h} = \frac{Ah}{h}$
$\frac{3V}{h} = A$

25. $P = a + b + c$
$P - (b + c) = a + b + c - (b + c)$
$P - b - c = a + b + c - b - c$
$P - b - c = a$

27. $S = 2\pi rh + 2\pi r^2$
$S - 2\pi r^2 = 2\pi rh + 2\pi r^2 - 2\pi r^2$
$S - 2\pi r^2 = 2\pi rh$
$\frac{S - 2\pi r^2}{2\pi r} = \frac{2\pi rh}{2\pi r}$
$\frac{S - 2\pi r^2}{2\pi r} = h$

29. a. $A = lw$
$A = 11.5(9)$
$A = 103.5$

$P = 2l + 2w$
$P = 2(11.5) + 2(9)$
$P = 23 + 18$
$P = 41$

The area is 103.5 square feet and the perimeter is 41 feet.

b. Baseboards have to do with perimeter because they are installed around the edges. Carpet has to do with area because it is installed in the middle of the room.

31. a. $A = \frac{1}{2}h(b_1 + b_2)$
$A = \frac{1}{2}(12)(56 + 24)$
$A = 6(80)$
$A = 480$

$P = l_1 + l_2 + l_3 + l_4$
$P = 24 + 20 + 56 + 20$
$P = 120$

The area is 480 square inches and the perimeter is 120 inches.

b. The frame has to do with perimeter because it surrounds the edge of the picture. The glass has to do with area because it covers the entire picture.

33. $A = 3990$ and $w = 57$.
$A = lw$
$3990 = l \cdot 57$
$\frac{3990}{57} = \frac{57l}{57}$
$70 = l$
The length is 70 feet.

35. Let $F = 14$.
$14 = \frac{9}{5}C + 32$
$5(14) = 5\left(\frac{9}{5}\right)C + 5(32)$
$70 = 9C + 160$
$70 - 160 = 9C + 160 - 160$
$-90 = 9C$
$\frac{-90}{9} = \frac{9C}{9}$
$-10 = C$
The equivalent temperature is −10°C.

37. Let $d = 25{,}000$ and $r = 4000$.
$d = rt$
$25{,}000 = 4000t$
$\frac{25{,}000}{4000} = \frac{4000t}{4000}$
$6.25 = t$
It will take 6.25 hours.

39. Let $P = 260$ and $w = \frac{2}{3}l$.

$$P = 2l + 2w$$
$$260 = 2l + 2\left(\frac{2}{3}l\right)$$
$$260 = \frac{10}{3}l$$
$$3(260) = 3\left(\frac{10}{3}l\right)$$
$$780 = 10l$$
$$\frac{780}{10} = \frac{10l}{10}$$
$$78 = l$$
$$w = \frac{2}{3}l = \frac{2}{3}(78) = 52$$

The width is 52 feet and the length is 78 feet.

41. Let $P = 102$, a = the length of the shortest side, $b = 2a$, and $c = a + 30$.

$$P = a + b + c$$
$$102 = a + 2a + a + 30$$
$$102 = 4a + 30$$
$$102 - 30 = 4a + 30 - 30$$
$$72 = 4a$$
$$\frac{72}{4} = \frac{4a}{4}$$
$$18 = a$$

$b = 2a = 2(18) = 36$
$c = a + 30 = 18 + 30 = 48$
The lengths are 18 feet, 36 feet, and 48 feet.

43. Let $d = 138$ and $t = 2.5$.

$$d = rt$$
$$138 = r \cdot 2.5$$
$$\frac{138}{2.5} = \frac{r \cdot 2.5}{2.5}$$
$$55.2 = r$$

The speed is 55.2 mph.

45. Let $l = 8$, $w = 6$, and $h = 3$.
$V = lwh$
$V = 8(6)(3) = 144$
Let x = number of piranha and volume per fish = 1.5.

$$144 = 1.5x$$
$$\frac{144}{1.5} = \frac{1.5x}{1.5}$$
$$96 = x$$

96 piranhas can be placed in the tank.

47. Let $h = 60$, $B = 130$, and $b = 70$.

$$A = \frac{1}{2}(B + b)h$$
$$A = \frac{1}{2}(130 + 70)60 = \frac{1}{2}(200)(60) = 6000$$

Let x = number of bags of fertilizer and the area per bag = 4000.

$$4000x = 6000$$
$$\frac{4000x}{4000} = \frac{6000}{4000}$$
$$x = 1.5$$

Two bags must be purchased.

49. Let $d = 16$, so $r = 8$.
$A = \pi r^2 = \pi(8)^2 = 64\pi$
Let $d = 10$, so $r = 5$.
$A = 2\pi r^2 = 2\pi(5)^2 = 50\pi$
One 16-inch pizza has more area and therefore gives more pizza for the price.

51.

$$x + x + x + 2.5x + 2.5x = 48$$
$$8x = 48$$
$$\frac{8x}{8} = \frac{48}{8}$$
$$x = 6$$

$2.5x = 2.5(6) = 15$
Three sides measure 6 meters and two sides measure 15 meters.

53. $r = 361$ and $d = 72.2$.

$$d = rt$$
$$72.2 = 361t$$
$$\frac{72.2}{361} = \frac{361t}{361}$$
$$0.2 = t$$

It will take 0.2 hour or 0.2(60) = 12 minutes.

55. Let x = the length of a side of the square and $x + 5$ = the length of a side of the triangle.

$$P(\text{triangle}) = P(\text{square}) + 7$$
$$3(x + 5) = 4x + 7$$
$$3x + 15 = 4x + 7$$
$$3x - 3x + 15 = 4x - 3x + 7$$
$$15 = x + 7$$
$$15 - 7 = x + 7 - 7$$
$$8 = x$$

$x + 5 = 8 + 5 = 13$
The side of the triangle is 13 inches.

57. Let $d = 135$ and $r = 60$.
$$d = rt$$
$$135 = 60t$$
$$\frac{135}{60} = \frac{60t}{60}$$
$$2.25 = t$$
It would take 2.25 hours.

59. Let $A = 1{,}813{,}500$ and $w = 150$.
$$A = lw$$
$$1{,}813{,}500 = l(150)$$
$$\frac{1{,}813{,}500}{150} = \frac{150l}{150}$$
$$12{,}090 = l$$
The length is 12,090 feet.

61. Let $F = 122$.
$$122 = \frac{9}{5}C + 32$$
$$5(122) = 5\left(\frac{9}{5}\right)C + 5(32)$$
$$610 = 9C + 160$$
$$610 - 160 = 9C + 160 - 160$$
$$450 = 9C$$
$$\frac{450}{9} = \frac{9C}{9}$$
$$50 = C$$
The equivalent temperature is 50°C.

63. Let $l = 199$, $w = 78.5$, and $h = 33$.
$$V = lwh$$
$$V = 199(78.5)(33) = 515{,}509.5$$
The volume must be 515,509.5 cubic inches.

65. Let $\pi \approx 3.14$ and $d = 9.5$ so $r = 4.75$.
$$V = \frac{4}{3}\pi r^3 \approx \frac{4}{3}(3.14)(4.75)^3 \approx 449$$
The volume is 449 cubic inches.

67. Let $C = 167$.
$$F = \frac{9}{5}C + 32$$
$$= \frac{9}{5}(167) + 32$$
$$= 300.6 + 32$$
$$= 332.6$$
The equivalent temperature is 332.6°F.

69. Nine divided by the sum of a number and five is $\frac{9}{x+5}$.

71. Three times the sum of a number and four is $3(x + 4)$.

73. Triple the difference of a number and twelve is $3(x - 12)$.

75. $\square - \bigcirc \cdot \square = \triangle$
$$-\bigcirc \cdot \square = \triangle - \square$$
$$\frac{-\bigcirc\square}{-\square} = \frac{\triangle - \square}{-\square}$$
$$\bigcirc = \frac{\square - \triangle}{\square}$$

77. Let $C = -78.5$.
$$F = \frac{9}{5}C + 32$$
$$= \frac{9}{5}(-78.5) + 32$$
$$= -141.3 + 32$$
$$= -109.3$$
The equivalent temperature is −109.3°F.

79. Let $d = 93{,}000{,}000$ and $r = 186{,}000$.
$$d = rt$$
$$93{,}000{,}000 = 186{,}000t$$
$$\frac{93{,}000{,}000}{186{,}000} = \frac{186{,}000t}{186{,}000}$$
$$500 = t$$
It will take 500 seconds or $8\frac{1}{3}$ minutes.

81. Let $t = 365$ and $r = 20$.
$$d = rt = 20(365) = 7300 \text{ inches}$$
$$\frac{7300 \text{ inches}}{1} \cdot \frac{1 \text{ foot}}{12 \text{ inch}} \approx 608.33 \text{ feet}$$
It moves about 608.33 feet.

83. Let $d = 2$ then $r = 1$.
$$15 \text{ feet} = \frac{15 \text{ feet}}{1} \cdot \frac{12 \text{ inches}}{1 \text{ foot}} = 180 \text{ inches, so}$$
$h = 180$.
$$V = \pi r^2 h$$
$$V = (\pi)(1)^2(180) = 180\pi \approx 565.5$$
The volume of the column is 565.5 cubic inches.

85. The original parallelogram has an area $V = bh$. The altered box has a base $2b$, a height $2h$, and a new area.
$A = 2b(2h) = 4bh$
The area is multiplied by 4.

Section 2.7

Practice Exercises

1. Let x = the unknown percent.

$$35 = x \cdot 56$$
$$\frac{35}{56} = \frac{56x}{56}$$
$$0.625 = x$$

The number 35 is 62.5% of 56.

2. Let x = the unknown number.

$$198 = 55\% \cdot x$$
$$198 = 0.55x$$
$$\frac{198}{0.55} = \frac{0.55x}{0.55}$$
$$360 = x$$

The number 198 is 55% of 360.

3. a. From the circle graph, 4% of trips made by American travelers are for combined business/pleasure.

b. From the circle graph,
17% + 66% + 4% = 87% of trips are for business, pleasure, or combined business/pleasure.

c. Since 4% are trips for business/pleasure, find 4% of 325.
$0.04 \cdot 325 = 13$
We can expect 13 of the Americans to be traveling for business/pleasure.

4. Let x = discount.

$$x = 85\% \cdot 480$$
$$x = 0.85 \cdot 480$$
$$x = 408$$

The discount is \$408.
New price = \$480 − \$408 = \$72

5. Increase = 299,800 − 198,900 = 100,900
Let x = percent increase.

$$100,900 = x \cdot 198,900$$
$$\frac{100,900}{198,900} = \frac{198,900x}{198,900}$$
$$0.507 \approx x$$

The percent increase is 50.7%.

6. Let x = number of new films in 2004.

$$x + 0.028x = 535$$
$$1.028x = 535$$
$$\frac{1.028x}{1.028} = \frac{535}{1.028}$$
$$x \approx 520$$

There were 520 new feature films released in 2004.

7. Let x = number of liters of 2% solution.

Eyewash	No. of gallons	· Acid Strength	= Amt. of Acid
2%	x	2%	$0.02x$
5%	$6-x$	5%	$0.05(6-x)$
Mix: 3%	6	3%	$0.03(6)$

$$\begin{aligned} 0.02x + 0.05(6-x) &= 0.03(6) \\ 0.02x + 0.3 - 0.05x &= 0.18 \\ -0.03x + 0.3 &= 0.18 \\ -0.03x + 0.3 - 0.3 &= 0.18 - 0.3 \\ -0.03x &= -0.12 \\ \frac{-0.03x}{-0.03} &= \frac{-0.12}{-0.03} \\ x &= 4 \end{aligned}$$

$6 - x = 6 - 4 = 2$

She should mix 4 liters of 2% eyewash with 2 liters of 5% eyewash.

Vocabulary and Readiness Check

1. No, 25% + 25% + 40% = 90% ≠ 100%.

2. No, 30% + 30% + 30% = 90% ≠ 100%.

3. Yes, 25% + 25% + 25% + 25% = 100%.

4. Yes, 40% + 50% + 10% = 100%.

Exercise Set 2.7

1. Let x = the number.
$x = 16\% \cdot 70$
$x = 0.16 \cdot 70$
$x = 11.2$
11.2 is 16% of 70.

3. Let x = the percent.
$28.6 = x \cdot 52$
$\frac{28.6}{52} = \frac{52x}{52}$
$0.55 = x$
28.6 is 55% of 52.

5. Let x = the number.
$45 = 25\% \cdot x$
$45 = 0.25x$
$\frac{45}{0.25} = \frac{0.25x}{0.25}$
$180 = x$
45 is 25% of 180.

7. Animal Feed = 51%
Ethanol = 18%
51% + 18% = 69%
69% of corn production is used for animal feed or ethanol.

9. 18% of 10,535 = 0.18(10,535) = 1896.3
1896.3 million bushels or 1,896,300,000 bushels were used to make ethanol.

11. Let x = amount of discount.
$x = 8\% \cdot 18{,}500$
$x = 0.08 \cdot 18{,}500$
$x = 1480$
New price $= 18{,}500 - 1480 = 17{,}020$
The discount was \$1480 and the new price is \$17,020.

13. Let x = tip.
$x = 15\% \cdot 40.50$
$x = 0.15 \cdot 40.5$
$x = 6.075 \approx 6.08$
Total = 40.50 + 6.08 = 46.58
The total cost is \$46.58.

15. Increase = 280 − 208 = 72
Let x = percent.
$72 = x \cdot 208$
$\frac{72}{208} = \frac{208x}{208}$
$35 \approx x$
The percent increase is 35%.

17. Decrease = 40 − 28 = 12
Let x = percent.
$12 = x \cdot 40$
$\frac{12}{40} = \frac{40x}{40}$
$0.3 = x$
The percent decrease is 30%.

19. Let x = the original price and $0.25x$ = the discount.
$x - 0.25x = 78$
$0.75x = 78$
$\frac{0.75x}{0.75} = \frac{78}{0.75}$
$x = 104$
The original price was \$104.

21. Let x = last year's salary, and $0.04x$ = pay raise.
$x + 0.04x = 44{,}200$
$1.04x = 44{,}200$
$\frac{1.04x}{1.04} = \frac{44{,}200}{1.04}$
$x = 42{,}500$
Last year's salary was \$42,500.

23. Let x = the amount of pure acid.

	No. of gallons	· Strength	= Amt. of Acid
100%	x	1.00	x
40%	2	0.4	2(0.4)
70%	$x+2$	0.7	$0.7(x+2)$

$x + 2(0.4) = 0.7(x+2)$
$x + 0.8 = 0.7x + 1.4$
$x - 0.7x + 0.8 = 0.7x - 0.7x + 1.4$
$0.3x + 0.8 = 1.4$
$0.3x + 0.8 - 0.8 = 1.4 - 0.8$
$0.3x = 0.6$
$\frac{0.3x}{0.3} = \frac{0.6}{0.3}$
$x = 2$
Mix 2 gallons of pure acid.

25. Let x = the number of pounds at \$7/lb.

	No. of lb	· Cost/lb	= Value
\$7/lb	x	7	$7x$
\$4/lb	14	4	4(14)
\$5/lb	$x + 14$	5	$5(x+14)$

$7x + 4(14) = 5(x+14)$
$7x + 56 = 5x + 70$
$7x - 5x + 56 = 5x - 5x + 70$
$2x + 56 = 70$
$2x + 56 - 56 = 70 - 56$
$2x = 14$
$\frac{2x}{2} = \frac{14}{2}$
$x = 7$
Add 7 pounds of \$7/pound coffee.

27. Let x = the number.
$x = 23\% \cdot 20$
$x = 0.23 \cdot 20$
$x = 4.6$
23% of 20 is 4.6.

29. Let x = the number.
$40 = 80\% \cdot x$
$40 = 0.8x$
$\frac{40}{0.8} = \frac{0.8x}{0.8}$
$50 = x$
40 is 80% of 50.

31. Let x = the percent.
$144 = x \cdot 480$
$\frac{144}{480} = \frac{480x}{480}$
$0.3 = x$
144 is 30% of 480.

33. From the graph, the height of the bar is 71. Therefore, 71% of the population in Fairbanks, Alaska, shop by catalog.

35. 65% of 275,043 $= 0.65 \cdot 275{,}043 \approx 178{,}778$
We predict 178,778 catalog shoppers live in Anchorage.

37.

Ford Motor Company Model Year 2006 Vehicle Sales Worldwide

	Thousands of Vehicles	Percent of Total (Rounded to Nearest Percent)
North America	3051	$\frac{3051}{6597} \approx 46\%$
Europe	1846	$\frac{1846}{6597} \approx 28\%$
Asia-Pacific-Africa	589	$\frac{589}{6597} \approx 9\%$
South America	381	$\frac{381}{6597} \approx 6\%$
Rest of the World	730	Example: $\frac{730}{6597} \approx 11\%$
Total	6597	100%

Source: Ford Motor Company

39. Let x = the decrease in price.
$x = 0.25(256) = 64$
The decrease in price is \$64.
The sale price is $256 - 64 = \$192$.

41. Increase $= 86 - 40 = 46$
Let x = the percent.
$46 = x \cdot 40$
$\frac{46}{40} = \frac{40x}{40}$
$1.15 = x$
The percent increase is 115%.

43. Let x = the number of cards (in millions) issued in 2001.
Increase $= 7.26x$
No. in 2001 + increase = no. in 2006
$x + 7.26 = 1900$
$8.26x = 1900$
$\frac{8.26x}{8.26} = \frac{1900}{8.26}$
$x = 230$
230 million cards were issued in 2006.

45. Let x = the amount of 20% alloy.

	No. of Oz	· Strength	= Amt. of Copper
50%	200	0.5	200(0.5)
20%	x	0.2	$0.2x$
Mix	$x + 200$	0.3	$0.3(x + 200)$

$$200(0.5)+0.2x=0.3(x+200)$$
$$100+0.2x=0.3x+60$$
$$100+0.2x-0.2x=0.3x-0.2x+60$$
$$100=0.1x+60$$
$$100-60=0.1x+60-60$$
$$40=0.1x$$
$$\frac{40}{0.1}=\frac{0.1x}{0.1}$$
$$400=x$$

Mix with 400 ounces of 20% alloy.

47. Let x = mark-up.
$x = 70\% \cdot 27$
$x = 0.7 \cdot 27$
$x = 18.9$
Adult price = 27 + 18.9 = 45.9
The mark-up is \$18.90 and the adult ticket price is \$45.90.

49. Increase = 144 − 36 = 108
Let x = percent.
$$108 = x \cdot 36$$
$$\frac{108}{36}=\frac{36x}{36}$$
$$3 = x$$
The percent increase is 300%.

51. Let x = the number of employees prior to layoff.
Decrease = $0.35x$
$$x-0.35x=78$$
$$0.65x=78$$
$$\frac{0.65x}{0.65}=\frac{78}{0.65}$$
$$x=120$$
There were 120 employees prior to the layoffs.

53. Let x = pounds of peanuts.

	pounds	price/lb	price
peanuts	x	5	$5x$
bites	10	2	2(10)
trail mix	$10 + x$	3	$3(10 + x)$

$$5x+2(10)=3(10+x)$$
$$5x+20=30+3x$$
$$5x+20-3x=30+3x-3x$$
$$20+2x=30$$
$$20+2x-20=30-20$$
$$2x=10$$
$$\frac{2x}{2}=\frac{10}{2}$$
$$x=5$$
Therefore, 5 pounds of chocolate-covered peanuts should be mixed.

55. Decrease = 2.19 − 2.09 = 0.10
Let x = percent.
$$0.10 = x \cdot 2.19$$
$$\frac{0.10}{2.19}=\frac{2.19x}{2.19}$$
$$0.046 \approx x$$
The percent decrease is 4.6%.

57. Let x = number of decisions in 1982–1983, then $0.457x$ = decrease.
$$x-0.457x=182$$
$$0.543x=182$$
$$\frac{0.543x}{0.543}=\frac{182}{0.543}$$
$$x \approx 335$$
There were 335 decisions in 1982–1983.

59. Let x = increase.
$x = 48\% \cdot 577$
$x = 0.48 \cdot 577$
$x \approx 277$
Naga pepper = 577 + 277 = 854
The Naga Jolokia pepper measures 854 thousand Scoville units.

61. 42% of 860 = $0.42 \cdot 860 \approx 361$
You would expect 361 college students to rank flexible hours as their top priority.

63. $-5 > -7$

65. $|-5| = -(-5)$

67. $(-3)^2 = 9; -3^2 = -9$
$(-3)^2 > -3^2$

69. No; answers may vary

71. No; answers may vary

73. 230 is x percent of 2400.

$x(2400) = 230$

$$x = \frac{230}{2400} \approx 0.096 = 9.6\%$$

This is about 9.6% of the daily value.

75. Let x = percent of calories from fat.

$x(130) = 35$

$$x = \frac{35}{130} \approx 0.269 = 26.9\%$$

This is less than 30% so it satisfies the recommendation.

77. Let x = percent of calories from protein.

$x(280) = 4(12)$

$$x = \frac{48}{280} \approx 0.171 = 17.1\%$$

About 17.1% of the calories in one serving come from protein.

Section 2.8

Practice Exercises

1. Let x = time down, then $x + 1$ = time up.

	Rate ·	Time =	Distance
Up	1.5	$x + 1$	$1.5(x + 1)$
Down	4	x	$4x$

$$d = d$$
$$1.5(x+1) = 4x$$
$$1.5x + 1.5 = 4x$$
$$1.5 = 2.5x$$
$$\frac{1.5}{2.5} = \frac{2.5x}{2.5}$$
$$0.6 = x$$

Total Time = $x + 1 + x = 0.6 + 1 + 0.6 = 2.2$

The entire hike took 2.2 hours.

2. Let x = speed of eastbound train, then $x - 10$ = speed of westbound train.

	r ·	t =	d
East	x	1.5	$1.5x$
West	$x - 10$	1.5	$1.5(x - 10)$

$$1.5x + 1.5(x - 10) = 171$$
$$1.5x + 1.5x - 15 = 171$$
$$3x - 15 = 171$$
$$3x = 186$$
$$\frac{3x}{3} = \frac{186}{3}$$
$$x = 62$$

$x - 10 = 62 - 10 = 52$

The eastbound train is traveling at 62 mph and the westbound train is traveling at 52 mph.

3. Let x = the number of \$20 bills, then $x + 47$ = number of \$5 bills.

Denomination	Number	Value
\$5 bills	$x + 47$	$5(x + 47)$
\$20 bills	x	$20x$

$$5(x+47) + 20x = 1710$$
$$5x + 235 + 20x = 1710$$
$$235 + 25x = 1710$$
$$25x = 1475$$
$$x = 59$$

$x + 47 = 59 + 47 = 106$

There are 106 \$5 bills and 59 \$20 bills.

4. Let x = amount invested at 11.5%, then
$30{,}000 - x$ = amount invested at 6%.

	Principal ·	Rate ·	Time =	Interest
11.5%	x	0.115	1	$x(0.115)(1)$
6%	$30{,}000 - x$	0.06	1	$0.06(30{,}000 - x)(1)$
Total	30,000			2790

$$0.115x + 0.06(30{,}000 - x) = 2790$$
$$0.115x + 1800 - 0.06x = 2790$$
$$1800 + 0.055x = 2790$$
$$0.055x = 990$$
$$\frac{0.055x}{0.055} = \frac{990}{0.055}$$
$$x = 18{,}000$$

$30{,}000 - x = 30{,}000 - 18{,}000 = 12{,}000$
She invested \$18,000 at 11.5% and \$12,000 at 6%.

Exercise Set 2.8

1. Let x = the time traveled by the jet plane.

	Rate ·	Time =	Distance
Jet	500	x	$500x$
Prop	200	$x + 2$	$200(x + 2)$

$$d = d$$
$$500x = 200(x + 2)$$
$$500x = 200x + 400$$
$$300x = 400$$
$$\frac{300x}{300} = \frac{400}{300}$$
$$x = \frac{4}{3}$$

The jet traveled for $\frac{4}{3}$ hours.

$$d = rt$$
$$d = 500\left(\frac{4}{3}\right) = 666\frac{2}{3}$$

The planes are $666\frac{2}{3}$ miles from the starting point.

3. Let x = the average speed on the winding road and $x + 20$ on the level.

	Rate	· Time	= Distance
Winding	x	4	$4x$
Level	$x + 20$	3	$3(x + 20)$
Total			305

$$4x+3(x+20)=305$$
$$4x+3x+60=305$$
$$7x+60=305$$
$$7x=245$$
$$\frac{7x}{7}=\frac{245}{7}$$
$$x=35$$
$x + 20 = 35 + 20 = 55$
The average speed on level road was 55 mph.

5. The value of y dimes is $0.10y$.

7. The value of $x + 7$ nickels is $0.05(x + 7)$.

9. The value of $4y$ \$20 bills is $20(4y)$ or $80y$.

11. The value of $35 - x$ \$50 bills is $50(35 - x)$.

13. Let x = number of \$10 bills, then $20 + x$ number of \$5 bills.

	Number of Bills	Value of Bills
\$5 bills	$20 + x$	$5(20 + x)$
\$10 bills	x	$10x$
Total		280

$$5(20+x)+10x=280$$
$$100+5x+10x=280$$
$$100+15x=280$$
$$15x=180$$
$$x=12$$
$20 + x = 32$
There are 12 \$10 bills and 32 \$5 bills.

15. Let x = the amount invested at 9% for one year.

	Principal	· Rate =	Interest
9%	x	0.09	$0.09x$
8%	$25{,}000 - x$	0.08	$0.08(25{,}000 - x)$
Total	25,000		2135

$$0.09x+0.08(25{,}000-x)=2135$$
$$0.09x+2000-0.08x=2135$$
$$0.01x+2000=2135$$
$$0.01x=135$$
$$\frac{0.01x}{0.01}=\frac{135}{0.01}$$
$$x=13{,}500$$
$25{,}000 - x = 25{,}000 - 13{,}500 = 11{,}500$
She invested \$11,500 at 8% and \$13,500 at 9%.

17. Let x = the amount invested at 11% for one year.

	Principal	· Rate =	Interest
11%	x	0.11	$0.11x$
4%	$10{,}000 - x$	-0.04	$-0.04(10{,}000 - x)$
Total	10,000		650

$$0.11x-0.04(10{,}000-x)=650$$
$$0.11x-400+0.04x=650$$
$$0.15x-400=650$$
$$0.15x=1050$$
$$\frac{0.15x}{0.15}=\frac{1050}{0.15}$$
$$x=7000$$
$10{,}000 - x = 10{,}000 - 7000 = 3000$
He invested \$7000 at 11% and \$3000 at 4%.

19. Let x = the number of adult tickets, then $500 - x$ = the number of child tickets.

	Number ·	Rate =	Cost
Adult	x	43	$43x$
Child	$500 - x$	28	$28(500 - x)$
Total	500		16,805

$$43x + 28(500 - x) = 16{,}805$$
$$43x + 14{,}000 - 28x = 16{,}805$$
$$14{,}000 + 15x = 16{,}805$$
$$15x = 2805$$
$$x = 187$$

$500 - x = 500 - 187 = 313$

Sales included 187 adult tickets and 313 child tickets.

21. Let x = the amount invested at 10% for one year.

	Principal ·	Rate =	Interest
10%	x	0.10	$0.10x$
8%	$54{,}000 - x$	0.08	$0.08(54{,}000 - x)$

$$0.10x = 0.08(54{,}000 - x)$$
$$0.10x = 4320 - 0.08x$$
$$0.18x = 4320$$
$$\frac{0.18x}{0.18} = \frac{4320}{0.18}$$
$$x = 24{,}000$$

$54{,}000 - x = 54{,}000 - 24{,}000 = 30{,}000$

Invest \$30,000 at 8% and \$24,000 at 10%.

23. Let x = the time they are able to talk.

	Rate ·	Time =	Distance
Alan	55	x	$55x$
Dave	65	$x - 1$	$65(x - 1)$
Total			250

$$55x + 65(x - 1) = 250$$
$$55x + 65x - 65 = 250$$
$$120x - 65 = 250$$
$$120x = 315$$
$$\frac{120x}{120} = \frac{315}{120}$$
$$x = 2\frac{5}{8}$$

They can talk for $2\frac{5}{8}$ hours or

2 hours $37\frac{1}{2}$ minutes.

25. Let x = number of nickels, then $3x$ = number of dimes.

	Number	Value
Nickels	x	$0.05x$
Dimes	$3x$	$0.10(3x)$
Total		56.35

$$0.05x + 0.10(3x) = 56.35$$
$$0.05x + 0.3x = 56.35$$
$$0.35x = 56.35$$
$$x = 161$$

$3x = 3(161) = 483$

They collected 161 nickels and 483 dimes.

27. Let x = the amount invested at 9% for one year.

	Principal ·	Rate =	Interest
9%	x	0.09	$0.09x$
6%	3000	0.06	$0.06(3000)$
Total			585

$$0.09x + 0.06(3000) = 585$$
$$0.09x + 180 = 585$$
$$0.09x = 405$$
$$\frac{0.09x}{0.09} = \frac{405}{0.09}$$
$$x = 4500$$

Should invest \$4500 at 9%.

29. Let x = the rate of hiker 1.

	Rate ·	Time =	Distance
Hiker 1	x	2	$2x$
Hiker 2	$x + 1.1$	2	$2(x + 1.1)$
Total			11

$$2x + 2(x + 1.1) = 11$$
$$2x + 2x + 2.2 = 11$$
$$4x + 2.2 = 11$$
$$4x = 8.8$$
$$\frac{4x}{4} = \frac{8.8}{4}$$
$$x = 2.2$$

$x + 1.1 = 2.2 + 1.1 = 3.3$

Hiker 1: 2.2 mph; Hiker 2: 3.3 mph

31. Let x = the time spent rowing upstream.

	Rate	· Time	= Distance
Upstream	5	x	$5x$
Downstream	11	$4-x$	$11(4-x)$

$$5x = 11(4-x)$$
$$5x = 44-11x$$
$$16x = 44$$
$$\frac{16x}{16} = \frac{44}{16}$$
$$x = 2.75$$

He rowed upstream for 2.75 hours.
$d = rt$
$d = 5(2.75) = 13.75$
He rowed 13.75 miles each way for a total of 27.5 miles.

33. $3 + (-7) = -4$

35. $\frac{3}{4}-\frac{3}{16}=\frac{4}{4}\cdot\frac{3}{4}-\frac{3}{16}=\frac{12}{16}-\frac{3}{16}=\frac{12-3}{16}=\frac{9}{16}$

37. $-5-(-1) = -5+1 = -4$

39. Let x = number of \$100 bills, then
$x + 46$ = number of \$50 bills, and
$7x$ = number of \$20 bills.

	Number	Value
\$100 bills	x	$100x$
\$50 bills	$x+46$	$50(x+46)$
\$20 bills	$7x$	$20(7x)$
Total		9550

$$100x+50(x+46)+20(7x) = 9550$$
$$100x+50x+2300+140x = 9550$$
$$290x+2300 = 9550$$
$$290x = 7250$$
$$x = 25$$

$x + 46 = 71$
$7x = 7(25) = 175$
There were 25 \$100 bills, 71 \$50 bills, and 175 \$20 bills.

41.
$$R = C$$
$$24x = 100+20x$$
$$4x = 100$$
$$\frac{4x}{x} = \frac{100}{4}$$
$$x = 25$$

Should sell 25 skateboards to break even.

43.
$$R = C$$
$$7.50x = 4.50x+2400$$
$$3x = 2400$$
$$\frac{3x}{3} = \frac{2400}{3}$$
$$x = 800$$

Should sell 800 books to break even.

45. Answers may vary

Section 2.9

Practice Exercises

1. $x < 5$
Place a parenthesis at 5 since the inequality symbol is <. Shade to the left of 5. The solution set is $(-\infty, 5)$.

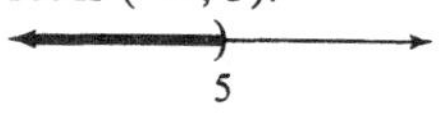

2.
$$x+11 \ge 6$$
$$x+11-11 \ge 6-11$$
$$x \ge -5$$

The solution set is $[-5, \infty)$.

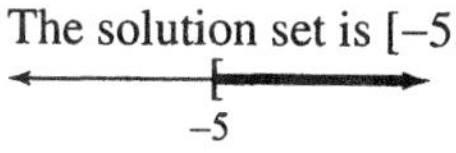

3.
$$-5x \ge -15$$
$$\frac{-5x}{-5} \le \frac{-15}{-5}$$
$$x \le 3$$

The solution set is $(-\infty, 3]$.

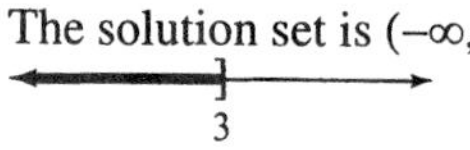

4.
$$3x > -9$$
$$\frac{3x}{3} > \frac{-9}{3}$$
$$x > -3$$

The solution set is $(-3, \infty)$.

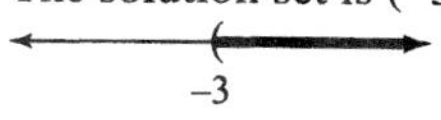

5.
$$45-7x \le -4$$
$$45-7x-45 \le -4-45$$
$$-7x \le -49$$
$$\frac{-7x}{-7} \ge \frac{-49}{-7}$$
$$x \ge 7$$
The solution set is $[7, \infty)$.

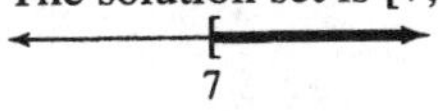

6.
$$3x+20 \le 2x+13$$
$$3x+20-2x \le 2x+13-2x$$
$$x+20 \le 13$$
$$x+20-20 \le 13-20$$
$$x \le -7$$
The solution set is $(-\infty, -7]$.

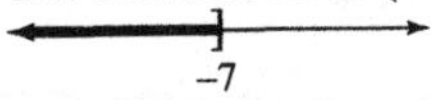

7.
$$6-5x > 3(x-4)$$
$$6-5x > 3x-12$$
$$6-5x-3x > 3x-12-3x$$
$$6-8x > -12$$
$$6-8x-6 > -12-6$$
$$-8x > -18$$
$$\frac{-8x}{-8} < \frac{-18}{-8}$$
$$x < \frac{9}{4}$$
The solution set is $\left(-\infty, \frac{9}{4}\right)$.

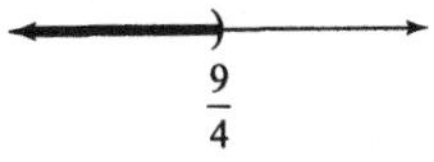

8.
$$3(x-4)-5 \le 5(x-1)-12$$
$$3x-12-5 \le 5x-5-12$$
$$3x-17 \le 5x-17$$
$$3x-17-5x \le 5x-17-5x$$
$$-2x-17 \le -17$$
$$-2x-17+17 \le -17+17$$
$$-2x \le 0$$
$$\frac{-2x}{-2} \ge \frac{0}{-2}$$
$$x \ge 0$$
The solution set is $[0, \infty)$.

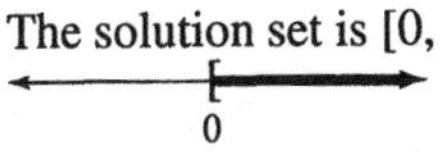

9. $-3 \le x < 1$

Graph all numbers greater than or equal to -3 and less than 1. Place a bracket at -3 and a parenthesis at 1.

The solution set is $[-3, 1)$.

10.
$$-4 < 3x+2 \le 8$$
$$-4-2 < 3x+2-2 \le 8-2$$
$$-6 < 3x \le 6$$
$$\frac{-6}{3} < \frac{3x}{3} \le \frac{6}{3}$$
$$-2 < x \le 2$$
The solution set is $(-2, 2]$.

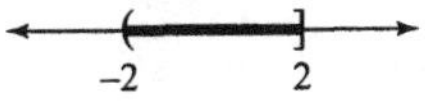

11.
$$1 < \frac{3}{4}x+5 < 6$$
$$4(1) < 4\left(\frac{3}{4}x+5\right) < 4(6)$$
$$4 < 3x+20 < 24$$
$$4-20 < 3x+20-20 < 24-20$$
$$-16 < 3x < 4$$
$$\frac{-16}{3} < \frac{3x}{3} < \frac{4}{3}$$
$$-\frac{16}{3} < x < \frac{4}{3}$$
The solution set is $\left(-\frac{16}{3}, \frac{4}{3}\right)$.

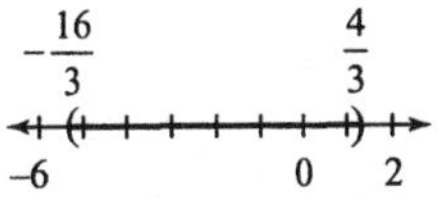

12. Let x = number of classes.
$$300+375x \le 1500$$
$$300+375x-300 \le 1500-300$$
$$375x \le 1200$$
$$\frac{375x}{375} \le \frac{1200}{375}$$
$$x \le 3.2$$
Kasonga can afford at most 3 community college classes this semester.

Vocabulary and Readiness Check

1. $6x - 7(x + 9)$ is an expression.
2. $6x = 7(x + 9)$ is an equation.
3. $6x < 7(x + 9)$ is an inequality.

4. $5y - 2 \geq -38$ is an inequality.

5. $\frac{9}{7} = \frac{x+2}{14}$ is an equation.

6. $\frac{9}{7} - \frac{x+2}{14}$ is an expression.

7. -5 is not a solution to $x \geq -3$.

8. $|-6| = 6$ is not a solution to $x < 6$.

9. 4.1 is not a solution to $x < 4.01$.

10. -4 is not a solution to $x \geq -3$.

Exercise Set 2.9

1. $[2, \infty), x \geq 2$

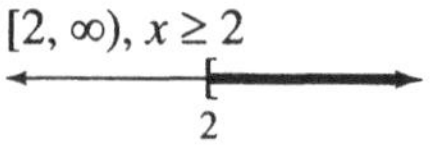

2

3. $(-\infty, -5), x < -5$

−5

5. $x \leq -1, (-\infty, -1]$

−1

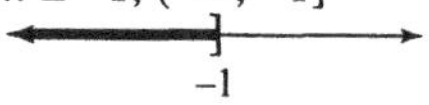

7. $x < \frac{1}{2}, \left(-\infty, \frac{1}{2}\right)$

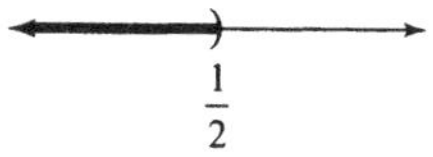

$\frac{1}{2}$

9. $y \geq 5, [5, \infty)$

5

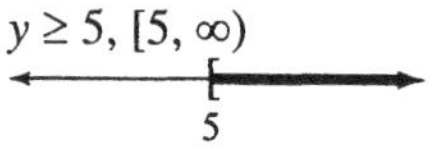

11. $2x < -6$

$x < -3, (-\infty, -3)$

−3

13. $x - 2 \geq -7$

$x \geq -5, [-5, \infty)$

−5

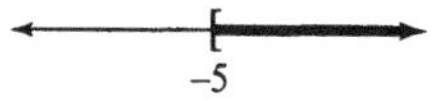

15. $-8x \leq 16$

$\frac{-8x}{-8} \geq \frac{16}{-8}$

$x \geq -2, [-2, \infty)$

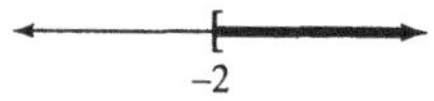

−2

17. $3x - 5 > 2x - 8$

$x - 5 > -8$

$x > -3, (-3, \infty)$

−3

19. $4x - 1 \leq 5x - 2x$

$4x - 1 \leq 3x$

$x - 1 \leq 0$

$x \leq 1, (-\infty, 1]$

1

21. $x - 7 < 3(x + 1)$

$x - 7 < 3x + 3$

$-2x - 7 < 3$

$-2x < 10$

$\frac{-2x}{-2} > \frac{10}{-2}$

$x > -5, (-5, \infty)$

−5

23. $-6x + 2 \geq 2(5 - x)$

$-6x + 2 \geq 10 - 2x$

$-4x + 2 \geq 10$

$-4x \geq 8$

$\frac{-4x}{-4} \leq \frac{8}{-4}$

$x \leq -2, (\infty, -2]$

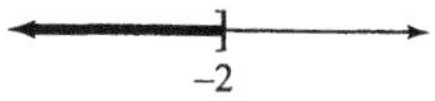

−2

25. $4(3x - 1) \leq 5(2x - 4)$

$12x - 4 \leq 10x - 20$

$2x - 4 \leq -20$

$2x \leq -16$

$x \leq -8, (-\infty, -8]$

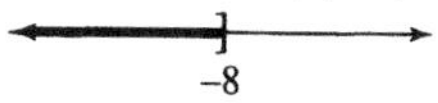

−8

27. $3(x + 2) - 6 > -2(x - 3) + 14$

$3x + 6 - 6 > -2x + 6 + 14$

$3x > -2x + 20$

$5x > 20$

$x > 4, (4, \infty)$

4

29. $-2x \le -40$
$\frac{-2x}{-2} \ge \frac{-40}{-2}$
$x \ge 20$, $[20, \infty)$

20

31. $-9 + x > 7$
$x > 16$, $(16, \infty)$

16

33. $3x - 7 < 6x + 2$
$-3x - 7 < 2$
$-3x < 9$
$\frac{-3x}{-3} > \frac{9}{-3}$
$x > -3$, $(-3, \infty)$

-3

35. $5x - 7x \ge x + 2$
$-2x \ge x + 2$
$-3x \ge 2$
$\frac{-3x}{-3} \le \frac{2}{-3}$
$x \le -\frac{2}{3}$, $\left(-\infty, -\frac{2}{3}\right]$

$-\frac{2}{3}$

37. $\frac{3}{4}x > 2$
$x > \frac{8}{3}$, $\left(\frac{8}{3}, \infty\right)$

$\frac{8}{3}$

39. $3(x - 5) < 2(2x - 1)$
$3x - 15 < 4x - 2$
$-x - 15 < -2$
$-x < 13$
$\frac{-x}{-1} > \frac{13}{-1}$
$x > -13$, $(-13, \infty)$

-13

41. $4(2x + 1) < 4$
$8x + 4 < 4$
$8x < 0$
$x < 0$, $(-\infty, 0)$

0

43. $-5x + 4 \ge -4(x - 1)$
$-5x + 4 \ge -4x + 4$
$-x + 4 \ge 4$
$-x \ge 0$
$\frac{-x}{-1} \le \frac{0}{-1}$
$x \le 0$, $(-\infty, 0]$

0

45. $-2(x - 4) - 3x < -(4x + 1) + 2x$
$-2x + 8 - 3x < -4x - 1 + 2x$
$-5x + 8 < -2x - 1$
$-3x + 8 < -1$
$-3x < -9$
$\frac{-3x}{-3} > \frac{-9}{-3}$
$x > 3$, $(3, \infty)$

3

47. $-3x + 6 \ge 2x + 6$
$-5x + 6 \ge 6$
$-5x \ge 0$
$\frac{-5x}{-5} \le \frac{0}{-5}$
$x \le 0$, $(-\infty, 0]$

0

49. Answers may vary

51. $-1 < x < 3$, $(-1, 3)$

-1 3

53. $0 \le y < 2$, $[0, 2)$

0 2

55. $-3 < 3x < 6$
$-1 < x < 2$, $(-1, 2)$

-1 2

57. $2 \le 3x - 10 \le 5$
$12 \le 3x \le 15$
$4 \le x \le 5$, [4, 5]

59. $-4 < 2(x-3) \le 4$
$-4 < 2x - 6 \le 4$
$2 < 2x \le 10$
$1 < x \le 5$, (1, 5]

61. $-2 < 3x - 5 < 7$
$3 < 3x < 12$
$1 < x < 4$, (1, 4)

63. $-6 < 3(x-2) \le 8$
$-6 < 3x - 6 \le 8$
$0 < 3x \le 14$
$0 < x \le \frac{14}{3}, \left(0, \frac{14}{3}\right]$

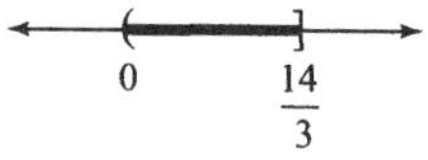

65. Answers may vary

67. $2x + 6 > -14$
$2x > -20$
$x > -10$

69. Let x = the number of people invited.
$34x + 50 \le 3000$
$34x \le 2950$
$x \le 86.8$
They may invite 86 people.

71. Let x = the length.
$2l + 2w = P$
$2x + 2(15) \le 100$
$2x + 30 \le 100$
$2x \le 70$
$x \le 35$
The length can be no greater than 35 cm.

73. Let x = the rate for \$5000 for one year.

	Principal ·	Rate =	Interest
11%	10,000	0.11	0.11(10,000)
?	5000	x	$5000x$
Total			1600

$0.11(10,000) + 5000x \ge 1600$
$1100 + 5000x \ge 1600$
$5000x \ge 500$
$x \ge 0.1$
Should invest the \$5000 at 10% or more.

75. Let x = his score on the third game.
$\frac{146 + 201 + x}{3} \ge 180$
$3\left(\frac{146 + 201 + x}{3}\right) \ge 3(180)$
$347 + x \ge 540$
$x \ge 193$
He must bowl at least 193.

77.

$x < 200$	recommended
$200 \le x \le 240$	borderline
$x > 240$	high

79. Let x = the unknown number.
$-5 < 2x + 1 < 7$
$-6 < 2x < 6$
$-3 < x < 3$
All numbers between −3 and 3

81. $-39 \le \frac{5}{9}(F - 32) \le 45$
$-351 \le 5(F - 32) \le 405$
$-351 \le 5F - 160 \le 405$
$-191 \le 5F \le 565$
$-38.2° \le F \le 113°$

83. $(2)^3 = (2)(2)(2) = 8$

85. $(1)^{12} = (1)(1)(1)(1)(1)(1)(1)(1)(1)(1)(1)(1) = 1$

87. $\left(\frac{4}{7}\right) = \left(\frac{4}{7}\right)\left(\frac{4}{7}\right) = \frac{16}{49}$

89. Read the value on the vertical axis corresponding to 2003; \$52.70

91. The greatest drop was in 2005.

93. $C = 3.14d$

$$2.9 \le 3.14d \le 3.1$$
$$0.924 \le d \le 0.987$$

The diameter must be between 0.924 cm and 0.987 cm.

95. $x(x+4) > x^2 - 2x + 6$

$$x^2 + 4x > x^2 - 2x + 6$$
$$4x > -2x + 6$$
$$6x > 6$$
$$x > 1,\ (1, \infty)$$

97. $x^2 + 6x - 10 < x(x-10)$

$$x^2 + 6x - 10 < x^2 - 10x$$
$$6x - 10 < -10x$$
$$16x - 10 < 0$$
$$16x < 10$$
$$x < \frac{10}{6}$$
$$x < \frac{5}{8},\ \left(-\infty, \frac{5}{8}\right)$$

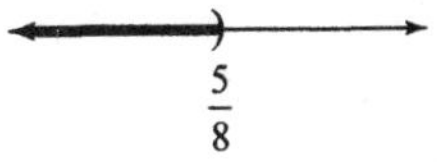

The Bigger Picture

1. $-5x = 15$

$$\frac{-5x}{-5} = \frac{15}{-5}$$
$$x = -3$$

The solution is −3.

2. $-5x > 15$

$$\frac{-5x}{-5} < \frac{15}{-5}$$
$$x < -3$$

The solution set is $(-\infty, -3)$.

3. $9y - 14 = -12$

$$9y - 14 + 14 = -12 + 14$$
$$9y = 2$$
$$\frac{9y}{9} = \frac{2}{9}$$
$$y = \frac{2}{9}$$

The solution is $\frac{2}{9}$.

4. $9x - 3 = 5x - 4$

$$9x - 3 - 5x = 5x - 4 - 5x$$
$$4x - 3 = -4$$
$$4x - 3 + 3 = -4 + 3$$
$$4x = -1$$
$$\frac{4x}{4} = \frac{-1}{4}$$
$$x = -\frac{1}{4}$$

The solution is $-\frac{1}{4}$.

5. $4(x-2) \le 5x + 7$

$$4x - 8 \le 5x + 7$$
$$4x - 8 - 5x \le 5x + 7 - 5x$$
$$-x - 8 \le 7$$
$$-x - 8 + 8 \le 7 + 8$$
$$-x \le 15$$
$$\frac{-x}{-1} \ge \frac{15}{-1}$$
$$x \ge -15$$

The solution set is $[-15, \infty)$.

6. $5(4x-1) = 2(10x-1)$

$$20x - 5 = 20x - 2$$
$$20x - 5 - 20x = 20x - 2 - 20x$$
$$-5 = -2$$

Since this is a false statement, there is no solution.

7. $-5.4 = 0.6x - 9.6$

$$-5.4 + 9.6 = 0.6x - 9.6 + 9.6$$
$$4.2 = 0.6x$$
$$\frac{4.2}{0.6} = \frac{0.6x}{0.6}$$
$$7 = x$$

The solution is 7.

8. $\frac{1}{3}(x-4) < \frac{1}{4}(x+7)$

$$12\left[\frac{1}{3}(x-4)\right] < 12\left[\frac{1}{4}(x+7)\right]$$
$$4(x-4) < 3(x+7)$$
$$4x - 16 < 3x + 21$$
$$4x - 16 - 3x < 3x + 21 - 3x$$
$$x - 16 < 21$$
$$x - 16 + 16 < 21 + 16$$
$$x < 37$$

The solution set is $(-\infty, 37)$.

9. $3y-5(y-4)=-2(y-10)$
$3y-5y+20=-2y+20$
$-2y+20=-2y+20$
$-2y+20+2y=-2y+20+2y$
$20=20$
All real numbers are solutions.

10. $\frac{7(x-1)}{3}=\frac{2(x+1)}{5}$
$15\left[\frac{7(x-1)}{3}\right]=15\left[\frac{2(x+1)}{5}\right]$
$35(x-1)=6(x+1)$
$35x-35=6x+6$
$35x-35-6x=6x+6-6x$
$29x-35=6$
$29x-35+35=6+35$
$29x=41$
$\frac{29x}{29}=\frac{41}{29}$
$x=\frac{41}{29}$
The solution is $\frac{41}{29}$.

Chapter 2 Vocabulary Check

1. Terms with the same variables raised to exactly the same powers are called like terms.
2. A linear equation in one variable can be written in the form $ax + b = c$.
3. Equations that have the same solution are called equivalent equations.
4. Inequalities containing two inequality symbols are called compound inequalities.
5. An equation that describes a known relationship among quantities is called a formula.
6. A linear inequality in one variable can be written in the form $ax + b < c$, (or $>$, $\le$, $\ge$).
7. The numerical coefficient of a term is its numerical factor.

Chapter 2 Review

1. $5x-x+2x=6x$
2. $0.2z-4.6x-7.4z=-4.6x-7.2z$
3. $\frac{1}{2}x+3+\frac{7}{2}x-5=\frac{8}{2}x-2=4x-2$
4. $\frac{4}{5}y+1+\frac{6}{5}y+2=\frac{10}{5}y+3=2y+3$
5. $2(n-4)+n-10=2n-8+n-10=3n-18$
6. $3(w+2)-(12-w)=3w+6-12+w=4w-6$
7. $(x+5)-(7x-2)=x+5-7x+2=-6x+7$
8. $(y-0.7)-(1.4y-3)=y-0.7-1.4y+3$
$=-0.4y+2.3$
9. Three times a number decreased by 7 is $3x-7$.
10. Twice the sum of a number and 2.8 added to 3 times a number is $2(x+2.8)+3x$.

11. $8x+4=9x$
$8x+4-8x=9x-8x$
$4=x$

12. $5y-3=6y$
$5y-3-5y=6y-5y$
$-3=y$

13. $\frac{2}{7}x+\frac{5}{7}x=6$
$\frac{7}{7}x=6$
$x=6$

14. $3x-5=4x+1$
$-5=x+1$
$-6=x$

15. $2x-6=x-6$
$x-6=-6$
$x=0$

16. $4(x+3)=3(1+x)$
$4x+12=3+3x$
$x+12=3$
$x=-9$

17. $6(3+n)=5(n-1)$
$18+6n=5n-5$
$18+n=-5$
$n=-23$

18. $5(2+x)-3(3x+2)=-5(x-6)+2$
$10+5x-9x-6=-5x+30+2$
$-4x+4=-5x+32$
$x+4=32$
$x=28$

19. $x-5=3$
$x-5+\underline{5}=3+\underline{5}$
$x=8$

20. $x+9=-2$
$x+9-\underline{9}=-2-\underline{9}$
$x=-11$

21. $10-x$; choice b.

22. $x-5$; choice a.

23. Complementary angles sum to 90°.
$(90-x)°$; choice b.

24. Supplementary angles sum to 180°.
$180-(x+5)=180-x-5=175-x$
$(175-x)°$; choice c.

25. $\frac{3}{4}x=-9$
$\frac{4}{3}\left(\frac{3}{4}x\right)=\frac{4}{3}(-9)$
$x=-12$

26. $\frac{x}{6}=\frac{2}{3}$
$6\cdot\frac{x}{6}=6\cdot\frac{2}{3}$
$x=4$

27. $-5x=0$
$\frac{-5x}{-5}=\frac{0}{-5}$
$x=0$

28. $-y=7$
$\frac{-y}{-1}=\frac{7}{-1}$
$y=-7$

29. $0.2x=0.15$
$\frac{0.2x}{0.2}=\frac{0.15}{0.2}$
$x=0.75$

30. $\frac{-x}{3}=1$
$-3\cdot\frac{-x}{3}=-3\cdot 1$
$x=-3$

31. $-3x+1=19$
$-3x=18$
$\frac{-3x}{-3}=\frac{18}{-3}$
$x=-6$

32. $5x+25=20$
$5x=-5$
$\frac{5x}{5}=\frac{-5}{5}$
$x=-1$

33. $7(x-1)+9=5x$
$7x-7+9=5x$
$7x+2=5x$
$2=-2x$
$\frac{2}{-2}=\frac{-2x}{-2}$
$-1=x$

34. $7x-6=5x-3$
$2x-6=-3$
$2x=3$
$\frac{2x}{2}=\frac{3}{2}$
$x=\frac{3}{2}$

35. $-5x+\frac{3}{7}=\frac{10}{7}$
$7\left(-5x+\frac{3}{7}\right)=7\cdot\frac{10}{7}$
$-35x+3=10$
$-35x=7$
$x=-\frac{7}{35}$
$x=-\frac{1}{5}$

36. $5x+x=9+4x-1+6$
$6x=4x+14$
$2x=14$
$x=7$

37. Let x = the first integer, then
$x + 1$ = the second integer, and
$x + 2$ = the third integer.
sum $= x + (x + 1) + (x + 2) = 3x + 3$

38. Let x = the first integer, then
$x + 2$ = the second integer
$x + 4$ = the third integer
$x + 6$ = the fourth integer.
sum $= x + (x + 6) = 2x + 6$

39.
$$\begin{aligned} \frac{5}{3}x + 4 &= \frac{2}{3}x \\ 3\left(\frac{5}{3}x + 4\right) &= 3\left(\frac{2}{3}x\right) \\ 5x + 12 &= 2x \\ 12 &= -3x \\ -4 &= x \end{aligned}$$

40.
$$\begin{aligned} \frac{7}{8}x + 1 &= \frac{5}{8}x \\ 8\left(\frac{7}{8}x + 1\right) &= 8\left(\frac{5}{8}x\right) \\ 7x + 8 &= 5x \\ 8 &= -2x \\ -4 &= x \end{aligned}$$

41.
$$\begin{aligned} -(5x + 1) &= -7x + 3 \\ -5x - 1 &= -7x + 3 \\ 2x - 1 &= 3 \\ 2x &= 4 \\ x &= 2 \end{aligned}$$

42.
$$\begin{aligned} -4(2x + 1) &= -5x + 5 \\ -8x - 4 &= -5x + 5 \\ -3x - 4 &= 5 \\ -3x &= 9 \\ x &= -3 \end{aligned}$$

43.
$$\begin{aligned} -6(2x - 5) &= -3(9 + 4x) \\ -12x + 30 &= -27 - 12x \\ 30 &= -27 \end{aligned}$$
There is no solution.

44.
$$\begin{aligned} 3(8y - 1) &= 6(5 + 4y) \\ 24y - 3 &= 30 + 24y \\ -3 &= 30 \end{aligned}$$
There is no solution.

45.
$$\begin{aligned} \frac{3(2 - z)}{5} &= z \\ 3(2 - z) &= 5z \\ 6 - 3z &= 5z \\ 6 &= 8z \\ \frac{6}{8} &= z \\ \frac{3}{4} &= z \end{aligned}$$

46.
$$\begin{aligned} \frac{4(n + 2)}{5} &= -n \\ 4(n + 2) &= -5n \\ 4n + 8 &= -5n \\ 8 &= -9n \\ -\frac{8}{9} &= n \end{aligned}$$

47.
$$\begin{aligned} 0.5(2n - 3) - 0.1 &= 0.4(6 + 2n) \\ 10[0.5(2n - 3) - 0.1] &= 10[0.4(6 + 2n)] \\ 5(2n - 3) - 1 &= 4(6 + 2n) \\ 10n - 15 - 1 &= 24 + 8n \\ 10n - 16 &= 24 + 8n \\ 2n - 16 &= 24 \\ 2n &= 40 \\ n &= 20 \end{aligned}$$

48.
$$\begin{aligned} -9 - 5a &= 3(6a - 1) \\ -9 - 5a &= 18a - 3 \\ -9 &= 23a - 3 \\ -6 &= 23a \\ -\frac{6}{23} &= a \end{aligned}$$

49.
$$\begin{aligned} \frac{5(c + 1)}{6} &= 2c - 3 \\ 5(c + 1) &= 6(2c - 3) \\ 5c + 5 &= 12c - 18 \\ -7c + 5 &= -18 \\ -7c &= -23 \\ c &= \frac{23}{7} \end{aligned}$$

50. $\frac{2(8-a)}{3}=4-4a$
$2(8-a)=3(4-4a)$
$16-2a=12-12a$
$10a+16=12$
$10a=-4$
$a=\frac{-4}{10}$
$a=-\frac{2}{5}$

51. $200(70x-3560)=-179(150x-19{,}300)$
$14{,}000x-712{,}000=-26{,}850x+3{,}454{,}700$
$40{,}850x-712{,}000=3{,}454{,}700$
$40{,}850x=4{,}166{,}700$
$x=102$

52. $1.72y-0.04y=0.42$
$1.68y=0.42$
$y=0.25$

53. Let x = length of a side of the square, then $50.5 + 10x$ = the height.
$x+(50.5+10x)=7327$
$11x+50.5=7327$
$11x=7276.5$
$x=661.5$
$50.5 + 10x = 50.5 + 10(661.5) = 6665.5$
The height is 6665.5 inches.

54. Let x = the length of the shorter piece and $2x$ = the length of the other.
$x+2x=12$
$3x=12$
$x=4$
$2x = 2(4) = 8$
The lengths are 4 feet and 8 feet.

55. Let x = number of Keebler plants, then $2x - 1$ = number of Kellogg plants.
$x+(2x-1)=53$
$3x-1=53$
$3x=54$
$x=18$
$2x - 1 = 2(18) - 1 = 35$
There were 18 Keebler plants and 35 Kellogg plants.

56. Let x = first integer, then
$x + 1$ = second integer, and
$x + 2$ = third integer.
$x+(x+1)+(x+2)=-114$
$3x+3=-114$
$3x=-117$
$x=-39$
$x + 1 = -39 + 1 = -38$
$x + 2 = -39 + 2 = -37$
The integers are −39, −38, −37.

57. Let x = the unknown number.
$\frac{x}{3}=x-2$
$3\cdot\frac{x}{3}=3(x-2)$
$x=3x-6$
$-2x=-6$
$x=3$
The number is 3.

58. Let x = the unknown number.
$2(x+6)=-x$
$2x+12=-x$
$12=-3x$
$-4=x$
The number is −4.

59. Let $P = 46$ and $l = 14$.
$P=2l+2w$
$46=2(14)+2w$
$46=28+2w$
$18=2w$
$9=w$

60. Let $V = 192$, $l = 8$, and $w = 6$.
$V=lwh$
$192=8(6)h$
$192=48h$
$4=h$

61. $y=mx+b$
$y-b=mx$
$\frac{y-b}{x}=m$

62. $r=vst-5$
$r+5=vst$
$\frac{r+5}{vt}=s$

63. $2y - 5x = 7$
$-5x = -2y + 7$
$x = \frac{-2y+7}{-5}$
$x = \frac{2y-7}{5}$

64. $3x - 6y = -2$
$-6y = -3x - 2$
$y = \frac{-3x-2}{-6}$
$y = \frac{3x+2}{6}$

65. $C = \pi D$
$\frac{C}{D} = \pi$

66. $C = 2\pi r$
$\frac{C}{2r} = \pi$

67. Let $V = 900$, $l = 20$, and $h = 3$.
$V = lwh$
$900 = 20w(3)$
$900 = 60w$
$15 = w$
The width is 15 meters.

68. Let x = width, then $x + 6$ = length.
$60 = 2x + 2(x + 6)$
$60 = 2x + 2x + 12$
$60 = 4x + 12$
$48 = 4x$
$12 = x$
$x + 6 = 12 + 6 = 18$
The dimensions are 18 feet by 12 feet.

69. Let $d = 10{,}000$ and $r = 125$.
$d = rt$
$10{,}000 = 125t$
$80 = t$
It will take 80 minutes or 1 hour and 20 minutes.

70. Let $F = 104$.
$C = \frac{5}{9}(F - 32)$
$= \frac{5}{9}(104 - 32)$
$= \frac{5}{9}(72)$
$= 40$
The temperature was 40°C.

71. Let x = the percent.
$9 = x \cdot 45$
$\frac{9}{45} = \frac{45x}{45}$
$0.2 = x$
9 is 20% of 45.

72. Let x = the percent.
$59.5 = x \cdot 85$
$\frac{59.5}{85} = \frac{85x}{85}$
$0.7 = x$
59.5 is 70% of 85.

73. Let x = the number.
$137.5 = 125\% \cdot x$
$137.5 = 1.25x$
$\frac{137.5}{1.25} = \frac{1.25x}{1.25}$
$110 = x$
137.5 is 125% of 110.

74. Let x = the number.
$768 = 60\% \cdot x$
$768 = 0.6x$
$\frac{768}{0.6} = \frac{0.6x}{0.6}$
$1280 = x$
768 is 60% of 1280.

75. Let x = mark-up.
$x = 11\% \cdot 1900$
$x = 0.11 \cdot 1900$
$x = 209$
New price = 1900 + 209 = 2109
The mark-up is $209 and the new price is $2109.

76. Find 66.9% of 76,000.
$0.669 \cdot 76{,}000 = 50{,}844$
We would expect 50,844 people to use the Internet.

77. Let x = gallons of 40% solution.

Strength	gallons	Concentration	
40%	x	0.4	$0.4x$
10%	$30-x$	0.1	$0.1(30-x)$
20%	30	0.2	$0.2(30)$

$$0.4x + 0.1(30-x) = 0.2(30)$$
$$0.4x + 3 - 0.1x = 6$$
$$0.3x + 3 = 6$$
$$0.3x = 3$$
$$x = 10$$

$30 - x = 30 - 10 = 20$

Mix 10 gallons of 40% acid solution with 20 gallons of 10% acid solution.

78. Increase $= 21.0 - 20.7 = 0.3$

Let x = percent.

$$0.3 = x \cdot 20.7$$
$$\frac{0.3}{20.7} = \frac{20.7x}{20.7}$$
$$0.0145 \approx x$$

The percent increase is 1.45%.

79. From the graph, the height of 'Almost hit a car' is 18%.

80. Choose the tallest graph. The most common effect is swerving into another lane.

81. Find 21% of 4600.

$0.21 \cdot 4600 = 966$

We would expect 966 customers to have cut someone off.

82. 46% + 41% + 21% + 18% = 126%

No; answers may vary

83. Let x = time up, then $3 - x$ = time down.

Rate · Time = Distance

	Rate	Time	Distance
Up	10	x	$10x$
Down	50	$3-x$	$50(3-x)$

$$d = d$$
$$10x = 50(3-x)$$
$$10x = 150 - 50x$$
$$60x = 150$$
$$x = 2.5$$

$$\begin{aligned}\text{Total distance} &= 10x + 50(3 - x)\\ &= 10(2.5) + 50(3 - 2.5)\\ &= 25 + 50(0.5)\\ &= 25 + 25\\ &= 50\end{aligned}$$

The distance traveled was 50 km.

84. Let x = the amount invested at 10.5% for one year.

	Principal ·	Rate =	Interest
10.5%	x	0.105	0.105
8.5%	50,000 − x	0.085	0.085(50,000 − x)
Total	50,000		4550

$$\begin{aligned}0.105x + 0.085(50,000 - x) &= 4550\\ 0.105x + 4250 - 0.085x &= 4550\\ 0.02x + 4250 &= 4550\\ 0.02x &= 300\\ x &= 15,000\end{aligned}$$

50,000 − x = 50,000 − 15,000 = 35,000

Invest \$35,000 at 8.5% and \$15,000 at 10.5%.

85. Let x = the number of dimes,
$2x$ = the number of quarters, and
$500 - x - 2x$ the number of nickels.

	No. of Coins ·	Value =	Amt. of Money
Dimes	x	0.1	$0.1x$
Quarters	$2x$	0.25	$0.25(2x)$
Nickels	$500 - 3x$	0.05	$0.05(500 - 3x)$
Total	500		88

$$\begin{aligned}0.1x + 0.25(2x) + 0.05(500 - 3x) &= 88\\ 0.1x + 0.5x + 25 - 0.15x &= 88\\ 0.45x + 25 &= 88\\ 0.45x &= 63\\ x &= 140\end{aligned}$$

$500 - 3x = 500 - 3(140) = 500 - 420 = 80$

There were 80 nickels in the pay phone.

86. Let x = the time traveled by the Amtrak train.

	Rate ·	Time =	Distance
Amtrak	60	x	$60x$
Freight	45	$x + 1.5$	$45(x + 1.5)$

$$d = d$$
$$60x = 45(x+1.5)$$
$$60x = 45x + 67.5$$
$$15x = 67.5$$
$$x = 4.5$$

It will take 4.5 hours.

87. $x > 0$, $(0, \infty)$

0

88. $x \le -2$, $(-\infty, -2]$

−2

89. $0.5 \le y < 1.5$, $[0.5, 1.5)$

0.5 1.5

90. $-1 < x < 1$, $(-1, 1)$

−1 1

91. $-3x > 12$
$$\frac{-3x}{-3} < \frac{12}{-3}$$
$$x < -4,\ (-\infty, -4)$$

−4

92. $-2x \ge -20$
$$\frac{-2x}{-2} \le \frac{-20}{-2}$$
$$x \le 10,\ (-\infty, 10]$$

10

93. $x + 4 \ge 6x - 16$
$$-5x + 4 \ge -16$$
$$-5x \ge -20$$
$$\frac{-5x}{-5} \le \frac{-20}{-5}$$
$$x \le 4,\ (-\infty, 4]$$

4

94. $5x - 7 > 8x + 5$
$$-3x - 7 > 5$$
$$-3x > 12$$
$$\frac{-3x}{-3} < \frac{12}{-3}$$
$$x < -4,\ (-\infty, -4)$$

−4

95. $-3 < 4x - 1 < 2$
$$-2 < 4x < 3$$
$$-\frac{1}{2} < x < \frac{3}{4},\ \left(-\frac{1}{2}, \frac{3}{4}\right)$$

$-\frac{1}{2}$ $\frac{3}{4}$

96. $2 \le 3x - 4 < 6$
$$6 \le 3x < 10$$
$$2 \le x < \frac{10}{3},\ \left[2, \frac{10}{3}\right)$$

2 $\frac{10}{3}$

97. $4(2x - 5) \le 5x - 1$
$$8x - 20 \le 5x - 1$$
$$3x - 20 \le -1$$
$$3x \le 19$$
$$x \le \frac{19}{3},\ \left(-\infty, \frac{19}{3}\right]$$

$\frac{19}{3}$

98. $-2(x - 5) > 2(3x - 2)$
$$-2x + 10 > 6x - 4$$
$$-8x + 10 > -4$$
$$-8x > -14$$
$$\frac{-8x}{-8} < \frac{-14}{-8}$$
$$x < \frac{7}{4},\ \left(-\infty, \frac{7}{4}\right)$$

$\frac{7}{4}$

99. Let x = the amount of sales then $0.05x$ = her commission.
$$175 + 0.05x \geq 300$$
$$0.05x \geq 125$$
$$x \geq 2500$$
Sales must be at least \$2500.

100. Let x = her score on the fourth round.
$$\frac{76+82+79+x}{4} < 80$$
$$237 + x < 320$$
$$x < 83$$
Her score must be less than 83.

101. $6x + 2x - 1 = 5x + 11$
$$8x - 1 = 5x + 11$$
$$3x - 1 = 11$$
$$3x = 12$$
$$x = 4$$

102. $2(3y - 4) = 6 + 7y$
$$6y - 8 = 6 + 7y$$
$$-8 = 6 + y$$
$$-14 = y$$

103. $4(3 - a) - (6a + 9) = -12a$
$$12 - 4a - 6a - 9 = -12a$$
$$3 - 10a = -12a$$
$$3 = -2a$$
$$-\frac{3}{2} = a$$

104. $\frac{x}{3} - 2 = 5$
$$\frac{x}{3} = 7$$
$$3 \cdot \frac{x}{3} = 3 \cdot 7$$
$$x = 21$$

105. $2(y + 5) = 2y + 10$
$$2y + 10 = 2y + 10$$
$$10 = 10$$
All real numbers are solutions.

106. $7x - 3x + 2 = 2(2x - 1)$
$$4x + 2 = 4x - 2$$
$$2 = -2$$
There is no solution.

107. Let x = the number.
$$6 + 2x = x - 7$$
$$6 + x = -7$$
$$x = -13$$
The number is -13.

108. Let x = length of shorter piece, then $4x + 3$ = length of longer piece.
$$x + (4x + 3) = 23$$
$$5x + 3 = 23$$
$$5x = 20$$
$$x = 4$$
$4x + 3 = 4(4) + 3 = 19$
The shorter piece is 4 inches and the longer piece is 19 inches.

109. $V = \frac{1}{3}Ah$
$$3 \cdot V = 3 \cdot \frac{1}{3}Ah$$
$$3V = Ah$$
$$\frac{3V}{A} = \frac{Ah}{A}$$
$$\frac{3V}{A} = h$$

110. Let x = the number.
$x = 26\% \cdot 85$
$x = 0.26 \cdot 85$
$x = 22.1$
22.1 is 26% of 85.

111. Let x = the number.
$$72 = 45\% \cdot x$$
$$72 = 0.45x$$
$$\frac{72}{0.45} = \frac{0.45x}{0.45}$$
$$160 = x$$
72 is 45% of 160.

112. Increase = 282 − 235 = 47
Let x = percent.
$$47 = x \cdot 235$$
$$\frac{47}{235} = \frac{235x}{235}$$
$$0.2 = x$$
The percent increase is 20%.

113. $4x - 7 > 3x + 2$
$$x - 7 > 2$$
$$x > 9, \ (9, \infty)$$

9

114. $-5x < 20$

$\frac{-5x}{-5} > \frac{20}{-5}$

$x > -4, (-4, \infty)$

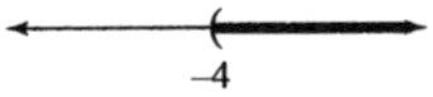

115. $-3(1+2x)+x \geq -(3-x)$

$-3-6x+x \geq -3+x$

$-3-5x \geq -3+x$

$-5x \geq x$

$-6x \geq 0$

$\frac{-6x}{-6} \leq \frac{0}{-6}$

$x \leq 0, (-\infty, 0]$

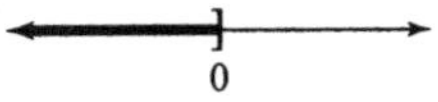

Chapter 2 Test

1. $2y - 6 - y - 4 = y - 10$

2. $2.7x + 6.1 + 3.2x - 4.9 = 5.9x + 1.2$

3. $4(x-2)-3(2x-6) = 4x-8-6x+18$
$= -2x+10$

4. $7 + 2(5y - 3) = 7 + 10y - 6 = 10y + 1$

5. $-\frac{4}{5}x = 4$

$-\frac{5}{4}\cdot\left(-\frac{4}{5}x\right) = -\frac{5}{4}\cdot 4$

$x = -5$

6. $4(n-5) = -(4-2n)$

$4n-20 = -4+2n$

$2n-20 = -4$

$2n = 16$

$n = 8$

7. $5y-7+y = -(y+3y)$

$6y-7 = -4y$

$-7 = -10y$

$\frac{7}{10} = y$

8. $4z+1-z = 1+z$

$3z+1 = 1+z$

$2z+1 = 1$

$2z = 0$

$z = 0$

9. $\frac{2(x+6)}{3} = x-5$

$2(x+6) = 3(x-5)$

$2x+12 = 3x-15$

$12 = x-15$

$27 = x$

10. $\frac{1}{2}-x+\frac{3}{2} = x-4$

$2\left(\frac{1}{2}-x+\frac{3}{2}\right) = 2(x-4)$

$1-2x+3 = 2x-8$

$-2x+4 = 2x-8$

$-4x+4 = -8$

$-4x = -12$

$x = 3$

11. $-0.3(x-4)+x = 0.5(3-x)$

$10[-0.3(x-4)+x] = 10[0.5(3-x)]$

$-3(x-4)+10x = 5(3-x)$

$-3x+12+10x = 15-5x$

$7x+12 = 15-5x$

$12x+12 = 15$

$12x = 3$

$x = \frac{3}{12} = \frac{1}{4} = 0.25$

12. $-4(a+1)-3a = -7(2a-3)$

$-4a-4-3a = -14a+21$

$-7a-4 = -14a+21$

$7a-4 = 21$

$7a = 25$

$a = \frac{25}{7}$

13. $-2(x-3) = x+5-3x$

$-2x+6 = -2x+5$

$6 = 5$

There is no solution.

14. Let x = the number.

$x+\frac{2}{3}x = 35$

$3\left(x+\frac{2}{3}x\right) = 3(35)$

$3x+2x = 105$

$5x = 105$

$x = 21$

The number is 21.

15. Let $l = 35$, and $w = 20$.
$2A = 2lw = 2(35)(20) = 1400$
Let x = the number of gallons needed at 200 square feet per gallon.
$1400 = 200x$
$7 = x$
7 gallons are needed.

16. Let x = one area code, then
$2x$ = other area code.
$x + 2x = 1203$
$3x = 1203$
$\frac{3x}{3} = \frac{1203}{3}$
$x = 401$
$2x = 2(401) = 802$
The area codes are 401 and 802.

17. Let x = the amount invested at 10% for one year.

Principal · Rate = Interest

10%	x	0.10	$0.1x$
12%	$2x$	0.12	$0.12(2x)$
Total			2890

$0.1x + 0.12(2x) = 2890$
$0.1x + 0.24x = 2890$
$0.34x = 2890$
$x = 8500$
$2x = 2(8500) = 17,000$
He invested \$8500 at 10% and \$17,000 at 12%.

18. Let x = the time they travel.

Rate · Time = Distance

Train 1	50	x	$50x$
Train 2	64	x	$64x$
Total			285

$50x + 64x = 285$
$114x = 285$
$x = 2\frac{1}{2}$

They must travel for $2\frac{1}{2}$ hours.

19. Let $y = -14$, $m = -2$, and $b = -2$.
$y = mx + b$
$-14 = -2x - 2$
$-12 = -2x$
$6 = x$

20. $V = \pi r^2 h$
$\frac{V}{\pi r^2} = \frac{\pi r^2 h}{\pi r^2}$
$\frac{V}{\pi r^2} = h$

21. $3x - 4y = 10$
$-4y = -3x + 10$
$y = \frac{-3x + 10}{-4}$
$y = \frac{3x - 10}{4}$

22. $3x - 5 \geq 7x + 3$
$-4x - 5 \geq 3$
$-4x \geq 8$
$\frac{-4x}{-4} \leq \frac{8}{-4}$
$x \leq -2$, $(-\infty, -2]$

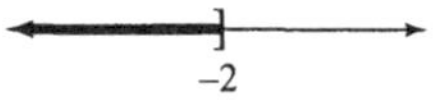

23. $x + 6 > 4x - 6$
$-3x + 6 > -6$
$-3x > -12$
$\frac{-3x}{-3} < \frac{-12}{-3}$
$x < 4$, $(-\infty, 4)$

24. $-2 < 3x + 1 < 8$
$-3 < 3x < 7$
$-1 < x < \frac{7}{3}$, $\left(-1, \frac{7}{3}\right)$

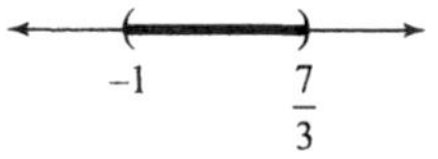

25. $\frac{2(5x+1)}{3} > 2$
$2(5x+1) > 6$
$10x + 2 > 6$
$10x > 4$
$x > \frac{4}{10} = \frac{2}{5}, \left(\frac{2}{5}, \infty\right)$

$\frac{2}{5}$

Cumulative Review Chapters 1–2

1. a. the natural numbers are 11 and 112.

b. The whole numbers are 0, 11, and 112.

c. The integers are –3, –2, 0, 11, and 112.

d. The rational numbers are –3, –2, –1.5, 0, $\frac{1}{4}$, 11, and 112.

e. The irrational number is $\sqrt{2}$.

f. All the numbers in the given set are real numbers.

2. a. The natural numbers are 2, 7, and 8.

b. The whole numbers are 0, 2, 7, and 8.

c. The integers are –185, 0, 2, 7, and 8.

d. The rational numbers are –185, $-\frac{1}{5}$, 0, 2, 7, and 8.

e. The irrational number is $\sqrt{3}$.

f. All the numbers in the given set are real numbers.

3. a. $|4| = 4$

b. $|-5| = 5$

c. $|0| = 0$

d. $\left|-\frac{1}{2}\right| = \frac{1}{2}$

e. $|5.6| = 5.6$

4. a. $|5| = 5$

b. $|-8| = 8$

c. $\left|-\frac{2}{3}\right| = \frac{2}{3}$

5. a. $40 = 2 \cdot 2 \cdot 2 \cdot 5$

b. $63 = 3 \cdot 3 \cdot 7$

6. a. $44 = 2 \cdot 2 \cdot 11$

b. $90 = 2 \cdot 3 \cdot 3 \cdot 5$

7. $\frac{2}{5} = \frac{2}{5} \cdot \frac{4}{4} = \frac{8}{20}$

8. $\frac{2}{3} = \frac{2}{3} \cdot \frac{8}{8} = \frac{16}{24}$

9. $3[4+2(10-1)] = 3[4+2(9)]$
$= 3[4+18]$
$= 3[22]$
$= 66$

10. $5[16-4(2+1)] = 5[16-4(3)]$
$= 5[16-12]$
$= 5[4]$
$= 20$

11. Let $x = 2$.
$3x + 10 = 8x$
$3(2) + 10 \stackrel{?}{=} 8(2)$
$6 + 10 \stackrel{?}{=} 16$
$16 = 16$
2 is a solution of the equation.

12. Let $x = 3$.
$5x - 2 = 4x$
$5(3) - 2 \stackrel{?}{=} 4(3)$
$15 - 2 \stackrel{?}{=} 12$
$13 \neq 12$
3 is not a solution of the equation.

13. $-1 + (-2) = -3$

14. $(-2) + (-8) = -10$

15. $-4 + 6 = 2$

16. $-3 + 10 = 7$

17. a. $-(-10) = 10$

b. $-\left(-\frac{1}{2}\right) = \frac{1}{2}$

c. $-(-2x) = 2x$

d. $-|-6| = -(6) = -6$

18. a. $-(-5) = 5$

b. $-\left(-\frac{2}{3}\right) = \frac{2}{3}$

c. $-(-a) = a$

d. $-|-3| = -(3) = -3$

19. a. $5.3 - (-4.6) = 5.3 + 4.6 = 9.9$

b. $$-\frac{3}{10} - \frac{5}{10} = -\frac{3}{10} + \left(-\frac{5}{10}\right) = \frac{-3-5}{10} = -\frac{8}{10} = -\frac{4}{5}$$

c. $$-\frac{2}{3} - \left(-\frac{4}{5}\right) = -\frac{2}{3}\cdot\frac{5}{5} + \frac{4}{5}\cdot\frac{3}{3} = -\frac{10}{15} + \frac{12}{15} = \frac{2}{15}$$

20. a. $-2.7 - 8.4 = -2.7 + (-8.4) = -11.1$

b. $-\frac{4}{5} - \left(-\frac{3}{5}\right) = -\frac{4}{5} + \frac{3}{5} = \frac{-4+3}{5} = -\frac{1}{5}$

c. $\frac{1}{4} - \left(-\frac{1}{2}\right) = \frac{1}{4} + \frac{1}{2}\cdot\frac{2}{2} = \frac{1}{4} + \frac{2}{4} = \frac{3}{4}$

21. a. $x = 90 - 38 = 90 + (-38) = 52$
The complementary angle is 52°.

b. $y = 180 - 62 = 180 + (-62) = 118$
The supplementary angle is 118°.

22. a. $x = 90 - 72 = 90 + (-72) = 18$
The complementary angle is 18°.

b. $y = 180 - 47 = 180 + (-47) = 133$
The supplementary angle is 133°.

23. a. $(-1.2)(0.05) = -0.06$

b. $\frac{2}{3}\cdot\left(-\frac{7}{10}\right) = -\frac{2\cdot 7}{3\cdot 10} = -\frac{14}{30} = -\frac{7}{15}$

c. $\left(-\frac{4}{5}\right)(-20) = \frac{4\cdot 20}{5} = \frac{80}{5} = 16$

24. a. $(4.5)(-0.08) = -0.36$

b. $-\frac{3}{4}\cdot -\frac{8}{17} = \frac{3\cdot 8}{4\cdot 17} = \frac{24}{68} = \frac{6}{17}$

25. a. $\frac{-24}{-4} = 6$

b. $\frac{-36}{3} = -12$

c. $\frac{2}{3} \div \left(-\frac{5}{4}\right) = \frac{2}{3}\left(-\frac{4}{5}\right) = -\frac{8}{15}$

d. $-\frac{3}{2} \div 9 = -\frac{3}{2} \div \frac{9}{1} = -\frac{3}{2}\cdot\frac{1}{9} = -\frac{3}{18} = -\frac{1}{6}$

26. a. $\frac{-32}{8} = -4$

b. $\frac{-108}{-12} = 9$

c. $-\frac{5}{7} \div \left(\frac{-9}{2}\right) = -\frac{5}{7}\left(-\frac{2}{9}\right) = \frac{10}{63}$

27. a. $x + 5 = 5 + x$

b. $3 \cdot x = x \cdot 3$

28. a. $y + 1 = 1 + y$

b. $y \cdot 4 = 4 \cdot y$

29. a. $8 \cdot 2 + 8 \cdot x = 8(2 + x)$

b. $7s + 7t = 7(s + t)$

30. **a.** $4 \cdot y + 4 \cdot \frac{1}{3} = 4\left(y + \frac{1}{3}\right)$

b. $0.10x + 0.10y = 0.10(x + y)$

31. $(2x-3)-(4x-2) = 2x-3-4x+2 = -2x-1$

32. $(-5x+1)-(10x+3) = -5x+1-10x-3$
$= -15x-2$

33. $\frac{1}{2} = x - \frac{3}{4}$
$4\left(\frac{1}{2}\right) = 4(x) - 4\left(\frac{3}{4}\right)$
$2 = 4x - 3$
$5 = 4x$
$\frac{5}{4} = x$

34. $\frac{5}{6} + x = \frac{2}{3}$
$6\left(\frac{5}{6}\right) + 6(x) = 6\left(\frac{2}{3}\right)$
$5 + 6x = 4$
$6x = -1$
$x = -\frac{1}{6}$

35. $6(2a-1)-(11a+6) = 7$
$12a-6-11a-6 = 7$
$a-12 = 7$
$a = 19$

36. $-3x+1-(-4x-6) = 10$
$-3x+1+4x+6 = 10$
$x+7 = 10$
$x = 3$

37. $\frac{y}{7} = 20$
$y = 140$

38. $\frac{x}{4} = 18$
$x = 72$

39. $4(2x-3)+7 = 3x+5$
$8x-12+7 = 3x+5$
$8x-5 = 3x+5$
$5x-5 = 5$
$5x = 10$
$x = 2$

40. $6x+5 = 4(x+4)-1$
$6x+5 = 4x+16-1$
$6x+5 = 4x+15$
$2x+5 = 15$
$2x = 10$
$x = 5$

41. Let x = a number.
$2(x+4) = 4x-12$
$2x+8 = 4x-12$
$8 = 2x-12$
$20 = 2x$
$10 = x$
The number is 10.

42. Let x = a number.
$x+4 = 3x-8$
$4 = 2x-8$
$12 = 2x$
$6 = x$
The number is 6.

43. $V = lwh$
$\frac{V}{wh} = \frac{lwh}{wh}$
$\frac{V}{wh} = l$

44. $C = 2\pi r$
$\frac{C}{2\pi} = \frac{2\pi r}{2\pi}$
$\frac{C}{2\pi} = r$

45. $x+4 \le -6$
$x \le -10,\ (-\infty, -10]$

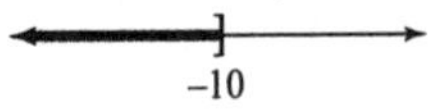

46. $x-3 > 2$
$x > 5,\ (5, \infty)$

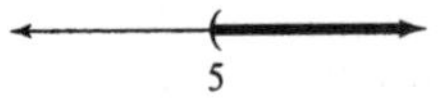

Chapter 3

Section 3.1

Practice Exercises

1. a. We look for the shortest bar, which is the bar representing Germany. We move from the right edge of this bar vertically downward to the Internet user axis. Germany has approximately 45 million Internet users.

b. India has approximately 50 million Internet users. Germany has approximately 45 million Internet users. We subtract $50 - 45 = 5$ or 5 million. India has 5 million more Internet users than Germany.

2. a. We locate the number 40 along the time axis and move vertically upward until the line is reached. From this point on the line, we move horizontally to the left until the pulse rate axis is reached. Reading the number of beats per minute, we find that the pulse rate is 70 beats per minute 40 minutes after a cigarette is lit.

b. The number 0 on the time axis corresponds to the time when the cigarette is being lit. We move vertically upward to the point on the line and then horizontally to the left to the pulse rate axis. The pulse rate is 60 beats per minute when the cigarette is being lit.

c. We find the highest point of the line graph, which represents the highest pulse rate. From this point, we move vertically downward to the time axis. We find the pulse rate is the highest at 5 minutes, which means 5 minutes after lighting a cigarette.

3. a. Point (4, –3) lies in quadrant IV.

b. Point (–3, 5) lies in quadrant II.

c. Point (0, 4) lies on an axis, so it is not in any quadrant.

d. Point (–6, 1) lies in quadrant II.

e. Point (–2, 0) lies on an axis, so it is not in any quadrant.

f. Point (5, 5) lies in quadrant I.

g. Point $\left(3\frac{1}{2}, 1\frac{1}{2}\right)$ lies in quadrant I.

h. Point (–4, –5) lies in quadrant III.

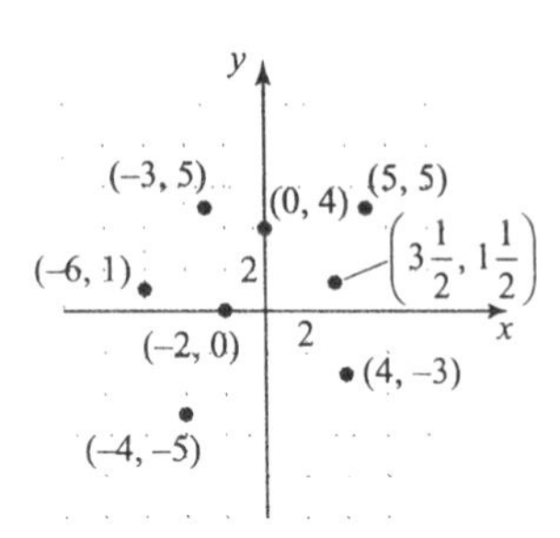

4. a. The ordered pairs are (2000, 92), (2001, 84), (2002, 73), (2003, 64), (2004, 65), (2005, 67), and (2006, 96).

b. We plot the ordered pairs. We label the horizontal axis "Year" and the vertical axis "Wildfires (in thousands)."

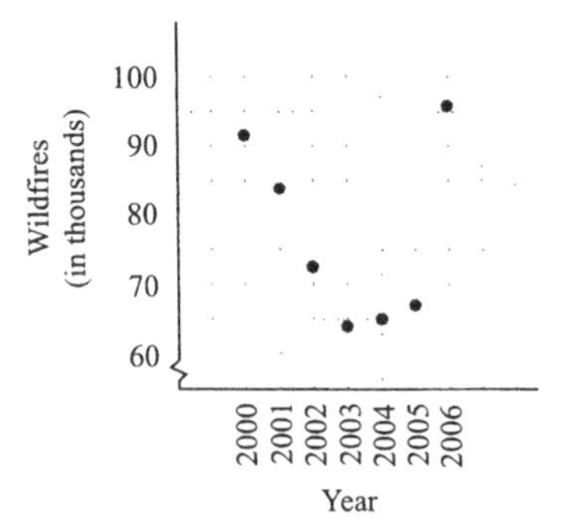

5. a. Let $x = 3$ and $y = 1$.

$$x + 3y = 6$$
$$3 + 3(1) = 6$$
$$3 + 3 = 6$$
$$6 = 6 \quad \text{true}$$

(3, 1) is a solution.

b. Let $x = 6$ and $y = 0$.

$$x + 3y = 6$$
$$6 + 3(0) = 6$$
$$6 + 0 = 6$$
$$6 = 6 \quad \text{true}$$

(6, 0) is a solution.

c. Let $x = -2$ and $y = \frac{2}{3}$.

$$x + 3y = 6$$
$$-2 + 3\left(\frac{2}{3}\right) = 6$$
$$-2 + 2 = 6$$
$$0 = 6 \quad \text{false}$$

$\left(-2, \frac{2}{3}\right)$ is not a solution.

6. a. Let $x = 0$ and solve for y.

$$2x - y = 8$$
$$2(0) - y = 8$$
$$0 - y = 8$$
$$-y = 8$$
$$y = -8$$

The ordered pair is $(0, -8)$.

b. Let $y = 4$ and solve for x.

$$2x - y = 8$$
$$2x - 4 = 8$$
$$2x = 12$$
$$x = 6$$

The ordered pair is $(6, 4)$.

c. Let $x = -3$ and solve for y.

$$2x - y = 8$$
$$2(-3) - y = 8$$
$$-6 - y = 8$$
$$-y = 14$$
$$y = -14$$

The ordered pair is $(-3, -14)$.

7. a. Replace x with -2 in the equation and solve for y.

$$y = -4x$$
$$y = -4(-2)$$
$$y = 8$$

The ordered pair is $(-2, 8)$.

b. Replace y with -12 in the equation and solve for x.

$$y = -4x$$
$$-12 = -4x$$
$$3 = x$$

The ordered pair is $(3, -12)$.

c. Replace x with 0 in the equation and solve for y.

$$y = -4x$$
$$y = -4(0)$$
$$y = 0$$

The ordered pair is $(0, 0)$.

The completed table is shown below.

x	y
−2	8
3	−12
0	0

8. a. Let $x = -10$.

$$y = \frac{1}{5}x - 2$$
$$y = \frac{1}{5}(-10) - 2$$
$$y = -2 - 2$$
$$y = -4$$

Ordered pair: $(-10, -4)$

b. Let $x = 0$.

$$y = \frac{1}{5}x - 2$$
$$y = \frac{1}{5}(0) - 2$$
$$y = 0 - 2$$
$$y = -2$$

Ordered pair: $(0, -2)$

c. Let $y = 0$.

$$y = \frac{1}{5}x - 2$$
$$0 = \frac{1}{5}x - 2$$
$$2 = \frac{1}{5}x$$
$$10 = x$$

Ordered pair: $(10, 0)$

The completed table is shown below.

x	y
−10	−4
0	−2
10	0

9. When $x = 0$,
$y = -1800x + 12{,}000$
$y = -1800 \cdot 0 + 12{,}000$
$y = 0 + 12{,}000$
$y = 12{,}000$

When $x = 1$,
$y = -1800x + 12{,}000$
$y = -1800 \cdot 1 + 12{,}000$
$y = -1800 + 12{,}000$
$y = 10{,}200$

When $x = 2$,
$y = -1800x + 12{,}000$
$y = -1800 \cdot 2 + 12{,}000$
$y = -3600 + 12{,}000$
$y = 8400$

When $x = 3$,
$y = -1800x + 12{,}000$
$y = -1800 \cdot 3 + 12{,}000$
$y = -5400 + 12{,}000$
$y = 6600$

When $x = 4$,
$y = -1800x + 12{,}000$
$y = -1800 \cdot 4 + 12{,}000$
$y = -7200 + 12{,}000$
$y = 4800$

The completed table is shown below.

x	0	1	2	3	4
y	12,000	10,200	8400	6600	4800

Vocabulary and Readiness Check

1. The horizontal axis is called the <u>x-axis</u>.

2. The vertical axis is called the <u>y-axis</u>.

3. The intersection of the horizontal axis and the vertical axis is a point called the <u>origin</u>.

4. The axes divide the plane into regions, called <u>quadrants</u>. There are <u>four</u> of these regions.

5. In the ordered pair of numbers (–2, 5), the number –2 is called the <u>x-coordinate</u> and the number 5 is called the <u>y-coordinate</u>.

6. Each ordered pair of numbers corresponds to <u>one</u> point in the plane.

7. An ordered pair is a <u>solution</u> of an equation in two variables if replacing the variables by the coordinates of the ordered pair results in a true statement.

Exercise Set 3.1

1. We look for the tallest bar, which is the bar representing France. France is the most popular tourist destination.

3. We look for bars extending above the horizontal line at 40 on the vertical axis, which are the bars representing France, U.S., Spain, and China. These countries have more than 40 million tourists per year.

5. The bar for the United Kingdom (U.K.) extends to the horizontal line at 30 on the vertical axis. The United Kingdom has approximately 30 million tourists per year.

7. From 2000 on the year axis, we move vertically up to the point on the line graph. This point is on the attendance axis at approximately 72,600. The Super Bowl attendance in 2000 was approximately 72,600.

9. We find the highest point on the line graph and move vertically downward to the year axis. The year with the greatest Super Bowl attendance was 2007. From the highest point on the graph, we move horizontally to the attendance axis. The highest attendance was approximately 104,000.

11. From 2002 on the year axis, we move vertically up to the point on the line graph. When we move horizontally to the vertical axis. The number of students per teacher was approximately 15.9 in 2002.

13. The number of students per teacher shows the greatest decrease between 1996 and 1998. Notice that the line graph is steepest between 1996 and 1998.

15. The points on the line graph for 1994, 1996, and 1998 lie above the horizontal line at 16 on the vertical axis. The point on the line graph for 2000 appears to lie on the horizontal line at 16 on the vertical axis. The point for 2002 is the first that lies below this horizontal line. The first year shown that the number of students per teacher fell below 16 was 2002.

17. a. Point (1, 5) lies in quadrant I.

b. Point (–5, –2) lies in quadrant III.

c. Point (–3, 0) lies on the x-axis, so it is not in any quadrant.

d. Point (0, –1) lies on the y-axis, so it is not in any quadrant.

e. Point (2, –4) lies in quadrant IV.

f. Point $\left(-1, 4\frac{1}{2}\right)$ lies in quadrant II.

g. (3.7, 2.2) lies in quadrant I.

h. Point $\left(\frac{1}{2}, -3\right)$ lies in quadrant IV.

19. Point A lies at the origin. Its coordinates are given by the ordered pair (0, 0).

21. Point C lies three units to the right and two units above the origin. Its coordinates are given by the ordered pair (3, 2).

23. Point E lies two units to the left and two units below the origin. Its coordinates are given by the ordered pair (–2, –2).

25. Point G lies two units to the right and one unit below the origin. Its coordinates are given by the ordered pair (2, –1).

27. Point B lies on the y-axis three units below the origin. Its coordinates are given by the ordered pair (0, –3).

29. Point D lies one unit to the right and three units above the origin. Its coordinates are given by the ordered pair (1, 3).

31. Point F lies three units to the left and one unit below the origin. Its coordinates are given by the ordered pair (–3, –1).

33. a. The ordered pairs are (2002, 12), (2003, 14), (2004, 14), (2005, 11), and (2006, 12).

b. We plot the ordered pairs. We label the horizontal axis "Year" and the vertical axis "Regular-Season Games Won by Super Bowl Winner."

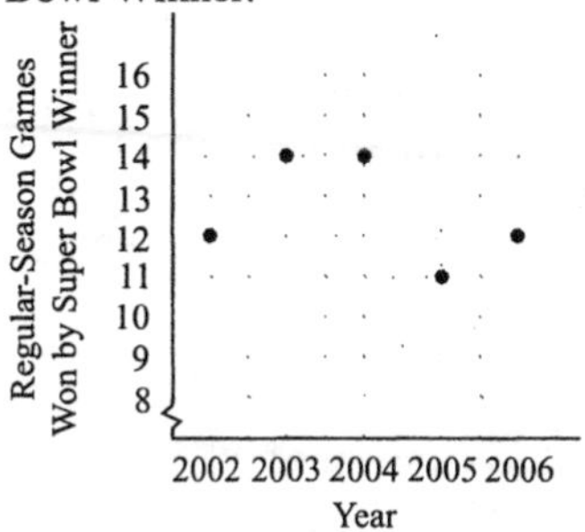

35. a. The ordered pairs are (2001, 1770), (2003, 2800), (2005, 3904), (2007, 7500), and (2009, 10,800).

b. We plot the ordered pairs. We label the horizontal axis "Year" and the vertical axis "Ethanol Fuel Production (in millions of gallons)."

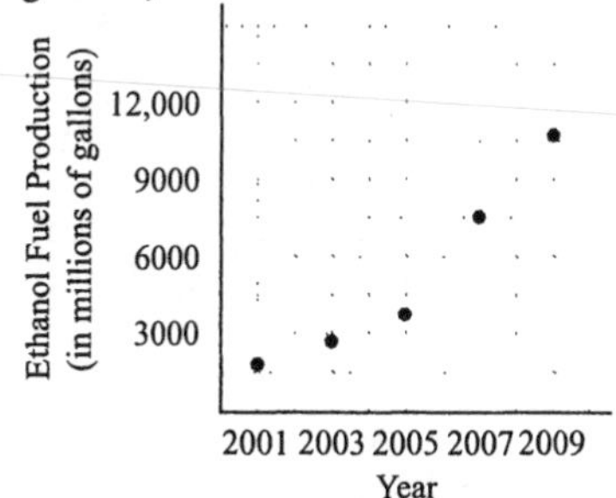

c. The ethanol production is increasing as the years increase.

37. a. The ordered pairs are (2313, 2), (2085, 1), (2711, 21), (2869, 39), (2920, 42), (4038, 99), (1783, 0), and (2493, 9).

b. We plot the ordered pairs. We label the horizontal axis "Distance from Equator (in miles)" and the vertical axis "Average Annual Snowfall (in inches)."

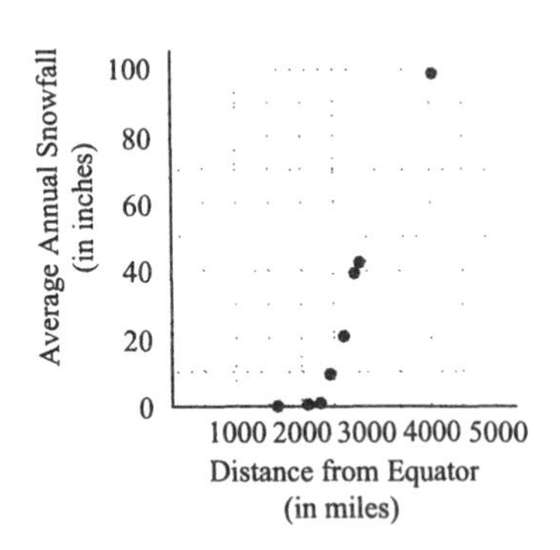

c. The farther from the equator, the more snowfall.

39. For (3, 1), let $x = 3$ and $y = 1$.

$2x + y = 7$
$2(3) + 1 = 7$
$6 + 1 = 7$
$7 = 7$ true

Yes, (3, 1) is a solution.

For (7, 0), let $x = 7$ and $y = 0$.

$2x + y = 7$
$2(7) + 0 = 7$
$14 + 0 = 7$
$14 = 7$ false

No, (7, 0) is not a solution.

For (0, 7), let $x = 0$ and $y = 7$.

$2x + y = 7$
$2(0) + 7 = 7$
$0 + 7 = 7$
$7 = 7$ true

Yes, (0, 7) is a solution.

41. For (0, 0), let $x = 0$ and $y = 0$.

$x = -\frac{1}{3}y$
$0 = -\frac{1}{3}(0)$
$0 = 0$ true

Yes, (0, 0) is a solution.

For (3, –9), let $x = 3$ and $y = -9$.

$x = -\frac{1}{3}y$
$3 = -\frac{1}{3}(-9)$
$3 = 3$ true

Yes, (3, –9) is a solution.

43. For (4, 5), let $x = 4$ and $y = 5$.

$x = 5$
$4 = 5$ false

No, (4, 5) is not a solution.

For (5, 4), let $x = 5$ and $y = 4$.

$x = 5$
$5 = 5$ true

Yes, (5, 4) is a solution.

For (5, 0), let $x = 5$ and $y = 0$.

$x = 5$
$5 = 5$ true

Yes, (5, 0) is a solution.

45. Replace y with –2 and solve for x.

$x - 4y = 4$
$x - 4(-2) = 4$
$x + 8 = 4$
$x = -4$

The ordered pair is (–4, –2).

Replace x with 4 and solve for y.

$x - 4y = 4$
$4 - 4y = 4$
$-4y = 0$
$y = 0$

The ordered pair is (4, 0).

47. Replace x with –8 and solve for y.

$y = \frac{1}{4}x - 3$
$y = \frac{1}{4}(-8) - 3$
$y = -2 - 3$
$y = -5$

The ordered pair is (–8, –5).

Replace y with 1 and solve for x.

$y = \frac{1}{4}x - 3$
$1 = \frac{1}{4}x - 3$
$4 = \frac{1}{4}x$
$16 = x$

The ordered pair is (16, 1).

49. Replace x with 0 and solve for y.
$y = -7x$
$y = -7(0)$
$y = 0$
The ordered pair is (0, 0).

Replace x with −1 and solve for y.
$y = -7x$
$y = -7(-1)$
$y = 7$
The ordered pair is (−1, 7).

Replace y with 2 and solve for x.
$y = -7x$
$2 = -7x$
$-\frac{2}{7} = x$

The ordered pair is $\left(-\frac{2}{7}, 2\right)$.

The completed table is shown below.

x	y
0	0
−1	7
$-\frac{2}{7}$	2

51. Replace x with 0 and solve for y.
$y = -x + 2$
$y = -0 + 2$
$y = 2$
The ordered pair is (0, 2).

Replace y with 0 and solve for x.
$y = -x + 2$
$0 = -x + 2$
$x = 2$
The ordered pair is (2, 0).

Replace x with −3 and solve for y.
$y = -x + 2$
$y = -(-3) + 2$
$y = 3 + 2$
$y = 5$
The ordered pair is (−3, 5).
The completed table is shown below.

x	y
0	2
2	0
−3	5

53. Replace x with 0 and solve for y.
$y = \frac{1}{2}x$
$y = \frac{1}{2}(0)$
$y = 0$
The ordered pair is (0, 0).

Replace x with −6 and solve for y.
$y = \frac{1}{2}x$
$y = \frac{1}{2}(-6)$
$y = -3$
The ordered pair is (−6, −3).

Replace y with 1 and solve for x.
$y = \frac{1}{2}x$
$1 = \frac{1}{2}x$
$2 = x$
The ordered pair is (2, 1).
The completed table is shown below.

x	y
0	0
−6	−3
2	1

55. Replace x with 0 and solve for y.
$x + 3y = 6$
$0 + 3y = 6$
$3y = 6$
$y = 2$
The ordered pair is (0, 2).

Replace y with 0 and solve for x.
$x + 3y = 6$
$x + 3(0) = 6$
$x + 0 = 6$
$x = 6$
The ordered pair is (6, 0).

Replace y with 1 and solve for x.
$x + 3y = 6$
$x + 3(1) = 6$
$x + 3 = 6$
$x = 3$

The ordered pair is (3, 1).
The completed table is shown below.

x	y
0	2
6	0
3	1

57. Replace x with 0 and solve for y.

$y = 2x - 12$
$y = 2(0) - 12$
$y = 0 - 12$
$y = -12$

The ordered pair is (0, −12).

Replace y with −2 and solve for x.

$y = 2x - 12$
$-2 = 2x - 12$
$10 = 2x$
$5 = x$

The ordered pair is (5, −2).

Replace x with 3 and solve for y.

$y = 2x - 12$
$y = 2(3) - 12$
$y = 6 - 12$
$y = -6$

The ordered pair is (3, −6).
The completed table is shown below.

x	y
0	−12
5	−2
3	−6

59. Replace x with 0 and solve for y.

$2x + 7y = 5$
$2(0) + 7y = 5$
$7y = 5$
$y = \frac{5}{7}$

The ordered pair is $\left(0, \frac{5}{7}\right)$.

Replace y with 0 and solve for x.

$2x + 7y = 5$
$2x + 7(0) = 5$
$2x = 5$
$x = \frac{5}{2}$

The ordered pair is $\left(\frac{5}{2}, 0\right)$.

Replace y with 1 and solve for x.

$2x + 7y = 5$
$2x + 7(1) = 5$
$2x + 7 = 5$
$2x = -2$
$x = -1$

The ordered pair is (−1, 1).
The completed table is shown below.

x	y
0	$\frac{5}{7}$
$\frac{5}{2}$	0
−1	1

61. Replace y with 0 and solve for x.

$x = -5y$
$x = -5(0)$
$x = 0$

The ordered pair is (0, 0).

Replace y with 1 and solve for x.

$x = -5y$
$x = -5(1)$
$x = -5$

The ordered pair is (−5, 1).

Replace x with 10 and solve for y.

$x = -5y$
$10 = -5y$
$-2 = y$

The ordered pair is (10, −2).
The completed table is shown below.

x	y
0	0
−5	1
10	−2

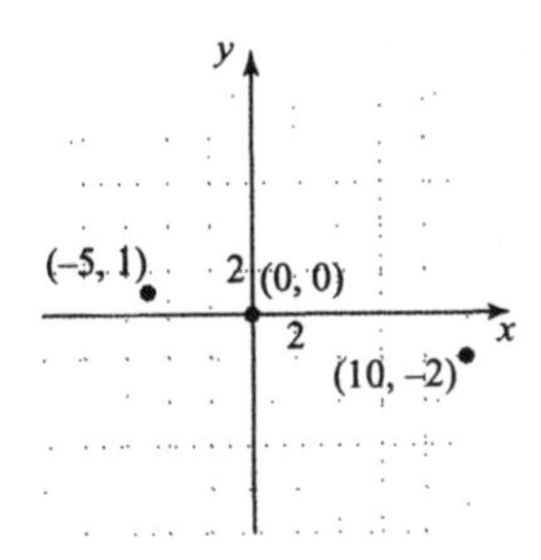

63. Replace x with 0 and solve for y.

$y = \frac{1}{3}x + 2$

$y = \frac{1}{3}(0) + 2$

$y = 2$

The ordered pair is (0, 2).

Replace x with −3 and solve for y.

$y = \frac{1}{3}x + 2$

$y = \frac{1}{3}(-3) + 2$

$y = -1 + 2$

$y = 1$

The ordered pair is (−3, 1).

Replace y with 0 and solve for x.

$y = \frac{1}{3}x + 2$

$0 = \frac{1}{3}x + 2$

$-\frac{1}{3}x = 2$

$x = -6$

The ordered pair is (−6, 0).
The completed table is shown below.

x	y
0	2
−3	1
−6	0

65. a. When $x = 100$,
$y = 80x + 5000$
$y = 80(100) + 5000$
$y = 8000 + 5000$
$y = 13,000$

When $x = 200$,
$y = 80x + 5000$
$y = 80(200) + 5000$
$y = 16,000 + 5000$
$y = 21,000$

When $x = 300$,
$y = 80x + 5000$
$y = 80(300) + 5000$
$y = 24,000 + 5000$
$y = 29,000$
The completed table is shown below.

x	100	200	300
y	13,000	21,000	29,000

b. Replace y with 8600 and solve for x.

$y = 80x + 5000$

$8600 = 80x + 5000$

$3600 = 80x$

$45 = x$

Thus, 45 computer desks can be produced for $8600.

67. a. When $x = 1$,
$y = -2.35x + 55.92$
$y = -2.35(1) + 55.92$
$y = -2.35 + 55.92$
$y = 53.57$

When $x = 3$,
$y = -2.35x + 55.92$
$y = -2.35(3) + 55.92$
$y = -7.05 + 55.92$
$y = 48.87$

When $x = 5$,
$y = -2.35x + 55.92$
$y = -2.35(5) + 55.92$
$y = -11.75 + 55.92$
$y = 44.17$
The completed table is shown below.

x	1	3	5
y	53.57	48.87	44.17

b. Replace y with 46 and solve for x.

$y = -2.35x + 55.92$
$46 = -2.35x + 55.92$
$-9.92 = -2.35x$
$4 \approx x$

The yearly average amount spent on recorded music was approximately \$46 in year 4 or 2005.

69. Four years after 2000, or in 2004, there were 1308 Target stores.

71. In year 0, there appear to be approximately 975 Target stores. In year 1, there appear to be approximately 1050 Target stores. The increase for year 1 is approximately
$1050 - 975 = 75$ stores.

In year 1, there appear to be approximately 1050 Target stores. In year 2, there appear to be approximately 1150 Target stores. The increase for year 2 is approximately
$1150 - 1050 = 100$ stores.

In year 2, there appear to be approximately 1150 Target stores. In year 3, there appear to be approximately 1225 Target stores. The increase for year 3 is approximately
$1225 - 1150 = 75$ stores.

73. The graph of the ordered pair (a, b) is the same as the graph of the ordered pair (b, a) when $a = b$.

75. Subtract x from each side.

$x + y = 5$
$y = 5 - x$

77. Subtract $2x$ from each side. Then divide each side by 4.

$2x + 4y = 5$
$4y = -2x + 5$
$y = -\frac{1}{2}x + \frac{5}{4}$

79. Divide each side by -5.

$10x = -5y$
$-2x = y$
$y = -2x$

81. Subtract x from each side. Then divide each side by -3.

$x - 3y = 6$
$-3y = -x + 6$
$y = \frac{1}{3}x - 2$

83. False; the point $(-1, 5)$ lies in quadrant II.

85. True

87. In quadrant III, both coordinates are negative: (negative, negative).

89. In quadrant IV, the x-coordinate is positive and the y-coordinate is negative: (positive, negative).

91. At the origin, both coordinates are zero: (0, 0).

93. A point of the form (0, number) is located on the y-axis.

95. No; answers may vary.

97. Answers may vary

99. The point four units to the right of the y-axis and seven units below the x-axis has ordered pair $(4, -7)$.

101.

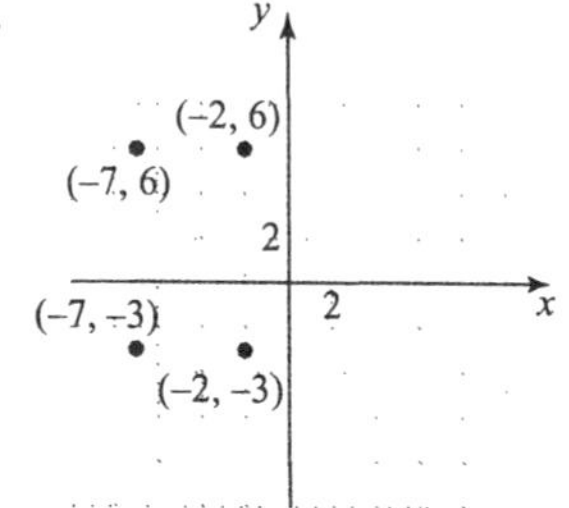

a. The fourth vertex is $(-2, 6)$. The rectangle is 9 units by 5 units.

b. The perimeter is $9 + 5 + 9 + 5 = 28$ units.

c. The area is $9 \times 5 = 45$ square units.

Section 3.2

Practice Exercises

1. a. $3x + 2.7y = -5.3$ is a linear equation in two variables because it is written in the form $Ax + By = C$ with $A = 3$, $B = 2.7$, and $C = -5.3$.

b. $x^2 + y = 8$ is not a linear equation in two variables because x is squared.

c. $y = 12$ is a linear equation in two variables because it can be written in the form $Ax + By = C$: $0x + y = 12$.

d. $5x = -3y$ is a linear equation in two variables because it can be written in the form $Ax + By = C$: $5x + 3y = 0$.

2. Find three ordered pair solutions.
Let $x = 0$.
$$x + 3y = 9$$
$$0 + 3y = 9$$
$$3y = 9$$
$$y = 3$$

Let $x = 3$.
$$x + 3y = 9$$
$$3 + 3y = 9$$
$$3y = 6$$
$$y = 2$$

Let $y = 1$.
$$x + 3y = 9$$
$$x + 3(1) = 9$$
$$x + 3 = 9$$
$$x = 6$$
The ordered pairs are (0, 3), (3, 2), and (6, 1).

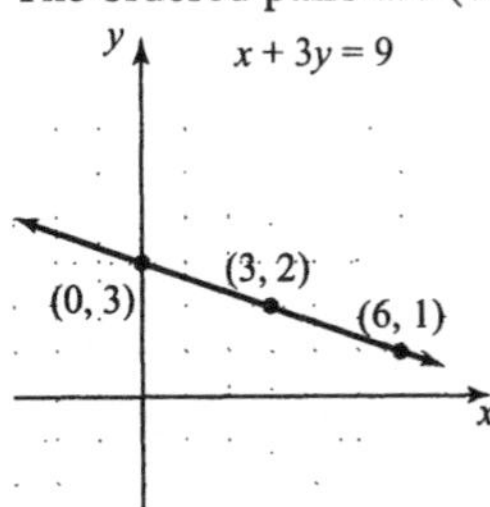

3. Find three ordered pair solutions.
Let $x = 0$.
$$3x - 4y = 12$$
$$3(0) - 4y = 12$$
$$-4y = 12$$
$$y = -3$$

Let $y = 0$.
$$3x - 4y = 12$$
$$3x - 4(0) = 12$$
$$3x = 12$$
$$x = 4$$

Let $x = 2$.
$$3x - 4y = 12$$
$$3(2) - 4y = 12$$
$$6 - 4y = 12$$
$$-4y = 6$$
$$y = -\frac{6}{4} = -\frac{3}{2}$$
The ordered pairs are (0, –3), (4, 0), and $\left(2, -\frac{3}{2}\right)$.

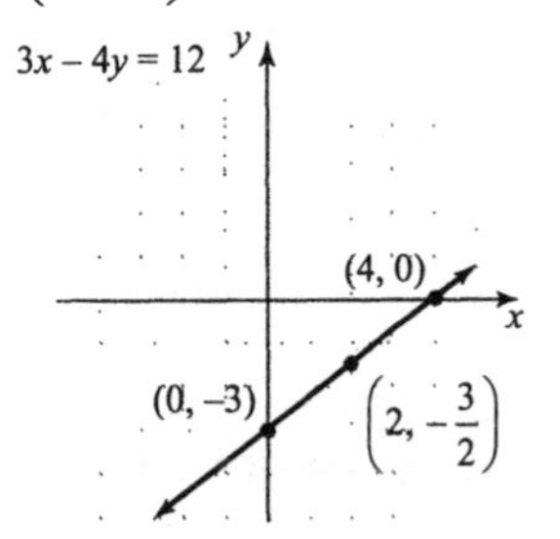

4. Find three ordered pair solutions.
If $x = 1$, $y = -2(1) = -2$.
If $x = 0$, $y = -2(0) = 0$.
If $x = -1$, $y = -2(-1) = 2$.

x	y
1	–2
0	0
–1	2

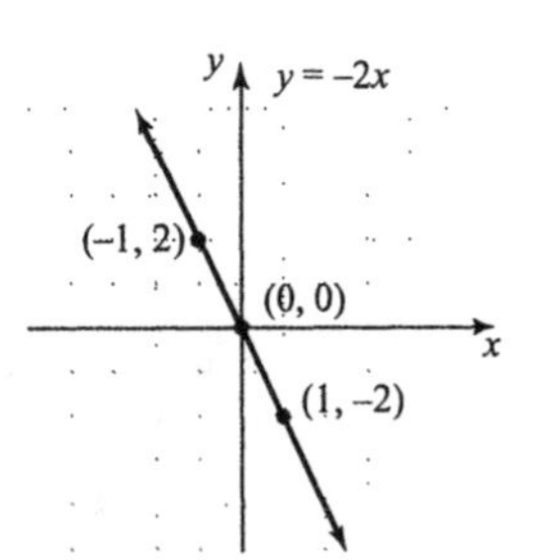

5. Find three ordered pair solutions.

If $x = 2$, $y = \frac{1}{2}(2) + 3 = 1 + 3 = 4$.

If $x = 0$, $y = \frac{1}{2}(0) + 3 = 0 + 3 = 3$.

If $x = -4$, $y = \frac{1}{2}(-4) + 3 = -2 + 3 = 1$.

x	y
2	4
0	3
−4	1

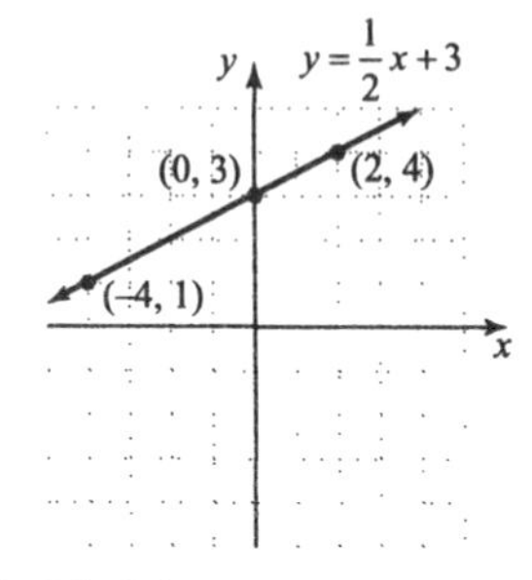

6. Find three ordered pair solutions.
If $x = 1$, $y = -2(1) + 3 = -2 + 3 = 1$.
If $x = 0$, $y = -2(0) + 3 = 0 + 3 = 3$.
If $x = 3$, $y = -2(3) + 3 = -6 + 3 = -3$.

x	y
1	1
0	0
3	−3

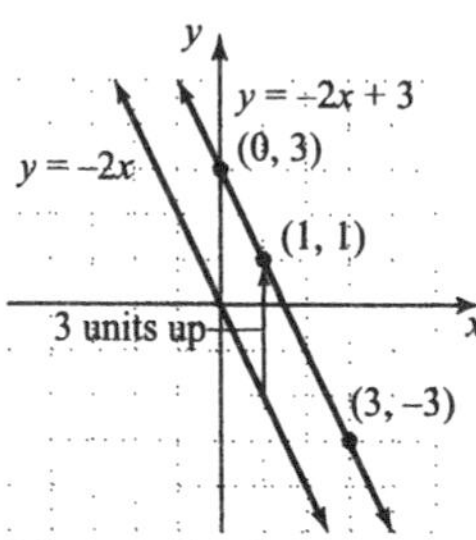

The graph of $y = -2x + 3$ is the same as the graph of $y = -2x$ except that the graph of $y = -2x + 3$ is moved three units upward.

7. a. Find three ordered pair solutions.
If $x = 0$, $y = 22.2(0) + 371 = 0 + 371 = 371$.
If $x = 6$,
$y = 22.2(6) + 371 = 133.2 + 371 = 504.2$.
If $x = 9$,
$y = 22.2(9) + 371 = 199.8 + 371 = 570.8$.

x	y
0	371
6	504.2
9	570.8

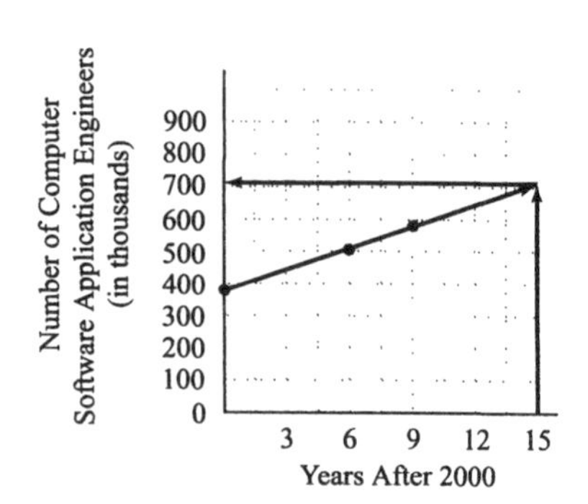

b. The graph shows that we predict approximately 700 thousand computer software application engineers in the year 2015.

Calculator Explorations

1. $y = -3x + 7$

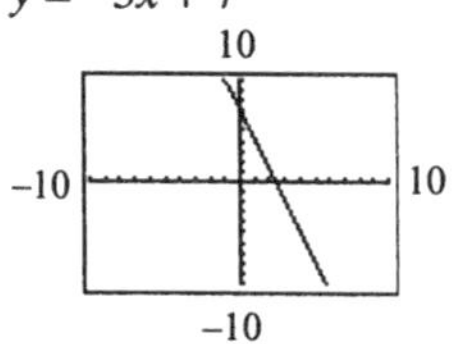

2. $y = -x + 5$

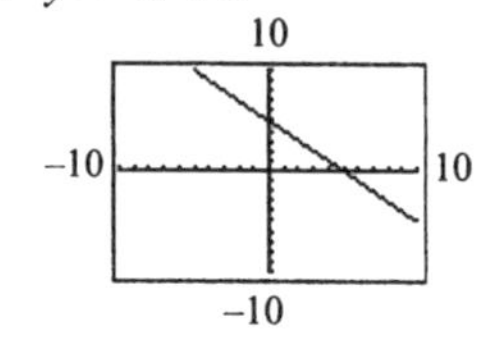

3. $y = 2.5x - 7.9$

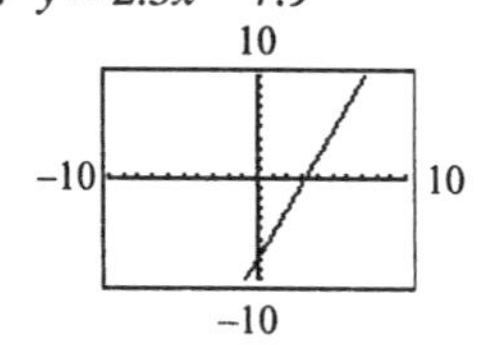

4. $y = -1.3x + 5.2$

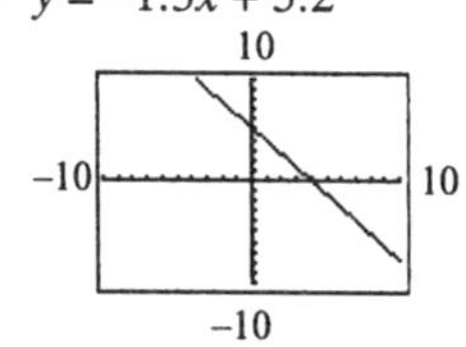

5. $y = -\frac{3}{10}x + \frac{32}{5}$

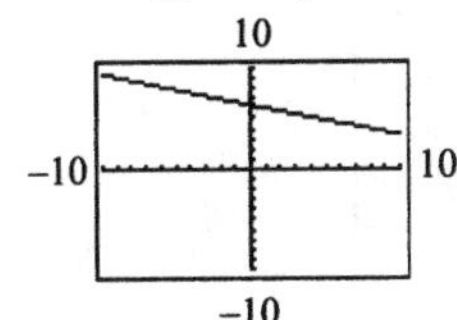

6. $y = \frac{2}{9}x - \frac{22}{3}$

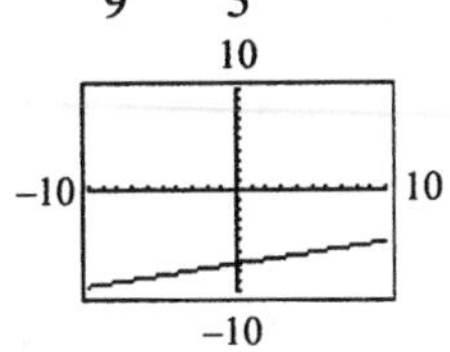

Exercise Set 3.2

1. Yes; it can be written in the form $Ax + By = C$.

3. Yes; it can be written in the form $Ax + By = C$.

5. No; x is squared.

7. Yes; it can be written in the form $Ax + By = C$.

9. Let $y = 0$.

$$\begin{aligned} x - y &= 6 \\ x - 0 &= 6 \\ x &= 6 \end{aligned}$$

Let $x = 4$.

$$\begin{aligned} x - y &= 6 \\ 4 - y &= 6 \\ -y &= 2 \\ y &= -2 \end{aligned}$$

Let $y = -1$

$$\begin{aligned} x - y &= 6 \\ x - (-1) &= 6 \\ x + 1 &= 6 \\ x &= 5 \end{aligned}$$

x	y
6	0
4	−2
5	−1

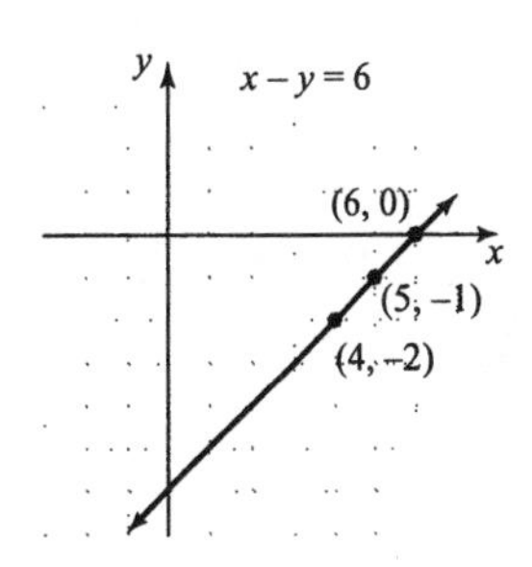

11. $y = -4x$

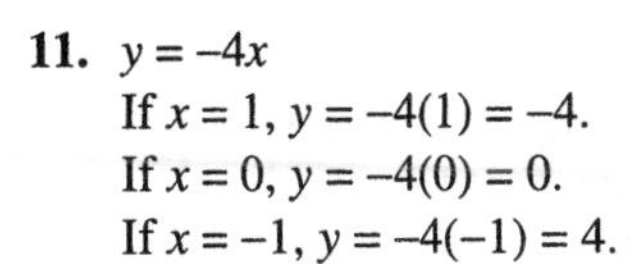

If $x = 1$, $y = -4(1) = -4$.
If $x = 0$, $y = -4(0) = 0$.
If $x = -1$, $y = -4(-1) = 4$.

x	y
1	−4
0	0
−1	4

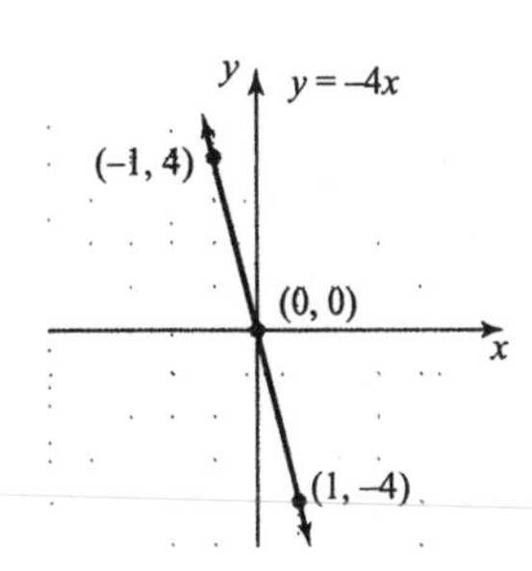

13. $y = \frac{1}{3}x$

If $x = 0$, $y = \frac{1}{3}(0) = 0$.

If $x = 6$, $y = \frac{1}{3}(6) = 2$.

If $x = -3$, $y = \frac{1}{3}(-3) = -1$.

x	y
0	0
6	2
−3	−1

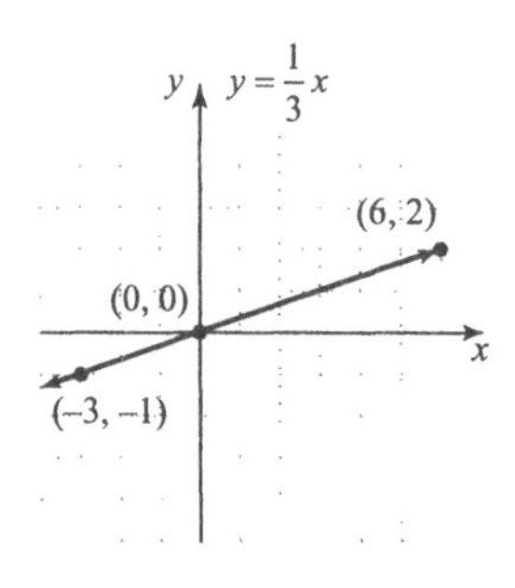

15. $y = -4x + 3$
If $x = 0$, $y = -4(0) + 3 = 0 + 3 = 3$.
If $x = 1$, $y = -4(1) + 3 = -4 + 3 = -1$.
If $x = 2$, $y = -4(2) + 3 = -8 + 3 = -5$.

x	y
0	3
1	−1
2	−5

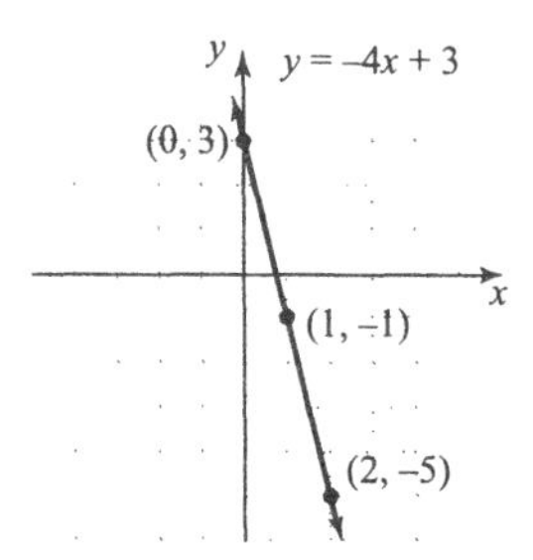

17. $x + y = 1$

x	y
0	1
1	0
2	−1

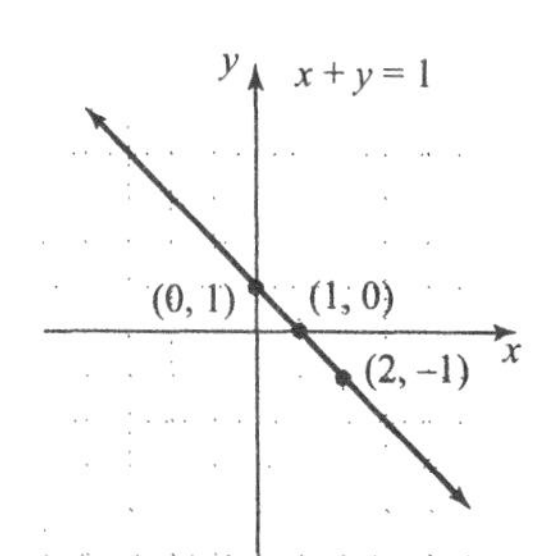

19. $x - y = -2$

x	y
−2	0
0	2
2	4

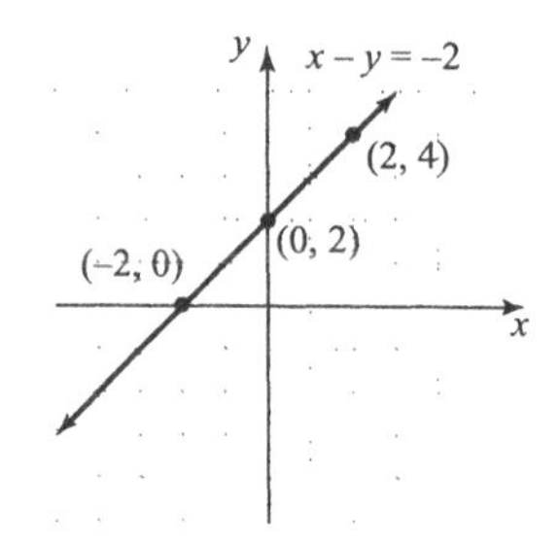

21. $x - 2y = 6$

x	y
−4	−5
0	−3
4	−1

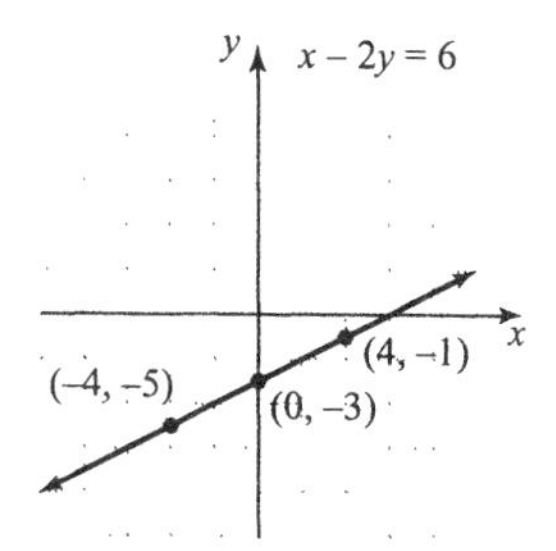

23. $y = 6x + 3$

x	y
−1	−3
0	3
1	9

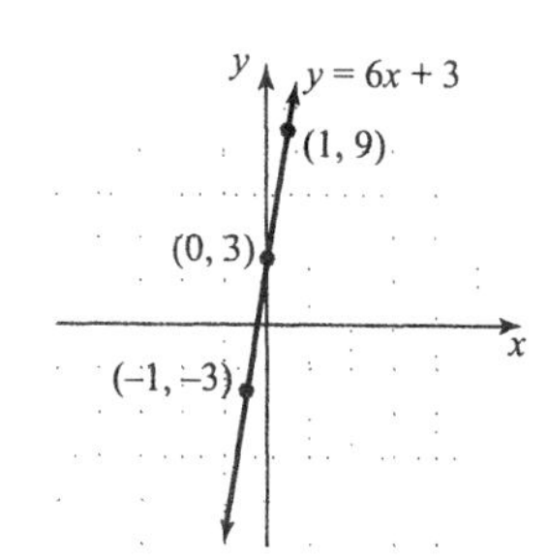

25. $x = -4$

x	y
–4	–1
–4	0
–4	2

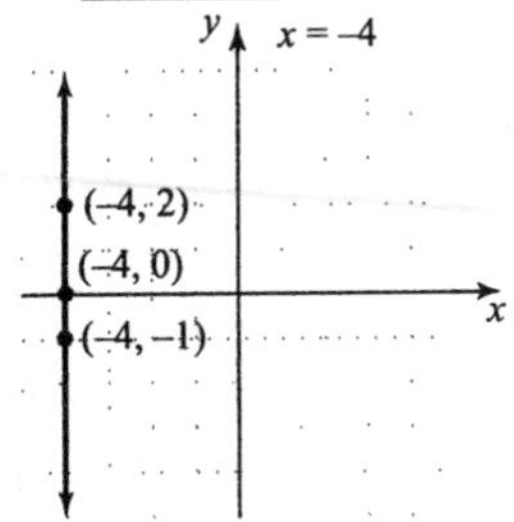

27. $y = 3$

x	y
–1	3
0	3
2	3

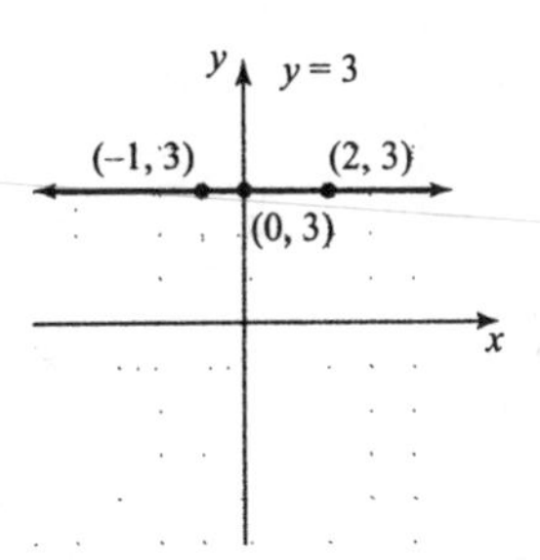

29. $y = x$

x	y
–1	–1
0	0
2	2

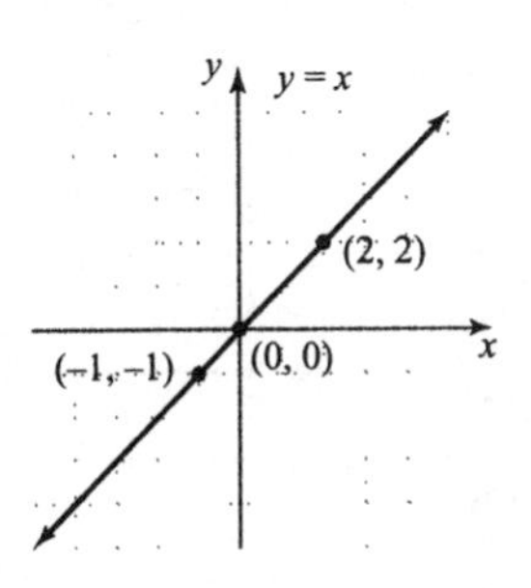

31. $x = -3y$

x	y
–6	2
0	0
6	–2

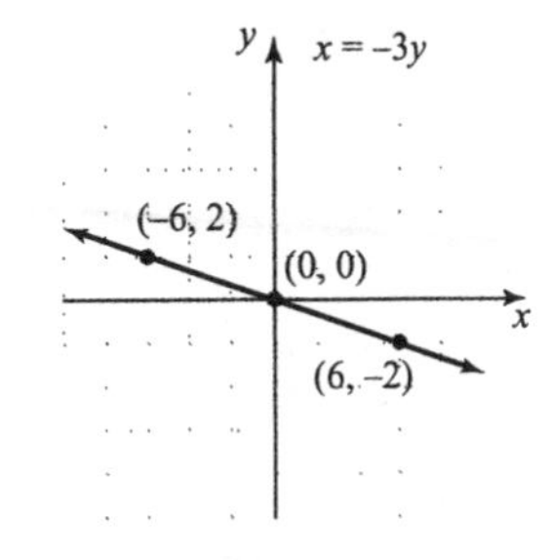

33. $x + 3y = 9$

x	y
–9	6
0	3
3	2

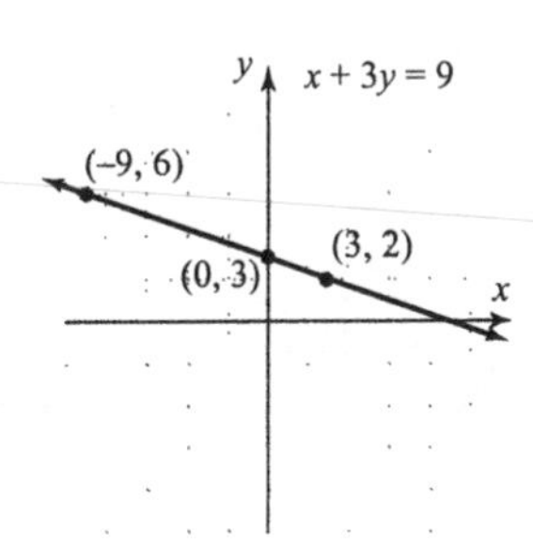

35. $y = \frac{1}{2}x + 2$

x	y
–4	0
0	2
4	4

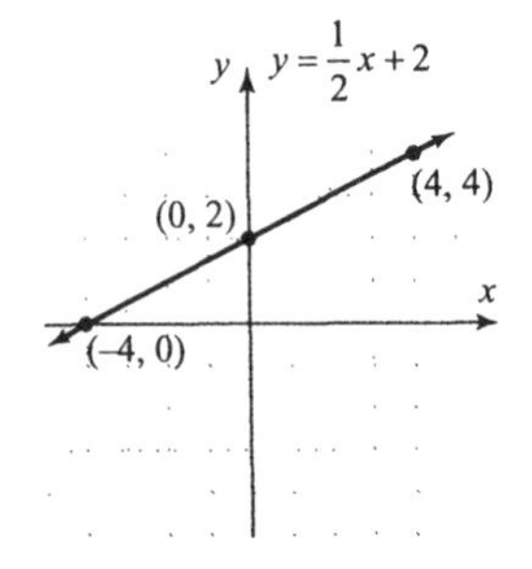

37. $3x - 2y = 12$

x	y
0	−6
2	−3
4	0

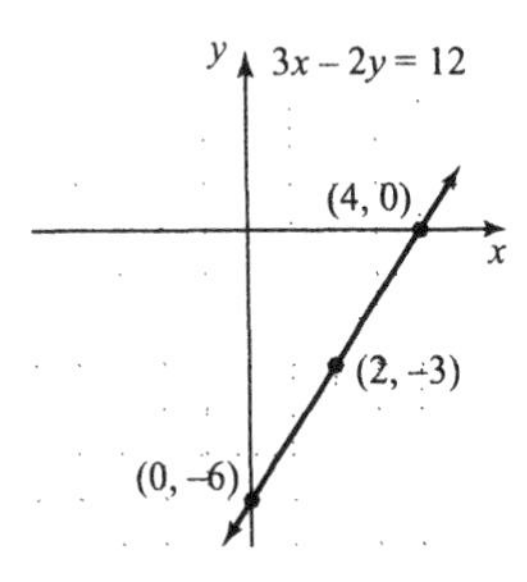

39. $y = -3.5x + 4$

x	y
0	4
1	0.5
2	−3

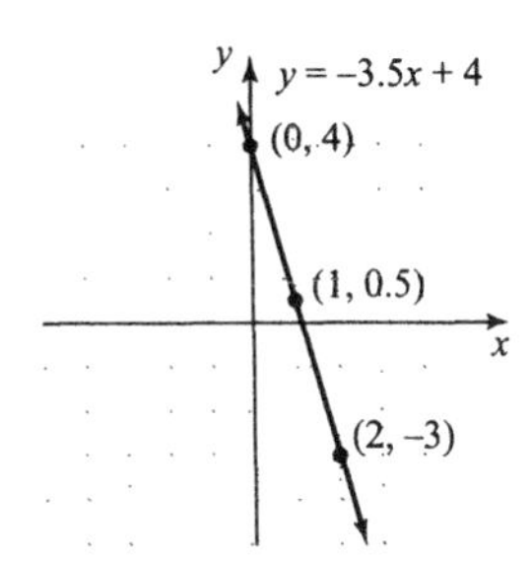

41. $y = 5x$

x	y
−1	−5
0	0
1	5

$y = 5x + 4$

x	y
−1	−1
0	4
1	9

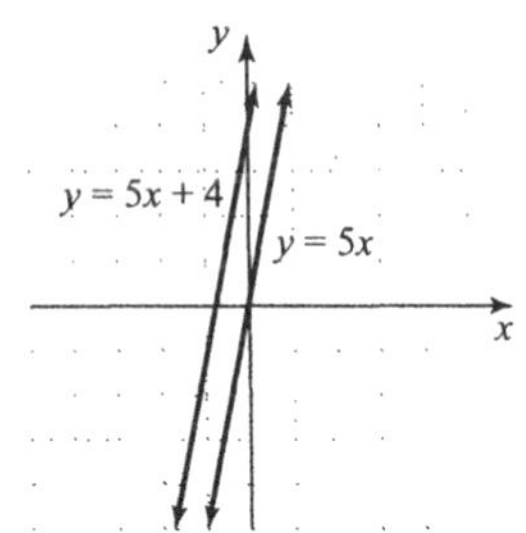

Answers may vary; possible answer: The graph of $y = 5x + 4$ is the same as the graph of $y = 5x$ except it is moved 4 units upward.

43. $y = -2x$

x	y
−2	4
0	0
2	−4

$y = -2x - 3$

x	y
−2	1
0	−3
2	−7

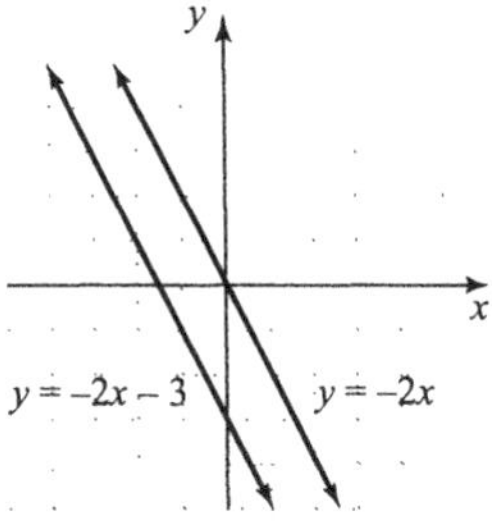

Answers may vary; possible answer: The graph of $y = -2x - 3$ is the same as the graph of $y = -2x$ except it is moved 3 units downward.

45. $y = \frac{1}{2}x$

x	y
−4	−2
0	0
4	2

$y = \frac{1}{2}x + 2$

x	y
−4	0
0	2
4	4

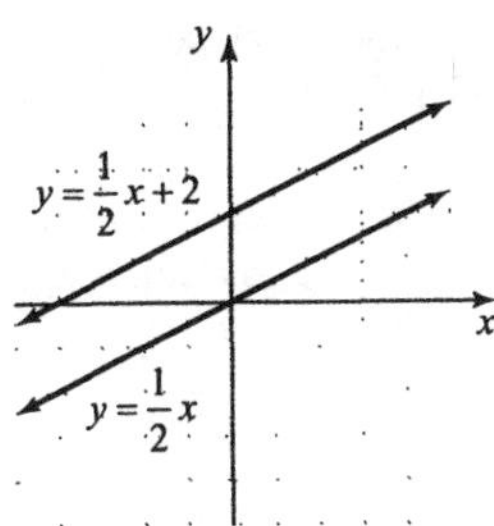

Answers may vary; possible answer: The graph of $y=\frac{1}{2}x+2$ is the same as the graph of $y=\frac{1}{2}x$ except it is moved 2 units upward.

47. Comparing $y = 5x + 5$ to $y = mx + b$, we see that $b = 5$. We see that graph c crosses the y-axis at (0, 5).

49. Comparing $y = 5x - 1$ to $y = mx + b$, we see that $b = -1$. We see that graph d crosses the y-axis at (0, −1).

51. a. Using the equation, let $x = 8$.
$y = 0.5x + 3$
$y = 0.5(8) + 3 = 4 + 3 = 7$
The ordered pair is (8, 7).

b. Eight years after 1997, in 2005, there were 7 million snowboarders.

c. The year 2012 is 15 years after 1997, so let $x = 15$.
$y = 0.5x + 3$
$y = 0.5(15) + 3 = 7.5 + 3 = 10.5$
If the trend continues, there will be 10.5 million snowboarders in 2012.

53. Let $x = 5$.
$y = 54x + 275$
$y = 54(5) + 275 = 270 + 275 = 545$
The expected minimum salary after 5 years' experience is $545 thousand.

55.

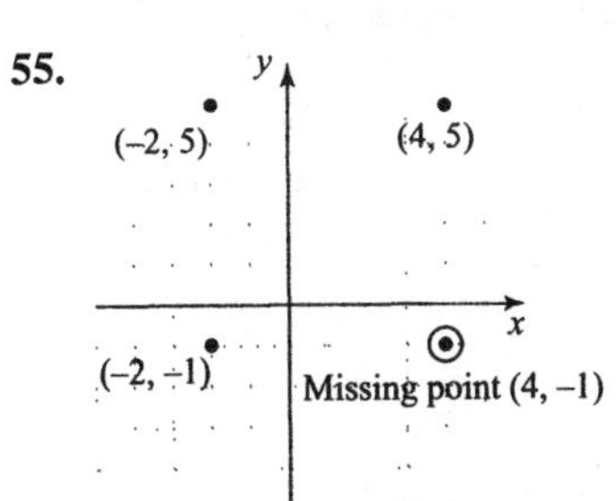

The fourth vertex is at (4, −1).

57.
$$\begin{aligned}3(x-2)+5x&=6x-16\\3x-6+5x&=6x-16\\8x-6&=6x-16\\2x-6&=-16\\2x&=-10\\x&=-5\end{aligned}$$

59.
$$\begin{aligned}3x+\frac{2}{5}&=\frac{1}{10}\\10(3x)+10\left(\frac{2}{5}\right)&=10\left(\frac{1}{10}\right)\\30x+4&=1\\30x&=-3\\x&=-\frac{1}{10}\end{aligned}$$

61. The equation is $y = x + 5$.

x	y
−2	3
0	5
2	7

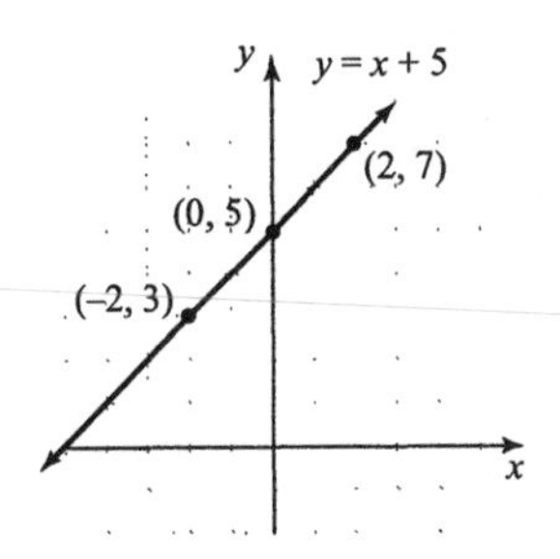

63. The equation is $2x + 3y = 6$.

x	y
3	0
0	2
−3	4

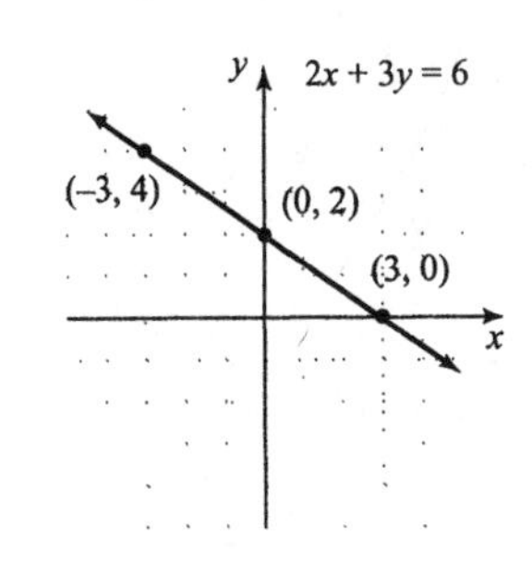

65. $x+y+5+5=22$
$x+y+10=22$
$x+y=12$
Let $x=3$.
$3+y=12$
$y=9$ centimeters

67. Answers may vary

69. $y=x^2$

x	y
0	0
1	1
−1	1
2	4
−2	4

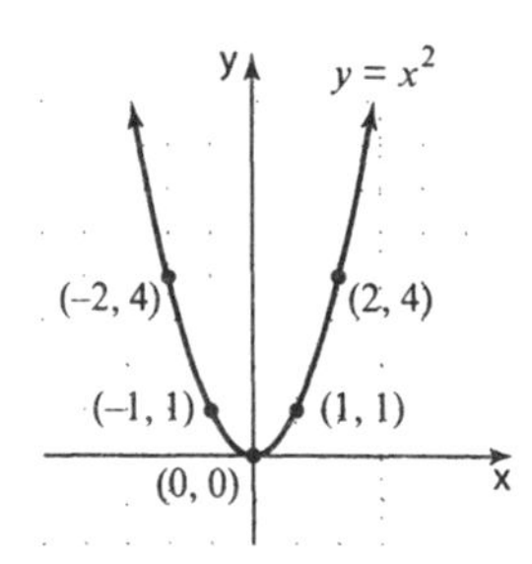

Section 3.3

Practice Exercises

1. The graph crosses the x-axis at the point (−4, 0). The x-intercept is (−4, 0).
The graph crosses the y-axis at the point (0, −6). The y-intercept is (0, −6).

2. The graph crosses the x-axis at the point (−1, 0) and at the point (−0.5, 0). The x-intercepts are (−1, 0) and (−0.5, 0).
The graph crosses the y-axis at the point (0, 1). The y-intercept is (0, 1).

3. The graph crosses both the x-axis and the y-axis at the point (0, 0). The x-intercept is (0, 0), and the y-intercept is (0, 0).

4. The graph does not cross the x-axis. There is no x-intercept. The graph crosses the y-axis at the point (0, 3). The y-intercept is (0, 3).

5. The graph crosses the x-axis at the point (−1, 0) and at the point (5, 0). The x-intercepts are (−1, 0) and (5, 0).
The graph crosses the y-axis at the point (0, −2) and at the point (0, 2). The y-intercepts are (0, −2) and (0, 2).

6. Let $y=0$.
$x+2y=-4$
$x+2(0)=-4$
$x+0=-4$
$x=-4$

Let $x=0$.
$x+2y=-4$
$0+2y=-4$
$2y=-4$
$y=-2$

The x-intercept is (−4, 0), and the y-intercept is (0, −2).
Let $x=2$.
$x+2y=-4$
$2+2y=-4$
$2y=-6$
$y=-3$

x	y
−4	0
0	−2
2	−3

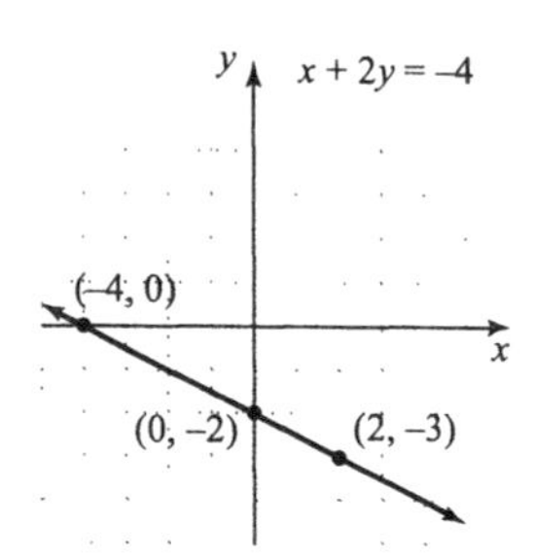

7. Let $y=0$.
$x=3y$
$x=3(0)$
$x=0$

Let $x=0$.
$x=3y$
$0=3y$
$0=y$

Both the x-intercept and the y-intercept are (0, 0).
Let $y=-1$
$x=3(-1)$
$x=-3$

Let $y=1$.
$x=3(1)$
$x=3$

x	y
0	0
3	1
−3	−1

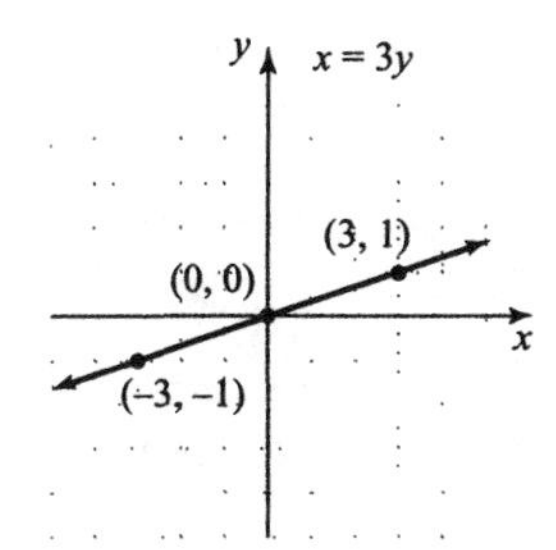

8. Let $y = 0$.

$3x = 2y + 4$
$3x = 2(0) + 4$
$3x = 4$
$x = \frac{4}{3}$

Let $x = 0$.

$3x = 2y + 4$
$3(0) = 2y + 4$
$-4 = 2y$
$-2 = y$

Let $x = 2$.

$3x = 2y + 4$
$3(2) = 2y + 4$
$6 = 2y + 4$
$2 = 2y$
$1 = y$

x	y
0	−2
$\frac{4}{3}$	0
2	1

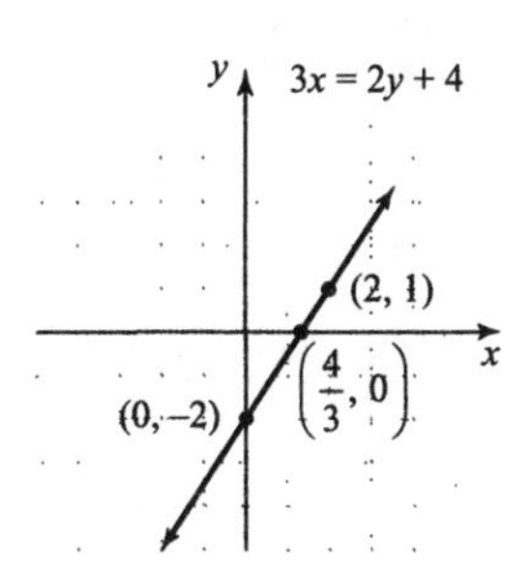

9. For any x-value chosen, notice that y is 2.

x	y
−5	2
0	2
5	2

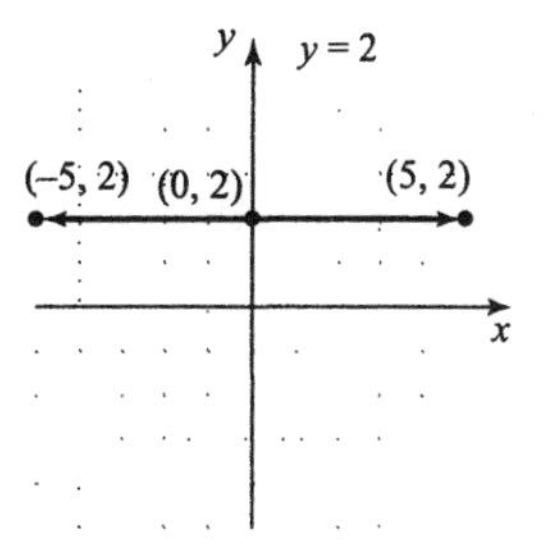

10. For any y-value chosen, notice that x is −2.

x	y
−2	−4
−2	0
−2	4

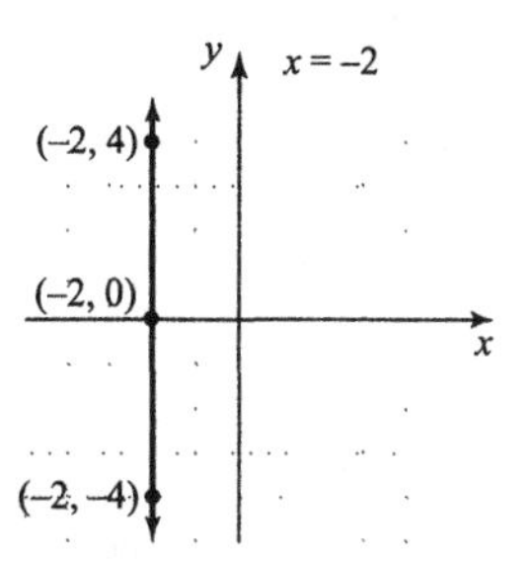

Calculator Explorations

1. $x = 3.78y$

$y = \frac{x}{3.78}$

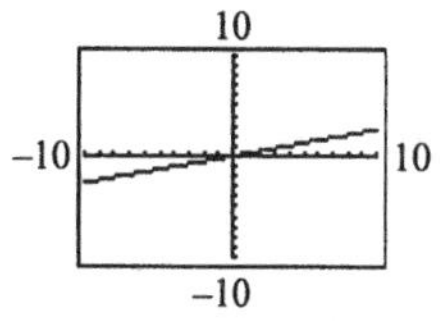

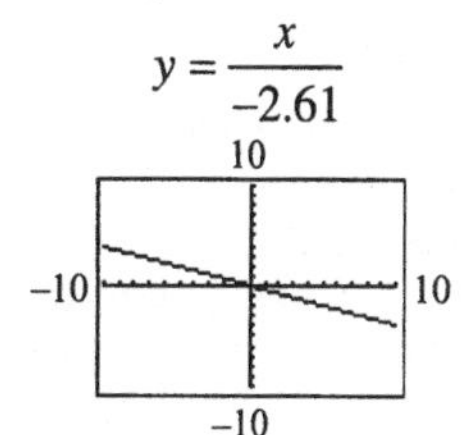

2. $-2.61y = x$

$y = \frac{x}{-2.61}$

3. $3x + 7y = 21$

$$7y = -3x + 21$$

$$y = -\frac{3}{7}x + 3$$

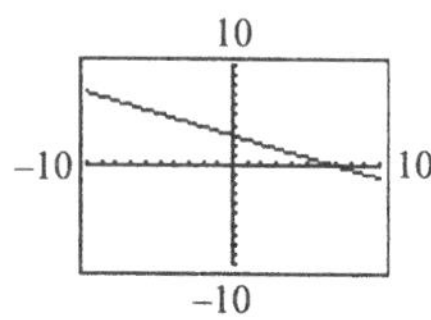

4. $-4x + 6y = 12$

$$6y = 4x + 12$$

$$y = \frac{2}{3}x + 2$$

5. $-2.2x + 6.8y = 15.5$

$$6.8y = 2.2x + 15.5$$

$$y = \frac{2.2}{6.8}x + \frac{15.5}{6.8}$$

6. $5.9x - 0.8y = -10.4$

$$-0.8y = -5.9x - 10.4$$

$$y = \frac{5.9}{0.8}x + \frac{10.4}{0.8}$$

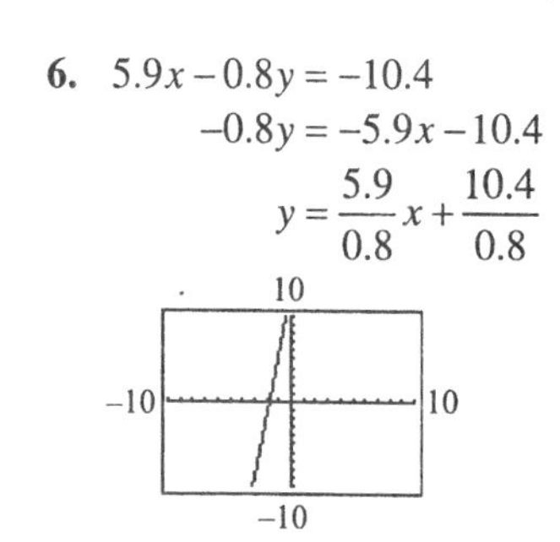

Vocabulary and Readiness Check

1. An equation that can be written in the form $Ax + By = C$ is called a linear equation in two variables.

2. The form $Ax + By = C$ is called standard form.

3. The graph of the equation $y = -1$ is a horizontal line.

4. The graph of the equation $x = 5$ is a vertical line.

5. A point where a graph crosses the y-axis is called a y-intercept.

6. A point where a graph crosses the x-axis is called a x-intercept.

7. Given an equation of a line, to find the x-intercept (if there is one), let $y = 0$ and solve for x.

8. Given an equation of a line, to find the y-intercept (if there is one), let $x = 0$ and solve for y.

9. False; for example, the horizontal line $y = 2$ does not have an x-intercept.

10. True

11. True

12. False; the graph of $y = 5x$ contains the point (1, 5) but not the point (5, 1).

Exercise Set 3.3

1. x-intercept: (−1, 0); y-intercept: (0, 1)

3. x-intercept: (−2, 0), (2, 0)

5. x-intercepts: (−2, 0), (1, 0), (3, 0)
y-intercept: (0, 3)

7. x-intercepts: (−1, 0), (1, 0)
y-intercepts: (0, 1), (0, −2)

9. Infinite; because the line could be vertical ($x = 0$) or horizontal ($y = 0$).

11. 0; because the circle could completely reside within one quadrant.

13. $x - y = 3$
$y = 0$: $x - 0 = 3$, $x = 3$
$x = 0$: $0 - y = 3$, $y = -3$
x-intercept: (3, 0); y-intercept: (0, −3)

x	y
3	0
0	−3

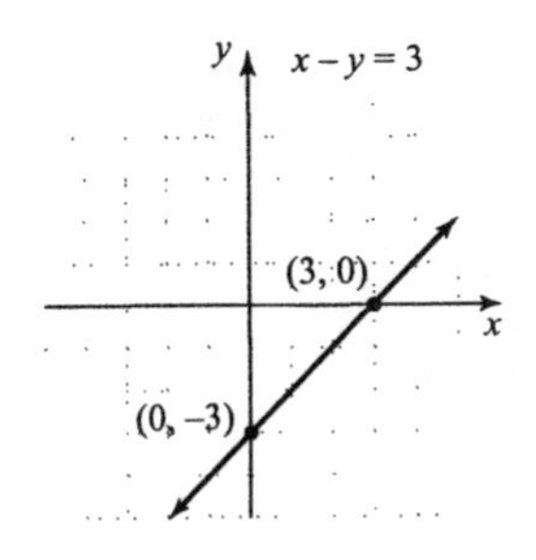

15. $x = 5y$
$y = 0$: $x = 5(0) = 0$
$x = 0$: $0 = 5y$, $y = 0$
x-intercept: (0, 0); y-intercept: (0, 0)
$y = 1$: $x = 5(1) = 5$

x	y
0	0
5	1

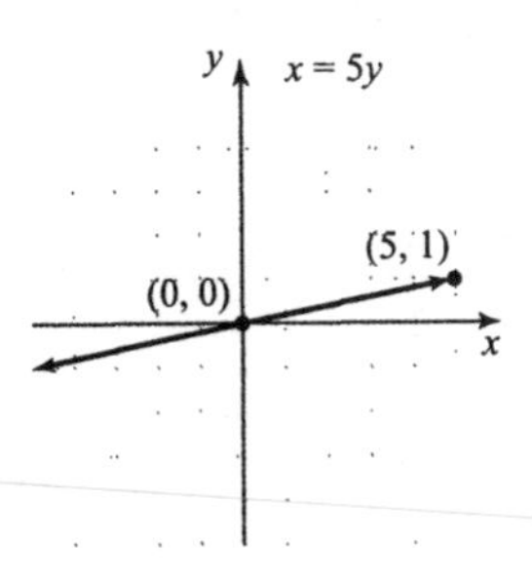

17. $-x + 2y = 6$
$y = 0$: $-x + 2(0) = 6$, $x = -6$
$x = 0$: $-0 + 2y = 6$, $y = 3$
x-intercept: (−6, 0); y-intercept: (0, 3)

x	y
−6	0
0	3

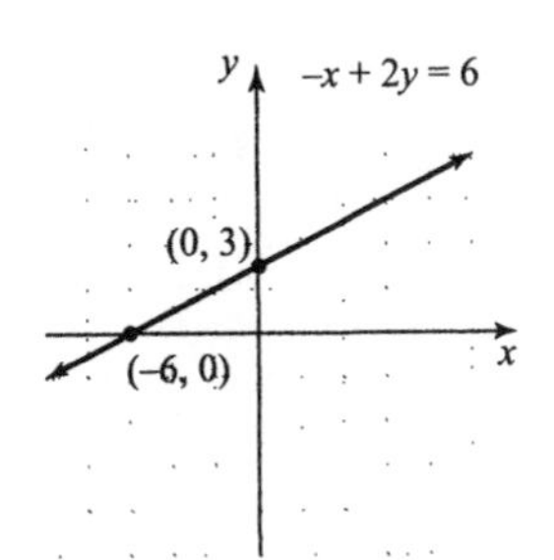

19. $2x - 4y = 8$
$y = 0$: $2x - 4(0) = 8$, $x = 4$
$x = 0$: $2(0) - 4y = 8$, $y = -2$
x-intercept: (4, 0); y-intercept: (0, −2)

x	y
4	0
0	−2

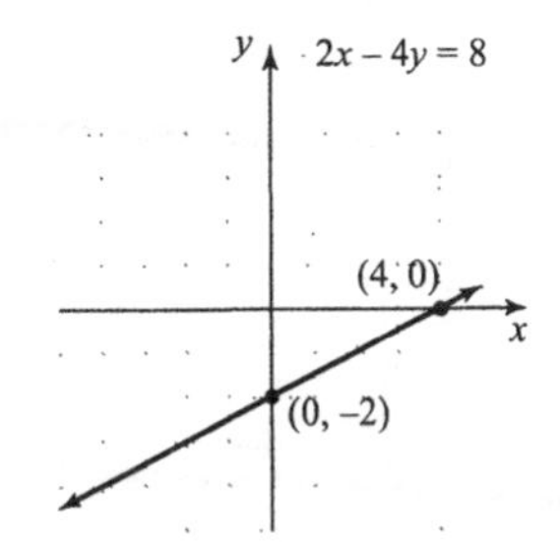

21. $y = 2x$
$y = 0$: $0 = 2x$, $0 = x$
$x = 0$: $y = 2(0)$, $y = 0$
x-intercept: (0, 0); y-intercept: (0, 0)
$x = 1$: $y = 2(1)$, $y = 2$

x	y
0	0
1	2

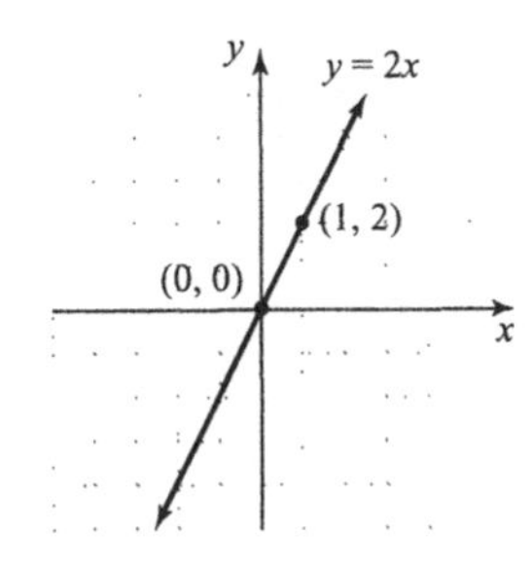

23. $y = 3x + 6$
$y = 0$: $0 = 3x + 6$, $-6 = 3x$, $-2 = x$
$x = 0$: $y = 3(0) + 6$, $y = 6$
x-intercept: (−2, 0); y-intercept: (0, 6)

x	y
−2	0
0	6

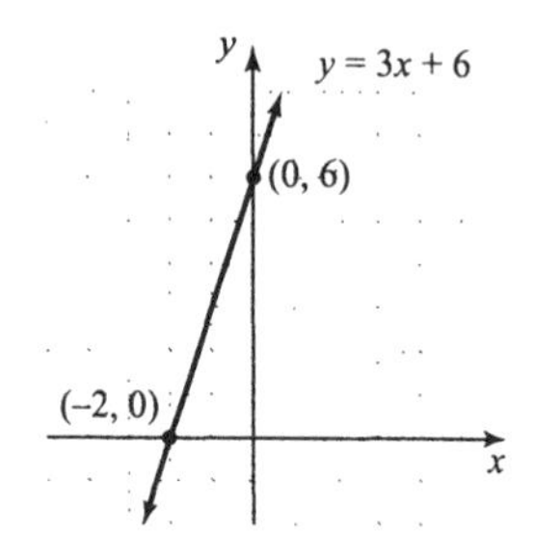

25. $x = -1$ for all values of y.

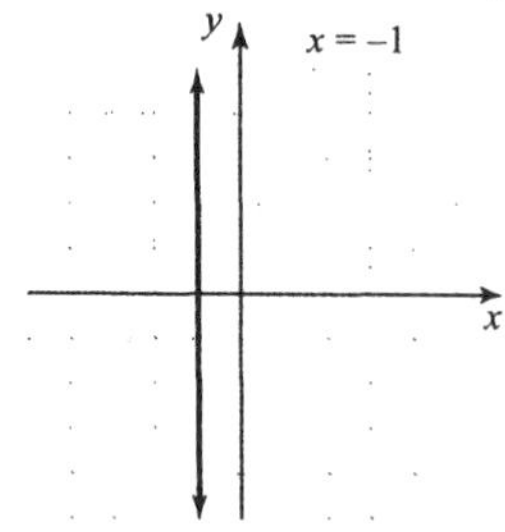

27. $y = 0$ for all values of x.

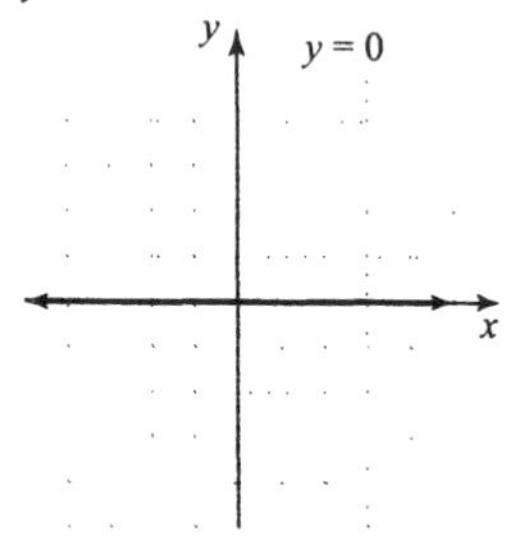

29. $y + 7 = 0$

$y = -7$ for all values of x.

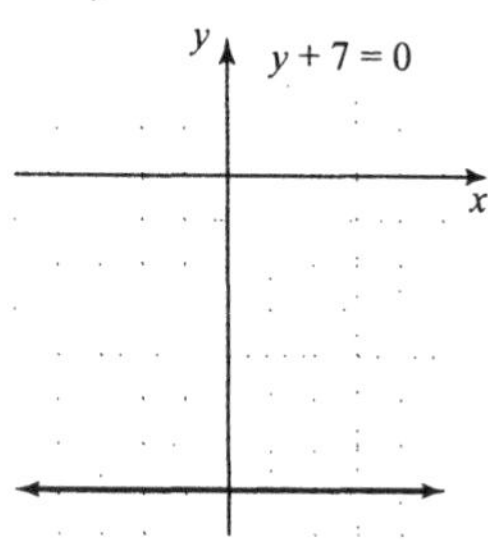

31. $x + 3 = 0$; $x = -3$ for all values of y.

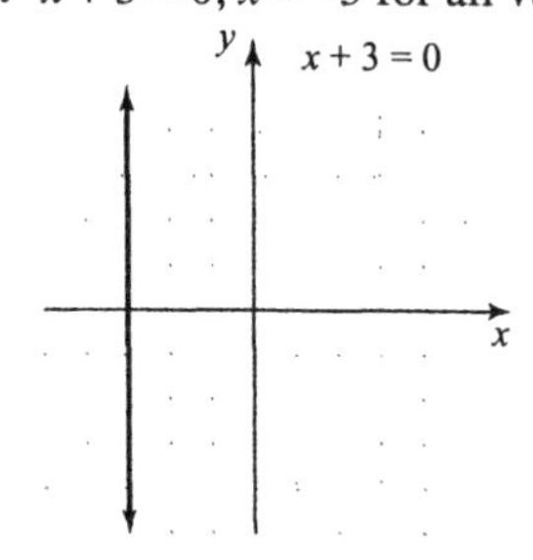

33. $x = y$

x-intercept: (0, 0); y-intercept: (0, 0)

Second point: (4, 4)

x	y
4	4
0	0

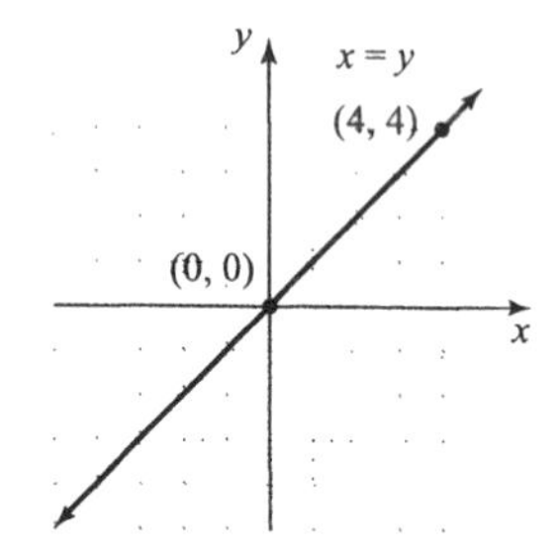

35. $x + 8y = 8$

x-intercept: (8, 0); y-intercept: (0, 1)

x	y
8	0
0	1

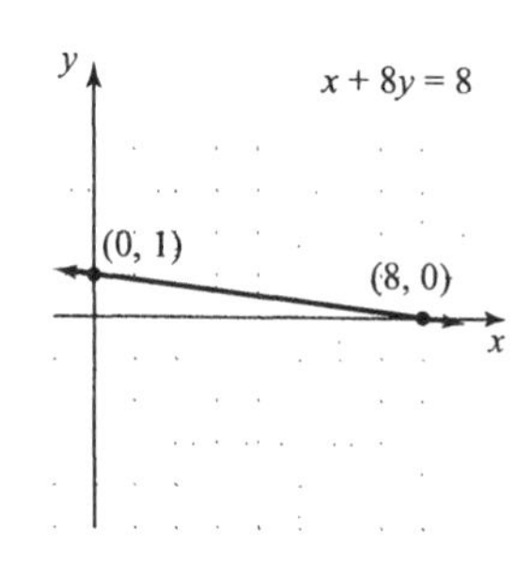

37. $5 = 6x - y$

x-intercept: $\left(\frac{5}{6}, 0\right)$; y-intercept: (0, −5)

x	y
$\frac{5}{6}$	0
0	−5

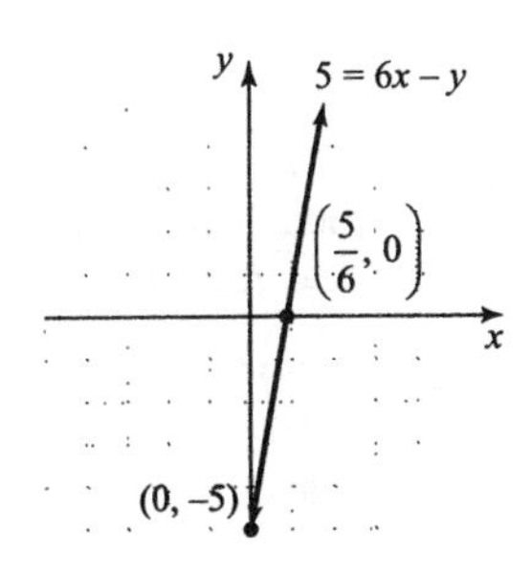

39. $-x + 10y = 11$

x-intercept: $(-11, 0)$; y-intercept: $\left(0, \frac{11}{10}\right)$

x	y
-11	0
0	$\frac{11}{10}$

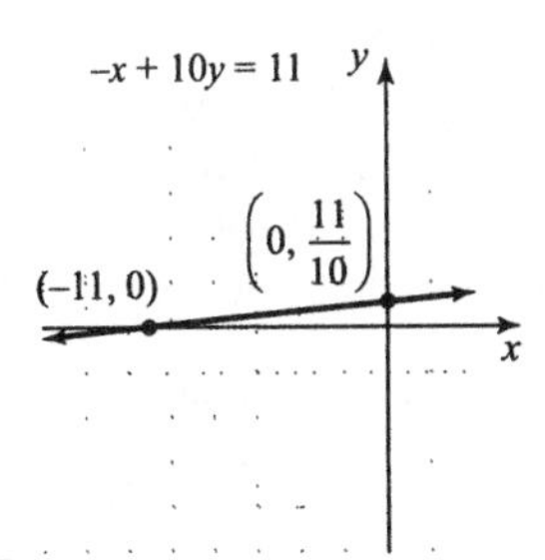

41. $x = -4\frac{1}{2}$ for all values of y.

x	y
$-4\frac{1}{2}$	0
$-4\frac{1}{2}$	3

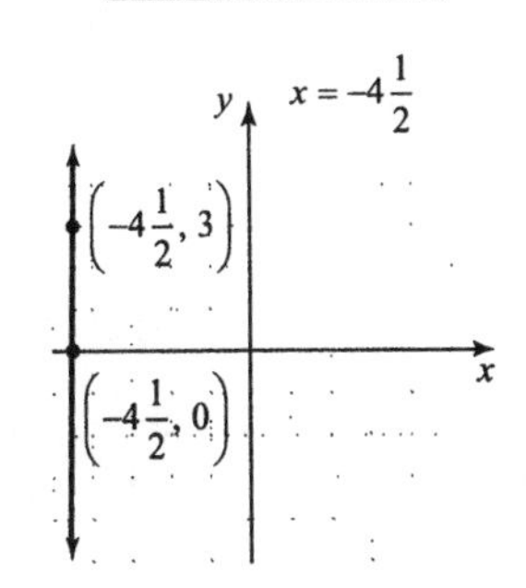

43. $y = 3\frac{1}{4}$ for all values of x.

x	y
0	$3\frac{1}{4}$
2	$3\frac{1}{4}$

45. $y = -\frac{2}{3}x + 1$

x-intercept: $\left(\frac{3}{2}, 0\right)$; y-intercept: $(0, 1)$

x	y
$\frac{3}{2}$	0
0	1

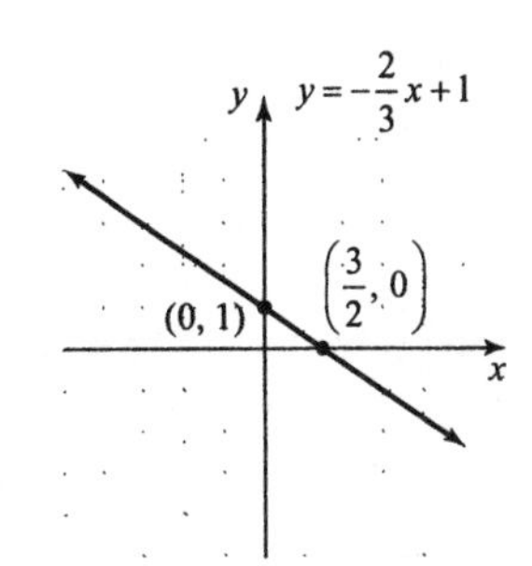

47. $4x - 6y + 2 = 0$

x-intercept: $\left(-\frac{1}{2}, 0\right)$; y-intercept: $\left(0, \frac{1}{3}\right)$

x	y
$-\frac{1}{2}$	0
0	$\frac{1}{3}$

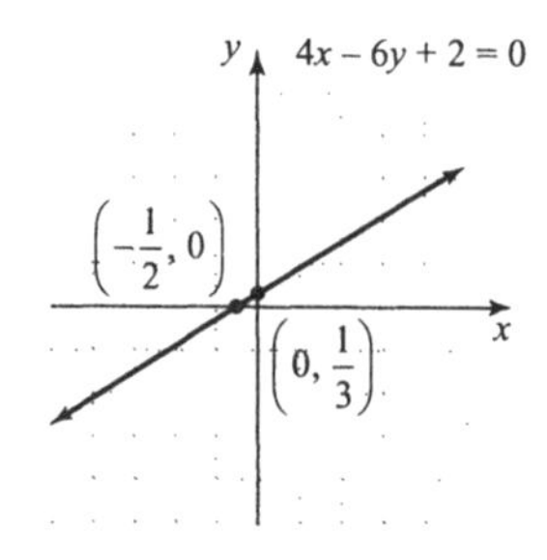

49. $y = 3$
The graph is a horizontal line with y-intercept (0, 3).
C

51. $x = -1$
The graph is a vertical line with x-intercept (−1, 0).
E

53. $y = 2x + 3$
The y-intercept is (0, 3) and the x-intercept is $\left(-\frac{3}{2}, 0\right)$.
B

55. $\frac{-6-3}{2-8} = \frac{-9}{-6} = \frac{3}{2}$

57. $\frac{-8-(-2)}{-3-(-2)} = \frac{-6}{-1} = 6$

59. $\frac{0-6}{5-0} = \frac{-6}{5} = -\frac{6}{5}$

61. $y = 1181x + 6505$

a. Let $x = 0$ and solve for y. The y-intercept is (0, 6505).

b. In 2003, the revenue for Disney Parks and Resorts was about $6505 million.

63. $y = -0.075x + 1.65$

a. $y = 0$:
$$0 = -0.075x + 1.65$$
$$0.075x = 1.65$$
$$x = 22$$
(22, 0)

b. 22 years after 2002 (2024); 0 people will attend movies at the theatre.

c. Answers may vary

65. $3x + 6y = 1200$

a. $x = 0$: $3(0) + 6y = 1200$, $y = 200$
(0, 200) corresponds to no chairs and 200 desks being manufactured.

b. $y = 0$: $3x + 6(0) = 1200$, $x = 400$
(400, 0) corresponds to 400 chairs and no desks being manufactured.

c.

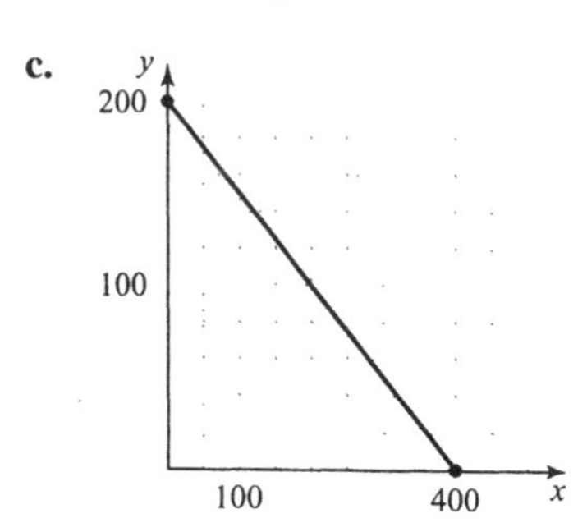

d. $y = 50$:
$$3x + 6(50) = 1200$$
$$3x + 300 = 1200$$
$$3x = 900$$
$$x = 300$$
300 chairs can be made.

67. Parallel to $y = -1$ is horizontal.
y-intercept is (0, −4), so $y = -4$ for all values of x. $y = -4$

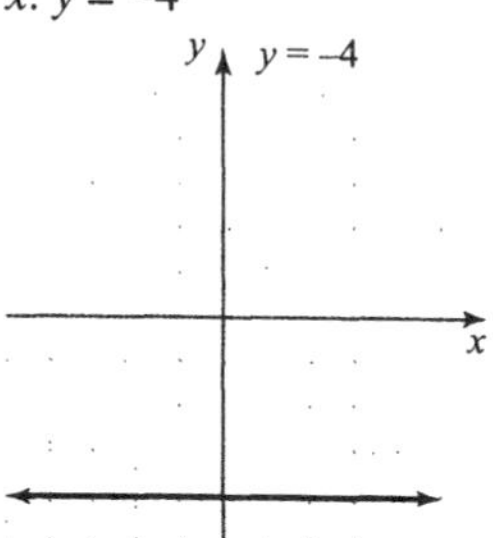

69. Answers may vary

71. Answers may vary

Section 3.4

Practice Exercises

1. If we let (x_1, y_1) be (−4, 11), then $x_1 = -4$ and $y_1 = 11$. Also, let (x_2, y_2) be (2, 5) so that $x_2 = 2$ and $y_2 = 5$.
$$m = \frac{y_2 - y_1}{x_2 - x_1} = \frac{5-11}{2-(-4)} = \frac{-6}{6} = -1$$
The slope of the line is −1.

2. Let (x_1, y_1) be (3, 1) and (x_2, y_2) be (−3, −1).

$$m = \frac{y_2 - y_1}{x_2 - x_1} = \frac{-1-1}{-3-3} = \frac{-2}{-6} = \frac{1}{3}$$

3. $y = \frac{2}{3}x - 2$

The equation is in slope-intercept form, $y = mx + b$. The coefficient of x, $\frac{2}{3}$, is the slope.

The constant term, −2, is the y-value of the y-intercept, (0, −2).

4. Write the equation in slope-intercept form by solving the equation for y.

$$6x - y = 5$$
$$-y = -6x + 5$$
$$y = 6x - 5$$

The coefficient of x, 6, is the slope. The constant term, −5, is the y-value of the y-intercept, (0, −5).

5. Write the equation in slope-intercept form by solving the equation for y.

$$5x + 2y = 8$$
$$2y = -5x + 8$$
$$\frac{2y}{2} = \frac{-5x}{2} + \frac{8}{2}$$
$$y = -\frac{5}{2}x + 4$$

The coefficient of x, $-\frac{5}{2}$, is the slope, and the y-intercept is (0, 4).

6. Recall that $y = 3$ is a horizontal line. Two ordered pair solutions of $y = 3$ and (1, 3) and (3, 3).

$$m = \frac{y_2 - y_1}{x_2 - x_1} = \frac{3-3}{3-1} = \frac{0}{2} = 0$$

The slope of the line $y = 3$ is 0.

7. Recall that the graph of $x = -4$ is a vertical line. Two ordered pair solutions of $x = -4$ and (−4, 1) and (−4, 3).

$$m = \frac{y_2 - y_1}{x_2 - x_1} = \frac{3-1}{-4-(-4)} = \frac{2}{0}$$

The slope of the vertical line $x = -4$ is undefined.

8. a. The slope of the line $y = -5x + 1$ is −5. We solve the second equation for y.

$$x - 5y = 10$$
$$-5y = -x + 10$$
$$\frac{-5y}{-5} = \frac{-x}{-5} + \frac{10}{-5}$$
$$y = \frac{1}{5}x - 2$$

The slope of the second line is $\frac{1}{5}$. Since the product of the slopes is $\frac{1}{5}(-5) = -1$, the lines are perpendicular.

b. Solve each equation for y.

$$x + y = 11 \qquad 2x + y = 11$$
$$y = -x + 11 \qquad y = -2x + 11$$

The slopes are −1 and −2. The slopes are not the same, and their product is not −1. Thus, the lines are neither parallel nor perpendicular.

c. Solve each equation for y.

$$2x + 3y = 21 \qquad 6y = -4x - 2$$
$$3y = -2x + 21 \qquad \frac{6y}{6} = \frac{-4x}{6} - \frac{2}{6}$$
$$\frac{3y}{3} = \frac{-2x}{3} + \frac{21}{3} \qquad y = -\frac{2}{3}x - \frac{1}{3}$$
$$y = -\frac{2}{3}x + 7$$

The slopes are $-\frac{2}{3}$ and $-\frac{2}{3}$. Since the lines have the same slope and different y-intercepts, they are parallel.

9. $\text{grade} = \frac{\text{rise}}{\text{run}} = \frac{1794}{7176} = 0.25 = 25\%$

The grade is 25%.

10. Use (2, 2) and (6, 5) to calculate slope.

$$m = \frac{5-2}{6-2} = \frac{3}{4} = \frac{0.75 \text{ dollar}}{1 \text{ pound}}$$

The Wash-n-Fold charges $0.75 per pound of laundry.

Calculator Explorations

1. $y_1 = 3.8x$
   $y_2 = 3.8x - 3$
   $y_3 = 3.8x + 9$

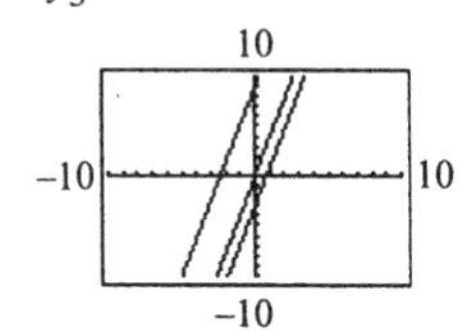

2. $y_1 = -4.9x$
   $y_2 = -4.9x + 1$
   $y_3 = -4.9x + 8$

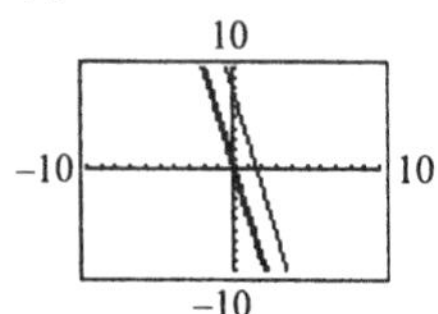

3. $y_1 = \frac{1}{4}x$
   $y_2 = \frac{1}{4}x + 5$
   $y_3 = \frac{1}{4}x - 8$

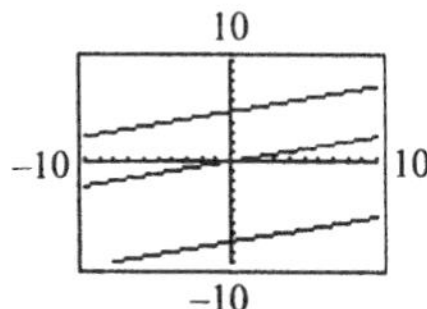

4. $y_1 = -\frac{3}{4}x$
   $y_2 = -\frac{3}{4}x - 5$
   $y_3 = -\frac{3}{4}x + 6$

10
−10 10
−10

Vocabulary and Readiness Check

1. The measure of the steepness or tilt of a line is called slope.
2. If an equation is written in the form $y = mx + b$, the value of the letter m is the value of the slope of the graph.
3. The slope of a horizontal line is 0.
4. The slope of a vertical line is undefined.
5. If the graph of a line moves upward from left to right, the line has positive slope.
6. If the graph of a line moves downward from left to right, the line has negative slope.
7. Given two points of a line, slope $= \frac{\text{change in } y}{\text{change in } x}$.
8. The line goes down. The slope is negative.
9. The line goes up. The slope is positive.
10. The line is vertical. The slope is undefined.
11. The line is horizontal. The slope is 0.
12. The slope is positive. The line is "upward."
13. The slope is negative. The line is "downward."
14. The slope is 0. The line is horizontal.
15. The slope is undefined. The line is vertical.

Exercise Set 3.4

1. $(x_1, y_1) = (-1, 5)$ and $(x_2, y_2) = (6, -2)$
   $$m = \frac{y_2 - y_1}{x_2 - x_1} = \frac{-2-5}{6-(-1)} = \frac{-7}{7} = -1$$

3. $(x_1, y_1) = (-4, 3)$ and $(x_2, y_2) = (-4, 5)$
   $$m = \frac{y_2 - y_1}{x_2 - x_1} = \frac{5-3}{-4-(-4)} = \frac{2}{0}$$
   The slope is undefined.

5. $(x_1, y_1) = (-2, 8)$ and $(x_2, y_2) = (1, 6)$
   $$m = \frac{y_2 - y_1}{x_2 - x_1} = \frac{6-8}{1-(-2)} = \frac{-2}{3} = -\frac{2}{3}$$

7. $(x_1, y_1) = (5, 1)$ and $(x_2, y_2) = (-2, 1)$
   $$m = \frac{y_2 - y_1}{x_2 - x_1} = \frac{1-1}{-2-5} = \frac{0}{-7} = 0$$

9. $(x_1, y_1) = (-1, 2)$ and $(x_2, y_2) = (2, -2)$

$$m = \frac{y_2 - y_1}{x_2 - x_1} = \frac{-2-2}{2-(-1)} = \frac{-4}{3} = -\frac{4}{3}$$

11. $(x_1, y_1) = (2, 3)$ and $(x_2, y_2) = (2, -1)$

$$m = \frac{y_2 - y_1}{x_2 - x_1} = \frac{-1-3}{2-2} = \frac{-4}{0}$$

The slope is undefined.

13. $(x_1, y_1) = (-3, -2)$ and $(x_2, y_2) = (-1, 3)$

$$m = \frac{y_2 - y_1}{x_2 - x_1} = \frac{3-(-2)}{-1-(-3)} = \frac{5}{2}$$

15. The slope of line 1 is positive, and the slope of line 2 is negative. Thus, line 1 has the greater slope.

17. Both line 1 and line 2 have positive slopes, but line 2 is steeper than line 1. Thus, line 2 has the greater slope.

19. (0, 0) and (2, 2)

$$m = \frac{y_2 - y_1}{x_2 - x_1} = \frac{2-0}{2-0} = \frac{2}{2} = 1$$

D

21. A vertical line has undefined slope.
B

23. (2, 0) and (4, −1)

$$m = \frac{y_2 - y_1}{x_2 - x_1} = \frac{-1-0}{4-2} = -\frac{1}{2}$$

E

25. $x = 6$ is a vertical line, so it has an undefined slope.

27. $y = -4$ is a horizontal line, so it has a slope $m = 0$.

29. $x = -3$ is a vertical line, so it has an undefined slope.

31. $y = 0$ is a horizontal line, so it has a slope $m = 0$.

33. $y = 5x - 2$
The equation is in slope-intercept form. The coefficient of x, 5, is the slope.

35. $y = -0.3x + 2.5$
The equation is in slope-intercept form. The coefficient of x, −0.3, is the slope.

37. Solve for y.

$$2x + y = 7$$
$$y = -2x + 7$$

The coefficient of x, −2, is the slope.

39. Solve for y.

$$2x - 3y = 10$$
$$-3y = -2x + 10$$
$$\frac{-3y}{-3} = \frac{-2x}{-3} + \frac{10}{-3}$$
$$y = \frac{2}{3}x - \frac{10}{3}$$

The coefficient of x, $\frac{2}{3}$, is the slope.

41. The graph of $x = 1$ is a vertical line. The slope is undefined.

43. Solve for y.

$$x = 2y$$
$$\frac{1}{2}x = y \text{ or } y = \frac{1}{2}x$$

The coefficient of x, $\frac{1}{2}$, is the slope.

45. The graph of $y = -3$ is a horizontal line. The slope is 0.

47. Solve for y.

$$-3x - 4y = 6$$
$$-4y = 3x + 6$$
$$\frac{-4y}{-4} = \frac{3x}{-4} + \frac{6}{-4}$$
$$y = -\frac{3}{4}x - \frac{3}{2}$$

The coefficient of x, $-\frac{3}{4}$, is the slope.

49. Solve for y.

$$20x - 5y = 1.2$$
$$-5y = -20x + 1.2$$
$$\frac{-5y}{-5} = \frac{-20x}{-5} + \frac{1.2}{-5}$$
$$y = 4x - 0.24$$

The coefficient of x, 4, is the slope.

51. $y = \frac{2}{9}x + 3$, $y = -\frac{2}{9}x$

The slopes are $\frac{2}{9}$ and $-\frac{2}{9}$. The slopes are not the same, and their product is not −1. The lines are neither parallel nor perpendicular.

53. The slope of $y = 3x - 9$ is 3. Solve the other equation for y.

$$x - 3y = -6$$
$$-3y = -x - 6$$
$$\frac{-3y}{-3} = -\frac{x}{-3} - \frac{6}{-3}$$
$$y = \frac{1}{3}x + 2$$

The slope is $\frac{1}{3}$. The slopes are not the same, and their product is not −1. The lines are neither parallel nor perpendicular.

55. Solve the equations for y.

$$6x = 5y + 1 \qquad -12x + 10y = 1$$
$$6x - 1 = 5y \qquad 10y = 12x + 1$$
$$\frac{6x}{5} - \frac{1}{5} = \frac{5y}{5} \qquad \frac{10y}{10} = \frac{12x}{10} + \frac{1}{10}$$
$$y = \frac{6}{5}x - \frac{1}{5} \qquad y = \frac{6}{5}x + \frac{1}{10}$$

The lines have the same slope, $\frac{6}{5}$, but different y-intercepts. The lines are parallel.

57. Solve the equations for y.

$$6 + 4x = 3y \qquad 3x + 4y = 8$$
$$\frac{6}{3} + \frac{4x}{3} = \frac{3y}{3} \qquad 4y = -3x + 8$$
$$y = \frac{4}{3}x + 2 \qquad \frac{4y}{4} = -\frac{3x}{4} + \frac{8}{4}$$
$$y = -\frac{3}{4}x + 2$$

The slopes are $\frac{4}{3}$ and $-\frac{3}{4}$. Their product is −1, so the lines are perpendicular.

59. $\text{pitch} = \frac{6}{10} = \frac{3}{5}$

61. $\text{grade} = \frac{\text{rise}}{\text{run}} = \frac{2}{16} = 0.125 = 12.5\%$

63. $\text{grade} = \frac{\text{rise}}{\text{run}} = \frac{2580}{6450} = 0.40 = 40\%$

65. $\text{grade} = \frac{\text{rise}}{\text{run}} = \frac{10}{12.66} = 0.79 = 79\%$

67. Use (2002, 74) and (2007, 89) to calculate slope.

$$m = \frac{89 - 74}{2007 - 2002} = \frac{15}{5} = \frac{3 \text{ million households}}{1 \text{ year}}$$

Every 1 year, there are/should be 3 million more U.S. households with personal computers.

69. Use (5000, 2100) and (20,000, 8400) to calculate slope.

$$m = \frac{8400 - 2100}{20{,}000 - 5000} = \frac{6300}{15{,}000} = \frac{0.42 \text{ dollar}}{1 \text{ mile}}$$

It costs \$0.42 per 1 mile to own and operate a compact car.

71.
$$y - (-6) = 2(x - 4)$$
$$y + 6 = 2x - 8$$
$$y = 2x - 14$$

73.
$$y - 1 = -6(x - (-2))$$
$$y - 1 = -6(x + 2)$$
$$y - 1 = -6x - 12$$
$$y = -6x - 11$$

75. (−3, −3) and (0, 0)

$$m = \frac{y_2 - y_1}{x_2 - x_1} = \frac{0 - (-3)}{0 - (-3)} = \frac{3}{3} = 1$$

a. $m = 1$

b. $m = -1$

77. (−8, −4) and (3, 5)

$$m = \frac{y_2 - y_1}{x_2 - x_1} = \frac{5 - (-4)}{3 - (-8)} = \frac{9}{11}$$

a. $m = \frac{9}{11}$

b. $m = -\frac{11}{9}$

79. (2, 1) and (0, 0): $m = \frac{0 - 1}{0 - 2} = \frac{-1}{-2} = \frac{1}{2}$

(2, 1) and (−2, −1): $m = \frac{-1 - 1}{-2 - 2} = \frac{-2}{-4} = \frac{1}{2}$

(2, 1) and (−4, −2): $m = \frac{-2 - 1}{-4 - 2} = \frac{-3}{-6} = \frac{1}{2}$

(0, 0) and (−2, −1): $m = \frac{-1 - 0}{-2 - 0} = \frac{-1}{-2} = \frac{1}{2}$

(0, 0) and (−4, −2): $m=\frac{-2-0}{-4-0}=\frac{-2}{-4}=\frac{1}{2}$

(−2, −1) and (−4, −2): $m=\frac{-2-(-1)}{-4-(-2)}=\frac{-1}{-2}=\frac{1}{2}$

Since the slope of the line between each pair of points is the same, the points lie on the same line.

81. Answers may vary

83. In 2001, the average fuel economy was approximately 28.5 miles per gallon.

85. The lowest point on the graph corresponds to the year 2000. The average fuel economy was approximately 28.1 miles per gallon.

87. The line segment from 2000 to 2001 is the steepest, so it has the greatest slope.

89. $\text{pitch}=\frac{\text{rise}}{\text{run}}$

$$\frac{1}{3}=\frac{x}{18}$$
$$3x=18$$
$$x=6$$

91. **a.** (2006, 1657) and (2001, 1132)

b. $m=\frac{y_2-y_1}{x_2-x_1}=\frac{1132-1657}{2001-2006}=\frac{-525}{-5}=105$

c. For the years 2001 through 2006, the price per acre of U.S. farmland rose approximately $105 per year.

93. (1, 1), (−4, 4) and (−3, 0)

$m_1=\frac{0-1}{-3-1}=\frac{1}{4},\ m_2=\frac{0-4}{-3-(-4)}=-4$

$m_1m_2=-1$, so the sides are perpendicular.

95. (2.1, 6.7) and (−8.3, 9.3)

$m=\frac{y_2-y_1}{x_2-x_1}=\frac{9.3-6.7}{-8.3-2.1}=\frac{2.6}{-10.4}=-0.25$

97. (2.3, 0.2) and (7.9, 5.1)

$m=\frac{y_2-y_1}{x_2-x_1}=\frac{5.1-0.2}{7.9-2.3}=\frac{4.9}{5.6}=0.875$

99. $y=-\frac{1}{3}x+2$

$y=-2x+2$

$y=-4x+2$

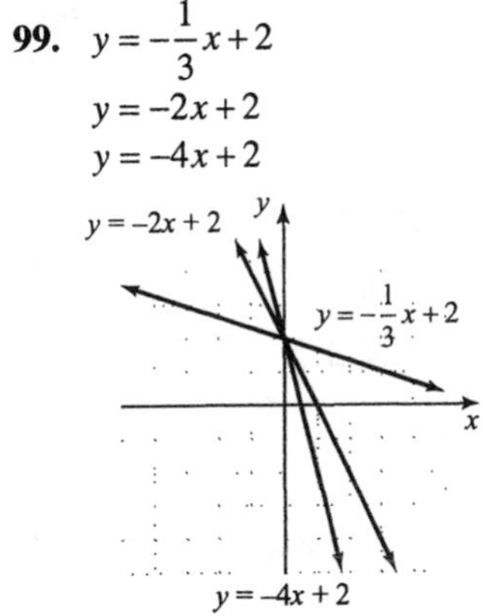

The line becomes steeper.

Integrated Review

1. (0, 0) and (2, 4)

$m=\frac{y_2-y_1}{x_2-x_1}=\frac{4-0}{2-0}=\frac{4}{2}=2$

2. Horizontal line, $m=0$

3. (0, 1) and (3, −1)

$m=\frac{y_2-y_1}{x_2-x_1}=\frac{-1-1}{3-0}=-\frac{2}{3}$

4. Vertical line, slope is undefined.

5. $y=-2x$

$m=-2,\ b=0$

x	y
0	0
1	−2
−1	2

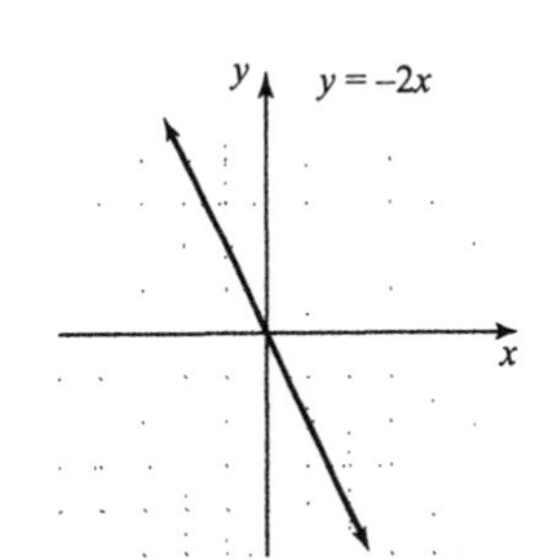

6. $x + y = 3$
$y = -x + 3$
$m = -1,\ b = 3$

x	y
0	3
3	0
1	2

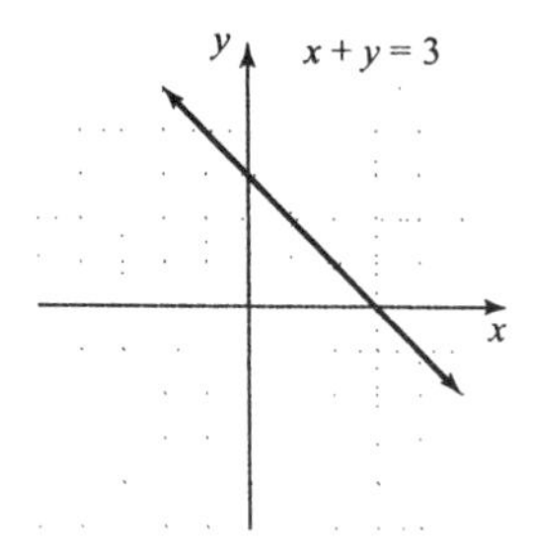

7. $x = -1$ for all values of y.
Vertical line; slope is undefined.

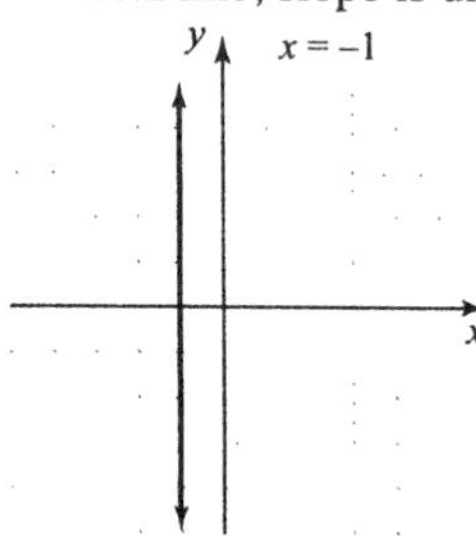

8. $y = 4$ for all values of x.
Horizontal line; $m = 0$

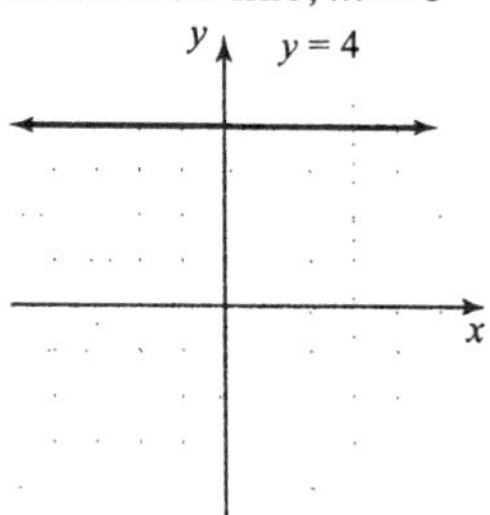

9. $x - 2y = 6$
$-2y = -x + 6$
$y = \frac{1}{2}x - 3$
$m = \frac{1}{2},\ b = -3$

x	y
0	−3
2	−2
4	−1

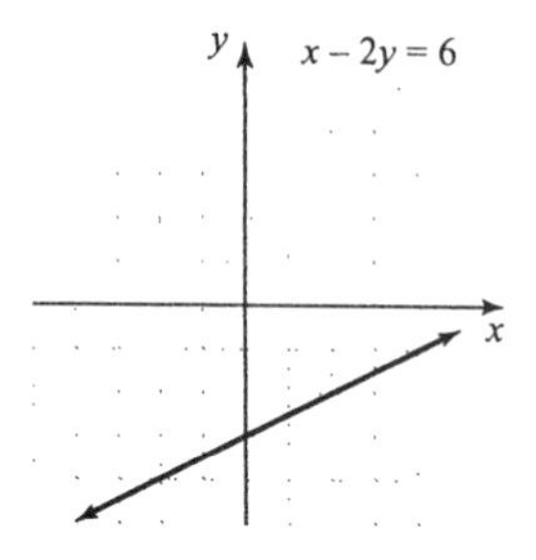

10. $y = 3x + 2$
$m = 3,\ b = 2$

x	y
0	2
−1	−1
−2	−4

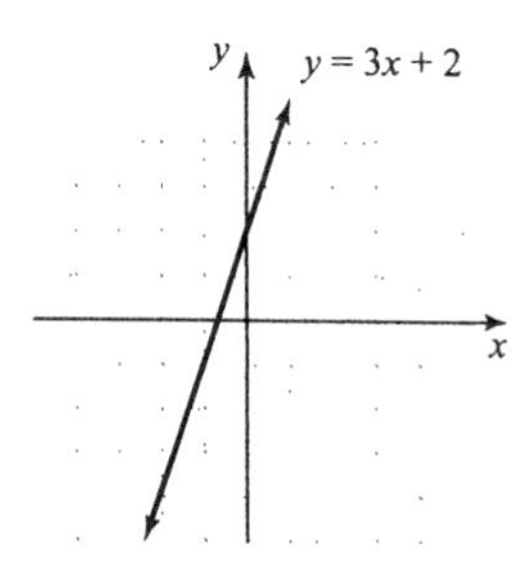

11. $5x + 3y = 15$

x	y
0	5
3	0

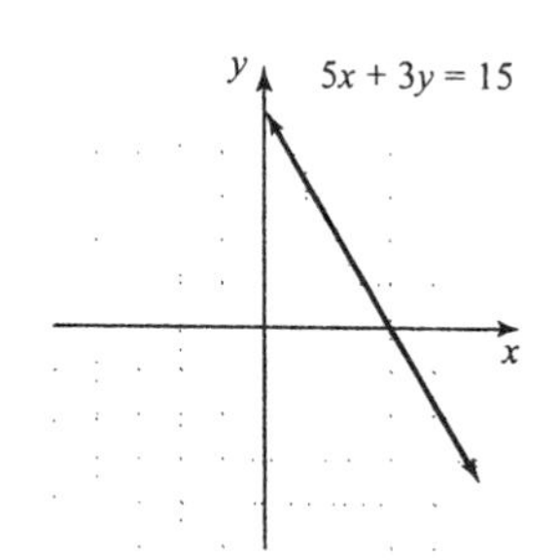

12. $2x - 4y = 8$

x	y
0	–2
4	0

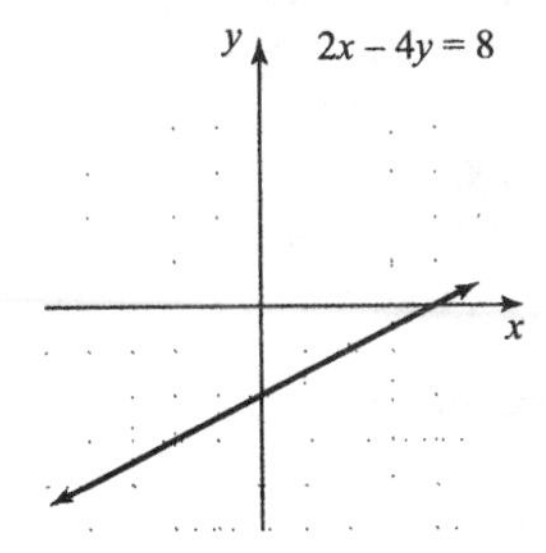

13. The slope of the first line is $-\frac{1}{5}$. Solve the second equation for y.

$$3x = -15y$$
$$\frac{3x}{-15} = \frac{-15y}{-15}$$
$$y = -\frac{1}{5}x$$

The slope of the second line is also $-\frac{1}{5}$. Since the lines have the same slope but different y-intercepts, the lines are parallel.

14. Solve the equations for y.

$$x - y = \frac{1}{2} \qquad 3x - y = \frac{1}{2}$$
$$-y = -x + \frac{1}{2} \qquad -y = -3x + \frac{1}{2}$$
$$y = x - \frac{1}{2} \qquad y = 3x - \frac{1}{2}$$

The slopes are 1 and 3. Since the slopes are not equal and their product is not –1, the lines are neither parallel nor perpendicular.

15. a. Let $x = 0$.
$y = -75(0) + 1650 = 1650$
The y-intercept is (0, 1650).

b. In 2002, there were 1650 million admissions to movie theaters in the United States.

c. The equation is in slope-intercept form. The coefficient of x, –75, is the slope.

d. For the years 2002 through 2005, the number of movie theater admissions decreased at a rate of 75 million per year.

16. a. Let $x = 9$.
$y = 3.3(9) - 3.1 = 29.7 - 3.1 = 26.6$
The ordered pair is (9, 26.6).

b. In 2009, the predicted revenue for online advertising is $26.6 billion.

Section 3.5

Practice Exercises

1. y-intercept: (0, 7); slope: $\frac{1}{2}$

Let $m = \frac{1}{2}$ and $b = 7$.
$$y = mx + b$$
$$y = \frac{1}{2}x + 7$$

2. $y = \frac{2}{3}x - 5$

The slope is $\frac{2}{3}$, and the y-intercept is (0, –5).

We plot (0, –5). From this point, we move up 2 units and then right 3 units. We stop at the point (3, –3).

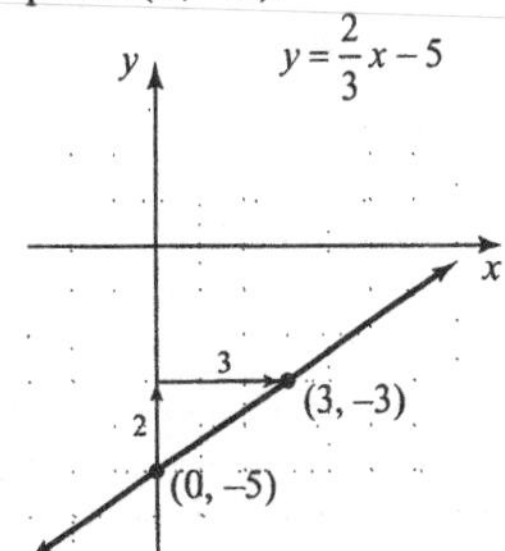

3. Solve the equation for y.
$$3x - y = 2$$
$$-y = -3x + 2$$
$$y = 3x - 2$$

The slope is 3, and the y-intercept is (0, –2). We plot (0, –2). From this point, we move up 3 units and then right 1 unit. We stop at the point (1, 1).

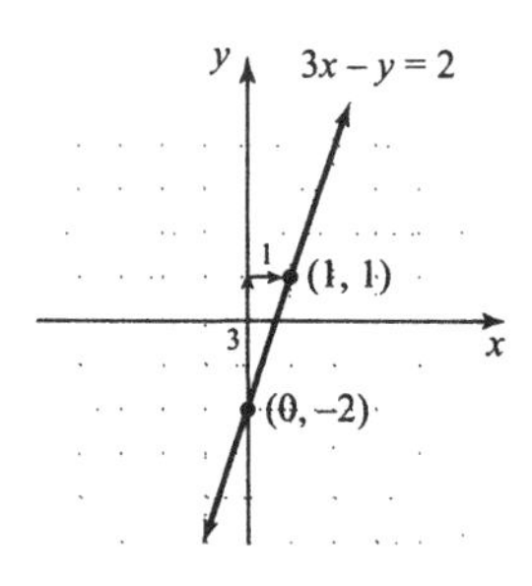

4. Line passing through (2, 3) with slope 4

$$y - y_1 = m(x - x_1)$$
$$y - 3 = 4(x - 2)$$
$$y - 3 = 4x - 8$$
$$-4x + y = -5$$
$$4x - y = 5$$

5. Line through (−1, 6) and (3, 1)

$$m = \frac{1-6}{3-(-1)} = \frac{-5}{4} = -\frac{5}{4}$$

Use the slope $-\frac{5}{4}$ and the point (3, 1).

$$y - y_1 = m(x - x_1)$$
$$y - 1 = -\frac{5}{4}(x - 3)$$
$$4(y-1) = 4\left(-\frac{5}{4}\right)(x - 3)$$
$$4y - 4 = -5(x - 3)$$
$$4y - 4 = -5x + 15$$
$$5x + 4y = 19$$

6. The equation of a vertical line can be written in the form $x = c$, so an equation for a vertical line passing through (3, −2) is $x = 3$.

7. Since the graph of $y = -2$ is a horizontal line, any line parallel to it is also vertical. The equation of a horizontal line can be written in the form $y = c$. An equation for the horizontal line passing through (4, 3) is $y = 3$.

8. a. Write two ordered pairs, (30, 150,000) and (50, 120,000).

$$m = \frac{120{,}000 - 150{,}000}{50 - 30} = \frac{-30{,}000}{20} = -1500$$

Use the slope −1500 and the point (30, 150,000).

$$y - y_1 = m(x - x_1)$$
$$y - 150{,}000 = -1500(x - 30)$$
$$y - 150{,}000 = -1500x + 45{,}000$$
$$y = -1500x + 195{,}000$$

b. Find y when $x = 60$.

$y = -1500x + 195{,}000$
$y = -1500(60) + 195{,}000$
$y = -90{,}000 + 195{,}000$
$y = 105{,}000$

To sell 60 condos per month, the price should be $105,000.

Calculator Explorations

1. $y_1 = x,\ y_2 = 6x,\ y_3 = -6x$

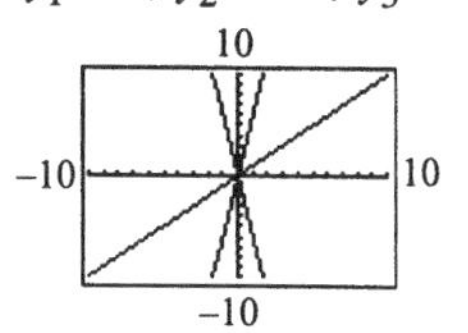

2. $y_1 = -x,\ y_2 = -5x,\ y_3 = -10x$

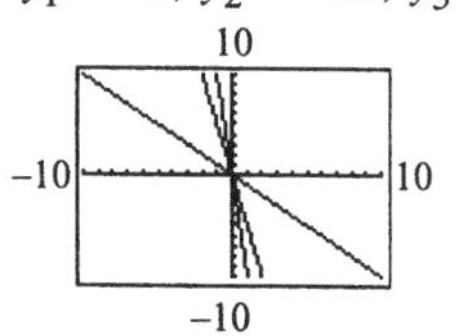

3. $y_1 = \frac{1}{2}x + 2,\ y_2 = \frac{3}{4}x + 2,\ y_3 = x + 2$

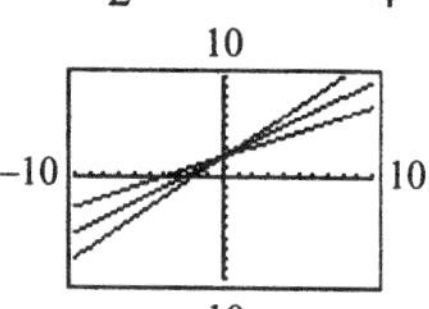

4. $y_1 = x + 1,\ y_2 = \frac{5}{4}x + 1,\ y_3 = \frac{5}{2}x + 1$

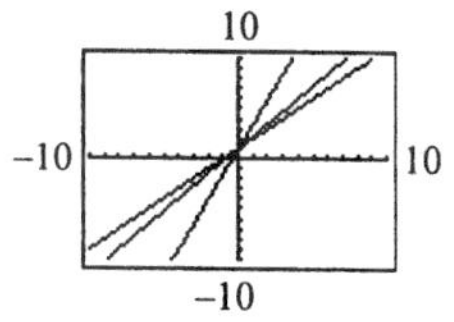

5. $y_1 = -7x + 5,\ y_2 = 7x + 5$

10

−10 10

−10

6. $y_1 = 3x - 1,\ y_2 = -3x - 1$

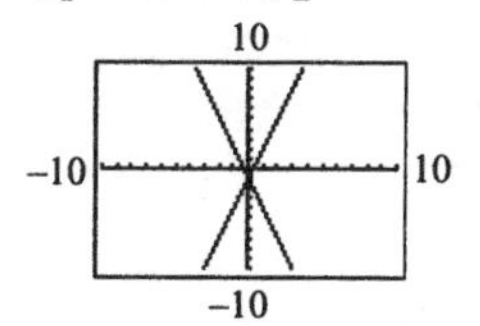

Vocabulary and Readiness Check

1. The form $y = mx + b$ is called <u>slope-intercept</u> form. When a linear equation in two variables is written in this form, <u>m</u> is the slope of its graph and (0, <u>b</u>) is its y-intercept.

2. The form $y - y_1 = m(x - x_1)$ is called <u>point-slope</u> form. When a linear equation in two variables is written in this form, <u>m</u> is the slope of its graph and <u>(x_1, y_1)</u> is a point on the graph.

3. $y - 7 = 4(x + 3)$; <u>point-slope</u> form

4. $5x - 9y = 11$; <u>standard</u> form

5. $y = \frac{1}{2}$; <u>horizontal</u> line

6. $x = -17$; <u>vertical</u> line

7. $y = \frac{3}{4}x - \frac{1}{3}$; <u>slope-intercept</u> form

Exercise Set 3.5

1. $m = 5, b = 3$
$y = mx + b$
$y = 5x + 3$

3. $m = -4,\ b = -\frac{1}{6}$
$y = mx + b$
$y = -4x + \left(-\frac{1}{6}\right)$
$y = -4x - \frac{1}{6}$

5. $m = \frac{2}{3},\ b = 0$
$y = mx + b$
$y = \frac{2}{3}x + 0$
$y = \frac{2}{3}x$

7. $m = 0, b = -8$
$y = mx + b$
$y = 0x + (-8)$
$y = -8$

9. $m = -\frac{1}{5},\ b = \frac{1}{9}$
$y = mx + b$
$y = -\frac{1}{5}x + \frac{1}{9}$

11. $y = 2x + 1$

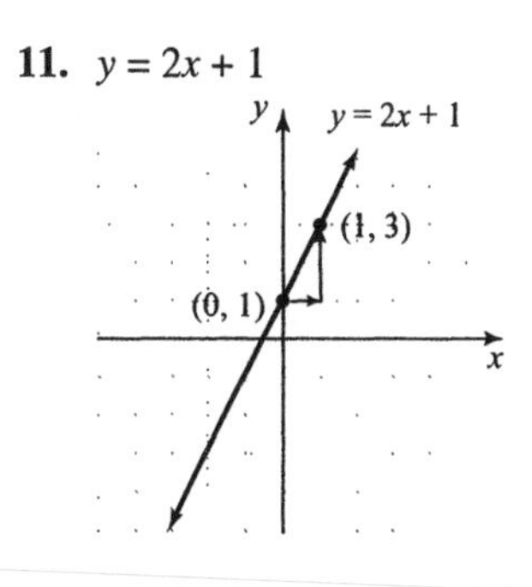

13. $y = \frac{2}{3}x + 5$

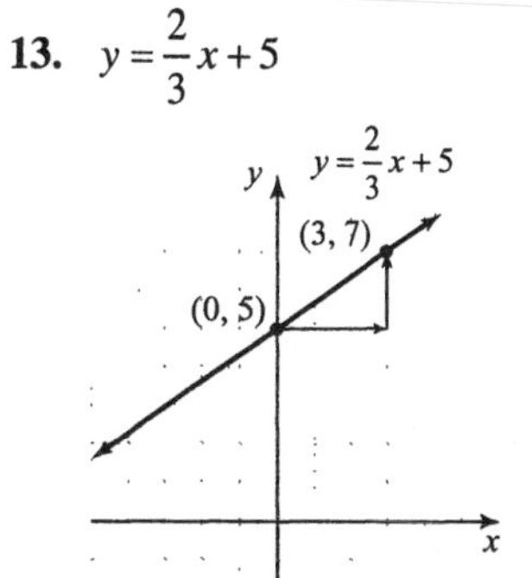

15. $y = -5x$

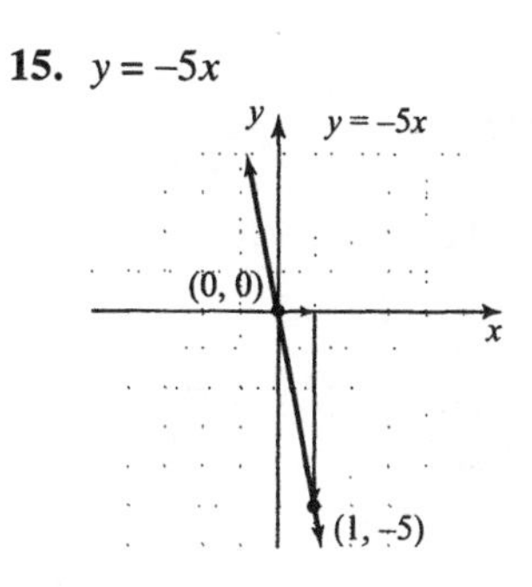

17. $4x+y=6$
$y=-4x+6$

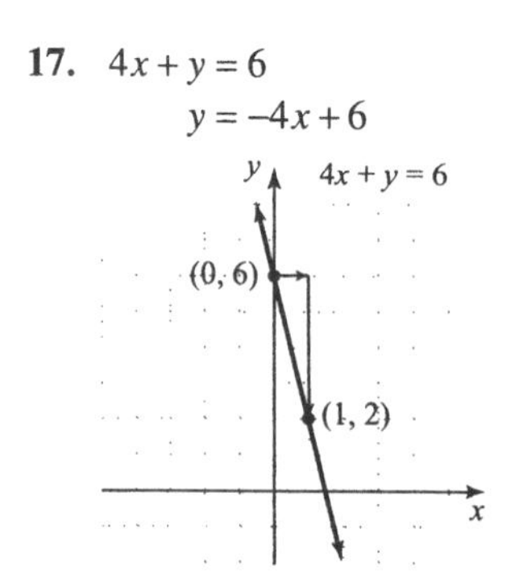

19. $4x-7y=-14$
$-7y=-4x-14$
$y=\frac{4}{7}x+2$

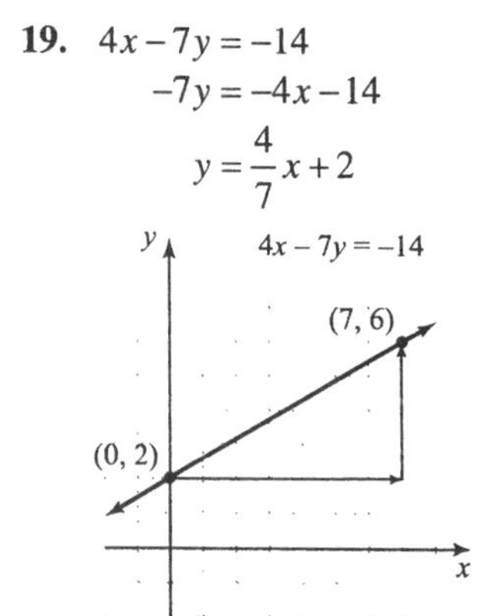

21. $x=\frac{5}{4}y$
$\frac{4}{5}x=y$
$y=\frac{4}{5}x$

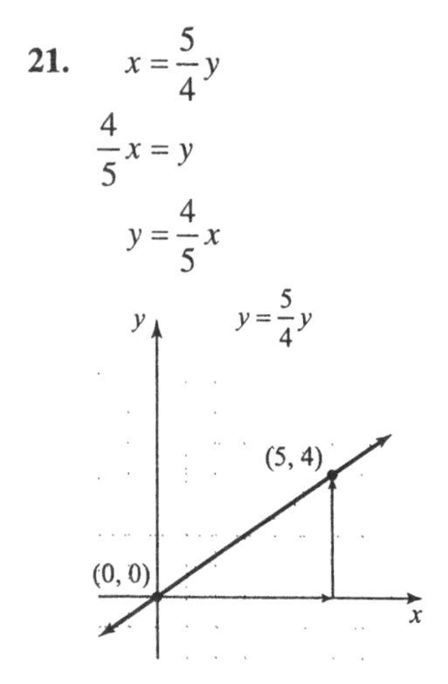

23. $m=6;\ (2, 2)$
$y-y_1=m(x-x_1)$
$y-2=6(x-2)$
$y-2=6x-12$
$-6x+y=-10$ or $6x-y=10$

25. $m=-8;\ (-1, -5)$
$y-y_1=m(x-x_1)$
$y-(-5)=-8(x-(-1))$
$y+5=-8x-8$
$8x+y=-13$

27. $m=\frac{3}{2};\ (5, -6)$
$y-y_1=m(x-x_1)$
$y-(-6)=\frac{3}{2}(x-5)$
$2(y+6)=3(x-5)$
$2y+12=3x-15$
$-3x+2y=-27$
$3x-2y=27$

29. $m=-\frac{1}{2};\ (-3, 0)$
$y-y_1=m(x-x_1)$
$y-0=-\frac{1}{2}(x-(-3))$
$y=-\frac{1}{2}(x+3)$
$-2y=x+3$
$-x-2y=3$
$x+2y=-3$

31. (3, 2) and (5, 6)
$m=\frac{y_2-y_1}{x_2-x_1}=\frac{6-2}{5-3}=\frac{4}{2}=2$
$m=2;\ (3, 2)$
$y-y_1=m(x-x_1)$
$y-2=2(x-3)$
$y-2=2x-6$
$-2x+y=-4$
$2x-y=4$

33. (−1, 3) and (−2, −5)
$m=\frac{y_2-y_1}{x_2-x_1}=\frac{-5-3}{-2-(-1)}=\frac{-8}{-1}=8$
$m=8;\ (-1, 3)$
$y-y_1=m(x-x_1)$
$y-3=8(x-(-1))$
$y-3=8x+8$
$-8x+y=11$
$8x-y=-11$

35. (2, 3) and (–1, –1)

$$m = \frac{y_2 - y_1}{x_2 - x_1} = \frac{-1-3}{-1-2} = \frac{-4}{-3} = \frac{4}{3}$$

$m = \frac{4}{3}$; (2, 3)

$$y - y_1 = m(x - x_1)$$
$$y - 3 = \frac{4}{3}(x-2)$$
$$3(y-3) = 4(x-2)$$
$$3y - 9 = 4x - 8$$
$$-4x + 3y = 1$$
$$4x - 3y = -1$$

37. (0, 0) and $\left(-\frac{1}{8}, \frac{1}{13}\right)$

$$m = \frac{\frac{1}{13} - 0}{-\frac{1}{8} - 0} = \frac{\frac{1}{13}}{-\frac{1}{8}} = \frac{1}{13}\left(-\frac{8}{1}\right) = -\frac{8}{13}$$

$m = -\frac{8}{13}$; (0, 0)

$$y - y_1 = m(x - x_1)$$
$$y - 0 = -\frac{8}{13}(x-0)$$
$$y = -\frac{8}{13}x$$
$$13y = -8x$$
$$8x + 13y = 0$$

39. Vertical line, point (0, 2)
$x = c$
$x = 0$

41. Horizontal line, point (–1, 3)
$y = c$
$y = 3$

43. Vertical line, point $\left(-\frac{7}{3}, -\frac{2}{5}\right)$

$x = c$

$x = -\frac{7}{3}$

45. $y = 5$ is horizontal.
Parallel to $y = 5$ is horizontal; $y = c$.
Point (1, 2)
$y = 2$

47. $x = -3$ is vertical.
Perpendicular to $x = -3$ is horizontal; $y = c$.
Point (–2, 5)
$y = 5$

49. $x = 0$ is vertical.
Parallel to $x = 0$ is vertical; $x = c$.
Point (6, –8)
$x = 6$

51. $m = -\frac{1}{2}$; $\left(0, \frac{5}{3}\right)$

$$y = mx + b$$
$$y = -\frac{1}{2}x + \frac{5}{3}$$

53. (10, 7) and (7, 10)

$$m = \frac{y_2 - y_1}{x_2 - x_1} = \frac{10-7}{7-10} = \frac{3}{-3} = -1$$

$m = -1$; (10, 7)

$$y - y_1 = m(x - x_1)$$
$$y - 7 = -1(x - 10)$$
$$y - 7 = -x + 10$$
$$y = -x + 17$$

55. Undefined slope, through $\left(-\frac{3}{4}, 1\right)$

A line with undefined slope is vertical. A vertical line has an equation of the form $x = c$.

$x = -\frac{3}{4}$

57. $m = 1$; (–7, 9)

$$y - y_1 = m(x - x_1)$$
$$y - 9 = 1[x - (-7)]$$
$$y - 9 = x + 7$$
$$y = x + 16$$

59. $m = -5$, $b = 7$
$y = mx + b$
$y = -5x + 7$

61. x-axis is horizontal.
Parallel to x-axis is horizontal; $y = c$.
Point (6, 7)
$y = 7$

63. (2, 3) and (0, 0)

$$m = \frac{y_2 - y_1}{x_2 - x_1} = \frac{3-0}{2-0} = \frac{3}{2}; \ b = 0$$
$$y = mx + b$$
$$y = \frac{3}{2}x + 0$$
$$y = \frac{3}{2}x$$

65. y-axis is vertical.
Perpendicular to y-axis is horizontal; $y = c$.
Point $(-2, -3)$
$y = -3$

67. $m = -\frac{4}{7}$; $(-1, -2)$

$$y - y_1 = m(x - x_1)$$
$$y - (-2) = -\frac{4}{7}[x - (-1)]$$
$$y + 2 = -\frac{4}{7}x - \frac{4}{7}$$
$$y = -\frac{4}{7}x - \frac{4}{7} - 2$$
$$y = -\frac{4}{7}x - \frac{18}{7}$$

69. a. $(1, 32)$ and $(3, 96)$

$$m = \frac{y_2 - y_1}{x_2 - x_1} = \frac{96 - 32}{3 - 1} = \frac{64}{2} = 32$$

$m = 32$; $(1, 32)$

$$s - s_1 = m(t - t_1)$$
$$s - 32 = 32(t - 1)$$
$$s - 32 = 32t - 32$$
$$s = 32t$$

b. If $t = 4$, then $s = 32(4) = 128$ ft/sec.

71. a. Use $(0, 29{,}000)$ and $(3, 71{,}000)$.

$$m = \frac{71{,}000 - 29{,}000}{3 - 0} = \frac{42{,}000}{3} = 14{,}000$$

$b = 29{,}000$
$y = mx + b$
$y = 14{,}000x + 29{,}000$

b. Let $x = 2010 - 2004 = 6$.

$$y = 14{,}000(6) + 29{,}000$$
$$= 84{,}000 + 29{,}000$$
$$= 113{,}000$$

We predict there will be 113,000 hybrids in 2010.

73. a. Use $(0, 79.6)$ and $(6, 85)$.

$$m = \frac{85 - 79.6}{6 - 0} = \frac{5.4}{6} = 0.9$$

$b = 79.6$
$y = mx + b$
$y = 0.9x + 79.6$

b. Let $x = 2010 - 2000 = 10$.
$y = 0.9(10) + 79.6 = 9 + 79.6 = 88.6$
We predict there will be 88.6 persons per square mile in 2010.

75. a. The ordered pairs are $(0, 14.7)$ and $(10, 14.14)$.

b. $m = \frac{14.14 - 14.7}{10 - 0} = \frac{-0.56}{10} = -0.056$

$b = 14.7$
$y = mx + b$
$y = -0.056x + 14.7$

c. Let $x = 2016 - 1996 = 20$.

$$y = -0.056(20) + 14.7$$
$$= -1.12 + 14.7$$
$$= 13.58$$

We predict that there will be 13.58 births per thousand population in 2016.

77. a. The ordered pairs are $(0, 5)$ and $(3, 20)$.

b. $m = \frac{20 - 5}{3 - 0} = \frac{15}{3} = 5$

$b = 5$
$y = mx + b$
$y = 5x + 5$

c. Let $x = 2012 - 2003 = 9$.
$y = 5(9) + 5 = 45 + 5 = 50$
We predict that the membership will be 50 thousand, or 50,000, in 2012.

79. If $x = 2$, then

$$x^2 - 3x + 1 = (2)^2 - 3(2) + 1 = 4 - 6 + 1 = -1$$

81. If $x = -1$, then

$$x^2 - 3x + 1 = (-1)^2 - 3(-1) + 1 = 1 + 3 + 1 = 5$$

83. No

85. Yes

87. Answers may vary

89. $y = 3x - 1$, $m_1 = 3$

a. Parallel: $m_2 = m_1 = 3$; $(-1, 2)$

$$y - y_1 = m_2(x - x_1)$$
$$y - 2 = 3(x - (-1))$$
$$y - 2 = 3x + 3$$
$$-3x + y = 5$$
$$3x - y = -5$$

b. Perpendicular: $m_2 = -\frac{1}{m_1} = -\frac{1}{3}$; $(-1, 2)$

$$\begin{aligned} y - y_1 &= m_2(x - x_1) \\ y - 2 &= -\frac{1}{3}(x - (-1)) \\ 3(y - 2) &= -1(x + 1) \\ 3y - 6 &= -x - 1 \\ x + 3y &= 5 \end{aligned}$$

91. $3x + 2y = 7$, $y = -\frac{3}{2}x + \frac{7}{2}$, $m_1 = -\frac{3}{2}$

a. Parallel: $m_2 = m_1 = -\frac{3}{2}$; $(3, -5)$

$$\begin{aligned} y - y_1 &= m_2(x - x_1) \\ y - (-5) &= -\frac{3}{2}(x - 3) \\ 2(y + 5) &= -3(x - 3) \\ 2y + 10 &= -3x + 9 \\ 3x + 2y &= -1 \end{aligned}$$

b. Perpendicular: $m_2 = -\frac{1}{m_1} = \frac{2}{3}$; $(3, -5)$

$$\begin{aligned} y - y_1 &= m_2(x - x_1) \\ y - (-5) &= \frac{2}{3}(x - 3) \\ 3(y + 5) &= 2(x - 3) \\ 3y + 15 &= 2x - 6 \\ 2x - 3y &= 21 \end{aligned}$$

Section 3.6

Practice Exercises

1. The domain is the set of all x-values $\{0, 1, 5\}$. The range is the set of all y-values: $\{-2, 0, 3, 4\}$.

2. a. $\{(4, 1), (3, -2), (8, 5), (-5, 3)\}$ Each x-value is assigned to only one y-value, so this set of ordered pairs is a function.

b. $\{(1, 2), (-4, 3), (0, 8), (1, 4)\}$ The x-value 1 is assigned to two y-values, 2 and 4, so this set of ordered pairs is not a function.

3. a. This is the graph of the relation $\{(-2, 1), (3, -3), (3, 2)\}$. The x-coordinate 3 is paired with two y-coordinates, -3 and 2, so this is not the graph of a function.

b. This is the graph of the relation $\{(-2, 1), (0, 1), (1, -3), (3, 2)\}$. Each x-coordinate has exactly one y-coordinate, so this is the graph of a function.

4. a. This is the graph of a function since no vertical line will intersect this graph more than once.

b. This is the graph of a function since no vertical line will intersect this graph more than once.

c. This is the graph of a function since no vertical line will intersect this graph more than once.

d. This is not the graph of a function. Vertical lines can be drawn that intersect the graph in two points. An example of one is shown.

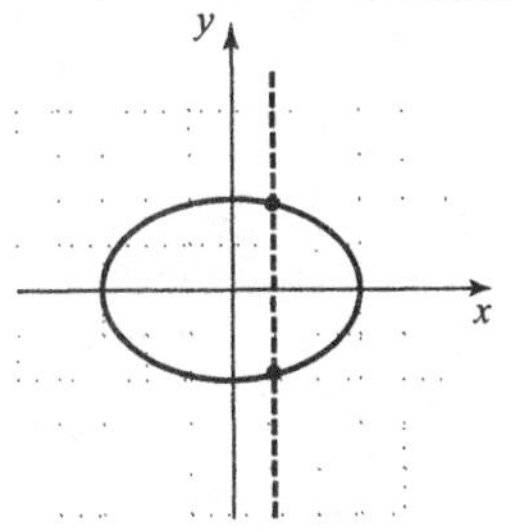

5. a. $y = 2x$ is a function because its graph is a nonvertical line.

b. $y = -3x - 1$ is a function because its graph is a nonvertical line.

c. $y = 8$ is a function because its graph is a nonvertical line.

d. $x = 2$ is not a function because its graph is a vertical line.

6. a. Since June is the sixth month, we look for 6 on the horizontal axis. From this point, we move vertically upward until the graph is reached. From the point on the graph, we move horizontally to the left to the vertical axis. The vertical axis there reads about 69°F.

b. We find 40°F on the temperature axis and move horizontally to the right. We eventually reach the point corresponding to 11, or November.

c. Yes, this is the graph of a function. It passes the vertical line test.

7. $h(x) = x^2 + 5$

a. $h(2) = 2^2 + 5 = 4 + 5 = 9$
(2, 9)

b. $h(-5) = (-5)^2 + 5 = 25 + 5 = 30$
(−5, 30)

c. $h(0) = 0^2 + 5 = 0 + 5 = 5$
(0, 5)

8. a. $h(x) = 6x + 3$
In this function, x can be any real number. The domain of $h(x)$ is the set of all real numbers, or $(-\infty, \infty)$ in interval notation.

b. $f(x) = \dfrac{1}{x^2}$
Recall that we cannot divide by 0 so that the domain of $f(x)$ is the set of all real numbers except 0. In interval notation, we write $(-\infty, 0) \cup (0, \infty)$.

9. a.

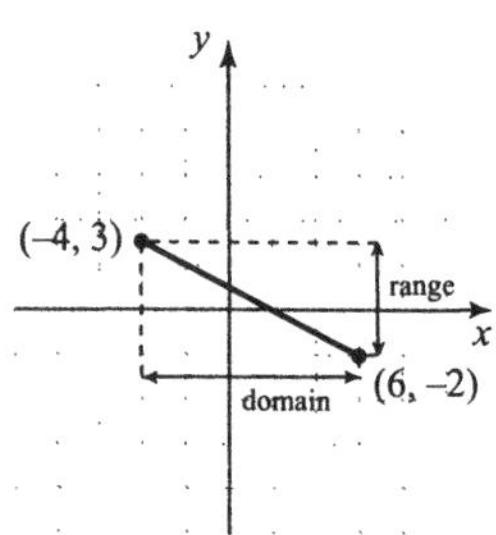

The domain is $[-4, 6]$.
The range is $[-2, 3]$.

b.

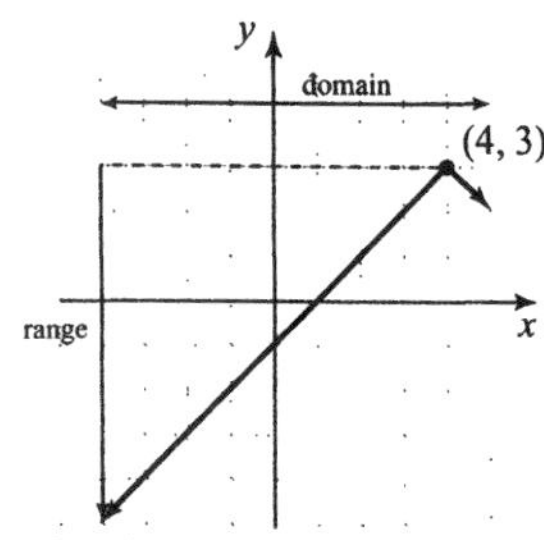

The domain is $(-\infty, \infty)$.
The range is $(-\infty, 3]$.

Vocabulary and Readiness Check

1. A set of ordered pairs is called a relation.

2. A set of ordered pairs that assigns to each x-value exactly one y-value is called a function.

3. The set of all y-coordinates of a relation is called the range.

4. The set of all x-coordinates of a relation is called the domain.

5. All linear equations are functions except those whose graphs are vertical lines.

6. All linear equations are functions except those whose equations are of the form $x = c$.

7. If $f(3) = 7$, the corresponding ordered pair is (3, 7).

8. The domain of $f(x) = x + 5$ is $(-\infty, \infty)$.

Exercise Set 3.6

1. {(2, 4), (0, 0), (−7, 10), (10, −7)}
Domain: {−7, 0, 2, 10}
Range: {−7, 0, 4, 10}

3. {(0, −2), (1, −2), (5, −2)}
Domain: {0, 1, 5}
Range: {−2}

5. Every point has a unique x-value: it is a function.

7. Two or more points have the same x-value: it is not a function.

9. No; two points have x-coordinate 1.

11. Yes; no two points have the same x-coordinate.

13. Yes; no vertical line can be drawn that intersects the graph more than once.

15. No; there are many vertical lines that intersect the graph twice, $x = 1$, for example.

17. Yes; $y = x + 1$ is a non-vertical line.

19. Yes; $y - x = 7$ is a non-vertical line.

21. Yes; $y = 6$ is a non-vertical line.

23. No; $x = -2$ is a vertical line.

25. No; does not pass the vertical line test.

27. The point on the graph above June corresponds to approximately 9:30 P.M. on the time axis.

29. The sunset is at approximately 3 P.M. twice, on January 1 and on December 1.

31. Yes; it passes the vertical line test.

33. \$4.25 per hour; the segment representing dates before October 1996 corresponds to 4.25 on the vertical axis.

35. 2009; the first line segment above 7.00 on the vertical axis represents dates beginning July 24, 2009.

37. Yes; answers may vary

39. $f(x) = 2x - 5$
$f(-2) = 2(-2) - 5 = -4 - 5 = -9$
$f(0) = 2(0) - 5 = -5$
$f(3) = 2(3) - 5 = 6 - 5 = 1$

41. $f(x) = x^2 + 2$
$f(-2) = (-2)^2 + 2 = 4 + 2 = 6$
$f(0) = (0)^2 + 2 = 2$
$f(3) = (3)^2 + 2 = 9 + 2 = 11$

43. $f(x) = 3x$
$f(-2) = 3(-2) = -6$
$f(0) = 3(0) = 0$
$f(3) = 3(3) = 9$

45. $f(x) = |x|$
$f(-2) = |-2| = 2$
$f(0) = |0| = 0$
$f(3) = |3| = 3$

47. $h(x) = -5x$
$h(-1) = -5(-1) = 5$
$h(0) = -5(0) = 0$
$h(4) = -5(4) = -20$

49. $h(x) = 2x^2 + 3$
$h(-1) = 2(-1)^2 + 3 = 2 + 3 = 5$
$h(0) = 2(0)^2 + 3 = 3$
$h(4) = 2(4)^2 + 3 = 2 \cdot 16 + 3 = 32 + 3 = 35$

51. $f(3) = 6$ corresponds to the ordered pair (3, 6).

53. $g(0) = -\frac{1}{2}$ corresponds to the ordered pair $\left(0, -\frac{1}{2}\right)$.

55. $h(-2) = 9$ corresponds to the ordered pair (−2, 9).

57. $(-\infty, \infty)$

59. $x + 5 \neq 0 \Rightarrow x \neq -5$, therefore $(-\infty, -5) \cup (-5, \infty)$ or all real numbers except −5.

61. $(-\infty, \infty)$

63. D: $(-\infty, \infty)$, R: $x \geq -4$, $[-4, \infty)$

65. D: $(-\infty, \infty)$, R: $(-\infty, \infty)$

67. D: $(-\infty, \infty)$, R: $\{2\}$

69. (−2, 1)

71. (−3, −1)

73. $f(-5) = 12$

75. (3, −4)

77. $f(5) = 0$

79. $H(x) = 2.59x + 47.24$

a. $H(46) = 2.59(46) + 47.24 = 166.38$ cm

b. $H(39) = 2.59(39) + 47.24 = 148.25$ cm

81. Answers may vary

83. $y = x + 7$
$f(x) = x + 7$

85. $g(x) = -3x + 12$

a. $g(s) = -3(s) + 12 = -3s + 12$

b. $g(r) = -3(r) + 12 = -3r + 12$

87. $f(x) = x^2 - 12$

a. $f(12) = (12)^2 - 12 = 132$

b. $f(a) = (a)^2 - 12 = a^2 - 12$

Chapter 3 Vocabulary Check

1. An ordered pair is a solution of an equation in two variables if replacing the variables by the coordinates of the ordered pair results in a true statement.

2. The vertical number line in the rectangle coordinate system is called the y-axis.

3. A linear equation can be written in the form $Ax + By = C$.

4. An x-intercept is a point of the graph where the graph crosses the x-axis.

5. The form $Ax + By = C$ is called standard form.

6. A y-intercept is a point of the graph where the graph crosses the y-axis.

7. The equation $y = 7x - 5$ is written in slope-intercept form.

8. The equation $y + 1 = 7(x - 2)$ is written in point-slope form.

9. To find an x-intercept of a graph, let $y = 0$.

10. The horizontal number line in the rectangular coordinate system is called the x-axis.

11. To find a y-intercept of a graph, let $x = 0$.

12. The slope of a line measures the steepness or tilt of a line.

13. A set of ordered pairs that assigns to each x-value exactly one y-value is called a function.

14. The set of all x-coordinates of a relation is called the domain of the relation.

15. The set of all y-coordinates of a relation is called the range of the relation.

16. A set of ordered pairs is called a relation.

Chapter 3 Review

1–6.

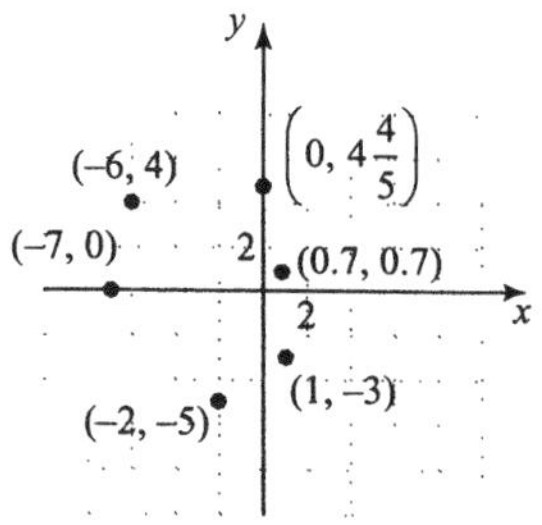

y
(−6, 4)
$\left(0, 4\frac{4}{5}\right)$
(−7, 0)
2
(0.7, 0.7)
2
x
(1, −3)
(−2, −5)

7. a. (8.00, 1), (7.50, 10), (6.50, 25), (5.00, 50), (2.00, 100)

b.

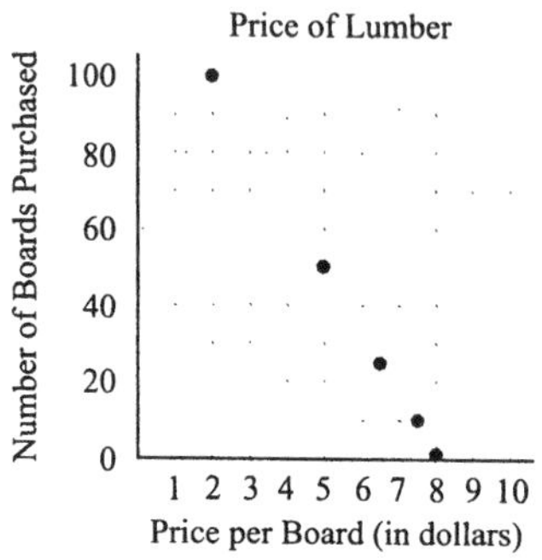

Price of Lumber
Number of Boards Purchased
100
80
60
40
20
0
1 2 3 4 5 6 7 8 9 10
Price per Board (in dollars)

8. a. (2001, 9.8), (2002, 15.1), (2003, 14.6), (2004, 14.0), (2005, 13.8), (2006, 13.6)

b.

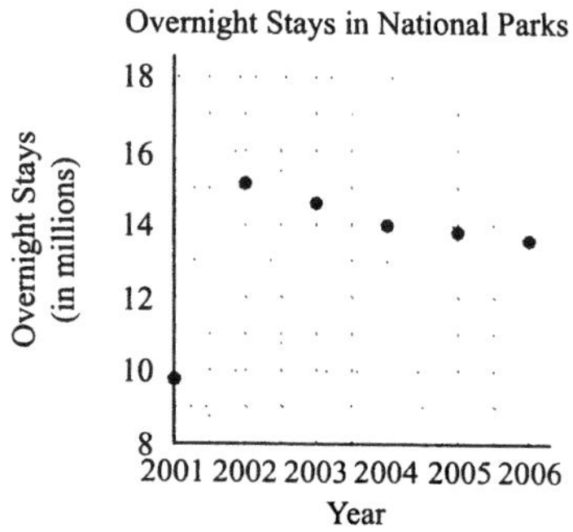

Overnight Stays in National Parks
Overnight Stays (in millions)
18
16
14
12
10
8
2001 2002 2003 2004 2005 2006
Year

9. $7x - 8y = 56$

(0, 56)

$7(0) - 8(56) \stackrel{?}{=} 56$

$-448 \neq 0$ No

(8, 0)

$7(8) - 8(0) \stackrel{?}{=} 56$

$56 = 56$ Yes

10. $-2x + 5y = 10$
$(-5, 0)$
$-2(-5) + 5(0) \stackrel{?}{=} 10$
$10 = 10$ Yes

$(1, 1)$
$-2(1) + 5(1) \stackrel{?}{=} 10$
$3 \neq 10$ No

11. $x = 13$
$(13, 5)$
$(13) \stackrel{?}{=} 13$
$13 = 13$ Yes

$(13, 13)$
$(13) \stackrel{?}{=} 13$
$13 = 13$ Yes

12. $y = 2$
$(7, 2)$
$(2) \stackrel{?}{=} 2$
$2 = 2$ Yes

$(2, 7)$
$(7) \stackrel{?}{=} 2$
$7 \neq 2$ No

13. $-2 + y = 6x,\ x = 7$
$-2 + y = 6(7)$
$-2 + y = 42$
$y = 44$
$(7, 44)$

14. $y = 3x + 5,\ y = -8$
$-8 = 3x + 5$
$-13 = 3x$
$-\frac{13}{3} = x$
$\left(-\frac{13}{3}, -8\right)$

15. $9 = -3x + 4y$
$y = 0$: $9 = -3x + 4(0)$, $9 = -3x$, $-3 = x$
$y = 3$: $9 = -3x + 4(3)$, $9 = -3x + 12$, $-3 = -3x$, $1 = x$
$x = 9$: $9 = -3(9) + 4y$, $9 = -27 + 4y$, $36 = 4y$, $9 = y$

x	y
–3	0
1	3
9	9

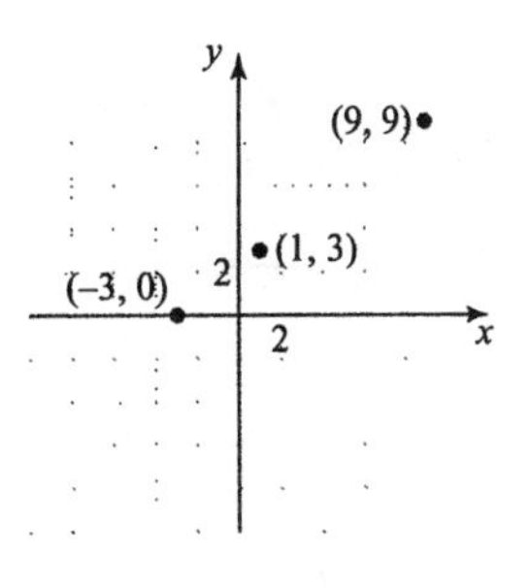

16. $y = 5$ for all values of x.

x	y
7	5
–7	5
0	5

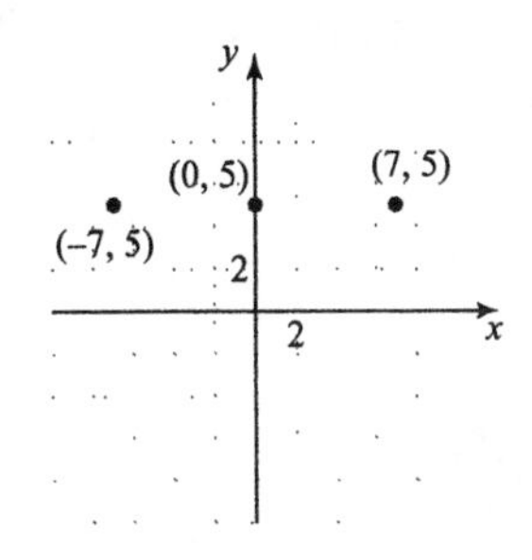

17. $x = 2y$
$y = 0$: $x = 2(0) = 0$
$y = 5$: $x = 2(5) = 10$
$y = -5$: $x = 2(-5) = -10$

x	y
0	0
10	5
–10	–5

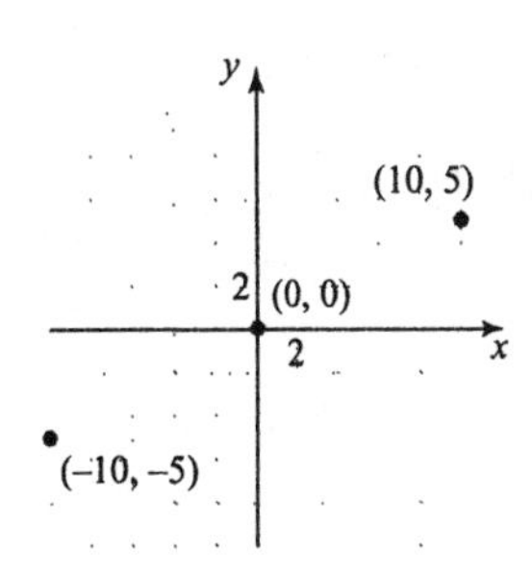

18. **a.** $y = 5x + 2000$
$x = 1$: $y = 5(1) + 2000 = 2005$
$x = 100$: $y = 5(100) + 2000 = 2500$
$x = 1000$: $y = 5(1000) + 2000 = 7000$

x	1	100	1000
y	2005	2500	7000

b. Let $y = 6430$.
$6430 = 5x + 2000$
$4430 = 5x$
$886 = x$
886 CD holders can be produced.

19. $x - y = 1$

x	y
1	0
0	−1

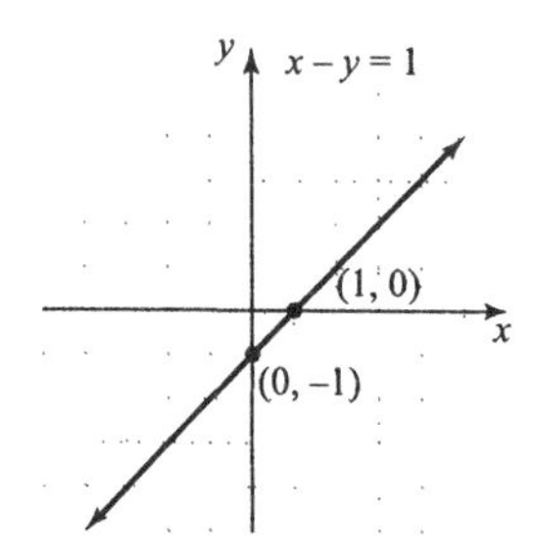

20. $x + y = 6$

x	y
6	0
0	6

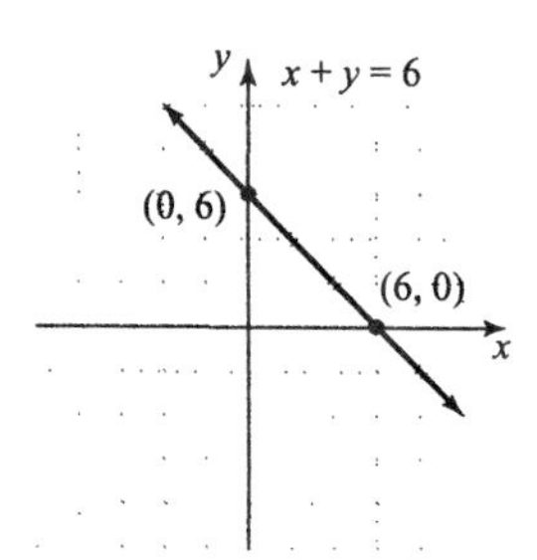

21. $x - 3y = 12$

x	y
12	0
0	−4

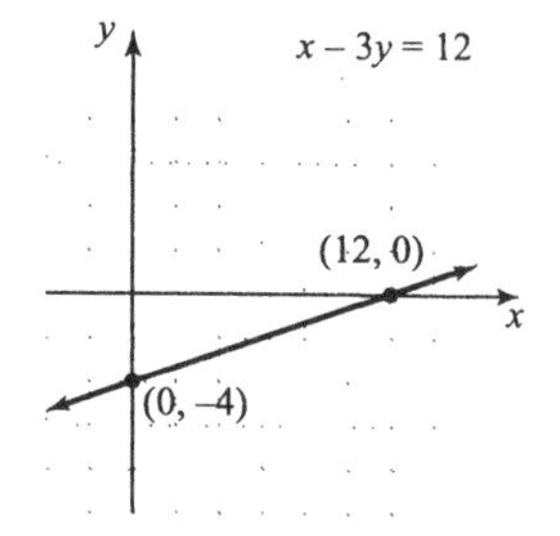

22. $5x - y = -8$

x	y
−2	−2
0	8

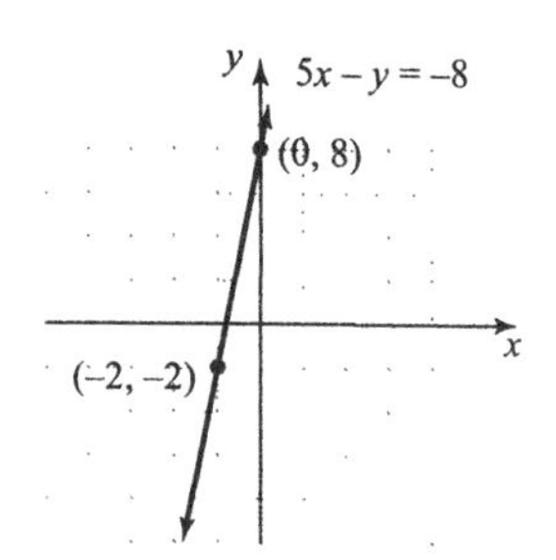

23. $x = 3y$

x	y
0	0
6	2

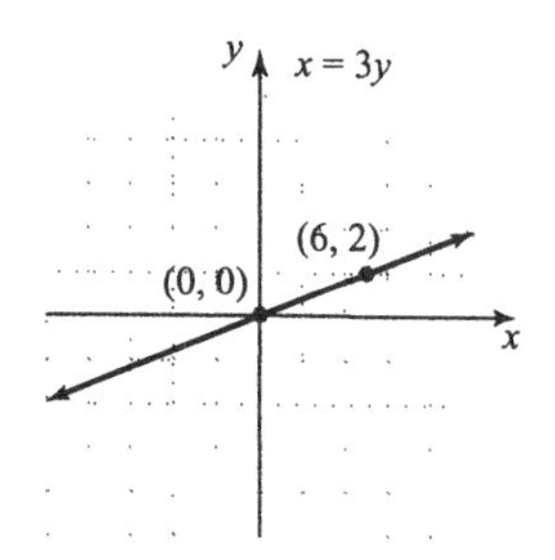

24. $y = -2x$

x	y
0	0
4	−8

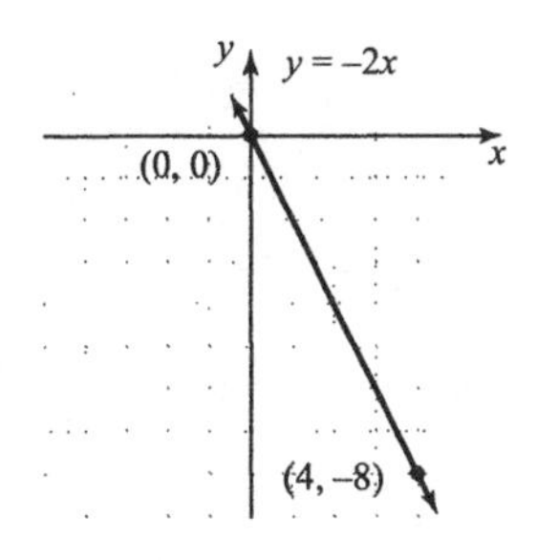

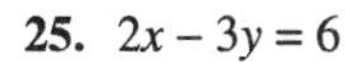
25. $2x - 3y = 6$

x	y
0	−2
3	0

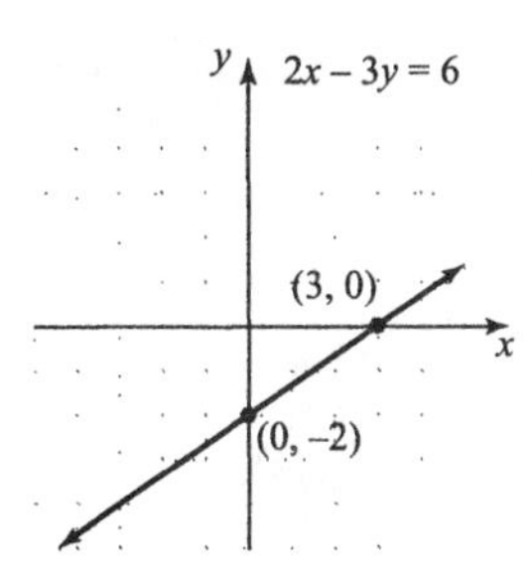

26. $4x - 3y = 12$

x	y
0	−4
3	0

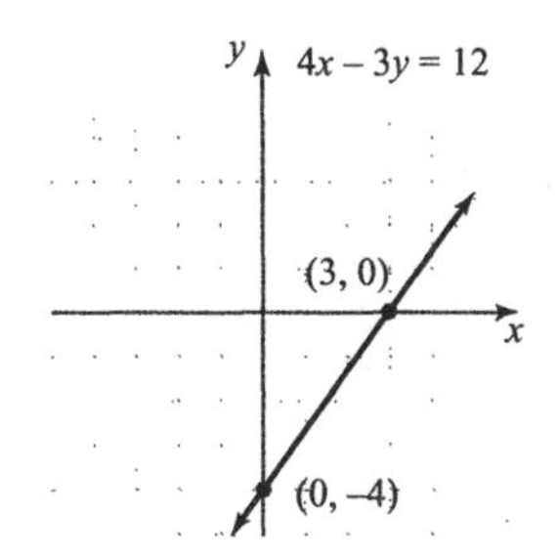

27. $y = 3x + 111$

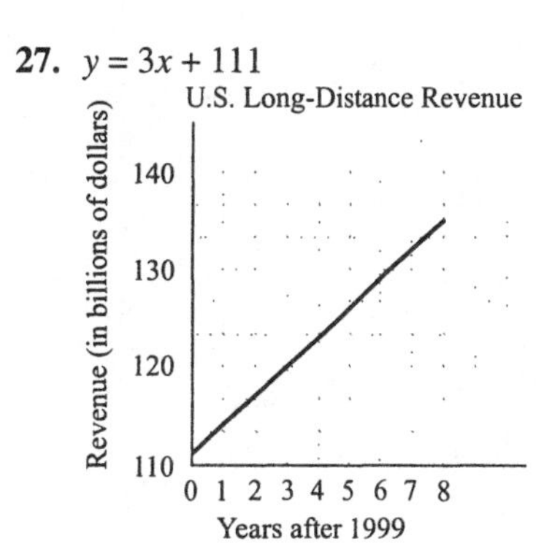

Expect a revenue of $135 billion in 2007.

28. x-intercept: (4, 0)
y-intercept: (0, −2)

29. y-intercept: (0, −3)

30. x-intercepts: (−2, 0), (2, 0)
y-intercepts: (0, 2), (0, −2)

31. x-intercepts: (−1, 0), (2, 0), (3, 0)
y-intercept: (0, −2)

32. $x - 3y = 12$

x	y
0	−4
12	0

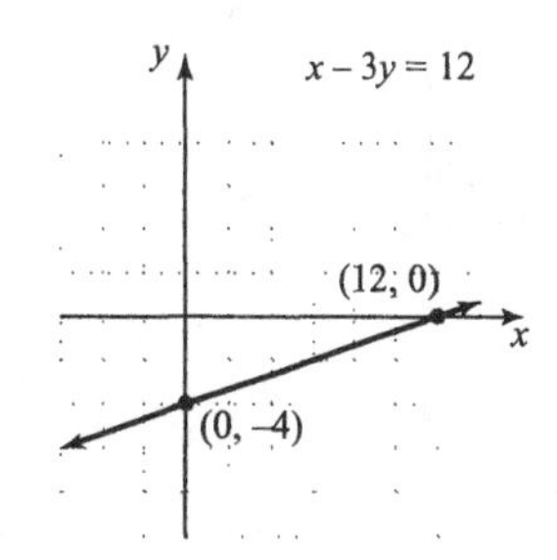

33. $-4x + y = 8$

x	y
0	8
−2	0

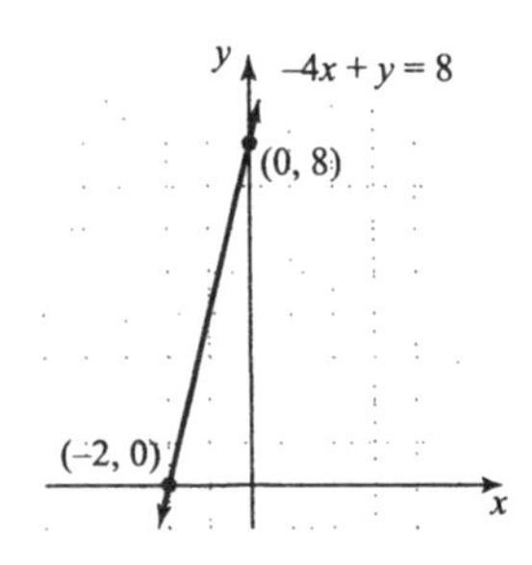

34. $y = -3$ for all x

x	y
0	−3

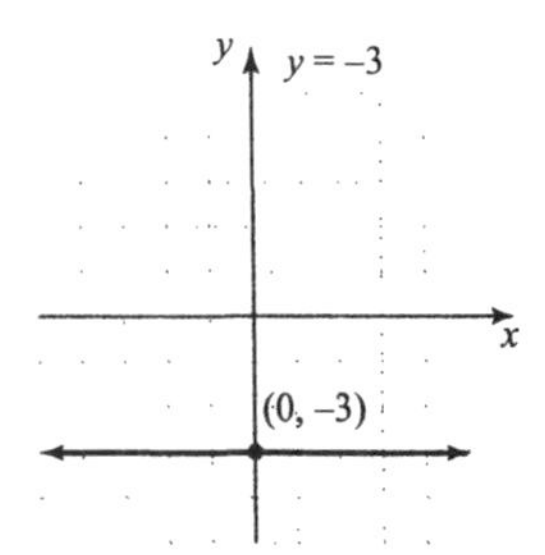

35. $x = 5$ for all y

x	y
5	0

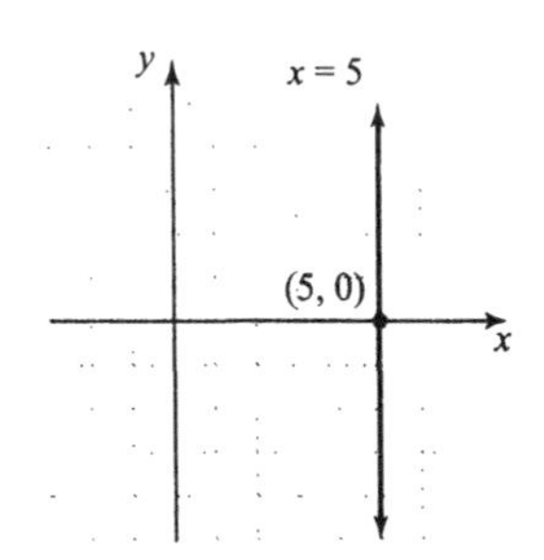

36. $y = -3x$
Find a second point.

x	y
0	0
3	−9

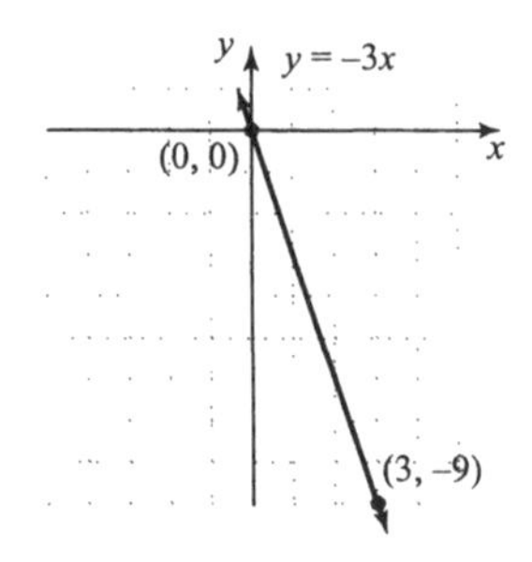

37. $x = 5y$
Find a second point.

x	y
0	0
5	1

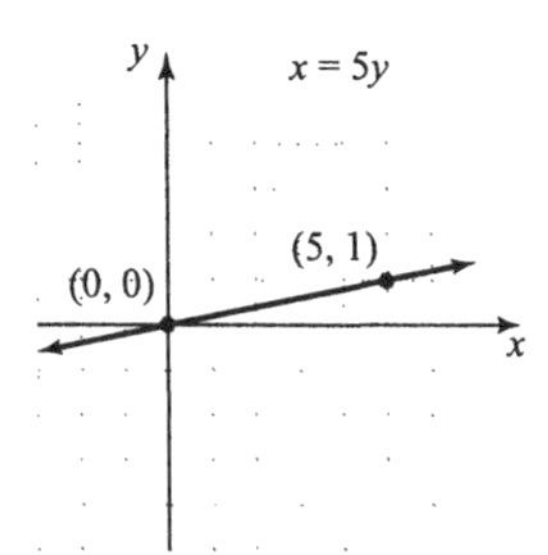

38. $x - 2 = 0$
$x = 2$ for all y

x	y
2	0

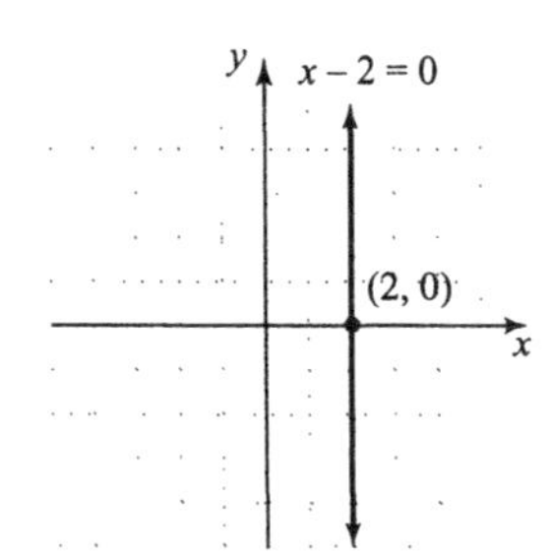

39. $y + 6 = 0$
$y = -6$ for all x

x	y
0	−6

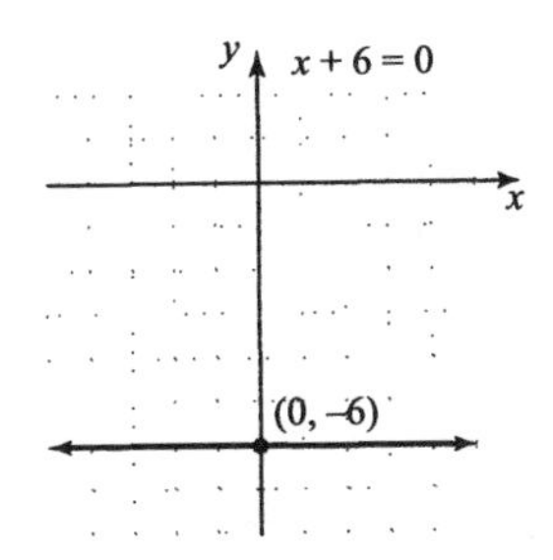

40. (–1, 2), and (3, –1)

$$m = \frac{y_2 - y_1}{x_2 - x_1} = \frac{-1-2}{3-(-1)} = -\frac{3}{4}$$

41. (–2, –2) and (3, –1)

$$m = \frac{y_2 - y_1}{x_2 - x_1} = \frac{-1-(-2)}{3-(-2)} = \frac{1}{5}$$

42. $m = 0$
d

43. $m = -1$
b

44. Slope is undefined.
c

45. $m = 3$
a

46. $m = \frac{2}{3}$
e

47. (2, 5) and (6, 8)

$$m = \frac{y_2 - y_1}{x_2 - x_1} = \frac{8-5}{6-2} = \frac{3}{4}$$

48. (4, 7) and (1, 2)

$$m = \frac{y_2 - y_1}{x_2 - x_1} = \frac{2-7}{1-4} = \frac{-5}{-3} = \frac{5}{3}$$

49. (1, 3) and (–2, –9)

$$m = \frac{y_2 - y_1}{x_2 - x_1} = \frac{-9-3}{-2-1} = \frac{-12}{-3} = 4$$

50. (–4, 1), and (3, –6)

$$m = \frac{y_2 - y_1}{x_2 - x_1} = \frac{-6-1}{3-(-4)} = \frac{-7}{7} = -1$$

51. $y = 3x + 7$
The equation is in slope-intercept form. The slope is the coefficient of x, or 3.

52. Solve for y.

$$x - 2y = 4$$
$$-2y = -x + 4$$
$$y = \frac{1}{2}x - 2$$

The slope is $\frac{1}{2}$.

53. $y = -2$
This is the equation of a horizontal line. The slope is 0.

54. $x = 0$
This is the equation of a vertical line. The slope is undefined.

55. Solve the equations for y.

$$x - y = 6 \qquad x + y = 3$$
$$-y = -x + 6 \qquad y = -x + 3$$
$$y = x - 6$$

The slopes are 1 and –1. Since their product is –1, the lines are perpendicular.

56. Solve the equations for y.

$$3x + y = 7 \qquad -3x - y = 10$$
$$y = -3x + 7 \qquad -y = 3x + 10$$
$$y = -3x - 10$$

The slopes are both –3. Since the lines have the same slope but different y-intercepts, they are parallel.

57. The first line, $y = 4x + \frac{1}{2}$, has slope 4. Solve the second equation for y.

$$4x + 2y = 1$$
$$2y = -4x + 1$$
$$y = -2x + \frac{1}{2}$$

The second line has slope –2. Since the slopes are not the same and their product is not –1, the lines are neither parallel nor perpendicular.

58. $x = 4$, $y = -2$
The first equation's graph is a vertical line, and the second equation's graph is a horizontal line. These lines are perpendicular.

59. Use the points (1985, 232) and (2006, 608).

$$m = \frac{608-232}{2006-1985} = \frac{376}{21} \approx \frac{17.90 \text{ dollars}}{1 \text{ year}}$$

Every 1 year, monthly daycare costs increase by \$17.90.

60. Use the points (2004, 46) and (2009, 56.5).

$$m = \frac{56.5-46}{2009-2004} = \frac{10.5}{5} \approx \frac{2.1 \text{ billion dollars}}{1 \text{ year}}$$

Every 1 year, \$2.1 billion more dollars are spent on technology.

61. $3x + y = 7$

$y = -3x + 7$

$y = mx + b$

$m = -3$, y-intercept = (0, 7)

62. $x - 6y = -1$

$-6y = -x - 1$

$y = \frac{1}{6}x + \frac{1}{6}$

$y = mx + b$

$m = \frac{1}{6}$, y-intercept $= \left(0, \frac{1}{6}\right)$

63. $y = 2$

$y = mx + b$

$m = 0$, y-intercept = (0, 2)

64. $x = -5$

$y = mx + b$

m is undefined.

There is no y-intercept.

65. $m = -5$, $b = \frac{1}{2}$

$y = mx + b$

$y = -5x + \frac{1}{2}$

66. $m = \frac{2}{3}$, $b = 6$

$y = mx + b$

$y = \frac{2}{3}x + 6$

67. $y = 3x - 1$

$y = mx + b$

$m = 3$, $b = -1$

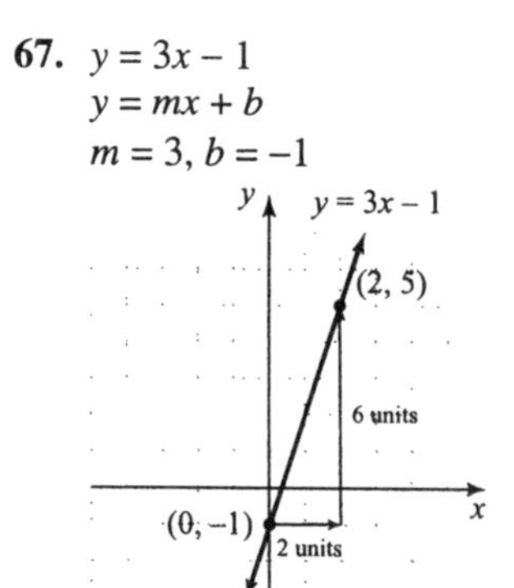

68. $y = -3x$

$y = mx + b$

$m = -3$, $b = 0$

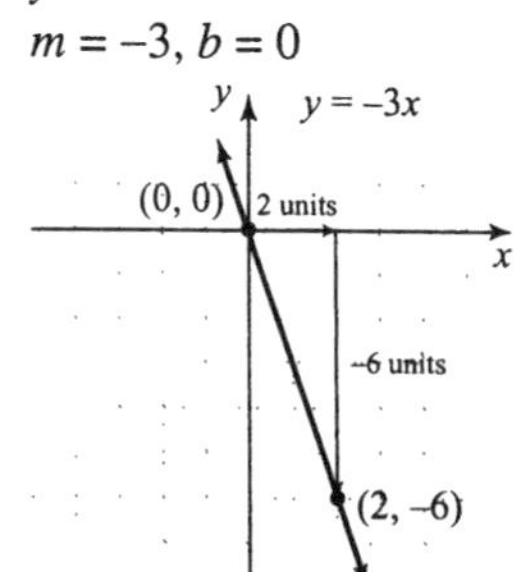

69. $5x - 3y = 15$

$-3y = -5x + 15$

$y = \frac{5}{3}x - 5$

$y = mx + b$

$m = \frac{5}{3}$, $b = -5$

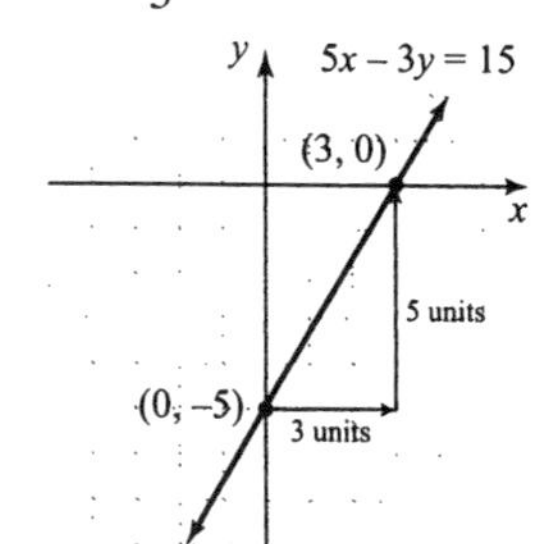

70. $-x + 2y = 8$

$2y = x + 8$

$y = \frac{1}{2}x + 4$

$y = mx + b$

$m = \frac{1}{2}$, $b = 4$

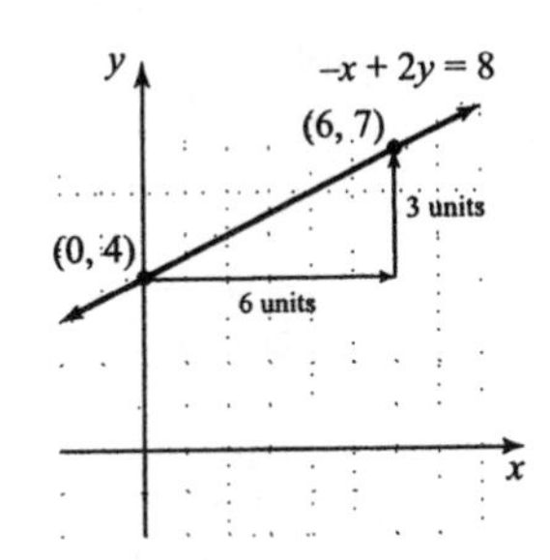

71. $y = -4x$
$m = -4, b = 0$
c

72. $y = -2x + 1$
$m = -2, b = 1$
d

73. $y = 2x - 1$
$m = 2, b = -1$
b

74. $y = 2x$
$m = 2, b = 0$
a

75. $m = -3$; $(0, -5)$
$$y = mx + b$$
$$y = -3x - 5$$
$$3x + y = -5$$

76. $m = \frac{1}{2}; \left(0, -\frac{7}{2}\right)$
$$y = mx + b$$
$$y = \frac{1}{2}x - \frac{7}{2}$$
$$2y = x - 7$$
$$x - 2y = 7$$

77. Horizontal line, point $(-2, -3)$
$y = c$
$y = -3$

78. Horizontal line, point $(0, 0)$
$y = c$
$y = 0$

79. $m = -6$; $(2, -1)$
$$y - y_1 = m(x - x_1)$$
$$y - (-1) = -6(x - 2)$$
$$y + 1 = -6x + 12$$
$$6x + y = 11$$

80. $m = 12; \left(\frac{1}{2}, 5\right)$
$$y - y_1 = m(x - x_1)$$
$$y - 5 = 12\left(x - \frac{1}{2}\right)$$
$$y - 5 = 12x - 6$$
$$12x - y = 1$$

81. $(0, 6)$ and $(6, 0)$
$$m = \frac{y_2 - y_1}{x_2 - x_1} = \frac{0-6}{6-0} = \frac{-6}{6} = -1$$
$m = -1$; $(0, 6)$
$$y - y_1 = m(x - x_1)$$
$$y - 6 = -1(x - 0)$$
$$y - 6 = -x$$
$$x + y = 6$$

82. $(0, -4)$ and $(-8, 0)$
$$m = \frac{y_2 - y_1}{x_2 - x_1} = \frac{0-(-4)}{-8-0} = \frac{4}{-8} = -\frac{1}{2}$$
$m = -\frac{1}{2}$; $(0, -4)$
$$y - y_1 = m(x - x_1)$$
$$y - (-4) = -\frac{1}{2}(x - 0)$$
$$y + 4 = -\frac{1}{2}x$$
$$2y + 8 = -x$$
$$x + 2y = -8$$

83. Vertical line, point $(5, 7)$
$x = c$
$x = 5$

84. Horizontal line, point $(-6, 8)$
$y = c$
$y = 8$

85. $y = 8$ is horizontal.
Perpendicular to $y = 8$ is vertical; $x = c$.
Point $(6, 0)$
$x = 6$

86. $x = -2$ is vertical.
Perpendicular to $x = -2$ is horizontal; $y = c$, point $(10, 12)$
$y = 12$

87. Two points have the same x-value: it is not a function.

88. Every point has a unique x-value: it is a function.

89. Yes; $7x - 6y = 1$ is a non-vertical line.

90. Yes; $y = 7$ is a non-vertical line.

91. No; $x = 2$ is a vertical line.

92. Yes; for each value of x there is only one value of y.

93. No; the graph does not pass the vertical line test.

94. Yes; the graph passes the vertical line test.

95. $f(x) = -2x + 6$

a. $f(0) = -2(0) + 6 = 6$

b. $f(-2) = -2(-2) + 6 = 4 + 6 = 10$

c. $f\left(\frac{1}{2}\right) = -2\left(\frac{1}{2}\right) + 6 = -1 + 6 = 5$

96. $h(x) = -5 - 3x$

a. $h(2) = -5 - 3(2) = -11$

b. $h(-3) = -5 - 3(-3) = 4$

c. $h(0) = -5 - 3(0) = -5$

97. $g(x) = x^2 + 12x$

a. $g(3) = (3)^2 + 12(3) = 45$

b. $g(-5) = (-5)^2 + 12(-5) = -35$

c. $g(0) = (0)^2 + 12(0) = 0$

98. $h(x) = 6 - |x|$

a. $h(-1) = 6 - |-1| = 6 - 1 = 5$

b. $h(1) = 6 - |1| = 6 - 1 = 5$

c. $h(-4) = 6 - |-4| = 6 - 4 = 2$

99. $(-\infty, \infty)$

100. $x - 2 \neq 0 \Rightarrow x \neq 2$, therefore $(-\infty, 2) \cup (2, \infty)$ or all real numbers except 2.

101. D: $[-3, 5]$, R: $[-4, 2]$

102. D: $(-\infty, \infty)$, R: $x \geq 0$, $[0, \infty)$

103. D: $\{3\}$, R: $(-\infty, \infty)$

104. D: $(-\infty, \infty)$, R: $x \leq 2$, $(-\infty, 2]$

105. $2x - 5y = 9$

Let $y = 1$.
$2x - 5(1) = 9$
$2x - 5 = 9$
$2x = 14$
$x = 7$

Let $x = 2$.
$2(2) - 5y = 9$
$4 - 5y = 9$
$-5y = 5$
$y = -1$

Let $y = -3$.
$2x - 5(-3) = 9$
$2x + 15 = 9$
$2x = -6$
$x = -3$

x	y
7	1
2	−1
−3	−3

106. $x = -3y$

Let $x = 0$.
$0 = -3y$
$0 = y$

Let $y = 1$.
$x = -3(1)$
$x = -3$

Let $x = 6$.
$6 = -3y$
$-2 = y$

x	y
0	0
−3	1
6	−2

107. $2x - 3y = 6$

Let $y = 0$.
$2x - 3(0) = 6$
$2x = 6$
$x = 3$

Let $x = 0$.
$2(0) - 3y = 6$
$-3y = 6$
$y = -2$

x-intercept: $(3, 0)$
y-intercept: $(0, -2)$

108. $-5x + y = 10$

Let $y = 0$.
$$-5x + 0 = 10$$
$$-5x = 10$$
$$x = -2$$

Let $x = 0$.
$$-5(0) + y = 10$$
$$y = 10$$

x-intercept: $(-2, 0)$
y-intercept: $(0, 10)$

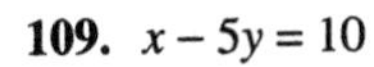

109. $x - 5y = 10$

x	y
10	0
0	−2

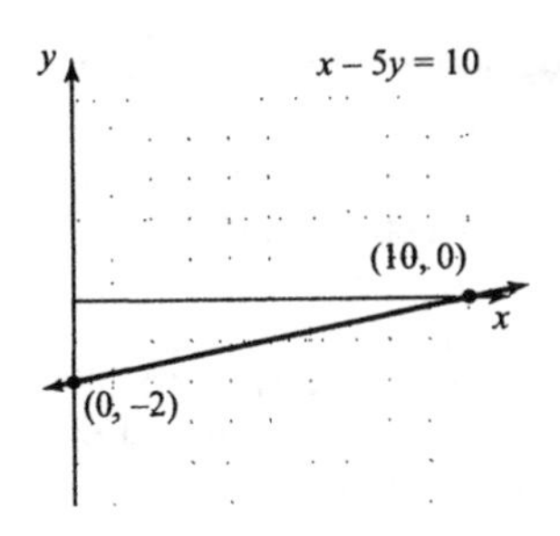

110. $x + y = 4$

x	y
4	0
0	4

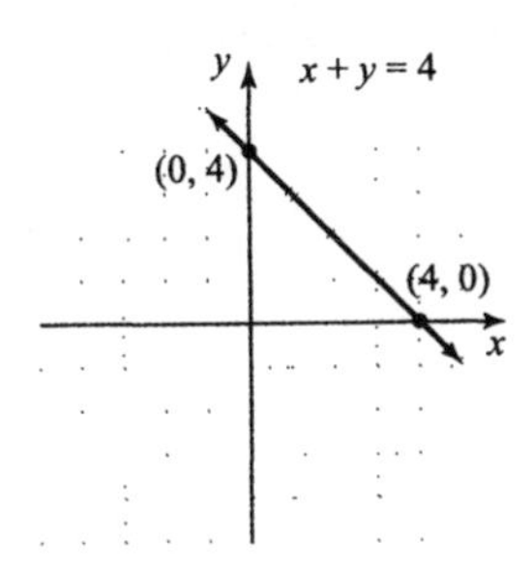

111. $y = -4x$

x	y
0	0
1	−4

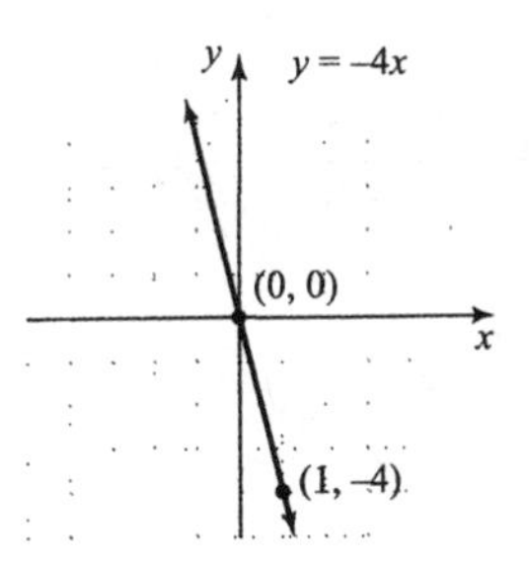

112. $2x + 3y = -6$

x	y
−3	0
0	−2

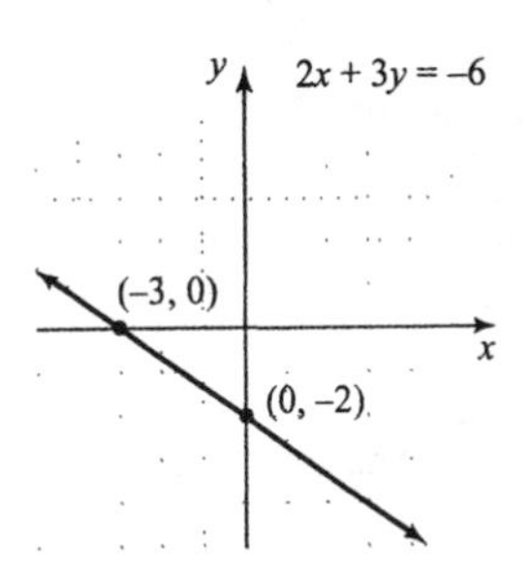

113. $x = 3$
This is the equation of a vertical line with x-intercept $(3, 0)$.

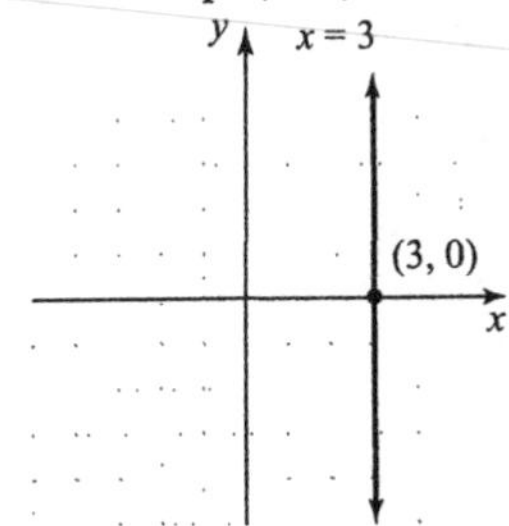

114. $y = -2$
This is the equation of a horizontal line with y-intercept $(0, -2)$.

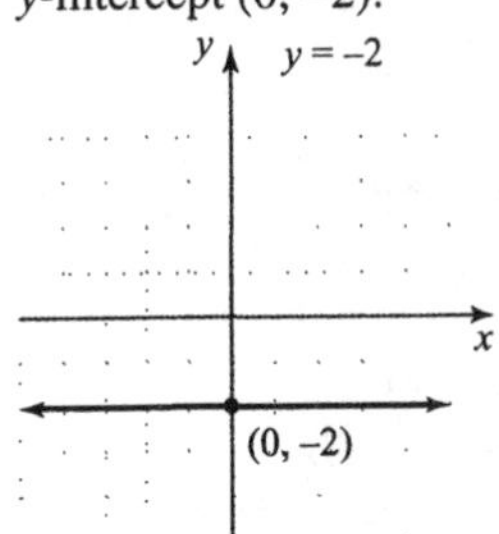

115. (3, –5) and (–4, 2)

$$m = \frac{y_2 - y_1}{x_2 - x_1} = \frac{2-(-5)}{-4-3} = \frac{7}{-7} = -1$$

116. (1, 3) and (–6, –8)

$$m = \frac{y_2 - y_1}{x_2 - x_1} = \frac{-8-3}{-6-1} = \frac{-11}{-7} = \frac{11}{7}$$

117. (0, –4) and (2, 0)

$$m = \frac{y_2 - y_1}{x_2 - x_1} = \frac{0-(-4)}{2-0} = \frac{4}{2} = 2$$

118. (0, 2) and (6, 0)

$$m = \frac{y_2 - y_1}{x_2 - x_1} = \frac{0-2}{6-0} = \frac{-2}{6} = -\frac{1}{3}$$

119. Solve for *y*.

$$-2x + 3y = -15$$
$$3y = 2x - 15$$
$$y = \frac{2}{3}x - 5$$

The slope is $\frac{2}{3}$. The *y*-intercept is (0, –5).

120. Solve for *y*.

$$6x + y - 2 = 0$$
$$y = -6x + 2$$

The slope is –6. The *y*-intercept is (0, 2).

121. $m = -5$; (3, –7)

$$y - y_1 = m(x - x_1)$$
$$y - (-7) = -5(x - 3)$$
$$y + 7 = -5x + 15$$
$$5x + y = 8$$

122. $m = 3$; (0, 6)

$$y = mx + b$$
$$y = 3x + 6$$
$$3x - y = -6$$

123. (–3, 9) and (–2, 5)

$$m = \frac{y_2 - y_1}{x_2 - x_1} = \frac{5-9}{-2-(-3)} = \frac{-4}{1} = -4$$

$m = -4$; (–2, 5)

$$y - y_1 = m(x - x_1)$$
$$y - 5 = -4(x - (-2))$$
$$y - 5 = -4(x + 2)$$
$$y - 5 = -4x - 8$$
$$4x + y = -3$$

124. (3, 1) and (5, –9)

$$m = \frac{y_2 - y_1}{x_2 - x_1} = \frac{-9-1}{5-3} = \frac{-10}{2} = -5$$

$m = -5$; (3, 1)

$$y - y_1 = m(x - x_1)$$
$$y - 1 = -5(x - 3)$$
$$y - 1 = -5x + 15$$
$$5x + y = 16$$

125. The highest point on the graph is above 2002 on the horizontal axis and corresponds to approximately 27.1 on the vertical axis, so the greatest beef production was 27.1 billion pounds in 2002.

126. The lowest point on the graph is above 2004 on the horizontal axis and corresponds to approximately 24.6 on the vertical axis, so the least beef production was 24.6 billion pounds in 2004.

127. The points for 2002, 2003, and 2006 lie above 25.0 on the vertical axis, so beef production was greater than 25 billion pounds in these years.

128. In 2003 and in 2004, there were decreases in beef production from the preceding years. In 2005 and in 2006, there were increases over the preceding years. The point for 2005 is only slightly higher than the point for 2004, denoting a small increase. However, the point for 2006 is much higher than the point for 2005, denoting a greater increase. The greatest increase occurred in 2006.

Chapter 3 Test

1. $y = \frac{1}{2}x$

$m = \frac{1}{2}$; $b = 0$

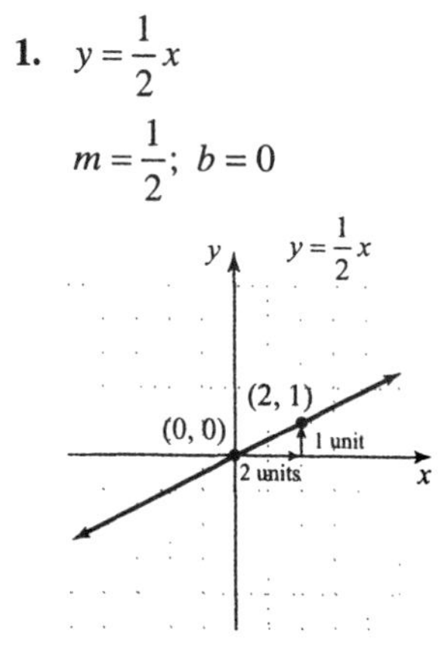

2. $2x + y = 8$

x	y
4	0
0	8

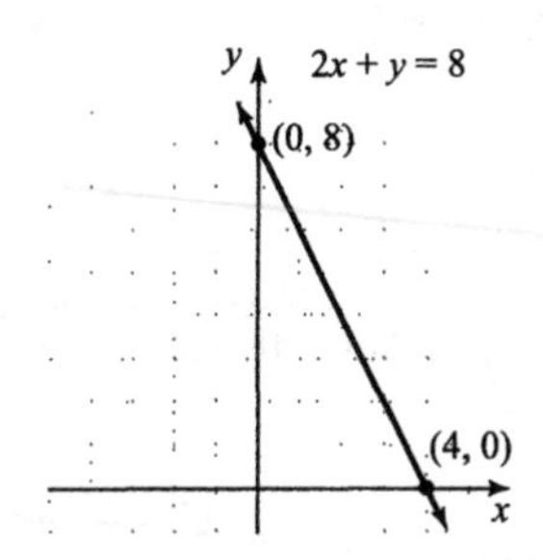

3. $5x - 7y = 10$

x	y
2	0
−5	−5

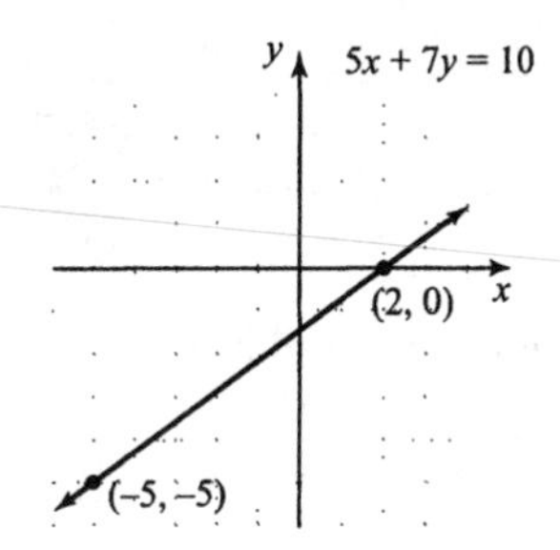

4. $y = -1$ for all values of x.

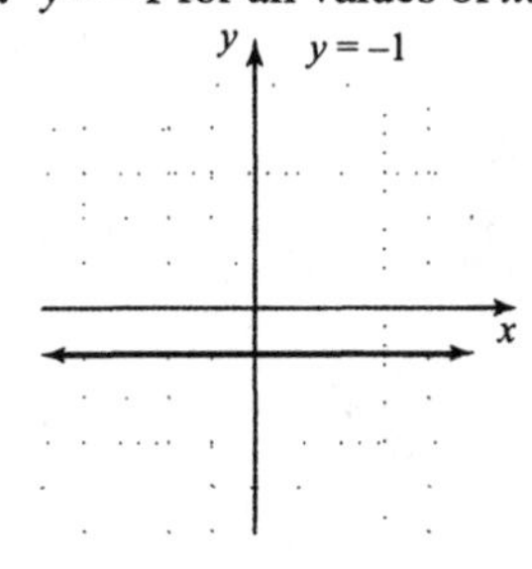

5. $x - 3 = 0$

$x = 3$ for all values of y.

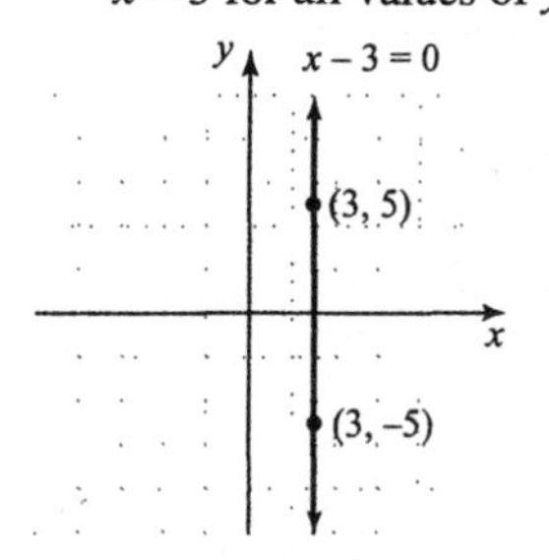

6. $(-1, -1)$ and $(4, 1)$

$$m = \frac{y_2 - y_1}{x_2 - x_1} = \frac{1-(-1)}{4-(-1)} = \frac{2}{5}$$

7. Horizontal line: $m = 0$

8. $(6, -5)$ and $(-1, 2)$

$$m = \frac{y_2 - y_1}{x_2 - x_1} = \frac{2-(-5)}{-1-6} = \frac{7}{-7} = -1$$

9. $-3x + y = 5$

$y = 3x + 5$

$y = mx + b$

$m = 3$

10. $x = 6$ is a vertical line. The slope is undefined.

11. $7x - 3y = 2$

$-3y = -7x + 2$

$y = \frac{7}{3}x - \frac{2}{3}$

$y = mx + b$

$m = \frac{7}{3},\ b = -\frac{2}{3},\ \left(0, -\frac{2}{3}\right)$

12. $y = 2x - 6,\ m_1 = 2$

$-4x = 2y,\ -2x = y$

$y = -2x,\ m_2 = -2$

$m_1 \neq m_2$ and $m_1 m_2 \neq -1$, neither

13. $m = -\frac{1}{4};\ (2, 2)$

$y - y_1 = m(x - x_1)$

$y - 2 = -\frac{1}{4}(x - 2)$

$4(y - 2) = -(x - 2)$

$4y - 8 = -x + 2$

$x + 4y = 10$

14. (0, 0) and (6, −7)

$$m = \frac{y_2 - y_1}{x_2 - x_1} = \frac{-7-0}{6-0} = -\frac{7}{6}$$

$$m = -\frac{7}{6};\ (0, 0)$$

$$y - y_1 = m(x - x_1)$$
$$y - 0 = -\frac{7}{6}(x - 0)$$
$$6y = -7x$$
$$7x + 6y = 0$$

15. (2, −5) and (1, 3)

$$m = \frac{y_2 - y_1}{x_2 - x_1} = \frac{3-(-5)}{1-2} = \frac{8}{-1} = -8$$

$m = -8$; (1, 3)

$$y - y_1 = m(x - x_1)$$
$$y - 3 = -8(x - 1)$$
$$y - 3 = -8x + 8$$
$$8x + y = 11$$

16. $x = 7$ is vertical.
Parallel to $x = 7$ is vertical;
$x = c$, point (−5, −1)
$x = -5$

17. $m = \frac{1}{8}$, $b = 12$

$$y = mx + b$$
$$y = \frac{1}{8}x + 12$$
$$8y = x + 96$$
$$x - 8y = -96$$

18. Yes; it passes the vertical line test.

19. No; it does not pass the vertical line test.

20. $h(x) = x^3 - x$

a. $h(-1) = (-1)^3 - (-1) = -1 + 1 = 0$

b. $h(0) = (0)^3 - (0) = 0$

c. $h(4) = (4)^3 - (4) = 64 - 4 = 60$

21. $x + 1 \neq 0 \Rightarrow x \neq -1$, therefore
$(-\infty, -1) \cup (-1, \infty)$ or all real numbers except −1.

22. D: $(-\infty, \infty)$, R: $x \leq 4$, $(-\infty, 4]$

23. D: $(-\infty, \infty)$, R: $(-\infty, \infty)$

24. $f(7) = 20$ corresponds to the ordered pair (7, 20).

25. The bar for Denmark extends to about 210 on the horizontal axis. The average water use per person per day in Denmark is approximately 210 liters.

26. The bar for Australia extends to about 490 on the horizontal axis. The average water use per person per day in Australia is approximately 490 liters.

27. The highest point on the graph corresponds to 7 on the horizontal axis, denoting July. The average high temperature is the greatest in July.

28. April corresponds to 4 on the horizontal axis. Moving horizontally to the left from the point on the graph above 4, we reach approximately 63 on the vertical axis. The average high temperature for April is approximately 63°F.

29. The points for months 1, 2, 3, 11, and 12 lie below 60 on the vertical axis. Thus, the average high temperature is below 60°F in January, February, March, November, and December.

Cumulative Review Chapters 1–3

1. a. $2 < 3$

b. $7 > 4$

c. $72 > 27$

2. $\frac{56}{64} = \frac{7 \cdot 8}{8 \cdot 8} = \frac{7}{8}$

3. $\frac{2}{15} \cdot \frac{5}{13} = \frac{2 \cdot 5}{3 \cdot 5 \cdot 13} = \frac{2}{39}$

4. $\frac{10}{3}+\frac{5}{21}=\frac{10\cdot 7}{3\cdot 7}+\frac{5}{21}$
$=\frac{70+5}{21}$
$=\frac{75}{21}$
$=\frac{3\cdot 25}{3\cdot 7}$
$=\frac{25}{7}$
$=3\frac{4}{7}$

5. $\frac{3+|4-3|+2^2}{6-3}=\frac{3+|1|+2^2}{6-3}=\frac{3+1+4}{6-3}=\frac{8}{3}$

6. $16-3\cdot 3+2^4=16-3\cdot 3+16$
$=16-9+16$
$=23$

7. a. $-8+(-11)=-19$

b. $-5+35=30$

c. $0.6+(-1.1)=-0.5$

d. $-\frac{7}{10}+\left(-\frac{1}{10}\right)=-\frac{8}{10}=-\frac{4}{5}$

e. $11.4+(-4.7)=6.7$

f. $-\frac{3}{8}+\frac{2}{5}=-\frac{3\cdot 5}{8\cdot 5}+\frac{2\cdot 8}{5\cdot 8}=\frac{-15+16}{40}=\frac{1}{40}$

8. $|9+(-20)|+|-10|=|-11|+|-10|=11+10=21$

9. a. $-14-8+10-(-6)=-14+(-8)+10+6$
$=-6$

b. $1.6-(-10.3)+(-5.6)=1.6+10.3+(-5.6)$
$=6.3$

10. $-9-(3-8)=-9-(-5)=-9+5=-4$

11. Let $x=-2$ and $y=-4$.

a. $5x-y=5(-2)-(-4)=-10+4=-6$

b. $x^4-y^2=(-2)^4-(-4)^2=16-16=0$

c. $\frac{3x}{2y}=\frac{3(-2)}{2(-4)}=\frac{-6}{-8}=\frac{3}{4}$

12. $\frac{x}{-10}=2$
Let $x=-20$.
$\frac{-20}{-10}\stackrel{?}{=}2$
$2=2$ True
-20 is a solution to the equation.

13. a. $10+(x+12)=10+x+12=x+22$

b. $-3(7x)=-21x$

14. $(12+x)-(4x-7)=12+x-4x+7=19-3x$

15. a. $-3y$: -3

b. $22z^4$: 22

c. $y=1y$: 1

d. $-x=-1x$: -1

e. $\frac{x}{7}=\frac{1}{7}x$: $\frac{1}{7}$

16. $-5(x-7)=-5x-(-5)(7)=-5x+35$

17. $y+0.6=-1.0$
$y=-1.6$

18. $5(3+z)-(8z+9)=-4$
$15+5z-8z-9=-4$
$-3z+6=-4$
$-3z=-10$
$z=\frac{10}{3}$

19. $-\frac{2}{3}x=-\frac{5}{2}$
$6\left(-\frac{2}{3}x\right)=6\left(-\frac{5}{2}\right)$
$-4x=-15$
$x=\frac{15}{4}$

20.
$$\frac{x}{4}-1=-7$$
$$4\left(\frac{x}{4}\right)-4(1)=4(-7)$$
$$x-4=-28$$
$$x=-24$$

21. Sum
= first integer + second integer + third integer
$$\text{Sum}=x+(x+1)+(x+2)$$
$$=x+x+1+x+2$$
$$=3x+3$$

22.
$$\frac{x}{3}-2=\frac{x}{3}$$
$$3\left(\frac{x}{3}\right)-3(2)=3\left(\frac{x}{3}\right)$$
$$x-6=x$$
$$-6=0$$
This is false. There is no solution.

23.
$$\frac{2(a+3)}{3}=6a+2$$
$$2(a+3)=3(6a+2)$$
$$2a+6=18a+6$$
$$-16a+6=6$$
$$-16a=0$$
$$a=0$$

24.
$$x+2y=6$$
$$x-x+2y=6-x$$
$$2y=6-x$$
$$\frac{2y}{2}=\frac{6-x}{2}$$
$$y=\frac{6-x}{2}$$

25. Let x = the number of Republican representatives and $x + 31$ = the number of Democratic representatives.
$$x+x+31=435$$
$$2x+31=435$$
$$2x=404$$
$$x=202$$
$$x+31=233$$
There were 202 Republican representatives and 233 Democratic.

26.
$$5(x+4)\geq 4(2x+3)$$
$$5x+20\geq 8x+12$$
$$-3x+20\geq 12$$
$$-3x\geq -8$$
$$\frac{-3x}{-3}\leq\frac{-8}{-3}$$
$$x\leq\frac{8}{3},\ \left(-\infty,\frac{8}{3}\right]$$

27. The perimeter of a rectangle is given by the formula $P = 2l + 2w$. Let l = the length of the garden.
$$P=2l+2w$$
$$140=2l+2w$$
$$140=2l+2(30)$$
$$140=2l+60$$
$$80=2l$$
$$40=l$$
The length of the garden is 40 feet.

28.
$$-3<4x-1\leq 2$$
$$-2<4x\leq 3$$
$$-\frac{1}{2}<x\leq\frac{3}{4},\ \left(-\frac{1}{2},\frac{3}{4}\right]$$

29.
$$y=mx+b$$
$$y-b=mx+b-b$$
$$y-b=mx$$
$$\frac{y-b}{m}=\frac{mx}{m}$$
$$\frac{y-b}{m}=x$$

30. $y=-5x$

x	y
0	0
−1	5
2	−10

31. Let x = the amount of 70% acid.
No. of liters · Strength = Amt of Acid

70%	x	0.7	$0.7x$
40%	$12-x$	0.4	$0.4(12-x)$
50%	12	0.5	0.5(12)

$$
\begin{aligned}
0.7x + 0.4(12 - x) &= 0.5(12) \\
0.7x + 4.8 - 0.4x &= 6 \\
0.3x + 4.8 &= 6 \\
0.3x &= 1.2 \\
x &= 4
\end{aligned}
$$

$12 - x = 12 - 4 = 8$

Mix 4 liters of 70% acid with 8 liters of 40% acid.

32. $y = -3x + 5$

x	y
−1	8
0	5
1	2

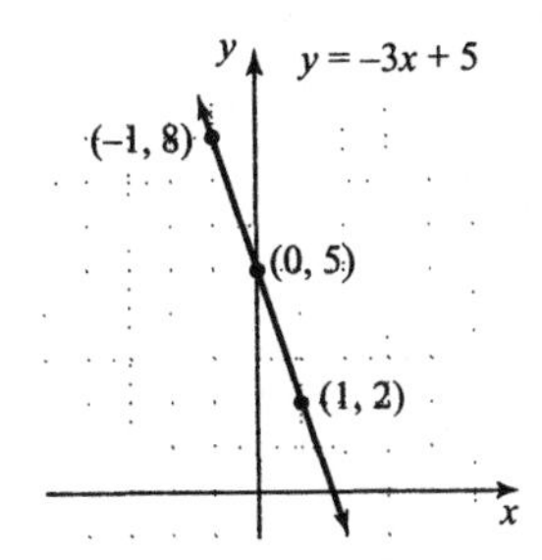

33. $x \geq -1$, $[-1, \infty)$

−1

34. $2x + 4y = -8$

x-intercept, $y = 0$

$2x + 4(0) = -8 \Rightarrow x = -4$: $(-4, 0)$

y-intercept, $x = 0$

$2(0) + 4y = -8 \Rightarrow y = -2$: $(0, -2)$

35.
$$
\begin{aligned}
-1 \leq 2x - 3 &< 5 \\
2 \leq 2x &< 8 \\
1 \leq x &< 4, \ [1, 4)
\end{aligned}
$$

1 4

36. $x = 2$

$x = 2$ for all values of y.

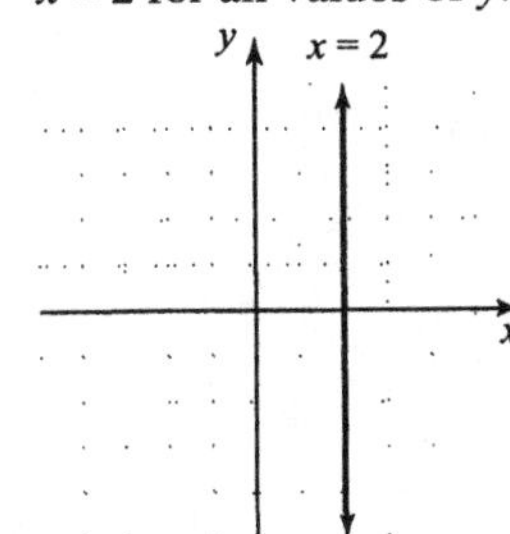

37. a. $x - 2y = 6$

$(6, 0)$

$(6) - 2(0) \stackrel{?}{=} 6$

$6 = 6$ Yes

b. $x - 2y = 6$

$(0, 3)$

$(0) - 2(3) \stackrel{?}{=} 6$

$-6 \neq 6$ No

c. $x - 2y = 6$

$\left(1, -\frac{5}{2}\right)$

$(1) - 2\left(-\frac{5}{2}\right) \stackrel{?}{=} 6$

$1 + 5 \stackrel{?}{=} 6$

$6 = 6$ Yes

38. $(0, 5)$ and $(-5, 4)$

$$m = \frac{y_2 - y_1}{x_2 - x_1} = \frac{4 - 5}{-5 - 0} = \frac{-1}{-5} = \frac{1}{5}$$

39. a. linear; because it can be written in the form $Ax + By = C$.

b. linear; because it can be written in the form $Ax + By = C$.

c. not linear; because y is squared.

d. linear; because it can be written in the form $Ax + By = C$.

40. $x = -10$ is a vertical line. The slope is undefined.

41. $y = -1$ is horizontal, slope is 0.

42.
$$
\begin{aligned}
2x - 5y &= 10 \\
-5y &= -2x + 10 \\
y &= \frac{2}{5}x - 2
\end{aligned}
$$

$y = mx + b$

$m = \frac{2}{5}$, $b = -2$

The slope is $\frac{2}{5}$.

The y-intercept is $(0, -2)$.

43. $m = \frac{1}{4};\ b = -3$

$y = mx + b$

$y = \frac{1}{4}x + (-3)$

$y = \frac{1}{4}x - 3$

44. (2, 3) and (0, 0)

$m = \frac{y_2 - y_1}{x_2 - x_1} = \frac{0-3}{0-2} = \frac{-3}{-2} = \frac{3}{2}$

Point: (0, 0)

$y - y_1 = m(x - x_1)$

$y - 0 = \frac{3}{2}(x - 0)$

$2y = 3x$

$3x - 2y = 0$

Chapter 4

Section 4.1

Practice Exercises

1. $\begin{cases} 4x - y = 2 \\ y = 3x \end{cases}$

$(4, 12)$
$4(4) - 12 \stackrel{?}{=} 2$
$16 - 12 \stackrel{?}{=} 2$
$4 = 2$ False
$(4, 12)$ is not a solution of the system.

2. $\begin{cases} x - 3y = -7 \\ 2x + 9y = 1 \end{cases}$

$(-4, 1)$
$-4 - 3(1) \stackrel{?}{=} -7$ $\quad$ $2(-4) + 9(1) \stackrel{?}{=} 1$
$-4 - 3 \stackrel{?}{=} -7$ $\quad$ $-8 + 9 \stackrel{?}{=} 1$
$-7 = -7$ True $\quad$ $1 = 1$ True
$(-4, 1)$ is a solution of the system.

3. $\begin{cases} x - y = 3 \\ x + 2y = 18 \end{cases}$

$x - y = 3$

x	y
−4	−7
0	−3
4	1

$x + 2y = 18$

x	y
−4	11
0	9
4	7

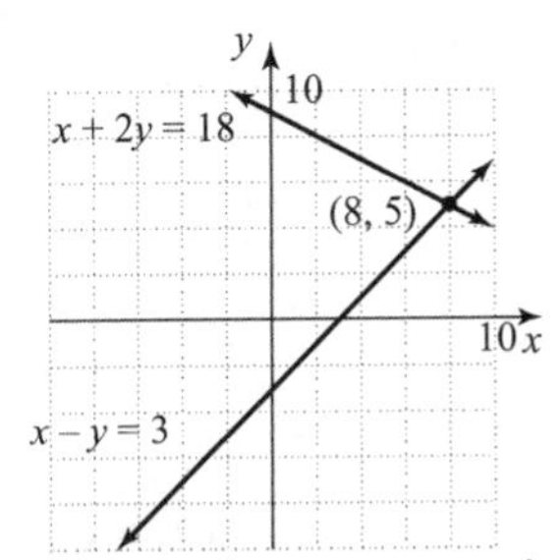

The two lines appear to intersect at $(8, 5)$.
$x - y = 3$ $\quad$ $x + 2y = 18$
$8 - 5 \stackrel{?}{=} 3$ $\quad$ $8 + 2(5) \stackrel{?}{=} 18$
$3 = 3$ True $\quad$ $8 + 10 \stackrel{?}{=} 18$
$18 = 18$ True
$(8, 5)$ is the solution of the system.

4. $\begin{cases} -4x + 3y = -3 \\ y = -5 \end{cases}$

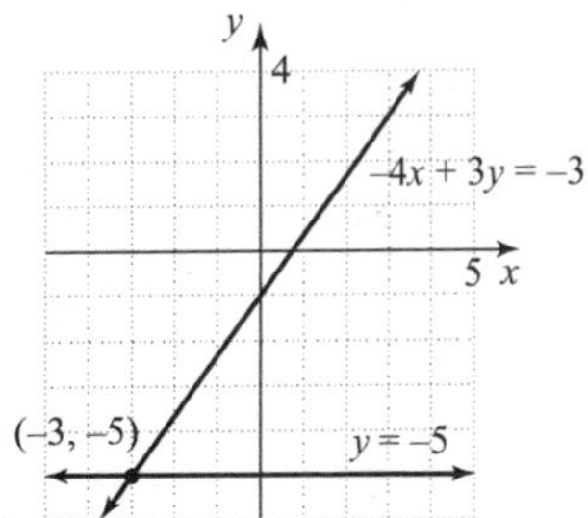

The two lines appear to intersect at $(-3, -5)$.
Check.
$-4x + 3y = -3$ $\quad$ $y = -5$
$-4(-3) + 3(-5) \stackrel{?}{=} -3$ $\quad$ $-5 = -5$ True
$12 - 15 \stackrel{?}{=} -3$
$-3 = -3$ True
$(-3, -5)$ is the solution of the system.

5. $\begin{cases} 3y = 9x \\ 6x - 2y = 12 \end{cases}$

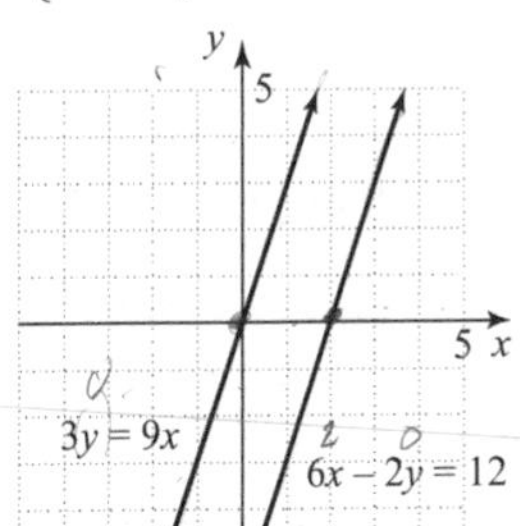

The lines appear to be parallel. To confirm this, write both equations in slope-intercept form.
$3y = 9x$ $\quad$ $6x - 2y = 12$
$y = 3x$ $\quad$ $-2y = -6x + 12$
$y = 3x - 6$

The slopes are the same, so the lines are parallel. Thus, there is no solution of the system and the system is inconsistent.

6. $\begin{cases} x - y = 4 \\ -2x + 2y = -8 \end{cases}$

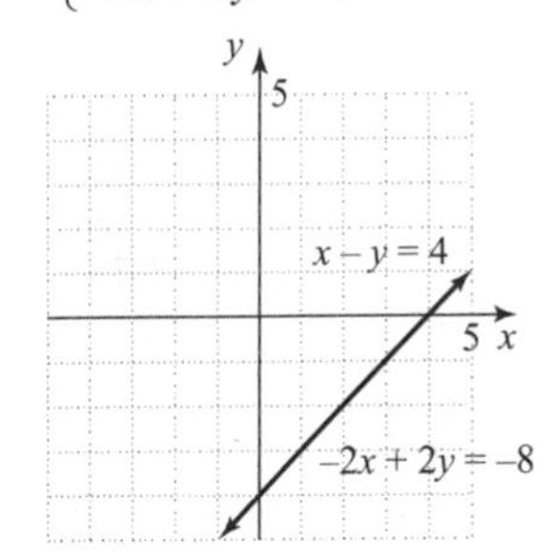

The graphs appear to be identical. To confirm this, write both equations in slope-intercept form.

$x-y=4$
$-y=-x+4$
$y=x-4$

$-2x+2y=-8$
$-x+y=-4$
$y=x-4$

The equations are identical. Thus, there is an infinite number of solutions of the system; the system is consistent; the equations are dependent.

7. $\begin{cases} 5x+4y=6 \\ x-y=3 \end{cases}$

Write each equation in slope-intercept form.

$5x+4y=6$
$4y=-5x+6$
$y=-\frac{5}{4}x+\frac{3}{2}$

$x-y=3$
$-y=-x+3$
$y=x-3$

The slopes are not equal, so the two lines are neither parallel nor identical and must intersect. Therefore, this system has one solution and is consistent.

8. $\begin{cases} -\frac{2}{3}x+y=6 \\ 3y=2x+5 \end{cases}$

Write each equation in slope-intercept form.

$-\frac{2}{3}x+y=6$
$y=\frac{2}{3}x+6$

$3y=2x+5$
$y=\frac{2}{3}x+\frac{5}{3}$

The slope of each line is $\frac{2}{3}$, but they have different y-intercepts. Therefore, the lines are parallel. The system has no solution and is inconsistent.

Calculator Explorations

1. $\begin{cases} y=-2.68x+1.21 \\ y=5.22x-1.68 \end{cases}$

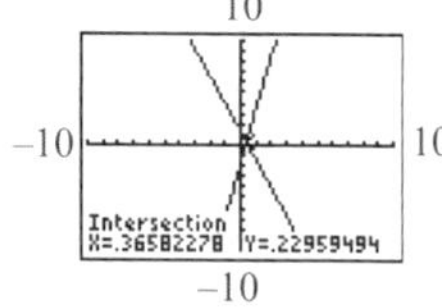

The approximate point of intersection is (0.37, 0.23).

2. $\begin{cases} y=4.25x+3.89 \\ y=-1.88x+3.21 \end{cases}$

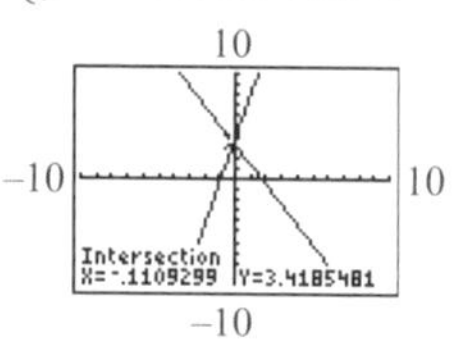

The approximate point of intersection is (−0.11, 3.42).

3. $\begin{cases} 4.3x-2.9y=5.6 \\ 8.1x+7.6y=-14.1 \end{cases}$

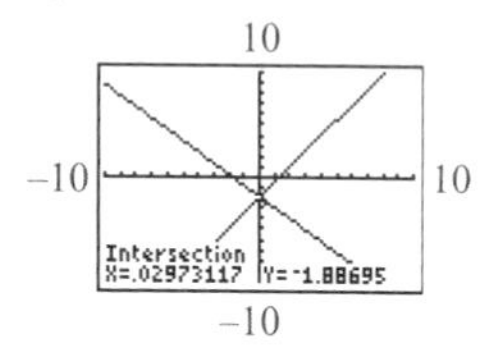

The approximate point of intersection is (0.03, −1.89).

4. $\begin{cases} -3.6x-8.6y=10 \\ -4.5x+9.6y=-7.7 \end{cases}$

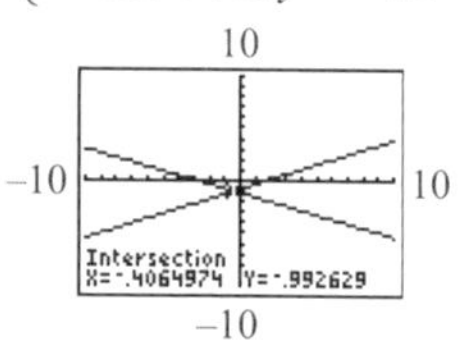

The approximate point of intersection is (−0.41, −0.99).

Vocabulary and Readiness Check

1. In a system of linear equations in two variables, if the graphs of the equations are the same, the equations are dependent equations.

2. Two or more linear equations are called a system of linear equations.

3. A system of equations that has at least one solution is called a consistent system.

4. A solution of a system of two equations in two variables is an ordered pair of numbers that is a solution of both equations in the system.

5. A system of equations that has no solution is called an inconsistent system.

6. In a system of linear equations in two variables, if the graphs of the equations are different, the equations are independent equations.

7. One solution, (−1, 3)

8. No solution

9. Infinite number of solutions

10. One solution, (3, 4)

Exercise Set 4.1

1. a. Let $x = 2$ and $y = 4$.

$x + y = 8$ | $3x + 2y = 21$
$2 + 4 \stackrel{?}{=} 8$ | $3(2) + 2(4) \stackrel{?}{=} 21$
$6 = 8$ False | $6 + 8 \stackrel{?}{=} 21$
 | $14 = 21$ False

(2, 4) is not a solution of the system.

b. Let $x = 5$ and $y = 3$.

$x + y = 8$ | $3x + 2y = 21$
$5 + 3 \stackrel{?}{=} 8$ | $3(5) + 2(3) \stackrel{?}{=} 21$
$8 = 8$ True | $15 + 6 \stackrel{?}{=} 21$
 | $21 = 21$ True

(5, 3) is a solution of the system.

3. a. Let $x = 3$ and $y = 4$.

$3x - y = 5$ | $x + 2y = 11$
$3(3) - 4 \stackrel{?}{=} 5$ | $3 + 2(4) \stackrel{?}{=} 11$
$9 - 4 \stackrel{?}{=} 5$ | $3 + 8 \stackrel{?}{=} 11$
$5 = 5$ True | $11 = 11$ True

(3, 4) is a solution of the system.

b. Let $x = 0$ and $y = -5$.

$3x - y = 5$ | $x + 2y = 11$
$3(0) - (-5) \stackrel{?}{=} 5$ | $0 + 2(-5) \stackrel{?}{=} 11$
$0 + 5 \stackrel{?}{=} 5$ | $0 - 10 \stackrel{?}{=} 11$
$5 = 5$ True | $-10 = 11$ False

(0, −5) is not a solution of the system.

5. a. Let $x = -3$ and $y = -3$.

$2y = 4x + 6$ | $2x - y = -3$
$2(-3) \stackrel{?}{=} 4(-3) + 6$ | $2(-3) - (-3) \stackrel{?}{=} -3$
$-6 \stackrel{?}{=} -12 + 6$ | $-6 + 3 \stackrel{?}{=} -3$
$-6 = -6$ True | $-3 = -3$ True

(−3, −3) is a solution of the system.

b. Let $x = 0$ and $y = 3$.

$2y = 4x + 6$ | $2x - y = -3$
$2(3) \stackrel{?}{=} 4(0) + 6$ | $2(0) - 3 \stackrel{?}{=} -3$
$6 \stackrel{?}{=} 0 + 6$ | $0 - 3 \stackrel{?}{=} -3$
$6 = 6$ True | $-3 = -3$ True

(0, 3) is a solution of the system.

7. a. Let $x = -2$ and $y = 0$.

$-2 = x - 7y$ | $6x - y = 13$
$-2 \stackrel{?}{=} -2 - 7(0)$ | $6(-2) - 0 \stackrel{?}{=} 13$
$-2 = -2$ True | $-12 = 13$ False

(−2, 0) is not a solution of the system.

b. Let $x = \frac{1}{2}$ and $y = \frac{5}{14}$.

$-2 = x - 7y$ | $6x - y = 13$
$-2 \stackrel{?}{=} \frac{1}{2} - 7\left(\frac{5}{14}\right)$ | $6\left(\frac{1}{2}\right) - \left(\frac{5}{14}\right) \stackrel{?}{=} 13$
$2 \stackrel{?}{=} \frac{1}{2} - \frac{5}{2} = -\frac{4}{2}$ | $3 - \frac{5}{14} \stackrel{?}{=} 13$
$-2 = -2$ True | $\frac{37}{14} = 13$ False

$\left(\frac{1}{2}, \frac{5}{14}\right)$ is not a solution of the system.

9. $\begin{cases} x + y = 4 \\ x - y = 2 \end{cases}$

The solution of the system is (3, 1), consistent and independent.

11. $\begin{cases} x + y = 6 \\ -x + y = -6 \end{cases}$

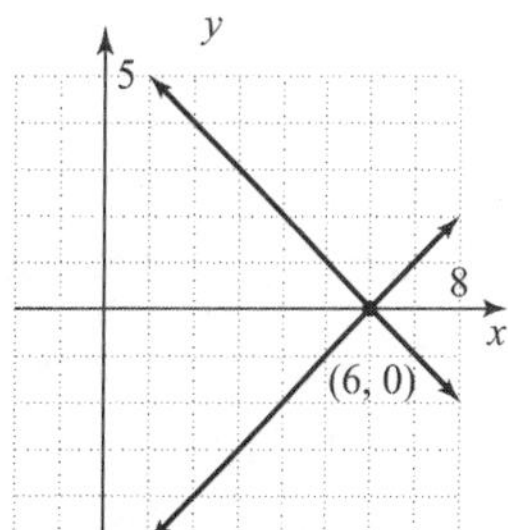

The solution of the system is (6, 0), consistent and independent.

13. $\begin{cases} y = 2x \\ 3x - y = -2 \end{cases}$

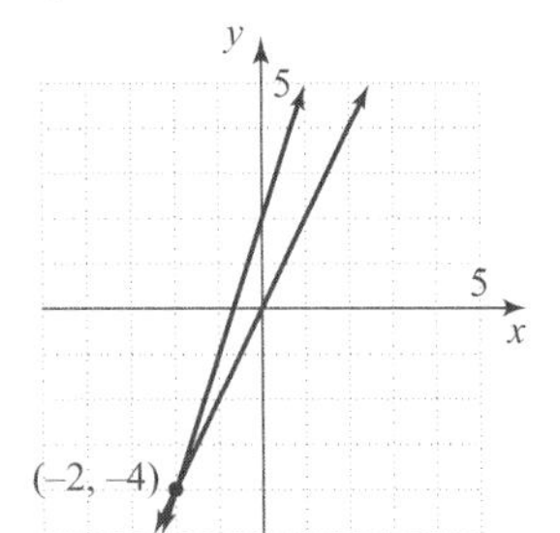

The solution of the system is (−2, −4), consistent and independent.

15. $\begin{cases} y = x + 1 \\ y = 2x - 1 \end{cases}$

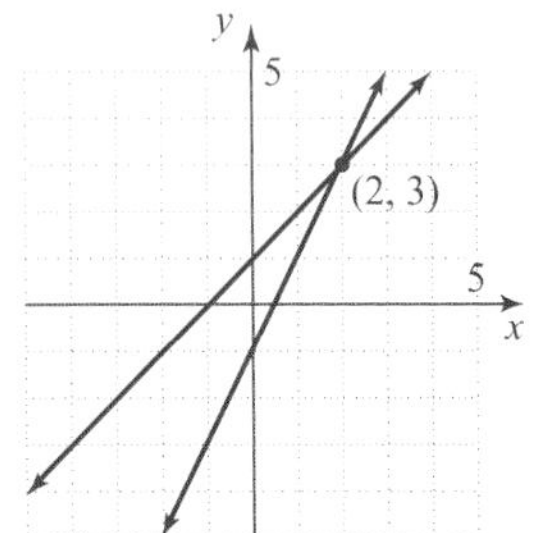

The solution of the system is (2, 3), consistent and independent.

17. $\begin{cases} 2x + y = 0 \\ 3x + y = 1 \end{cases}$

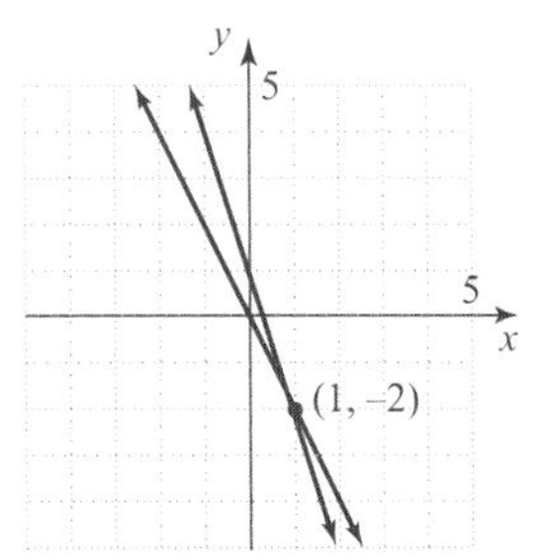

The solution of the system is (1, −2), consistent and independent.

19. $\begin{cases} y = -x - 1 \\ y = 2x + 5 \end{cases}$

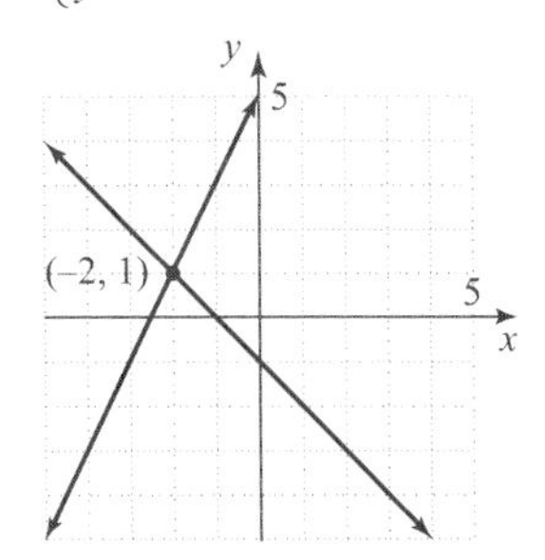

The solution of the system is (−2, 1), consistent and independent.

21. $\begin{cases} x + y = 5 \\ x + y = 6 \end{cases}$

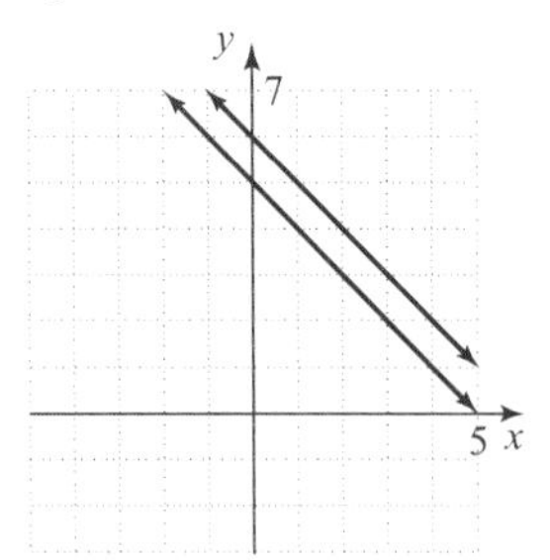

There is no solution, inconsistent and independent.

23. $\begin{cases} 2x - y = 6 \\ \quad\quad y = 2 \end{cases}$

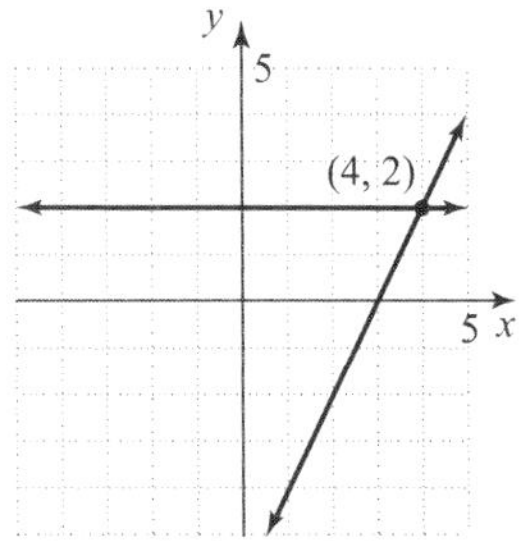

The solution of the system is (4, 2), consistent and independent.

25. $\begin{cases} x - 2y = 2 \\ 3x + 2y = -2 \end{cases}$

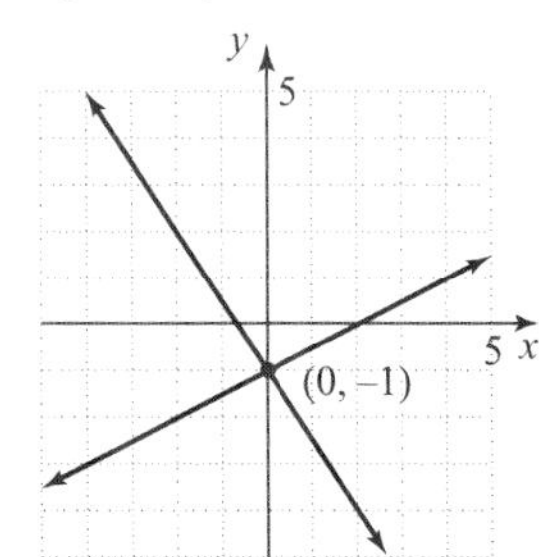

The solution of the system is (0, −1), consistent and independent.

27. $\begin{cases} 2x+y=4 \\ 6x=-3y+6 \end{cases}$

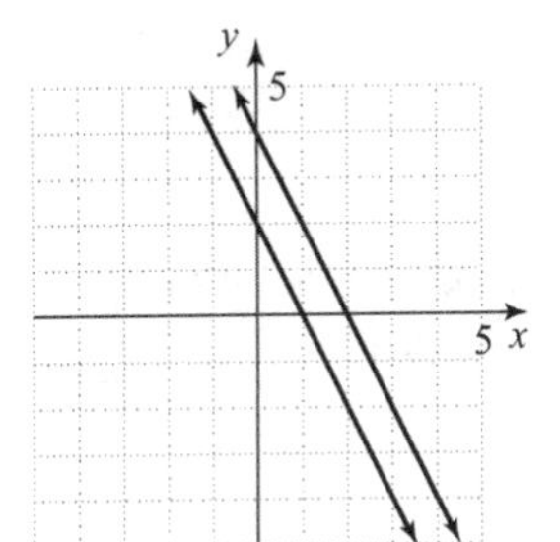

There is no solution, inconsistent and independent.

29. $\begin{cases} y-3x=-2 \\ 6x-2y=4 \end{cases}$

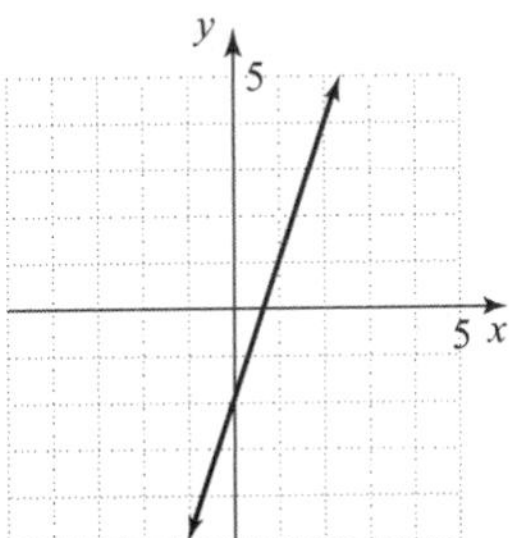

There is an infinite number of solutions, consistent and dependent.

31. $\begin{cases} x=3 \\ y=-1 \end{cases}$

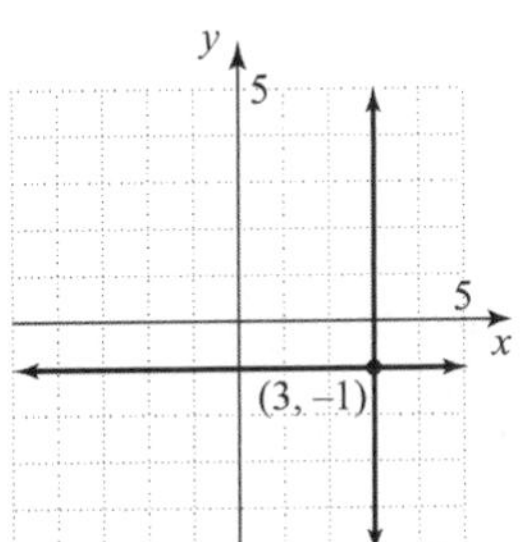

The solution of the system is (3, −1), consistent and independent.

33. $\begin{cases} y=x-2 \\ y=2x+3 \end{cases}$

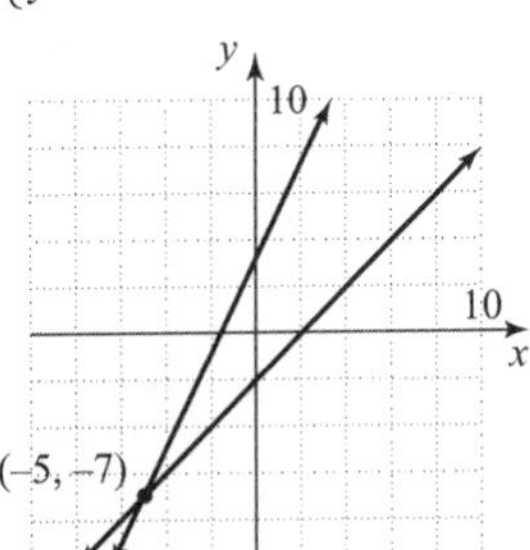

The solution of the system is (−5, −7), consistent and independent.

35. $\begin{cases} 2x-3y=-2 \\ -3x+5y=5 \end{cases}$

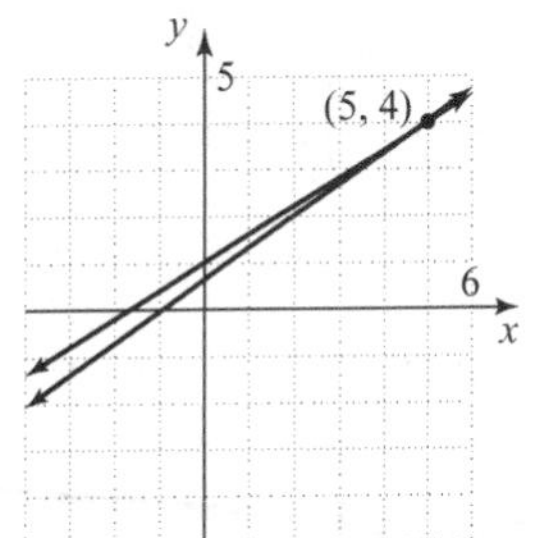

The solution of the system is (5, 4), consistent and independent.

37. $\begin{cases} 6x-y=4 \\ \frac{1}{2}y=-2+3x \end{cases}$

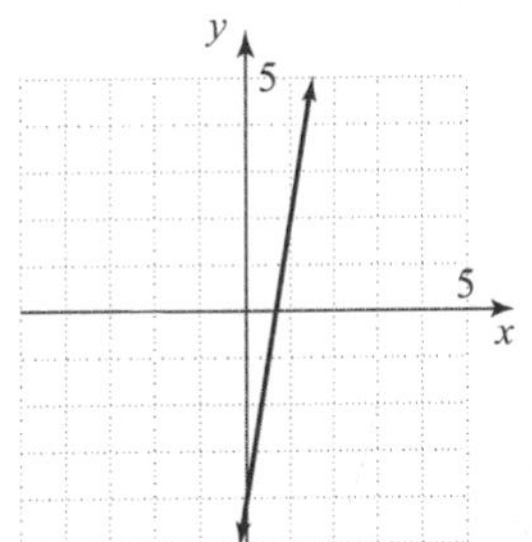

There is an infinite number of solutions, consistent and dependent.

39. $\begin{cases} 4x+y=24 \\ x+2y=2 \end{cases} \rightarrow \begin{cases} y=-4x+24 \\ y=-\frac{1}{2}x+1 \end{cases}$

The lines are intersecting; there is one solution.

41. $\begin{cases} 2x+y=0 \\ 2y=6-4x \end{cases} \rightarrow \begin{cases} y=-2x \\ y=-2x+3 \end{cases}$

The lines are parallel; there is no solution.

43. $\begin{cases} 6x-y=4 \\ \frac{1}{2}y=-2+3x \end{cases} \rightarrow \begin{cases} y=6x-4 \\ y=6x-4 \end{cases}$

The lines are identical; there is an infinite number of solutions.

45. $\begin{cases} x=5 \\ y=-2 \end{cases}$

The lines are intersecting; there is one solution.

47. $\begin{cases} 3y-2x=3 \\ x+2y=9 \end{cases} \to \begin{cases} y=\frac{2}{3}x+1 \\ y=-\frac{1}{2}x+\frac{9}{2} \end{cases}$

The lines are intersecting; there is one solution.

49. $\begin{cases} 6y+4x=6 \\ 3y-3=-2x \end{cases} \to \begin{cases} y=-\frac{2}{3}x+1 \\ y=-\frac{2}{3}x+1 \end{cases}$

The lines are identical; there is an infinite number of solutions.

51. $\begin{cases} x+y=4 \\ x+y=3 \end{cases} \to \begin{cases} y=-x+4 \\ y=-x+3 \end{cases}$

The lines are parallel; there is no solution.

53. $5(x-3)+3x=1$
$5x-15+3x=1$
$8x-15=1$
$8x=16$
$x=2$
The solution is 2.

55. $4\left(\frac{y+1}{2}\right)+3y=0$
$2(y+1)+3y=0$
$2y+2+3y=0$
$5y+2=0$
$5y=-2$
$y=-\frac{2}{5}$
The solution is $-\frac{2}{5}$.

57. $8a-2(3a-1)=6$
$8a-6a+2=6$
$2a+2=6$
$2a=4$
$a=2$
The solution is 2.

59. Answers may vary

61. Answers may vary

63. Answers may vary

65. The graph for fish is above the graph for shellfish for the years 2000, 2001, and 2002.

67. The graph for Toyota is lower than the graph for GM for the years 2001, 2002, and 2003.

69. Answers may vary

71. Answers may vary

73. **a.** Each table includes the point (4, 9). Therefore (4, 9) is a solution of the system.

b.

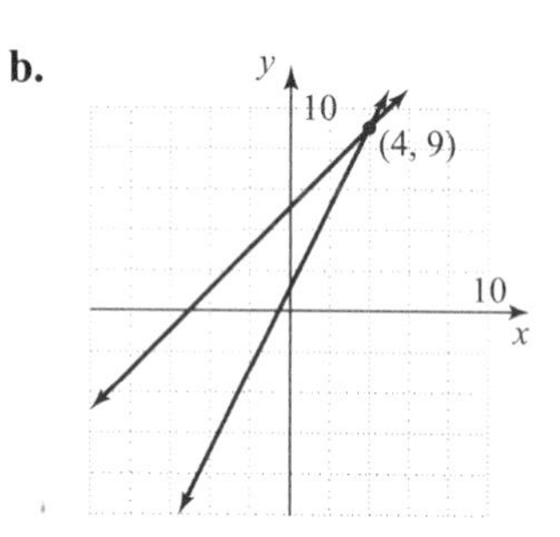

c. Yes

75. Answers may vary

Section 4.2

Practice Exercises

1. $\begin{cases} 2x-y=9 \\ x=y+1 \end{cases}$

Substitute $y + 1$ for x in the first equation.
$2x-y=9$
$2(y+1)-y=9$
$2y+2-y=9$
$y+2=9$
$y=7$
Let $y = 7$ in the second equation.
$x = y + 1 = 7 + 1 = 8$
The solution of the system is (8, 7).
Check.

$2x-y=9$	$x=y+1$
$2(8)-7 \stackrel{?}{=} 9$	$8 \stackrel{?}{=} 7+1$
$16-7 \stackrel{?}{=} 9$	$8=8$ True
$9=9$ True	

The solution of the system is (8, 7).

2. $\begin{cases} 7x-y=-15 \\ y=2x \end{cases}$

Substitute $2x$ for y in the first equation.
$7x-y=-15$
$7x-2x=-15$
$5x=-15$
$x=-3$

Let $x = -3$ in the second equation.
$y = 2x = 2(-3) = -6$
The solution of the system is $(-3, -6)$.

3. $\begin{cases} x+3y=6 \\ 2x+3y=10 \end{cases}$

Solve the first equation for x.

$$x+3y=6$$
$$x=-3y+6$$

Substitute $-3y + 6$ for x in the second equation.

$$2x+3y=10$$
$$2(-3y+6)+3y=10$$
$$-6y+12+3y=10$$
$$-3y+12=10$$
$$-3y=-2$$
$$y=\frac{2}{3}$$

Let $y=\frac{2}{3}$ in the equation for x.

$$x=3y+6=-3\left(\frac{2}{3}\right)+6=-2+6=4$$

The solution of the system is $\left(4, \frac{2}{3}\right)$.

4. $\begin{cases} 5x+3y=-9 \\ -2x+y=8 \end{cases}$

Solve the second equation for y.

$$-2x+y=8$$
$$y=2x+8$$

Substitute $2x + 8$ for y in the first equation.

$$5x+3y=-9$$
$$5x+3(2x+8)=-9$$
$$5x+6x+24=-9$$
$$11x+24=-9$$
$$11x=-33$$
$$x=-3$$

Let $x = -3$ in the equation for y.
$y = 2x + 8 = 2(-3) + 8 = -6 + 8 = 2$
The solution of the system is $(-3, 2)$.

5. $\begin{cases} \frac{1}{4}x-y=2 \\ x=4y+8 \end{cases}$

Substitute $4y + 8$ for x in the first equation.

$$\frac{1}{4}x-y=2$$
$$\frac{1}{4}(4y+8)-y=2$$
$$y+2-y=2$$
$$2=2$$

The two linear equations are equivalent. Thus, the system has an infinite number of solutions.

6. $\begin{cases} 4x-3y=12 \\ -8x+6y=-30 \end{cases}$

Solve the first equation for x.

$$4x-3y=12$$
$$4x=3y+12$$
$$x=\frac{3}{4}y+3$$

Substitute $\frac{3}{4}y+3$ for x in the second equation.

$$-8x+6y=-30$$
$$-8\left(\frac{3}{4}y+3\right)+6y=-30$$
$$-6y-24+6y=-30$$
$$-24=-30$$

The false statement $-24 = -30$ indicates that the system has no solution and is inconsistent.

Vocabulary and Readiness Check

1. Since $x = 1$, $y = 4x = 4(1) = 4$ and the solution is $(1, 4)$.
2. There is no solution, since $0 = 34$ is a false statement.
3. There is an infinite number of solutions, since the statement $0 = 0$ is true for all values of the variables.
4. Since $y = 0$, $x = y + 5 = 0 + 5 = 5$ and the solution is $(5, 0)$.
5. Since $x = 0$ and $x + y = 0$, $y = -x = -0 = 0$ and the solution is $(0, 0)$.
6. There is an infinite number of solutions, since the statement $0 = 0$ is true for all values of the variables.

Exercise Set 4.2

1. $\begin{cases} x+y=3 \\ x=2y \end{cases}$

Substitute $2y$ for x in the first equation.

$2y+y=3$
$3y=3$
$y=1$

Let $y = 1$ in the second equation.

$x = 2(1) = 2$

The solution is (2, 1).

3. $\begin{cases} x+y=6 \\ y=-3x \end{cases}$

Substitute $-3x$ for y in the first equation.

$x+(-3x)=6$
$-2x=6$
$x=-3$

Let $x = -3$ in the second equation.

$y = -3(-3) = 9$

The solution is (−3, 9).

5. $\begin{cases} y=3x+1 \\ 4y-8x=12 \end{cases}$

Substitute $3x + 1$ for y in the second equation.

$4(3x+1)-8x=12$
$12x+4-8x=12$
$4x+4=12$
$4x=8$
$x=2$

Let $x = 2$ in the first equation.

$y = 3(2) + 1 = 7$

The solution is (2, 7).

7. $\begin{cases} y=2x+9 \\ y=7x+10 \end{cases}$

Substitute $2x + 9$ for y in the second equation.

$2x+9=7x+10$
$-5x+9=10$
$-5x=1$
$x=-\frac{1}{5}$

Let $x=-\frac{1}{5}$ in the first equation.

$y=2\left(-\frac{1}{5}\right)+9=-\frac{2}{5}+\frac{45}{5}=\frac{43}{5}$

The solution is $\left(-\frac{1}{5}, \frac{43}{5}\right)$.

9. $\begin{cases} 3x-4y=10 \\ y=x-3 \end{cases}$

Substitute $x - 3$ for y in the first equation.

$3x-4(x-3)=10$
$3x-4x+12=10$
$-x=-2$
$x=2$

Let $x = 2$ in the second equation

$y = 2 - 3 = -1$

The solution is (2, −1).

11. $\begin{cases} x+2y=6 \\ 2x+3y=8 \end{cases}$

Solve the first equation for x.

$x = 6 - 2y$

Substitute $6 - 2y$ for x in the second equation.

$2(6-2y)+3y=8$
$12-4y+3y=8$
$-y=-4$
$y=4$

Let $y = 4$ in $x = 6 - 2y$.

$x = 6 - 2(4) = -2$

The solution is (−2, 4).

13. $\begin{cases} 3x+2y=16 \\ x=3y-2 \end{cases}$

Substitute $3y - 2$ for x in the first equation.

$3(3y-2)+2y=16$
$9y-6+2y=16$
$11y=22$
$y=2$

Let $y = 2$ in the second equation.

$x = 3(2) - 2 = 4$

The solution is (4, 2).

15. $\begin{cases} 2x-5y=1 \\ 3x+y=-7 \end{cases}$

Solve the second equation for y.

$y = -7 - 3x$

Substitute $-7 - 3x$ for y in the first equation.

$2x-5(-7-3x)=1$
$2x+35+15x=1$
$17x=-34$
$x=-2$

Let $x = -2$ in $y = -7 - 3x$.

$y = -7 - 3(-2) = -1$

The solution is (−2, −1).

17. $\begin{cases} 4x+2y=5 \\ -2x=y+4 \end{cases}$

Solve the second equation for y.

$y = -2x - 4$

Substitute $-2x - 4$ for y in the first equation.

$$4x+2(-2x-4)=5$$
$$4x-4x-8=5$$
$$-8=5 \quad \text{False}$$

The system has no solution.

19. $\begin{cases} 4x+y=11 \\ 2x+5y=1 \end{cases}$

Solve the first equation for y.

$y = 11 - 4x$

Substitute $11 - 4x$ for y in the second equation.

$$2x+5(11-4x)=1$$
$$2x+55-20x=1$$
$$-18x=-54$$
$$x=3$$

Let $x = 3$ in $y = 11 - 4x$.

$y = 11 - 4(3) = -1$

The solution is $(3, -1)$.

21. $\begin{cases} x+2y+5=-4+5y-x \\ 2x+9=3y \\ \\ 2x+x=y+4 \\ -4+3x=y \end{cases}$

Substitute $3x - 4$ for y in $2x + 9 = 3y$.

$$2x+9=3(3x-4)$$
$$2x+9=9x-12$$
$$21=7x$$
$$3=x$$

Let $x = 3$ in $3x - 4 = y$.

$y = 3(3) - 4 = 5$

The solution of the system is $(3, 5)$.

23. $\begin{cases} 6x-3y=5 \\ x+2y=0 \end{cases}$

Solve the second equation for x.

$x = -2y$

Substitute $-2y$ for x in the first equation.

$$6(-2y)-3y=5$$
$$-12y-3y=5$$
$$-15y=5$$
$$y=-\frac{1}{3}$$

Let $y=-\frac{1}{3}$ in $x = -2y$.

$$x=-2\left(-\frac{1}{3}\right)=\frac{2}{3}$$

The solution is $\left(\frac{2}{3}, -\frac{1}{3}\right)$.

25. $\begin{cases} 3x-y=1 \\ 2x-3y=10 \end{cases}$

Solve the first equation for y.

$y = 3x - 1$

Substitute $3x - 1$ for y in the second equation.

$$2x-3(3x-1)=10$$
$$2x-9x+3=10$$
$$-7x=7$$
$$x=-1$$

Let $x = -1$ in $y = 3x - 1$.

$y = 3(-1) - 1 = -4$

The solution is $(-1, -4)$.

27. $\begin{cases} -x+2y=10 \\ -2x+3y=18 \end{cases}$

Solve the first equation for x.

$x = 2y - 10$

Substitute $2y - 10$ for x in the second equation.

$$-2(2y-10)+3y=18$$
$$-4y+20+3y=18$$
$$-y=-2$$
$$y=2$$

Let $y = 2$ in $x = 2y - 10$.

$x = 2(2) - 10 = -6$

The solution is $(-6, 2)$.

29. $\begin{cases} 5x+10y=20 \\ 2x+6y=10 \end{cases}$

Solve the first equation for x.

$$x+2y=4$$
$$x=4-2y$$

Substitute $4 - 2y$ for x in the second equation.

$$2(4-2y)+6y=10$$
$$8-4y+6y=10$$
$$2y=2$$
$$y=1$$

Let $y = 1$ in $x = 4 - 2y$.

$x = 4 - 2(1) = 2$

The solution is $(2, 1)$.

31. $\begin{cases} 3x+6y=9 \\ 4x+8y=16 \end{cases}$

Solve the first equation for x.

$x+2y=3$
$x=3-2y$

Substitute $3-2y$ for x in the second equation.

$4(3-2y)+8y=16$
$12-8y+8y=16$
$12=16$ False

The system has no solution.

33. $\begin{cases} \frac{1}{3}x-y=2 \\ x-3y=6 \end{cases}$

Solve the second equation for x.

$x=6+3y$

Substitute $6+3y$ for x in the first equation.

$\frac{1}{3}(6+3y)-y=2$
$2+y-y=2$
$2=2$

The equations in the original system are equivalent and there are an infinite number of solutions.

35. $\begin{cases} x=\frac{3}{4}y-1 \\ 8x-5y=-6 \end{cases}$

Substitute $\frac{3}{4}y-1$ for x in the second equation.

$8\left(\frac{3}{4}y-1\right)-5y=-6$
$6y-8-5y=-6$
$y=2$

Let $y=2$ in the first equation.

$x=\frac{3}{4}(2)-1=\frac{1}{2}$

The solution of the system is $\left(\frac{1}{2}, 2\right)$.

37. $\begin{cases} -5y+6y=3x+2(x-5)-3x+5 \\ y=3x+2x-10-3x+5 \\ y=2x-5 \\ \\ 4(x+y)-x+y=-12 \\ 4x+4y-x+y=-12 \\ 3x+5y=-12 \end{cases}$

Substitute $2x-5$ for y in the second equation.

$3x+5(2x-5)=-12$
$3x+10x-25=-12$
$13x=13$
$x=1$

Let $x=1$ in $y=2x-5$.

$y=2(1)-5=-3$

The solution is $(1, -3)$.

39. $3x+2y=6$
$-2(3x+2y)=-2(6)$
$-6x-4y=-12$

41. $-4x+y=3$
$3(-4x+y)=3(3)$
$-12x+3y=9$

43. $$\begin{array}{r} 3n+6m \\ 2n-6m \\ \hline 5n \end{array}$$

45. $$\begin{array}{r} -5a-7b \\ 5a-8b \\ \hline -15b \end{array}$$

47. Answers may vary

49. No; answers may vary.

51. c; answers may vary.

53. a. $\begin{cases} y=3.9x+443 \\ y=14.2x+314 \end{cases}$

Substitute $14.2x+314$ for y in the first equation.

$14.2x+314=3.9x+443$
$10.3x=129$
$x\approx 12.52$

Let $x=12.52$ in $y=3.9x+443$.

$y\approx 3.9(12.52)+443\approx 491.828$

The solution is $(13, 492)$.

b. In $1970+13=1983$, the number of men and the number of women receiving bachelor's degrees was the same.

c.

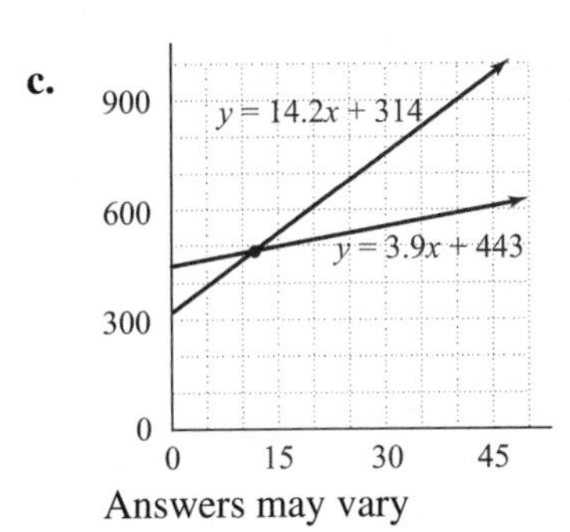

Answers may vary

55. $\begin{cases} y = 5.1x + 14.56 \\ y = -2x - 3.9 \end{cases}$

Substitute $-2x - 3.9$ for y in the first equation.

$$-2x - 3.9 = 5.1x + 14.56$$
$$-7.1x = 18.46$$
$$x = -2.6$$

Let $x = -2.6$ in $y = -2x - 3.9$.

$y = -2(-2.6) - 3.9 = 1.3$

The solution is $(-2.6, 1.3)$.

57. $\begin{cases} 3x + 2y = 14.05 \\ 5x + y = 18.5 \end{cases}$

Solve the second equation for y.

$y = -5x + 18.5$

Substitute $-5x + 18.5$ for y in the first equation.

$$3x + 2(-5x + 18.5) = 14.05$$
$$3x - 10x + 37 = 14.05$$
$$-7x = -22.95$$
$$x \approx 3.279$$

Let $x = 3.279$ in $y = -5x + 18.5$.

$y \approx -5(3.279) + 18.5 \approx 2.105$

The solution is approximately $(3.28, 2.11)$.

Section 4.3

Practice Exercises

1. $\begin{cases} x - y = 2 \\ x + y = 8 \end{cases}$

Add the left sides of the equations together and the right sides of the equations together.

$$\begin{array}{r} x - y = 2 \\ x + y = 8 \\ \hline 2x \quad = 10 \\ x = 5 \end{array}$$

Let $x = 5$ in the first equation.

$$x - y = 2$$
$$5 - y = 2$$
$$3 = y$$

The solution is $(5, 3)$.

Check.

$x - y = 2$ $x + y = 8$

$5 - 3 \stackrel{?}{=} 2$ $5 + 3 \stackrel{?}{=} 8$

$2 = 2$ True $8 = 8$ True

The solution of the system is $(5, 3)$.

2. $\begin{cases} x - 2y = 11 \\ 3x - y = 13 \end{cases}$

Multiply both sides of the first equation by -3 and add to the second equation.

$$\begin{array}{r} -3x + 6y = -33 \\ 3x - y = 13 \\ \hline 5y = -20 \\ y = -4 \end{array}$$

Let $y = -4$ in the first equation.

$$x - 2y = 11$$
$$x - 2(-4) = 11$$
$$x + 8 = 11$$
$$x = 3$$

The solution of the system is $(3, -4)$.

3. $\begin{cases} x - 3y = 5 \\ 2x - 6y = -3 \end{cases}$

Multiply both sides of the first equation by -2 and add to the second equation.

$$\begin{array}{r} -2x + 6y = -10 \\ 2x - 6y = -3 \\ \hline 0 = -13 \end{array} \quad \text{False}$$

The system has no solution.

4. $\begin{cases} 4x - 3y = 5 \\ -8x + 6y = -10 \end{cases}$

Multiply the first equation by 2 and add to the second equation.

$$\begin{array}{r} 8x - 6y = 10 \\ -8x + 6y = -10 \\ \hline 0 = 0 \end{array} \quad \text{True}$$

The equations are equivalent, so the system has an infinite number of solutions.

5. $\begin{cases} 4x + 3y = 14 \\ 3x - 2y = 2 \end{cases}$

Multiply the first equation by 2 and the second equation by 3 and add.

$$\begin{array}{r} 8x + 6y = 28 \\ 9x - 6y = 6 \\ \hline 17x \quad = 34 \\ x = 2 \end{array}$$

Let $x = 2$ in the second equation.

$$\begin{aligned} 3x - 2y &= 2 \\ 3(2) - 2y &= 2 \\ 6 - 2y &= 2 \\ -2y &= -4 \\ y &= 2 \end{aligned}$$

The solution of the system is (2, 2).

6. $\begin{cases} -2x + \dfrac{3y}{2} = 5 \\ -\dfrac{x}{2} - \dfrac{y}{4} = \dfrac{1}{2} \end{cases}$

Clear fractions by multiplying the first equation by 2 and the second by 4.

$\begin{cases} -4x + 3y = 10 \\ -2x - y = 2 \end{cases}$

Multiply the second simplified equation by 3 and add.

$$\begin{aligned} -4x + 3y &= 10 \\ -6x - 3y &= 6 \\ \hline -10x \quad &= 16 \end{aligned}$$

$$x = -\frac{16}{10} = -\frac{8}{5}$$

Now multiply the second simplified equation by -2 and add.

$$\begin{aligned} -4x + 3y &= 10 \\ 4x + 2y &= -4 \\ \hline 5y &= 6 \\ y &= \frac{6}{5} \end{aligned}$$

The solution of the system is $\left(-\frac{8}{5}, \frac{6}{5}\right)$.

Exercise Set 4.3

1. $\begin{cases} 3x + y = 5 \\ 6x - y = 4 \end{cases}$

$$\begin{aligned} 3x + y &= 5 \\ 6x - y &= 4 \\ \hline 9x \quad &= 9 \\ x &= 1 \end{aligned}$$

Let $x = 1$ in the first equation.

$$\begin{aligned} 3(1) + y &= 5 \\ 3 + y &= 5 \\ y &= 2 \end{aligned}$$

The solution of the system is (1, 2).

3. $\begin{cases} x - 2y = 8 \\ -x + 5y = -17 \end{cases}$

$$\begin{aligned} x - 2y &= 8 \\ -x + 5y &= -17 \\ \hline 3y &= -9 \\ y &= -3 \end{aligned}$$

Let $y = -3$ in the first equation.

$$\begin{aligned} x - 2(-3) &= 8 \\ x + 6 &= 8 \\ x &= 2 \end{aligned}$$

The solution of the system is (2, –3).

5. $\begin{cases} 3x + y = -11 \\ 6x - 2y = -2 \end{cases}$

Multiply the first equation by 2.

$$\begin{aligned} 6x + 2y &= -22 \\ 6x - 2y &= -2 \\ \hline 12x \quad &= -24 \\ x &= -2 \end{aligned}$$

Let $x = -2$ in the first equation.

$$\begin{aligned} 3(-2) + y &= -11 \\ -6 + y &= -11 \\ y &= -5 \end{aligned}$$

The solution of the system is (–2, –5).

7. $\begin{cases} 3x + 2y = 11 \\ 5x - 2y = 29 \end{cases}$

$$\begin{aligned} 3x + 2y &= 11 \\ 5x - 2y &= 29 \\ \hline 8x \quad &= 40 \\ x &= 5 \end{aligned}$$

Let $x = 5$ in the first equation.

$$\begin{aligned} 3(5) + 2y &= 11 \\ 15 + 2y &= 11 \\ 2y &= -4 \\ y &= -2 \end{aligned}$$

The solution of the system is (5, –2).

9. $\begin{cases} x + 5y = 18 \\ 3x + 2y = -11 \end{cases}$

Multiply the first equation by -3.

$$\begin{aligned} -3x - 15y &= -54 \\ 3x + 2y &= -11 \\ \hline -13y &= -65 \\ y &= 5 \end{aligned}$$

Let $y = 5$ in the first equation.

$x+5(5)=18$
$x+25=18$
$x=-7$

The solution of the system is (−7, 5).

11. $\begin{cases} x+y=6 \\ x-y=6 \end{cases}$

$x+y=6$
$x-y=6$
$2x=12$
$x=6$

Let $x = 6$ in the first equation.

$6+y=6$
$y=0$

The solution of the system is (6, 0).

13. $\begin{cases} 2x+3y=0 \\ 4x+6y=3 \end{cases}$

Multiply the first equation by −2.

$-4x-6y=0$
$4x+6y=3$
$0=3$ False

The system has no solution.

15. $\begin{cases} -x+5y=-1 \\ 3x-15y=3 \end{cases}$

Multiply the first equation by 3.

$-3x+15y=-3$
$3x-15y=3$
$0=0$

There are an infinite number of solutions.

17. $\begin{cases} 3x-2y=7 \\ 5x+4y=8 \end{cases}$

Multiply the first equation by 2.

$6x-4y=14$
$5x+4y=8$
$11x=22$
$x=2$

Let $x = 2$ in the first equation.

$3(2)-2y=7$
$6-2y=7$
$-2y=1$
$y=-\frac{1}{2}$

The solution of the system is $\left(2, -\frac{1}{2}\right)$.

19. $\begin{cases} 8x=-11y-16 \\ 2x+3y=-4 \end{cases}$

Add $11y$ to both sides of the first equation and multiply the second equation by −4, then add.

$8x+11y=-16$
$-8x-12y=16$
$-y=0$
$y=0$

Let $y = 0$ in the first equation.

$8x=-11(0)-16$
$8x=-16$
$x=-2$

The solution of the system is (−2, 0).

21. $\begin{cases} 4x-3y=7 \\ 7x+5y=2 \end{cases}$

Multiply the first equation by 5 and the second equation by 3.

$20x-15y=35$
$21x+15y=6$
$41x=41$
$x=1$

Let $x = 1$ in the first equation.

$4x-3y=7$
$4(1)-3y=7$
$4-3y=7$
$-3y=3$
$y=-1$

The solution of the system is (1, −1).

23. $\begin{cases} 4x-6y=8 \\ 6x-9y=12 \end{cases}$

Multiply the first equation by 3 and the second equation by −2.

$12x-18y=24$
$-12x+18y=-24$
$0=0$

The equations in the original system are equivalent and there is an infinite number of solutions.

25. $\begin{cases} 2x-5y=4 \\ 3x-2y=4 \end{cases}$

Multiply the first equation by −3 and the second equation by 2.

$$\begin{aligned} -6x+15y&=-12 \\ 6x-4y&=8 \\ \hline 11y&=-4 \\ y&=-\frac{4}{11} \end{aligned}$$

Multiply the first equation by −2 and the second equation by 5.

$$\begin{aligned} -4x+10y&=-8 \\ 15x-10y&=20 \\ \hline 11x\quad&=12 \\ x&=\frac{12}{11} \end{aligned}$$

The solution of the system is $\left(\frac{12}{11}, -\frac{4}{11}\right)$.

27. $\begin{cases} \frac{x}{3}+\frac{y}{6}=1 \\ \frac{x}{2}-\frac{y}{4}=0 \end{cases}$

Multiply the first equation by 6 and the second equation by 4.

$$\begin{aligned} 2x+y&=6 \\ 2x-y&=0 \\ \hline 4x\quad&=6 \\ x&=\frac{3}{2} \end{aligned}$$

Multiply the second equation of the simplified system by −1.

$$\begin{aligned} 2x+y&=6 \\ -2x+y&=0 \\ \hline 2y&=6 \\ y&=3 \end{aligned}$$

The solution of the system is $\left(\frac{3}{2}, 3\right)$.

29. $\begin{cases} \frac{10}{3}x+4y=-4 \\ 5x+6y=-6 \end{cases}$

Multiply the first equation by 3 and the second equation by −2.

$$\begin{aligned} 10x+12y&=-12 \\ -10x-12y&=12 \\ \hline 0&=0 \end{aligned}$$

The system has an infinite number of solutions.

31. $\begin{cases} x-\frac{y}{3}=-1 \\ -\frac{x}{2}+\frac{y}{8}=\frac{1}{4} \end{cases}$

Multiply the first equation by 3 and the second equation by 8.

$$\begin{aligned} 3x-y&=-3 \\ -4x+y&=2 \\ \hline -x\quad&=-1 \\ x&=1 \end{aligned}$$

Multiply the first equation of the simplified system by 4 and the second equation by 3.

$$\begin{aligned} 12x-4y&=-12 \\ -12x+3y&=6 \\ \hline -y&=-6 \\ y&=6 \end{aligned}$$

The solution of the system is (1, 6).

33. $\begin{cases} -4(x+2)=3y \\ 2x-2y=3 \end{cases} \rightarrow \begin{cases} -4x-8=3y \\ 2x-2y=3 \end{cases}$

$\rightarrow \begin{cases} -4x-3y=8 \\ 2x-2y=3 \end{cases}$

Multiply the second equation by 2.

$$\begin{aligned} -4x-3y&=8 \\ 4x-4y&=6 \\ \hline -7y&=14 \\ y&=-2 \end{aligned}$$

Let $y=-2$ in the second equation.

$$\begin{aligned} 2x-2(-2)&=3 \\ 2x+4&=3 \\ 2x&=-1 \\ x&=-\frac{1}{2} \end{aligned}$$

The solution of the system is $\left(-\frac{1}{2}, -2\right)$.

35. $\begin{cases} \frac{x}{3}-y=2 \\ -\frac{x}{2}+\frac{3y}{2}=-3 \end{cases}$

Multiply the first equation by 3 and the second equation by 2.

$$\begin{aligned} x-3y&=6 \\ -2x+3y&=-6 \\ \hline 0&=0 \end{aligned}$$

The equations of the original system are equivalent and there is an infinite number of solutions.

37. $\begin{cases} \frac{3}{5}x - y = -\frac{4}{5} \\ 3x + \frac{y}{2} = -\frac{9}{5} \end{cases}$

Multiply the first equation by 5 and the second equation by 10.

$$\begin{array}{l} 3x - 5y = -4 \\ \underline{30x + 5y = -18} \\ 33x \quad = -22 \end{array}$$

$$x = -\frac{2}{3}$$

Let $x = -\frac{2}{3}$ in $30x + 5y = -18$.

$$30\left(-\frac{2}{3}\right) + 5y = -18$$
$$-20 + 5y = -18$$
$$5y = 2$$
$$y = \frac{2}{5}$$

The solution of the system is $\left(-\frac{2}{3}, \frac{2}{5}\right)$.

39. $\begin{cases} 3.5x + 2.5y = 17 \\ -1.5x - 7.5y = -33 \end{cases}$

Multiply the first equation by 6 and the second equation by 2.

$$\begin{array}{l} 21x + 15y = 102 \\ \underline{-3x - 15y = -66} \\ 18x \quad = 36 \end{array}$$

$$x = 2$$

Let $x = 2$ in $-3x - 15y = -66$.

$$-3(2) - 15y = -66$$
$$-6 - 15y = -66$$
$$-15y = -60$$
$$y = 4$$

The solution of the system is (2, 4).

41. $\begin{cases} 0.02x + 0.04y = 0.09 \\ -0.1x + 0.3y = 0.8 \end{cases}$

Multiply the first equation by 100 and the second equation by 20.

$$\begin{array}{l} 2x + 4y = 9 \\ \underline{-2x + 6y = 16} \\ \quad 10y = 25 \end{array}$$

$$y = \frac{5}{2} = 2.5$$

Let $y = 2.5$ in $2x + 4y = 9$.

$$2x + 4(2.5) = 9$$
$$2x + 10 = 9$$
$$2x = -1$$
$$x = -\frac{1}{2} = -0.5$$

The solution of the system is (–0.5, 2.5).

43. $\begin{cases} 2x - 3y = -11 \\ y = 4x - 3 \end{cases}$

Substitute $4x - 3$ for y in the first equation.

$$2x - 3(4x - 3) = -11$$
$$2x - 12x + 9 = -11$$
$$-10x = -20$$
$$x = 2$$

Let $x = 2$ in the second equation.

$y = 4(2) - 3 = 5$

The solution of the system is (2, 5).

45. $\begin{cases} x + 2y = 1 \\ 3x + 4y = -1 \end{cases}$

Multiply the first equation by –2.

$$\begin{array}{l} -2x - 4y = -2 \\ \underline{3x + 4y = -1} \\ x \quad = -3 \end{array}$$

Let $x = -3$ in the first equation.

$$-3 + 2y = 1$$
$$2y = 4$$
$$y = 2$$

The solution is (–3, 2).

47. $\begin{cases} 2y = x + 6 \\ 3x - 2y = -6 \end{cases}$

Subtract x from both sides of the first equation.

$$\begin{array}{l} -x + 2y = 6 \\ \underline{3x - 2y = -6} \\ 2x \quad = 0 \end{array}$$

$$x = 0$$

Let $x = 0$ in the first equation.

$$2y = 0 + 6$$
$$2y = 6$$
$$y = 3$$

The solution of the system is (0, 3).

49. $\begin{cases} y = 2x - 3 \\ y = 5x - 18 \end{cases}$

Substitute $5x - 18$ for y in the first equation.

$$5x - 18 = 2x - 3$$
$$3x = 15$$
$$x = 5$$

Let $x = 5$ in the second equation.
$y = 5(5) - 18 = 7$
The solution of the system is (5, 7).

51. $\begin{cases} x + \frac{1}{6}y = \frac{1}{2} \\ 3x + 2y = 3 \end{cases}$

Multiply the first equation by −12.

$$\begin{array}{r} -12x - 2y = -6 \\ 3x + 2y = 3 \\ \hline -9x \quad = -3 \\ x = \frac{1}{3} \end{array}$$

Substitute $\frac{1}{3}$ for x in the second equation.

$$3\left(\frac{1}{3}\right) + 2y = 3$$
$$1 + 2y = 3$$
$$2y = 2$$
$$y = 1$$

The solution of the system is $\left(\frac{1}{3}, 1\right)$.

53. $\begin{cases} \frac{x+2}{2} = \frac{y+11}{3} \\ \frac{x}{2} = \frac{2y+16}{6} \end{cases}$

Multiply the first equation by 6 and the second equation by −6.

$$\begin{cases} 3(x+2) = 2(y+11) \\ 3x + 6 = 2y + 22 \\ 3x - 2y = 16 \\ \\ -3x = -2y - 16 \\ -3x + 2y = -16 \end{cases}$$

Add the two equations.

$$\begin{array}{r} 3x - 2y = 16 \\ -3x + 2y = -16 \\ \hline 0 = 0 \end{array}$$

There is an infinite number of solutions.

55. $\begin{cases} 2x + 3y = 14 \\ 3x - 4y = -69.1 \end{cases}$

Multiply the first equation by 3 and the second equation by −2.

$$\begin{array}{r} 6x + 9y = 42 \\ -6x + 8y = 138.2 \\ \hline 17y = 180.2 \\ y = 10.6 \end{array}$$

Let $y = 10.6$ in the first equation.

$$2x + 3(10.6) = 14$$
$$2x + 31.8 = 14$$
$$2x = -17.8$$
$$x = -8.9$$

The solution of the system is (−8.9, 10.6).

57. Let x = a number.
$2x + 6 = x - 3$

59. Let x = a number.
$20 - 3x = 2$

61. Let n = a number.
$4(n + 6) = 2n$

63. $\begin{cases} 4x + 2y = -7 \\ 3x - y = -12 \end{cases}$

To eliminate y, multiply the second equation by 2.
$6x - 2y = -24$

65. $\begin{cases} 3x + 8y = -5 \\ 2x - 4y = 3 \end{cases} = \begin{cases} 3x + 8y = -5 \\ 4x - 8y = 6 \end{cases}$

The correct answer is **b**; answers may vary

67. Answers may vary

69. $\begin{cases} x + y = 5 \\ 3x + 3y = b \end{cases}$

Multiply the first equation by −3.

$$\begin{array}{r} -3x - 3y = -15 \\ 3x + 3y = b \\ \hline 0 = b - 15 \end{array}$$

a. The system has an infinite number of solutions if this statement is true.
$b = 15$

b. The system has no solution if this statement is false. b = any real number except 15.

71. $\begin{cases} 1.2x + 3.4y = 27.6 \\ 7.2x - 1.7y = -46.56 \end{cases}$

Multiply the second equation by 2.

$$\begin{array}{r} 1.2x+3.4y=27.6 \\ 14.4x-3.4y=-93.12 \\ \hline 15.6x \quad\quad =-65.52 \\ x=-4.2 \end{array}$$

Let $x = -4.2$ in the first equation.

$1.2(-4.2)+3.4y=27.6$
$-5.04+3.4y=27.6$
$3.4y=32.64$
$y=9.6$

The solution of the system is (−4.2, 9.6).

73. a. $\begin{cases} 7.4x-y=-258 \\ 12.6x-y=-231 \end{cases}$

Multiply the first equation by −1.

$$\begin{array}{r} -7.4x+y=258 \\ 12.6x-y=-231 \\ \hline 5.2x \quad\quad =27 \\ x\approx 5 \end{array}$$

Let $x = 5$ in the first equation.

$7.4(5)-y=-258$
$37-y=-258$
$-y=-295$
$y=295$

The solution of the system is approximately (5, 295) [or (5, 294) or (5, 296); answers may vary].

b. In 2009 (2004 + 5), the number of pharmacy technician jobs equals the number of network and data analyst jobs.

c. There will be approximately 294–296 thousand jobs.

Integrated Review

1. $\begin{cases} 2x-3y=-11 \\ y=4x-3 \end{cases}$

Substitute $4x - 3$ for y in the first equation.

$2x-3(4x-3)=-11$
$2x-12x+9=-11$
$-10x=-20$
$x=2$

Let $x = 2$ in the second equation.

$y = 4(2) - 3 = 5$

The solution of the system is (2, 5).

2. $\begin{cases} 4x-5y=6 \\ y=3x-10 \end{cases}$

Substitute $3x - 10$ for y in the first equation.

$4x-5(3x-10)=6$
$4x-15x+50=6$
$-11x=-44$
$x=4$

Let $x = 4$ in the second equation.

$y = 3(4) - 10 = 2$

The solution of the system is (4, 2).

3. $\begin{cases} x+y=3 \\ x-y=7 \end{cases}$

$$\begin{array}{r} x+y=3 \\ x-y=7 \\ \hline 2x \quad\quad =10 \\ x=5 \end{array}$$

Let $x = 5$ in the first equation.

$5+y=3$
$y=-2$

The solution of the system is (5, −2).

4. $\begin{cases} x-y=20 \\ x+y=-8 \end{cases}$

$$\begin{array}{r} x-y=20 \\ x+y=-8 \\ \hline 2x \quad\quad =12 \\ x=6 \end{array}$$

Let $x = 6$ in the second equation.

$6+y=-8$
$y=-14$

The solution of the system is (6, −14).

5. $\begin{cases} x+2y=1 \\ 3x+4y=-1 \end{cases}$

Solve the first equation for x.

$x = 1 - 2y$

Substitute $1 - 2y$ for x in the second equation.

$3(1-2y)+4y=-1$
$3-6y+4y=-1$
$-2y=-4$
$y=2$

Let $y = 2$ in $x = 1 - 2y$.

$x = 1 - 2(2) = -3$

The solution is (−3, 2).

6. $\begin{cases} x+3y=5 \\ 5x+6y=-2 \end{cases}$

Solve the first equation for x.

$x = 5 - 3y$

Substitute $5 - 3y$ for x in the second equation.

$$5(5-3y)+6y=-2$$
$$25-15y+6y=-2$$
$$-9y=-27$$
$$y=3$$

Let $y = 3$ in $x = 5 - 3y$.
$x = 5 - 3(3) = -4$
The solution is (−4, 3).

7. $\begin{cases} y=x+3 \\ 3x-2y=-6 \end{cases}$

Substitute $x + 3$ for y in the second equation.

$$3x-2(x+3)=-6$$
$$3x-2x-6=-6$$
$$x=0$$

Let $x = 0$ in the first equation.
$y = 0 + 3 = 3$
The solution is (0, 3).

8. $\begin{cases} y=-2x \\ 2x-3y=-16 \end{cases}$

Substitute $-2x$ for y in the second equation.

$$2x-3(-2x)=-16$$
$$2x+6x=-16$$
$$8x=-16$$
$$x=-2$$

Let $x = -2$ in the first equation.
$y = -2(-2) = 4$
The solution is (−2, 4).

9. $\begin{cases} y=2x-3 \\ y=5x-18 \end{cases}$

Substitute $5x - 18$ for y in the first equation.

$$5x-18=2x-3$$
$$3x=15$$
$$x=5$$

Let $x = 5$ in the second equation.
$y = 5(5) - 18 = 7$
The solution is (5, 7).

10. $\begin{cases} y=6x-5 \\ y=4x-11 \end{cases}$

Substitute $6x - 5$ for y in the second equation.

$$6x-5=4x-11$$
$$2x=-6$$
$$x=-3$$

Let $x = -3$ in the first equation.
$y = 6(-3) - 5 = -23$
The solution is (−3, −23).

11. $\begin{cases} x+\frac{1}{6}y=\frac{1}{2} \\ 3x+2y=3 \end{cases}$

Multiply the first equation by 6.

$\begin{cases} 6x+y=3 \\ 3x+2y=3 \end{cases}$

Multiply the first equation of the simplified system by −2.

$$\begin{array}{r} -12x-2y=-6 \\ 3x+2y=3 \\ \hline -9x \quad\quad =-3 \end{array}$$
$$x=\frac{1}{3}$$

Multiply the second equation of the simplified system by −2.

$$\begin{cases} 6x\ +y=3 \\ -6x-4y=-6 \end{cases}$$
$$-3y=-3$$
$$y=1$$

The solution of the system is $\left(\frac{1}{3}, 1\right)$.

12. $\begin{cases} x+\frac{1}{3}y=\frac{5}{12} \\ 8x+3y=4 \end{cases}$

Multiply the first equation by 12.

$\begin{cases} 12x+4y=5 \\ 8x+3y=4 \end{cases}$

Multiply the first equation of the simplified system by 2 and the second equation by −3.

$$\begin{array}{r} 24x+8y=10 \\ -24x-9y=-12 \\ \hline -y=-2 \end{array}$$
$$y=2$$

Multiply the first equation of the simplified system by 3 and the second equation by −4.

$$\begin{array}{r} 36x+12y=15 \\ -32x-12y=-16 \\ \hline 4x \quad\quad =-1 \end{array}$$
$$x=-\frac{1}{4}$$

The solution of the system is $\left(-\frac{1}{4}, 2\right)$.

13. $\begin{cases} x-5y=1 \\ -2x+10y=3 \end{cases}$

Multiply the first equation by 2.

$$\begin{array}{r} 2x-10y=2 \\ -2x+10y=3 \\ \hline 0=5 \end{array} \quad \text{False}$$

The system has no solution.

14. $\begin{cases} -x+2y=3 \\ 3x-6y=-9 \end{cases}$

Multiply the first equation by 3.

$$\begin{array}{r} -3x+6y=9 \\ 3x-6y=-9 \\ \hline 0=0 \end{array}$$

The equations in the original system are equivalent and there is an infinite number of solutions.

15. $\begin{cases} 0.2x-0.3y=-0.95 \\ 0.4x+0.1y=0.55 \end{cases}$

Multiply both equations by 10.

$\begin{cases} 2x-3y=-9.5 \\ 4x+y=5.5 \end{cases}$

Multiply the first equation of the simplified system by −2.

$$\begin{array}{r} -4x+6y=19 \\ 4x+y=5.5 \\ \hline 7y=24.5 \\ y=3.5 \end{array}$$

Multiply the second equation of the simplified system by 3.

$$\begin{array}{r} 2x-3y=-9.5 \\ 12x+3y=16.5 \\ \hline 14x=7 \\ x=0.5 \end{array}$$

The solution of the system is (0.5, 3.5).

16. $\begin{cases} 0.08x-0.04y=-0.11 \\ 0.02x-0.06y=-0.09 \end{cases}$

Multiply both equations by 100.

$\begin{cases} 8x-4y=-11 \\ 2x-6y=-9 \end{cases}$

Multiply the second equation of the simplified system by −4.

$$\begin{array}{r} 8x-4y=-11 \\ -8x+24y=36 \\ \hline 20y=25 \\ y=1.25 \end{array}$$

Multiply the first equation of the simplified system by −3 and the second equation by 2.

$$\begin{array}{r} -24x+12y=33 \\ 4x-12y=-18 \\ \hline -20x=15 \\ x=-0.75 \end{array}$$

The solution of the system is (−0.75, 1.25).

17. $\begin{cases} x=3y-7 \\ 2x-6y=-14 \end{cases}$

Substitute $3y-7$ for x in the second equation.

$$\begin{aligned} 2(3y-7)-6y&=-14 \\ 6y-14-6y&=-14 \\ -14&=-14 \end{aligned}$$

The equations in the original system are equivalent and there is an infinite number of solutions.

18. $\begin{cases} y=\frac{x}{2}-3 \\ 2x-4y=0 \end{cases}$

Substitute $\frac{x}{2}-3$ for y in the second equation.

$$\begin{aligned} 2x-4\left(\frac{x}{2}-3\right)&=0 \\ 2x-2x+12&=0 \\ 12&=0 \quad \text{False} \end{aligned}$$

There is no solution.

19. $\begin{cases} 2x+5y=-1 \\ 3x-4y=33 \end{cases}$

Multiply the first equation by 4 and the second equation by 5.

$$\begin{array}{r} 8x+20y=-4 \\ 15x-20y=165 \\ \hline 23x=161 \\ x=7 \end{array}$$

Let $x = 7$ in the first equation.

$$\begin{aligned} 2(7)+5y&=-1 \\ 14+5y&=-1 \\ 5y&=-15 \\ y&=-3 \end{aligned}$$

The solution of the system is (7, −3).

20. $\begin{cases} 7x-3y=2 \\ 6x+5y=-21 \end{cases}$

Multiply the first equation by 5 and the second equation by 3.

$$\begin{aligned} 35x - 15y &= 10 \\ 18x + 15y &= -63 \\ \hline 53x \quad &= -53 \\ x &= -1 \end{aligned}$$

Let $x = -1$ in the first equation.

$$\begin{aligned} 7(-1) - 3y &= 2 \\ -7 - 3y &= 2 \\ -3y &= 9 \\ y &= -3 \end{aligned}$$

The solution of the system is (–1, –3).

21. Answers may vary

22. Answers may vary

Section 4.4

Practice Exercises

1. Let x be one number and y be the other.

The system is $\begin{cases} x + y = 30 \\ x - y = 6 \end{cases}$.

Add.

$$\begin{aligned} x + y &= 30 \\ x - y &= 6 \\ \hline 2x \quad &= 36 \\ x &= 18 \end{aligned}$$

Let $x = 18$ in the first equation.

$$\begin{aligned} x + y &= 30 \\ 18 + y &= 30 \\ y &= 12 \end{aligned}$$

The numbers are 18 and 12.

2. Let x = price for adult admission and y = price per child admission.

$$\begin{cases} 3x + 3y = 75 \\ 2x + 4y = 62 \end{cases}$$

Multiply the first equation by 2 and the second equation by –3.

$$\begin{aligned} 6x + 6y &= 150 \\ -6x - 12y &= -186 \\ \hline -6y &= -36 \\ y &= 6 \end{aligned}$$

Let $y = 6$ in the second equation.

$$\begin{aligned} 2x + 4y &= 62 \\ 2x + 4(6) &= 62 \\ 2x + 24 &= 62 \\ 2x &= 38 \\ x &= 19 \end{aligned}$$

a. $x = 19$, so the adult price is \$19.

b. $y = 6$, so the child price is \$6.

c. $5(19) + 15(6) = 95 + 90 = 185 < 200$
No, the regular rates are less than the group rate.

3. Let x and y be the speed of the hikers. Let the slower hiker be y; then $y = x - 2$.

r	$\cdot\ t$	$= d$
x	4	$4x$
y	4	$4y$

The total distance is 22 miles, so the system is:

$$\begin{cases} 4x + 4y = 22 \\ y = x - 2 \end{cases} \rightarrow \begin{cases} 2x + 2y = 11 \\ -x + y = -2 \end{cases}$$

Multiply $-x + y = -2$ by 2.

$$\begin{aligned} 2x + 2y &= 11 \\ -2x + 2y &= -4 \\ \hline 4y &= 7 \\ y &= \frac{7}{4} = 1.75 \end{aligned}$$

Let $y = 1.75$ in $-x + y = -2$.

$$\begin{aligned} -x + 1.75 &= -2 \\ -x &= -3.75 \\ x &= 3.75 \end{aligned}$$

The speeds are 1.75 miles per hour and 3.75 miles per hour.

4. Let x = pounds of Kona and y = pounds of Blue.
Then, $x + y = 20$ pounds of mix.
$20x$ is the total price of the Kona.
$28y$ is the total price of the Blue.
22(20) is the total price of the blend.
Then, $20x + 28y = 440$.

The system is $\begin{cases} x + y = 20 \\ 20x + 28y = 440 \end{cases}$.

Multiply the first equation by –20.

$$\begin{aligned} -20x - 20y &= -400 \\ 20x + 28y &= 440 \\ \hline 8y &= 40 \\ y &= 5 \end{aligned}$$

Let $y = 5$ in the first equation.

$$\begin{aligned} x + y &= 20 \\ x + 5 &= 20 \\ x &= 15 \end{aligned}$$

Jemima should use 15 pounds of Kona and 5 pounds of Blue Mountain.

Exercise Set 4.4

1. a. $l - w = 8 - 5 = 3$
$$\begin{aligned} P &= 2l + 2w \\ &= 2(8) + 2(5) \\ &= 13 + 10 \\ &= 23 \neq 30 \end{aligned}$$

b. $l - w = 8 - 7 = 1 \neq 3$

c. $l - w = 9 - 6 = 3$
$P = 2l + 2w = 2(9) + 2(6) = 18 + 12 = 30$

Choice **c** is correct.

3. a. $2d + 3n = 2(3) + 3(4) = 6 + 12 = 18 \neq 17$

b. $2d + 3n = 2(4) + 3(3) = 8 + 9 = 17$
$5d + 4n = 5(4) + 4(3) = 20 + 12 = 32$

c. $2d + 3n = 2(2) + 3(5) = 4 + 15 = 19 \neq 17$

Choice **b** is correct.

5. a. $80 + 20 = 100$
$$\begin{aligned} 80d + 20q &= 80(0.10) + 20(0.25) \\ &= 8 + 5 \\ &= 13 \end{aligned}$$

b. $20 + 44 = 64 \neq 100$

c. $60 + 40 = 100$
$$\begin{aligned} 60d + 40q &= 60(0.10) + 40(0.25) \\ &= 6 + 10 \\ &= 16 \neq 13 \end{aligned}$$

Choice **a** is correct.

7. Let x = the larger number and y = the smaller number.
$$\begin{cases} x + y = 15 \\ x - y = 7 \end{cases}$$

9. Let x = the amount invested in the larger account and y = the amount invested in the smaller account.
$$\begin{cases} x + y = 6500 \\ x = y + 800 \end{cases}$$

11. Let x = the first number and y = the second number.
$$\begin{cases} x + y = 83 \\ x - y = 17 \end{cases}$$

$$\begin{array}{r} x + y = 83 \\ x - y = 17 \\ \hline 2x \quad = 100 \\ x = 50 \end{array}$$

Let $x = 50$ in the first equation.
$$\begin{aligned} 50 + y &= 83 \\ y &= 33 \end{aligned}$$

The numbers are 50 and 33.

13. Let x = the first number and y = the second number.
$$\begin{cases} x + 2y = 8 \\ 2x + y = 25 \end{cases}$$

Multiply the first equation by -2.
$$\begin{array}{r} -2x - 4y = -16 \\ 2x \;\; + y = 25 \\ \hline -3y = 9 \\ y = -3 \end{array}$$

Let $y = -3$ in the first equation.
$$\begin{aligned} x + 2(-3) &= 8 \\ x - 6 &= 8 \\ x &= 14 \end{aligned}$$

The numbers are 14 and -3.

15. Let x = Taurasi's points and y = Augustus's points.
$$\begin{cases} x - y = 116 \\ x + y = 1604 \end{cases}$$

$$\begin{array}{r} x - y = 116 \\ x + y = 1604 \\ \hline 2x \quad = 1720 \\ x = 860 \end{array}$$

Let $x = 860$ in the second equation.
$$\begin{aligned} x + y &= 1604 \\ 860 + y &= 1604 \\ y &= 744 \end{aligned}$$

Taurasi scored 860 points and Augustus scored 744 points.

17. Let x = the price of an adult's ticket and y = the price of a child's ticket.
$$\begin{cases} 3x + 4y = 159 \\ 2x + 3y = 112 \end{cases}$$

Multiply the first equation by -2 and the second equation by 3.
$$\begin{array}{r} -6x - 8y = -318 \\ 6x + 9y = 336 \\ \hline y = 18 \end{array}$$

Let $y = 18$ in the first equation.

$$\begin{aligned} 3x+4(18)&=159\\ 3x+72&=159\\ 3x&=87\\ x&=29 \end{aligned}$$

An adult's ticket is $29 and a child's ticket is $18.

19. Let x = the number of quarters and y = the number of nickels.

$$\begin{cases} x+y=80\\ 0.25x+0.05y=14.6 \end{cases}$$

Solve the first equation for y.

$y = 80 - x$

Substitute $80 - x$ for y in the second equation.

$$\begin{aligned} 0.25x+0.05(80-x)&=14.6\\ 0.25x+4-0.05x&=14.6\\ 0.20x&=10.6\\ x&=53 \end{aligned}$$

Let $x = 53$ in $y = 80 - x$.

$y = 80 - 53$

$y = 27$

There are 53 quarters and 27 nickels.

21. Let x = price of Apple stock and y = price of Microsoft stock.

$$\begin{cases} 50x+60y=6035.90\\ x-y=60.68 \end{cases}$$

Multiply the second equation by 60.

$$\begin{aligned} 50x+60y&=6035.90\\ 60x-60y&=3640.80\\ \hline 110x\qquad&=9676.70\\ x&=87.97 \end{aligned}$$

Let $x = 87.97$ in the second equation.

$$\begin{aligned} x-y&=60.68\\ 87.97-y&=60.68\\ -y&=-27.29\\ y&=27.29 \end{aligned}$$

The price of Apple's stock was $87.97 and the price of Microsoft stock was $27.29.

23. Let x = the daily fee and y = the mileage charge.

$$\begin{cases} 4x+450y=240.50\\ 3x+200y=146.00 \end{cases}$$

Multiply the first equation by 3 and the second equation by –4.

$$\begin{aligned} 12x+1350y&=721.50\\ -12x-800y&=-584.00\\ \hline 550y&=137.5\\ y&=0.25 \end{aligned}$$

Let $y = 0.25$ in the second equation.

$$\begin{aligned} 3x+200(0.25)&=146.00\\ 3x+50&=146.00\\ 3x&=96.00\\ x&=32.00 \end{aligned}$$

The daily fee is $32 and the mileage charge is $0.25 per mile.

25.

	d	= r	· t
Downstream	18	$x+y$	2
Upstream	18	$x-y$	$4\frac{1}{2}$

$$\begin{cases} 2(x+y)=18\\ \frac{9}{2}(x-y)=18 \end{cases}$$

Multiply the first equation by $\frac{1}{2}$ and the second equation by $\frac{2}{9}$.

$$\begin{aligned} x+y&=9\\ x-y&=4\\ \hline 2x\qquad&=13\\ x&=6.5 \end{aligned}$$

Let $x = 6.5$ in $x + y = 9$.

$6.5+y=9$

$y=2.5$

Pratap can row 6.5 miles per hour in still water. The rate of the current is 2.5 miles per hour.

27.

	d	= r	· t
With the wind	780	$x+y$	$1\frac{1}{2}$
Into the wind	780	$x-y$	2

$$\begin{cases} \frac{3}{2}(x+y)=780\\ 2(x-y)=780 \end{cases}$$

Multiply the first equation by $\frac{2}{3}$ and the second equation by $\frac{1}{2}$.

$$\begin{aligned} x+y&=520\\ x-y&=390\\ \hline 2x\qquad&=910\\ x&=455 \end{aligned}$$

Let $x = 455$ in $x + y = 520$.

$455 + y = 520$
$y = 65$

The plane can fly 455 miles per hour in still air. The speed of the wind is 65 miles per hour.

29. Let x = the time spent walking and y = the time spent on the bicycle.

	r	$\cdot\ t$	$= d$
Walking	4	x	$4x$
Biking	20	y	$20y$

$$\begin{cases} x + y = 6 \\ 4x + 20y = 96 \end{cases}$$

Multiply the first equation by –4.

$-4x - 4y = -24$
$4x + 20y = 96$
$16y = 72$
$y = 4.5$

He spent $4\frac{1}{2}$ hours on the bicycle.

31. Let x = ounces of 4% solution and y = ounces of 12% solution.

Concentration Rate	Ounces of Solution	Ounces of Pure Acid
0.04	x	$0.04x$
0.12	y	$0.12y$
0.09	12	0.09(12)

$$\begin{cases} x + y = 12 \\ 0.04x + 0.12y = 0.09(12) \end{cases}$$

Multiply the first equation by –4 and the second equation by 100.

$-4x - 4y = -48$
$4x + 12y = 108$
$8y = 60$
$y = 7.5$

Let $y = 7.5$ in the first equation.

$x + 7.5 = 12$
$x = 4.5$

$4\frac{1}{2}$ ounces of 4% solution and $7\frac{1}{2}$ ounces of 12% solution should be mixed.

33. Let x = pounds of \$4.95 per pound beans and y = pounds of \$2.65 per pound beans.

	Cost Rate	Pounds of Beans	Dollars Cost
High Quality	4.95	x	$4.95x$
Low Quality	2.65	y	$2.65y$
Mixture	3.95	200	3.95(200)

$$\begin{cases} x + y = 200 \\ 4.95x + 2.65y = 3.95(200) \end{cases}$$

Solve the first equation for y.

$y = 200 - x$

Substitute $200 - x$ for y in the second equation.

$4.95x + 2.65(200 - x) = 3.95(200)$
$4.95x + 530 - 2.65x = 790$
$2.30x = 260$
$x \approx 113.04$

Let $x = 113.04$ in the first equation.

$113.04 + y = 200$
$y \approx 86.96$

He needs 113 pounds of \$4.95 per pound beans and 87 pounds of \$2.65 per pound beans.

35. Let x = the first angle and y = the second angle.

$$\begin{cases} x + y = 90 \\ x = 2y \end{cases}$$

Substitute $2y$ for x in the first equation.

$2y + y = 90$
$3y = 90$
$y = 30$

Let $y = 30$ in the second equation.

$x = 2(30) = 60$

The angles are 60° and 30°.

37. Let x = the first angle and y = the second angle.

$$\begin{cases} x + y = 90 \\ x = 3y + 10 \end{cases}$$

Substitute $3y + 10$ for x in the first equation.

$3y + 10 + y = 90$
$4y = 80$
$y = 20$

Let $y = 20$ in the second equation.

$x = 3(20) + 10 = 70$

The angles are 70° and 20°.

39. Let x = the number sold at \$9.50 and y = the number sold at \$7.50.

$$\begin{cases} x+y=90 \\ 9.5x+7.5y=721 \end{cases}$$

Solve the first equation for y.

$y = 90 - x$

Substitute $90 - x$ for y in the second equation.

$$9.5x+7.5(90-x)=721$$
$$9.5x+675-7.5x=721$$
$$2x=46$$
$$x=23$$

Let $x = 23$ in $y = 90 - x$.

$y = 90 - 23 = 67$

They sold 23 at \$9.50 and 67 at \$7.50.

41. Let x = the rate of the faster group and y = the rate of the slower group.

	r	$\cdot\ t$	$=\ d$
Faster group	x	240	$240x$
Slower group	y	240	$240y$

$$\begin{cases} x=y+\frac{1}{2} \\ 240x+240y=1200 \end{cases}$$

Substitute $y+\frac{1}{2}$ for x in the second equation.

$$240\left(y+\frac{1}{2}\right)+240y=1200$$
$$240y+120+240y=1200$$
$$480y=1080$$
$$y=\frac{1080}{480}=2\frac{1}{4}$$

Let $y=2\frac{1}{4}$ in the first equation.

$$x=2\frac{1}{4}+\frac{1}{2}=2\frac{3}{4}$$

The rate of the faster group is $2\frac{3}{4}$ miles per hour. The rate of the slower group is $2\frac{1}{4}$ miles per hour.

43. Let x = gallons of 30% solution and y = gallons of 60% solution.

Concentration Rate	Gallons of Solution	Gallons of Pure Fertilizer
0.30	x	$0.30x$
0.60	y	$0.60y$
0.50	150	0.50(150)

$$\begin{cases} x+y=150 \\ 0.30x+0.60y=0.50(150) \end{cases}$$

Multiply the first equation by –3 and the second equation by 10.

$$-3x-3y=-450$$
$$3x+6y=750$$
$$3y=300$$
$$y=100$$

Let $y = 100$ in the first equation.

$$x+100=150$$
$$x=50$$

50 gallons of 30% solution and 100 gallons of 60% solution.

45. Let x = the width and y = the length.

$$\begin{cases} 2x+2y=144 \\ y=x+12 \end{cases}$$

Substitute $x + 12$ for y in the first equation.

$$2x+2(x+12)=144$$
$$2x+2x+24=144$$
$$4x=120$$
$$x=30$$

Let $x = 30$ in the second equation.

$y = 30 + 12 = 42$

The width is 30 inches and the length is 42 inches.

47. $-3x<-9$

$$\frac{-3x}{-3}>\frac{-9}{-3}$$
$$x>3,\ (3,\ \infty)$$

49. $4(2x-1)\ge 0$

$$8x-4\ge 0$$
$$8x\ge 4$$
$$x\ge \frac{1}{2},\ \left[\frac{1}{2},\ \infty\right)$$

51. The minimum price is \$0.49.
The maximum price is \$0.65.
$0.72 > 0.65$ Impossible
$0.29 < 0.49$ Impossible
$0.49 < 0.58 < 0.65$ Possible
The answer is **a**.

53. Let x = the width and y = the length.
$$\begin{cases} 2x + y = 33 \\ y = 2x - 3 \end{cases}$$
Substitute $2x - 3$ for y in the first equation.
$$2x + 2x - 3 = 33$$
$$4x = 36$$
$$x = 9$$
Let $x = 9$ in the second equation.
$y = 2(9) - 3 = 15$
The width is 9 feet and the length is 15 feet.

55. a. $$\begin{cases} y = 0.82x + 17.2 \\ y = 0.33x + 30.5 \end{cases}$$
Substitute $0.82x + 17.2$ for y in the second equation.
$$0.82x + 17.2 = 0.33x + 30.5$$
$$0.49x = 13.3$$
$$x \approx 27.14$$
Let $x = 27.14$ in $y = 0.82x + 17.2$.
$y = 0.82(27.14) + 17.2$
$y \approx 39.4548$
The solution is approximately (27.1, 39.5).

b. For viewers who are 27.1 years over 18 (or 45.1 years of age) the percent who watch cable news and network news is the same, 39.5%.

c. Answers may vary

Section 4.5

Practice Exercises

1.

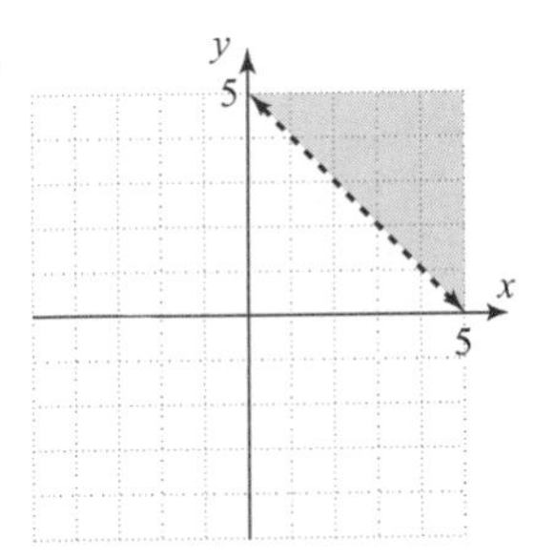

2.

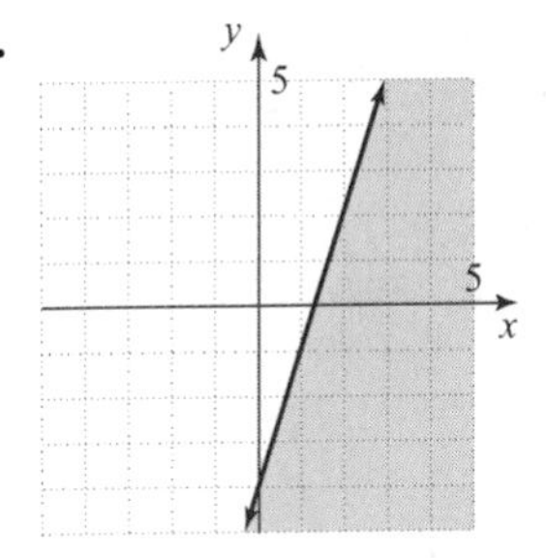

3.

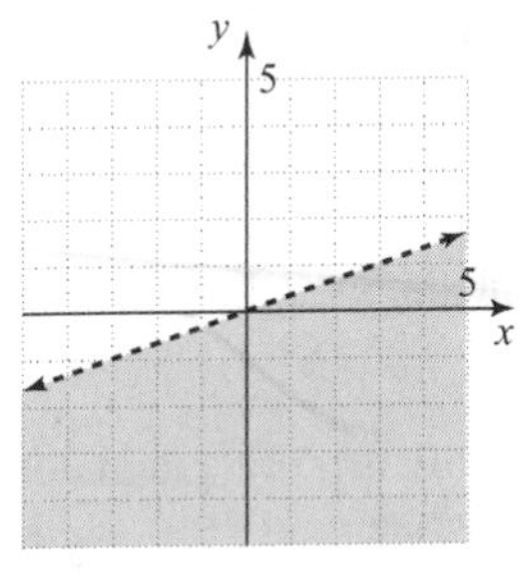

4.

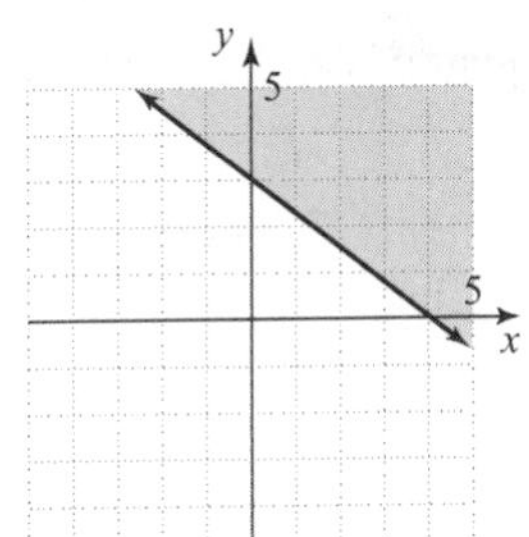

5.

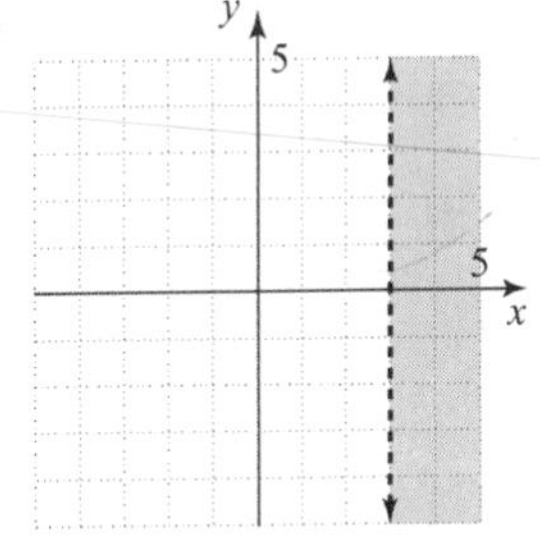

Vocabulary and Readiness Check

1. The statement $5x - 6y < 7$ is an example of a linear inequality in two variables.

2. A boundary line divides a plane into two regions called half-planes.

3. The graph of $5x - 6y < 7$ does not include its corresponding boundary line. The statement is false.

4. When graphing a linear inequality to determine which side of the boundary line to shade, choose a point *not* on the boundary line. The statement is true.

5. The boundary line for the inequality $5x - 6y < 7$ is the graph of $5x - 6y = 7$. The statement is true.

6. The graph of $\underline{y < 3}$ is

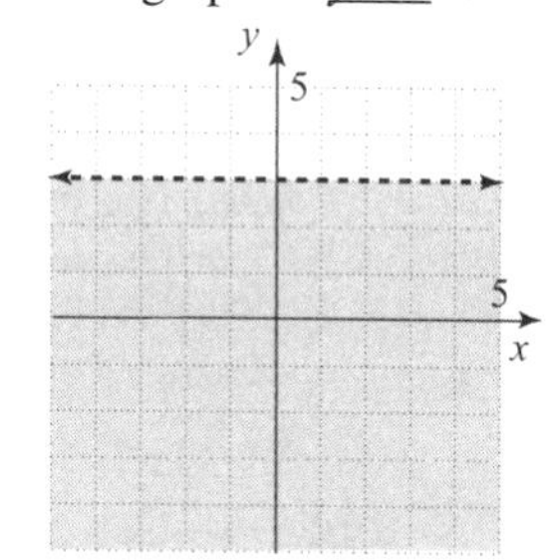

7. Yes, since the inequality is $\geq$, the graph includes the boundary line.

8. No, since the inequality is >, the graph does not include the boundary line.

9. Yes, since the inequality is $\geq$, the graph includes the boundary line.

10. No, since the inequality is >, the graph does not include the boundary line.

11. $x + y > -5$, (0, 0)

$$0 + 0 \overset{?}{>} -5$$
$$0 \overset{?}{>} -5$$

Yes, (0, 0) is a solution.

12. $2x + 3y < 10$, (0, 0)

$$2(0) + 3(0) \overset{?}{<} 10$$
$$0 \overset{?}{<} 10$$

Yes, (0, 0) is a solution.

13. $x - y \leq -1$, (0,)

$$0 - 0 \overset{?}{\leq} -1$$
$$0 \overset{?}{\leq} -1$$

No, (0, 0) is not a solution.

14. $\frac{2}{3}x + \frac{5}{6}y > 4$, (0, 0)

$$\frac{2}{3}(0) + \frac{5}{6}(0) \overset{?}{>} 4$$
$$0 \overset{?}{>} 4$$

No, (0, 0) is not a solution.

Exercise Set 4.5

1. $x - y > 3$

$$(2, -1),\ 2 - (-1) \overset{?}{>} 3$$
$$2 + 1 \overset{?}{>} 3$$
$$3 \overset{?}{>} 3, \text{ False}$$

(2, −1) is not a solution.

$$(5, 1),\ 5 - 1 \overset{?}{>} 3$$
$$4 \overset{?}{>} 3, \quad \text{True}$$

(5, 1) is a solution.

3. $3x - 5y \leq -4$

$$(-1, -1),\ 3(-1) - 5(-1) \overset{?}{\leq} -4$$
$$-3 + 5 \overset{?}{\leq} -4$$
$$2 \overset{?}{\leq} -4, \quad \text{False}$$

(−1, −1) is not a solution.

$$(4, 0),\ 3(4) - 5(0) \overset{?}{\leq} -4$$
$$12 - 0 \overset{?}{\leq} -4$$
$$12 \overset{?}{\leq} -4, \text{ False}$$

(4, 0) is not a solution.

5. $x < -y$

$$(0, 2),\ 0 \overset{?}{<} -2, \text{ False}$$

(0, 2) is not a solution.

$$(-5, 1),\ -5 \overset{?}{<} -1, \text{ True}$$

(−5, 1) is a solution.

7. $x + y \leq 1$

Test (0, 0)

$$0 + 0 \overset{?}{\leq} 1, \text{ True}$$

Shade below.

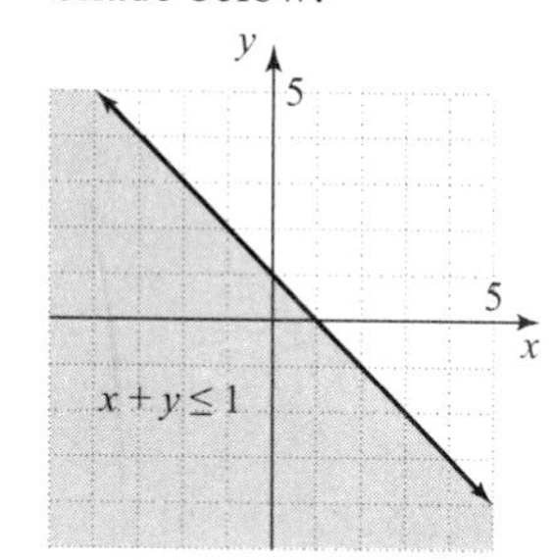

9. $2x + y > -4$
Test (0, 0)

$2(0)+0 \overset{?}{>} -4$

True
Shade above.

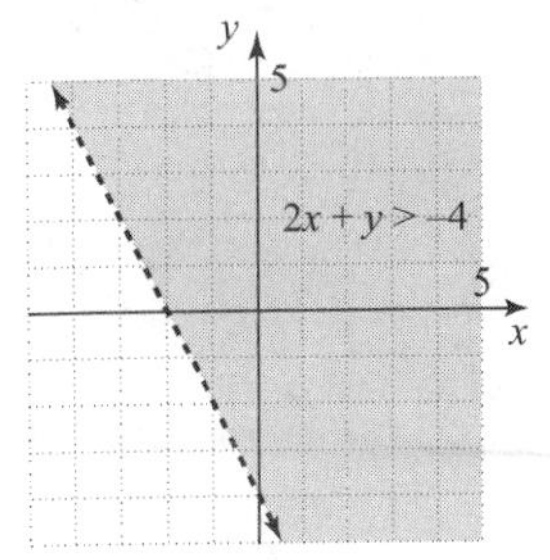

11. $x + 6y \le -6$
Test (0, 0)

$0+6(0) \overset{?}{\le} -6$

False
Shade below.

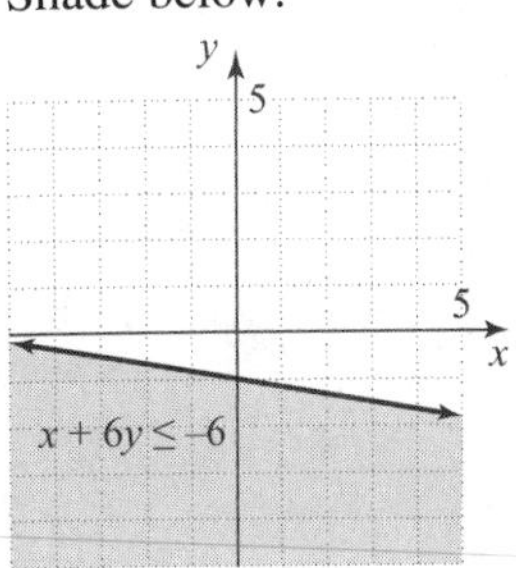

13. $2x + 5y > -10$
Test (0, 0)

$2(0)+5(0) \overset{?}{>} -10$

True
Shade above.

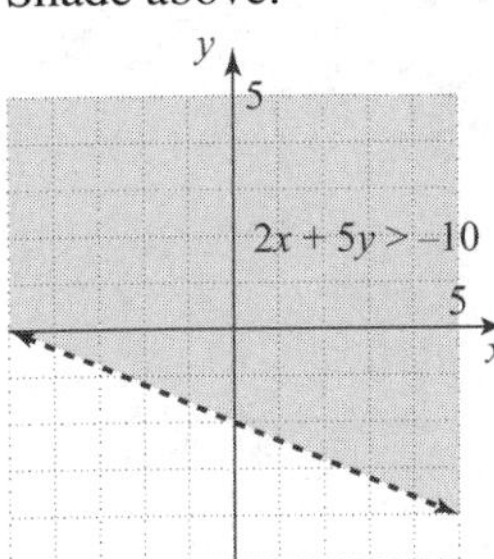

15. $x + 2y \le 3$
Test (0, 0)

$0+2(0) \overset{?}{\le} 3$

True
Shade below.

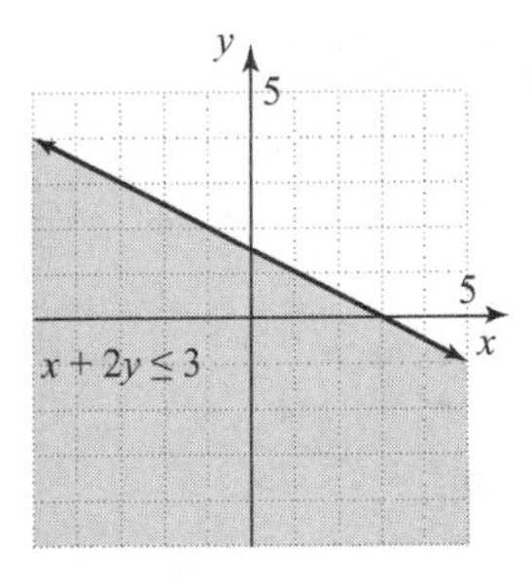

17. $2x + 7y > 5$
Test (0, 0)

$2(0)+7(0) \overset{?}{>} 5$

False
Shade above.

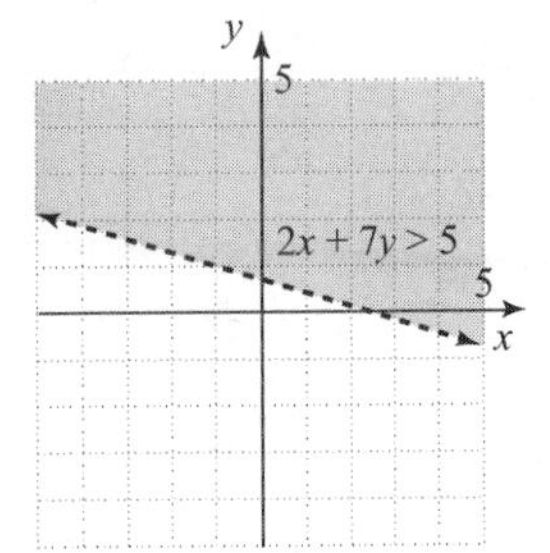

19. $x - 2y \ge 3$
Test (0, 0)

$(0)-2(0) \overset{?}{\ge} 3$

False
Shade below.

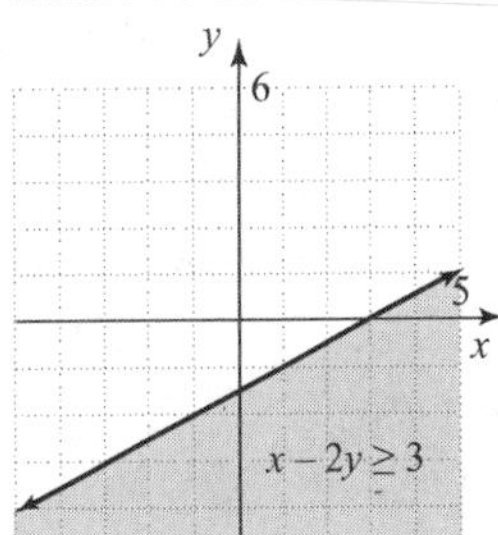

21. $5x + y < 3$
Test (0, 0)

$5(0)+0 \overset{?}{<} 3$

True
Shade below.

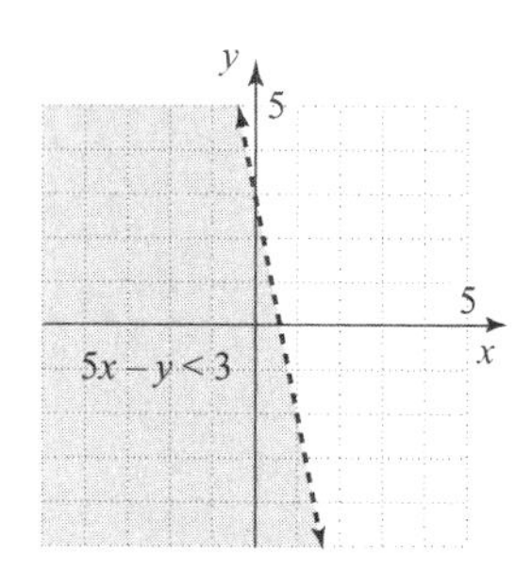

23. $4x + y < 8$
Test (0, 0)

$4(09) + 0 \overset{?}{<} 8$

True
Shade below.

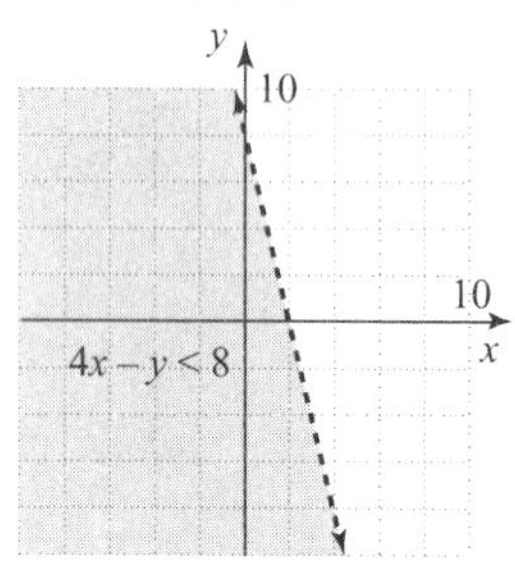

25. $y \geq 2x$
Test (1, 0)

$0 \overset{?}{\geq} 2(1)$

False
Shade above.

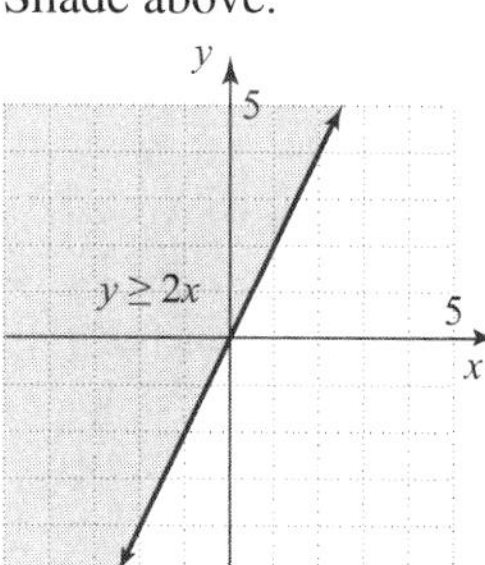

27. $x \geq 0$
Shade right.

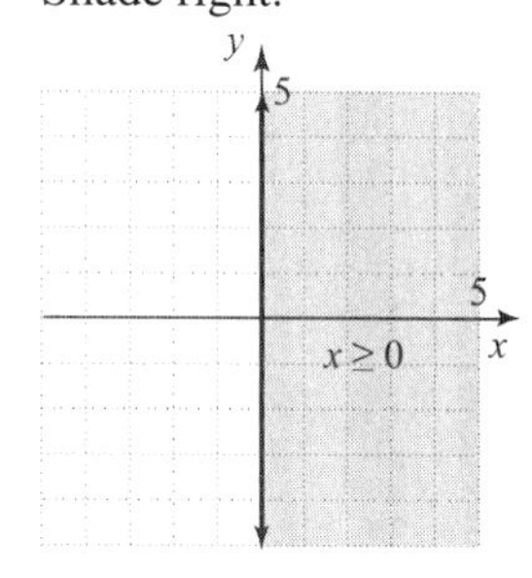

29. $y \leq -3$
Shade below.

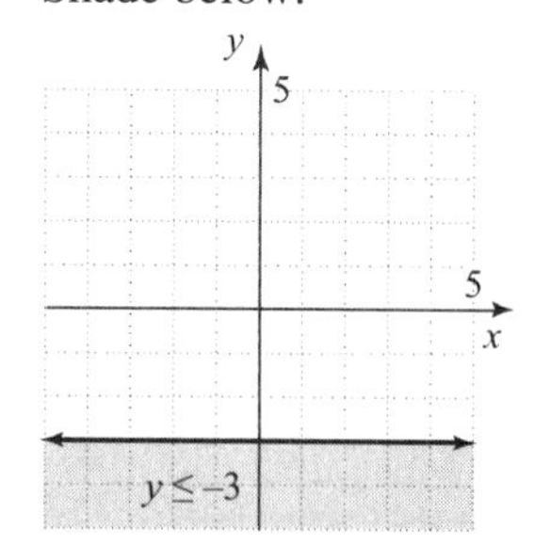

31. $2x - 7y > 0$
Test (1, 0)

$2(1) - 7(0) \overset{?}{>} 0$

True
Shade below.

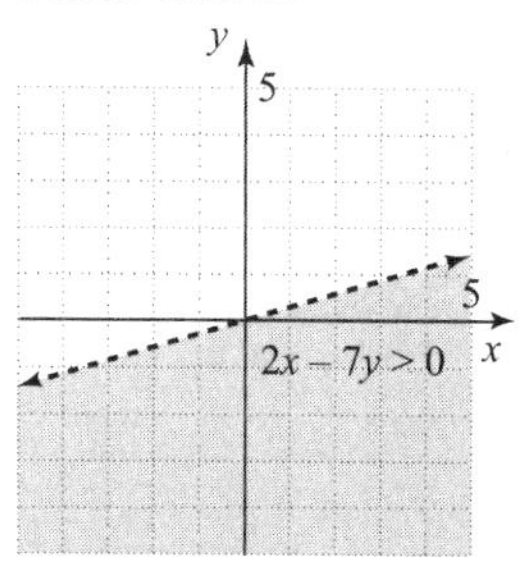

33. $3x - 7y \geq 0$
Test (1, 0)

$3(1) - 7(0) \overset{?}{\geq} 0$

True
Shade below.

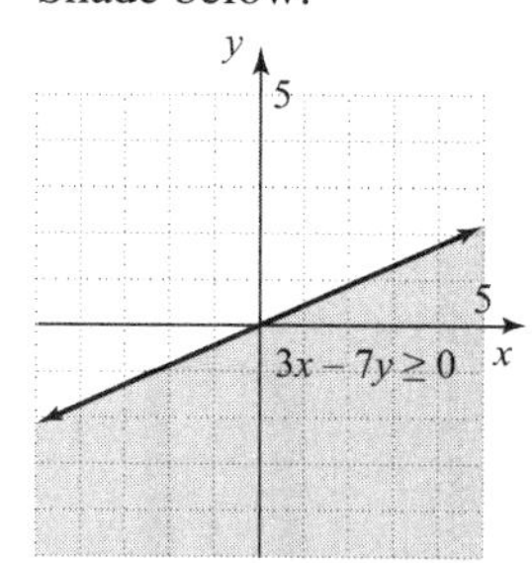

35. $x > y$
Test (0, 1)

$0 \overset{?}{>} 1$

False
Shade below.

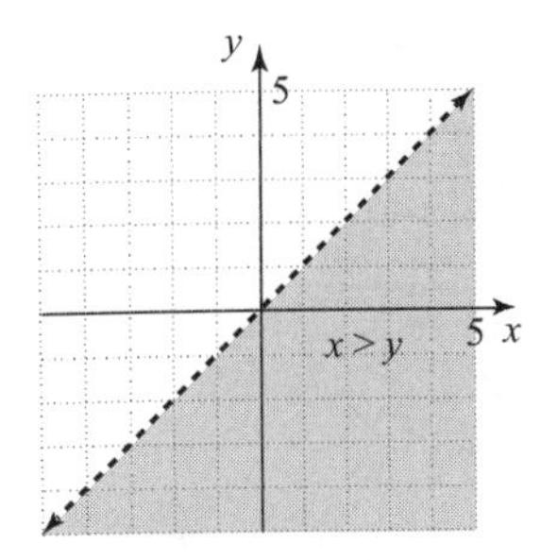

37. $x - y \leq 6$
Test (0, 0)
$0 - 0 \overset{?}{\leq} 6$
True
Shade above.

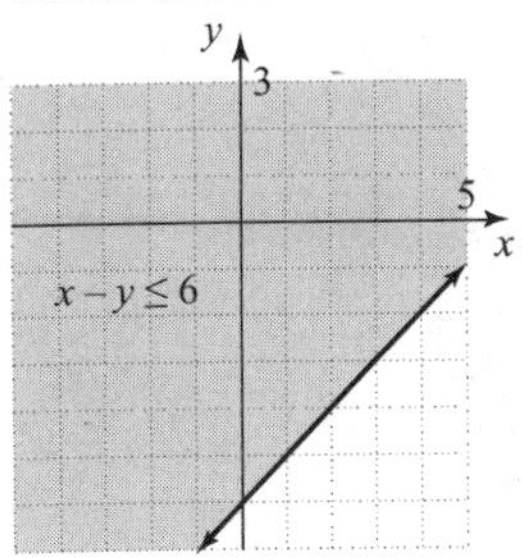

39. $-\frac{1}{4}y + \frac{1}{3}x > 1$
Test (0, 0)
$-\frac{1}{4}(0) + \frac{1}{3}(0) \overset{?}{>} 1$
False
Shade below.

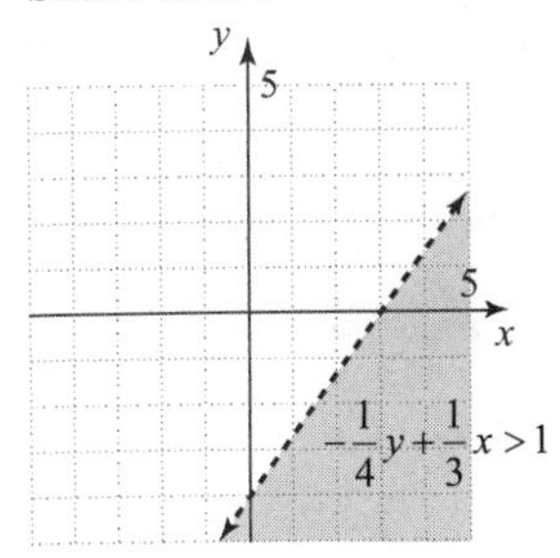

41. $-x < 0.4y$
Test (1, 0)
$-(1) \overset{?}{<} 0$
True
Shade above.

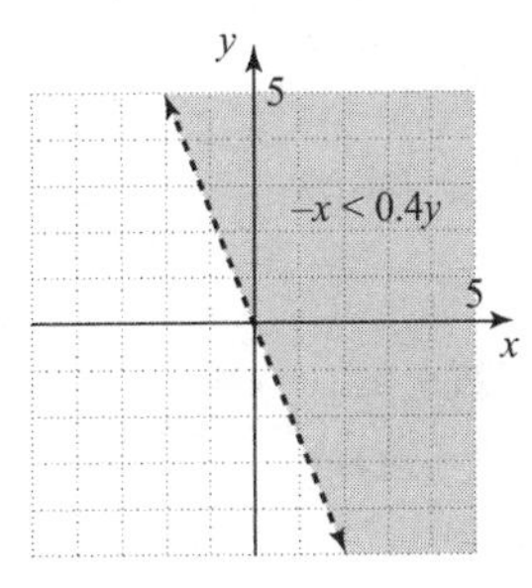

43. e

45. c

47. f

49. $2^3 = 2 \cdot 2 \cdot 2 = 8$

51. $(-2)^5 = (-2)(-2)(-2)(-2)(-2) = -32$

53. $3 \cdot 4^2 = 3 \cdot 4 \cdot 4 = 48$

55. Let $x = -5$.
$x^2 = (-5)(-5) = 25$

57. Let $x = -1$.
$2x^3 = 2(-1)(-1)(-1) = -2$

59. $3x + 4y < 8$; (1,1)
$3(1) + 4(1) < 8$
$3 + 4 < 8$
$7 < 8$ True
(1, 1) is included in the graph.

61. $y \geq -\frac{1}{2}x$; (1, 1)
$y \geq -\frac{1}{2}(1)$
$y \geq -\frac{1}{2}$ True
(1, 1) is included in the graph.

63. The inequality is $x + y \geq 13$.

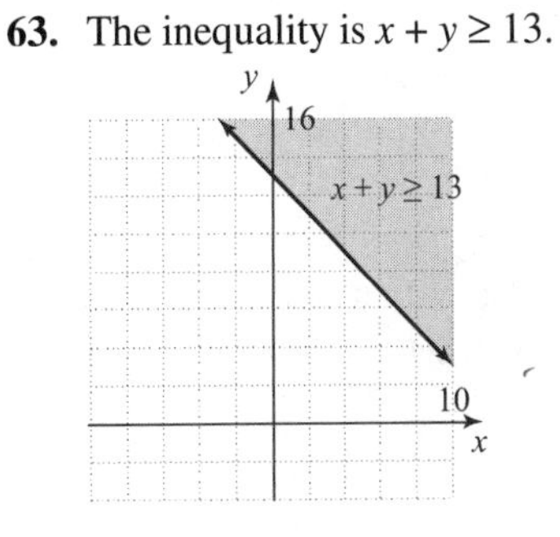

65. Answers may vary

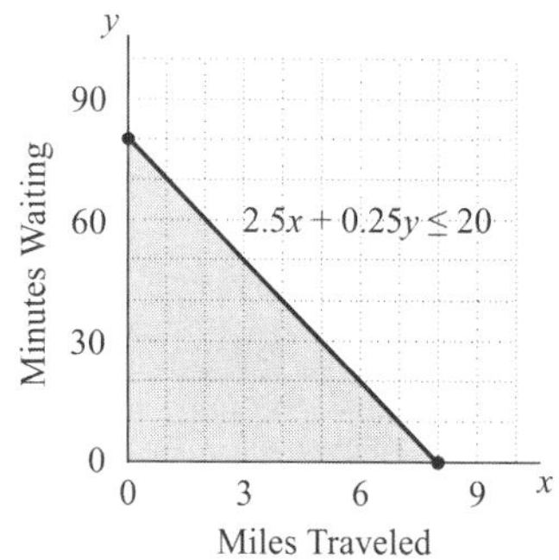

69. Answers may vary

71. a. $30x + 0.15y \le 500$

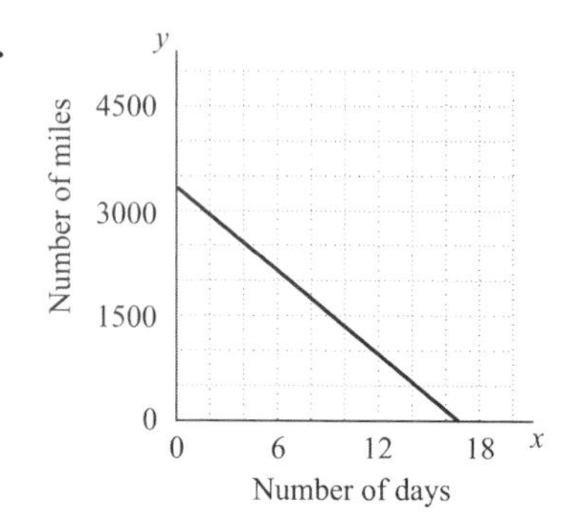

c. Answers may vary

Section 4.6

Practice Exercises

1. $\begin{cases} 4x \le y \\ x + 3y \ge 9 \end{cases}$

Graph $4x \le y$ with a solid line.
Test (1, 0)

$4(1) \overset{?}{\le} 0$

False
Shade above.
Graph $x + 3y \ge 9$ with a solid line.
Test (0, 0)

$0 + 3(0) \overset{?}{\ge} 9$

False
Shade above.
The solution of the system is the darker shaded region and includes parts of both boundary lines.

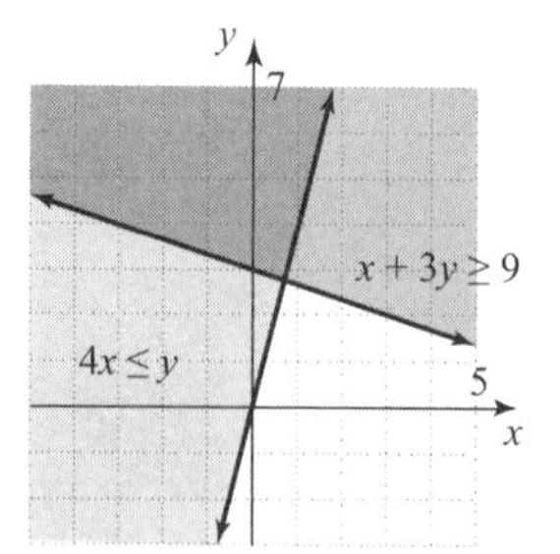

2. $\begin{cases} x - y > 4 \\ x + 3y < -4 \end{cases}$

Graph both inequalities using dashed lines. The solution of the system is the darker shaded region which does not include any of the boundary lines.

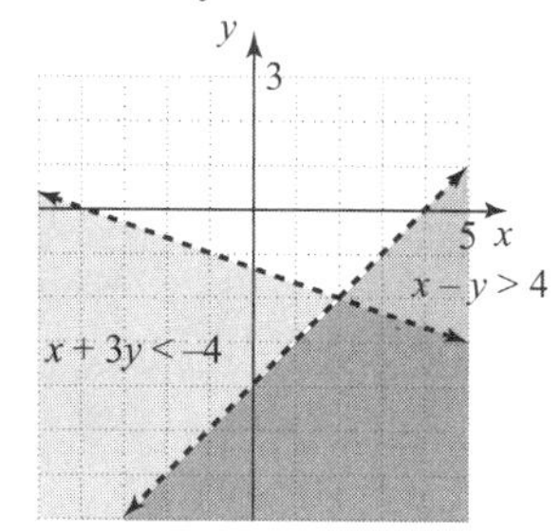

3. $\begin{cases} y \le 6 \\ -2x + 5y > 10 \end{cases}$

Graph both inequalities. The solution of the system is the darker shaded region.

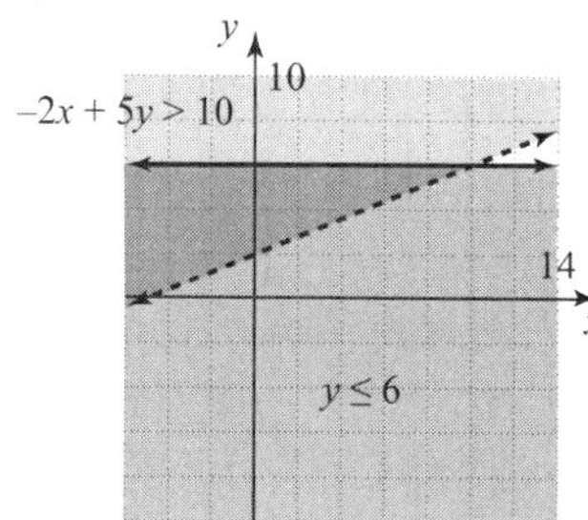

Exercise Set 4.6

1. $\begin{cases} y \ge x + 1 \\ y \ge 3 - x \end{cases}$

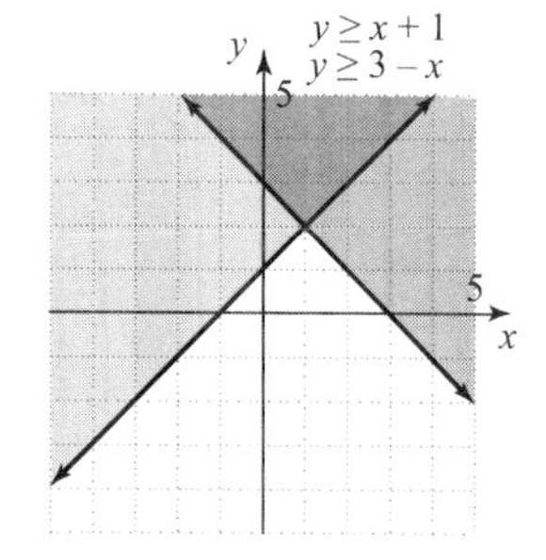

3. $\begin{cases} y < 3x - 4 \\ y \le x + 2 \end{cases}$

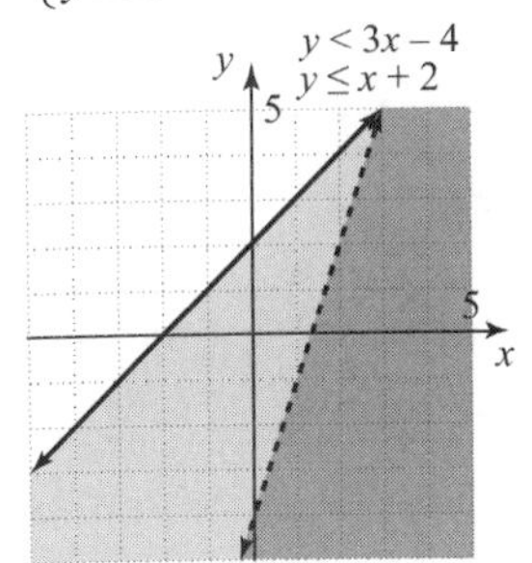

5. $\begin{cases} y \le -2x - 2 \\ y \ge x + 4 \end{cases}$

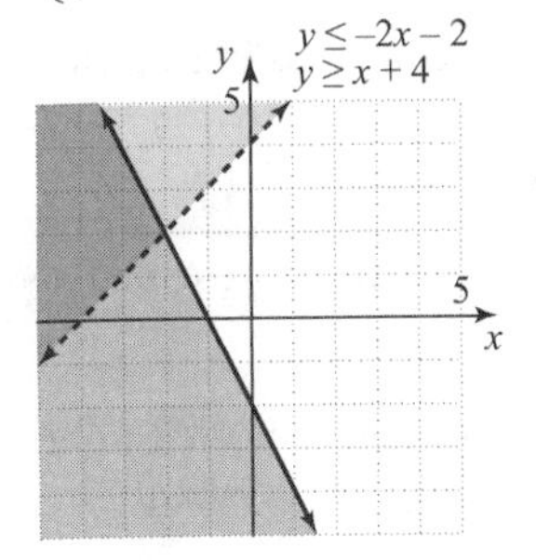

7. $\begin{cases} y \ge -x + 2 \\ y \le 2x + 5 \end{cases}$

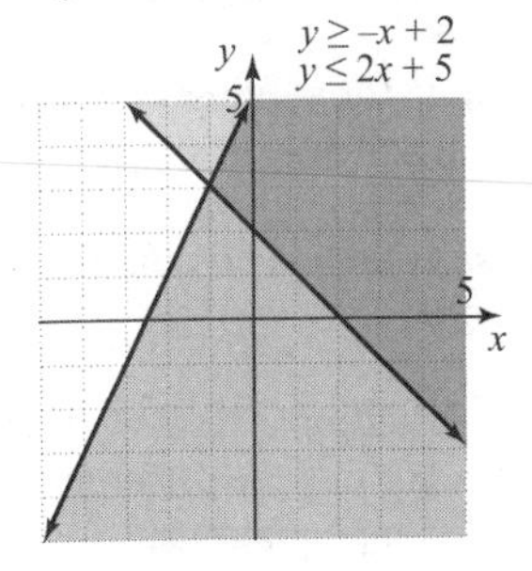

9. $\begin{cases} x \ge 3y \\ x + 3y \le 6 \end{cases}$

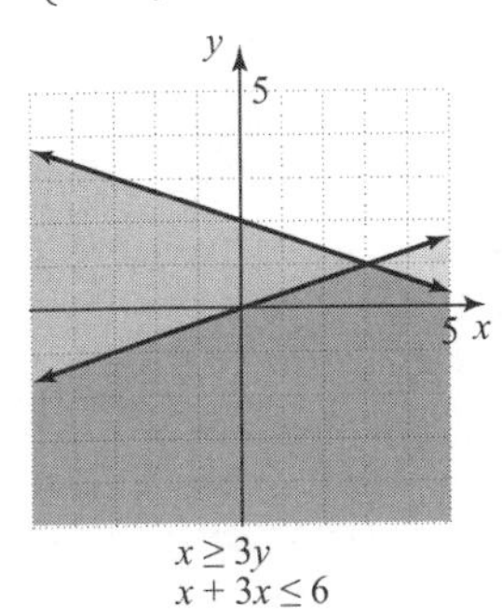

11. $\begin{cases} y + 2x \ge 0 \\ 5x - 3y \le 12 \end{cases}$

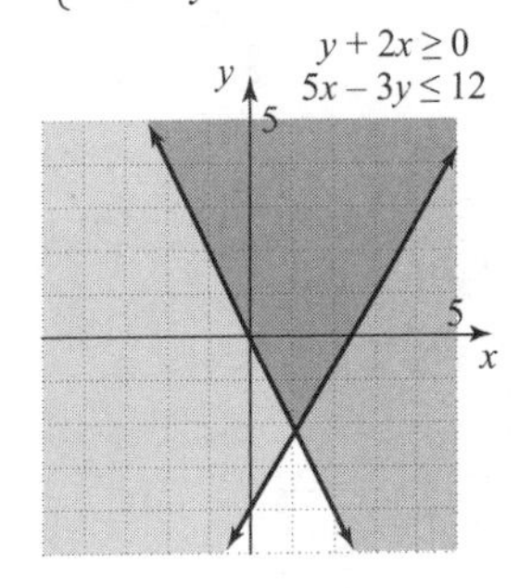

13. $\begin{cases} 3x - 4y \ge -6 \\ 2x + y \le 7 \end{cases}$

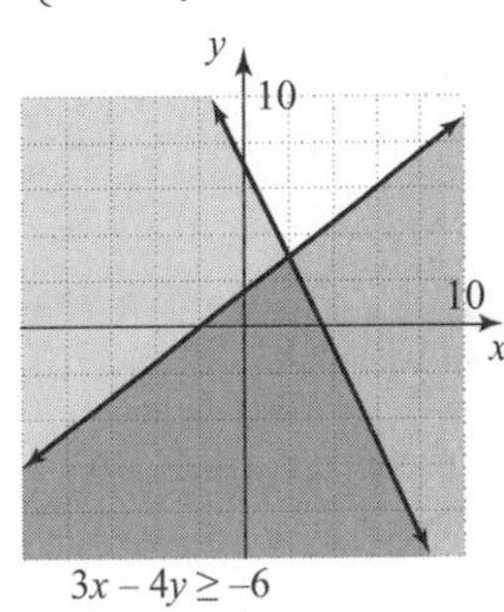

15. $\begin{cases} x \le 2 \\ y \ge -3 \end{cases}$

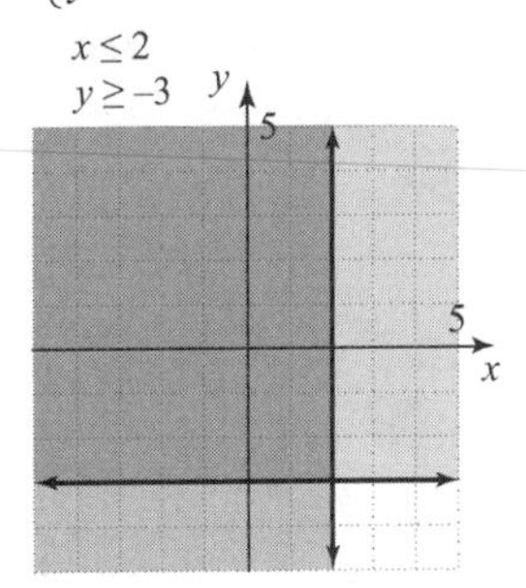

17. $\begin{cases} y \ge 1 \\ x < -3 \end{cases}$

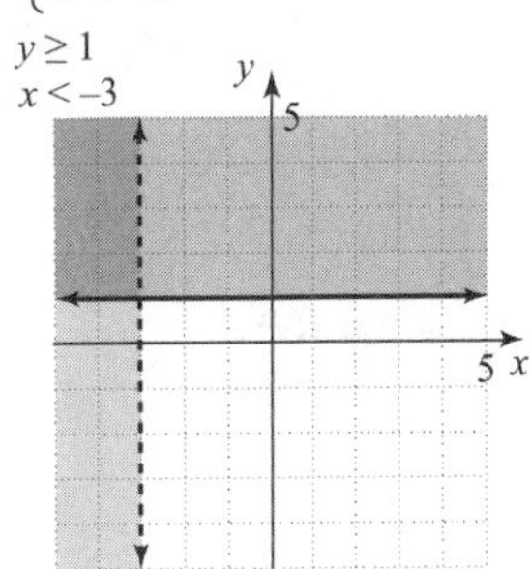

19. $\begin{cases} 2x+3y<-8 \\ x \geq -4 \end{cases}$

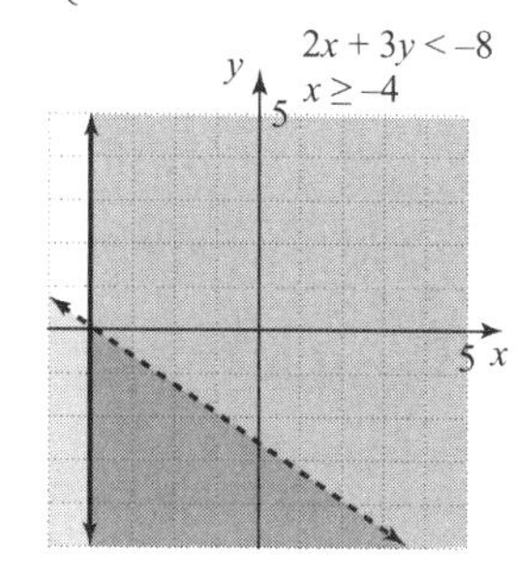

21. $\begin{cases} 2x-5y \leq 9 \\ y \leq -3 \end{cases}$

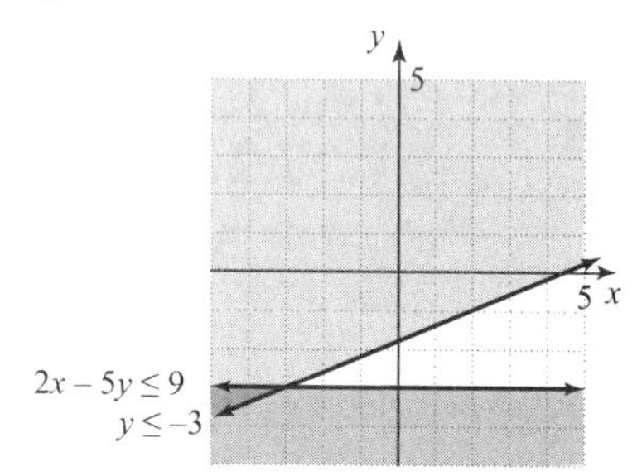

23. $\begin{cases} y \geq \frac{1}{2}x+2 \\ y \leq \frac{1}{2}x-3 \end{cases}$

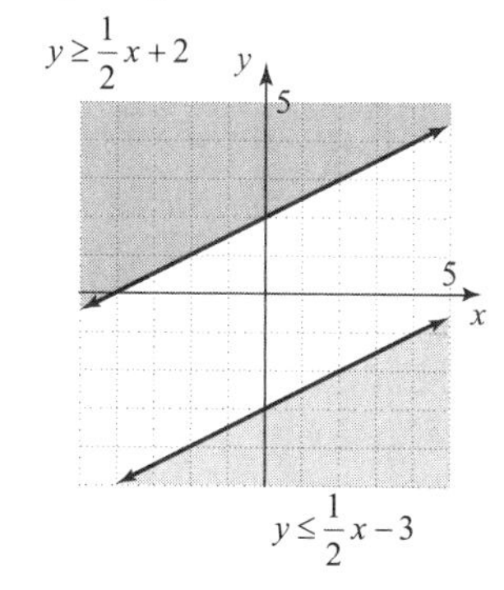

25. $4^2 = (4)(4) = 16$

27. $(6x)^2 = (6x)(6x) = 36x^2$

29. $(10y^3)^2 = (10y^3)(10y^3) = 100y^6$

31. C

33. D

35. Answers may vary

37. $\begin{cases} 2x-y \leq 6 \\ x \geq 3 \\ y > 2 \end{cases}$

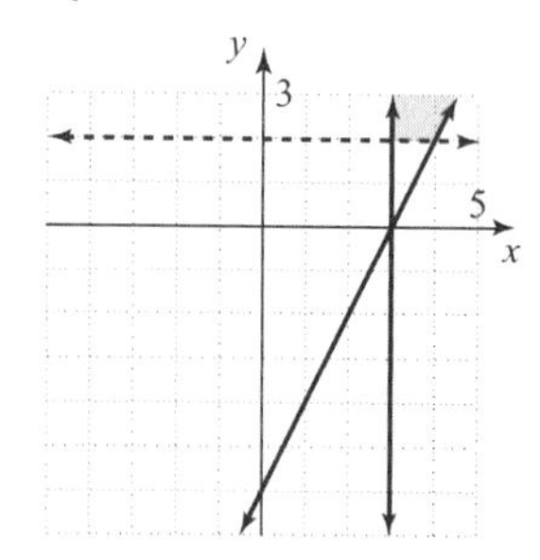

Chapter 4 Vocabulary Check

1. In a system of linear equations in two variables, if the graphs of the equations are the same, the equations are dependent equations.
2. Two or more linear equations are called a system of linear equations.
3. A system of equations that has at least one solution is called a consistent system.
4. A solution of a system of two equations in two variables is an ordered pair of numbers that is a solution of both equations in the system.
5. Two algebraic methods for solving systems of equations are addition and substitution.
6. A system of equations that has no solution is called an inconsistent system.
7. In a system of linear equations in two variables, if the graphs of the equations are different, the equations are independent equations.
8. Two or more linear inequalities are called a system of linear inequalities.

Chapter 4 Review

1. a. Let $x = 12$ and $y = 4$.

$$2x-3y=12$$
$$2(12)-3(4) \stackrel{?}{=} 12$$
$$24-12 \stackrel{?}{=} 12$$
$$12=12 \quad \text{True}$$

$$3x+4y=1$$
$$3(12)+4(4) \stackrel{?}{=} 1$$
$$36+16 \stackrel{?}{=} 1$$
$$52=1 \quad \text{False}$$

(12, 4) is not a solution of the system.

b. Let $x = 3$ and $y = -2$.

$$\begin{aligned} 2x - 3y &= 12 \\ 2(3) - 3(-2) &\stackrel{?}{=} 12 \\ 6 + 6 &\stackrel{?}{=} 12 \\ 2 &= 12 \quad \text{True} \end{aligned}$$

$$\begin{aligned} 3x + 4y &= 1 \\ 3(3) + 4(-2) &\stackrel{?}{=} 1 \\ 9 - 8 &\stackrel{?}{=} 1 \\ 1 &= 1 \quad \text{True} \end{aligned}$$

$(3, -2)$ is a solution of the system.

c. Let $x = -3$ and $y = 6$.

$$\begin{aligned} 2x - 3y &= 12 \\ 2(-3) - 3(6) &\stackrel{?}{=} 12 \\ -6 - 18 &\stackrel{?}{=} 12 \\ -24 &= 12 \quad \text{False} \end{aligned}$$

$$\begin{aligned} 3x + 4y &= 1 \\ 3(-3) + 4(6) &\stackrel{?}{=} 1 \\ -9 + 24 &\stackrel{?}{=} 1 \\ 15 &= 1 \quad \text{False} \end{aligned}$$

$(-3, 6)$ is not a solution of the system.

2. a. Let $x = \frac{3}{4}$ and $y = -3$.

$$\begin{aligned} 4x + y &= 0 \\ 4\left(\frac{3}{4}\right) - 3 &\stackrel{?}{=} 0 \\ 3 - 3 &\stackrel{?}{=} 0 \\ 0 &= 0 \quad \text{True} \end{aligned}$$

$$\begin{aligned} -8x - 5y &= 9 \\ -8\left(\frac{3}{4}\right) - 5(-3) &\stackrel{?}{=} 9 \\ -6 + 15 &\stackrel{?}{=} 9 \\ 9 &= 9 \quad \text{True} \end{aligned}$$

$\left(\frac{3}{4}, -3\right)$ is a solution of the system.

b. Let $x = -2$ and $y = 8$.

$$\begin{aligned} 4x + y &= 0 \\ 4(-2) + 8 &\stackrel{?}{=} 0 \\ -8 + 8 &\stackrel{?}{=} 0 \\ 0 &= 0 \quad \text{True} \end{aligned}$$

$$\begin{aligned} -8x - 5y &= 9 \\ -8(-2) - 5(8) &\stackrel{?}{=} 9 \\ 16 - 40 &\stackrel{?}{=} 9 \\ -24 &= 9 \quad \text{False} \end{aligned}$$

$(-2, 8)$ is not a solution of the system.

c. Let $x = \frac{1}{2}$ and $y = -2$.

$$\begin{aligned} 4x + y &= 0 \\ 4\left(\frac{1}{2}\right) - 2 &\stackrel{?}{=} 0 \\ 2 - 2 &\stackrel{?}{=} 0 \\ 0 &= 0 \quad \text{True} \end{aligned}$$

$$\begin{aligned} -8x - 5y &= 9 \\ -8\left(\frac{1}{2}\right) - 5(-2) &\stackrel{?}{=} 9 \\ -4 + 10 &\stackrel{?}{=} 9 \\ 6 &= 9 \quad \text{False} \end{aligned}$$

$\left(\frac{1}{2}, -2\right)$ is not a solution of the system.

3. a. Let $x = -6$ and $y = -8$.

$$\begin{aligned} 5x - 6y &= 18 \\ 5(-6) - 6(-8) &\stackrel{?}{=} 18 \\ -30 + 48 &\stackrel{?}{=} 18 \\ 18 &= 18 \quad \text{True} \end{aligned}$$

$$\begin{aligned} 2y - x &= -4 \\ 2(-8) - (-6) &\stackrel{?}{=} -4 \\ -16 + 6 &\stackrel{?}{=} -4 \\ -10 &= -4 \quad \text{False} \end{aligned}$$

$(-6, -8)$ is not a solution of the system.

b. Let $x = 3$ and $y = \frac{5}{2}$.

$$\begin{aligned} 5x - 6y &= 18 \\ 5(3) - 6\left(\frac{5}{2}\right) &\stackrel{?}{=} 18 \\ 15 - 15 &\stackrel{?}{=} 18 \\ 0 &= 18 \quad \text{False} \end{aligned}$$

$$\begin{aligned} 2y - x &= -4 \\ 2\left(\frac{5}{2}\right) - 3 &\stackrel{?}{=} -4 \\ 5 - 3 &\stackrel{?}{=} -4 \\ 2 &= -4 \quad \text{False} \end{aligned}$$

$\left(3, \frac{5}{2}\right)$ is not a solution of the system.

c. Let $x = 3$ and $y = -\frac{1}{2}$.

$$5x - 6y = 18$$
$$5(3) - 6\left(-\frac{1}{2}\right) \stackrel{?}{=} 18$$
$$15 + 3 \stackrel{?}{=} 18$$
$$18 = 18 \quad \text{True}$$

$$2y - x = -4$$
$$2\left(-\frac{1}{2}\right) - 3 \stackrel{?}{=} -4$$
$$-1 - 3 \stackrel{?}{=} -4$$
$$-4 = -4 \quad \text{True}$$

$\left(3, -\frac{1}{2}\right)$ is a solution of the system.

4. a. Let $x = 2$ and $y = 2$.

$$2x + 3y = 1 \qquad 3y - x = 4$$
$$2(2) + 3(2) \stackrel{?}{=} 1 \qquad 3(2) - 2 \stackrel{?}{=} 4$$
$$4 + 6 \stackrel{?}{=} 1 \qquad 6 - 2 \stackrel{?}{=} 4$$
$$10 = 1 \quad \text{False} \qquad 4 = 4 \quad \text{True}$$

(2, 2) is not a solution of the system.

b. Let $x = -1$ and $y = 1$.

$$2x + 3y = 1 \qquad 3y - x = 4$$
$$2(-1) + 3(1) \stackrel{?}{=} 1 \qquad 3(1) - (-1) \stackrel{?}{=} 4$$
$$-2 + 3 \stackrel{?}{=} 1 \qquad 3 + 1 \stackrel{?}{=} 4$$
$$1 = 1 \quad \text{True} \qquad 4 = 4 \quad \text{True}$$

(−1, 1) is a solution of the system.

c. Let $x = 2$ and $y = -1$.

$$2x + 3y = 1$$
$$2(2) + 3(-1) \stackrel{?}{=} 1$$
$$4 - 3 \stackrel{?}{=} 1$$
$$1 = 1 \quad \text{True}$$

$$3y - x = 4$$
$$3(-1) - 2 \stackrel{?}{=} 4$$
$$-3 - 2 \stackrel{?}{=} 4$$
$$-5 = 4 \quad \text{False}$$

(2, −1) is not a solution of the system.

5. $\begin{cases} x + y = 5 \\ x - 1 = y \end{cases}$

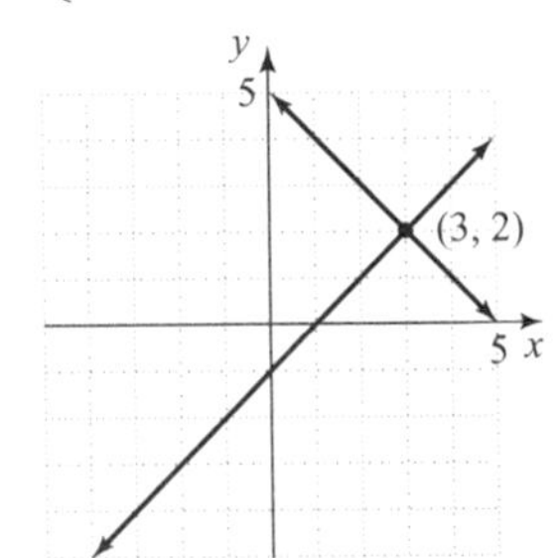

The solution of the system is (3, 2).

6. $\begin{cases} x + y = 3 \\ x - y = -1 \end{cases}$

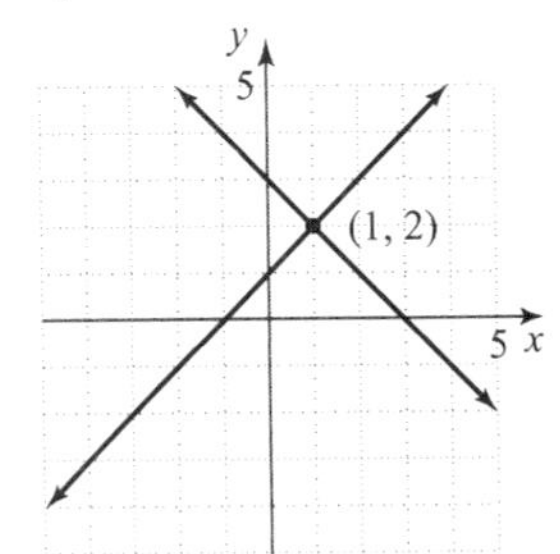

The solution of the system is (1, 2).

7. $\begin{cases} x = 5 \\ y = -1 \end{cases}$

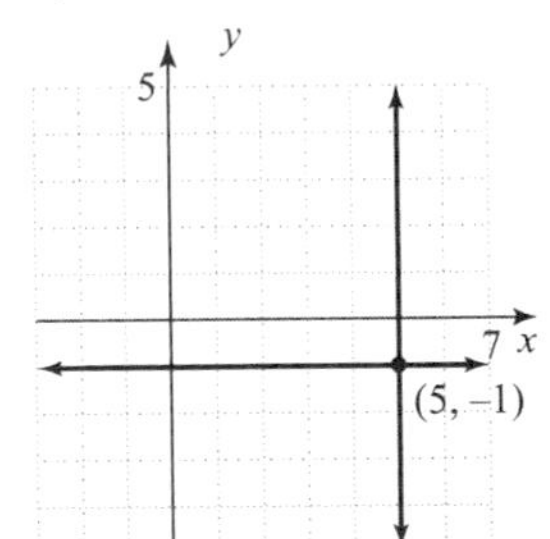

The solution of the system is (5, −1).

8. $\begin{cases} x = -3 \\ y = 2 \end{cases}$

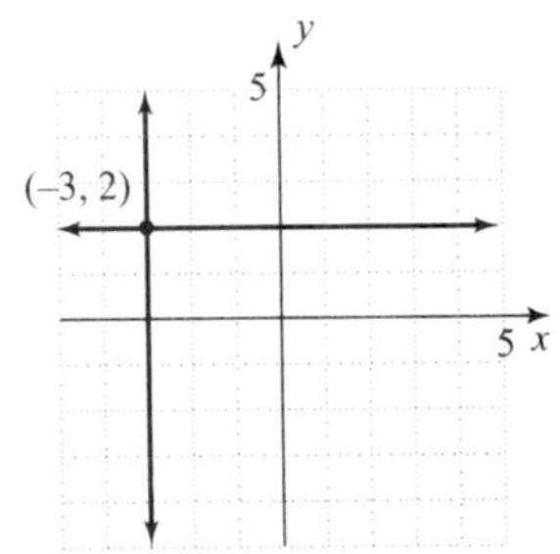

The solution of the system is (−3, 2).

9. $\begin{cases} 2x + y = 5 \\ x = -3y \end{cases}$

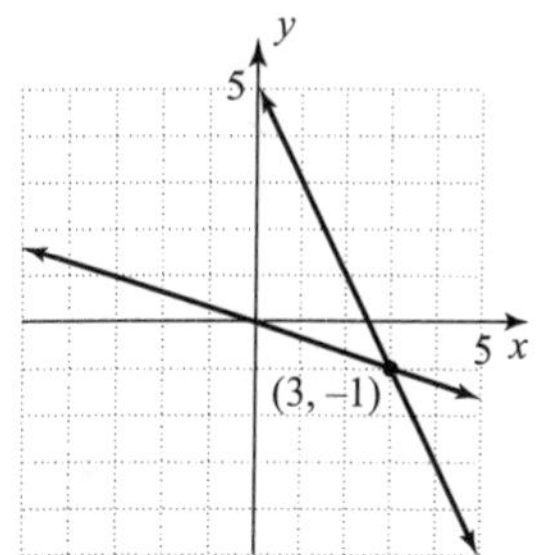

The solution of the system is (3, −1).

10. $\begin{cases} 3x + y = -2 \\ y = -5x \end{cases}$

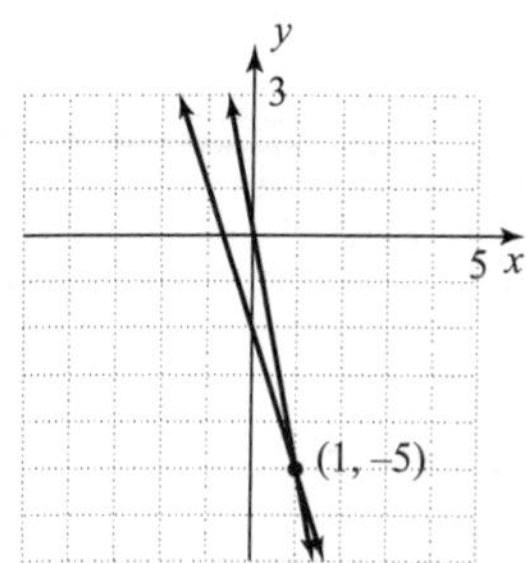

The solution of the system is (1, −5).

11. $\begin{cases} y = 3x \\ -6x + 2y = 6 \end{cases}$

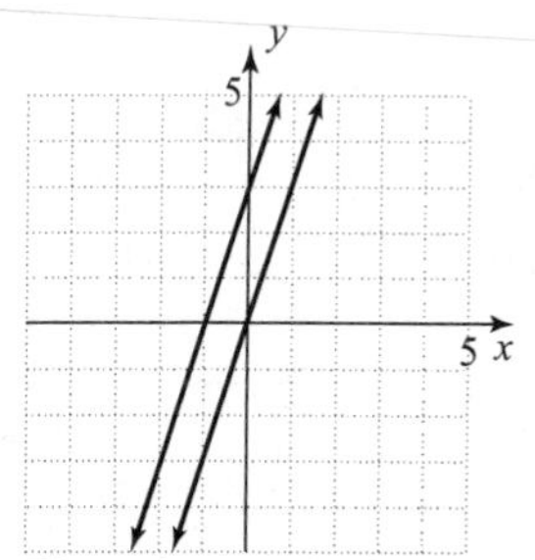

There is no solution.

12. $\begin{cases} x - 2y = 2 \\ -2x + 4y = -4 \end{cases}$

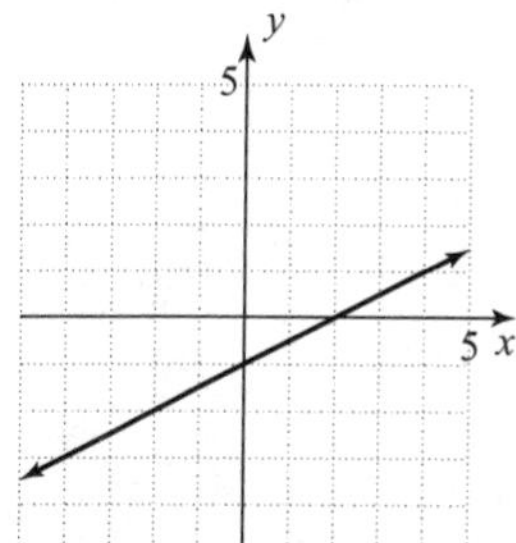

There is an infinite number of solutions.

13. $\begin{cases} y = 2x + 6 \\ 3x - 2y = -11 \end{cases}$

Substitute $2x + 6$ for y in the second equation.

$$\begin{aligned} 3x - 2(2x + 6) &= -11 \\ 3x - 4x - 12 &= -11 \\ -x &= 1 \\ x &= -1 \end{aligned}$$

Let $x = -1$ in the first equation.
$y = 2(-1) + 6 = 4$
The solution is (−1, 4).

14. $\begin{cases} y = 3x - 7 \\ 2x - 3y = 7 \end{cases}$

Substitute $3x - 7$ for y in the second equation.

$$\begin{aligned} 2x - 3(3x - 7) &= 7 \\ 2x - 9x + 21 &= 7 \\ -7x &= -14 \\ x &= 2 \end{aligned}$$

Let $x = 2$ in the first equation.
$y = 3(2) - 7 = -1$
The solution is (2, −1).

15. $\begin{cases} x + 3y = -3 \\ 2x + y = 4 \end{cases}$

Solve the first equation for x.
$x = -3y - 3$
Substitute $-3y - 3$ for x in the second equation.

$$\begin{aligned} 2(-3y - 3) + y &= 4 \\ -6y - 6 + y &= 4 \\ -5y &= 10 \\ y &= -2 \end{aligned}$$

Let $y = -2$ in $x = -3y - 3$.
$x = -3(-2) - 3 = 3$
The solution is (3, −2).

16. $\begin{cases} 3x + y = 11 \\ x + 2y = 12 \end{cases}$

Solve the first equation for y.
$y = 11 - 3x$
Substitute $11 - 3x$ for y in the second equation.

$$\begin{aligned} x + 2(11 - 3x) &= 12 \\ x + 22 - 6x &= 12 \\ -5x &= -10 \\ x &= 2 \end{aligned}$$

Let $x = 2$ in $y = 11 - 3x$.
$y = 11 - 3(2) = 5$
The solution is (2, 5).

17. $\begin{cases} 4y = 2x + 6 \\ x - 2y = -3 \end{cases}$

Solve the second equation for x.

$x = 2y - 3$

Substitute $2y - 3$ for x in the first equation.

$$4y = 2(2y - 3) + 6$$
$$4y = 4y - 6 + 6$$
$$0 = 0$$

The system has an infinite number of solutions.

18. $\begin{cases} 9x = 6y + 3 \\ 6x - 4y = 2 \end{cases}$

Solve the first equation for y.

$$9x = 6y + 3$$
$$9x - 3 = 6y$$
$$\frac{3}{2}x - \frac{1}{2} = y$$

Substitute $\frac{3}{2}x - \frac{1}{2}$ for y in the second equation.

$$6x - 4\left(\frac{3}{2}x - \frac{1}{2}\right) = 2$$
$$6x - 6x + 2 = 2$$
$$2 = 2$$

The system has an infinite number of solutions.

19. $\begin{cases} x + y = 6 \\ y = -x - 4 \end{cases}$

Substitute $-x - 4$ for y in the first equation.

$$x + (-x - 4) = 6$$
$$x - x - 4 = 6$$
$$-4 = 6 \quad \text{False}$$

There is no solution.

20. $\begin{cases} -3x + y = 6 \\ y = 3x + 2 \end{cases}$

Substitute $3x + 2$ for y in the first equation.

$$-3x + (3x + 2) = 6$$
$$-3x + 3x + 2 = 6$$
$$2 = 6 \quad \text{False}$$

There is no solution.

21. $\begin{cases} 2x + 3y = -6 \\ x - 3y = -12 \end{cases}$

$$\begin{array}{r} 2x + 3y = -6 \\ x - 3y = -12 \\ \hline 3x \quad\quad = -18 \end{array}$$
$$x = -6$$

Let $x = -6$ in the first equation.

$$2(-6) + 3y = -6$$
$$-12 + 3y = -6$$
$$3y = 6$$
$$y = 2$$

The solution of the system is $(-6, 2)$.

22. $\begin{cases} 4x + y = 15 \\ -4x + 3y = -19 \end{cases}$

$$\begin{array}{r} 4x \ + y = 15 \\ -4x + 3y = -19 \\ \hline 4y = -4 \end{array}$$
$$y = -1$$

Let $y = -1$ in the first equation.

$$4x + (-1) = 15$$
$$4x - 1 = 15$$
$$4x = 16$$
$$x = 4$$

The solution of the system is $(4, -1)$.

23. $\begin{cases} 2x - 3y = -15 \\ x + 4y = 31 \end{cases}$

Multiply the second equation by -2.

$$\begin{array}{r} 2x - 3y = -15 \\ -2x - 8y = -62 \\ \hline -11y = -77 \end{array}$$
$$y = 7$$

Let $y = 7$ in the second equation.

$$x + 4(7) = 31$$
$$x + 28 = 31$$
$$x = 3$$

The solution of the system is $(3, 7)$.

24. $\begin{cases} x - 5y = -22 \\ 4x + 3y = 4 \end{cases}$

Multiply the first equation by -4.

$$\begin{array}{r} -4x + 20y = 88 \\ 4x \ + 3y = 4 \\ \hline 23y = 92 \end{array}$$
$$y = 4$$

Let $y = 4$ in the first equation.

$$x - 5(4) = -22$$
$$x - 20 = -22$$
$$x = -2$$

The solution of the system is $(-2, 4)$.

25. $\begin{cases} 2x - 6y = -1 \\ -x + 3y = \frac{1}{2} \end{cases}$

Multiply the second equation by 2.

$$\begin{array}{r} 2x-6y=-1 \\ -2x+6y=1 \\ \hline 0=0 \end{array}$$

There is an infinite number of solutions.

26. $\begin{cases} 0.6x-0.3y=-1.5 \\ 0.04x-0.02y=-0.1 \end{cases}$

Multiply the first equation by 20 and the second equation by −300.

$$\begin{array}{r} 12x-6y=-30 \\ -12x+6y=30 \\ \hline 0=0 \end{array}$$

There are an infinite number of solutions.

27. $\begin{cases} \frac{3}{4}x+\frac{2}{3}y=2 \\ x+\frac{y}{3}=6 \end{cases}$

Multiply the first equation by 12 and the second equation by 3.

$\begin{cases} 9x+8y=24 \\ 3x+y=18 \end{cases}$

Multiply the second equation in the simplified system by −3.

$$\begin{array}{r} 9x+8y=24 \\ -9x-3y=-54 \\ \hline 5y=-30 \\ y=-6 \end{array}$$

Let $y = -6$ in $3x + y = 18$.

$$3x+(-6)=18$$
$$3x=24$$
$$x=8$$

The solution of the system is (8, −6).

28. $\begin{cases} 10x+2y=0 \\ 3x+5y=33 \end{cases}$

Multiply the first equation by −5 and the second equation by 2.

$$\begin{array}{r} -50x-10y=0 \\ 6x+10y=66 \\ \hline -44x\qquad=66 \\ x=-\frac{3}{2} \end{array}$$

Let $x=-\frac{3}{2}$ in the first equation.

$$10\left(-\frac{3}{2}\right)+2y=0$$
$$-15+2y=0$$
$$2y=15$$
$$y=\frac{15}{2}$$

The solution is $\left(-\frac{3}{2}, \frac{15}{2}\right)$.

29. Let x = the larger number and y = the smaller number.

$\begin{cases} x+y=16 \\ 3x-y=72 \end{cases}$

$$\begin{array}{r} x+y=16 \\ 3x-y=72 \\ \hline 4x\qquad=88 \\ x=22 \end{array}$$

Let $x = 22$ in the first equation.

$$22+y=16$$
$$y=-6$$

The numbers are −6 and 22.

30. Let x = the number of orchestra seats and y = the number of balcony seats.

$\begin{cases} x+y=360 \\ 45x+35y=15,150 \end{cases}$

Solve the first equation for x.

$x = 360 - y$

Substitute $360 - y$ for x in the second equation.

$$45(360-y)+35y=15,150$$
$$16,200-45y+35y=15,150$$
$$-10y=-1050$$
$$y=105$$

Let $y = 105$ in $x = 360 - y$.

$x = 360 - 105 = 255$

There are 255 orchestra seats and 105 balcony seats.

31. Let x = the riverboat's speed in still water and y = the rate of the current.

	d =	r ·	t
Downriver	340	$x+y$	14
Upriver	340	$x-y$	19

$\begin{cases} 14(x+y)=340 \\ 19(x-y)=340 \end{cases}$

Multiply the first equation by $\frac{1}{14}$ and the second equation by $\frac{1}{19}$.

$$\begin{aligned} x+y &= \frac{340}{14} \approx 24.29 \\ x-y &= \frac{340}{19} \approx 17.89 \\ \hline 2x &\approx 42.18 \\ x &\approx 21.09 \end{aligned}$$

Multiply the second equation of the simplified system by −1.

$$\begin{aligned} x+y &\approx 24.29 \\ -x+y &\approx -17.89 \\ \hline 2y &\approx 6.4 \\ y &\approx 3.2 \end{aligned}$$

The riverboat's speed in still water is 21.1 miles per hour. The rate of the current is 3.2 miles per hour.

32. Let x = amount of 6% solution and y = amount of 14% solution.

Concentration Rate	Amount of Solution	Amount of Pure Acid
0.06	x	$0.06x$
0.14	y	$0.14y$
0.12	50	0.12(50)

$$\begin{cases} x+y=50 \\ 0.06x+0.14y=0.12(50) \end{cases}$$

Multiply the first equation by −6 and the second equation by 100.

$$\begin{aligned} -6x-6y &= -300 \\ 6x+14y &= 600 \\ \hline 8y &= 300 \\ y &= 37.5 \end{aligned}$$

Let $y = 37.5$ in the first equation.

$$\begin{aligned} x+37.5 &= 50 \\ x &= 12.5 \end{aligned}$$

$12\frac{1}{2}$ cc of 6% solution and $37\frac{1}{2}$ cc of 14% solution.

33. Let x = the cost of an egg and y = the cost of a strip of bacon.

$$\begin{cases} 3x+4y=3.80 \\ 2x+3y=2.75 \end{cases}$$

Multiply the first equation by −2 and the second equation by 3.

$$\begin{aligned} -6x-8y &= -7.60 \\ 6x+9y &= 8.25 \\ \hline y &= 0.65 \end{aligned}$$

Let $y = 0.65$ in the first equation.

$$\begin{aligned} 3x+4(0.65) &= 3.80 \\ 3x+2.60 &= 3.80 \\ 3x &= 1.20 \\ x &= 0.40 \end{aligned}$$

An egg costs 40¢ and a strip of bacon costs 65¢.

34. Let x = the time spent walking and y = the time spent jogging.

	r	$\cdot\ t$	$= d$
Walking	4	x	$4x$
Jogging	7.5	y	$7.5y$

$$\begin{cases} x+y=3 \\ 4x+7.5y=15 \end{cases}$$

Multiply the first equation by −4.

$$\begin{aligned} -4x-4y &= -12 \\ 4x+7.5y &= 15 \\ \hline 3.5y &= 3 \\ y &\approx 0.857 \end{aligned}$$

Let $y = 0.857$ in the first equation.

$$\begin{aligned} x+0.857 &= 3 \\ x &\approx 2.143 \end{aligned}$$

He spent 2.14 hours walking and 0.86 hours jogging.

35. $5x + 4y < 20$

Test (0, 0)

$0+0 \stackrel{?}{<} 20$

True

Shade below.

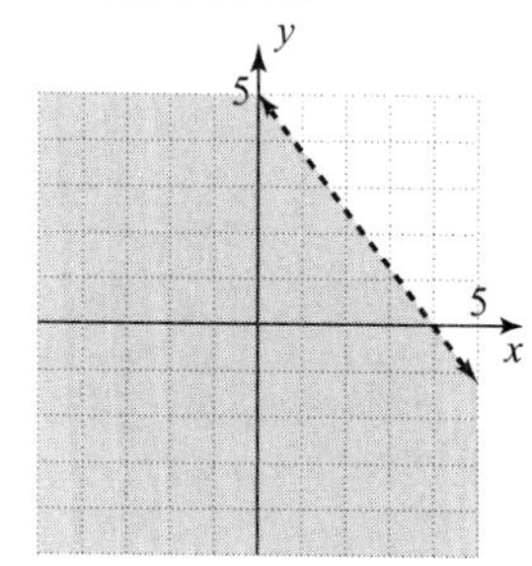

36. $x + 3y > 4$
Test (0, 0)

$0 + 0 \overset{?}{>} 4$

False
Shade above.

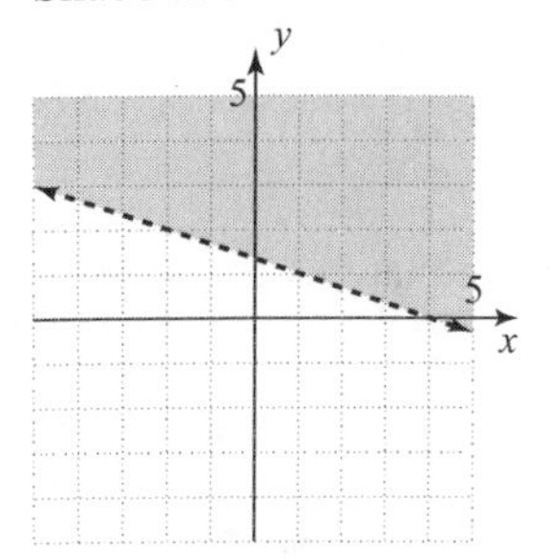

37. $y \geq -7$
Shade above.

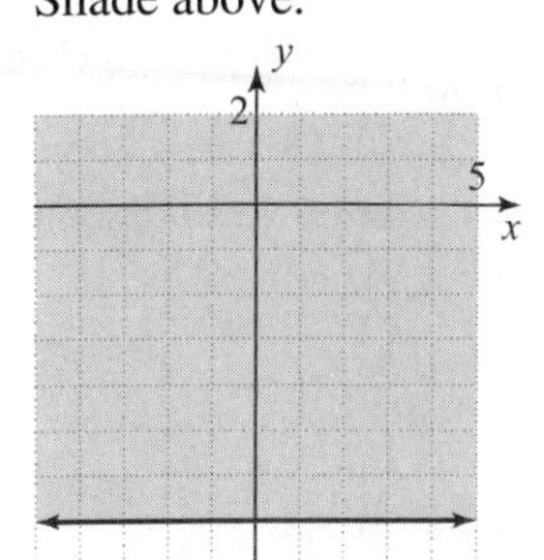

38. $y \leq -4$
Shade below.

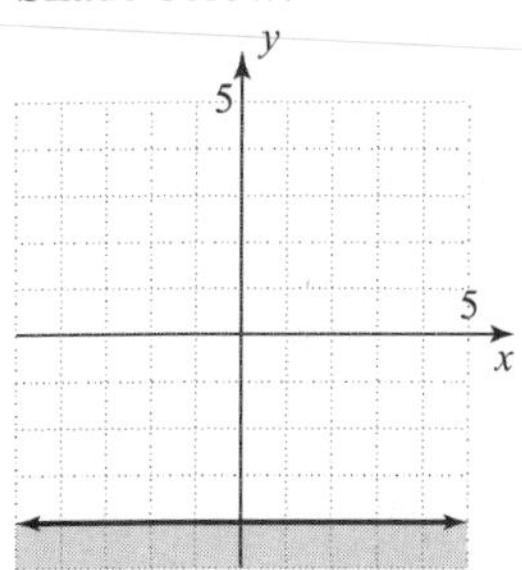

39. $-x \leq y$
Test (1, 0)

$-1 \overset{?}{\leq} 0$

True
Shade above.

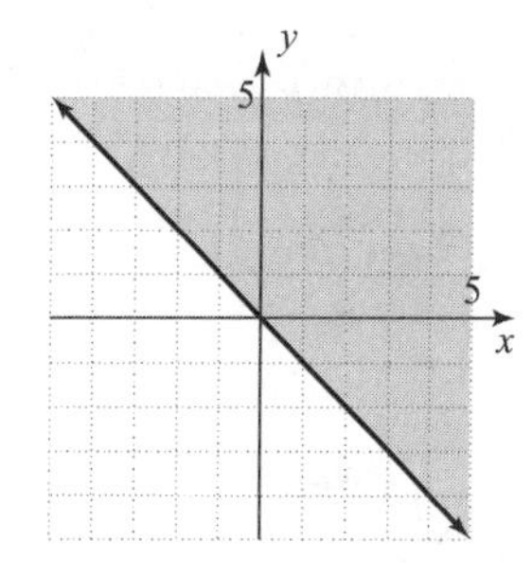

40. $x \geq -y$
Test (1, 0)

$1 \overset{?}{\geq} 0$

True
Shade above.

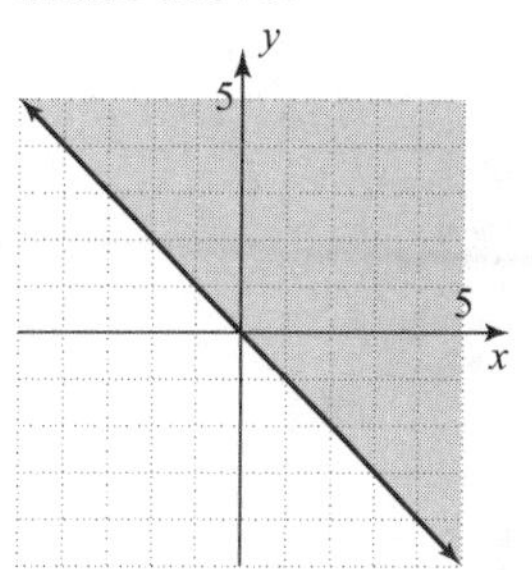

41. $\begin{cases} y \geq 2x - 3 \\ y \leq -2x + 1 \end{cases}$

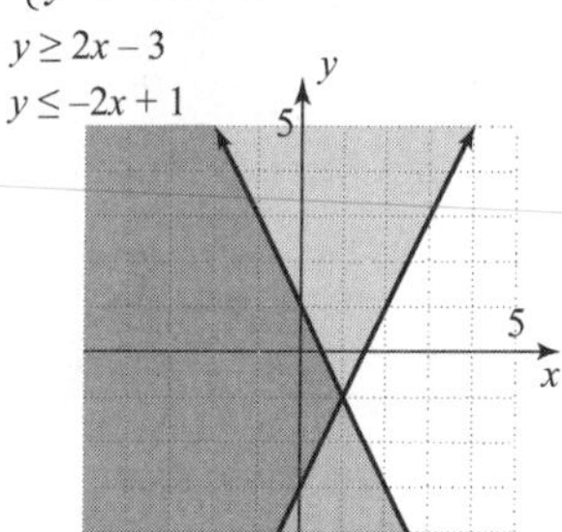

42. $\begin{cases} y \leq -3x - 3 \\ y \leq 2x + 7 \end{cases}$

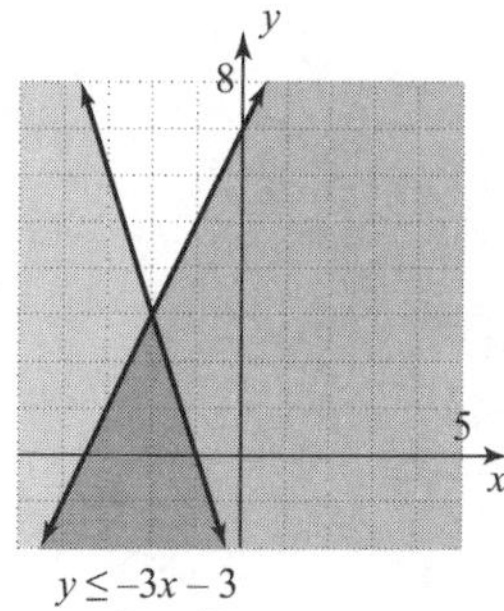

43. $\begin{cases} -3x+2y>-1 \\ y<-2 \end{cases}$

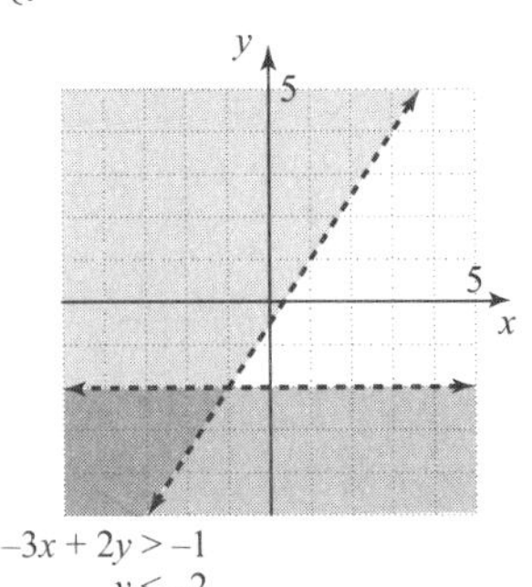

44. $\begin{cases} -2x+3y>-7 \\ x\ge -2 \end{cases}$

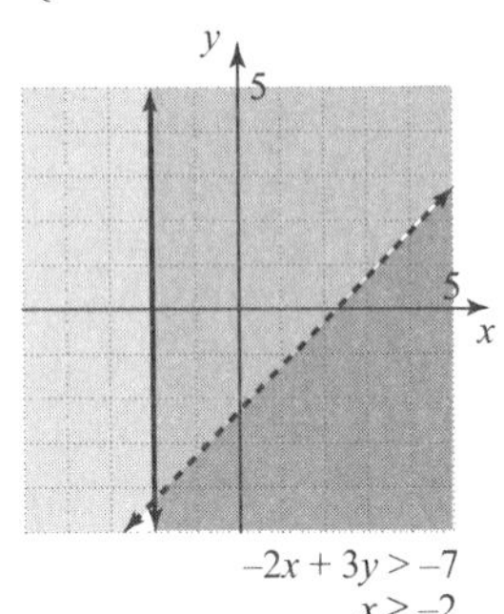

45. $\begin{cases} x-2y=1 \\ 2x+3y=-12 \end{cases}$

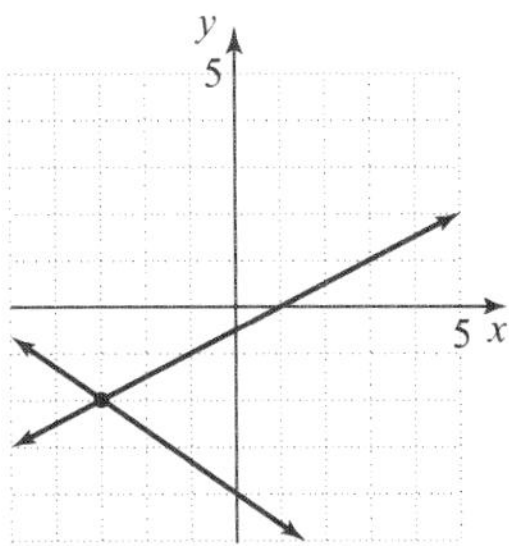

The solution is (−3, −2).

46. $\begin{cases} 3x-y=-4 \\ 6x-2y=-8 \end{cases}$

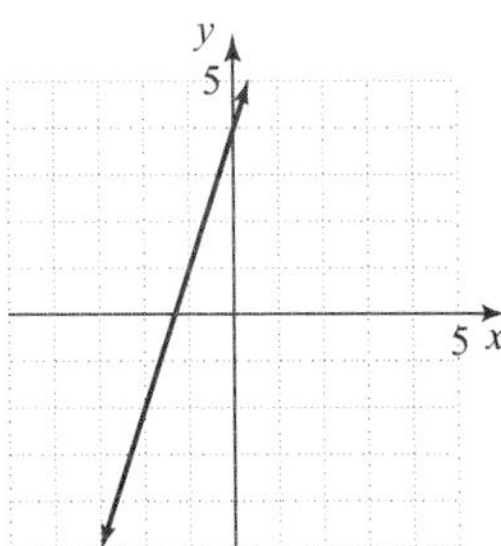

There is an infinite number of solutions.

47. $\begin{cases} x+4y=11 \\ 5x-9y=-3 \end{cases}$

Solve the first equation for x.

$x = 11 - 4y$

Substitute $11 - 4y$ for x in the second equation.

$$\begin{aligned} 5(11-4y)-9y &= -3 \\ 55-20y-9y &= -3 \\ -29y &= -58 \\ y &= 2 \end{aligned}$$

Let $y = 2$ in the first equation.

$$\begin{aligned} x+4(2) &= 11 \\ x+8 &= 11 \\ x &= 3 \end{aligned}$$

The solution is (3, 2).

48. $\begin{cases} x+9y=16 \\ 3x-8y=13 \end{cases}$

Solve the first equation for x.

$x = 16 - 9y$

Substitute $16 - 9y$ for x in the second equation.

$$\begin{aligned} 3(16-9y)-8y &= 13 \\ 48-27y-8y &= 13 \\ -35y &= -35 \\ y &= 1 \end{aligned}$$

Let $y = 1$ in the first equation.

$$\begin{aligned} x+9(1) &= 16 \\ x+9 &= 16 \\ x &= 7 \end{aligned}$$

The solution is (7, 1).

49. $\begin{cases} y=-2x \\ 4x+7y=-15 \end{cases}$

Substitute $-2x$ for y in the second equation.

$$\begin{aligned} 4x+7(-2x) &= -15 \\ 4x-14x &= -15 \\ -10x &= -15 \\ x &= \frac{3}{2} = 1\frac{1}{2} \end{aligned}$$

Let $x = \frac{3}{2}$ in the first equation.

$$y = -2\left(\frac{3}{2}\right) = -3$$

The solution is $\left(1\frac{1}{2}, -3\right)$.

50. $\begin{cases} 3y=2x+15 \\ -2x+3y=21 \end{cases}$

Solve the first equation for x.

$$3y = 2x + 15$$
$$3y - 15 = 2x$$
$$\frac{3}{2}y - \frac{15}{2} = x$$

Substitute $\frac{3}{2}y - \frac{15}{2}$ for x in the second equation.

$$-2\left(\frac{3}{2}y - \frac{15}{2}\right) + 3y = 21$$
$$-3y + 15 + 3y = 21$$
$$15 = 21 \quad \text{False}$$

The system has no solution.

51. $\begin{cases} 3x - y = 4 \\ 4y = 12x - 16 \end{cases}$

Solve the first equation for y.

$3x - 4 = y$

Substitute $3x - 4$ for y in the second equation.

$$4(3x - 4) = 12x - 16$$
$$12x - 16 = 12x - 16$$
$$0 = 0$$

There is an infinite number of solutions.

52. $\begin{cases} x + y = 19 \\ x - y = -3 \end{cases}$

$$x + y = 19$$
$$\underline{x - y = -3}$$
$$2x \quad = 16$$
$$x = 8$$

Let $x = 8$ in the first equation.

$$8 + y = 19$$
$$y = 11$$

The solution is (8, 11).

53. $\begin{cases} x - 3y = -11 \\ 4x + 5y = -10 \end{cases}$

Solve the first equation for x.

$x = 3y - 11$

Substitute $3y - 11$ for x in the second equation.

$$4(3y - 11) + 5y = -10$$
$$12y - 44 + 5y = -10$$
$$17y = 34$$
$$y = 2$$

Let $y = 2$ in the first equation.

$$x - 3(2) = -11$$
$$x - 6 = -11$$
$$x = -5$$

The solution is (–5, 2).

54. $\begin{cases} -x - 15y = 44 \\ 2x + 3y = 20 \end{cases}$

Solve the first equation for x.

$$-x - 15y = 44$$
$$-x = 15y + 44$$
$$x = -15y - 44$$

Substitute $-15y - 44$ for x in the second equation.

$$2(-15y - 44) + 3y = 20$$
$$-30y - 88 + 3y = 20$$
$$-27y = 108$$
$$y = -4$$

Let $y = -4$ in $x = -15y - 44$.

$x = -15(-4) - 44 = 60 - 44 = 16$

The solution is (16, –4).

55. $\begin{cases} 2x + y = 3 \\ 6x + 3y = 9 \end{cases}$

Solve the first equation for y.

$y = -2x + 3$

Substitute $-2x + 3$ for y in the second equation.

$$6x + 3(-2x + 3) = 9$$
$$6x - 6x + 9 = 9$$
$$9 = 9$$

There is an infinite number of solutions.

56. $\begin{cases} -3x + y = 5 \\ -3x + y = -2 \end{cases}$

Multiply the first equation by –1.

$$3x - y = -5$$
$$\underline{-3x + y = -2}$$
$$0 = -7 \quad \text{False}$$

There is no solution.

57. Let x = the larger number and y = the smaller number.

$\begin{cases} x + y = 12 \\ x + 3y = 20 \end{cases}$

Multiply the first equation by –1.

$$-x - y = -12$$
$$\underline{x + 3y = 20}$$
$$2y = 8$$
$$y = 4$$

Let $y = 4$ in the first equation.

$$x + 4 = 12$$
$$x = 8$$

The numbers are 4 and 8.

58. Let x = the smaller number and y = the larger number.

$$\begin{cases} x - y = -18 \\ 2x - y = -23 \end{cases}$$

Multiply the first equation by −1.

$$\begin{array}{r} -x + y = 18 \\ \underline{2x - y = -23} \\ x \quad = -5 \end{array}$$

Let $x = -5$ in the first equation.

$$-5 - y = -18$$
$$-y = -13$$
$$y = 13$$

The numbers are −5 and 13.

59. Let x = the number of nickels and y = the number of dimes.

$$\begin{cases} x + y = 65 \\ 0.05x + 0.10y = 5.30 \end{cases}$$

Multiply the first equation by −5 and the second equation by 100.

$$\begin{array}{r} -5x - 5y = -325 \\ \underline{5x + 10y = 530} \\ 5y = 205 \\ y = 41 \end{array}$$

Let $y = 41$ in the first equation.

$$x + 41 = 65$$
$$x = 24$$

There are 24 nickels and 41 dimes.

60. Let x = the number of 13¢ stamps and y = the number of 22¢ stamps.

$$\begin{cases} x + y = 26 \\ 0.13x + 0.22y = 4.19 \end{cases}$$

Multiply the first equation by −13 and the second equation by 100.

$$\begin{array}{r} -13x - 13y = -338 \\ \underline{13x + 22y = 419} \\ 9y = 81 \\ y = 9 \end{array}$$

Let $y = 9$ in the first equation.

$$x + 9 = 26$$
$$x = 17$$

They purchased 17 13¢ stamps and 9 22¢ stamps.

61. $x + 6y < 6$

Test (0, 0)

$$0 + 6(0) \stackrel{?}{<} 6$$

True

Shade below.

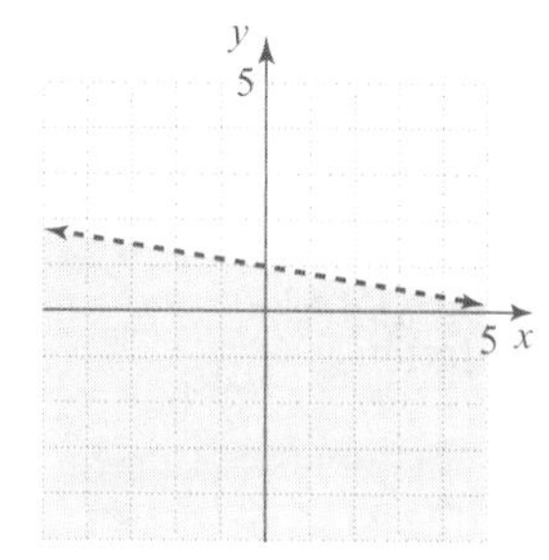

62. $x + y > -2$

Test (0, 0)

$$0 + 0 \stackrel{?}{>} -2$$

True

Shade above.

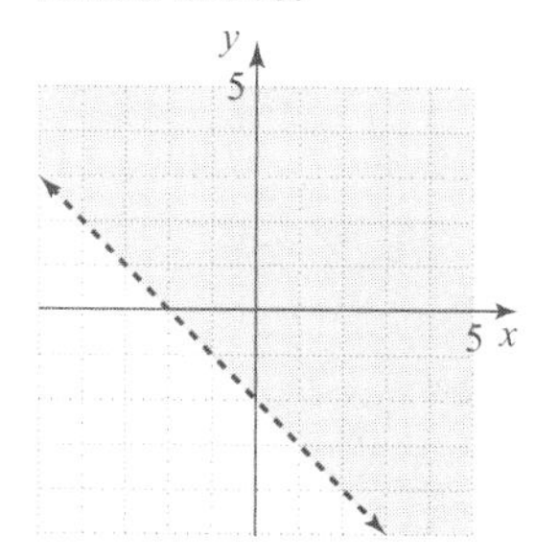

Chapter 4 Test

1. False; one solution, infinitely many solutions, or no solutions are the only possibilities.

2. False; a solution of a system of equations must be a solution of each equation in the system.

3. True

4. False; $x = 0$ is part of the solution.

5. Let $x = 1$ and $y = -1$.

$$\begin{array}{rl} 2x - 3y = 5 & \\ 2(1) - 3(-1) \stackrel{?}{=} 5 & \\ 2 + 3 \stackrel{?}{=} 5 & \\ 5 = 5 & \text{True} \end{array} \qquad \begin{array}{rl} 6x + y = 1 & \\ 6(1) + (-1) \stackrel{?}{=} 1 & \\ 6 - 1 \stackrel{?}{=} 1 & \\ 5 = 1 & \text{False} \end{array}$$

(1, −1) is not a solution of the system.

6. Let $x = 3$ and $y = -4$.

$$4x - 3y = 24$$
$$4(3) - 3(-4) \stackrel{?}{=} 24$$
$$12 + 12 \stackrel{?}{=} 24$$
$$24 = 24 \quad \text{True}$$

$$4x + 5y = -8$$
$$4(3) + 5(-4) \stackrel{?}{=} -8$$
$$12 - 20 \stackrel{?}{=} -8$$
$$-8 = -8 \quad \text{True}$$

(3, –4) is a solution of the system.

7. $\begin{cases} x - y = 2 \\ 3x - y = -2 \end{cases}$

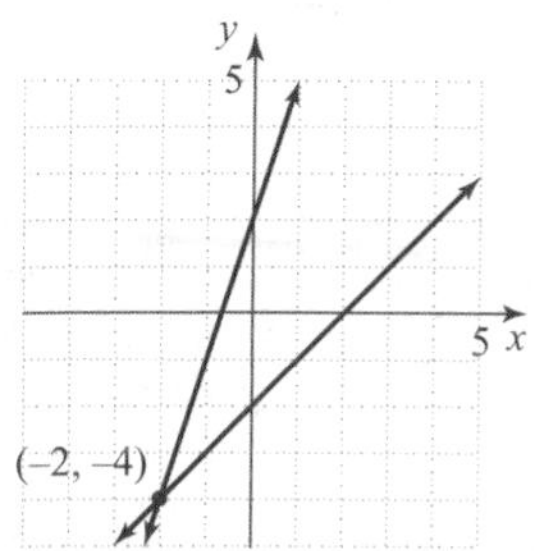

The solution of the system is (–2, –4).

8. $\begin{cases} y = -3x \\ 3x + y = 6 \end{cases}$

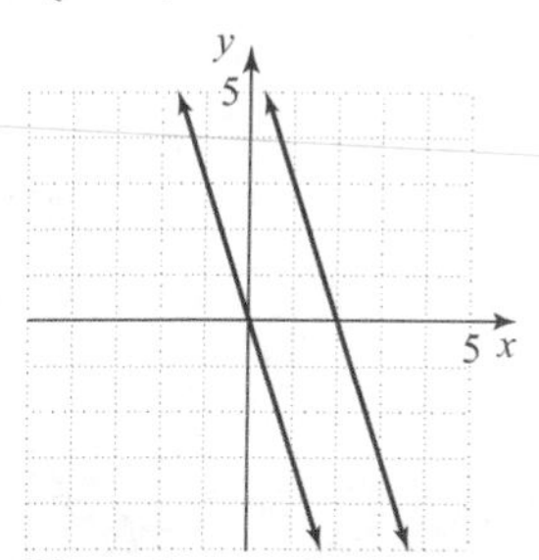

There is no solution.

9. $\begin{cases} 3x - 2y = -14 \\ y = x + 5 \end{cases}$

Substitute $x + 5$ for y in the first equation.

$$3x - 2(x + 5) = -14$$
$$3x - 2x - 10 = -14$$
$$x = -4$$

Let $x = -4$ in $y = x + 5$.

$y = -4 + 5 = 1$

The solution is (–4, 1).

10. $\begin{cases} \frac{1}{2}x + 2y = -\frac{15}{4} \\ 4x = -y \end{cases}$

Solve the second equation for y.

$y = -4x$

Substitute $-4x$ for y in the first equation.

$$\frac{1}{2}x + 2(-4x) = -\frac{15}{4}$$
$$\frac{1}{2}x - 8x = -\frac{15}{4}$$
$$-\frac{15}{2}x = -\frac{15}{4}$$
$$x = \frac{1}{2}$$

Let $x = \frac{1}{2}$ in the equation $y = -4x$.

$$y = -4\left(\frac{1}{2}\right) = -2$$

The solution is $\left(\frac{1}{2}, -2\right)$.

11. $\begin{cases} x + y = 28 \\ x - y = 12 \end{cases}$

$$\begin{aligned} x + y &= 28 \\ x - y &= 12 \\ \hline 2x \quad &= 40 \\ x &= 20 \end{aligned}$$

Let $x = 20$ in the first equation.

$$20 + y = 28$$
$$y = 8$$

The solution is (20, 8).

12. $\begin{cases} 4x - 6y = 7 \\ -2x + 3y = 0 \end{cases}$

Multiply the second equation by 2.

$$\begin{aligned} 4x - 6y &= 7 \\ -4x + 6y &= 0 \\ \hline 0 &= 7 \end{aligned}$$

The system is inconsistent. There is no solution.

13. $\begin{cases} 3x + y = 7 \\ 4x + 3y = 1 \end{cases}$

Solve the first equation for y.

$y = 7 - 3x$

Substitute $7 - 3x$ for y in the second equation.

$$4x + 3(7 - 3x) = 1$$
$$4x + 21 - 9x = 1$$
$$-5x = -20$$
$$x = 4$$

Let $x = 4$ in $y = 7 - 3x$.

$y = 7 - 3(4) = -5$

The solution is (4, –5).

14. $\begin{cases} 3(2x+y)=4x+20 \\ 6x+3y=4x+20 \\ 2x+3y=20 \\ \\ x-2y=3 \end{cases}$

Multiply the second equation by −2.

$$\begin{array}{r} 2x+3y=20 \\ \underline{-2x+4y=-6} \\ 7y=14 \\ y=2 \end{array}$$

Let $y = 2$ in the second equation.

$$\begin{aligned} x-2(2)&=3 \\ x-4&=3 \\ x&=7 \end{aligned}$$

The solution of the system is (7, 2).

15. $\begin{cases} \dfrac{x-3}{2}=\dfrac{2-y}{4} \\ \dfrac{7-2x}{3}=\dfrac{y}{2} \end{cases}$

Multiply the first equation by 4 and the second equation by 6.

$\begin{cases} 2(x-3)=2-y \\ 2x-6=2-y \\ 2x+y=8 \\ \\ 2(7-2x)=3y \\ 14-4x=3y \\ 4x+3y=14 \end{cases}$

Multiply the first equation by −3.

$$\begin{array}{r} -6x-3y=-24 \\ \underline{4x+3y=14} \\ -2x \qquad =-10 \\ x=5 \end{array}$$

Let $x = 5$ in the first equation.

$$\begin{aligned} 2(5)+y&=8 \\ 10+y&=8 \\ y&=-2 \end{aligned}$$

The solution of the system is (5, −2).

16. $\begin{cases} 8x-4y=12 \\ y=2x-3 \end{cases}$

Substitute $2x - 3$ for y in the first equation.

$$\begin{aligned} 8x-4(2x-3)&=12 \\ 8x-8x+12&=12 \\ 12&=12 \end{aligned}$$

There is an infinite number of solutions.

17. $\begin{cases} 0.01x-0.06y=-0.23 \\ 0.2x+0.4y=0.2 \end{cases}$

Multiply the first equation by 100 and the second equation by −5.

$$\begin{array}{r} x-6y=-23 \\ \underline{-x-2y=-1} \\ -8y=-24 \\ y=3 \end{array}$$

Let $y = 3$ in $x - 6y = -23$.

$$\begin{aligned} x-6(3)&=-23 \\ x-18&=-23 \\ x&=-5 \end{aligned}$$

The solution is (−5, 3).

18. $\begin{cases} x-\dfrac{2}{3}y=3 \\ -2x+3y=10 \end{cases}$

Multiply the first equation by 9 and the second equation by 2.

$$\begin{array}{r} 9x-6y=27 \\ \underline{-4x+6y=20} \\ 5x \qquad =47 \\ x=\dfrac{47}{5} \end{array}$$

Let $x=\dfrac{47}{5}$ in the first equation.

$$\begin{aligned} \frac{47}{5}-\frac{2}{3}y&=3 \\ 141-10y&=45 \\ -10y&=-96 \\ y&=\frac{48}{5} \end{aligned}$$

The solution is $\left(\dfrac{47}{5}, \dfrac{48}{5}\right)$.

19. Let x = the larger number and y = the smaller number.

$\begin{cases} x+y=124 \\ x-y=32 \end{cases}$

$$\begin{array}{r} x+y=124 \\ \underline{x-y=32} \\ 2x \qquad =156 \\ x=78 \end{array}$$

Let $x = 78$ in the first equation.

$$\begin{aligned} 78+y&=124 \\ y&=46 \end{aligned}$$

The numbers are 78 and 46.

20. Let x = cc's of 12% solution and y = cc's of 16% solution.

Concentration Rate	cc's of Solution	cc's of salt
12%	x	$0.12x$
22%	80	$0.22(80)$
16%	y	$0.16y$

$$\begin{cases} x+80=y \\ 0.12x+0.22(80)=0.16y \end{cases}$$

Multiply the first equation by −16 and the second equation by 100.

$$\begin{aligned} -16x-1280&=-16y \\ 12x+1760&=16y \\ \hline -4x+480&=0 \\ -4x&=-480 \\ x&=120 \end{aligned}$$

Should add 120 cc's of 12% solution

21. Let x = the number of thousands of farms in Texas and y = the number of thousands of farms in Missouri.

$$\begin{cases} x+y=336 \\ x-y=116 \end{cases}$$

$$\begin{aligned} x+y&=336 \\ x-y&=116 \\ \hline 2x\quad&=452 \\ x&=226 \end{aligned}$$

Let $x = 226$ in the first equation.

$$\begin{aligned} 226+y&=336 \\ y&=110 \end{aligned}$$

There are 226,000 farms in Texas and 110,000 farms in Missouri.

22. Let x = the speed of the faster hiker and y = the speed of the slower hiker.

	r ·	t =	d
Faster	x	4	$4x$
Slower	y	4	$4y$

$$\begin{cases} 4x+4y=36 \\ x=2y \end{cases}$$

Substitute $2y$ for x in the first equation.

$$\begin{aligned} 4(2y)+4y&=36 \\ 8y+4y&=36 \\ 12y&=36 \\ y&=3 \end{aligned}$$

Let $y = 3$ in the second equation.

$x = 2(3) = 6$

The speeds are 3 miles per hour and 6 miles per hour.

23. $x-y \ge -2$

Test (0, 0)

$0-0 \ge -2$

True

Shade below.

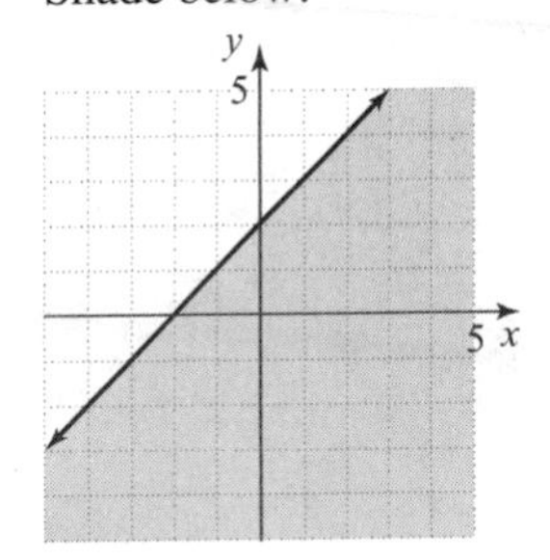

24. $y \ge -4x$

Test (1, 0)

$0 \overset{?}{\ge} -4(1)$

True

Shade above.

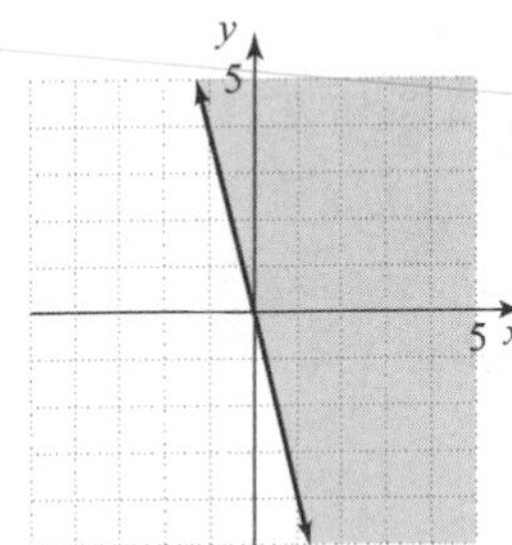

25. $2x-3y > -6$

Test (0, 0)

$2(0)-3(0) \overset{?}{>} -6$

True

Shade below.

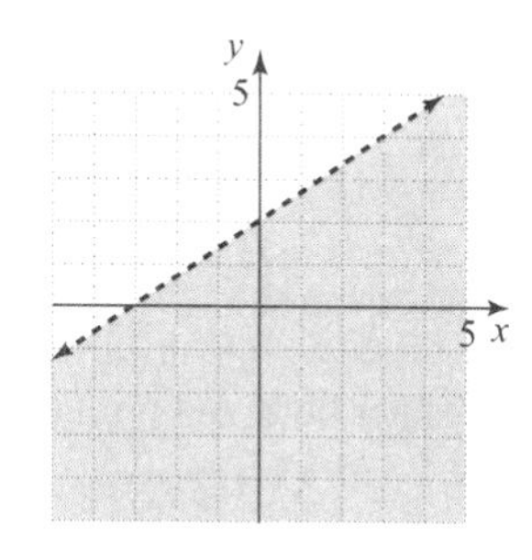

26. $\begin{cases} y+2x \le 4 \\ y \ge 2 \end{cases}$

$y+2x \le 4$
$y \ge 2$

27. $\begin{cases} 2y-x \ge 1 \\ x+y \ge -4 \end{cases}$

$2y-x \ge 1$
$x+y \ge -4$

Cumulative Review Chapters 1–4

1. a. $-1 < 0$

b. $7 = \frac{14}{2}$

c. $-5 > -6$

2. a. $5^2 = 5 \cdot 5 = 25$

b. $2^5 = 2 \cdot 2 \cdot 2 \cdot 2 \cdot 2 = 32$

3. a. commutative property of multiplication

b. associative property of addition

c. identity element for addition

d. commutative property of multiplication

e. multiplicative inverse property

f. additive inverse property

g. commutative and associative properties of multiplication

4. Let $x = 8$, $y = 5$.
$y^2 - 3x = 5^2 - 3(8) = 25 - 24 = 1$

5. $(2x-3)-(4x-2) = 2x-3-4x+2 = -2x-1$

6. $7-12+(-5)-2+(-2)$
$= 7+(-12)+(-5)+(-2)+(-2)$
$= 7+(-21)$
$= -14$

7. $5t-5 = 6t+2$
$-t-5 = 2$
$-t = 7$
$t = -7$

8. Let $x = -7$, $y = -3$.
$2y^2 - x^2 = 2(-3)^2 - (-7)^2$
$= 2(9) - 49$
$= 18 - 49$
$= -31$

9. $\frac{5}{2}x = 15$
$\frac{2}{5} \cdot \frac{5}{2}x = \frac{2}{5} \cdot 15$
$x = 6$

10. $0.4y - 6.7 + y - 0.3 - 2.6y$
$= 0.4y + y + (-2.6y) + (-6.7) + (-0.3)$
$= -1.2y - 7$

11. $\frac{x}{2} - 1 = \frac{2}{3}x - 3$
$6\left(\frac{x}{2} - 1\right) = 6\left(\frac{2}{3}x - 3\right)$
$3x - 6 = 4x - 18$
$-x - 6 = -18$
$-x = -12$
$x = 12$

12. $7(x-2)-6(x+1)=20$
$7x-14-6x-6=20$
$x-20=20$
$x=40$

13. Let x = the number.
$2(x+4)=4x-12$
$2x+8=4x-12$
$-2x+8=-12$
$-2x=-20$
$x=10$
The number is 10.

14. $5(y-5)=5y+10$
$5y-25=5y+10$
$-25=10$
False statement; there is no solution.

15. $y=mx+b$
$y-b=mx+b-b$
$y-b=mx$
$\frac{y-b}{m}=\frac{mx}{m}$
$\frac{y-b}{m}=x$

16. Let x = the number.
$5(x-1)=6x$
$5x-5=6x$
$-x-5=0$
$-x=5$
$x=-5$
The number is -5.

17. $-2x\le-4$
$\frac{-2x}{-2}\ge\frac{-4}{-2}$
$x\ge2$, $[2, \infty)$

18. $P=a+b+c$
$P-a-c=a+b+c-a-c$
$P-a-c=b$

19. $x=-2y$

x	y
0	0
–4	2

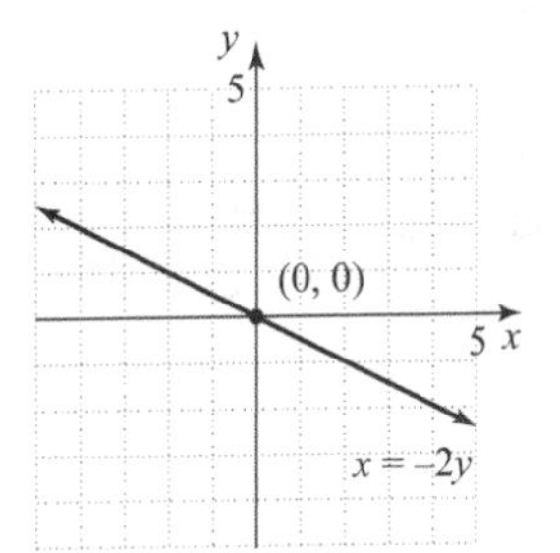

20. $3x+7\ge x-9$
$2x+7\ge-9$
$2x\ge-16$
$x\ge-8$, $[-8, \infty)$

21. $(-1, 5)$ and $(2, -3)$
$m=\frac{y_2-y_1}{x_2-x_1}=\frac{-3-5}{2-(-1)}=\frac{-8}{3}=-\frac{8}{3}$

22. $x-3y=3$

x	y
0	–1
3	0
9	2

23. $y=\frac{3}{4}x+6$
$y=mx+b$
$m=\frac{3}{4}$

24. $(-1, 3)$ and $(2, -8)$
$m=\frac{y_2-y_1}{x_2-x_1}=\frac{-8-3}{2-(-1)}=-\frac{11}{3}$
A parallel line has the same slope.
Slope is $-\frac{11}{3}$.

25. $3x-4y=4$
$-4y=-3x+4$
$y=\frac{-3x}{-4}+\frac{4}{-4}$
$y=\frac{3}{4}x-1$
$y=mx+b$

$m = \frac{3}{4},\ b = -1$

Slope is $\frac{3}{4}$, y-intercept is $(0, -1)$.

26. $y = 7x + 0$
$y = mx + b$
$m = 7, b = 0$
Slope is 7, y-intercept is $(0, 0)$.

27. $m = -2$, with point $(-1, 5)$

$$y - y_1 = m(x - x_1)$$
$$y - 5 = -2[x - (-1)]$$
$$y - 5 = -2x - 2$$
$$2x + y = 3$$

28. Line: $y = 4x - 5 \Rightarrow m_1 = 4$

Line 2: $-4x + y = 7 \Rightarrow y = 4x + 7 \Rightarrow m_2 = 4$

$m_2 = m_1$

The lines are parallel.

29. A vertical line has an equation $x = c$.
Point, $(-1, 5)$
$x = -1$

30. $m = -5$, with point $(-2, 3)$

$$y - y_1 = m(x - x_1)$$
$$y - 3 = -5[x - (-2)]$$
$$y - 3 = -5x - 10$$
$$y = -5x - 7$$

31. Domain is $\{-1, 0, 3\}$
Range is $\{-2, 0, 2, 3\}$

32. $f(x) = 5x^2 - 6$

$f(0) = 5(0)^2 - 6 = -6$

$f(-2) = 5(-2)^2 - 6 = 5(4) - 6 = 14$

33. **a.** function

b. not a function

34. **a.** not a function

b. function

c. not a function

35.
$$\begin{cases} 3x - y = 4 \\ y = 3x - 4,\ m = 3 \\ \\ x + 2y = 8 \\ y = -\frac{1}{2}x + 4,\ m = -\frac{1}{2} \end{cases}$$

Because they have different slopes, there is only one solution.

36. **a.** Let $x = 1$ and $y = -4$.

$$2x - y = 6$$
$$2(1) - (-4) \stackrel{?}{=} 6$$
$$2 + 4 \stackrel{?}{=} 6$$
$$6 = 6 \quad \text{True}$$

$$3x + 2y = -5$$
$$3(1) + 2(-4) \stackrel{?}{=} -5$$
$$3 - 8 \stackrel{?}{=} -5$$
$$-5 = -5 \quad \text{True}$$

$(1, -4)$ is a solution of the system.

b. Let $x = 0$ and $y = 6$.

$$2x - y = 6$$
$$2(0) - (6) \stackrel{?}{=} 6$$
$$0 - 6 \stackrel{?}{=} 6$$
$$-6 = 6 \quad \text{False}$$

$3x + 2y = -5$
Test not needed

$(0, 6)$ is not a solution of the system.

c. Let $x = 3$ and $y = 0$.

$$2x - y = 6$$
$$2(3) - (0) \stackrel{?}{=} 6$$
$$6 - 0 \stackrel{?}{=} 6$$
$$6 = 6 \quad \text{True}$$

$$3x + 2y = -5$$
$$3(3) + 2(0) \stackrel{?}{=} -5$$
$$9 + 0 \stackrel{?}{=} -5$$
$$9 = -5 \quad \text{False}$$

$(3, 0)$ is not a solution of the system.

37.
$$\begin{cases} x + 2y = 7 \\ 2x + 2y = 13 \end{cases}$$

Solve the first equation for x.
$x = 7 - 2y$
Substitute $7 - 2y$ for x in the second equation.

$$2(7 - 2y) + 2y = 13$$
$$14 - 4y + 2y = 13$$
$$-2y = -1$$
$$y = \frac{1}{2}$$

Let $y = \frac{1}{2}$ in $x = 7 - 2y$.

$x = 7 - 2\left(\frac{1}{2}\right) = 6$

The solution is $\left(6, \frac{1}{2}\right)$.

38. $\begin{cases} 3x - 4y = 10 \\ y = 2x \end{cases}$

Substitute $2x$ for y in the first equation.

$3x - 4(2x) = 10$
$3x - 8x = 10$
$-5x = 10$
$x = -2$

Let $x = -2$ in the second equation.

$y = 2(-2) = -4$

The solution is $(-2, -4)$.

39. $\begin{cases} x + y = 7 \\ x - y = 5 \end{cases}$

$x + y = 7$
$x - y = 5$
$2x = 12$
$x = 6$

Let $x = 6$ in the first equation.

$6 + y = 7$
$y = 1$

The solution to the system is $(6, 1)$.

40. $\begin{cases} x = 5y - 3 \\ x = 8y + 4 \end{cases}$

Substitute $8y + 4$ for x in the first equation.

$8y + 4 = 5y - 3$
$3y + 4 = -3$
$3y = -7$
$y = -\frac{7}{3}$

Let $y = -\frac{7}{3}$ in the second equation.

$x = 8\left(-\frac{7}{3}\right) + 4$

$x = -\frac{56}{3} + \frac{12}{3}$

$x = -\frac{44}{3}$

The solution is $\left(-\frac{44}{3}, -\frac{7}{3}\right)$.

41. Let x = the first number and y = the second number.

$\begin{cases} x + y = 37 \\ x - y = 21 \end{cases}$

$x + y = 37$
$x - y = 21$
$2x = 58$
$x = 29$

Let $x = 29$ in the first equation.

$29 + y = 37$
$y = 8$

The numbers are 29 and 8.

42. $x > 1$
Shade right.

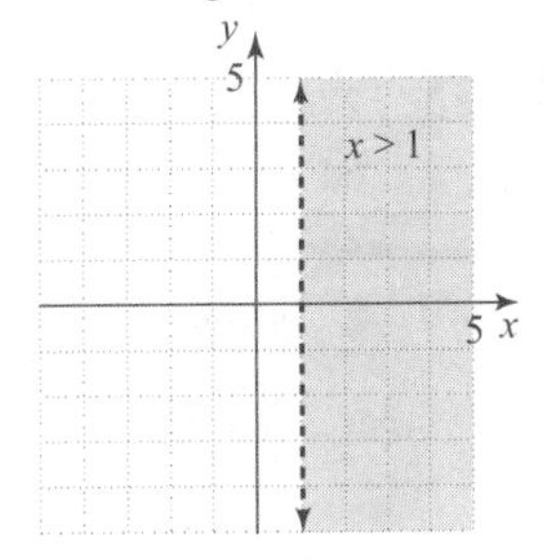

43. $2x - y \geq 3$
Test (0, 0)

$2(0) - 0 \overset{?}{\geq} 3$

False
Shade below.

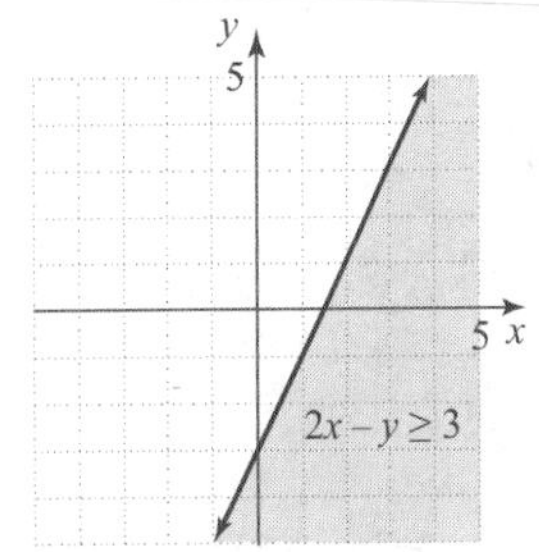

44. $\begin{cases} 2x + 3y < 6 \\ y < 2 \end{cases}$

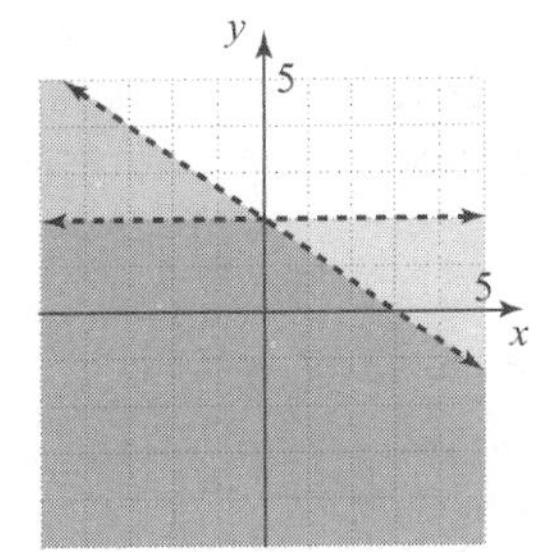

$2x + 3x < 6$
$y < 2$

Chapter 5

Section 5.1

Practice Exercises

1. a. $3^3 = 3 \cdot 3 \cdot 3 = 27$

b. Use 4 as a factor once, $4^1 = 4$

c. $(-8)^2 = (-8)(-8) = 64$

d. $-8^2 = -(8 \cdot 8) = -64$

e. $\left(\frac{3}{4}\right)^3 = \frac{3}{4} \cdot \frac{3}{4} \cdot \frac{3}{4} = \frac{27}{64}$

f. $(0.3)^4 = (0.3)(0.3)(0.3)(0.3) = 0.0081$

g. $3 \cdot 5^2 = 3 \cdot 25 = 75$

2. a. If x is 3, $3x^4 = 3 \cdot (3)^4$
$= 3 \cdot (3 \cdot 3 \cdot 3 \cdot 3)$
$= 3 \cdot 81$
$= 243$

b. If x is -4, $\frac{6}{x^2} = \frac{6}{(-4)^2} = \frac{6}{(-4)(-4)} = \frac{6}{16} = \frac{3}{8}$

3. a. $3^4 \cdot 3^6 = 3^{4+6} = 3^{10}$

b. $y^3 \cdot y^2 = y^{3+2} = y^5$

c. $z \cdot z^4 = z^1 \cdot z^4 = z^{1+4} = z^5$

d. $x^3 \cdot x^2 \cdot x^6 = x^{3+2+6} = x^{11}$

e. $(-2)^5 \cdot (-2)^3 = (-2)^{5+3} = (-2)^8$

f. $b^3 \cdot t^5$, cannot be simplified because b and t are different bases.

4. $(-5y^3)(-3y^4) = -5 \cdot y^3 \cdot -3 \cdot y^4$
$= -5 \cdot -3 \cdot y^3 \cdot y^4$
$= 15y^7$

5. a. $(y^7z^3)(y^5z) = (y^7 \cdot y^5) \cdot (z^3 \cdot z^1)$
$= y^{12} \cdot z^4$ or $y^{12}z^4$

b. $(-m^4n^4)(7mn^{10})$
$= (-1 \cdot 7) \cdot (m^4 \cdot m^1) \cdot (n^4 \cdot n^{10})$
$= (-7) \cdot (m^5) \cdot (n^{14})$ or $-7m^5n^{14}$

6. a. $(x^4)^3 = x^{4 \cdot 3} = x^{12}$

b. $(z^3)^7 = z^{3 \cdot 7} = z^{21}$

c. $[(-2)^3]^5 = (-2)^{3 \cdot 5} = (-2)^{15}$

7. a. $(pr)^5 = p^5 \cdot r^5 = p^5r^5$

b. $(6b)^2 = 6^2 \cdot b^2 = 36b^2$

c. $\left(\frac{1}{4}x^2y\right)^3 = \left(\frac{1}{4}\right)^3 \cdot (x^2)^3 \cdot y^3$
$= \frac{1}{64} \cdot x^6 \cdot y^3$
$= \frac{1}{64}x^6y^3$

d. $(-3a^3b^4c)^4 = (-3)^4 \cdot (a^3)^4 \cdot (b^4)^4 \cdot c^4$
$= 81a^{12}b^{16}c^4$

8. a. $\left(\frac{x}{y^2}\right)^5 = \frac{x^5}{(y^2)^5} = \frac{x^5}{y^{10}}$, $y \neq 0$

b. $\left(\frac{2a^4}{b^3}\right)^5 = \frac{2^5 \cdot (a^4)^5}{(b^3)^5} = \frac{32a^{20}}{b^{15}}$, $b \neq 0$

9. a. $\frac{z^8}{z^4} = z^{8-4} = z^4$

b. $\frac{(-5)^5}{(-5)^3} = (-5)^{5-3} = (-5)^2 = 25$

c. $\frac{8^8}{8^6} = 8^{8-6} = 8^2 = 64$

d. $\frac{q^5}{t^2}$ cannot be simplified because q and t are different bases.

e. Begin by grouping common bases.
$\frac{6x^3y^7}{xy^5} = 6 \cdot \frac{x^3}{x} \cdot \frac{y^7}{y^5} = 6 \cdot x^{3-1} \cdot y^{7-5} = 6x^2y^2$

10. a. $-3^0 = -1 \cdot 3^0 = -1 \cdot 1 = -1$

b. $(-3)^0 = 1$

c. $8^0 = 1$

d. $(0.2)^0 = 1$

e. $(xz)^0 = x^0 \cdot z^0 = 1 \cdot 1 = 1$

11. a. This is a quotient raised to a power, so we use the power of a quotient rule.
$\left(\frac{5}{xz}\right)^3 = \frac{5^3}{x^3z^3} = \frac{125}{x^3z^3}$

b. This is a product raised to a power, so we use the power of a product rule.
$(2z^8x^5)^4 = 2^4(z^8)^4(x^5)^4 = 16z^{32}x^{20}$

c. Use the power of a product or quotient rule; then use the power rule for exponents.
$\left(\frac{-3x^3}{y^4}\right)^3 = \frac{(-3)^3(x^3)^3}{(y^4)^3} = -\frac{27x^9}{y^{12}}$

Vocabulary and Readiness Check

1. Repeated multiplication of the same factor can be written using an <u>exponent</u>.

2. In 5^2, the 2 is called the <u>exponent</u> and the 5 is called the <u>base</u>.

3. To simplify $x^2 \cdot x^7$, keep the base and <u>add</u> the exponents.

4. To simplify $(x^3)^6$, keep the base and <u>multiply</u> the exponents.

5. The understood exponent on the term y is <u>1</u>.

6. If $x^{\square} = 1$, the exponent is <u>0</u>.

7. In $\underline{3^2}$, the base is <u>3</u> and the exponent is <u>2</u>.

8. In $\underline{(-3)^6}$, the base is <u>−3</u> and the exponent is <u>6</u>.

9. In $\underline{-4^2}$, the base is <u>4</u> and the exponent is <u>2</u>.

10. In $5 \cdot 3^4$, the base 5 has exponent 1 and the base 3 has exponent 4.

11. In $5x^2$, the base 5 has exponent 1 and the base x has exponent 2.

12. In $\underline{(5x)^2}$, the base is <u>5x</u> and the exponent is <u>2</u>.

Exercise Set 5.1

1. $7^2 = 7 \cdot 7 = 49$

3. $(-5)^1 = -5$

5. $-2^4 = -2 \cdot 2 \cdot 2 \cdot 2 = -16$

7. $(-2)^4 = (-2)(-2)(-2)(-2) = 16$

9. $(0.1)^5 = (0.1)(0.1)(0.1)(0.1)(0.1) = 0.00001$

11. $\left(\frac{1}{3}\right)^4 = \left(\frac{1}{3}\right)\left(\frac{1}{3}\right)\left(\frac{1}{3}\right)\left(\frac{1}{3}\right) = \frac{1}{81}$

13. $7 \cdot 2^5 = 7 \cdot 2 \cdot 2 \cdot 2 \cdot 2 \cdot 2 = 224$

15. $-2 \cdot 5^3 = -2 \cdot 5 \cdot 5 \cdot 5 = -250$

17. Answers may vary

19. $x^2 = (-2)^2 = (-2)(-2) = 4$

21. $5x^3 = 5(3)^3 = 5 \cdot 3 \cdot 3 \cdot 3 = 135$

23. $2xy^2 = 2(3)(5)^2 = 2(3)(5)(5) = 150$

25. $\frac{2z^4}{5} = \frac{2(-2)^4}{5} = \frac{2(-2)(-2)(-2)(-2)}{5} = \frac{32}{5}$

27. $x^2 \cdot x^5 = x^{2+5} = x^7$

29. $(-3)^3 \cdot (-3)^9 = (-3)^{3+9} = (-3)^{12}$

31. $(5y^4)(3y) = 5(3)y^{4+1} = 15y^5$

33. $(x^9y)(x^{10}y^5) = x^{9+10}y^{1+5} = x^{19}y^6$

35. $(-8mn^6)(9m^2n^2) = --8(9)m^{1+2}n^{6+2}$
$= -72m^3n^8$

37. $(4z^{10})(-6z^7)(z^3) = 4(-6)z^{10+7+3} = -24z^{20}$

39. $A = (4x^2)\cdot(5x^3)$
$= (4\cdot 5)\cdot(x^2 \cdot x^3)$
$= 20x^{2+3}$
$= 20x^5$

The area is $20x^5$ square feet.

41. $(x^9)^4 = x^{9\cdot 4} = x^{36}$

43. $(pq)^8 = p^8q^8$

45. $(2a^5)^3 = 2^3 \cdot (a^5)^3 = 8 \cdot a^{5\cdot 3} = 8a^{15}$

47. $(x^2y^3)^5 = (x^2)^5 \cdot (y^3)^5 = x^{2\cdot 5} \cdot y^{3\cdot 5} = x^{10}y^{15}$

49. $(-7a^2b^5c)^2 = (-7)^2 \cdot (a^2)^2 \cdot (b^5)^2 \cdot c^2$
$= 49a^{2\cdot 2}b^{5\cdot 2}c^2$
$= 49a^4b^{10}c^2$

51. $\left(\frac{r}{s}\right)^9 = \frac{r^9}{s^9}$

53. $\left(\frac{mp}{n}\right)^5 = \frac{(mp)^5}{n^5} = \frac{m^5 \cdot p^5}{n^5} = \frac{m^5p^5}{n^5}$

55. $\left(\frac{-2xz}{y^5}\right)^2 = \frac{(-2)^2x^2z^2}{y^{5\cdot 2}} = \frac{4x^2z^2}{y^{10}}$

57. $A = (8z^5)^2 = 8^2 \cdot (z^5)^2 = 64 \cdot z^{5\cdot 2} = 64z^{10}$

The area is $64z^{10}$ square decimeters.

59. $V = (3y^4)^3 = 3^3y^{4\cdot 3} = 27y^{12}$

The volume is $27y^{12}$ cubic feet.

61. $\frac{x^3}{x} = \frac{x^3}{x^1} = x^{3-1} = x^2$

63. $\frac{(-4)^6}{(-4)^3} = (-4)^{6-3} = (-4)^3 = -64$

65. $\frac{p^7q^{20}}{pq^{15}} = p^{7-1}q^{20-15} = p^6q^5$

67. $\frac{7x^2y^6}{14x^2y^3} = \frac{7}{14}x^{2-2}y^{6-3} = \frac{1}{2}x^0y^3 = \frac{y^3}{2}$

69. $7^0 = 1$

71. $(2x)^0 = 1$

73. $-7x^0 = -7(1) = -7$

75. $5^0 + y^0 = 1 + 1 = 2$

77. $-9^2 = -9\cdot 9 = -81$

79. $\left(\frac{1}{4}\right)^3 = \frac{1}{4}\cdot\frac{1}{4}\cdot\frac{1}{4} = \frac{1}{64}$

81. $\left(\frac{9}{qr}\right)^2 = \frac{9^2}{(qr)^2} = \frac{81}{q^2r^2}$

83. $a^2a^3a = a^{2+3+1} = a^6$

85. $(2x^3)(-8x^4) = 2(-8)x^{3+4} = -16x^7$

87. $(a^7b^{12})(a^4b^8) = a^{7+4}b^{12+8} = a^{11}b^{20}$

89. $(-2mn^6)(-13m^8n) = -2(-13)m^{1+8}n^{6+1}$
$= 26m^9n^7$

91. $(z^4)^{10} = z^{4\cdot 10} = z^{40}$

93. $(-6xyz^3)^2 = (-6)^2x^2y^2z^{3\cdot 2} = 36x^2y^2z^6$

95. $\frac{3x^5}{x^4} = 3x^{5-4} = 3x$

97. $(9xy)^2 = 9^2x^2y^2 = 81x^2y^2$

99. $2^0 + 2^5 = 1 + 32 = 33$

101. $\left(\frac{3y^5}{6x^4}\right)^3 = \frac{3^3y^{5\cdot3}}{6^3x^{4\cdot3}} = \frac{27y^{15}}{216x^{12}} = \frac{y^{15}}{8x^{12}}$

103. $\frac{2x^3y^2z}{xyz} = 2x^{3-1}y^{2-1}z^{1-1} = 2x^2y^1z^0 = 2x^2y$

105. $y - 10 + y = y + y - 10 = 2y - 10$

107. $7x + 2 - 8x - 6 = 7x - 8x + 2 - 6 = -x - 4$

109. $2(x-5) + 3(5-x) = 2x - 10 + 15 - 3x = -x + 5$

111. $(x^{14})^{23} = x^{14\cdot23} = x^{322}$
Multiply the exponents; choice c.

113. $x^{14} + x^{23}$ cannot be simplified further; choice e.

115. Answers may vary

117. Answers may vary

119. $V = x^3 = 7^3 = 7\cdot7\cdot7 = 343$
The volume is 343 cubic meters.

121. Volume; volume measures capacity.

123. Answers may vary

125. $x^{5a}x^{4a} = x^{5a+4a} = x^{9a}$

127. $(a^b)^5 = a^{b\cdot5} = a^{5b}$

129. $\frac{x^{9a}}{x^{4a}} = x^{9a-4a} = x^{5a}$

131. $A = P\left(1 + \frac{r}{12}\right)^6$

$$A = 1000\left(1 + \frac{0.09}{12}\right)^6$$
$$= 1000(1.0075)^6$$
$$\approx 1045.85$$

You need $1045.85 to pay off the loan.

Section 5.2

Practice Exercises

1. a. The exponent on y is 3, so the degree of $5y^3$ is 3.

b. $-3a^2b^5c$ can be written as $-3a^2b^5c^1$. The degree of the term is the sum of the exponents, so the degree is 2 + 5 + 1 or 8.

c. The constant, 8, can be written as $8x^0$ (since $x^0 = 1$). The degree of 8 or $8x^0$ is 0.

2. a. The degree of the trinomial $5b^2 - 3b + 7$ is 2, the greatest degree of any of its terms.

b. Rewrite the binomial as $7t^1 + 3$, the degree is 1.

c. The degree of the polynomial $5x^2 + 3x - 6x^3 + 4$ is 3.

3.

Term	numerical coefficient	degree of term
$-3x^3y^2$	-3	5
$4xy^2$	4	3
$-y^2$	-1	2
$3x$	3	1
-2	-2	0

4. $2x^2 - 5x + 3 = 2(-3)^2 - 5(-3) + 3$

$$= 2(9) + 15 + 3$$
$$= 18 + 15 + 3$$
$$= 36$$

5. To find each height, we evaluate the polynomial when $t = 1$ and when $t = 2$.

$$\begin{aligned}-16t^2 + 130 &= -16(1)^2 + 130\\ &= -16 + 130\\ &= 114\end{aligned}$$

The height of the camera at 1 second is 114 feet.

$$\begin{aligned}-16t^2 + 130 &= -16(2)^2 + 130\\ &= -16(4) + 130\\ &= -64 + 130\\ &= 66\end{aligned}$$

The height of the camera at 2 seconds is 66 feet.

6. a. $-4y + 2y = (-4 + 2)y = -2y$

b. These terms cannot be combined because z and $5z^3$ are not like terms.

c.
$$\begin{aligned}7a^2 - 5 - 3a^2 - 7 &= 7a^2 - 3a^2 - 5 - 7\\ &= 4a^2 - 12\end{aligned}$$

d.
$$\begin{aligned}&\frac{3}{8}x^3 - x^2 + \frac{5}{6}x^4 + \frac{1}{12}x^3 - \frac{1}{2}x^4\\ &= \left(\frac{5}{6} - \frac{1}{2}\right)x^4 + \left(\frac{3}{8} + \frac{1}{12}\right)x^3 - x^2\\ &= \left(\frac{5}{6} - \frac{3}{6}\right)x^4 + \left(\frac{9}{24} + \frac{2}{24}\right)x^3 - x^2\\ &= \frac{2}{6}x^4 + \frac{11}{24}x^3 - x^2\\ &= \frac{1}{3}x^4 + \frac{11}{24}x^3 - x^2\end{aligned}$$

7.
$$\begin{aligned}9xy - 3x^2 - 4yx + 5y^2 &= -3x^2 + (9 - 4)xy + 5y^2\\ &= -3x^2 + 5xy + 5y^2\end{aligned}$$

8.
$$\begin{aligned}&x \cdot x + 2 \cdot x + 2 \cdot 2 + 5 \cdot x + x \cdot 3x\\ &= x^2 + 2x + 4 + 5x + 3x^2\\ &= 4x^2 + 7x + 4\end{aligned}$$

9.
$$\begin{aligned}&(-3x^2 - 4x + 9) + (2x^2 - 2x)\\ &= -3x^2 - 4x + 9 + 2x^2 - 2x\\ &= (-3x^2 + 2x^2) + (-4x - 2x) + 9\\ &= -x^2 - 6x + 9\end{aligned}$$

10.
$$\begin{aligned}&(-3x^3 + 7x^2 + 3x - 4) + (3x^2 - 9x)\\ &= -3x^3 + 7x^2 + 3x - 4 + 3x^2 - 9x\\ &= -3x^3 + (7x^2 + 3x^2) + (3x - 9x) - 4\\ &= -3x^3 + 10x^2 - 6x - 4\end{aligned}$$

11.
$$\begin{array}{r}5z^3 + 3z^2 + 4z\\ \underline{5z^2 + 4z}\\ 5z^3 + 8z^2 + 8z\end{array}$$

12.
$$\begin{aligned}(8x - 7) - (3x - 6) &= (8x - 7) + [-(3x - 6)]\\ &= (8x - 7) + (-3x + 6)\\ &= (8x - 3x) + (-7 + 6)\\ &= 5x - 1\end{aligned}$$

13. First, change the sign of each term of the second polynomial and then add.

$$\begin{aligned}&(3x^3 - 5x^2 + 4x) - (x^3 - x^2 + 6)\\ &= (3x^3 - 5x^2 + 4x) + (-x^3 + x^2 - 6)\\ &= 3x^3 - x^3 - 5x^2 + x^2 + 4x - 6\\ &= 2x^3 - 4x^2 + 4x - 6\end{aligned}$$

14. Arrange the polynomials in vertical format, lining up like terms.

$$\begin{array}{r}-2z^2 - 8z + 5\\ \underline{-(6z^2 + 3z - 6)}\end{array} \qquad \begin{array}{r}-2z^2 - 8z + 5\\ \underline{-6z^2 - 3z + 6}\\ -8z^2 - 11z + 11\end{array}$$

15.
$$\begin{aligned}&[(8x - 11) + (2x + 5)] - (3x + 5)\\ &= 8x - 11 + 2x + 5 - 3x - 5\\ &= 8x + 2x - 3x - 11 + 5 - 5\\ &= 7x - 11\end{aligned}$$

16. a.
$$\begin{aligned}&(3a^2 - 4ab + 7b^2) + (-8a^2 + 3ab - b^2)\\ &= 3a^2 - 4ab + 7b^2 - 8a^2 + 3ab - b^2\\ &= -5a^2 - ab + 6b^2\end{aligned}$$

b.
$$\begin{aligned}&(5x^2y^2 - 6xy - 4xy^2)\\ &\qquad - (2x^2y^2 + 4xy - 5 + 6y^2)\\ &= 5x^2y^2 - 6xy - 4xy^2 - 2x^2y^2\\ &\qquad - 4xy + 5 - 6y^2\\ &= 3x^2y^2 - 10xy - 4xy^2 - 6y^2 + 5\end{aligned}$$

Vocabulary and Readiness Check

1. A binomial is a polynomial with exactly 2 terms.
2. A monomial is a polynomial with exactly one term.
3. A trinomial is a polynomial with exactly three terms.

4. The numerical factor of a term is called the coefficient.

5. A number term is also called a constant.

6. The degree of a polynomial is the greatest degree of any term of the polynomial.

7. $-9y - 5y = (-9 - 5)y = -14y$

8. $6m^5 + 7m^5 = (6+7)m^5 = 13m^5$

9. $x + 6x = (1 + 6)x = 7x$

10. $7z - z = (7 - 1)z = 6z$

11. $5m^2 + 2m$ Not like terms.

12. $8p^3 + 3p^2$ Not like terms.

Exercise Set 5.2

1. $x + 2$ is a binomial because it has two terms. The degree is 1 since x is x^1.

3. $9m^3 - 5m^2 + 4m - 8$ is neither a monomial, a binomial, nor a trinomial because it has more than three terms. The degree is 3, the greatest degree of any of its terms.

5. $12x^4y - x^2y^2 - 12x^2y^4$ is a trinomial because it has three terms. The degree is 6, the greatest degree of any of its terms.

7. $3zx - 5x^2$ is a binomial because it has two terms. The degree is 2 because 2 is the degree of the term with the highest degree.

	Polynomial	*Degree*
9.	$3xy^2 - 4$	3
11.	$5a^2 - 2a + 1$	2

13. a. $x + 6 = 0 + 6 = 6$

b. $x + 6 = -1 + 6 = 5$

15. a. $x^2 - 5x - 2 = 0^2 - 5(0) - 2 = -2$

b. $x^2 - 5x - 2 = (-1)^2 - 5(-1) - 2$
$= 1 + 5 - 2$
$= 4$

17. a. $x^3 - 15 = 0^3 - 15 = -15$

b. $x^3 - 15 = (-1)^3 - 15 = -1 - 15 = -16$

19. $-16t^2 + 1150$
$t = 1$; $-16(1)^2 + 1150 = -16 + 1150 = 1134$
After 1 second, the height is 1134 feet.

21. $-16t^2 + 1150$
$t = 3$; $-16(3)^2 + 1150 = -144 + 1150 = 1006$
After 3 seconds, the height is 1006 feet.

23. $14x^2 + 9x^2 = (14+9)x^2 = 23x^2$

25. $15x^2 - 3x^2 - y = (15-3)x^2 - y = 12x^2 - y$

27. $8s - 5s + 4s = (8 - 5 + 4)s = 7s$

29. $0.1y^2 - 1.2y^2 + 6.7 - 1.9$
$= (0.1 - 1.2)y^2 + (6.7 - 1.9)$
$= -1.1y^2 + 4.8$

31. $\frac{2}{5}x^2 - \frac{1}{3}x^3 + x^2 - \frac{1}{4}x^3 + 6$
$= \left(-\frac{1}{3} - \frac{1}{4}\right)x^3 + \left(\frac{2}{5} + 1\right)x^2 + 6$
$= \left(-\frac{4}{12} - \frac{3}{12}\right)x^3 + \left(\frac{2}{5} + \frac{5}{5}\right)x^2 + 6$
$= -\frac{7}{12}x^3 + \frac{7}{5}x^2 + 6$

33. $6a^2 - 4ab + 7b^2 - a^2 - 5ab + 9b^2$
$= (6-1)a^2 + (-4-5)ab + (7+9)b^2$
$= 5a^2 - 9ab + 16b^2$

35. $(3x+7) + (9x+5) = 3x + 7 + 9x + 5$
$= (3x + 9x) + (7 + 5)$
$= 12x + 12$

37. $(-7x+5)+(-3x^2+7x+5)$
$=-7x+5-3x^2+7x+5$
$=-3x^2+(-7x+7x)+(5+5)$
$=-3x^2+10$

39. $(2x^2+5)-(3x^2-9)=2x^2+5-3x^2+9$
$=(2x^2-3x^2)+(5+9)$
$=-x^2+14$

41. $3x-(5x-9)=3x-5x+9$
$=(3x-5x)+9$
$=-2x+9$

43. $(2x^2+3x-9)-(-4x+7)$
$=2x^2+3x-9+4x-7$
$=2x^2+(3x+4x)+(-9-7)$
$=2x^2+7x-16$

45.
$$\begin{array}{r} 3t^2+4 \\ +\ 5t^2-8 \\ \hline 8t^2-4 \end{array}$$

47.
$$\begin{array}{r} 4z^2-8z+3 \\ -\ (6z^2+8z-3) \\ \hline \end{array} \qquad \begin{array}{r} 4z^2-8z+3 \\ +\ (-6z^2-8z+3) \\ \hline -2z^2-16z+6 \end{array}$$

49.
$$\begin{array}{r} 5x^3-4x^2+6x-2 \\ -\ (3x^3-2x^2-x-4) \\ \hline \end{array} \qquad \begin{array}{r} 5x^3-4x^2+6x-2 \\ +\ (-3x^3+2x^2+x+4) \\ \hline 2x^3-2x^2+7x+2 \end{array}$$

51. $(81x^2+10)-(19x^2+5)=81x^2+10-19x^2-5$
$=62x^2+5$

53. $[(8x+1)+(6x+3)]-(2x+2)$
$=8x+1+6x+3-2x-2$
$=8x+6x-2x+1+3-2$
$=12x+2$

55. $(-3y^2-4y)+(2y^2+y-1)$
$=-3y^2-4y+2y^2+y-1$
$=-y^2-3y-1$

57. $(5x+8)-(-2x^2-6x+8)$
$=5x+8+2x^2+6x-8$
$=2x^2+11x$

59. $(-8x^4+7x)+(-8x^4+x+9)$
$=-8x^4+7x-8x^4+x+9$
$=-16x^4+8x+9$

61. $(3x^2+5x-8)+(5x^2+9x+12)-(x^2-14)$
$=3x^2+5x-8+5x^2+9x+12-x^2+14$
$=7x^2+14x+18$

63. $(7x-3)-4x=7x-3-4x=3x-3$

65. $(7x^2+3x+9)-(5x+7)=7x^2+3x+9-5x-7$
$=7x^2-2x+2$

67. $[(8y^2+7)+(6y+9)]-(4y^2-6y-3)$
$=8y^2+7+6y+9-4y^2+6y+3$
$=4y^2+12y+19$

69. $[(-x^2-2x)+(5x^2+x+9)]-(-2x^2+4x-12)$
$=-x^2-2x+5x^2+x+9+2x^2-4x+12$
$=6x^2-5x+21$

71. $2x\cdot 2x+x\cdot 7+x\cdot x+x\cdot 5=4x^2+7x+x^2+5x$
$=5x^2+12x$

73. $9x+10+3x+12+4x+15+2x+7$
$=(9x+3x+4x+2x)+(10+12+15+7)$
$=18x+44$

75. $(-x^2+3x)+(2x^2+5)+(4x-1)$
$=-x^2+3x+2x^2+5+4x-1$
$=x^2+7x+4$
The perimeter is (x^2+7x+4) feet.

77. $(4y^2+4y+1)-(y^2-10)$
$=4y^2+4y+1-y^2+10$
$=3y^2+4y+11$
The length of the remaining piece is $(3y^2+4y+11)$ meters.

79. $(9a+6b-5)+(-11a-7b+6)$
$=9a+6b-5-11a-7b+6$
$=-2a-b+1$

81. $(4x^2+y^2+3)-(x^2+y^2-2)$
$=4x^2+y^2+3-x^2-y^2+2$
$=3x^2+5$

83. $(x^2+2xy-y^2)+(5x^2-4xy+20y^2)$
$=x^2+2xy-y^2+5x^2-4xy+20y^2$
$=6x^2-2xy+19y^2$

85. $(11r^2s+16rs-3-2r^2s^2)-(3sr^2+5-9r^2s^2)$
$=11r^2s+16rs-3-2r^2s^2-3sr^2-5+9r^2s^2$
$=8r^2s+16rs-8+7r^2s^2$

87. $7.75x+9.16x^2-1.27-14.58x^2-18.34$
$=(9.16-14.58)x^2+7.75x+(-1.27-18.34)$
$=-5.42x^2+7.75x-19.61$

89. $[(7.9y^4-6.8y^3+3.3y)+(6.1y^3-5)]$
$-(4.2y^4+1.1y-1)$
$=7.9y^4-6.8y^3+3.3y+6.1y^3-5-4.2y^4$
$-1.1y+1$
$=3.7y^4-0.7y^3+2.2y-4$

91. $3x(2x)=3\cdot 2\cdot x\cdot x=6x^2$

93. $(12x^3)(-x^5)=(12x^3)(-1x^5)$
$=(12)(-1)(x^3)(x^5)$
$=-12x^8$

95. $10x^2(20xy^2)=10\cdot 20x^2\cdot x\cdot y^2=200x^3y^2$

97. Answers may vary

99. Answers may vary

101. $10y-6y^2-y=(10-1)y-6y^2=9y-6y^2$
choice b

103. $(5x-3)+(5x-3)=(5x+5x)+(-3-3)$
$=(5+5)x-6$
$=10x-6$
choice e

105. a. $z+3z=(1+3)z=4z$

b. $z\cdot 3z=3\cdot z\cdot z=3\cdot z^{1+1}=3z^2$

c. $-z-3z=(-1-3)z=-4z$

d. $(-z)(-3z)=(-1\cdot -3)\cdot z\cdot z=3\cdot z^{1+1}=3z^2$

107. Answers may vary

109. $x^2+x^2+xy+xy+xy+xy=2x^2+4xy$

111. $6.4x^2+37.9x+2856.8$
$=6.4(26)^2+37.9(26)+2856.8$
$=4326.4+985.4+2856.8$
$=8168.6$
Costs are predicted to be \$8169 in 2010.

113. $(2.13x^2+21.89x+1190)+(8.71x^2-1.46x+2095)$
$=(2.13+8.71)x^2+(21.89-1.46)x+(1190+2095)$
$=10.84x^2+20.43x+3285$

Section 5.3

Practice Exercises

1. $5y\cdot 2y=(5\cdot 2)(y\cdot y)=10y^2$

2. $(5z^3)\cdot(-0.4z^5)=(5\cdot -0.4)(z^3\cdot z^5)=-2z^8$

3. $\left(-\frac{1}{9}b^6\right)\left(-\frac{7}{8}b^3\right)=\left(-\frac{1}{9}\cdot -\frac{7}{8}\right)(b^6\cdot b^3)=\frac{7}{72}b^9$

4. a. $3x(5x^5+5)=3x(5x^5)+3x(5)=15x^6+15x$

b. $-5x^3(2x^2-9x+2)$
$=-5x^3(2x^2)+(-5x^3)(-9x)+(-5x^3)(2)$
$=-10x^5+45x^4-10x^3$

5. Multiply each term of the first binomial by each term of the second.
$(5x-2)(2x+3)$
$=5x(2x)+5x(3)+(-2)(2x)+(-2)(3)$
$=10x^2+15x-4x-6$
$=10x^2+11x-6$

6. Recall that $a^2 = a \cdot a$, so $(5x-3y)^2 = (5x-3y)(5x-3y)$. Multiply each term of the first binomial by each term of the second.

$$\begin{aligned}&(5x-3y)(5x-3y)\\&= 5x(5x)+5x(-3y)+(-3y)(5x)+(-3y)(-3y)\\&= 25x^2-15xy-15xy+9y^2\\&= 25x^2-30xy+9y^2\end{aligned}$$

7. Multiply each term of the first polynomial by each term of the second.

$$\begin{aligned}&(y+4)(2y^2-3y+5)\\&= y(2y^2)+y(-3y)+y(5)+4(2y^2)\\&\quad +4(-3y)+4(5)\\&= 2y^3-3y^2+5y+8y^2-12y+20\\&= 2y^3+5y^2-7y+20\end{aligned}$$

8. Write $(s+2t)^3$ as $(s + 2t)(s + 2t)(s + 2t)$.

$$\begin{aligned}&(s+2t)(s+2t)(s+2t)\\&= (s^2+2st+2st+4t^2)(s+2t)\\&= (s^2+4st+4t^2)(s+2t)\\&= (s^2+4st+4t^2)s+(s^2+4st+4t^2)(2t)\\&= s^3+4s^2t+4st^2+2s^2t+8st^2+8t^3\\&= s^3+6s^2t+12st^2+8t^3\end{aligned}$$

9.

$$\begin{array}{r}5x^2-3x+5\\ \times \qquad\qquad x-4\\ \hline -20x^2+12x-20\\ 5x^3-3x^2+5x\qquad\\ \hline 5x^3-23x^2+17x-20\end{array}$$

10.

$$\begin{array}{r}x^3-2x^2+1\\ \times \qquad\qquad x^2+2\\ \hline 2x^3-4x^2+2\\ x^5-2x^4\qquad\quad +x^2\qquad\\ \hline x^5-2x^4+2x^3-3x^2+2\end{array}$$

11.

$$\begin{array}{r}5x^2+2x-2\\ x^2-x+3\\ \hline 15x^2+6x-6\\ -5x^3-2x^2+2x\qquad\\ 5x^4+2x^3-2x^2\qquad\qquad\\ \hline 5x^4-3x^3+11x^2+8x-6\end{array}$$

Vocabulary and Readiness Check

1. The expression $5x(3x + 2)$ equals $5x \cdot 3x + 5x \cdot 2$ by the distributive property.

2. The expression $(x + 4)(7x - 1)$ equals $x(7x - 1) + 4(7x - 1)$ by the distributive property.

3. The expression $(5y-1)^2$ equals $(5y - 1)(5y - 1)$.

4. The expression $9x \cdot 3x$ equals $27x^2$.

5. $x^3 \cdot x^5 = x^{3+5} = x^8$

6. $x^2 \cdot x^6 = x^{2+6} = x^8$

7. $x^3 + x^5$ cannot be simplified.

8. $x^2 + x^6$ cannot be simplified.

9. $x^7 \cdot x^7 = x^{7+7} = x^{14}$

10. $x^{11} \cdot x^{11} = x^{11+11} = x^{22}$

11. $x^7 + x^7 = (1+1)x^7 = 2x^7$

12. $x^{11} + x^{11} = (1+1)x^{11} = 2x^{11}$

Exercise Set 5.3

1. $-4n^3 \cdot 7n^7 = (-4 \cdot 7)(n^3 \cdot n^7) = -28n^{10}$

3. $(-3.1x^3)(4x^9) = (-3.1 \cdot 4)(x^3 \cdot x^9) = -12.4x^{12}$

5. $\left(-\frac{1}{3}y^2\right)\left(\frac{2}{5}y\right) = \left(-\frac{1}{3} \cdot \frac{2}{5}\right)(y^2 \cdot y) = -\frac{2}{15}y^3$

7. $(2x)(-3x^2)(4x^5) = (2 \cdot -3 \cdot 4)(x \cdot x^2 \cdot x^5) = -24x^8$

9. $3x(2x+5) = 3x(2x)+3x(5) = 6x^2+15x$

11. $-2a(a+4) = -2a(a)+(-2a)(4) = -2a^2-8a$

13. $3x(2x^2-3x+4) = 3x(2x^2)+3x(-3x)+3x(4)$
$= 6x^3-9x^2+12x$

15. $-2a^2(3a^2-2a+3)$
$=-2a^2(3a^2)+(-2a^2)(-2a)+(-2a^2)(3)$
$=-6a^4+4a^3-6a^2$

17. $-y(4x^3-7x^2y+xy^2+3y^3)$
$=-y(4x^3)+(-y)(-7x^2y)+(-y)(xy^2)$
$\quad+(-y)(3y^3)$
$=-4x^3y+7x^2y^2-xy^3-3y^4$

19. $\frac{1}{2}x^2(8x^2-6x+1)$
$=\frac{1}{2}x^2(8x^2)+\frac{1}{2}x^2(-6x)+\frac{1}{2}x^2(1)$
$=4x^4-3x^3+\frac{1}{2}x^2$

21. $(x+4)(x+3)=x(x)+x(3)+4(x)+4(3)$
$=x^2+3x+4x+12$
$=x^2+7x+12$

23. $(a+7)(a-2)=a(a)+a(-2)+7(a)+7(-2)$
$=a^2-2a+7a-14$
$=a^2+5a-14$

25. $\left(x+\frac{2}{3}\right)\left(x-\frac{1}{3}\right)$
$=x(x)+x\left(-\frac{1}{3}\right)+\frac{2}{3}(x)+\frac{2}{3}\left(-\frac{1}{3}\right)$
$=x^2-\frac{1}{3}x+\frac{2}{3}x-\frac{2}{9}$
$=x^2+\frac{1}{3}x-\frac{2}{9}$

27. $(3x^2+1)(4x^2+7)$
$=3x^2(4x^2)+3x^2(7)+1(4x^2)+1(7)$
$=12x^4+21x^2+4x^2+7$
$=12x^4+25x^2+7$

29. $(2y-4)^2$
$=(2y-4)(2y-4)$
$=2y(2y)+2y(-4)+(-4)(2y)+(-4)(-4)$
$=4y^2-8y-8y+16$
$=4y^2-16y+16$

31. $(4x-3)(3x-5)$
$=4x(3x)+4x(-5)+(-3)(3x)+(-3)(-5)$
$=12x^2-20x-9x+15$
$=12x^2-29x+15$

33. $(3x^2+1)^2=(3x^2+1)(3x^2+1)$
$=3x^2(3x^2)+3x^2(1)+1(3x^2)+1(1)$
$=9x^4+3x^2+3x^2+1$
$=9x^4+6x^2+1$

35. **a.** $(3x+5)+(3x+7)=(3x+3x)+(5+7)$
$=6x+12$

b. $(3x+5)(3x+7)$
$=3x(3x)+3x(7)+5(3x)+5(7)$
$=9x^2+21x+15x+35$
$=9x^2+36x+35$

c. Answers may vary

37. $(x-2)(x^2-3x+7)$
$=x(x^2)+x(-3x)+x(7)+(-2)(x^2)$
$\quad+(-2)(-3x)+(-2)(7)$
$=x^3-3x^2+7x-2x^2+6x-14$
$=x^3-5x^2+13x-14$

39. $(x+5)(x^3-3x+4)$
$=x(x^3)+x(-3x)+x(4)+5(x^3)+5(-3x)+5(4)$
$=x^4-3x^2+4x+5x^3-15x+20$
$=x^4+5x^3-3x^2-11x+20$

41. $(2a-3)(5a^2-6a+4)$
$=2a(5a^2)+2a(-6a)+2a(4)+(-3)(5a^2)$
$\quad+(-3)(-6a)+(-3)(4)$
$=10a^3-12a^2+8a-15a^2+18a-12$
$=10a^3-27a^2+26a-12$

43. $(x+2)^3=(x+2)(x+2)(x+2)$
$=(x^2+2x+2x+4)(x+2)$
$=(x^2+4x+4)(x+2)$
$=(x^2+4x+4)x+(x^2+4x+4)2$
$=x^3+4x^2+4x+2x^2+8x+8$
$=x^3+6x^2+12x+8$

45. $(2y-3)^3$
$= (2y-3)(2y-3)(2y-3)$
$= (4y^2-6y-6y+9)(2y-3)$
$= (4y^2-12y+9)(2y-3)$
$= (4y^2-12y+9)2y+(4y^2-12y+9)(-3)$
$= 8y^3-24y^2+18y-12y^2+36y-27$
$= 8y^3-36y^2+54y-27$

47.
$$\begin{array}{r} 2x-11 \\ \times \quad 6x+1 \\ \hline 2x-11 \\ 12x^2-66x \quad \\ \hline 12x^2-64x-11 \end{array}$$

49.
$$\begin{array}{r} 2x^2+4x-1 \\ \times \quad 5x+1 \\ \hline 2x^2+4x-1 \\ 10x^3+20x^2-5x \quad \\ \hline 10x^3+22x^2-x-1 \end{array}$$

51.
$$\begin{array}{r} 2x^2-7x-9 \\ \times \quad x^2+5x-7 \\ \hline -14x^2+49x+63 \\ 10x^3-35x^2-45x \quad \\ 2x^4-7x^3-9x^2 \quad \quad \\ \hline 2x^4+3x^3-58x^2+4x+63 \end{array}$$

53. $-1.2y(-7y^6) = -1.2(-7)(y \cdot y^6) = 8.4y^7$

55. $-3x(x^2+2x-8)$
$= -3x(x^2)+(-3x)(2x)+(-3x)(-8)$
$= -3x^3-6x^2+24x$

57. $(x+19)(2x+1) = x(2x)+x(1)+19(2x)+19(1)$
$= 2x^2+x+38x+19$
$= 2x^2+39x+19$

59. $\left(x+\frac{1}{7}\right)\left(x-\frac{3}{7}\right)$
$= x(x)+x\left(-\frac{3}{7}\right)+\frac{1}{7}(x)+\frac{1}{7}\left(-\frac{3}{7}\right)$
$= x^2-\frac{3}{7}x+\frac{1}{7}(x)-\frac{3}{49}$
$= x^2-\frac{2}{7}x-\frac{3}{49}$

61. $(3y+5)^2 = (3y+5)(3y+5)$
$= 3y(3y)+3y(5)+5(3y)+5(5)$
$= 9y^2+15y+15y+25$
$= 9y^2+30y+25$

63. $(a+4)(a^2-6a+6)$
$= a(a^2)+a(-6a)+a(6)+4(a^2)+4(-6a)+4(6)$
$= a^3-6a^2+6a+4a^2-24a+24$
$= a^3-2a^2-18a+24$

65. $(2x-5)^3$
$= (2x-5)(2x-5)(2x-5)$
$= (4x^2-10x-10x+25)(2x-5)$
$= (4x^2-20x+25)(2x-5)$
$= (4x^2-20x+25)2x+(4x^2-20x+25)(-5)$
$= 8x^3-40x^2+50x-20x^2+100x-125$
$= 8x^3-60x^2+150x-125$

67. $(4x+5)(8x^2+2x-4)$
$= 4x(8x^2)+4x(2x)+4x(-4)+5(8x^2)$
$\quad +5(2x)+5(-4)$
$= 32x^3+8x^2-16x+40x^2+10x-20$
$= 32x^3+48x^2-6x-20$

69.
$$\begin{array}{r} 3x^2+2x-4 \\ \times \quad 2x^2-4x+3 \\ \hline 9x^2+6x-12 \\ -12x^3-8x^2+16x \quad \\ 6x^4+4x^3-8x^2 \quad \quad \\ \hline 6x^4-8x^3-7x^2+22x-12 \end{array}$$

71. $(2x-5)(2x+5)$
$= 2x(2x)+2x(5)+(-5)(2x)+(-5)(5)$
$= 4x^2+10x-10x-25$
$= 4x^2-25$
The area is $(4x^2-25)$ square yards.

73. $\frac{1}{2}(3x-2)(4x) = 2x(3x-2)$
$= 2x(3x)+2x(-2)$
$= 6x^2-4x$
The area is $(6x^2-4x)$ square inches.

75. $(5x)^2 = (5x)(5x) = (5\cdot 5)(x\cdot x) = 25x^2$

77. $(-3y^3)^2 = (-3y^3)(-3y^3)$
$= (-3\cdot -3)(y^3\cdot y^3)$
$= 9y^6$

79. left rectangle: $x\cdot x = x^2$
right rectangle: $x\cdot 3 = 3x$
left rectangle + right rectangle: x^2+3x

81. top left rectangle: $x\cdot x = x^2$
top right rectangle: $x\cdot 3 = 3x$
bottom left rectangle: $2\cdot x = 2x$
bottom right rectangle: $2\cdot 3 = 6$
entire figure: $x^2+3x+2x+6 = x^2+5x+6$

83. $5a+6a = (5+6)a = 11a$

85. $(5x)^2+(2y)^2 = (5x)(5x)+(2y)(2y)$
$= 25x^2+4y^2$

87. $(3x-1)+(10x-6) = (3x+10x)+(-1-6)$
$= 13x-7$

89. $(3x-1)(10x-6)$
$= 3x(10x)+3x(-6)+(-1)(10x)+(-1)(-6)$
$= 30x^2-18x-10x+6$
$= 30x^2-28x+6$

91. $(3x-1)-(10x-6) = 3x-1-10x+6$
$= (3x-10x)+(-1+6)$
$= -7x+5$

93. a. $(a+b)(a-b) = a(a)+a(-b)+b(a)+b(-b)$
$= a^2-ab+ab-b^2$
$= a^2-b^2$

b. $(2x+3y)(2x-3y)$
$= 2x(2x)+2x(-3y)+3y(2x)+3y(-3y)$
$= 4x^2-6xy+6xy-9y^2$
$= 4x^2-9y^2$

c. $(4x+7)(4x-7)$
$= 4x(4x)+4x(-7)+7(4x)+7(-7)$
$= 16x^2-28x+28x-49$
$= 16x^2-49$

d. Answers may vary

95. larger square: $(x+3)^2 = (x+3)(x+3)$
$= x(x)+x(3)+3(x)+3(3)$
$= x^2+3x+3x+9$
$= x^2+6x+9$
smaller square: $2^2 = 2\cdot 2 = 4$
shaded region: $x^2+6x+9-4 = x^2+6x+5$
The area of the shaded region is
(x^2+6x+5) square units.

Section 5.4

Practice Exercises

1. $(x+2)(x-5)$
$= (x)(x)+(x)(-5)+(2)(x)+(2)(-5)$
$= x^2-5x+2x-10$
$= x^2-3x-10$

2. $(4x-9)(x-1)$
$= 4x(x)+4x(-1)+(-9)(x)+(-9)(-1)$
$= 4x^2-4x-9x+9$
$= 4x^2-13x+9$

3. $3(x+5)(3x-1) = 3(3x^2-x+15x-5)$
$= 3(3x^2+14x-5)$
$= 9x^2+42x-15$

4. $(4x-1)^2$
$= (4x-1)(4x-1)$
$= (4x)(4x) + (4x)(-1) + (-1)(4x) + (-1)(-1)$
$= 16x^2 - 4x - 4x + 1$
$= 16x^2 - 8x + 1$

5. a. $(b+3)^2 = b^2 + 2(b)(3) + 3^2 = b^2 + 6b + 9$

b. $(x-y)^2 = x^2 - 2(x)(y) + y^2 = x^2 - 2xy + y^2$

c. $(3y+2)^2 = (3y)^2 + 2(3y)(2) + 2^2$
$= 9y^2 + 12y + 4$

d. $(a^2 - 5b)^2 = (a^2)^2 - 2(a^2)(5b) + (5b)^2$
$= a^4 - 10a^2b + 25b^2$

6. a. $3(x+5)(x-5) = 3(x^2 - 5^2)$
$= 3x(x^2 - 25)$
$= 3x^2 - 75$

b. $(4b-3)(4b+3) = (4b)^2 - 3^2 = 16b^2 - 9$

c. $\left(x+\frac{2}{3}\right)\left(x-\frac{2}{3}\right) = x^2 - \left(\frac{2}{3}\right)^2 = x^2 - \frac{4}{9}$

d. $(5s+t)(5s-t) = (5s)^2 - t^2 = 25s^2 - t^2$

e. $(2y - 3z^2)(2y + 3z^2) = (2y)^2 - (3z^2)^2$
$= 4y^2 - 9z^4$

7. a. $(4x+3)(x-6) = 4x^2 - 24x + 3x - 18$
$= 4x^2 - 21x - 18$

b. $(7b-2)^2 = (7b)^2 - 2(7b)(2) + 2^2$
$= 49b^2 - 28b + 4$

c. $(x+0.4)(x-0.4) = x^2 - (0.4)^2 = x^2 - 0.16$

d. $(x^2 - 3)(3x^4 + 2) = 3x^6 + 2x^2 - 9x^4 - 6$

e. $(x+1)(x^2 + 5x - 2)$
$= x(x^2 + 5x - 2) + 1(x^2 + 5x - 2)$
$= x^3 + 5x^2 - 2x + x^2 + 5x - 2$
$= x^3 + 6x^2 + 3x - 2$

Vocabulary and Readiness Check

1. $(x+4)^2 = x^2 + 2(x)(4) + 4^2$
$= x^2 + 8x + 16 \neq x^2 + 16$
The statement is false.

2. $(x+6)(2x-1) = 2x^2 - x + 12x - 6$
$= 2x^2 + 11x - 6$
The statement is true.

3. $(x+4)(x-4) = x^2 - 4^2 = x^2 - 16 \neq x^2 + 16$
The statement is false.

4. $(x-1)(x^3 + 3x - 1)$
$= x(x^3 + 3x - 1) - 1(x^3 + 3x - 1)$
$= x^4 + 3x^2 - x - x^3 - 3x + 1$
$= x^4 - x^3 + 3x^2 - 4x + 1$
This is a polynomial of degree 4; the statement is false.

Exercise Set 5.4

1. $(x+3)(x+4) = x^2 + 4x + 3x + 12 = x^2 + 7x + 12$

3. $(x-5)(x+10) = x^2 + 10x - 5x - 50$
$= x^2 + 5x - 50$

5. $(5x-6)(x+2) = 5x^2 + 10x - 6x - 12$
$= 5x^2 + 4x - 12$

7. $(y-6)(4y-1) = 4y^2 - 1y - 24y + 6$
$= 4y^2 - 25y + 6$

9. $(2x+5)(3x-1) = 6x^2 - 2x + 15x - 5$
$= 6x^2 + 13x - 5$

11. $(x-2)^2 = x^2 - 2(x)(2) + 2^2 = x^2 - 4x + 4$

13. $(2x-1)^2 = (2x)^2 - 2(2x)(1) + (1)^2$
$= 4x^2 - 4x + 1$

15. $(3a-5)^2 = (3a)^2 - 2(3a)(5) + 5^2$
$= 9a^2 - 30a + 25$

17. $(5x+9)^2 = (5x)^2 + 2(5x)(9) + 9^2$
$= 25x^2 + 90x + 81$

19. Answers may vary

21. $(a-7)(a+7)=a^2-7^2=a^2-49$

23. $(3x-1)(3x+1)=(3x)^2-1^2=9x^2-1$

25. $\left(3x-\frac{1}{2}\right)\left(3x+\frac{1}{2}\right)=(3x)^2-\left(\frac{1}{2}\right)^2=9x^2-\frac{1}{4}$

27. $(9x+y)(9x-y)=(9x)^2-y^2=81x^2-y^2$

29. $(2x+0.1)(2x-0.1)=(2x)^2-(0.1)^2=4x^2-0.01$

31. $(a+5)(a+4)=a^2+4a+5a+20=a^2+9a+20$

33. $(a+7)^2=a^2+2(a)(7)+7^2=a^2+14a+49$

35. $(4a+1)(3a-1)=12a^2-4a+3a-1$
$=12a^2-a-1$

37. $(x+2)(x-2)=x^2-2^2=x^2-4$

39. $(3a+1)^2=(3a)^2+2(3a)(1)+1^2=9a^2+6a+1$

41. $(x^2+y)(4x-y^4)=4x^3-x^2y^4+4xy-y^5$

43. $(x+3)(x^2-6x+1)$
$=x(x^2-6x+1)+3(x^2-6x+1)$
$=x^3-6x^2+x+3x^2-18x+3$
$=x^3-3x^2-17x+3$

45. $(2a-3)^2=(2a)^2-2(2a)(3)+(3)^2$
$=4a^2-12a+9$

47. $(5x-6z)(5x+6z)=(5x)^2-(6z)^2=25x^2-36z^2$

49. $(x^5-3)(x^5-5)=x^{10}-5x^5-3x^5+15$
$=x^{10}-8x^5+15$

51. $\left(x-\frac{1}{3}\right)\left(x+\frac{1}{3}\right)=x^2-\left(\frac{1}{3}\right)^2=x^2-\frac{1}{9}$

53. $(a^3+11)(a^4-3)=a^7-3a^3+11a^4-33$

55. $3(x-2)^2=3[x^2-2(x)(2)+2^2]$
$=3(x^2-4x+4)$
$=3x^2-12x+12$

57. $(3b+7)(2b-5)=6b^2-15b+14b-35$
$=6b^2-b-35$

59. $(7p-8)(7p+8)=(7p)^2-(8)^2=49p^2-64$

61. $\left(\frac{1}{3}a^2-7\right)\left(\frac{1}{3}a^2+7\right)=\left(\frac{1}{3}a^2\right)^2-(7)^2$
$=\frac{1}{9}a^4-49$

63. $5x^2(3x^2-x+2)=5x^2(3x^2)+5x^2(-x)+5x^2(2)$
$=15x^4-5x^3+10x^2$

65. $(2r-3s)(2r+3s)=(2r)^2-(3s)^2=4r^2-9s^2$

67. $(3x-7y)^2=(3x)^2-2(3x)(7y)+(7y)^2$
$=9x^2-42xy+49y^2$

69. $(4x+5)(4x-5)=(4x)^2-5^2=16x^2-25$

71. $(8x+4)^2=(8x)^2+2(8x)(4)+(4)^2$
$=64x^2+64x+16$

73. $\left(a-\frac{1}{2}y\right)\left(a+\frac{1}{2}y\right)=a^2-\left(\frac{1}{2}y\right)^2=a^2-\frac{1}{4}y^2$

75. $\left(\frac{1}{5}x-y\right)\left(\frac{1}{5}x+y\right)=\left(\frac{1}{5}x\right)^2-y^2=\frac{1}{25}x^2-y^2$

77. $(a+1)(3a^2-a+1)$
$=a(3a^2-a+1)+1(3a^2-a+1)$
$=3a^3-a^2+a+3a^2-a+1$
$=3a^3+2a^2+1$

79. $(2x+1)^2=(2x)^2+2(2x)(1)+1^2=4x^2+4x+1$
The area is $(4x^2+4x+1)$ square feet.

81. $\frac{50b^{10}}{70b^5}=\frac{50}{70}b^{10-5}=\frac{5b^5}{7}$

83. $\frac{8a^{17}b^{15}}{-4a^7b^{10}} = \frac{8}{-4}a^{17-7}b^{15-10} = -2a^{10}b^5$

85. $\frac{2x^4y^{12}}{3x^4y^4} = \frac{2}{3}x^{4-4}y^{12-4} = \frac{2y^8}{3}$

87. (−1, 1) and (2, 2)
$m = \frac{y_2 - y_1}{x_2 - x_1} = \frac{2-1}{2-(-1)} = \frac{1}{3}$

89. (−1, −2) and (1, 0)
$m = \frac{y_2 - y_1}{x_2 - x_1} = \frac{0-(-2)}{1-(-1)} = \frac{2}{2} = 1$

91. $(a-b)^2 = a^2 - 2ab + b^2$
Choice c.

93. $(a+b)^2 = a^2 + 2ab + b^2$
Choice d.

95. From FOIL, the first term in the result is $(x^{\square})^2 = x^{2\square}$. Thus, $2\square = 4$ so $\square = 2$.

97. $\frac{1}{2}(5a+b)(5a-b) = \frac{1}{2}(25a^2 - b^2) = \frac{25a^2}{2} - \frac{b^2}{2}$
The area is $\left(\frac{25a^2}{2} - \frac{b^2}{2}\right)$ square units.

99. $(5x-3)^2 - (x+1)^2$
$= (25x^2 - 30x + 9) - (x^2 + 2x + 1)$
$= 25x^2 - 30x + 9 - x^2 - 2x - 1$
$= (24x^2 - 32x + 8)$
The shaded area is
$(24x^2 - 32x + 8)$ square meters.

101. $(x+5)(x+5) = (x+5)^2$
$= x^2 + 2(x)(5) + 5^2$
$= x^2 + 10x + 25$
The area is $(x^2 + 10x + 25)$ square units.

103. Answers may vary

105. $[(x+y)-3][(x+y)+3] = (x+y)^2 - 3^2$
$= x^2 + 2xy + y^2 - 9$

107. $[(a-3)+b][(a-3)-b] = (a-3)^2 - b^2$
$= a^2 - 6a + 9 - b^2$

Integrated Review

1. $(5x^2)(7x^3) = (5\cdot 7)(x^2 \cdot x^3) = 35x^5$

2. $(4y^2)(8y^7) = (4\cdot 8)(y^2 \cdot y^7) = 32y^9$

3. $-4^2 = -(4\cdot 4) = -16$

4. $(-4)^2 = (-4)(-4) = 16$

5. $(x-5)(2x+1) = 2x^2 + x - 10x - 5$
$= 2x^2 - 9x - 5$

6. $(3x-2)(x+5) = 3x^2 + 15x - 2x - 10$
$= 3x^2 + 13x - 10$

7. $(x-5)+(2x+1) = x - 5 + 2x + 1 = 3x - 4$

8. $(3x-2)+(x+5) = 3x - 2 + x + 5 = 4x + 3$

9. $\frac{7x^9y^{12}}{x^3y^{10}} = 7x^{9-3}y^{12-10} = 7x^6y^2$

10. $\frac{20a^2b^8}{14a^2b^2} = \frac{20}{14}a^{2-2}b^{8-2} = \frac{10b^6}{7}$

11. $(12m^7n^6)^2 = 12^2m^{7\cdot 2}n^{6\cdot 2} = 144m^{14}n^{12}$

12. $(4y^9z^{10})^3 = 4^3y^{9\cdot 3}z^{10\cdot 3} = 64y^{27}z^{30}$

13. $3(4y-3)(4y+3) = 3[(4y)^2 - 3^2]$
$= 3(16y^2 - 9)$
$= 48y^2 - 27$

14. $2(7x-1)(7x+1) = 2[(7x)^2 - 1^2]$
$= 2(49x^2 - 1)$
$= 98x^2 - 2$

15. $(x^7y^5)^9 = x^{7\cdot 9}y^{5\cdot 9} = x^{63}y^{45}$

16. $(3^1x^9)^3 = 3^{1\cdot 3}x^{9\cdot 3} = 3^3x^{27} = 27x^{27}$

17. $(7x^2-2x+3)-(5x^2+9)$
$=7x^2-2x+3-5x^2-9$
$=2x^2-2x-6$

18. $(10x^2+7x-9)-(4x^2-6x+2)$
$=10x^2+7x-9-4x^2+6x-2$
$=6x^2+13x-11$

19. $0.7y^2-1.2+1.8y^2-6y+1=2.5y^2-6y-0.2$

20. $7.8x^2-6.8x+3.3+0.6x^2-9$
$=8.4x^2-6.8x-5.7$

21. $(x+4y)^2=x^2+2(x)(4y)+(4y)^2$
$=x^2+8xy+16y^2$

22. $(y-9z)^2=y^2-2(y)(9z)+(9z)^2$
$=y^2-18yz+81z^2$

23. $(x+4y)+(x+4y)=x+4y+x+4y=2x+8y$

24. $(y-9z)+(y-9z)=y-9z+y-9z=2y-18z$

25. $7x^2-6xy+4(y^2-xy)=7x^2-6xy+4y^2-4xy$
$=7x^2-10xy+4y^2$

26. $5a^2-3ab+6(b^2-a^2)=5a^2-3ab+6b^2-6a^2$
$=-a^2-3ab+6b^2$

27. $(x-3)(x^2+5x-1)$
$=x(x^2+5x-1)-3(x^2+5x-1)$
$=x^3+5x^2-x-3x^2-15x+3$
$=x^3+2x^2-16x+3$

28. $(x+1)(x^2-3x-2)$
$=x(x^2-3x-2)+1(x^2-3x-2)$
$=x^3-3x^2-2x+x^2-3x-2$
$=x^3-2x^2-5x-2$

29. $(2x^3-7)(3x^2+10)$
$=2x^3(3x^2)+2x^3(10)-7(3x^2)-7(10)$
$=6x^5+20x^3-21x^2-70$

30. $(5x^3-1)(4x^4+5)$
$=5x^3(4x^4)+5x^3(5)-1(4x^4)-1(5)$
$=20x^7+25x^3-4x^4-5$

31. $(2x-7)(x^2-6x+1)$
$=2x(x^2-6x+1)-7(x^2-6x+1)$
$=2x^3-12x^2+2x-7x^2+42x-7$
$=2x^3-19x^2+44x-7$

32. $(5x-1)(x^2+2x-3)$
$=5x(x^2+2x-3)-1(x^2+2x-3)$
$=5x^3+10x^2-15x-x^2-2x+3$
$=5x^3+9^2-17x+3$

33. $5x^3+5y^3$ cannot be simplified.

34. $(5x^3)(5y^3)=5\cdot 5x^3y^3=25x^3y^3$

35. $(5x^3)^3=5^3x^{3\cdot 3}=125x^9$

36. $\dfrac{5x^3}{5y^3}=\dfrac{x^3}{y^3}$

37. $x+x=2x$

38. $x\cdot x=x^2$

Section 5.5

Practice Exercises

1. a. $5^{-3}=\dfrac{1}{5^3}=\dfrac{1}{125}$

b. $3y^{-4}=3\cdot\dfrac{1}{y^4}=\dfrac{3}{y^4}$

c. $3^{-1}+2^{-1}=\dfrac{1}{3}+\dfrac{1}{2}=\dfrac{2}{6}+\dfrac{3}{6}=\dfrac{5}{6}$

d. $(-5)^{-2}=\dfrac{1}{(-5)^2}=\dfrac{1}{(-5)(-5)}=\dfrac{1}{25}$

e. $\dfrac{1}{x^{-5}}=\dfrac{1}{\frac{1}{x^5}}=x^5$

f. $\frac{1}{4^{-3}} = \frac{1}{\frac{1}{4^3}} = \frac{4^3}{1} = 64$

2. a. $\frac{1}{s^{-5}} = \frac{s^5}{1} = s^5$

b. $\frac{1}{2^{-3}} = \frac{2^3}{1} = 8$

c. $\frac{x^{-7}}{y^{-5}} = \frac{y^5}{x^7}$

d. $\frac{4^{-3}}{3^{-2}} = \frac{3^2}{4^3} = \frac{9}{64}$

3. a. $\frac{x^{-3}}{x^2} = x^{-3-2} = x^{-5} = \frac{1}{x^5}$

b. $\frac{5}{y^{-7}} = 5 \cdot \frac{1}{y^{-7}} = 5 \cdot y^7 = 5y^7$

c. $\frac{z}{z^{-4}} = \frac{z^1}{z^{-4}} = z^{1-(-4)} = z^5$

4. a. $\left(\frac{3}{4}\right)^{-2} = \frac{3^{-2}}{4^{-2}} = \frac{4^2}{3^2} = \frac{16}{9}$

b.
$$\begin{aligned}\frac{x^2(x^5)^3}{x^7} &= \frac{x^2 \cdot x^{15}}{x^7}\\ &= \frac{x^{2+15}}{x^7}\\ &= \frac{x^{17}}{x^7}\\ &= x^{17-7}\\ &= x^{10}\end{aligned}$$

c.
$$\begin{aligned}\left(\frac{5p^8}{q}\right)^{-2} &= \frac{5^{-2}(p^8)^{-2}}{q^{-2}}\\ &= \frac{5^{-2}p^{-16}}{q^{-2}}\\ &= \frac{q^2}{5^2p^{16}}\\ &= \frac{q^2}{25p^{16}}\end{aligned}$$

d.
$$\begin{aligned}\frac{6^{-2}x^{-4}y^{-7}}{6^{-3}x^3y^{-9}} &= 6^{-2-(-3)}x^{-4-3}y^{-7-(-9)}\\ &= 6^1x^{-7}y^2\\ &= \frac{6y^2}{x^7}\end{aligned}$$

e. $(a^4b^{-3})^{-5} = a^{-20}b^{15} = \frac{b^{15}}{a^{20}}$

f.
$$\begin{aligned}\left(\frac{-3x^4y}{x^2y^{-2}}\right)^3 &= \frac{(-3)^3x^{12}y^3}{x^2y^{-2}}\\ &= \frac{-27x^{12}y^3}{x^6y^{-6}}\\ &= -27x^{12-6}y^{3-(-6)}\\ &= -27x^6y^9\end{aligned}$$

5. a. $0.000007 = 7 \times 10^{-6}$
The decimal point is moved 6 places, and the original number is less than 1, so the count is –6.

b. $20,700,000 = 2.07 \times 10^7$
The decimal point is moved 7 places, and the original number is 10 or greater, so the count is 7.

c. $0.0043 = 4.3 \times 10^{-3}$
The decimal point is moved 3 places, and the original number is less than 1, so the count is –3.

d. $812,000,000 = 8.12 \times 10^8$
The decimal point is moved 8 places, and the original number is 10 or greater, so the count is 8.

6. a. Move the decimal point 4 places to the left.
$3.67\times10^{-4} = 0.000367$

b. Move the decimal point 6 places to the right.
$8.954\times10^{6} = 8{,}954{,}000$

c. Move the decimal point 5 places to the left.
$2.009\times10^{-5} = 0.00002009$

d. Move the decimal point 3 places to the right.
$4.054\times10^{3} = 4054$

7. a. $(5\times10^{-4})(8\times10^{6}) = (5\cdot8)\times(10^{-4}\cdot10^{6})$
$= 40\times10^{2}$
$= 4000$

b. $\dfrac{64\times10^{3}}{32\times10^{-7}} = \dfrac{64}{32}\times10^{3-(-7)}$
$= 2\times10^{10}$
$= 20{,}000{,}000{,}000$

Calculator Explorations

1. $5.31\times10^{3} = 5.31$ EE 3
2. $-4.8\times10^{14} = -4.8$ EE 14
3. $6.6\times10^{-9} = 6.6$ EE -9
4. $-9.9811\times10^{-2} = -9.9811$ EE -2
5. $3{,}000{,}000\times5{,}000{,}000 = 1.5\times10^{13}$
6. $230{,}000\times1000 = 2.3\times10^{8}$
7. $(3.26\times10^{6})(2.5\times10^{13}) = 8.15\times10^{19}$
8. $(8.76\times10^{-4})(1.237\times10^{9}) = 1.083612\times10^{6}$

Vocabulary and Readiness Check

1. The expression x^{-3} equals $\underline{\dfrac{1}{x^3}}$.
2. The expression 5^{-4} equals $\underline{\dfrac{1}{625}}$.
3. The number 3.021×10^{-3} is written in <u>scientific notation</u>.
4. The number 0.0261 is written in <u>standard form</u>.
5. $5x^{-2} = 5\cdot\dfrac{1}{x^2} = \dfrac{5}{x^2}$
6. $3x^{-3} = 3\cdot\dfrac{1}{x^3} = \dfrac{3}{x^3}$
7. $\dfrac{1}{y^{-6}} = \dfrac{1}{\frac{1}{y^6}} = \dfrac{y^6}{1} = y^6$
8. $\dfrac{1}{x^{-3}} = \dfrac{1}{\frac{1}{x^3}} = \dfrac{x^3}{1} = x^3$
9. $\dfrac{4}{y^{-3}} = \dfrac{4}{\frac{1}{y^3}} = 4\cdot\dfrac{y^3}{1} = 4y^3$
10. $\dfrac{16}{y^7} = \dfrac{16}{\frac{1}{y^7}} = 16\cdot\dfrac{y^7}{1} = 16y^7$

Exercise Set 5.5

1. $4^{-3} = \dfrac{1}{4^3} = \dfrac{1}{64}$

3. $(-2)^{-4} = \dfrac{1}{(-2)^4} = \dfrac{1}{16}$

5. $7x^{-3} = 7\cdot\dfrac{1}{x^3} = \dfrac{7}{x^3}$

7. $\left(\dfrac{1}{2}\right)^{-5} = \dfrac{1^{-5}}{2^{-5}} = \dfrac{2^5}{1^5} = 32$

9. $\left(-\dfrac{1}{4}\right)^{-3} = \dfrac{(-1)^{-3}}{(4)^{-3}} = \dfrac{4^3}{(-1)^3} = \dfrac{64}{-1} = -64$

11. $3^{-1} + 2^{-1} = \dfrac{1}{3} + \dfrac{1}{2} = \dfrac{2}{6} + \dfrac{3}{6} = \dfrac{5}{6}$

13. $\frac{1}{p^{-3}} = p^3$

15. $\frac{p^{-5}}{q^{-4}} = \frac{q^4}{p^5}$

17. $\frac{x^{-2}}{x} = x^{-2-1} = x^{-3} = \frac{1}{x^3}$

19. $\frac{z^{-4}}{z^{-7}} = z^{-4-(-7)} = z^3$

21. $3^{-2} + 3^{-1} = \frac{1}{3^2} + \frac{1}{3} = \frac{1}{9} + \frac{1}{3} = \frac{1}{9} + \frac{3}{9} = \frac{4}{9}$

23. $\frac{-1}{p^{-4}} = -1(p^4) = -p^4$

25. $-2^0 - 3^0 = -1(1) - 1 = -1 - 1 = -2$

27. $\frac{x^2 x^5}{x^3} = x^{2+5-3} = x^4$

29. $\frac{p^2 p}{p^{-1}} = p^{2+1-(-1)} = p^4$

31. $\frac{(m^5)^4 m}{m^{10}} = \frac{m^{20} m}{m^{10}} = m^{20+1-10} = m^{11}$

33. $\frac{r}{r^{-3} r^{-2}} = r^{1-(-3)-(-2)} = r^6$

35. $(x^5 y^3)^{-3} = x^{5(-3)} y^{3(-3)} = x^{-15} y^{-9} = \frac{1}{x^{15} y^9}$

37. $\frac{(x^2)^3}{x^{10}} = \frac{x^6}{x^{10}} = x^{6-10} = x^{-4} = \frac{1}{x^4}$

39. $\frac{(a^5)^2}{(a^3)^4} = \frac{a^{10}}{a^{12}} = a^{10-12} = a^{-2} = \frac{1}{a^2}$

41. $\frac{8k^4}{2k} = \frac{8}{2} \cdot k^{4-1} = 4k^3$

43. $\frac{-6m^4}{-2m^3} = \frac{-6}{-2} \cdot m^{4-3} = 3m$

45. $\frac{-24a^6 b}{6ab^2} = \frac{-24}{6} \cdot a^{6-1} b^{1-2} = -4a^5 b^{-1} = -\frac{4a^5}{b}$

47.
$$\begin{aligned}(-2x^3 y^{-4})(3x^{-1} y) &= -2(3) x^{3+(-1)} y^{-4+1} \\ &= -6x^2 y^{-3} \\ &= -\frac{6x^2}{y^3}\end{aligned}$$

49. $(a^{-5} b^2)^{-6} = a^{-5(-6)} b^{2(-6)} = a^{30} b^{-12} = \frac{a^{30}}{b^{12}}$

51.
$$\begin{aligned}\left(\frac{x^{-2} y^4}{x^3 y^7}\right)^2 &= \frac{x^{-2(2)} y^{4(2)}}{x^{3(2)} y^{7(2)}} \\ &= \frac{x^{-4} y^8}{x^6 y^{14}} \\ &= x^{-4-6} y^{8-14} \\ &= x^{-10} y^{-6} \\ &= \frac{1}{x^{10} y^6}\end{aligned}$$

53. $\frac{4^2 z^{-3}}{4^3 z^{-5}} = 4^{2-3} z^{-3-(-5)} = 4^{-1} z^2 = \frac{z^2}{4}$

55.
$$\begin{aligned}\frac{2^{-3} x^{-4}}{2^2 x} &= 2^{-3-2} x^{-4-1} \\ &= 2^{-5} x^{-5} \\ &= \frac{1}{2^5 x^5} \\ &= \frac{1}{32x^5}\end{aligned}$$

57.
$$\begin{aligned}\frac{7ab^{-4}}{7^{-1} a^{-3} b^2} &= 7^{1-(-1)} a^{1-(-3)} b^{-4-2} \\ &= 7^2 a^4 b^{-6} \\ &= \frac{49a^4}{b^6}\end{aligned}$$

59. $\left(\frac{a^{-5}b}{ab^3}\right)^{-4} = \frac{a^{-5(-4)}b^{-4}}{a^{-4}b^{3(-4)}}$
$= \frac{a^{20}b^{-4}}{a^{-4}b^{-12}}$
$= a^{20-(-4)}b^{-4-(-12)}$
$= a^{24}b^8$

61. $\frac{(xy^3)^5}{(xy)^{-4}} = \frac{x^5y^{3(5)}}{x^{-4}y^{-4}}$
$= \frac{x^5y^{15}}{x^{-4}y^{-4}}$
$= x^{5-(-4)}y^{15-(-4)}$
$= x^9y^{19}$

63. $\frac{(-2xy^{-3})^{-3}}{(xy^{-1})^{-1}} = \frac{(-2)^{-3}x^{-3}y^9}{x^{-1}y^1}$
$= (-2)^{-3}x^{-3-(-1)}y^{9-1}$
$= (-2)^{-3}x^{-2}y^8$
$= \frac{y^8}{(-2)^3x^2}$
$= -\frac{y^8}{8x^2}$

65. $\frac{6x^2y^3}{-7xy^5} = -\frac{6}{7}x^{2-1}y^{3-5} = -\frac{6}{7}x^1y^{-2} = -\frac{6x}{7y^2}$

67. $\frac{(a^4b^{-7})^{-5}}{(5a^2b^{-1})^{-2}} = \frac{(a^4)^{-5}(b^{-7})^{-5}}{5^{-2}(a^2)^{-2}(b^{-1})^{-2}}$
$= \frac{a^{-20}b^{35}}{5^{-2}a^{-4}b^2}$
$= 5^2a^{-20-(-4)}b^{35-2}$
$= 5^2a^{-16}b^{33}$
$= \frac{25b^{33}}{a^{16}}$

69. $78,000 = 7.8\times10^4$

71. $0.00000167 = 1.67\times10^{-6}$

73. $0.00635 = 6.35\times10^{-3}$

75. $1,160,000 = 1.16\times10^6$

77. $2,000,000,000 = 2\times10^9$

79. $1,212,000,000 = 1.212\times10^9$

81. $8.673\times10^{-10} = 0.0000000008673$

83. $3.3\times10^{-2} = 0.033$

85. $2.032\times10^4 = 20,320$

87. $7.0\times10^8 = 700,000,000$

89. $9.460\times10^{12} = 9,460,000,000,000$

91. The longest bar corresponds to Yahoo! sites, with approximately 130,000,000 or 1.3×10^8 visits.

93. $1,000,000,000 = 1\times10^9$

95. $5.7\times10^7 = 57,000,000$

97. $(1.2\times10^{-3})(3\times10^{-2}) = (1.2\cdot3)\times(10^{-3}\cdot10^{-2})$
$= 3.6\times10^{-5}$
$= 0.000036$

99. $(4\times10^{-10})(7\times10^{-9}) = (4\cdot7)\times(10^{-10}\cdot10^{-9})$
$= 28\times10^{-19}$
$= 2.8\times10^{-18}$
$= 0.0000000000000000028$

101. $\frac{8\times10^{-1}}{16\times10^5} = \frac{8}{16}\times10^{-1-5}$
$= 0.5\times10^{-6}$
$= 5\times10^{-7}$
$= 0.0000005$

103. $\frac{1.4\times10^{-2}}{7\times10^{-8}} = \frac{1.4}{7}\times10^{-2-(-8)}$
$= 0.2\times10^6$
$= 2.0\times10^5$
$= 200,000$

105. $\frac{5x^7}{3x^4} = \frac{5}{3}\cdot x^{7-4} = \frac{5x^3}{3}$

107. $\frac{15z^4y^3}{21zy} = \frac{15}{21}z^{4-1}y^{3-1} = \frac{5z^3y^2}{7}$

109. $\frac{1}{y}(5y^2 - 6y + 5) = \frac{1}{y}(5y^2) + \frac{1}{y}(-6y) + \frac{1}{y}(5)$
$= 5y - 6 + \frac{5}{y}$

111. $\left(\frac{3x^{-2}}{z}\right)^3 = \frac{3^3x^{-6}}{z^3} = \frac{27}{x^6z^3}$

The volume is $\frac{27}{x^6z^3}$ cubic inches.

113. $(2a^3)^3a^4 + a^5a^8 = 2^3(a^3)^3a^4 + a^{5+8}$
$= 8a^9a^4 + a^{13}$
$= 8a^{13} + a^{13}$
$= 9a^{13}$

115. $x^{-5} = \frac{1}{x^5}$

117. Answers may vary

119. a. $9.7 \times 10^{-2} = 0.097$
$1.3 \times 10^1 = 130$
1.3×10^1 is larger.

b. $8.6 \times 10^5 = 860,000$
$4.4 \times 10^7 = 44,000,000$
4.4×10^7 is larger.

c. $6.1 \times 10^{-2} = 0.061$
$5.6 \times 10^{-4} = 0.00056$
6.1×10^{-2} is larger.

121. a. $5^{-1} = \frac{1}{5}$
$5^{-2} = \frac{1}{5^2} = \frac{1}{25}$
$\frac{1}{5} > \frac{1}{25}$
The statement is false.

b. $\left(\frac{1}{5}\right)^{-1} = \frac{1^{-1}}{5^{-1}} = \frac{5}{1} = 5$
$\left(\frac{1}{5}\right)^{-2} = \frac{1^{-2}}{5^{-2}} = \frac{5^2}{1^2} = 25$
$5 < 25$
The statement is true.

c. The statement is false, since the statement in part a is false.

123. $(x^{-3s})^3 = x^{-3s\cdot 3} = x^{-9s} = \frac{1}{x^{9s}}$

125. $a^{4m+1} \cdot a^4 = a^{4m+1+4} = a^{4m+5}$

127. $(6.785 \times 10^{-4})(4.68 \times 10^{10}) = 31,753,800$

129. $t = \frac{d}{r}$
$t = \frac{93,000,000}{1.86 \times 10^5}$
$= \frac{9.3 \times 10^7}{1.86 \times 10^5}$
$= \frac{9.3}{1.86} \times \frac{10^7}{10^5}$
$= 5 \times 10^2$
$= 500$
It takes the light of the sun 500 seconds to reach Earth.

Section 5.6

Practice Exercises

1. $\frac{8t^3 + 4t^2}{4t^2} = \frac{8t^3}{4t^2} + \frac{4t^2}{4t^2} = 2t + 1$

Check: $4t^2(2t+1) = 4t^2(2t) + 4t^2(1)$
$= 8t^3 + 4t^2$

2. $\dfrac{16x^6+20x^3-12x}{4x^2}=\dfrac{16x^6}{4x^2}+\dfrac{20x^3}{4x^2}-\dfrac{12x}{4x^2}$
$=4x^4+5x-\dfrac{3}{x}$

Check: $4x^2\left(4x^4+5x-\dfrac{3}{x}\right)$
$=4x^2(4x^4)+4x^2(5x)-4x^2\left(\dfrac{3}{x}\right)$
$=16x^6+20x^3-12x$

3. $\dfrac{15x^4y^4-10xy+y}{5xy}=\dfrac{15x^4y^4}{5xy}-\dfrac{10xy}{5xy}+\dfrac{y}{5xy}$
$=3x^3y^3-2+\dfrac{1}{5x}$

Check: $5xy\left(3x^3y^3-2+\dfrac{1}{5x}\right)$
$=5xy(3x^3y^3)-5xy(2)+5xy\left(\dfrac{1}{5x}\right)$
$=15x^4y^4-10xy+y$

4.

$$\begin{array}{r|l} & x+3 \\ x+2 & x^2+5x+6 \\ & \underline{x^2+2x} \\ & \quad 3x+6 \\ & \quad \underline{3x+6} \\ & \qquad 0 \end{array}$$

Check: $(x+2)\cdot(x+3)+0=x^2+5x+6$
The quotient checks.

5.

$$\begin{array}{r|l} & 2x+3 \\ 2x+1 & 4x^2+8x-7 \\ & \underline{4x^2+2x} \\ & \quad 6x-7 \\ & \quad \underline{6x+3} \\ & \quad -10 \end{array}$$

$\dfrac{4x^2+8x-7}{2x+1}=2x+3+\dfrac{-10}{2x+1}$

Check:
$(2x+1)(2x+3)+(-10)=(4x^2+8x+3)-10$
$=4x^2+8x-7$
The quotient checks.

6. Rewrite $11x-3+9x^3$ as $9x^3+0x^2+11x-3$.

$$\begin{array}{r|l} & 3x^2-2x+5 \\ 3x+2 & 9x^3+0x^2+11x-3 \\ & \underline{9x^3+6x^2} \\ & \quad -6x^2+11x \\ & \quad \underline{-6x^2-4x} \\ & \qquad 15x-3 \\ & \qquad \underline{15x+10} \\ & \qquad -13 \end{array}$$

$\dfrac{11x-3+9x^3}{3x+2}=3x^2-2x+5+\dfrac{-13}{3x+2}$

7. Rewrite x^2+2 as x^2+0x+2.

$$\begin{array}{r|l} & 3x^2-2x-9 \\ x^2+0x+2 & 3x^4-2x^3-3x^2+x+4 \\ & \underline{3x^4+0x^3+6x^2} \\ & \quad -2x^3-9x^2+x \\ & \quad \underline{-2x^3+0x^2-4x} \\ & \qquad -9x^2+5x+4 \\ & \qquad \underline{-9x^2+0x-18} \\ & \qquad 5x+22 \end{array}$$

$\dfrac{3x^4-2x^3-3x^2+x+4}{x^2+2}=3x^2-2x-9+\dfrac{5x+22}{x^2+2}$

Vocabulary and Readiness Check

1. In $6\overline{)18}$ with quotient 3, the 18 is the dividend, the 3 is the quotient and the 6 is the divisor.

2. In $x+1\overline{)x^2+3x+2}$ with quotient $x+2$, the $x+1$ is the divisor, the x^2+3x+2 is the dividend and the $x+2$ is the quotient.

3. $\dfrac{a^6}{a^4}=a^{6-4}=2$

4. $\dfrac{p^8}{p^3}=p^{8-3}=p^5$

5. $\dfrac{y^2}{y}=\dfrac{y^2}{y^1}=y^{2-1}=y$

6. $\frac{a^3}{a} = \frac{a^3}{a^1} = a^{3-1} = a^2$

Exercise Set 5.6

1. $\frac{12x^4 + 3x^2}{x} = \frac{12x^4}{x} + \frac{3x^2}{x} = 12x^3 + 3x$

3. $\frac{20x^3 - 30x^2 + 5x + 5}{5} = \frac{20x^3}{5} - \frac{30x^2}{5} + \frac{5x}{5} + \frac{5}{5}$
$= 4x^3 - 6x^2 + x + 1$

5. $\frac{15p^3 + 18p^2}{3p} = \frac{15p^3}{3p} + \frac{18p^2}{3p} = 5p^2 + 6p$

7. $\frac{-9x^4 + 18x^5}{6x^5} = \frac{-9x^4}{6x^5} + \frac{18x^5}{6x^5} = -\frac{3}{2x} + 3$

9. $\frac{-9x^5 + 3x^4 - 12}{3x^3} = \frac{-9x^5}{3x^3} + \frac{3x^4}{3x^3} - \frac{12}{3x^3}$
$= -3x^2 + x - \frac{4}{x^3}$

11. $\frac{4x^4 - 6x^3 + 7}{-4x^4} = \frac{4x^4}{-4x^4} - \frac{6x^3}{-4x^4} + \frac{7}{-4x^4}$
$= -1 + \frac{3}{2x} - \frac{7}{4x^4}$

13.
$$\begin{array}{r} x+1 \\ x+3\overline{)x^2+4x+3} \\ \underline{x^2+3x} \\ x+3 \\ \underline{x+3} \\ 0 \end{array}$$

$\frac{x^2 + 4x + 3}{x + 3} = x + 1$

15.
$$\begin{array}{r} 2x+3 \\ x+5\overline{)2x^2+13x+15} \\ \underline{2x^2+10x} \\ 3x+15 \\ \underline{3x+15} \\ 0 \end{array}$$

$\frac{2x^2 + 13x + 15}{x + 5} = 2x + 3$

17.
$$\begin{array}{r} 2x+1 \\ x-4\overline{)2x^2-7x+3} \\ \underline{2x^2-8x} \\ x+3 \\ \underline{x-4} \\ 7 \end{array}$$

$\frac{2x^2 - 7x + 3}{x - 4} = 2x + 1 + \frac{7}{x - 4}$

19.
$$\begin{array}{r} 3a^2-3a+1 \\ 3a+2\overline{)9a^3-3a^2-3a+4} \\ \underline{9a^3+6a^2} \\ -9a^2-3a \\ \underline{-9a^2-6a} \\ 3a+4 \\ \underline{3a+2} \\ 2 \end{array}$$

$\frac{9a^3 - 3a^2 - 3a + 4}{3a + 2} = 3a^2 - 3a + 1 + \frac{2}{3a + 2}$

21.
$$\begin{array}{r} 4x+3 \\ 2x+1\overline{)8x^2+10x+1} \\ \underline{8x^2+4x} \\ 6x+1 \\ \underline{6x+3} \\ -2 \end{array}$$

$\frac{8x^2 + 10x + 1}{2x + 1} = 4x + 3 - \frac{2}{2x + 1}$

23.
$$\begin{array}{r} 2x^2+6x-5 \\ x-2\overline{)2x^3+2x^2-17x+8} \\ \underline{2x^3-4x^2} \\ 6x^2-17x \\ \underline{6x^2-12x} \\ -5x+8 \\ \underline{-5x+10} \\ -2 \end{array}$$

$\frac{2x^3 + 2x^2 - 17x + 8}{x - 2} = 2x^2 + 6x - 5 - \frac{2}{x - 2}$

25. Rewrite x^2-36 as $x^2+0x-36$.

$$\begin{array}{r} x+6 \\ x-6\overline{)x^2+0x-36} \\ \underline{x^2-6x} \\ 6x-36 \\ \underline{6x-36} \\ 0 \end{array}$$

$$\frac{x^2-36}{x-6}=x+6$$

27. Rewrite x^3-27 as $x^3+0x^2+0x-27$.

$$\begin{array}{r} x^2+3x+9 \\ x-3\overline{)x^3+0x^2+0x-27} \\ \underline{x^3-3x^2} \\ 3x^2+0x \\ \underline{3x^2-9x} \\ 9x-27 \\ \underline{9x-27} \\ 0 \end{array}$$

$$\frac{x^3-27}{x-3}=x^2+3x+9$$

29. Rewrite $1-3x^2$ as $-3x^2+0x+1$.

$$\begin{array}{r} -3x+6 \\ x+2\overline{)-3x^2+0x+1} \\ \underline{-3x^2-6x} \\ 6x+1 \\ \underline{6x+12} \\ -11 \end{array}$$

$$\frac{1-3x^2}{x+2}=-3x+6-\frac{11}{x+2}$$

31. Rewrite $-4b+4b^2-5$ as $4b^2-4b-5$.

$$\begin{array}{r} 2b-1 \\ 2b-1\overline{)4b^2-4b-5} \\ \underline{4b^2-2b} \\ -2b-5 \\ \underline{-2b+1} \\ -6 \end{array}$$

$$\frac{-4b+4b^2-5}{2b-1}=2b-1-\frac{6}{2b-1}$$

33. $\dfrac{a^2b^2-ab^3}{ab}=\dfrac{a^2b^2}{ab}-\dfrac{ab^3}{ab}=ab-b^2$

35.

$$\begin{array}{r} 4x+9 \\ 2x-3\overline{)8x^2+6x-27} \\ \underline{8x^2-12x} \\ 18x-27 \\ \underline{18x-27} \\ 0 \end{array}$$

$$\frac{8x^2+6x-27}{2x-3}=4x+9$$

37. $\dfrac{2x^2y+8x^2y^2-xy^2}{2xy}=\dfrac{2x^2y}{2xy}+\dfrac{8x^2y^2}{2xy}-\dfrac{xy^2}{2xy}$

$=x+4xy-\dfrac{y}{2}$

39.

$$\begin{array}{r} 2b^2+b+2 \\ b+4\overline{)2b^3+9b^2+6b-4} \\ \underline{2b^3+8b^2} \\ b^2+6b \\ \underline{b^2+4b} \\ 2b-4 \\ \underline{2b+8} \\ -12 \end{array}$$

$$\frac{2b^3+9b^2+6b-4}{b+4}=2b^2+b+2-\frac{12}{b+4}$$

41.

$$\begin{array}{r} 5x-2 \\ x+6\overline{)5x^2+28x-10} \\ \underline{5x^2+30x} \\ -2x-10 \\ \underline{-2x-12} \\ 2 \end{array}$$

$$\frac{5x^2+28x-10}{x+6}=5x-2+\frac{2}{x+6}$$

43. $\dfrac{10x^3-24x^2-10x}{10x}=\dfrac{10x^3}{10x}-\dfrac{24x^2}{10x}-\dfrac{10x}{10x}$

$=x^2-\dfrac{12x}{5}-1$

45.
$$\begin{array}{r} 6x-1 \\ x+3\overline{)6x^2+17x-4} \\ \underline{6x^2+18x} \\ -x-4 \\ \underline{-x-3} \\ -1 \end{array}$$

$$\frac{6x^2+17x-4}{x+3}=6x-1-\frac{1}{x+3}$$

47.
$$\begin{array}{r} 6x-1 \\ 5x-2\overline{)30x^2-17x+2} \\ \underline{30x^2-12x} \\ -5x+2 \\ \underline{-5x+2} \\ 0 \end{array}$$

$$\frac{30x^2-17x+2}{5x-2}=6x-1$$

49.
$$\frac{3x^4-9x^3+12}{-3x}=\frac{3x^4}{-3x}-\frac{9x^3}{-3x}+\frac{12}{-3x}$$
$$=-x^3+3x^2-\frac{4}{x}$$

51.
$$\begin{array}{r} x^2+3x+9 \\ x+3\overline{)x^3+6x^2+18x+27} \\ \underline{x^3+3x^2} \\ 3x^2+18x \\ \underline{3x^2+9x} \\ 9x+27 \\ \underline{9x+27} \\ 0 \end{array}$$

$$\frac{x^3+6x^2+18x+27}{x+3}=x^2+3x+9$$

53. Rewrite y^3+3y^2+4 as y^3+3y^2+0y+4.
$$\begin{array}{r} y^2+5y+10 \\ y-2\overline{)y^3+3y^2+0y+4} \\ \underline{y^3-2y^2} \\ 5y^2+0y \\ \underline{5y^2-10y} \\ 10y+4 \\ \underline{10y-20} \\ 24 \end{array}$$

$$\frac{y^3+3y^2+4}{y-2}=y^2+5y+10+\frac{24}{y-2}$$

55. Rewrite $5-6x^2$ as $-6x^2+0x+5$.
$$\begin{array}{r} -6x-12 \\ x-2\overline{)-6x^2+0x+5} \\ \underline{-6x^2+12x} \\ -12x+5 \\ \underline{-12x+24} \\ -19 \end{array}$$

$$\frac{5-6x^2}{x-2}=-6x-12-\frac{19}{x-2}$$

57. Rewrite x^5+x^2 as $x^5+0x^4+0x^3+x^2$.
$$\begin{array}{r} x^3-x^2+x \\ x^2+x\overline{)x^5+0x^4+0x^3+x^2} \\ \underline{x^5+x^4} \\ -x^4+0x^3 \\ \underline{-x^4-x^3} \\ x^3+x^2 \\ \underline{x^3+x^2} \\ 0 \end{array}$$

$$\frac{x^5+x^2}{x^2+x}=x^3-x^2+x$$

59. $2a(a^2+1)=2a(a^2)+2a(1)=2a^3+2a$

61.
$$2x(x^2+7x-5)=2x(x^2)+2x(7x)+2x(-5)$$
$$=2x^3+14x^2-10x$$

63.
$$-3xy(xy^2+7x^2y+8)$$
$$=-3xy(xy^2)-3xy(7x^2y)-3xy(8)$$
$$=-3x^2y^3-21x^3y^2-24xy$$

65. $9ab(ab^2c+4bc-8)$
$=9ab(ab^2c)+9ab(4bc)+9ab(-8)$
$=9a^2b^3c+36ab^2c-72ab$

67. The longest guitar corresponds to The Rolling Stones (2005).

69. The 2005 concert tour of U2 grossed approximately $139 million.

71. $P=4s,\ s=\dfrac{P}{4}$

$$\frac{12x^3+4x-16}{4}=\frac{12x^3}{4}+\frac{4x}{4}-\frac{16}{4}=3x^3+x-4$$

Each side is $(3x^3+x-4)$ feet.

73. $\dfrac{a+7}{7}=\dfrac{a}{7}+\dfrac{7}{7}=\dfrac{a}{7}+1;$
choice c

75. Answers may vary

77. $A=l\cdot w,\ w=\dfrac{A}{l}$

$$w=\frac{49x^2+70x-200}{7x+20}$$

$$\begin{array}{r} 7x-10 \\ 7x+20\overline{)\,49x^2+70x-200} \\ \underline{49x^2+140x} \\ -70x-200 \\ \underline{-70x-200} \\ 0 \end{array}$$

The width is $(7x-10)$ inches.

79. $$\frac{25y^{11b}+5y^{6b}-20y^{3b}+100y^b}{5y^b}=\frac{25y^{11b}}{5y^b}+\frac{5y^{6b}}{5y^b}-\frac{20y^{3b}}{5y^b}+\frac{100y^b}{5y^b}=5y^{10b}+y^{5b}-4y^{2b}+20$$

The Bigger Picture

1. $-5.7+(-0.23)=-5.93$

2. $\dfrac{1}{2}-\dfrac{9}{10}=\dfrac{5}{10}-\dfrac{9}{10}=\dfrac{-4}{10}=-\dfrac{2}{5}$

3. $(-5x^2y^3)(-x^7y)=(-5\cdot-1)(x^2\cdot x^7)(y^3\cdot y)=5x^9y^4$

4. $2^{-3}a^{-7}a^3=\dfrac{1}{2^3}\cdot\dfrac{1}{a^7}\cdot a^3=\dfrac{1}{8}\cdot\dfrac{a^3}{a^7}=\dfrac{1}{8}\cdot a^{-4}=\dfrac{1}{8a^4}$

5. $(7y^3-6y+2)-(y^3+2y^2+2)$
$=7y^3-6y+2-y^3-2y^2-2$
$=6y^3-2y^2-6y$

6. $(9y^2-3y)-(y^2+7)=9y^2-3y-y^2-7$
$=8y^2-3y-7$

7. $(x-3)(4x^2-x+7)$
$=x(4x^2)+x(-x)+x(7)-3(4x^2)-3(-x)-3(7)$
$=4x^3-x^2+7x-12x^2+3x-21$
$=4x^3-13x^2+10x-21$

8. $(6m-5)^2=(6m)^2-2(6m)(5)+5^2$
$=36m^2-60m+25$

9. $\dfrac{20n^2-5n+10}{5n}=\dfrac{20n^2}{5n}-\dfrac{5n}{5n}+\dfrac{10}{5n}=4n-1+\dfrac{2}{n}$

10.
$$\begin{array}{r} 2x-6 \\ 3x-1\overline{)\,6x^2-20x+20} \\ \underline{6x^2-2x} \\ -18x+20 \\ \underline{-18x+6} \\ 14 \end{array}$$

$$\frac{6x^2-20x+20}{3x-1}=2x-6+\frac{14}{3x-1}$$

11. $-6x=3.6$
$x=\dfrac{3.6}{-6}$
$x=-0.6$

12. $-6x<3.6$
$x>\dfrac{3.6}{-6}$
$x>-0.6$
$(-0.6,\infty)$

13. $6x+6 \ge 8x+2$
$6x \ge 8x-4$
$-2x \ge -4$
$x \le \frac{-4}{-2}$
$x \le 2$
$(-\infty, 2]$

14. $7y+3(y-1)=4(y+1)-3$
$7y+3y-3=4y+4-3$
$10y-3=4y+1$
$6y-3=1$
$6y=4$
$\frac{6y}{6}=\frac{4}{6}$
$y=\frac{2}{3}$

Chapter 5 Review

1. In 7^9, the base is 7 and the exponent is 9.

2. In $(-5)^4$, the base is -5 and the exponent is 4.

3. In -5^4, the base is 5 and the exponent is 4.

4. In x^6, the base is x and the exponent is 6.

5. $8^3 = 8 \cdot 8 \cdot 8 = 512$

6. $(-6)^2 = (-6)(-6) = 36$

7. $-6^2 = -6 \cdot 6 = -36$

8. $-4^3 - 4^0 = -64 - 1 = -65$

9. $(3b)^0 = 1$

10. $\frac{8b}{8b} = 1$

11. $y^2 \cdot y^7 = y^{2+7} = y^9$

12. $x^9 \cdot x^5 = x^{9+5} = x^{14}$

13. $(2x^5)(-3x^6) = (2 \cdot -3)(x^5 \cdot x^6) = -6x^{11}$

14.

15. $(x^4)^2 = x^{4 \cdot 2} = x^8$

16. $(y^3)^5 = y^{3 \cdot 5} = y^{15}$

17. $(3y^6)^4 = 3^4(y^6)^4 = 81y^{24}$

18. $2^3(x^3)^3 = 8x^9$

19. $\frac{x^9}{x^4} = x^{9-4} = x^5$

20. $\frac{z^{12}}{z^5} = z^{12-5} = z^7$

21. $\frac{a^5b^4}{ab} = a^{5-1}b^{4-1} = a^4b^3$

22. $\frac{x^4y^6}{xy} = x^{4-1}y^{6-1} = x^3y^5$

23. $\frac{12xy^6}{3x^4y^{10}} = \frac{12}{3}x^{1-3}y^{6-10} = 4x^{-2}y^{-4} = \frac{4}{x^2y^4}$

24. $\frac{2x^7y^8}{8xy^2} = \frac{2}{8}x^{7-1}y^{8-2} = \frac{x^6y^6}{4}$

25. $5a^7(2a^4)^3 = 5a^7(2^3)(a^4)^3$
$= (5 \cdot 8)(a^7 \cdot a^{12})$
$= 40a^{19}$

26. $(2x)^2(9x) = (2^2 \cdot x^2)(9x)$
$= (4 \cdot 9)(x^2 \cdot x)$
$= 36x^3$

27. $(-5a)^0 + 7^0 + 8^0 = 1 + 1 + 1 = 3$

28. $8x^0 + 9^0 = 8(1) + 1 = 9$

29. $\left(\frac{3x^4}{4y}\right)^3 = \frac{3^3x^{4\cdot3}}{4^3y^3} = \frac{27x^{12}}{64y^3}$, choice b.

30. $\left(\frac{5a^6}{b^3}\right)^2 = \frac{5^2a^{6\cdot2}}{b^{3\cdot2}} = \frac{25a^{12}}{b^6}$, choice c.

31. The degree of $-5x^4y^3$ is 4 + 3 = 7.

32. The degree of $10x^3y^2z$ is 3 + 2 + 1 = 6.

33. The degree of $35a^5bc^2$ is 5 + 1 + 2 = 8.

34. The degree of 95*xyz* is 1 + 1 + 1 = 3.

35. The degree is 5 because y^5 is the term with the highest degree.

36. The degree is 2 because $9y^2$ is the term with the highest degree.

37. The degree is 5 because $-28x^2y^3$ is the term with the highest degree.

38. The degree is 6 because $6x^2y^2z^2$ is the term with the highest degree.

39. a.

Term	Numerical Coefficient	Degree of Term
x^2y^2	1	4
$5x^2$	5	2
$-7y^2$	−7	2
$11xy$	11	2
−1	−1	0

b. The degree is 4.

40. $2x^2 + 20x$:

$x = 1$: $2(1)^2 + 20(1) = 22$

$x = 3$: $2(3)^2 + 20(3) = 78$

$x = 5.1$: $2(5.1)^2 + 20(5.1) = 154.02$

$x = 10$: $2(10)^2 + 20(10) = 400$

41. $7a^2 - 4a^2 - a^2 = (7 - 4 - 1)a^2 = 2a^2$

42. $9y + y - 14y = (9 + 1 - 14)y = -4y$

43. $6a^2 + 4a + 9a^2 = (6 + 9)a^2 + 4a = 15a^2 + 4a$

44. $21x^2 + 3x + x^2 + 6 = (21 + 1)x^2 + 3x + 6$
$= 22x^2 + 3x + 6$

45. $4a^2b - 3b^2 - 8q^2 - 10a^2b + 7q^2$
$= (4a^2b - 10a^2b) - 3b^2 + (-8q^2 + 7q^2)$
$= -6a^2b - 3b^2 - q^2$

46. $2s^{14} + 3s^{13} + 12s^{12} - s^{10}$ cannot be combined.

47. $(3x^2 + 2x + 6) + (5x^2 + x)$
$= 3x^2 + 2x + 6 + 5x^2 + x$
$= 8x^2 + 3x + 6$

48. $(2x^5 + 3x^4 + 4x^3 + 5x^2) + (4x^2 + 7x + 6)$
$= 2x^5 + 3x^4 + 4x^3 + 5x^2 + 4x^2 + 7x + 6$
$= 2x^5 + 3x^4 + 4x^3 + 9x^2 + 7x + 6$

49. $(-5y^2 + 3) - (2y^2 + 4) = -5y^2 + 3 - 2y^2 - 4$
$= -7y^2 - 1$

50. $(3x^2 - 7xy + 7y^2) - (4x^2 - xy + 9y^2)$
$= 3x^2 - 7xy + 7y^2 - 4x^2 + xy - 9y^2$
$= -x^2 - 6xy - 2y^2$

51. $(-9x^2 + 6x + 2) + (4x^2 - x - 1)$
$= -9x^2 + 6x + 2 + 4x^2 - x - 1$
$= -5x^2 + 5x + 1$

52. $(7x - 14y) - (3x - y) = 7x - 14y - 3x + y$
$= 4x - 13y$

53. $[(x^2 + 7x + 9) + (x^2 + 4)] - (4x^2 + 8x - 7)$
$= x^2 + 7x + 9 + x^2 + 4 - 4x^2 - 8x + 7$
$= -2x^2 - x + 20$

54. Let $x = 8$.
$754x^2 - 228x + 80{,}134$
$= 754(8)^2 - 228(8) + 80{,}134$
$= 48{,}256 - 1824 + 80{,}134$
$= 126{,}566$
Revenues from software sales in 2009 are predicted to be $126,566 million.

55. $4(2a + 7) = 4(2a) + 4(7) = 8a + 28$

56. $9(6a - 3) = 9(6a) - 9(3) = 54a - 27$

57. $-7x(x^2+5)=-7(x^2)-7x(5)=-7x^3-35x$

58. $-8y(4y^2-6)=-8y(4y^2)-8y(-6)$
$=-32y^3+48y$

59. $(3a^3-4a+1)(-2a)$
$=3a^3(-2a)-4a(-2a)+1(-2a)$
$=-6a^4+8a^2-2a$

60. $(6b^3-4b+2)(7b)=6b^3(7b)-4b(7b)+2(7b)$
$=42b^4-28b^2+14b$

61. $(2x+2)(x-7)=2x^2-14x+2x-14$
$=2x^2-12x-14$

62. $(2x-5)(3x+2)=6x^2+4x-15x-10$
$=6x^2-11x-10$

63. $(x-9)^2=(x-9)(x-9)$
$=x^2-9x-9x+81$
$=x^2-18x+81$

64. $(x-12)^2=(x-12)(x-12)$
$=x^2-12x-12x+144$
$=x^2-24x+144$

65. $(4a-1)(a+7)=4a^2+28a-a-7$
$=4a^2+27a-7$

66. $(6a-1)(7a+3)=42a^2+18a-7a-3$
$=42a^2+11a-3$

67. $(5x+2)^2=(5x+2)(5x+2)$
$=25x^2+10x+10x+4$
$=25x^2+20x+4$

68. $(3x+5)^2=(3x+5)(3x+5)$
$=9x^2+15x+15x+25$
$=9x^2+30x+25$

69. $(x+7)(x^3+4x-5)$
$=x(x^3+4x-5)+7(x^3+4x-5)$
$=x^4+4x^2-5x+7x^3+28x-35$
$=x^4+7x^3+4x^2+23x-35$

70. $(x+2)(x^5+x+1)=x(x^5+x+1)+2(x^5+x+1)$
$=x^6+x^2+x+2x^5+2x+2$
$=x^6+2x^5+x^2+3x+2$

71. $(x^2+2x+4)(x^2+2x-4)$
$=x^2(x^2+2x-4)+2x(x^2+2x-4)$
$+4(x^2+2x-4)$
$=x^4+2x^3-4x^2+2x^3+4x^2-8x$
$+4x^2+8x-16$
$=x^4+4x^3+4x^2-16$

72. $(x^3+4x+4)(x^3+4x-4)$
$=x^3(x^3+4x-4)+4x(x^3+4x-4)$
$+4(x^3+4x-4)$
$=x^6+4x^4-4x^3+4x^4+16x^2-16x+4x^3$
$+16x-16$
$=x^6+8x^4+16x^2-16$

73. $(x+7)^3=(x+7)(x+7)(x+7)$
$=(x^2+7x+7x+49)(x+7)$
$=(x^2+14x+49)(x+7)$
$=(x^2+14x+49)x+(x^2+14x+49)7$
$=x^3+14x^2+49x+7x^2+98x+343$
$=x^3+21x^2+147x+343$

74. $(2x-5)^3$
$=(2x-5)(2x-5)(2x-5)$
$=(4x^2-10x-10x+25)(2x-5)$
$=(4x^2-20x+25)(2x-5)$
$=(4x^2-20x+25)(2x)+(4x^2-20x+25)(-5)$
$=8x^3-40x^2+50x-20x^2+100x-125$
$=8x^3-60x^2+150x-125$

75. $(x+7)^2=x^2+2(x)(7)+7^2=x^2+14x+49$

76. $(x-5)^2=x^2-2(x)(5)+5^2=x^2-10x+25$

77. $(3x-7)^2=(3x)^2-2(3x)(7)+7^2$
$=9x^2-42x+49$

78. $(4x+2)^2=(4x)^2+2(4x)(2)+2^2$
$=16x^2+16x+4$

79. $(5x-9)^2 = (5x)^2 - 2(5x)(9) + 9^2$
$= 25x^2 - 90x + 81$

80. $(5x+1)(5x-1) = (5x)^2 - 1^2 = 25x^2 - 1$

81. $(7x+4)(7x-4) = (7x)^2 - 4^2 = 49x^2 - 16$

82. $(a+2b)(a-2b) = a^2 - (2b)^2 = a^2 - 4b^2$

83. $(2x-6)(2x+6) = (2x)^2 - 6^2 = 4x^2 - 36$

84. $(4a^2-2b)(4a^2+2b) = (4a^2)^2 - (2b)^2$
$= 16a^4 - 4b^2$

85. $(3x-1)^2 = (3x)^2 - 2(3x)(1) + 1^2$
$= 9x^2 - 6x + 1$
The area is $(9x^2 - 6x + 1)$ square meters.

86. $(5x+2)(x-1) = 5x^2 - 5x + 2x - 2$
$= 5x^2 - 3x - 2$
The area is $(5x^2 - 3x - 2)$ square miles.

87. $7^{-2} = \frac{1}{7^2} = \frac{1}{49}$

88. $-7^{-2} = -\frac{1}{7^2} = -\frac{1}{49}$

89. $2x^{-4} = \frac{2}{x^4}$

90. $(2x)^{-4} = \frac{1}{(2x)^4} = \frac{1}{16x^4}$

91. $\left(\frac{1}{5}\right)^{-3} = \frac{1^{-3}}{5^{-3}} = \frac{5^3}{1^3} = 125$

92. $\left(\frac{-2}{3}\right)^{-2} = \frac{(-2)^{-2}}{3^{-2}} = \frac{3^2}{(-2)^2} = \frac{9}{4}$

93. $2^0 + 2^{-4} = 1 + \frac{1}{2^4} = \frac{16}{16} + \frac{1}{16} = \frac{17}{16}$

94. $6^{-1} - 7^{-1} = \frac{1}{6} - \frac{1}{7} = \frac{7}{42} - \frac{6}{42} = \frac{1}{42}$

95. $\frac{x^5}{x^{-3}} = x^{5-(-3)} = x^8$

96. $\frac{z^4}{z^{-4}} = z^{4-(-4)} = z^8$

97. $\frac{r^{-3}}{r^{-4}} = r^{-3-(-4)} = r$

98. $\frac{y^{-2}}{y^{-5}} = y^{-2-(-5)} = y^3$

99. $\left(\frac{bc^{-2}}{bc^{-3}}\right)^4 = \frac{b^4c^{-8}}{b^4c^{-12}} = b^{4-4}c^{-8-(-12)} = c^4$

100. $\left(\frac{x^{-3}y^{-4}}{x^{-2}y^{-5}}\right)^{-3} = \frac{x^9y^{12}}{x^6y^{15}}$
$= x^{9-6}y^{12-15}$
$= x^3y^{-3}$
$= \frac{x^3}{y^3}$

101. $\frac{x^{-4}y^{-6}}{x^2y^7} = x^{-4-2}y^{-6-7}$
$= x^{-6}y^{-13}$
$= \frac{1}{x^6y^{13}}$

102. $\frac{a^5b^{-5}}{a^{-5}b^5} = a^{5-(-5)}b^{-5-5} = a^{10}b^{-10} = \frac{a^{10}}{b^{10}}$

103. $a^{6m}a^{5m} = a^{6m+5m} = a^{11m}$

104. $\frac{(x^{5+h})^3}{x^5} = \frac{x^{3(5+h)}}{x^5}$
$= \frac{x^{15+3h}}{x^5}$
$= x^{15+3h-5}$
$= x^{10+3h}$

105. $(3xy^{2z})^3 = 3^3x^3y^{2z(3)} = 27x^3y^{6z}$

106. $a^{m+2}a^{m+3} = a^{(m+2)+(m+3)} = a^{2m+5}$

107. $0.00027 = 2.7 \times 10^{-4}$

108. $0.8868 = 8.868 \times 10^{-1}$

109. $80{,}800{,}000 = 8.08 \times 10^{7}$

110. $868{,}000 = 8.68 \times 10^{5}$

111. $91{,}000{,}000 = 9.1 \times 10^{7}$

112. $150{,}000 = 1.5 \times 10^{5}$

113. $8.67 \times 10^{5} = 867{,}000$

114. $3.86 \times 10^{-3} = 0.00386$

115. $8.6 \times 10^{-4} = 0.00086$

116. $8.936 \times 10^{5} = 893{,}600$

117. $1.43128 \times 10^{15} = 1{,}431{,}280{,}000{,}000{,}000$

118. $1 \times 10^{-10} = 0.0000000001$

119.
$$\begin{aligned}(8\times 10^{4})(2\times 10^{-7}) &= (8\cdot 2)\times(10^{4}\cdot 10^{-7})\\ &= 16\times 10^{-3}\\ &= 0.016\end{aligned}$$

120.
$$\begin{aligned}\frac{8\times 10^{4}}{2\times 10^{-7}} &= \frac{8}{2}\times 10^{4-(-7)}\\ &= 4\times 10^{11}\\ &= 400{,}000{,}000{,}000\end{aligned}$$

121.
$$\begin{aligned}\frac{x^2+21x+49}{7x^2} &= \frac{x^2}{7x^2}+\frac{21x}{7x^2}+\frac{49}{7x^2}\\ &= \frac{1}{7}+\frac{3}{x}+\frac{7}{x^2}\end{aligned}$$

122.
$$\begin{aligned}\frac{5a^3b-15ab^2+20ab}{-5ab} &= \frac{5a^3b}{-5ab}-\frac{15ab^2}{-5ab}+\frac{20ab}{-5ab}\\ &= -a^2+3b-4\end{aligned}$$

123.
$$\begin{array}{r} a+1 \\ a-2\,\overline{)\,a^2 - a + 4} \\ \underline{a^2 - 2a} \\ a + 4 \\ \underline{a - 2} \\ 6 \end{array}$$

$$(a^2-a+4)\div(a-2) = a+1+\frac{6}{a-2}$$

124.
$$\begin{array}{r} 4x \\ x+5\,\overline{)\,4x^2 + 20x + 7} \\ \underline{4x^2 + 20x} \\ 7 \end{array}$$

$$(4x^2+20x+7)\div(x+5) = 4x+\frac{7}{x+5}$$

125.
$$\begin{array}{r} a^2 + 3a + 8 \\ a-2\,\overline{)\,a^3 + a^2 + 2a + 6} \\ \underline{a^3 - 2a^2} \\ 3a^2 + 2a \\ \underline{3a^2 - 6a} \\ 8a + 6 \\ \underline{8a - 16} \\ 22 \end{array}$$

$$\frac{a^3+a^2+2a+6}{a-2} = a^2+3a+8+\frac{22}{a-2}$$

126.
$$\begin{array}{r} 3b^2 - 4b \\ 3b-2\,\overline{)\,9b^3 - 18b^2 + 8b - 1} \\ \underline{9b^3 - 6b^2} \\ -12b^2 + 8b \\ \underline{-12b^2 + 8b} \\ -1 \end{array}$$

$$\frac{9b^3-18b^2+8b-1}{3b-2} = 3b^2-4b-\frac{1}{3b-2}$$

127.

$$\begin{array}{r|l} & 2x^3 - x^2 + 2 \\ 2x-1 & 4x^4 - 4x^3 + x^2 + 4x - 3 \\ & \underline{4x^4 - 2x^3} \\ & -2x^3 + x^2 \\ & \underline{-2x^3 + x^2} \\ & 4x - 3 \\ & \underline{4x - 2} \\ & -1 \end{array}$$

$$\frac{4x^4 - 4x^3 + x^2 + 4x - 3}{2x-1} = 2x^3 - x^2 + 2 - \frac{1}{2x-1}$$

128. Rewrite $-10x^2 - x^3 - 21x + 18$ as $-x^3 - 10x^2 - 21x + 18$.

$$\begin{array}{r|l} & -x^2 - 16x - 117 \\ x-6 & -x^3 - 10x^2 - 21x + 18 \\ & \underline{-x^3 + 6x^2} \\ & -16x^2 - 21x \\ & \underline{-16x^2 + 96x} \\ & -117x + 18 \\ & \underline{-117x + 702} \\ & -684 \end{array}$$

$$\frac{-10x^2 - x^3 - 21x + 18}{x-6} = -x^2 - 16x - 117 - \frac{684}{x-6}$$

129. $\frac{15x^3 - 3x^2 + 60}{3x^2} = \frac{15x^3}{3x^2} - \frac{3x^2}{3x^2} + \frac{60}{3x^2}$
$= 5x - 1 + \frac{20}{x^2}$

The width is $\left(5x - 1 + \frac{20}{x^2}\right)$ feet.

130. $\frac{21a^3b^6 + 3a - 3}{3} = \frac{21a^3b^6}{3} + \frac{3a}{3} - \frac{3}{3}$
$= 7a^3b^6 + a - 1$

The length of a side is $(7a^3b^6 + a - 1)$ units.

131. $\left(-\frac{1}{2}\right)^3 = \left(-\frac{1}{2}\right)\left(-\frac{1}{2}\right)\left(-\frac{1}{2}\right) = -\frac{1}{8}$

132. $(4xy^2)(x^3y^5) = 4(x \cdot x^3)(y^2 \cdot y^5)$
$= 4x^{1+3}y^{2+5}$
$= 4x^4y^7$

133. $\frac{18x^9}{27x^3} = \frac{18}{27}x^{9-3} = \frac{2x^6}{3}$

134. $\left(\frac{3a^4}{b^2}\right)^3 = \frac{3^3(a^4)^3}{(b^2)^3} = \frac{27a^{12}}{b^6}$

135. $(2x^{-4}y^3)^{-4} = 2^{-4}(x^{-4})^{-4}(y^3)^{-4}$
$= \frac{1}{2^4}x^{16}y^{-12}$
$= \frac{x^{16}}{16y^{12}}$

136. $\frac{a^{-3}b^6}{9^{-1}a^{-5}b^{-2}} = 9a^{-3-(-5)}b^{6-(-2)} = 9a^2b^8$

137. $(6x + 2) + (5x - 7) = 6x + 2 + 5x - 7 = 11x - 5$

138. $(-y^2 - 4) + (3y^2 - 6) = -y^2 - 4 + 3y^2 - 6$
$= 2y^2 - 10$

139. $(8y^2 - 3y + 1) - (3y^2 + 2) = 8y^2 - 3y^2 - 3y + 1 - 2$
$= 5y^2 - 3y - 1$

140. $(5x^2 + 2x - 6) - (-x - 4) = 5x^2 + 2x - 6 + x + 4$
$= 5x^2 + 3x - 2$

141. $4x(7x^2 + 3) = 4x(7x^2) + 4x(3)$
$= 28x^3 + 12x$

142. $(2x + 5)(3x - 2) = 6x^2 - 4x + 15x - 10$
$= 6x^2 + 11x - 10$

143. $(x - 3)(x^2 + 4x - 6)$
$= x(x^2 + 4x - 6) - 3(x^2 + 4x - 6)$
$= x^3 + 4x^2 - 6x - 3x^2 - 12x + 18$
$= x^3 + x^2 - 18x + 18$

144. $(7x - 2)(4x - 9) = 28x^2 - 63x - 8x + 18$
$= 28x^2 - 71x + 18$

145. $(5x + 4)^2 = (5x)^2 + 2(5x)(4) + 4^2$
$= 25x^2 + 40x + 16$

146. $(6x + 3)(6x - 3) = (6x)^2 - (3)^2 = 36x^2 - 9$

147. $\frac{8a^4-2a^3+4a-5}{2a^3} = \frac{8a^4}{2a^3} - \frac{2a^3}{2a^3} + \frac{4a}{2a^3} - \frac{5}{2a^3}$
$= 4a - 1 + \frac{2}{a^2} - \frac{5}{2a^3}$

148.
$$\begin{array}{r} x-3 \\ x+5\overline{)x^2+2x+10} \\ \underline{x^2+5x} \\ -3x+10 \\ \underline{-3x-15} \\ 25 \end{array}$$

$\frac{x^2+2x+10}{x+5} = x-3+\frac{25}{x+5}$

149.
$$\begin{array}{r} 2x^2+7x+5 \\ 2x-3\overline{)4x^3+8x^2\ -11x+4} \\ \underline{4x^3-6x^2} \\ 14x^2-11x \\ \underline{14x^2-21x} \\ 10x\ +4 \\ \underline{10x-15} \\ 19 \end{array}$$

$\frac{4x^3+8x^2-11x+4}{2x-3} = 2x^2+7x+5+\frac{19}{2x-3}$

Chapter 5 Test

1. $2^5 = 2\cdot2\cdot2\cdot2\cdot2 = 32$

2. $(-3)^4 = (-3)(-3)(-3)(-3) = 81$

3. $-3^4 = -3\cdot3\cdot3\cdot3 = -81$

4. $4^{-3} = \frac{1}{4^3} = \frac{1}{64}$

5. $(3x^2)(-5x^9) = (3)(-5)(x^2\cdot x^9) = -15x^{11}$

6. $\frac{y^7}{y^2} = y^{7-2} = y^5$

7. $\frac{r^{-8}}{r^{-3}} = r^{-8-(-3)} = r^{-5} = \frac{1}{r^5}$

8. $\left(\frac{x^2y^3}{x^3y^{-4}}\right)^2 = \frac{x^4y^6}{x^6y^{-8}}$
$= x^{4-6}y^{6-(-8)}$
$= x^{-2}y^{14}$
$= \frac{y^{14}}{x^2}$

9. $\left(\frac{6^2x^{-4}y^{-1}}{6^3x^{-3}y^7}\right) = 6^{2-3}x^{-4-(-3)}y^{-1-7}$
$= 6^{-1}x^{-1}y^{-8}$
$= \frac{1}{6xy^8}$

10. $563,000 = 5.63\times10^5$

11. $0.0000863 = 8.63\times10^{-5}$

12. $1.5\times10^{-3} = 0.0015$

13. $6.23\times10^4 = 62,300$

14. $(1.2\times10^5)(3\times10^{-7}) = (1.2)(3)\times10^{5-7}$
$= 3.6\times10^{-2}$
$= 0.036$

15. **a.**

Term	Numerical Coefficient	Degree of Term
$4xy^2$	4	3
$7xyz$	7	3
x^3y	1	4
-2	-2	0

b. The degree is 4.

16. $5x^2+4xy-7x^2+11+8xy$
$= (5x^2-7x^2)+(4xy+8xy)+11$
$= -2x^2+12xy+11$

17. $(8x^3+7x^2+4x-7)+(8x^3-7x-6)$
$= 8x^3+7x^2+4x-7+8x^3-7x-6$
$= 16x^3+7x^2-3x-13$

18.
$$\begin{array}{r} 5x^3 + x^2 + 5x - 2 \\ -(8x^3 - 4x^2 + x - 7) \\ \hline \end{array}$$

$$\begin{array}{r} 5x^3 + x^2 + 5x - 2 \\ -8x^3 + 4x^2 - x + 7 \\ \hline -3x^3 + 5x^2 + 4x + 5 \end{array}$$

19. $[(8x^2 + 7x + 5) + (x^3 - 8)] - (4x + 2)$
$= 8x^2 + 7x + 5 + x^3 - 8 - 4x - 2$
$= x^3 + 8x^2 + 3x - 5$

20. $(3x + 7)(x^2 + 5x + 2)$
$= 3x(x^2 + 5x + 2) + 7(x^2 + 5x + 2)$
$= 3x^3 + 15x^2 + 6x + 7x^2 + 35x + 14$
$= 3x^3 + 22x^2 + 41x + 14$

21. $3x^2(2x^2 - 3x + 7)$
$= 3x^2(2x^2) + 3x^2(-3x) + 3x^2(7)$
$= 6x^4 - 9x^3 + 21x^2$

22. $(x + 7)(3x - 5) = 3x^2 - 5x + 21x - 35$
$= 3x^2 + 16x - 35$

23. $\left(3x - \frac{1}{5}\right)\left(3x + \frac{1}{5}\right) = (3x)^2 - \left(\frac{1}{5}\right)^2 = 9x^2 - \frac{1}{25}$

24. $(4x - 2)^2 = (4x)^2 - 2(4x)(2) + 2^2$
$= 16x^2 - 16x + 4$

25. $(8x + 3)^2 = (8x)^2 + 2(8x)(3) + (3)^2$
$= 64x^2 + 48x + 9$

26. $(x^2 - 9b)(x^2 + 9b) = (x^2)^2 - (9b)^2 = x^4 - 81b^2$

27. $-16t^2 + 1001$
$t = 0$: $-16(0)^2 + 1001 = 1001$ ft
$t = 1$: $-16(1)^2 + 1001 = 985$ ft
$t = 3$: $-16(3)^2 + 1001 = 857$ ft
$t = 5$: $-16(5)^2 + 1001 = 601$ ft

28. $(2x + 3)(2x - 3) = (2x)^2 - (3)^2 = 4x^2 - 9$
The area is $(4x^2 - 9)$ square inches.

29. $\frac{4x^2 + 24xy - 7x}{8xy} = \frac{4x^2}{8xy} + \frac{24xy}{8xy} - \frac{7x}{8xy}$
$= \frac{x}{2y} + 3 - \frac{7}{8y}$

30.
$$\begin{array}{r} x + 2 \\ x + 5 \overline{)\, x^2 + 7x + 10} \\ \underline{x^2 + 5x} \\ 2x + 10 \\ \underline{2x + 10} \\ 0 \end{array}$$

$\frac{x^2 + 7x + 10}{x + 5} = x + 2$

31. Rewrite $27x^3 - 8$ as $27x^3 + 0x^2 + 0x - 8$.
$$\begin{array}{r} 9x^2 - 6x + 4 \\ 3x + 2 \overline{)\, 27x^3 + 0x^2 + 0x - 8} \\ \underline{27x^3 + 18x^2} \\ -18x^2 + 0x \\ \underline{-18x^2 - 12x} \\ 12x - 8 \\ \underline{12x + 8} \\ -16 \end{array}$$

$\frac{27x^3 - 8}{3x + 2} = 9x^2 - 6x + 4 - \frac{16}{3x + 2}$

Cumulative Review Chapters 1–5

1. a. $8 \geq 8$ is true since $8 = 8$.

b. $8 \leq 8$ is true since $8 = 8$.

c. $23 \leq 0$ is false.

d. $23 \geq 0$ is true

2. a. $|-7.2| = 7.2$

b. $|0| = 0$

c. $\left|-\frac{1}{2}\right| = \frac{1}{2}$

3. a. $\frac{4}{5} \div \frac{5}{16} = \frac{4}{5} \cdot \frac{16}{5} = \frac{64}{25}$

b. $\frac{7}{10} \div 14 = \frac{7}{10} \div \frac{14}{1} = \frac{7}{10} \cdot \frac{1}{14} = \frac{7}{10 \cdot 7 \cdot 2} = \frac{1}{20}$

c. $\frac{3}{8} \div \frac{3}{10} = \frac{3}{8} \cdot \frac{10}{3} = \frac{3 \cdot 2 \cdot 5}{2 \cdot 4 \cdot 3} = \frac{5}{4}$

4. a. $\frac{3}{4} \cdot \frac{7}{21} = \frac{3 \cdot 7}{4 \cdot 3 \cdot 7} = \frac{1}{4}$

b. $\frac{1}{2} \cdot 4\frac{5}{6} = \frac{1}{2} \cdot \frac{29}{6} = \frac{29}{12} = 2\frac{5}{12}$

5. a. $3^2 = 3 \cdot 3 = 9$

b. $5^3 = 5 \cdot 5 \cdot 5 = 125$

c. $2^4 = 2 \cdot 2 \cdot 2 \cdot 2 = 16$

d. $7^1 = 7$

e. $\left(\frac{3}{7}\right)^2 = \left(\frac{3}{7}\right)\left(\frac{3}{7}\right) = \frac{9}{49}$

6. Let $x = 5$ and $y = 1$.

$$\begin{aligned} \frac{2x - 7y}{x^2} &= \frac{2(5) - 7(1)}{5^2} \\ &= \frac{10 - 7}{25} \\ &= \frac{3}{25} \end{aligned}$$

7. a. $-3 + (-7) = -10$

b. $-1 + (-20) = -21$

c. $-2 + (-10) = -12$

8.
$$\begin{aligned} 8 + 3(2 \cdot 6 - 1) &= 8 + 3(12 - 1) \\ &= 8 + 3(11) \\ &= 8 + 33 \\ &= 41 \end{aligned}$$

9. $-4 - 8 = -4 + (-8) = -12$

10. $x = 1$

$$\begin{aligned} 5x^2 + 2 &= x - 8 \\ 5(1)^2 + 2 &\stackrel{?}{=} 1 - 8 \\ 5 + 2 &\stackrel{?}{=} -7 \\ 7 &\stackrel{?}{=} -7 \quad \text{False} \end{aligned}$$

$x = 1$ is not a solution.

11. a. The reciprocal of 22 is $\frac{1}{22}$.

b. The reciprocal of $\frac{3}{16}$ is $\frac{16}{3}$.

c. The reciprocal of -10 is $-\frac{1}{10}$.

d. The reciprocal of $-\frac{9}{13}$ is $-\frac{13}{9}$.

12. a. $7 - 40 = 7 + (-40) = -33$

b. $-5 - (-10) = -5 + 10 = 5$

13. a. $5 + (4 + 6) = (5 + 4) + 6$

b. $(-1 \cdot 2) \cdot 5 = -1 \cdot (2 \cdot 5)$

14. $\frac{4(-3) + (-8)}{5 + (-5)} = \frac{-12 + (-8)}{0}$ is undefined.

15. a.
$$\begin{aligned} 10 + (x + 12) &= 10 + (12 + x) \\ &= (10 + 12) + x \\ &= 22 + x \end{aligned}$$

b. $-3(7x) = (-3 \cdot 7)x = -21x$

16.
$$\begin{aligned} -2(x + 3y - z) &= -2(x) + (-2)(3y) - (-2)(z) \\ &= -2x - 6y + 2z \end{aligned}$$

17. a. $5(x + 2) = 5x + 5(2) = 5x + 10$

b.
$$\begin{aligned} &-2(y + 0.3z - 1) \\ &= -2(y) + (-2)(0.3z) - (-2)(1) \\ &= -2y - 0.6z + 2y \end{aligned}$$

c.
$$\begin{aligned} &-(x + y - 2z + 6) \\ &= -1(x + y - 2z + 6) \\ &= -1(x) + (-1)(y) - (-1)(2z) + (-1)(6) \\ &= -x - y + 2z - 6 \end{aligned}$$

18.
$$\begin{aligned} 2(6x - 1) - (x - 7) &= 12x - 2 - x + 7 \\ &= 11x + 5 \end{aligned}$$

19.
$$\begin{aligned} x - 7 &= 10 \\ x - 7 + 7 &= 10 + 7 \\ x &= 17 \end{aligned}$$

20. Let x = a number.

$(x + 7) - 2x$

21.
$$\frac{5}{2}x = 15$$
$$\frac{2}{5}\cdot\frac{5}{2}x = \frac{2}{5}\cdot 15$$
$$x = 6$$

22.
$$2x + \frac{1}{8} = x - \frac{3}{8}$$
$$x + \frac{1}{8} = -\frac{3}{8}$$
$$x = -\frac{4}{8}$$
$$x = -\frac{1}{2}$$

23. Let x = a number.
$$7 + 2x = x - 3$$
$$7 + x = -3$$
$$x = -10$$
The number is −10.

24.
$$10 = 5j - 2$$
$$12 = 5j$$
$$\frac{12}{5} = j$$

25. Let x = a number.
$$2(x+4) = 4x - 12$$
$$2x + 8 = 4x - 12$$
$$-2x + 8 = -12$$
$$-2x = -20$$
$$x = 10$$
The number is 10.

26.
$$\frac{7x+5}{3} = x + 3$$
$$3\left(\frac{7x+5}{3}\right) = 3(x+3)$$
$$7x + 5 = 3x + 9$$
$$4x + 5 = 9$$
$$4x = 4$$
$$x = 1$$

27. Let x = the width and $3x - 2$ = the length.
$$2L + 2W = P$$
$$2(3x-2) + 2x = 28$$
$$6x - 4 + 2x = 28$$
$$8x - 4 = 28$$
$$8x = 32$$
$$x = 4$$
$3x - 2 = 3(4) - 2 = 10$
The width is 4 feet and the length is 10 feet.

28. $x < 5$, $(-\infty, 5)$

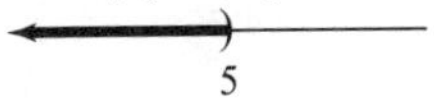

29.
$$F = \frac{9}{5}C + 32$$
$$F - 32 = \frac{9}{5}C$$
$$\frac{5}{9}(F - 32) = C$$
$$\frac{5F - 160}{9} = C$$

30. a. $x = -1$ is a vertical line and the slope is undefined.

b. $y = 7$ is a horizontal line and the slope is zero.

31. $2 < x \le 4$

32. $m = \frac{y_2 - y_1}{x_2 - x_1} = \frac{2}{20} = \frac{1}{10}\cdot 100\% = 10\%$

33. $3x + y = 12$

a. (0,): $3(0) + y = 12$
$$y = 12,\ (0, 12)$$

b. (, 6): $3x + 6 = 12$
$$3x = 6$$
$$x = 2,\ (2, 6)$$

c. (−1,): $3(-1) + y = 12$
$$-3 + y = 12$$
$$y = 15,\ (-1, 15)$$

34.
$$\begin{cases} 3x + 2y = -8 \\ 2x - 6y = -9 \end{cases}$$
Multiply the first equation by 3 and add.
$$9x + 6y = -24$$
$$\underline{2x - 6y = -9}$$
$$11x \quad = -33$$
$$x = -3$$
Replace x with −3 in the first equation.

$$3(-3)+2y=-8$$
$$-9+2y=-8$$
$$2y=1$$
$$y=\frac{1}{2}$$

The solution to the system is $\left(-3, \frac{1}{2}\right)$.

35. $2x + y = 5$

x	y
0	5
$\frac{5}{2}$	0

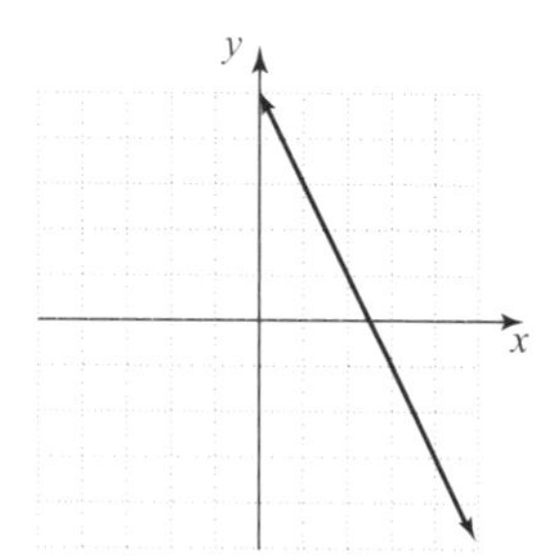

36. $\begin{cases} x=-3y+3 \\ 2x+9y=5 \end{cases}$

Replace x with $-3y + 3$ in the second equation.

$$2(-3y+3)+9y=5$$
$$-6y+6+9y=5$$
$$3y+6=5$$
$$3y=-1$$
$$y=-\frac{1}{3}$$

Replace y with $-\frac{1}{3}$ in the first equation.

$$x=-3\left(-\frac{1}{3}\right)+3=1+3=4$$

The solution to the system is $\left(4, -\frac{1}{3}\right)$.

37.

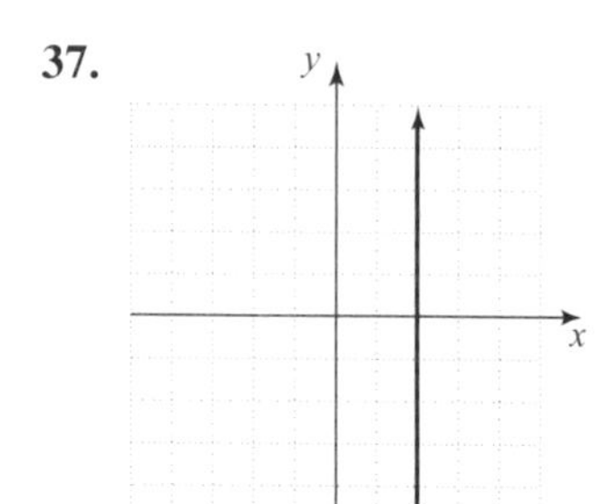

38. a. $(-5)^2=(-5)(-5)=25$

b. $-5^2=-(5)(5)=-25$

c. $2\cdot 5^2=2\cdot 5\cdot 5=50$

39. $x = 5$ is a vertical line and the slope is undefined.

40. $\frac{(z^2)^3\cdot z^7}{z^9}=\frac{z^6\cdot z^7}{z^9}=z^{6+7-9}=z^4$

41. $x+y<7$

Test (0, 0)

$0+0\overset{?}{<}7$

True

Shade below.

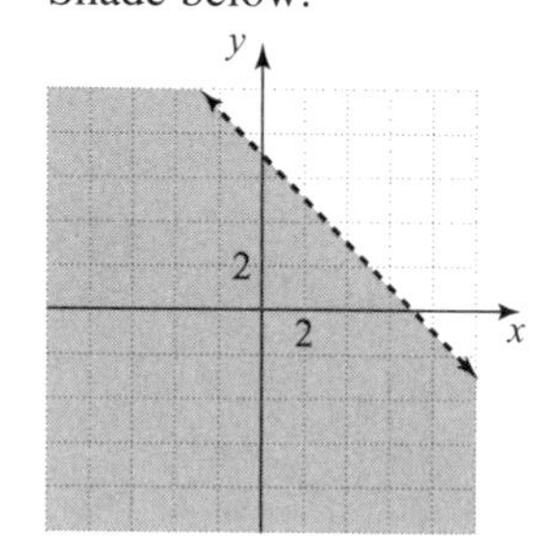

42. $(5y^2-6)-(y^2+2)=5y^2-6-y^2-2=4y^2-8$

43. $(2x^2)(-3x^5)=(2\cdot -3)(x^2\cdot x^5)=-6x^{2+5}=-6x^7$

44. $-x^2$

a. $-(2)^2=-4$

b. $-(-2)^2=-4$

45. $(-2x^2+5x-1)+(-2x^2+x+3)$
$=-2x^2+5x-1-2x^2+x+3$
$=-4x^2+6x+2$

46. $(10x^2-3)(10x^2+3)=(10x^2)^2-3^2$
$=100x^4-9$

47. $(2x-y)^2=(2x)^2-2(2x)(y)+(y)^2$
$=4x^2-4xy+y^2$

48. $(10x^2+3)^2=(10x^2)^2+2(10x^2)(3)+3^2$
$=100x^4+60x^2+9$

49. $\frac{6m^2+2m}{2m} = \frac{6m^2}{2m} + \frac{2m}{2m} = 3m+1$

50. **a.** $5^{-1} = \frac{1}{5}$

b. $7^{-2} = \frac{1}{7^2} = \frac{1}{49}$

Chapter 6

Section 6.1

Practice Exercises

1. a. $36 = 2 \cdot 2 \cdot 3 \cdot 3 = 2^2 \cdot 3^2$
$42 = 2 \cdot 3 \cdot 7$
$\text{GCF} = 2 \cdot 3 = 6$

b. $35 = 5 \cdot 7$
$44 = 2 \cdot 2 \cdot 11$
$\text{GCF} = 1$

c. $12 = 2 \cdot 2 \cdot 3 = 2^2 \cdot 3$
$16 = 2 \cdot 2 \cdot 2 \cdot 2 = 2^4$
$40 = 2 \cdot 2 \cdot 2 \cdot 5 = 2^3 \cdot 5$
$\text{GCF} = 2^2 = 4$

2. a. The GCF is y^4 since 4 is the smallest exponent to which y is raised.

b. The GCF is x^1 or x, since 1 is the smallest exponent on x.

3. a. $5y^4 = 5 \cdot y^4$
$15y^2 = 3 \cdot 5 \cdot y^2$
$-20y^3 = -1 \cdot 2 \cdot 2 \cdot 5 \cdot y^3$
$\text{GCF} = 5 \cdot y^2 = 5y^2$

b. $4x^2 = 2 \cdot 2 \cdot x^2$
$x^3 = x^3$
$3x^8 = 3 \cdot x^8$
$\text{GCF} = x^2$

c. The GCF of a^4, a^3, and a^2 is a^2.
The GCF of b^2, b^5, and b^3 is b^2.
Thus, the GCF of a^4b^2, a^3b^5, and a^2b^3 is a^2b^2.

4. a. $4t + 12$; GCF $= 4$
$4t + 12 = 4 \cdot t + 4 \cdot 3 = 4(t + 3)$

b. $y^8 + y^4$; GCF $= y^4$
$y^8 + y^4 = y^4 \cdot y^4 + y^4 \cdot 1 = y^4(y^4 + 1)$

5. $-8b^6 + 16b^4 - 8b^2$
$= -8b^2(b^4) - 8b^2(-2b^2) - 8b^2(1)$
$= -8b^2(b^4 - 2b^2 + 1)$ or $8b^2(-b^4 + 2b^2 - 1)$

6. $5x^4 - 20x = 5x(x^3 - 4)$

7. $\frac{5}{9}z^5 + \frac{1}{9}z^4 - \frac{2}{9}z^3 = \frac{1}{9}z^3(5z^2 + z - 2)$

8. $8a^2b^4 - 20a^3b^3 + 12ab^3 = 4ab^3(2ab - 5a^2 + 3)$

9. $8(y - 2) + x(y - 2) = (y - 2)(8 + x)$

10. $7xy^3(p+q) - (p+q) = 7xy^3(p+q) - 1(p+q)$
$= (p+q)(7xy^3 - 1)$

11. $xy + 3y + 4x + 12 = (xy + 3y) + (4x + 12)$
$= y(x+3) + 4(x+3)$
$= (x+3)(y+4)$
Check: $(x + 3)(y + 4) = xy + 3y + 4x + 12$

12. $2xy + 3y^2 - 2x - 3y = (2xy + 3y^2) + (-2x - 3y)$
$= y(2x + 3y) - 1(2x + 3y)$
$= (2x + 3y)(y - 1)$

13. $7a^3 + 5a^2 + 7a + 5 = (7a^3 + 5a^2) + (7a + 5)$
$= a^2(7a + 5) + 1(7a + 5)$
$= (7a + 5)(a^2 + 1)$

14. $4xy + 15 - 12x - 5y = 4xy - 12x - 5y + 15$
$= (4xy - 12x) + (-5y + 15)$
$= 4x(y - 3) - 5(y - 3)$
$= (y - 3)(4x - 5)$

15. $9y - 18 + y^3 - 4y^2 = 9(y - 2) + y^2(y - 4)$
There is no common binomial factor, so it cannot be factored by grouping.

16. $3xy - 3ay - 6ax + 6a^2 = 3(xy - ay - 2ax + 2a^2)$
$= 3[y(x - a) - 2a(x - a)]$
$= 3(x - a)(y - 2a)$

Vocabulary and Readiness Check

1. Since $5 \cdot 4 = 20$, the numbers 5 and 4 are called factors of 20.

2. The greatest common factor of a list of integers is the largest integer that is a factor of all the integers in the list.

3. The greatest common factor of a list of common variables raised to powers is the variable raised to the least exponent in the list.

4. The process of writing a polynomial as a product is called factoring.

5. $7(x + 3) + y(x + 3)$ is a sum, not a product. The statement is false.

6. $3x^3 + 6x + x^2 + 2 = 3x(x^2 + 2) + (x^2 + 2)$
$= (x^2 + 2)(3x + 1)$
The statement is false.

7. $14 = 2 \cdot 7$

8. $15 = 3 \cdot 5$

9. The GCF of 18 and 3 is 3.

10. The GCF of 7 and 35 is 7.

11. The GCF of 20 and 15 is 5.

12. The GCF of 6 and 15 is 3.

Exercise Set 6.1

1. $32 = 2 \cdot 2 \cdot 2 \cdot 2 \cdot 2 = 2^5$
$36 = 2 \cdot 2 \cdot 3 \cdot 3 = 2^2 \cdot 3^2$
$\text{GCF} = 2 \cdot 2 = 4$

3. $18 = 2 \cdot 3 \cdot 3 = 2 \cdot 3^2$
$42 = 2 \cdot 3 \cdot 7$
$84 = 2 \cdot 2 \cdot 3 \cdot 7 = 2^2 \cdot 3 \cdot 7$
$\text{GCF} = 2 \cdot 3 = 6$

5. $24 = 2 \cdot 2 \cdot 2 \cdot 3 = 2^3 \cdot 3$
$14 = 2 \cdot 7$
$21 = 3 \cdot 7$
$\text{GCF} = 1$

7. The GCF of y^2, y^4, and y^7 is y^2.

9. The GCF of z^7, z^9, and z^{11} is z^7.

11. The GCF of x^{10}, x, and x^3 is x.
The GCF of y^2, y^2, and y^3 is y^2.
Thus the GCF of $x^{10}y^2$, xy^2, and x^3y^3 is xy^2.

13. $14x = 2 \cdot 7 \cdot x$
$21 = 3 \cdot 7$
$\text{GCF} = 7$

15. $12y^4 = 2 \cdot 2 \cdot 3 \cdot y^4$
$20y^3 = 2 \cdot 2 \cdot 5 \cdot y^3$
$\text{GCF} = 2 \cdot 2 \cdot y^3 = 4y^3$

17. $-10x^2 = -1 \cdot 2 \cdot 5 \cdot x^2$
$15x^3 = 3 \cdot 5 \cdot x^3$
$\text{GCF} = 5 \cdot x^2 = 5x^2$

19. $12x^3 = 2 \cdot 2 \cdot 3 \cdot x^3$
$-6x^4 = -1 \cdot 2 \cdot 3 \cdot x^4$
$3x^5 = 3 \cdot x^5$
$\text{GCF} = 3 \cdot x^3 = 3x^3$

21. $-18x^2y = -1 \cdot 2 \cdot 3 \cdot 3 \cdot x^2 \cdot y$
$9x^3y^3 = 3 \cdot 3 \cdot x^3 \cdot y^3$
$36x^3y = 2 \cdot 2 \cdot 3 \cdot 3 \cdot x^3 \cdot y$
$\text{GCF} = 3 \cdot 3 \cdot x^2 \cdot y = 9x^2y$

23. $20a^6b^2c^8 = 2 \cdot 2 \cdot 5 \cdot a^6 \cdot b^2 \cdot c^8$
$50a^7b = 2 \cdot 5 \cdot 5 \cdot a^7 \cdot b$
$\text{GCF} = 2 \cdot 5 \cdot a^6 \cdot b = 10a^6b$

25. $3a + 6 = 3(a + 2)$

27. $30x - 15 = 15(2x - 1)$

29. $x^3 + 5x^2 = x^2(x + 5)$

31. $6y^4 + 2y^3 = 2y^3(3y + 1)$

33. $4x - 8y + 4 = 4(x - 2y + 1)$

35. $6x^3 - 9x^2 + 12x = 3x(2x^2 - 3x + 4)$

37. $a^7b^6 - a^3b^2 + a^2b^5 - a^2b^2$
$= a^2b^2(a^5b^4 - a + b^3 - 1)$

39. $8x^5+16x^4-20x^3+12=4(2x^5+4x^4-5x^3+3)$

41. $\frac{1}{3}x^4+\frac{2}{3}x^3-\frac{4}{3}x^5+\frac{1}{3}x$
$=\frac{1}{3}x(x^3+2x^2-4x^4+1)$

43. $y(x^2+2)+3(x^2+2)=(x^2+2)(y+3)$

45. $z(y+4)-3(y+4)=(y+4)(z-3)$

47. $r(z^2-6)+(z^2-6)=r(z^2-6)+1(z^2-6)$
$=(z^2-6)(r+1)$

49. $-2x-14=-2(x+7)$

51. $-2x^5+x^7=-x^5(2-x^2)$

53. $-6a^4+9a^3-3a^2=-3a^2(2a^2-3a+1)$

55. $x^3+2x^2+5x+10=x^2(x+2)+5(x+2)$
$=(x+2)(x^2+5)$

57. $5x+15+xy+3y=5(x+3)+y(x+3)$
$=(x+3)(5+y)$

59. $6x^3-4x^2+15x-10=2x^2(3x-2)+5(3x-2)$
$=(3x-2)(2x^2+5)$

61. $5m^3+6mn+5m^2+6n$
$=m(5m^2+6n)+1(5m^2+6n)$
$=(5m^2+6n)(m+1)$

63. $2y-8+xy-4x=2(y-4)+x(y-4)$
$=(y-4)(2+x)$

65. $2x^3-x^2+8x-4=x^2(2x-1)+4(2x-1)$
$=(2x-1)(x^2+4)$

67. $4x^2-8xy-3x+6y=4x(x-2y)-3(x-2y)$
$=(x-2y)(4x-3)$

69. $5q^2-4pq-5q+4p=q(5q-4p)-1(5q-4p)$
$=(5q-4p)(q-1)$

71. $2x^4+5x^3+2x^2+5x=x(2x^3+5x^2+2x+5)$
$=x[x^2(2x+5)+1(2x+5)]$
$=x(2x+5)(x^2+1)$

73. $12x^2y-42x^2-4y+14$
$=2(6x^2y-21x^2-2y+7)$
$=2[3x^2(2y-7)-1(2y-7)]$
$=2(2y-7)(3x^2-1)$

75. $32xy^2-18x^2=2x(16y-9x)$

77. $y(x+2)-3(x+2)=(x+2)(y-3)$

79. $14x^3y+7x^2y-7xy=7xy(2x^2+x-1)$

81. $28x^3-7x^2+12x-3=7x^2(4x-1)+3(4x-1)$
$=(4x-1)(7x^2+3)$

83. $-40x^8y^6-16x^9y^5=-8x^8y^5(5y+2x)$

85. $6a^2+9ab^2+6ab+9b^3$
$=3(2a^2+3ab^2+2ab+3b^3)$
$=3[a(2a+3b^2)+b(2a+3b^2)]$
$=3(2a+3b^2)(a+b)$

87. $(x+2)(x+5)=x^2+5x+2x+10=x^2+7x+10$

89. $(b+1)(b-4)=b^2-4b+b-4=b^2-3b-4$

	Two Numbers	Their Product	Their Sum
91.	2, 6	12	8
93.	−1, −8	8	−9
95.	−2, 5	−10	3

97. a. $8\cdot a-24=8a-24$

b. $8(a-3)=8a-24$

c. $4(2a-12)=8a-48$

d. $8\cdot a-2\cdot 12=8a-24$

The answer is b.

99. $(x+5)(x+y)$ is factored.

101. $3x(a + 2b) + 2(a + 2b)$ is not factored.

103. Answers may vary

105. Answers may vary

107. a. $-8x^2 + 50x + 3020 = -8(2)^2 + 50(2) + 3020$
$= -32 + 100 + 3020$
$= 3088$
In 2005, 3088 thousand, or 3,088,000, students graduated from U.S. high schools.

b. Let $x = 2007 - 2003 = 4$
$-8x^2 + 50x + 3020 = -8(4)^2 + 50(4) + 3020$
$= -128 + 200 + 3020$
$= 3092$
We predict 3092 thousand, or 3,092,000, students to graduate from U.S. high schools in 2007.

c. $-8x^2 + 50x + 3020 = -2(4x^2 - 25x - 1510)$

109. length of side of square = $2x$
shaded area = square's area − circle's area
$= (2x)^2 - \pi x^2$
$= 4x^2 - \pi x^2$
$= x^2(4 - \pi)$

111. Area $= 5x^5 - 5x^2 = 5x^2(x^3 - 1)$
Since the width is $5x^2$ units, the length is $(x^3 - 1)$ units.

113. $x^{2n} + 6x^n + 10x^n + 60 = x^n(x^n + 6) + 10(x^n + 6)$
$= (x^n + 6)(x^n + 10)$

115. $12x^{2n} - 10x^n - 30x^n + 25$
$= 2x^n(6x^n - 5) - 5(6x^n - 5)$
$= (6x^n - 5)(2x^n - 5)$

Section 6.2

Practice Problems

1.

Positive Factors of 6	Sum of Factors
1, 6	7
2, 3	5

$x^2 + 5x + 6 = (x + 2)(x + 3)$

2.

Negative Factors of 70	Sum of Factors
−1, −70	−71
−2, −35	−37
−5, −14	−19
−7, −10	−17

$x^2 - 17x + 70 = (x - 7)(x - 10)$

3.

Factors of −14	Sum of Factors
−1, 14	13
1, −14	−13
−2, 7	5
2, −7	−5

$x^2 + 5x - 14 = (x - 2)(x + 7)$

4. The first term of each binomial is p. Then look for two numbers whose product is −63 and whose sum is −2.
$p^2 - 2p - 63 = (p - 9)(p + 7)$

5. The first term of each binomial is b. Then look for two numbers whose product is 1 and whose sum is 5. There are no such numbers.
$b^2 + 5b + 1$ is a prime polynomial.

6. The first term of each polynomial is x. Then look for two terms whose product is $12y^2$ and whose sum is $7y$.
$x^2 + 7xy + 12y^2 = (x + 3y)(x + 4y)$

7. The first term of each polynomial is x^2. Then look for two numbers whose product is 12 and whose sum is 13.
$x^4 + 13x^2 + 12 = (x^2 + 1)(x^2 + 12)$

8. $48 - 14x + x^2 = x^2 - 14x + 48$
The first term of each binomial is x. Then look for two factors whose product is 48 and whose sum is −14.
$x^2 - 14x + 48 = (x - 6)(x - 8)$

9. $4x^2 - 24x + 36 = 4(x^2 - 6x + 9)$
The first term of each binomial is x. Then look for two factors whose product is 9 and whose sum is -6.
$4(x^2 - 6x + 9) = 4(x-3)(x-3)$ or $4(x-3)^2$

10. $3y^4 - 18y^3 - 21y^2 = 3y^2(y^2 - 6y - 7)$
The first term of each binomial is y. Then look for two factors whose product is -7 and whose sum is -6.
$3y^2(y^2 - 6y - 7) = 3y^2(y-7)(y+1)$

Vocabulary and Readiness Check

1. The statement is true.

2. The statement is true.

3. Since $4x - 12 = 4(x - 3)$, the statement is false.

4. $(x+2y)^2 = (x+2y)(x+2y) \neq (x+2y)(x+y)$
The statement is false.

5. $x^2 + 9x + 20 = (x+4)(x\underline{+5})$

6. $x^2 + 12x + 35 = (x+5)(x\underline{+7})$

7. $x^2 - 7x + 12 = (x-4)(x\underline{-3})$

8. $x^2 - 13x + 22 = (x-2)(x\underline{-11})$

9. $x^2 + 4x + 4 = (x+2)(x\underline{+2})$

10. $x^2 + 10x + 24 = (x+6)(x\underline{+4})$

Exercise Set 6.2

1. $x^2 + 7x + 6 = (x+6)(x+1)$

3. $y^2 - 10y + 9 = (y-9)(y-1)$

5. $x^2 - 6x + 9 = (x-3)(x-3)$ or $(x-3)^2$

7. $x^2 - 3x - 18 = (x-6)(x+3)$

9. $x^2 + 3x - 70 = (x+10)(x-7)$

11. $x^2 + 5x + 2$ is a prime polynomial.

13. $x^2 + 8xy + 15y^2 = (x+5y)(x+3y)$

15. $a^4 - 2a^2 - 15 = (a^2 - 5)(a^2 + 3)$

17. $13 + 14m + m^2 = m^2 + 14m + 13 = (m+13)(m+1)$

19. $10t - 24 + t^2 = t^2 + 10t - 24 = (t-2)(t+12)$

21. $a^2 - 10ab + 16b^2 = (a-2b)(a-8b)$

23. $2z^2 + 20z + 32 = 2(z^2 + 10z + 16)$
$= 2(z+8)(z+2)$

25. $2x^3 - 18x^2 + 40x = 2x(x^2 - 9x + 20)$
$= 2x(x-5)(x-4)$

27. $x^2 - 3xy - 4y^2 = (x-4y)(x+y)$

29. $x^2 + 15x + 36 = (x+12)(x+3)$

31. $x^2 - x - 2 = (x-2)(x+1)$

33. $r^2 - 16r + 48 = (r-12)(r-4)$

35. $x^2 + xy - 2y^2 = (x+2y)(x-y)$

37. $3x^2 + 9x - 30 = 3(x^2 + 3x - 10) = 3(x+5)(x-2)$

39. $3x^2 - 60x + 108 = 3(x^2 - 20x + 36)$
$= 3(x-18)(x-2)$

41. $x^2 - 18x - 144 = (x-24)(x+6)$

43. $r^2 - 3r + 6$ is a prime polynomial.

45. $x^2 - 8x + 15 = (x-5)(x-3)$

47. $6x^3 + 54x^2 + 120x = 6x(x^2 + 9x + 20)$
$= 6x(x+4)(x+5)$

49. $4x^2y + 4xy - 12y = 4y(x^2 + x - 3)$

51. $x^2 - 4x - 21 = (x-7)(x+3)$

53. $x^2+7xy+10y^2=(x+5y)(x+2y)$

55. $64+24t+2t^2=2t^2+24t+64$
$=2(t^2+12t+32)$
$=2(t+8)(t+4)$

57. $x^3-2x^2-24x=x(x^2-2x-24)$
$=x(x-6)(x+4)$

59. $2t^5-14t^4+24t^3=2t^3(t^2-7t+12)$
$=2t^3(t-4)(t-3)$

61. $5x^3y-25x^2y^2-120xy^3=5xy(x^2-5xy-24y^2)$
$=5xy(x-8y)(x+3y)$

63. $162-45m+3m^2=3m^2-45m+162$
$=3(m^2-15m+54)$
$=3(m-9)(m-6)$

65. $-x^2+12x-11=-1(x^2-12x+11)$
$=-1(x-11)(x-1)$

67. $\frac{1}{2}y^2-\frac{9}{2}y-11=\frac{1}{2}(y^2-9y-22)$
$=\frac{1}{2}(y-11)(y+2)$

69. $x^3y^2+x^2y-20x=x(x^2y^2+xy-20)$
$=x(xy-4)(xy+5)$

71. $(2x+1)(x+5)=2x^2+10x+x+5$
$=2x^2+11x+5$

73. $(5y-4)(3y-1)=15y^2-5y-12y+4$
$=15y^2-17y+4$

75. $(a+3b)(9a-4b)=9a^2-4ab+27ab-12b^2$
$=9a^2+23ab-12b^2$

77. $(x-3)(x+8)=x^2+8x-3x-3(8)=x^2+5x-24$

79. Answers may vary

81. $P=2l+2w$
$l=x^2+10x$ and $w=4x+33$, so
$P=2(x^2+10x)+2(4x+33)$
$=2x^2+20x+8x+66$
$=2x^2+28x+66$
$=2(x^2+14x+33)$
$=2(x+11)(x+3)$
The perimeter of the rectangle is given by the polynomial $2x^2+28x+66$ which factors as $2(x+11)(x+3)$.

83. $-16t^2+64t+80=-16(t^2-4t-5)$
$=-16(t-5)(t+1)$

85. $x^2+\frac{1}{2}x+\frac{1}{16}=\left(x+\frac{1}{4}\right)\left(x+\frac{1}{4}\right)$ or $\left(x+\frac{1}{4}\right)^2$

87. $z^2(x+1)-3z(x+1)-70(x+1)$
$=(x+1)(z^2-3z-70)$
$=(x+1)(z-10)(z+7)$

89. $x^{2n}+8x^n-20=(x^n+10)(x^n-2)$

91. c must be the product of positive numbers that sum to 8.
$8=1+7;\ 1\cdot 7=7$
$8=2+6;\ 2\cdot 6=12$
$8=3+5;\ 3\cdot 5=15$
$8=4+4;\ 4\cdot 4=16$
t^2+8t+c is factorable when c is 7, 12, 15, or 16.

93. c must be the product of negative numbers that sum to -16.
$-16=-1+(-15);\ -1\cdot -15=15$
$-16=-2+(-14);\ -2\cdot -14=28$
$-16=-3+(-13);\ -3\cdot -13=39$
$-16=-4+(-12);\ -4\cdot -12=48$
$-16=-5+(-11);\ -5\cdot -11=55$
$-16=-6+(-10);\ -6\cdot -10=60$
$-16=-7+(-9);\ -7\cdot -9=63$
$-16=-8+(-8);\ -8\cdot -8=64$
$n^2-16n+c$ is factorable when c is 15, 28, 39, 48, 55, 60, 63, or 64.

95. b must be the sum of positive numbers which have a product of 20.
$20 = 1 \cdot 20;\ 1 + 20 = 21$
$20 = 2 \cdot 10;\ 2 + 10 = 12$
$20 = 4 \cdot 5;\ 4 + 5 = 9$
$y^2 + by + 20$ is factorable when b is 9, 12, or 21.

97. b must be the positive sum of a positive number and a negative number which have a product of -14.
$-14 = 14 \cdot -1;\ 14 + (-1) = 13$
$-14 = 7 \cdot -2;\ 7 + (-2) = 5$
$x^2 + bx - 14$ is factorable when b is 5 or 13.

Section 6.3

Practice Exercises

1. Factors of $2x^2$: $2x^2 = 2x \cdot x$
   Factors of 15: $15 = 1 \cdot 15,\ 15 = 3 \cdot 5$
   Try possible combinations.
   Factored form: $2x^2 + 11x + 15 = (2x + 5)(x + 3)$

2. Factors of $15x^2$: $15x^2 = 15x \cdot x,\ 15x^2 = 5x \cdot 3x$
   Factors of 8: $8 = -1 \cdot -8,\ 8 = -2 \cdot -4$
   Try possible combinations.
   Factored form: $15x^2 - 22x + 8 = (5x - 4)(3x - 2)$

3. Factors of $4x^2$: $4x^2 = 4x \cdot x,\ 4x^2 = 2x \cdot 2x$
   Factors of -3: $-3 = -1 \cdot 3,\ -3 = 1 \cdot -3$
   Try possible combinations.
   Factored form: $4x^2 + 11x - 3 = (4x - 1)(x + 3)$

4. Factors of $21x^2$: $21x^2 = 21x \cdot x,\ 21x^2 = 3x \cdot 7x$
   Factors of
   $-2y^2$: $-2y^2 = -2y \cdot y,\ -2y^2 = 2y \cdot -y$
   Try possible combinations.
   Factored form:
   $21x^2 + 11xy - 2y^2 = (7x - y)(3x + 2y)$

5. Factors of $2x^4$: $2x^4 = 2x^2 \cdot x^2$
   Factors of -7: $-7 = -7 \cdot 1,\ -7 = 7 \cdot -1$
   Try possible combinations.
   $2x^4 - 5x^2 - 7 = (2x^2 - 7)(x^2 + 1)$

6. $3x^3 + 17x^2 + 10x = x(3x^2 + 17x + 10)$
   Factors of $3x^2$: $3x^2 = 3x \cdot x$
   Factors of 10: $10 = 1 \cdot 10,\ 10 = 2 \cdot 5$
   Try possible combinations:
   $$3x^3 + 17x^2 + 10x = x(3x^2 + 17x + 10) = x(3x + 2)(x + 5)$$

7. $$-8x^2 + 2x + 3 = -1(8x^2 - 2x - 3) = -1(4x - 3)(2x + 1)$$

8. $x^2 = (x)^2$ and $49 = 7^2$
   Is $2 \cdot x \cdot 7 = 14x$ the middle term? Yes.
   $x^2 + 14x + 49 = (x + 7)^2$

9. $4x^2 = (2x)^2$ and $9y^2 = (3y)^2$
   Is $2 \cdot 2x \cdot 3y = 12xy$ the middle term? No.
   Try other possibilities.
   $4x^2 + 20xy + 9y^2 = (2x + 9y)(2x + y)$

10. $36n^4 = (6n^2)^2$ and $1 = 1^2$
    Is $2 \cdot 6n^2 \cdot 1 = 12n^2$ the middle term? Yes, the opposite of the middle term.
    $36n^4 - 12n^2 + 1 = (6n^2 - 1)^2$

11. $$12x^3 - 84x^2 + 147x = 3x(4x^2 - 28x + 49) = 3x[(2x)^2 - 2 \cdot 2x \cdot 7 + 7^2] = 3x(2x - 7)^2$$

Vocabulary and Readiness Check

1. A perfect square trinomial is a trinomial that is the square of a binomial.

2. The term $25y^2$ written as a square is $(5y)^2$.

3. The expression $x^2 + 10xy + 25y^2$ is called a perfect square trinomial.

4. The factorization $(x + 5y)(x + 5y)$ may also be written as $(x + 5y)^2$.

5. no

6. yes

7. $64 = 8^2$

8. $9 = 3^2$

9. $121a^2 = (11a)^2$

10. $81b^2 = (9b)^2$

11. $36p^4 = (6p^2)^2$

12. $4q^4 = (2q^2)^2$

Exercise Set 6.3

1. $5x^2 + 22x + 8 = (5x + 2)(x + 4)$

3. $50x^2 + 15x - 2 = (5x + 2)(10x - 1)$

5. $25x^2 - 20x + 4 = (5x - 2)(5x - 2)$

7. $2x^2 + 13x + 15 = (2x + 3)(x + 5)$

9. $8y^2 - 17y + 9 = (y - 1)(8y - 9)$

11. $2x^2 - 9x - 5 = (2x + 1)(x - 5)$

13. $20r^2 + 27r - 8 = (4r - 1)(5r + 8)$

15. $10x^2 + 31x + 3 = (10x + 1)(x + 3)$

17. $2m^2 + 17m + 10$ is prime.

19. $6x^2 - 13xy + 5y^2 = (3x - 5y)(2x - y)$

21. $15m^2 - 16m - 15 = (3m - 5)(5m + 3)$

23. $12x^3 + 11x^2 + 2x = x(12x^2 + 11x + 2)$
$= x(3x + 2)(4x + 1)$

25. $21b^2 - 48b - 45 = 3(7b^2 - 16b - 15)$
$= 3(7b + 5)(b - 3)$

27. $7z + 12z^2 - 12 = 12z^2 + 7z - 12 = (3z + 4)(4z - 3)$

29. $6x^2y^2 - 2xy^2 - 60y^2 = 2y^2(3x^2 - x - 30)$
$= 2y^2(3x - 10)(x + 3)$

31. $4x^2 - 8x - 21 = (2x - 7)(2x + 3)$

33. $-x^2 + 2x + 24 = -1(x^2 - 2x - 24)$
$= -1(x - 6)(x + 4)$

35. $4x^3 - 9x^2 - 9x = x(4x^2 - 9x - 9)$
$= x(4x + 3)(x - 3)$

37. $24x^2 - 58x + 9 = (4x - 9)(6x - 1)$

39. $x^2 + 22x + 121 = x^2 + 2 \cdot x \cdot 11 + 11^2 = (x + 11)^2$

41. $x^2 - 16x + 64 = x^2 - 2 \cdot x \cdot 8 + 8^2 = (x - 8)^2$

43. $16a^2 - 24a + 9 = (4a)^2 - 2 \cdot 4a \cdot 3 + 3^2 = (4a - 3)^2$

45. $x^4 + 4x^2 + 4 = (x^2)^2 + 2 \cdot x^2 \cdot 2 + 2^2 = (x^2 + 2)^2$

47. $2n^2 - 28n + 98 = 2(n^2 - 14n + 49)$
$= 2(n^2 - 2 \cdot n \cdot 7 + 7^2)$
$= 2(n - 7)^2$

49. $16y^2 + 40y + 25 = (4y)^2 + 2 \cdot 4y \cdot 5 + 5^2$
$= (4y + 5)^2$

51. $2x^2 - 7x - 99 = (2x + 11)(x - 9)$

53. $24x^2 + 41x + 12 = (8x + 3)(3x + 4)$

55. $3a^2 + 10ab + 3b^2 = (3a + b)(a + 3b)$

57. $-9x + 20 + x^2 = x^2 - 9x + 20 = (x - 4)(x - 5)$

59. $p^2 + 12pq + 36q^2 = p^2 + 2 \cdot p \cdot 6q + (6q)^2$
$= (p + 6q)^2$

61. $x^2y^2 - 10xy + 25 = (xy)^2 - 2 \cdot xy \cdot 5 + 5^2$
$= (xy - 5)^2$

63. $40a^2b + 9ab - 9b = b(40a^2 + 9a - 9)$
$= b(8a - 3)(5a + 3)$

65. $30x^3 + 38x^2 + 12x = 2x(15x^2 + 19x + 6)$
$= 2x(3x + 2)(5x + 3)$

67. $6y^3 - 8y^2 - 30y = 2y(3y^2 - 4y - 15)$
$= 2y(3y + 5)(y - 3)$

69. $10x^4 + 25x^3y - 15x^2y^2 = 5x^2(2x^2 + 5xy - 3y^2)$
$= 5x^2(2x - y)(x + 3y)$

71. $-14x^2 + 39x - 10 = -1(14x^2 - 39x + 10)$
$= -1(2x - 5)(7x - 2)$

73. $16p^4 - 40p^3 + 25p^2 = p^2(16p^2 - 40p + 25)$
$= p^2[(4p)^2 - 2 \cdot 4p \cdot 5 + 5^2]$
$= p^2(4p - 5)^2$

75. $x + 3x^2 - 2 = 3x^2 + x - 2 = (3x - 2)(x + 1)$

77. $8x^2 + 6xy - 27y^2 = (4x + 9y)(2x - 3y)$

79. $1 + 6x^2 + x^4 = x^4 + 6x^2 + 1$ is prime.

81. $9x^2 - 24xy + 16y^2 = (3x)^2 - 2 \cdot 3x \cdot 4y + (4y)^2$
$= (3x - 4y)^2$

83. $18x^2 - 9x - 14 = (6x - 7)(3x + 2)$

85. $-27t + 7t^2 - 4 = 7t^2 - 27t - 4 = (7t + 1)(t - 4)$

87. $49p^2 - 7p - 2 = (7p + 1)(7p - 2)$

89. $m^3 + 18m^2 + 81m = m(m^2 + 18m + 81)$
$= m(m^2 + 2 \cdot m \cdot 9 + 9^2)$
$= m(m + 9)^2$

91. $5x^2y^2 + 20xy + 1$ is prime.

93. $6a^5 + 37a^3b^2 + 6ab^4 = a(6a^4 + 37a^2b^2 + 6b^4)$
$= a(6a^2 + b^2)(a^2 + 6b^2)$

95. $(x - 2)(x + 2) = x^2 + 2x - 2x - 4 = x^2 - 4$

97. $(a + 3)(a^2 - 3a + 9)$
$= a^3 - 3a^2 + 9a + 3a^2 - 9a + 27$
$= a^3 + 27$

99. Look for the tallest graph. The income range is \$75,000 and above.

101. Answers may vary

103. no

105. Answers may vary

107. $P = (3x^2 + 1) + (6x + 4) + (x^2 + 15x)$
$= 3x^2 + 1 + 6x + 4 + x^2 + 15x$
$= 4x^2 + 21x + 5$
$= (4x + 1)(x + 5)$

109. $4x^2 + 2x + \frac{1}{4} = (2x)^2 + 2 \cdot 2x \cdot \frac{1}{2} + \left(\frac{1}{2}\right)^2$
$= \left(2x + \frac{1}{2}\right)^2$

111. $4x^2(y - 1)^2 + 10x(y - 1)^2 + 25(y - 1)^2$
$= (y - 1)^2(4x^2 + 10x + 25)$

113. $16 = 4^2$; $2 \cdot x \cdot 4 = 8x$; 8

115. $(a + b)^2 = a^2 + 2ab + b^2$

117. $b = 2$: $3x^2 + 2x - 5 = (3x + 5)(x - 1)$
$b = 14$: $3x^2 + 14x - 5 = (3x - 1)(x + 5)$

119. $c = 2$: $5x^2 + 7x + 2 = (5x + 2)(x + 1)$

121. $-12x^3y^2 + 3x^2y^2 + 15xy^2$
$= -3xy^2(4x^2 - x - 5)$
$= -3xy^2(4x - 5)(x + 1)$

123. $4x^2(y - 1)^2 + 20x(y - 1)^2 + 25(y - 1)^2$
$= (y - 1)^2(4x^2 + 20x + 25)$
$= (y - 1)^2[(2x)^2 + 2 \cdot 2x \cdot 5 + 5^2]$
$= (y - 1)^2(2x + 5)^2$

125. $3x^{2n} + 17x^n + 10 = (3x^n + 2)(x^n + 5)$

127. Answers may vary

Section 6.4

Practice Exercises

1.

Factors of $ac = 60$	Sum of Factors
1, 60	61 ← correct sum $b = 61.$
2, 30	32
3, 20	23
4, 15	19
5, 12	17
6, 10	16

$$\begin{aligned}5x^2+61x+12 &= 5x^2+1x+60x+12\\ &= x(5x+1)+12(5x+1)\\ &= (5x+1)(x+12)\end{aligned}$$

2.

Factors of $ac = 60$	Sum of Factors
−1, −60	−61
−2, −30	−32
−3, −20	−23
−4, −15	−19 ← Correct sum $b = -19$
−5, −12	−17
−6, −10	−60

$$\begin{aligned}12x^2-19x+5 &= 12x^2-15x-4x+5\\ &= 3x(4x-5)-1(4x-5)\\ &= (4x-5)(3x-1)\end{aligned}$$

3. $30x^2-14x-4 = 2(15x^2-7x-2)$
Find two numbers whose product is $ac = 15(-2) = -30$ and whose sum is b, −7. The numbers are −10 and 3.
$$\begin{aligned}2(15x^2-7x-2) &= 2(15x^2-10x+3x-2)\\ &= 2[5x(3x-2)+1(3x-2)]\\ &= 2(3x-2)(5x+1)\end{aligned}$$

4. $40m^4+5m^3-35m^2 = 5m^2(8m^2+m-7)$
Find two numbers whose product is $ac = 8(-7) = -56$ and whose sum is b, 1. The numbers are 8 and −7.
$$\begin{aligned}5m^2(8m^2+m-7) &= 5m^2(8m^2+8m-7m-7)\\ &= 5m^2[8m(m+1)-7(m+1)]\\ &= 5m^2(m+1)(8m-7)\end{aligned}$$

5. Find two numbers whose product is $ac = 16 \cdot 9 = 144$ and whose sum is b, 24. The numbers are 12 and 12.
$$\begin{aligned}16x^2+24x+9 &= 16x^2+12x+12x+9\\ &= 4x(4x+3)+3(4x+3)\\ &= (4x+3)(4x+3)\\ &= (4x+3)^2\end{aligned}$$

Exercise Set 6.4

1. $x^2+3x+2x+6 = x(x+3)+2(x+3)$
$= (x+3)(x+2)$

3. $y^2+8y-2y-16 = y(y+8)-2(y+8)$
$= (y+8)(y-2)$

5. $8x^2-5x-24x+15 = x(8x-5)-3(8x-5)$
$= (8x-5)(x-3)$

7. $5x^4-3x^2+25x^2-15 = x^2(5x^2-3)+5(5x^2-3)$
$= (5x^2-3)(x^2+5)$

9. **a.** $9 \cdot 2 = 18$; $9 + 2 = 11$; 9, 2

   b. $11x = 9x + 2x$

   c. $$\begin{aligned}6x^2+11x+3 &= 6x^2+9x+2x+3\\ &= 3x(2x+3)+1(2x+3)\\ &= (3x+1)(2x+3)\end{aligned}$$

11. **a.** $-20 \cdot (-3) = 60$; $-20 + (-3) = -23$; −20, −3

    b. $-23x = -20x - 3x$

    c. $$\begin{aligned}15x^2-23x+4 &= 15x^2-20x-3x+4\\ &= 5x(3x-4)-1(3x-4)\\ &= (3x-4)(5x-1)\end{aligned}$$

13. $ac = 21 \cdot 2 = 42$; $b = 17$; two numbers: 14, 3

$$\begin{aligned} 21y^2 + 17y + 2 &= 21y^2 + 14y + 3y + 2 \\ &= 7y(3y+2) + 1(3y+2) \\ &= (3y+2)(7y+1) \end{aligned}$$

15. $ac = 7 \cdot (-11) = -77$; $b = -4$;
two numbers: -11, 7

$$\begin{aligned} 7x^2 - 4x - 11 &= 7x^2 - 11x + 7x - 11 \\ &= x(7x-11) + 1(7x-11) \\ &= (7x-11)(x+1) \end{aligned}$$

17. $ac = 10 \cdot 2 = 20$; $b = -9$; two numbers: -4, -5

$$\begin{aligned} 10x^2 - 9x + 2 &= 10x^2 - 4x - 5x + 2 \\ &= 2x(5x-2) - 1(5x-2) \\ &= (5x-2)(2x-1) \end{aligned}$$

19. $ac = 2 \cdot 5 = 10$; $b = -7$; two numbers: -5, -2

$$\begin{aligned} 2x^2 - 7x + 5 &= 2x^2 - 5x - 2x + 5 \\ &= x(2x-5) - 1(2x-5) \\ &= (2x-5)(x-1) \end{aligned}$$

21. $12x + 4x^2 + 9 = 4x^2 + 12x + 9$
$ac = 4 \cdot 9 = 36$; $b = 12$; two numbers: 6, 6

$$\begin{aligned} 4x^2 + 12x + 9 &= 4x^2 + 6x + 6x + 9 \\ &= 2x(2x+3) + 3(2x+3) \\ &= (2x+3)(2x+3) \\ &= (2x+3)^2 \end{aligned}$$

23. $ac = 4 \cdot (-21) = -84$; $b = -8$;
two numbers: 6, -14

$$\begin{aligned} 4x^2 - 8x - 21 &= 4x^2 + 6x - 14x - 21 \\ &= 2x(2x+3) - 7(2x+3) \\ &= (2x+3)(2x-7) \end{aligned}$$

25. $ac = 10 \cdot 12 = 120$; $b = -23$;
two numbers: -8, -15

$$\begin{aligned} 10x^2 - 23x + 12 &= 10x^2 - 8x - 15x + 12 \\ &= 2x(5x-4) - 3(5x-4) \\ &= (5x-4)(2x-3) \end{aligned}$$

27. $2x^3 + 13x^2 + 15x = x(2x^2 + 13x + 15)$
$ac = 2 \cdot 15 = 30$; $b = 13$; two numbers: 3, 10

$$\begin{aligned} x(2x^2 + 13x + 15) &= x(2x^2 + 3x + 10x + 15) \\ &= x[x(2x+3) + 5(2x+3)] \\ &= x(2x+3)(x+5) \end{aligned}$$

29. $16y^2 - 34y + 18 = 2(8y^2 - 17y + 9)$
$ac = 8(9) = 72$; $b = -17$; two numbers: -9, -8

$$\begin{aligned} 2(8y^2 - 17y + 9) &= 2(8y^2 - 9y - 8y + 9) \\ &= 2[y(8y-9) - 1(8y-9)] \\ &= 2(8y-9)(y-1) \end{aligned}$$

31. $-13x + 6 + 6x^2 = 6x^2 - 13x + 6$
$ac = 6 \cdot 6 = 36$; $b = -13$; two numbers: -9, -4

$$\begin{aligned} 6x^2 - 13x + 6 &= 6x^2 - 9x - 4x + 6 \\ &= 3x(2x-3) - 2(2x-3) \\ &= (2x-3)(3x-2) \end{aligned}$$

33. $54a^2 - 9a - 30 = 3(18a^2 - 3a - 10)$
$ac = 18(-10) = -180$; $b = -3$;
two numbers: 12, -15

$$\begin{aligned} 3(18a^2 - 3a - 10) &= 3(18a^2 + 12a - 15a - 10) \\ &= 3[6a(3a+2) - 5(3a+2)] \\ &= 3(3a+2)(6a-5) \end{aligned}$$

35. $20a^3 + 37a^2 + 8a = a(20a^2 + 37a + 8)$
$ac = 20(8) = 160$; $b = 37$; two numbers: 5, 32

$$\begin{aligned} a(20a^2 + 37a + 8) &= a(20a^2 + 5a + 32a + 8) \\ &= a[5a(4a+1) + 8(4a+1)] \\ &= a(4a+1)(5a+8) \end{aligned}$$

37. $12x^3 - 27x^2 - 27x = 3x(4x^2 - 9x - 9)$
$ac = 4(-9) = -36$; $b = -9$; two numbers: 3, -12

$$\begin{aligned} 3x(4x^2 - 9x - 9) &= 3x(4x^2 + 3x - 12x - 9) \\ &= 3x[x(4x+3) - 3(4x+3)] \\ &= 3x(4x+3)(x-3) \end{aligned}$$

39. $3x^2y + 4xy^2 + y^3 = y(3x^2 + 4xy + y^2)$
$ac = 3 \cdot 1 = 3$; $b = 4$; two numbers: 1, 3

$$\begin{aligned} y(3x^2 + 4xy + y^2) &= y(3x^2 + xy + 3xy + y^2) \\ &= y[x(3x+y) + y(3x+y)] \\ &= y(3x+y)(x+y) \end{aligned}$$

41. $ac = 20 \cdot 1 = 20$; $b = 7$; there are no two numbers.

$20z^2 + 7z + 1$ is prime.

43. $5x^2+50xy+125y^2=5(x^2+10xy+25y^2)$
$ac = 1 \cdot 25 = 25$; $b = 10$; two numbers: 5, 5
$$\begin{aligned}5(x^2+10xy+25y^2) &= 5(x^2+5xy+5xy+25y^2)\\ &= 5[x(x+5y)+5y(x+5y)]\\ &= 5(x+5y)(x+5y)\\ &= 5(x+5y)^2\end{aligned}$$

45. $24a^2-6ab-30b^2=6(4a^2-ab-5b^2)$
$ac = 4 \cdot (-5) = -20$; $b = -1$; two numbers: 4, –5
$$\begin{aligned}6(4a^2-ab-5b^2) &= 6(4a^2+4ab-5ab-5b^2)\\ &= 6[4a(a+b)-5b(a+b)]\\ &= 6(a+b)(4a-5b)\end{aligned}$$

47. $15p^4+31p^3q+2p^2q^2=p^2(15p^2+31pq+2q^2)$
$ac = 15(2) = 30$; $b = 31$; two numbers: 1, 30
$$\begin{aligned}&p^2(15p^2+31pq+2q^2)\\ &= p^2(15p^2+pq+30pq+2q^2)\\ &= p^2[p(15p+q)+2q(15p+q)]\\ &= p^2(15p+q)(p+2q)\end{aligned}$$

49. $162a^4-72a^2+8=2(81a^4-36a^2+4)$
$ac = 81 \cdot 4 = 324$; $b = -36$;
two numbers: –18, –18
$$\begin{aligned}&2(81a^4-36a^2+4)\\ &= 2(81a^4-18a^2-18a^2+4)\\ &= 2[9a^2(9a^2-2)-2(9a^2-2)]\\ &= 2(9a^2-2)(9a^2-2)\\ &= 2(9a^2-2)^2\end{aligned}$$

51. $35+12x+x^2=x^2+12x+35$
$ac = 1 \cdot 35 = 35$; $b = 12$; two numbers: 5, 7
$$\begin{aligned}x^2+12x+35 &= x^2+5x+7x+35\\ &= x(x+5)+7(x+5)\\ &= (x+5)(x+7)\end{aligned}$$

53. $6-11x+5x^2=5x^2-11x+6$
$ac = 5 \cdot 6 = 30$; $b = -11$; two numbers: –6, –5
$$\begin{aligned}5x^2-11x+6 &= 5x^2-6x-5x+6\\ &= x(5x-6)-1(5x-6)\\ &= (5x-6)(x-1)\end{aligned}$$

55. $(x-2)(x+2)=x^2-2^2=x^2-4$

57. $(y+4)(y+4)=y^2+2\cdot y\cdot 4+4^2=y^2+8y+16$

59. $(9z+5)(9z-5)=(9z)^2-5^2=81z^2-25$

61. $(x-3)(x^2+3x+9)=x^3-3^3=x^3-27$

63. $5(2x^2+9x+9)=10x^2+45x+45$
$ac = 2 \cdot 9 = 18$; $b = 9$; two numbers: 3, 6
$$\begin{aligned}5(2x^2+9x+9) &= 5(2x^2+3x+6x+9)\\ &= 5[x(2x+3)+3(2x+3)]\\ &= 15(2x+3)(x+3)\end{aligned}$$

65. $$\begin{aligned}x^{2n}+2x^n+3x^n+6 &= x^n(x^n+2)+3(x^n+2)\\ &= (x^n+2)(x^n+3)\end{aligned}$$

67. $ac = 3 \cdot (-35) = -105$; $b = 16$;
two numbers: –5, 21
$$\begin{aligned}3x^{2n}+16x^n-35 &= 3x^{2n}-5x^n+21x^n-35\\ &= x^n(3x^n-5)+7(3x^n-5)\\ &= (3x^n-5)(x^n+7)\end{aligned}$$

69. Answers may vary

Section 6.5

Practice Exercises

1. $x^2-81=x^2-9^2=(x+9)(x-9)$

2. a. $9x^2-1=(3x)^2-1^2=(3x+1)(3x-1)$

b. $$\begin{aligned}36a^2-49b^2 &= (6a)^2-(7b)^2\\ &= (6a+7b)(6a-7b)\end{aligned}$$

c. $p^2-\frac{25}{36}=p^2-\left(\frac{5}{6}\right)^2=\left(p+\frac{5}{6}\right)\left(p-\frac{5}{6}\right)$

3. $p^4-q^{10}=(p^2)^2-(q^5)^2=(p^2+q^5)(p^2-q^5)$

4. a. $$\begin{aligned}z^4-81 &= (z^2)^2-9^2\\ &= (z^2+9)(z^2-9)\\ &= (z^2+9)(z+3)(z-3)\end{aligned}$$

b. m^2+49 is a prime polynomial.

5. $$\begin{aligned}36y^3-25y &= y(36y^2-25)\\ &= y[(6y)^2-5^2]\\ &= y(6y+5)(6y-5)\end{aligned}$$

6. $80y^4 - 5 = 5(16y^2 - 1)$
$= 5[(4y)^2 - 1^2]$
$= 5(4y+1)(4y-1)$

7. $-9x^2 + 100 = -1(9x^2 - 100)$
$= -1[(3x)^2 - 10^2]$
$= -1(3x+10)(3x-10)$

8. $x^3 + 64 = x^3 + 4^3$
$= (x+4)(x^2 - x \cdot 4 + 4^2)$
$= (x+4)(x^2 - 4x + 16)$

9. $x^3 - 125 = x^3 - 5^3$
$= (x-5)(x^2 + x \cdot 5 + 5^2)$
$= (x-5)(x^2 + 5x + 25)$

10. $27y^3 + 1 = (3y)^3 + 1^3$
$= (3y+1)[(3y)^2 - 3y \cdot 1 + 1^2]$
$= (3y+1)(9y^2 - 3y + 1)$

11. $32x^3 - 500y^3$
$= 4(8x^3 - 125y^3)$
$= 4[(2x)^3 - (5y)^3]$
$= 4(2x - 5y)[(2x)^2 + 2x \cdot 5y + (5y)^2]$
$= 4(2x - 5y)(4x^2 + 10xy + 25y^2)$

Calculator Explorations

x	$x^2 - 2x + 1$	$x^2 - 2x - 1$	$(x-1)^2$
5	16	14	16
−3	16	14	16
2.7	2.89	0.89	2.89
−12.1	171.61	169.61	171.61
0	1	−1	1

Vocabulary and Readiness Check

1. The expression $x^3 - 27$ is called a difference of two cubes.

2. The expression $x^2 - 49$ is called a difference of two squares.

3. The expression $z^3 + 1$ is called a sum of two cubes.

4. The binomial $y^2 + 9$ is prime. The statement is false.

5. $64 = 8^2$

6. $100 = 10^2$

7. $49x^2 = (7x)^2$

8. $25y^4 = (5y^2)^2$

9. $64 = 4^3$

10. $1 = 1^3$

11. $8y^3 = (2y)^3$

12. $x^6 = (x^2)^3$

Exercise Set 6.5

1. $x^2 - 4 = x^2 - 2^2 = (x+2)(x-2)$

3. $81p^2 - 1 = (9p)^2 - 1^2 = (9p+1)(9p-1)$

5. $25y^2 - 9 = (5y)^2 - 3^2 = (5y+3)(5y-3)$

7. $121m^2 - 100n^2 = (11m)^2 - (10n)^2$
$= (11m + 10n)(11m - 10n)$

9. $x^2y^2 - 1 = (xy)^2 - 1^2 = (xy+1)(xy-1)$

11. $x^2 - \frac{1}{4} = x^2 - \left(\frac{1}{2}\right)^2 = \left(x + \frac{1}{2}\right)\left(x - \frac{1}{2}\right)$

13. $-4r^2 + 1 = -1(4r^2 - 1)$
$= -1[(2r)^2 - 1^2]$
$= -1(2r+1)(2r-1)$

15. $16r^2 + 1$ is the sum of two squares, $(4r)^2 + 1^2$, not the difference of two squares. $16r^2 + 1$ is a prime polynomial.

17. $-36+x^2=-1(36-x^2)$
$=-1(6^2-x^2)$
$=-1(6+x)(6-x)$ or $(-6+x)(6+x)$

19. $m^4-1=(m^2)^2-1^2$
$=(m^2+1)(m^2-1)$
$=(m^2+1)(m+1)(m-1)$

21. $m^4-n^{18}=(m^2)^2-(n^9)^2$
$=(m^2+n^9)(m^2-n^9)$

23. $x^3+125=x^3+5^3$
$=(x+5)(x^2-x\cdot 5+5^2)$
$=(x+5)(x^2-5x+25)$

25. $8a^3-1=(2a)^3-1^3$
$=(2a-1)[(2a)^2+2a\cdot 1+1^2]$
$=(2a-1)(4a^2+2a+1)$

27. $m^3+27n^3=m^3+(3n)^3$
$=(m+3n)[m^2-m\cdot 3n+(3n)^2]$
$=(m+3n)(m^2-3mn+9n^2)$

29. $5k^3+40=5(k^3+8)$
$=5(k^3+2^3)$
$=5(k+2)[k^2-k\cdot 2+2^2]$
$=5(k+2)(k^2-2k+4)$

31. $x^3y^3-64=(xy)^3-4^3$
$=(xy-4)[(xy)^2+xy\cdot 4+4^2]$
$=(xy-4)(x^2y^2+4xy+16)$

33. $250r^3-128t^3=2(125r^3-64t^3)$
$=2[(5r)^3-(4t)^3]$
$=2(5r-4t)[(5r)^2+5r\cdot 4t+(4t)^2]$
$=2(5r-4t)(25r^2+20rt+16t^2)$

35. $r^2-64=r^2-8^2=(r+8)(r-8)$

37. $x^2-169y^2=x^2-(13y)^2=(x+13y)(x-13y)$

39. $27-t^3=3^3-t^3$
$=(3-t)(3^2+3\cdot t+t^2)$
$=(3-t)(9+3t+t^2)$

41. $18r^2-8=2(9r^2-4)$
$=2[(3r)^2-2^2]$
$=2(3r+2)(3r-2)$

43. $9xy^2-4x=x(9y^2-4)$
$=x[(3y)^2-2^2]$
$=x(3y+2)(3y-2)$

45. $8m^3+64=8(m^3+8)$
$=8(m^3+2^3)$
$=8(m+2)(m^2-m\cdot 2+2^2)$
$=8(m+2)(m^2-2m+4)$

47. $xy^3-9xyz^2=xy(y^2-9z^2)$
$=xy[y^2-(3z)^2]$
$=xy(y+3z)(y-3z)$

49. $36x^2-64y^2=4(9x^2-16y^2)$
$=4[(3x)^2-(4y)^2]$
$=4(3x+4y)(3x-4y)$

51. $144-81x^2=9(16-9x^2)$
$=9[4^2-(3x)^2]$
$=9(4+3x)(4-3x)$

53. $x^3y^3-z^6=(xy)^3-(z^2)^3$
$=(xy-z^2)[(xy)^2+xy\cdot z^2+(z^2)^2]$
$=(xy-z^2)(x^2y^2+xyz^2+z^4)$

55. $49-\frac{9}{25}m^2=7^2-\left(\frac{3}{5}m\right)^2=\left(7+\frac{3}{5}m\right)\left(7-\frac{3}{5}m\right)$

57. $t^3+343=t^3+7^3$
$=(t+7)(t^2-t\cdot 7+7^2)$
$=(t+7)(t^2-7t+49)$

59. $n^3-49n=n(n^2+49)$

61. $x^6 - 81x^2 = x^2(x^4 - 81)$
$= x^2[(x^2)^2 - 9^2]$
$= x^2(x^2 + 9)(x^2 - 9)$
$= x^2(x^2 + 9)(x + 3)(x - 3)$

63. $64p^3q - 81pq^3 = pq(64p^2 - 81q^2)$
$= pq[(8p)^2 - (9q)^2]$
$= pq(8p + 9q)(8p - 9q)$

65. $27x^2y^3 + xy^2 = xy^2(27xy + 1)$

67. $125a^4 - 64ab^3$
$= a(125a^3 - 64b^3)$
$= a[(5a)^3 - (4b)^3]$
$= a(5a - 4b)[(5a)^2 + 5a \cdot 4b + (4b)^2]$
$= a(5a - 4b)(25a^2 + 20ab + 16b^2)$

69. $16x^4 - 64x^2 = 16x^2(x^2 - 4)$
$= 16x^2(x^2 - 2^2)$
$= 16x^2(x + 2)(x - 2)$

71. $x - 6 = 0$
$x - 6 + 6 = 0 + 6$
$x = 6$

73. $2m + 4 = 0$
$2m + 4 - 4 = 0 - 4$
$2m = -4$
$\frac{2m}{2} = \frac{-4}{2}$
$m = -2$

75. $5z - 1 = 0$
$5z - 1 + 1 = 0 + 1$
$5z = 1$
$\frac{5z}{5} = \frac{1}{5}$
$z = \frac{1}{5}$

77. Let $x = 2003 - 2000 = 3$.
$-1.2x^2 + 4x + 80 = -1.2(3)^2 + 4(3) + 80$
$= -10.8 + 12 + 80$
$= 81.2$
81.2% of college students had credit cards in 2003.

79. $-1.2x^2 + 4x + 80 = -4(0.3x^2 - x - 20)$

81. $(x + 2)^2 - y^2 = (x + 2 + y)(x + 2 - y)$

83. $a^2(b - 4) - 16(b - 4) = (b - 4)(a^2 - 16)$
$= (b - 4)(a^2 - 4^2)$
$= (b - 4)(a + 4)(a - 4)$

85. $(x^2 + 6x + 9) - 4y^2 = (x + 3)^2 - 4y^2$
$= (x + 3)^2 - (2y)^2$
$= [(x + 3) + 2y][(x + 3) - 2y]$
$= (x + 3 + 2y)(x + 3 - 2y)$

87. $x^{2n} - 100 = (x^n)^2 - 10^2 = (x^n + 10)(x^n - 10)$

89. $x + 6$ since
$(x + 6)(x - 6) = x^2 - 6x + 6x - 36$
$= x^2 - 36$
$= x^2 - 6^2$

91. Answers may vary

93. **a.** Let $t = 2$.
$841 - 16t^2 = 841 - 16(2)^2$
$= 841 - 16(4)$
$= 841 - 64$
$= 777$
After 2 seconds, the height of the object is 777 feet.

b. Let $t = 5$.
$841 - 16t^2 = 841 - 16(5)^2$
$= 841 - 16(25)$
$= 841 - 400$
$= 441$
After 5 seconds the height of the object is 441 feet.

c. When the object hits the ground, its height is zero feet. Thus, to find the time, t, when the object's height is zero feet above the ground, we set the expression $841 - 16t^2$ equal to 0 and solve for t.

$$\begin{aligned} 841 - 16t^2 &= 0 \\ 841 - 16t^2 + 16t^2 &= 0 + 16t^2 \\ 841 &= 16t^2 \\ \frac{841}{16} &= \frac{16t^2}{16} \\ 52.5625 &= t^2 \\ \sqrt{52.5625} &= \sqrt{t^2} \\ 7.25 &= t \end{aligned}$$

Thus, the object will hit the ground after approximately 7 seconds.

d. $841 - 16t^2 = 29^2 - (4t)^2 = (29 + 4t)(29 - 4t)$

95. a. Let $t = 3$.
$1600 - 16t^2 = 1600 - 16(3)^2 = 1456$
After 3 seconds the height is 1456 feet.

b. Let $t = 7$.
$1600 - 16t^2 = 1600 - 16(7)^2 = 816$
After 7 seconds the height is 816 feet.

c. When it hits the ground, the height is 0.
Let $0 = 1600 - 16t^2$.

$$\begin{aligned} 16t^2 &= 1600 \\ t^2 &= 100 \\ t &= \sqrt{100} \\ t &= 10 \end{aligned}$$

Thus, it will hit the ground after 10 seconds.

d. $$\begin{aligned} 1600 - 16t^2 &= 16(100 - t^2) \\ &= 16(10^2 - t^2) \\ &= 16(10 + t)(10 - t) \end{aligned}$$

Integrated Review

Practice Exercises

1. $6x^2 - 11x + 3$
$ac = 6 \cdot 3 = 18$; $b = -11$; two numbers: -2, -9

$$\begin{aligned} 6x^2 - 11x + 3 &= 6x^2 - 2x - 9x + 3 \\ &= 2x(3x - 1) - 3(3x - 1) \\ &= (3x - 1)(2x - 3) \end{aligned}$$

2. $$\begin{aligned} 3x^3 + x^2 - 12x - 4 &= (3x^3 + x^2) + (-12x - 4) \\ &= x^2(3x + 1) - 4(3x + 1) \\ &= (3x + 1)(x^2 - 4) \\ &= (3x + 1)(x + 2)(x - 2) \end{aligned}$$

3. $$\begin{aligned} 27x^2 - 3y^2 &= 3(9x^2 - y^2) \\ &= 3[(3x)^2 - y^2] \\ &= 3(3x + y)(3x - y) \end{aligned}$$

4. $$\begin{aligned} 8a^3 + b^3 &= (2a)^3 + b^3 \\ &= (2a + b)[(2a)^2 - 2a \cdot b + b^2] \\ &= (2a + b)(4a^2 - 2ab + b^2) \end{aligned}$$

5. $$\begin{aligned} &60x^3y^2 - 66x^2y^2 - 36xy^2 \\ &= 6xy^2(10x^2 - 11x - 6) \\ &= 6xy^2(5x + 2)(2x - 3) \end{aligned}$$

Integrated Review

1. $x^2 + 2xy + y^2 = (x + y)(x + y) = (x + y)^2$

2. $x^2 - 2xy + y^2 = (x - y)(x - y) = (x - y)^2$

3. $a^2 + 11a - 12 = (a + 12)(a - 1)$

4. $a^2 - 11a + 10 = (a - 10)(a - 1)$

5. $a^2 - a - 6 = (a - 3)(a + 2)$

6. $a^2 - 2a + 1 = (a - 1)(a - 1) = (a - 1)^2$

7. $x^2 + 2x + 1 = (x + 1)(x + 1) = (x + 1)^2$

8. $x^2 + x - 2 = (x + 2)(x - 1)$

9. $x^2 + 4x + 3 = (x + 3)(x + 1)$

10. $x^2 + x - 6 = (x + 3)(x - 2)$

11. $x^2 + 7x + 12 = (x + 4)(x + 3)$

12. $x^2 + x - 12 = (x + 4)(x - 3)$

13. $x^2 + 3x - 4 = (x + 4)(x - 1)$

14. $x^2 - 7x + 10 = (x - 5)(x - 2)$

15. $x^2 + 2x - 15 = (x + 5)(x - 3)$

16. $x^2 + 11x + 30 = (x + 6)(x + 5)$

17. $x^2-x-30=(x-6)(x+5)$

18. $x^2+11x+24=(x+8)(x+3)$

19. $2x^2-98=2(x^2-49)$
$=2(x^2-7^2)$
$=2(x+7)(x-7)$

20. $3x^2-75=3(x^2-25)$
$=3(x^2-5^2)$
$=3(x+5)(x-5)$

21. $x^2+3x+xy+3y=x(x+3)+y(x+3)$
$=(x+3)(x+y)$

22. $3y-21+xy-7x=3(y-7)+x(y-7)$
$=(y-7)(3+x)$

23. $x^2+6x-16=(x+8)(x-2)$

24. $x^2-3x-28=(x-7)(x+4)$

25. $4x^3+20x^2-56x=4x(x^2+5x-14)$
$=4x(x+7)(x-2)$

26. $6x^3-6x^2-120x=6x(x^2-x-20)$
$=6x(x-5)(x+4)$

27. $12x^2+34x+24=2(6x^2+17x+12)$
$=2(6x^2+9x+8x+12)$
$=2[3x(2x+3)+4(2x+3)]$
$=2(2x+3)(3x+4)$

28. $8a^2+6ab-5b^2=8a^2+10ab-4ab-5b^2$
$=2a(4a+5b)-b(4a+5b)$
$=(4a+5b)(2a-b)$

29. $4a^2-b^2=(2a)^2-b^2=(2a+b)(2a-b)$

30. $28-13x-6x^2=28-21x+8x-6x^2$
$=7(4-3x)+2x(4-3x)$
$=(4-3x)(7+2x)$

31. $20-3x-2x^2=20-8x+5x-2x^2$
$=4(5-2x)+x(5-2x)$
$=(5-2x)(4+x)$

32. x^2-2x+4 is a prime polynomial.

33. a^2+a-3 is a prime polynomial.

34. $6y^2+y-15=6y^2+10y-9y-15$
$=2y(3y+5)-3(3y+5)$
$=(3y+5)(2y-3)$

35. $4x^2-x-5=4x^2-5x+4x-5$
$=x(4x-5)+1(4x-5)$
$=(4x-5)(x+1)$

36. $x^2y-y^3=y(x^2-y^2)=y(x-y)(x+y)$

37. $4t^2+36=4(t^2+9)$

38. $x^2+x+xy+y=x(x+1)+y(x+1)$
$=(x+1)(x+y)$

39. $ax+2x+a+2=x(a+2)+1(a+2)$
$=(a+2)(x+1)$

40. $18x^3-63x^2+9x=9x(2x^2-7x+1)$

41. $12a^3-24a^2+4a=4a(3a^2-6a+1)$

42. $x^2+14x-32=(x+16)(x-2)$

43. $x^2-14x-48$ is prime.

44. $16a^2-56ab+49b^2=(4a)^2-2(4a)(7b)+(7b)^2$
$=(4a-7b)^2$

45. $25p^2-70pq+49q^2=(5p)^2-2(5p)(7q)+(7q)^2$
$=(5p-7q)^2$

46. $7x^2+24xy+9y^2=7x^2+3xy+21xy+9y^2$
$=x(7x+3y)+3y(7x+3y)$
$=(7x+3y)(x+3y)$

47. $125-8y^3=5^3-(2y)^3$
$=(5-2y)[5^2+5\cdot 2y+(2y)^2]$
$=(5-2y)(25+10y+4y^2)$

48. $64x^3+27=(4x)^3+3^3$
$=(4x+3)[(4x)^2-4x\cdot 3+3^2]$
$=(4x+3)(16x^2-12x+9)$

49. $-x^2-x+30=-1(x^2+x-30)=-(x+6)(x-5)$

50. $-x^2+6x-8=-1(x^2-6x+8)=-(x-2)(x-4)$

51. $14+5x-x^2=(7-x)(2+x)$

52. $3-2x-x^2=(3+x)(1-x)$

53. $3x^4y+6x^3y-72x^2y=3x^2y(x^2+2x-24)$
$=3x^2y(x+6)(x-4)$

54. $2x^3y+8x^2y^2-10xy^3=2xy(x^2+4xy-5y^2)$
$=2xy(x+5y)(x-y)$

55. $5x^3y^2-40x^2y^3+35xy^4=5xy^2-8xy+7y^2)$
$=5xy^2(x-7y)(x-y)$

56. $4x^4y-8x^3y-60x^2y=4x^2y(x^2-2x-15)$
$=4x^2y(x-5)(x+3)$

57. $12x^3y+243xy=3xy(4x^2+81)$

58. $6x^3y^2+8xy^2=2xy^2(3x^2+4)$

59. $4-x^2=2^2-x^2=(2+x)(2-x)$

60. $9-y^2=3^2-y^2=(3+y)(3-y)$

61. $3rs-s+12r-4=s(3r-1)+4(3r-1)$
$=(3r-1)(s+4)$

62. $x^3-2x^2+3x-6=x^2(x-2)+3(x-2)$
$=(x-2)(x^2+3)$

63. $4x^2-8xy-3x+6y=4x(x-2y)-3(x-2y)$
$=(x-2y)(4x-3)$

64. $4x^2-2xy-7yz+14xz$
$=2x(2x-y)+7z(-y+2x)$
$=(2x-y)(2x+7z)$

65. $6x^2+18xy+12y^2=6(x^2+3xy+2y^2)$
$=6(x+2)(x+y)$

66. $12x^2+46xy-8y^2=2(6x^2+23xy-4y^2)$
$=2(6x^2+24xy-xy-4y^2)$
$=2[6x(x+4y)-y(x+4y)]$
$=2(x+4y)(6x-y)$

67. $xy^2-4x+3y^2-12=x(y^2-4)+3(y^2-4)$
$=(y^2-4)(x+3)$
$=(y^2-2^2)(x+3)$
$=(y+2)(y-2)(x+3)$

68. $x^2y^2-9x^2+3y^2-27=x^2(y^2-9)+3(y^2-9)$
$=(y^2-9)(x^2+3)$
$=(y^2-3^2)(x^2+3)$
$=(y-3)(y+3)(x^2+3)$

69. $5(x+y)+x(x+y)=(x+y)(5+x)$

70. $7(x-y)+y(x-y)=(x-y)(7+y)$

71. $14t^2-9t+1=14t^2-7t-2t+1$
$=7t(2t-1)-1(2t-1)$
$=(2t-1)(7t-1)$

72. $3t^2-5t+1$ is a prime polynomial.

73. $3x^2+2x-5=3x^2+5x-3x-5$
$=x(3x+5)-1(3x+5)$
$=(3x+5)(x-1)$

74. $7x^2+19x-6=7x^2+21x-2x-6$
$=7x(x+3)-2(x+3)$
$=(x+3)(7x-2)$

75. $x^2+9xy-36y^2=(x+12y)(x-3y)$

76. $3x^2+10xy-8y^2=3x^2-2xy+12xy-8y^2$
$=x(3x-2y)+4y(3x-2y)$
$=(3x-2y)(x+4y)$

77. $1-8ab-20a^2b^2=1-10ab+2ab-20a^2b^2$
$=1(1-10ab)+2ab(1-10ab)$
$=(1-10ab)(1+2ab)$

78. $1-7ab-60a^2b^2 = 1-12ab+5ab-60a^2b^2$
$= 1(1-12ab)+5ab(1-12ab)$
$= (1-12ab)(1+5ab)$

79. $9-10x^2+x^4 = (9-x^2)(1-x^2)$
$= (3^2-x^2)(1^2-x^2)$
$= (3+x)(3-x)(1+x)(1-x)$

80. $36-13x^2+x^4 = (9-x^2)(4-x^2)$
$= (3^2-x^2)(2^2-x^2)$
$= (3+x)(3-x)(2+x)(2-x)$

81. $x^4-14x^2-32 = (x^2+2)(x^2-16)$
$= (x^2+2)(x^2-4^2)$
$= (x^2+2)(x+4)(x-4)$

82. $x^4-22x^2-75 = (x^2+3)(x^2-25)$
$= (x^2+3)(x^2-5^2)$
$= (x^2+3)(x+5)(x-5)$

83. $x^2-23x+120 = (x-15)(x-8)$

84. $y^2+22y+96 = (y+16)(y+6)$

85. $6x^3-28x^2+16x = 2x(3x^2-14x+8)$
$= 2x(3x-2)(x-4)$

86. $6y^3-8y^2-30y = 2y(3y^2-4y-15)$
$= 2y(3y+5)(y-3)$

87. $27x^3-125y^3 = (3x)^3-(5y)^3$
$= (3x-5y)[(3x)^2+3x\cdot 5y+(5y)^2]$
$= (3x-5y)(9x^2+15xy+25y^2)$

88. $216y^3-z^3 = (6y)^3-z^3$
$= (6y-z)[(6y)^2+6y\cdot z+z^2]$
$= (6y-z)(36y^2+6yz+z^2)$

89. $x^3y^3+8z^3 = (xy)^3+(2z)^3$
$= (xy+2z)[(xy)^2-xy\cdot 2z+(2z)^2]$
$= (xy+2z)(x^2y^2-2xyz+4z^2)$

90. $27a^3b^3+8 = (3ab)^3+2^3$
$= (3ab+2)[(3ab)^2-3ab\cdot 2+2^2]$
$= (3ab+2)(9a^2b^2-6ab+4)$

91. $2xy-72x^3y = 2xy(1-36x^2)$
$= 2xy[1^2-(6x)^2]$
$= 2xy(1+6x)(1-6x)$

92. $2x^3-18x = 2x(x^2-9)$
$= 2x(x^2-3^2)$
$= 2x(x+3)(x-3)$

93. $x^3+6x^2-4x-24 = x^2(x+6)-4(x+6)$
$= (x+6)(x^2-4)$
$= (x+6)(x^2-2^2)$
$= (x+6)(x+2)(x-2)$

94. $x^3-2x^2-36x+72 = x^2(x-2)-36(x-2)$
$= (x-2)(x^2-36)$
$= (x-2)(x^2-6^2)$
$= (x-2)(x+6)(x-6)$

95. $6a^3+10a^2 = 2a^2(3a+5)$

96. $4n^2-6n = 2n(2n-3)$

97. $a^2(a+2)+2(a+2) = (a+2)(a^2+2)$

98. $a-b+x(a-b) = (a-b)(1+x)$

99. $x^3-28+7x^2-4x = x^3+7x^2-28-4x$
$= x^2(x+7)-4(7+x)$
$= (x+7)(x^2-4)$
$= (x+7)(x^2-2^2)$
$= (x+7)(x+2)(x-2)$

100. $a^3-45-9a+5a^2 = a^3+5a^2-9a-45$
$= a^2(a+5)-9(a+5)$
$= (a+5)(a^2-9)$
$= (a+5)(a^2-3^2)$
$= (a+5)(a+3)(a-3)$

101. $(x-y)^2-z^2 = (x-y+z)(x-y-z)$

102. $(x+2y)^2-9=(x+2y)^2-3^2$
$=(x+2y+3)(x+2y-3)$

103. $81-(5x+1)^2=9^2-(5x+1)^2$
$=[9+(5x+1)][9-(5x+1)]$
$=(9+5x+1)(9-5x-1)$

104. $b^2-(4a+c)^2$
$=[b+(4a+c)][b-(4a+c)]$
$=(b+4a+c)(b-4a-c)$

105. Answers may vary

106. Yes; $9x^2+81y^2=9(x^2+9y^2)$

107. a, c

Section 6.6

Practice Exercises

1. $(x+4)(x-5)=0$
$x+4=0$ or $x-5=0$
$x=-4$ $\quad x=5$
Check:
Let $x=-4$.
$(x+4)(x-5)=0$
$(-4+4)(-4-5)\stackrel{?}{=}0$
$0(-9)=0$ True
Let $x=5$.
$(x+4)(x-5)=0$
$(5+4)(5-5)\stackrel{?}{=}0$
$9(0)=0$ True
The solutions are −4 and 5.

2. $x(7x-6)=0$
$x=0$ or $7x-6=0$
$7x=6$
$x=\frac{6}{7}$
Check:
Let $x=0$.
$x(7x-6)=0$
$0(7\cdot 0-6)\stackrel{?}{=}0$
$0(-6)=0$ True
Let $x=\frac{6}{7}$.
$x(7x-6)=0$
$\frac{6}{7}\left(7\cdot\frac{6}{7}-6\right)\stackrel{?}{=}0$
$\frac{6}{7}(6-6)\stackrel{?}{=}0$
$\frac{6}{7}(0)=0$ True
The solutions are 0 and $\frac{6}{7}$.

3. $x^2-8x-48=0$
$(x+4)(x-12)=0$
$x+4=0$ or $x-12=0$
$x=-4$ $\quad x=12$
Check:
Let $x=-4$.
$x^2-8x-48=0$
$(-4)^2-8(-4)-48\stackrel{?}{=}0$
$16+32-48\stackrel{?}{=}0$
$48-48\stackrel{?}{=}0$
$0=0$ True
Let $x=12$.
$x^2-8x-48=0$
$12^2-8\cdot 12-48\stackrel{?}{=}0$
$144-96-48\stackrel{?}{=}0$
$48-48\stackrel{?}{=}0$
$0=0$ True
The solutions are −4 and 12.

4. $9x^2-24x=-16$
$9x^2-24x+16=0$
$(3x-4)(3x-4)=0$
$3x-4=0$
$3x=4$
$x=\frac{4}{3}$
The solution is $\frac{4}{3}$.

5. $x(3x+7)=6$
$3x^2+7x=6$
$3x^2+7x-6=0$
$(3x-2)(x+3)=0$
$3x-2=0$ or $x+3=0$
$3x=2$ $\quad x=-3$
$x=\frac{2}{3}$

The solutions are $\frac{2}{3}$ and -3.

6. $-3x^2-6x+72=0$
$-3(x^2+2x-24)=0$
$-3(x+6)(x-4)=0$
$x+6=0$ or $x-4=0$
$x=-6$ $\quad x=4$
The solutions are -6 and 4.

7. $7x^3-63x=0$
$7x(x^2-9)=0$
$7x(x+3)(x-3)=0$
$7x=0$ or $x+3=0$ or $x-3=0$
$x=0$ $\quad x=-3$ $\quad x=3$
The solutions are 0, -3, and 3.

8. $(3x-2)(2x^2-13x+15)=0$
$(3x-2)(2x-3)(x-5)=0$
$3x-2=0$ or $2x-3=0$ or $x-5=0$
$3x=2$ $\quad 2x=3$ $\quad x=5$
$x=\frac{2}{3}$ $\quad x=\frac{3}{2}$

The solutions are $\frac{2}{3}, \frac{3}{2}$, and 5.

9. $5x^3+5x^2-30x=0$
$5x(x^2+x-6)=0$
$5x(x+3)(x-2)=0$
$5x=0$ or $x+3=0$ or $x-2=0$
$x=0$ $\quad x=-3$ $\quad x=2$
The solutions are 0, -3, and 2.

10. $y=x^2-6x+8$
$0=x^2-6x+8$
$0=(x-4)(x-2)$
$x-4=0$ or $x-2=0$
$x=4$ $\quad x=2$

The x-intercepts of the graph of $y=x^2-6x+8$ are (2, 0) and (4, 0).

Calculator Explorations

1. -0.9, 2.2

2. -2.5, 3.5

3. no real solution

4. no real solution

5. -1.8, 2.8

6. -0.9, 0.3

Vocabulary and Readiness Check

1. An equation that can be written in the form $ax^2+bx+c=0$, (with $a \neq 0$), is called a quadratic equation.

2. If the product of two numbers is 0, then at least one of the numbers must be 0.

3. The solutions to $(x-3)(x+5)=0$ are 3, -5.

4. If $a \cdot b = 0$, then $a=0$ or $b=0$.

5. 3, 7

6. 5, 2

7. -8, -6

8. -2, -3

9. -1, 3

10. 1, -2

Exercise Set 6.6

1. $(x-2)(x+1)=0$
$x-2=0$ or $x+1=0$
$x=2$ $\quad x=-1$
The solutions are 2 and -1.

3. $(x+9)(x+17)=0$
$x+9=0$ or $x+17=0$
$x=-9$ $\quad x=-17$
The solutions are -9 and -17.

5. $x(x+6)=0$
$x=0$ or $x+6=0$
$x=-6$
The solutions are 0 and -6.

7. $3x(x-8)=0$
$3x=0$ or $x-8=0$
$x=0$ $\quad x=8$
The solutions are 0 and 8.

9. $(2x+3)(4x-5)=0$
$2x+3=0$ or $4x-5=0$
$2x=-3$ $\quad 4x=5$
$x=-\frac{3}{2}$ $\quad x=\frac{5}{4}$
The solutions are $-\frac{3}{2}$ and $\frac{5}{4}$.

11. $(2x-7)(7x+2)=0$
$2x-7=0$ or $7x+2=0$
$2x=7$ $\quad 7x=-2$
$x=\frac{7}{2}$ $\quad x=-\frac{2}{7}$
The solutions are $\frac{7}{2}$ and $-\frac{2}{7}$.

13. $\left(x-\frac{1}{2}\right)\left(x+\frac{1}{3}\right)=0$
$x-\frac{1}{2}=0$ or $x+\frac{1}{3}=0$
$x=\frac{1}{2}$ $\quad x=-\frac{1}{3}$
The solutions are $\frac{1}{2}$ and $-\frac{1}{3}$.

15. $(x+0.2)(x+1.5)=0$
$x+0.2=0$ or $x+1.5=0$
$x=-0.2$ $\quad x=-1.5$
The solutions are -0.2 and -1.5

17. Answers may vary. Possible answer:
If $x=6$ and $x=-1$ are the solutions, then
$x=6$ or $x=-1$
$x-6=0$ $\quad x+1=0$
$(x-6)(x+1)=0$

19. $x^2-13x+36=0$
$(x-9)(x-4)=0$
$x-9=0$ or $x-4=0$
$x=9$ $\quad x=4$
The solutions are 9 and 4.

21. $x^2+2x-8=0$
$(x+4)(x-2)=0$
$x+4=0$ or $x-2=0$
$x=-4$ $\quad x=2$
The solutions are -4 and 2.

23. $x^2-7x=0$
$x(x-7)=0$
$x=0$ or $x-7=0$
$x=7$
The solutions are 0 and 7.

25. $x^2-4x=32$
$x^2-4x-32=0$
$(x-8)(x+4)=0$
$x-8=0$ or $x+4=0$
$x=8$ $\quad x=-4$
The solutions are 8 and -4.

27. $x^2=16$
$x^2-16=0$
$(x+4)(x-4)=0$
$x+4=0$ or $x-4=0$
$x=-4$ $\quad x=4$
The solutions are -4 and 4.

29. $(x+4)(x-9)=4x$
$x^2-5x-36=4x$
$x^2-9x-36=0$
$(x-12)(x+3)=0$
$x-12=0$ or $x+3=0$
$x=12$ $\quad x=-3$
The solutions are 12 and -3.

31. $x(3x-1)=14$
$3x^2-x=14$
$3x^2-x-14=0$
$(3x-7)(x+2)=0$
$3x-7=0$ or $x+2=0$
$3x=7$ $\quad x=-2$
$x=\frac{7}{3}$
The solutions are $\frac{7}{3}$ and -2.

33.
$$\begin{aligned} -3x^2+75&=0\\ -3(x^2-25)&=0\\ -3(x+5)(x-5)&=0 \end{aligned}$$
$x+5=0$ or $x-5=0$
$x=-5$ $\quad x=5$
The solutions are −5 and 5.

35.
$$\begin{aligned} 24x^2+44x&=8\\ 24x^2+44x-8&=0\\ 4(6x^2+11x-2)&=0\\ 4(6x-1)(x+2)&=0 \end{aligned}$$
$6x-1=0$ or $x+2=0$
$6x=1$ $\quad x=-2$
$x=\frac{1}{6}$
The solutions are $\frac{1}{6}$ and −2.

37. $x^3-12x^2+32x=0$
$$\begin{aligned} x(x^2-12x+32)&=0\\ x(x-8)(x-4)&=0 \end{aligned}$$
$x=0$ or $x-8=0$ or $x-4=0$
$x=8$ $\quad x=4$
The solutions are 0, 8, and 4.

39. $(4x-3)(16x^2-24x+9)=0$
$$\begin{aligned} (4x-3)(4x-3)^2&=0\\ (4x-3)^3&=0\\ 4x-3&=0\\ 4x&=3\\ x&=\frac{3}{4} \end{aligned}$$
The solution is $\frac{3}{4}$.

41.
$$\begin{aligned} 4x^3-x&=0\\ x(4x^2-1)&=0\\ x(2x+1)(2x-1)&=0 \end{aligned}$$
$x=0$ or $2x+1=0$ or $2x-1=0$
$2x=-1$ $\quad 2x=1$
$x=-\frac{1}{2}$ $\quad x=\frac{1}{2}$
The solutions are 0, $-\frac{1}{2}$, and $\frac{1}{2}$.

43. $32x^3-4x^2-6x=0$
$$\begin{aligned} 2x(16x^2-2x-3)&=0\\ 2x(2x-1)(8x+3)&=0 \end{aligned}$$
$2x=0$ or $2x-1=0$ or $8x+3=0$
$x=0$ $\quad 2x=1$ $\quad 8x=-3$
$x=\frac{1}{2}$ $\quad x=-\frac{3}{8}$
The solutions are 0, $\frac{1}{2}$, and $-\frac{3}{8}$.

45. $(x+3)(x-2)=0$
$x+3=0$ or $x-2=0$
$x=-3$ $\quad x=2$
The solutions are −3 and 2.

47. $x^2+20x=0$
$x(x+20)=0$
$x=0$ or $x+20=0$
$x=-20$
The solutions are 0 and −20.

49. $4(x-7)=6$
$$\begin{aligned} 4x-28&=6\\ 4x&=34\\ x&=\frac{34}{4}\\ x&=\frac{17}{2} \end{aligned}$$
The solution is $\frac{17}{2}$.

51.
$$\begin{aligned} 4y^2-1&=0\\ (2y+1)(2y-1)&=0 \end{aligned}$$
$2y+1=0$ or $2y-1=0$
$2y=-1$ $\quad 2y=1$
$y=-\frac{1}{2}$ $\quad y=\frac{1}{2}$
The solutions are $-\frac{1}{2}$ and $\frac{1}{2}$.

53. $(2x+3)(2x^2-5x-3)=0$
$(2x+3)(2x+1)(x-3)=0$
$2x+3=0$ or $2x+1=0$ or $x-3=0$
$2x=-3$ $\quad 2x=-1$ $\quad x=3$
$x=-\frac{3}{2}$ $\quad x=-\frac{1}{2}$
The solutions are $-\frac{3}{2}$, $-\frac{1}{2}$, and 3.

55.
$$x^2 - 15 = -2x$$
$$x^2 + 2x - 15 = 0$$
$$(x+5)(x-3) = 0$$
$$x+5=0 \quad \text{or} \quad x-3=0$$
$$x=-5 \qquad\qquad x=3$$
The solutions are –5 and 3.

57.
$$30x^2 - 11x - 30 = 0$$
$$(6x+5)(5x-6) = 0$$
$$6x+5=0 \quad \text{or} \quad 5x-6=0$$
$$6x=-5 \qquad\qquad 5x=6$$
$$x=-\frac{5}{6} \qquad\qquad x=\frac{6}{5}$$
The solutions are $-\frac{5}{6}$ and $\frac{6}{5}$.

59.
$$5x^2 - 6x - 8 = 0$$
$$(5x+4)(x-2) = 0$$
$$5x+4=0 \quad \text{or} \quad x-2=0$$
$$5x=-4 \qquad\qquad x=2$$
$$x=-\frac{4}{5}$$
The solutions are $-\frac{4}{5}$ and 2.

61.
$$6y^2 - 22y - 40 = 0$$
$$2(3y^2 - 11y - 20) = 0$$
$$2(3y+4)(y-5) = 0$$
$$3y+4=0 \quad \text{or} \quad y-5=0$$
$$3y=-4 \qquad\qquad y=5$$
$$y=-\frac{4}{3}$$
The solutions are $-\frac{4}{3}$ and 5.

63.
$$(y-2)(y+3) = 6$$
$$y^2 + y - 6 = 6$$
$$y^2 + y - 12 = 0$$
$$(y+4)(y-3) = 0$$
$$y+4=0 \quad \text{or} \quad y-3=0$$
$$y=-4 \qquad\qquad y=3$$
The solutions are –4 and 3.

65.
$$3x^3 + 19x^2 - 72x = 0$$
$$x(3x^2 + 19x - 72) = 0$$
$$x(3x-8)(x+9) = 0$$
$$x=0 \quad \text{or} \quad 3x-8=0 \quad \text{or} \quad x+9=0$$
$$3x=8 \qquad\qquad x=-9$$
$$x=\frac{8}{3}$$
The solutions are 0, $\frac{8}{3}$, and –9.

67.
$$x^2 + 14x + 49 = 0$$
$$(x+7)^2 = 0$$
$$x+7=0$$
$$x=-7$$
The solution is –7.

69.
$$12y = 8y^2$$
$$0 = 8y^2 - 12y$$
$$0 = 4y(2y-3)$$
$$4y=0 \quad \text{or} \quad 2y-3=0$$
$$y=0 \qquad\qquad 2y=3$$
$$y=\frac{3}{2}$$
The solutions are 0 and $\frac{3}{2}$.

71.
$$7x^3 - 7x = 0$$
$$7x(x^2 - 1) = 0$$
$$7x(x+1)(x-1) = 0$$
$$7x=0 \quad \text{or} \quad x+1=0 \quad \text{or} \quad x-1=0$$
$$x=0 \qquad\qquad x=-1 \qquad\qquad x=1$$
The solutions are 0, –1, and 1.

73.
$$3x^2 + 8x - 11 = 13 - 6x$$
$$3x^2 + 14x - 24 = 0$$
$$(3x-4)(x+6) = 0$$
$$3x-4=0 \quad \text{or} \quad x+6=0$$
$$3x=4 \qquad\qquad x=-6$$
$$x=\frac{4}{3}$$
The solutions are $\frac{4}{3}$ and –6.

75. $3x^2 - 20x = -4x^2 - 7x - 6$
$7x^2 - 13x + 6 = 0$
$(7x - 6)(x - 1) = 0$
$7x - 6 = 0$ or $x - 1 = 0$
$7x = 6$ $\quad x = 1$
$x = \frac{6}{7}$

The solutions are $\frac{6}{7}$ and 1.

77. Let $y = 0$ and solve for x.
$y = (3x + 4)(x - 1)$
$0 = (3x + 4)(x - 1)$
$3x + 4 = 0$ or $x - 1 = 0$
$3x = -4$ $\quad x = 1$
$x = -\frac{4}{3}$

The intercepts are $\left(-\frac{4}{3}, 0\right)$ and (1, 0).

79. Let $y = 0$ and solve for x.
$y = x^2 - 3x - 10$
$0 = x^2 - 3x - 10$
$0 = (x - 5)(x + 2)$
$x - 5 = 0$ or $x + 2 = 0$
$x = 5$ $\quad x = -2$
The x-intercepts are (5, 0) and (−2, 0).

81. Let $y = 0$ and solve for x.
$y = 2x^2 + 11x - 6$
$0 = 2x^2 + 11x - 6$
$0 = (2x - 1)(x + 6)$
$2x - 1 = 0$ or $x + 6 = 0$
$2x = 1$ $\quad x = -6$
$x = \frac{1}{2}$

The x-intercepts are $\left(\frac{1}{2}, 0\right)$ and (−6, 0).

83. e; x-intercepts are (−2, 0), (1, 0)

85. b; x-intercepts are (0, 0), (−3, 0)

87. c; $y = 2x^2 - 8 = 2(x - 2)(x + 2)$
x-intercepts are (2, 0), (−2, 0).

89.
$$\frac{3}{5} + \frac{4}{9} = \frac{3 \cdot 9}{5 \cdot 9} + \frac{4 \cdot 5}{9 \cdot 5} = \frac{27}{45} + \frac{20}{45} = \frac{27 + 20}{45} = \frac{47}{45}$$

91.
$$\frac{7}{10} - \frac{5}{12} = \frac{7 \cdot 6}{10 \cdot 6} - \frac{5 \cdot 5}{12 \cdot 5} = \frac{42}{60} - \frac{25}{60} = \frac{42 - 25}{60} = \frac{17}{60}$$

93. $\frac{7}{8} \div \frac{7}{15} = \frac{7}{8} \cdot \frac{15}{7} = \frac{15}{8}$

95. $\frac{4}{5} \cdot \frac{7}{8} = \frac{4 \cdot 7}{5 \cdot 8} = \frac{4 \cdot 7}{5 \cdot 2 \cdot 4} = \frac{7}{10}$

97. Didn't write the equation in standard form; standard form should be:
$x(x - 2) = 8$
$x^2 - 2x = 8$
$x^2 - 2x - 8 = 0$
$(x - 4)(x + 2) = 0$
$x - 4 = 0$ or $x + 2 = 0$
$x = 4$ $\quad x = -2$

99. Answers may vary. Possible answer: If the solutions are $x = 5$ and $x = 7$, then, by the zero factor property,
$x = 5$ or $x = 7$
$x - 5 = 0$ $\quad x - 7 = 0$
$(x - 5)(x - 7) = 0$
$x^2 - 7x - 5x + 35 = 0$
$x^2 - 12x + 35 = 0$

101. $y = -16x^2 + 20x + 300$

a.

time x	0	1	2	3	4	5	6
height y	300	304	276	216	124	0	−156

b. The compass strikes the ground after 5 seconds, when the height, y, is zero feet.

c. The maximum height was approximately 304 feet.

d.

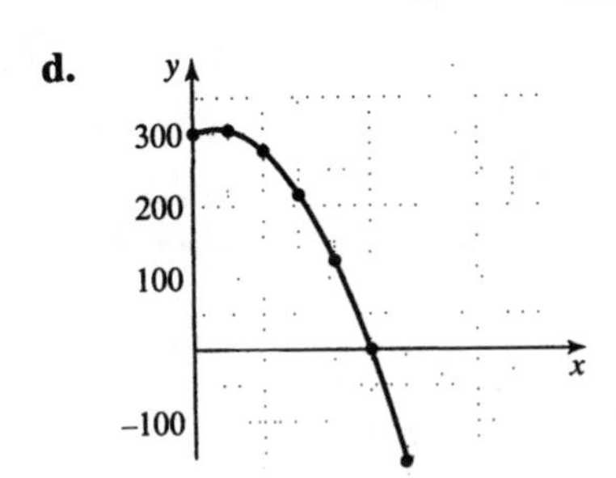

103. $(x-3)(3x+4) = (x+2)(x-6)$

$3x^2 - 5x - 12 = x^2 - 4x - 12$

$2x^2 - x = 0$

$x(2x-1) = 0$

$2x - 1 = 0$ or $x = 0$

$x = \frac{1}{2}$

The solutions are $\frac{1}{2}$ and 0.

105. $(2x-3)(x+8) = (x-6)(x+4)$

$2x^2 + 13x - 24 = x^2 - 2x - 24$

$x^2 + 15x = 0$

$x(x+15) = 0$

$x + 15 = 0$ or $x = 0$

$x = -15$

The solutions are −15 and 0.

The Bigger Picture

1. $-7 + (-27) = -34$

2. $\frac{(x^3)^4}{(x^{-2})^5} = \frac{x^{12}}{x^{-10}} = x^{12-(-10)} = x^{22}$

3. $(x^3-6x^2+2)-(5x^3-6)$
$=x^3-6x^2+2-5x^3+6$
$=x^3-5x^3-6x^2+2+6$
$=-4x^3-6x^2+8$

4. $\dfrac{3y^3-3y^2+9}{3y^2}=\dfrac{3y^3}{3y^2}-\dfrac{3y^2}{3y^2}+\dfrac{9}{3y^2}=y-1+\dfrac{3}{y^2}$

5. $10x^3-250x=10x(x^2-25)$
$=10x(x^2-5^2)$
$=10x(x+5)(x-5)$

6. $x^2-36x+35=(x-1)(x-35)$

7. $6xy+15x-6y-15=3(2xy+5x-2y-5)$
$=3[x(2y+5)-1(2y+5)]$
$=3(2y+5)(x-1)$

8. $5xy^2-2xy-7x=x(5y^2-2y-7)$
$=x(5y-7)(y+1)$

9. $(x-5)(2x+1)=0$
$x-5=0$ or $2x+1=0$
$x=5$ $\quad x=-\dfrac{1}{2}$
The solutions are 5 and $-\dfrac{1}{2}$.

10. $5x-5=0$
$5x=5$
$x=1$
The solution is 1.

11. $x(x-12)=28$
$x^2-12x=28$
$x^2-12x-28=0$
$(x+2)(x-14)=0$
$x+2=0$ or $x-14=0$
$x=-2$ $\quad x=14$
The solutions are –2 and 14.

12. $7(x-3)+2(5x+1)=14$
$7x-21+10x+2=14$
$17x-19=14$
$17x=33$
$x=\dfrac{33}{17}$
The solution is $\dfrac{33}{17}$.

Section 6.7

Practice Exercises

1. Find t when $h=0$.
$h=-16t^2+64$
$0=-16t^2+64$
$0=-16(t^2-4)$
$0=-16(t-2)(t+2)$
$t-2=0$ or $t+2=0$
$t=2$ $\quad t=-2$
Since time cannot be negative, the diver will reach the pool in 2 seconds.

2. Let x = the number.
$x^2-8x=48$
$x^2-8x-48=0$
$(x-12)(x+4)=0$
$x-12=0$ or $x+4=0$
$x=12$ $\quad x=-4$
There are two numbers. They are –4 and 12.

3. Let x = height, then $3x-1$ = base.
$A=\dfrac{1}{2}bh$
$210=\dfrac{1}{2}(3x-1)(x)$
$420=(3x-1)(x)$
$420=3x^2-x$
$0=3x^2-x-420$
$0=(3x+35)(x-12)$
$3x+35=0$ or $x-12=0$
$x=-\dfrac{35}{3}$ $\quad x=12$
Since height cannot be negative, the height is 12 feet and the base is $3(12)-1=35$ feet.

4. Let x = first integer, then
$x + 1$ = next consecutive integer.

$$x(x+1) = x + (x+1) + 41$$
$$x^2 + x = 2x + 42$$
$$x^2 - x - 42 = 0$$
$$(x-7)(x+6) = 0$$
$$x - 7 = 0 \quad \text{or} \quad x + 6 = 0$$
$$x = 7 \qquad\qquad x = -6$$

The numbers are 7 and 8 or −6 and −5.

5. Let x = first leg, then $2x - 1$ = second leg, and $2x + 1$ = hypotenuse.

$$x^2 + (2x-1)^2 = (2x+1)^2$$
$$x^2 + 4x^2 - 4x + 1 = 4x^2 + 4x + 1$$
$$x^2 - 8x = 0$$
$$x(x-8) = 0$$
$$x = 0 \quad \text{or} \quad x - 8 = 0$$
$$x = 8$$

Since the length cannot be 0, the legs have lengths 8 units and $2(8) - 1 = 15$ units and the hypotenuse has length $2(8) + 1 = 17$ units.

Exercise Set 6.7

1. Let x = the width, then $x + 4$ = the length.

3. Let x = the first odd integer, then $x + 2$ = the next consecutive odd integer.

5. Let x = the base, then $4x + 1$ = the height.

7. Let x = the length of one side.

$$A = x^2$$
$$121 = x^2$$
$$0 = x^2 - 121$$
$$0 = x^2 - 11^2$$
$$0 = (x+11)(x-11)$$
$$x + 11 = 0 \quad \text{or} \quad x - 11 = 0$$
$$x = -11 \qquad\qquad x = 11$$

Since the length cannot be negative, the sides are 11 units long.

9. The perimeter is the sum of the lengths of the sides.

$$120 = (x+5) + (x^2 - 3x) + (3x - 8) + (x + 3)$$
$$120 = x + 5 + x^2 - 3x + 3x - 8 + x + 3$$
$$120 = x^2 + 2x$$
$$0 = x^2 + 2x - 120$$
$$x^2 + 2x - 120 = 0$$
$$(x+12)(x-10) = 0$$
$$x + 12 = 0 \quad \text{or} \quad x - 10 = 0$$
$$x = -12 \qquad\qquad x = 10$$

Since the dimensions cannot be negative, the lengths of the sides are:
$10 + 5 = 15$ cm, $10^2 - 3(10) = 70$ cm, $3(10) - 8 = 22$ cm, and $10 + 3 = 13$ cm.

11. $x + 5$ = the base and $x - 5$ = the height.

$$A = bh$$
$$96 = (x+5)(x-5)$$
$$96 = x^2 - 25$$
$$0 = x^2 - 121$$
$$x^2 - 121 = 0$$
$$(x+11)(x-11) = 0$$
$$x + 11 = 0 \quad \text{or} \quad x - 11 = 0$$
$$x = -11 \qquad\qquad x = 11$$

Since the dimensions cannot be negative, $x = 11$. The base is $11 + 5 = 16$ miles, and the height is $11 - 5 = 6$ miles.

13. Find t when $h = 0$.

$$h = -16t^2 + 64t + 80$$
$$0 = -16t^2 + 64t + 80$$
$$0 = -16(t^2 - 4t - 5)$$
$$0 = -16(t-5)(t+1)$$
$$t - 5 = 0 \quad \text{or} \quad t + 1 = 0$$
$$t = 5 \qquad\qquad t = -1$$

Since the time t cannot be negative, the object hits the ground after 5 seconds.

15. Let x = the width then $2x - 7$ = the length.

$$A = lw$$
$$30 = (2x-7)(x)$$
$$30 = 2x^2 - 7x$$
$$0 = 2x^2 - 7x - 30$$
$$0 = (2x+5)(x-6)$$

$2x+5=0$ or $x-6=0$

$x=-\frac{5}{2}$ $x=6$

Since the dimensions cannot be negative, the width is 6 cm and the length is 2(6) – 7 = 5 cm.

17. Let $n = 12$.

$$D=\frac{1}{2}n(n-3)$$

$$D=\frac{1}{2}\cdot 12(12-3)=6(9)=54$$

A polygon with 12 sides has 54 diagonals.

19. Let $D = 35$ and solve for n.

$$D=\frac{1}{2}n(n-3)$$

$$35=\frac{1}{2}n(n-3)$$

$$70=n^2-3n$$

$$0=n^2-3n-70$$

$$0=(n-10)(n+7)$$

$n-10=0$ or $n+7=0$

$n=10$ $n=-7$

The polygon has 10 sides.

21. Let x = the unknown number.

$$x+x^2=132$$

$$x^2+x-132=0$$

$$(x+12)(x-11)=0$$

$x+12=0$ or $x-11=0$

$x=-12$ $x=11$

The two numbers are –12 and 11.

23. Let x = the first room number, then $x + 1$ = next room number.

$$x(x+1)=210$$

$$x^2+x=210$$

$$x^2+x-210=0$$

$$(x-14)(x+15)=0$$

$x-14=0$ or $x+15=0$

$x=14$ $x=-15$

Since the room number is not negative, the room numbers are 14 and 15.

25. Let x = hypotenuse, then $x - 1$ = height.

$$a^2+b^2=c^2$$

$$5^2+(x-1)^2=x^2$$

$$25+x^2-2x+1=x^2$$

$$26-2x=0$$

$$26=2x$$

$$13=x$$

The length of the ladder is 13 feet.

27. Let x = the length of a side of the original square. Then $x + 3$ = the length of a side of the larger square.

$$64=(x+3)^2$$

$$64=x^2+6x+9$$

$$0=x^2+6x-55$$

$$0=(x+11)(x-5)$$

$x+11=0$ or $x-5=0$

$x=-11$ $x=5$

Since the length cannot be negative, the sides of the original square are 5 inches long.

29. Let x = the length of the shorter leg. Then $x + 4$ = the length of the longer leg and $x + 8$ = the length of the hypotenuse.
By the Pythagorean theorem,

$$x^2+(x+4)^2=(x+8)^2$$

$$x^2+x^2+8x+16=x^2+16x+64$$

$$x^2-8x-48=0$$

$$(x-12)(x+4)=0$$

$x-12=0$ or $x+4=0$

$x=12$ $x=-4$

Since the length cannot be negative, the sides of the triangle are 12 mm, 12 + 4 = 16 mm, and 12 + 8 = 20 mm.

31. Let x = the height of the triangle, then $2x$ = the base.

$$A=\frac{1}{2}bh$$

$$100=\frac{1}{2}(2x)(x)$$

$$100=x^2$$

$$0=x^2-100$$

$$0=(x+10)(x-10)$$

$x+10=0$ or $x-10=0$

$x=-10$ $x=10$

Since the height cannot be negative, the height of the triangle is 10 km.

33. Let x = the length of the shorter leg, then
$x + 12$ = the length of the longer leg and
$2x - 12$ = the length of the hypotenuse.
By the Pythagorean theorem,

$$x^2 + (x+12)^2 = (2x-12)^2$$
$$x^2 + x^2 + 24x + 144 = 4x^2 - 48x + 144$$
$$0 = 2x^2 - 72x$$
$$0 = 2x(x-36)$$
$$2x = 0 \quad \text{or} \quad x - 36 = 0$$
$$x = 0 \qquad\qquad x = 36$$

Since the length cannot be zero feet, the shorter leg is 36 feet long.

35. Find t when $h = 0$.

$$h = -16t^2 + 1444$$
$$0 = -16t^2 + 1444$$
$$0 = -4(4t^2 - 361)$$
$$0 = -4(2t-19)(2t+19)$$
$$2t - 19 = 0 \quad \text{or} \quad 2t + 19 = 0$$
$$t = \frac{19}{2} \qquad\qquad t = -\frac{19}{2}$$

Since time cannot be negative, the object reaches the ground in $\frac{19}{2} = 9.5$ seconds.

37. Let $P = 100$ and $A = 144$.

$$A = P(1+r)^2$$
$$144 = 100(1+r)^2$$
$$144 = 100 + 200r + 100r^2$$
$$0 = 100r^2 + 200r - 44$$
$$0 = 4(25r^2 + 50r - 11)$$
$$0 = 4(5r-1)(5r+11)$$
$$5r - 1 = 0 \quad \text{or} \quad 5r + 11 = 0$$
$$5r = 1 \qquad\qquad 5r = -11$$
$$r = \frac{1}{5} \qquad\qquad r = -\frac{11}{5}$$
$$r = 0.2 \qquad\qquad r = -2.2$$

Since the interest rate cannot be negative $r = 0.2$ and the rate is 20%.

39. Let x = the length and $x - 7$ = the width.

$$A = lw$$
$$120 = (x-7)(x)$$
$$120 = x^2 - 7x$$
$$0 = x^2 - 7x - 120$$
$$0 = (x+8)(x-15)$$
$$x + 8 = 0 \quad \text{or} \quad x - 15 = 0$$
$$x = -8 \qquad\qquad x = 15$$

Since the length cannot be negative, the length is 15 miles. The width is $15 - 7 = 8$ miles.

41. Let $C = 9500$.

$$C = x^2 - 15x + 50$$
$$9500 = x^2 - 15x + 50$$
$$0 = x^2 - 15x - 9450$$
$$0 = (x+90)(x-105)$$
$$x + 90 = 0 \quad \text{or} \quad x - 105 = 0$$
$$x = -90 \qquad\qquad x = 105$$

Since the number of units cannot be negative the solution is 105 units.

43. In 1940, the size of the average farm was about 175 acres.

45. In 1940, there were approximately 6.25 million farms.

47. The lines appear to intersect in 1966.

49. Answers may vary

51. $\frac{24}{32} = \frac{2 \cdot 2 \cdot 2 \cdot 3}{2 \cdot 2 \cdot 2 \cdot 2 \cdot 2} = \frac{3}{4}$

53. $\frac{15}{27} = \frac{3 \cdot 5}{3 \cdot 3 \cdot 3} = \frac{5}{9}$

55. $\frac{45}{50} = \frac{3 \cdot 3 \cdot 5}{2 \cdot 5 \cdot 5} = \frac{9}{10}$

57. Let x = the length of a side of the square. Then
x = the width of the rectangle and
$x + 6$ = the length of the rectangle.
The area of the square is x^2. The area of the rectangle is $x(x+6) = x^2 + 6x$.

$$176 = x^2 + (x^2 + 6x)$$
$$176 = 2x^2 + 6x$$
$$0 = 2x^2 + 6x - 176$$
$$0 = 2(x^2 + 3x - 88)$$
$$0 = 2(x+11)(x-8)$$
$$x + 11 = 0 \quad \text{or} \quad x - 8 = 0$$
$$x = -11 \qquad\qquad x = 8$$

Since the length cannot be negative, the side of the square is 8 meters.

59. Let x = the first number, then $25 - x$ = the other number.

$$x^2 + (25 - x)^2 = 325$$
$$x^2 + 625 - 50x + x^2 = 325$$
$$2x^2 - 50x + 625 = 325$$
$$2x^2 - 50x + 300 = 0$$
$$2(x^2 - 25x + 150) = 0$$
$$2(x - 15)(x - 10) = 0$$
$$x - 15 = 0 \quad \text{or} \quad x - 10 = 0$$
$$x = 15 \qquad\qquad x = 10$$

The numbers are 15 and 10.

61. Pool: width = x and length = $x + 6$
Total Area: width = $x + 8$ and length = $x + 14$

$$\text{Total area} = 576 + \text{Pool area}$$
$$(x + 14)(x + 8) = 576 + (x + 6)(x)$$
$$x^2 + 22x + 112 = 576 + x^2 + 6x$$
$$16x + 112 = 576$$
$$16x = 464$$
$$x = 29$$

$x + 6 = 29 + 6 = 35$
The pool has length 35 meters and width 29 meters.

63. Answers may vary

Chapter 6 Vocabulary Check

1. An equation that can be written in the form $ax^2 + bx + c = 0$ (with a not 0) is called a quadratic equation.

2. Factoring is the process of writing an expression as a product.

3. The greatest common factor of a list of terms is the product of all common factors.

4. A trinomial that is the square of some binomial is called a perfect square trinomial.

5. The expression $a^2 - b^2$ is called a difference of two squares.

6. The expression $a^3 - b^3$ is called a difference of two cubes.

7. The expression $a^3 + b^3$ is called a sum of two cubes.

8. By the zero factor property, if the product of two numbers is 0, then at least one of the numbers must be 0.

Chapter 6 Review

1. $6x^2 - 15x = 3x(2x - 5)$

2. $2x^3y - 6x^2y^2 - 8xy^3 = 2xy(x^2 - 3xy - 4y^2)$
$= 2xy(x - 4y)(x + y)$

3. $20x^2 + 12x = 4x(5x + 3)$

4. $6x^2y^2 - 3xy^3 = 3xy^2(2x - y)$

5. $-8x^3y + 6x^2y^2 = -2x^2y(4x - 3y)$

6. $3x(2x + 3) - 5(2x + 3) = (2x + 3)(3x - 5)$

7. $5x(x + 1) - (x + 1) = (x + 1)(5x - 1)$

8. $3x^2 - 3x + 2x - 2 = 3x(x - 1) + 2(x - 1)$
$= (x - 1)(3x + 2)$

9. $6x^2 + 10x - 3x - 5 = 2x(3x + 5) - 1(3x + 5)$
$= (3x + 5)(2x - 1)$

10. $3a^2 + 9ab + 3b^2 + ab = 3a(a + 3b) + b(3b + a)$
$= (a + 3b)(3a + b)$

11. $x^2 + 6x + 8 = (x + 4)(x + 2)$

12. $x^2 - 11x + 24 = (x - 8)(x - 3)$

13. $x^2 + x + 2$ is a prime polynomial.

14. $x^2 - 5x - 6 = (x - 6)(x + 1)$

15. $x^2 + 2x - 8 = (x + 4)(x - 2)$

16. $x^2 + 4xy - 12y^2 = (x + 6y)(x - 2y)$

17. $x^2 + 8xy + 15y^2 = (x + 5y)(x + 3y)$

18. $3x^2y + 6xy^2 + 3y^3 = 3y(x^2 + 2xy + y^2)$
$= 3y(x + y)(x + y)$
$= 3y(x + y)^2$

19. $72-18x-2x^2=2(36-9x-x^2)$
$=2(3-x)(12+x)$

20. $32+12x-4x^2=4(8+3x-x^2)$

21. $2x^2+11x-6=(2x-1)(x+6)$

22. $4x^2-7x+4$ is a prime polynomial.

23. $4x^2+4x-3=4x^2+6x-2x-3$
$=2x(2x+3)-1(2x+3)$
$=(2x+3)(2x-1)$

24. $6x^2+5xy-4y^2=6x^2+8xy-3xy-4y^2$
$=2x(3x+4y)-y(3x+4y)$
$=(3x+4y)(2x-y)$

25. $6x^2-25xy+4y^2=(6x-y)(x-4y)$

26. $18x^2-60x+50=2(9x^2-30x+25)$
$=2[(3x)^2-2\cdot 3x\cdot 5+5^2]$
$=2(3x-5)^2$

27. $2x^2-23xy-39y^2=2x^2-26xy+3xy-39y^2$
$=2x(x-13y)+3y(x-13y)$
$=(x-13y)(2x+3y)$

28. $4x^2-28xy+49y^2=[(2x)^2-2\cdot 2x\cdot 7y+(7y)^2]$
$=(2x-7y)^2$

29. $18x^2-9xy-20y^2=18x^2-24xy+15xy-20y^2$
$=6x(3x-4y)+5y(3x-4y)$
$=(3x-4y)(6x+5y)$

30. $36x^3y+24x^2y^2-45xy^3$
$=3xy(12x^2+8xy-15y^2)$
$=3xy(12x^2+18xy-10xy-15y^2)$
$=3xy[6x(2x+3y)-5y(2x+3y)]$
$=3xy(2x+3y)(6x-5y)$

31. $4x^2-9=(2x)^2-3^2=(2x+3)(2x-3)$

32. $9t^2-25s^2=(3t)^2-(5s)^2=(3t+5s)(3t-5s)$

33. $16x^2+y^2$ is a prime polynomial.

34. $x^3-8y^3=x^3-(2y)^3$
$=(x-2y)[x^2+x\cdot 2y+(2y)^2]$
$=(x-2y)(x^2+2xy+4y^2)$

35. $8x^3+27=(2x)^3+3^3$
$=(2x+3)[(2x)^2-2x\cdot 3+3^2]$
$=(2x+3)(4x^2-6x+9)$

36. $2x^3+8x=2x(x^2+4)$

37. $54-2x^3y^3=2(27-x^3y^3)$
$=2[3^3-(xy)^3]$
$=2(3-xy)[3^2+3\cdot xy+(xy)^2]$
$=2(3-xy)(9+3xy+x^2y^2)$

38. $9x^2-4y^2=(3x)^2-(2y)^2=(3x-2y)(3x+2y)$

39. $16x^4-1=(4x^2)^2-1^2$
$=(4x^2+1)(4x^2-1)$
$=(4x^2+1)[(2x)^2-1^2]$
$=(4x^2+1)(2x+1)(2x-1)$

40. x^4+16 is a prime polynomial.

41. $(x+6)(x-2)=0$
$x+6=0$ or $x-2=0$
$x=-6$ $\quad x=2$
The solutions are –6 and 2.

42. $3x(x+1)(7x-2)=0$
$3x=0$ or $x+1=0$ or $7x-2=0$
$x=0$ $\quad x=-1$ $\quad 7x=2$
$x=\frac{2}{7}$

The solutions are 0, –1, and $\frac{2}{7}$.

43. $4(5x+1)(x+3)=0$
$5x+1=0$ or $x+3=0$
$5x=-1$ $\quad x=-3$
$x=-\frac{1}{5}$

The solutions are $-\frac{1}{5}$ and –3.

44. $x^2+8x+7=0$
$(x+7)(x+1)=0$
$x+7=0$ or $x+1=0$
$x=-7$ $x=-1$
The solutions are –7 and –1.

45. $x^2-2x-24=0$
$(x-6)(x+4)=0$
$x-6=0$ or $x+4=0$
$x=6$ $x=-4$
The solutions are 6 and –4.

46. $x^2+10x=-25$
$x^2+10x+25=0$
$(x+5)(x+5)=0$
$x+5=0$ or $x+5=0$
$x=-5$ $x=-5$
The solution is –5.

47. $x(x-10)=-16$
$x^2-10x=-16$
$x^2-10x+16=0$
$(x-8)(x-2)=0$
$x-8=0$ or $x-2=0$
$x=8$ $x=2$
The solutions are 8 and 2.

48. $(3x-1)(9x^2+3x+1)=0$
$3x-1=0$ or $9x^2+3x+1=0$
$9x^2+3x+1$ is a prime polynomial.
$3x-1=0$
$3x=1$
$x=\frac{1}{3}$
The solution is $\frac{1}{3}$.

49. $56x^2-5x-6=0$
$56x^2+16x-21x-6=0$
$8x(7x+2)-3(7x+2)=0$
$(7x+2)(8x-3)=0$
$7x+2=0$ or $8x-3=0$
$7x=-2$ $8x=3$
$x=-\frac{2}{7}$ $x=\frac{3}{8}$
The solutions are $-\frac{2}{7}$ and $\frac{3}{8}$.

50. $20x^2-7x-6=0$
$(4x-3)(5x+2)=0$
$4x-3=0$ or $5x+2=0$
$4x=3$ $5x=-2$
$x=\frac{3}{4}$ $x=-\frac{2}{5}$
The solutions are $\frac{3}{4}$ and $-\frac{2}{5}$.

51. $5(3x+2)=4$
$15x+10=4$
$15x=-6$
$x=-\frac{6}{15}=-\frac{2}{5}$
The solution is $-\frac{2}{5}$.

52. $6x^2-3x+8=0$
The equation has no real solution.

53. $12-5t=-3$
$-5t=-15$
$t=3$
The solution is 3.

54. $5x^3+20x^2+20x=0$
$5x(x^2+4x+4)=0$
$5x(x+2)(x+2)=0$
$x+2=0$ or $5x=0$
$x=-2$ $x=0$
The solutions are –2 and 0.

55. $4t^3-5t^2-21t=0$
$t(4t^2-5t-21)=0$
$t(4t+7)(t-3)=0$
$t=0$ or $4t+7=0$ or $t-3=0$
$4t=-7$ $t=3$
$t=-\frac{7}{4}$
The solutions are 0, $-\frac{7}{4}$, and 3.

56. Answers may vary. Possible answer:
$(x-4)(x-5)=0$
$x^2-9x+20=0$

57. a. $7\neq 2\cdot 5$

b. $10 = 2 \cdot 5$

$$\begin{aligned} P &= 2l + 2w \\ &= 2(10) + 2(5) \\ &= 20 + 10 \\ &= 30 \neq 24 \end{aligned}$$

c. $8 = 2 \cdot 4$

$P = 2l + 2w = 2(8) + 2(4) = 16 + 8 = 24$

d. $10 \neq 2 \cdot 2$

Choice **c** gives the correct dimensions.

58. a. $3 \cdot 8 + 1 = 25 \neq 10$

b. $3 \cdot 4 + 1 = 13$

$A = lw = 13(4) = 52 \neq 80$

c. $3 \cdot 4 + 1 = 13 \neq 20$

d. $3 \cdot 5 + 1 = 16$

$A = lw = 5(16) = 80$

Choice **d** gives the correct dimensions.

59.

$$\begin{aligned} x^2 &= 81 \\ x^2 - 81 &= 0 \\ (x-9)(x+9) &= 0 \end{aligned}$$

$x - 9 = 0$ or $x + 9 = 0$

$x = 9$ $\quad x = -9$

Since length is not negative, the length of the side is 9 units.

60. $(2x+3)+(3x+1)+(x^2-3x)+(x+3) = 47$

$$\begin{aligned} x^2 + 3x + 7 &= 47 \\ x^2 + 3x - 40 &= 0 \\ (x-5)(x+8) &= 0 \end{aligned}$$

$x - 5 = 0$ or $x + 8 = 0$

$x = 5$ $\quad x = -8$

Length is not negative, so $x = 5$. The lengths are:

$x + 3 = 5 + 3 = 8$ units

$2x + 3 = 2(5) + 3 = 13$ units

$3x + 1 = 3(5) + 1 = 16$ units

$x^2 - 3x = 5^2 - 3(5) = 10$ units

61. Let x = the width of the flag. Then $2x - 15$ = the length of the flag.

$$\begin{aligned} A &= lw \\ 500 &= (2x-15)(x) \\ 500 &= 2x^2 - 15x \\ 0 &= 2x^2 - 15x - 500 \\ 0 &= (2x+25)(x-20) \end{aligned}$$

$2x + 25 = 0$

$2x = -25$ or $x - 20 = 0$

$x = -\frac{25}{2}$ $\quad x = 20$

Since the dimensions cannot be negative, the width is 20 inches and the length is $2(20) - 15 = 25$ inches.

62. Let x = the height of the sail, then $4x$ = the base of the sail.

$$\begin{aligned} A &= \frac{1}{2}bh \\ 162 &= \frac{1}{2}(4x)(x) \\ 162 &= 2x^2 \\ 0 &= 2x^2 - 162 \\ 0 &= 2(x^2 - 81) \\ 0 &= 2(x+9)(x-9) \end{aligned}$$

$x + 9 = 0$ or $x - 9 = 0$

$x = -9$ $\quad x = 9$

Since the dimensions cannot be negative, the height is 9 yards and the base is $4 \cdot 9 = 36$ yards.

63. Let x = the first integer. Then $x + 1$ = the next consecutive integer.

$$\begin{aligned} x(x+1) &= 380 \\ x^2 + x &= 380 \\ x^2 + x - 380 &= 0 \\ (x+20)(x-19) &= 0 \end{aligned}$$

$x + 20 = 0$ or $x - 19 = 0$

$x = -20$ $\quad x = 19$

The integers are 19 and 20.

64. a. Let $h = 2800$ and solve for t.

$$\begin{aligned} h &= -16t^2 + 440t \\ 2800 &= -16t^2 + 440t \\ 0 &= -16t^2 + 440t - 2800 \\ 0 &= -8(2t^2 - 55t + 350) \\ 0 &= -8(2t - 35)(t - 10) \end{aligned}$$

$$2t-35=0 \quad \text{or} \quad t-10=0$$
$$2t=35 \qquad t=10$$
$$t=\frac{35}{2}$$
$$t=17.5$$

The solutions are 17.5 sec and 10 sec. There are two answers because the rocket reaches a height of 2800 feet on its way up and on its way back down.

b. Let $h = 0$ and solve for t.

$$h=-16t^2+440t$$
$$0=-16t^2+440t$$
$$0=-8t(2t-55)$$
$$-8t=0 \quad \text{or} \quad 2t-55=0$$
$$t=0 \qquad 2t=55$$
$$t=\frac{55}{2}$$
$$t=27.5$$

$t = 0$ is when the rocket is launched, so it reaches the ground again after 27.5 seconds.

65. Find t when $h = 0$.

$$h=-16t^2+625$$
$$0=-16t^2+625$$
$$0=-1(16t^2-625)$$
$$0=-1[(4t)^2-(25)^2]$$
$$0=-1(4t+25)(4t-25)$$
$$4t+25=0 \quad \text{or} \quad 4t-25=0$$
$$t=-\frac{25}{4} \qquad t=\frac{25}{4}$$

Since time cannot be negative, the object reaches the ground after $\frac{25}{4}=6.25$ seconds.

66. Let x = the length of the longer leg, then $x - 8$ = the length of the shorter leg and $x + 8$ = the length of the hypotenuse.
By the Pythagorean theorem,

$$x^2+(x-8)^2=(x+8)^2$$
$$x^2+x^2-16x+64=x^2+16x+64$$
$$x^2-32x=0$$
$$x(x-32)=0$$
$$x=0 \quad \text{or} \quad x-32=0$$
$$x=32$$

Since the length cannot be 0 cm, the length of the longer leg is 32 cm.

67. $7x - 63 = 7(x - 9)$

68. $11x(4x-3)-6(4x-3)=(4x-3)(11x-6)$

69. $m^2-\frac{4}{25}=m^2-\left(\frac{2}{5}\right)^2=\left(m+\frac{2}{5}\right)\left(m-\frac{2}{5}\right)$

70. $3x^3-4x^2+6x-8=x^2(3x-4)+2(3x-4)$
$=(3x-4)(x^2+2)$

71. $xy+2x-y-2=x(y+2)-1(y+2)$
$=(y+2)(x-1)$

72. $2x^2+2x-24=2(x^2+x-12)=2(x+4)(x-3)$

73. $3x^3-30x^2+27x=3x(x^2-10x+9)$
$=3x(x-9)(x-1)$

74. $4x^2-81=(2x)^2-9^2=(2x+9)(2x-9)$

75. $2x^2-18=2(x^2-9)$
$=2(x^2-3^2)$
$=2(x+3)(x-3)$

76. $16x^2-24x+9=(4x)^2-2\cdot 4x\cdot 3+3^2$
$=(4x-3)^2$

77. $5x^2+20x+20=5(x^2+4x+4)$
$=5(x^2+2\cdot x\cdot 2+2^2)$
$=5(x+2)^2$

78. $2x^2+5x-12=(2x-3)(x+4)$

79. $4x^2y-6xy^2=2xy(2x-3y)$

80. $8x^2-15x-x^3=-x(-8x+15+x^2)$
$=-x(x^2-8x+15)$
$=-x(x-5)(x-3)$

81. $125x^3+27=(5x)^3+3^3$
$=(5x+3)[(5x)^2-5x\cdot 3+3^2]$
$=(5x+3)(25x^2-15x+9)$

82. $24x^2-3x-18=3(8x^2-x-6)$

83. $(x+7)^2-y^2=[(x+7)+y][(x+7)-y]$
$=(x+7+y)(x+7-y)$

84. $x^2(x+3)-4(x+3)=(x+3)(x^2-4)$
$=(x+3)(x^2-2^2)$
$=(x+3)(x-2)(x+2)$

85. $54a^3b-2b=2b(27a^3-1)$
$=2b[(3a)^3-1^3]$
$=2b(3a-1)[(3a)^2+3a\cdot 1+1^2]$
$=2b(3a-1)(9a^2+3a+1)$

86. To factor $x^2+2x-48$, think of two numbers whose product is $\underline{-48}$ and whose sum if $\underline{2}$.

87. The first step is to factor out the GCF, 3.

88. $(x^2-2)+(x^2-4x)+(3x^2-5x)$
$=x^2+x^2+3x^2-4x-5x-2$
$=5x^2-9x-2$
$=(5x+1)(x-2)$

89. $2(2x^2+3)+2(6x^2-14x)$
$=4x^2+6+12x^2-28x$
$=16x^2-28x+6$
$=2(8x^2-14x+3)$
$=2(4x-1)(2x-3)$

90. $2x^2-x-28=0$
$(2x+7)(x-4)=0$
$2x+7=0$ or $x-4=0$
$x=-\frac{7}{2}$ $\quad x=4$
The solutions are $-\frac{7}{2}$ and 4.

91. $x^2-2x=15$
$x^2-2x-15=0$
$(x+3)(x-5)=0$
$x+3=0$ or $x-5=0$
$x=-3$ $\quad x=5$
The solutions are −3 and 5.

92. $2x(x+7)(x+4)=0$
$2x=0$ or $x+7=0$ or $x+4=0$
$x=0$ $\quad x=-7$ $\quad x=-4$
The solutions are 0, −7, and −4.

93. $x(x-5)=-6$
$x^2-5x=-6$
$x^2-5x+6=0$
$(x-3)(x-2)=0$
$x-3=0$ or $x-2=0$
$x=3$ $\quad x=2$
The solutions are 3 and 2.

94. $x^2=16x$
$x^2-16x=0$
$x(x-16)=0$
$x=0$ or $x-16=0$
$x=16$
The solutions are 0 and 16.

95. $(x^2+3)+(4x+5)+2x=48$
$x^2+6x+8=48$
$x^2+6x-40=0$
$(x-4)(x+10)=0$
$x-4=0$ or $x+10=0$
$x=4$ $\quad x=-10$
Since the length cannot be negative, $x=4$.
The lengths are:
$x^2+3=4^2+3=19$ inches
$4x+5=4\cdot 4+5=21$ inches
$2x=2\cdot 4=8$ inches

96. Let x = length, then $x-4$ = width.
$A=lw$
$12=x(x-4)$
$12=x^2-4x$
$0=x^2-4x-12$
$0=(x-6)(x+2)$
$x-6=0$ or $x+2=0$
$x=6$ $\quad x=-2$
Since length cannot be negative, the length is 6 inches and the width is $6-4=2$ inches.

97. Find t when $h=0$.
$h=-16t^2+729$
$0=-16t^2+729$
$0=-(16t^2-729)$
$0=-[(4t)^2-(27)^2]$
$0=-(4t+27)(4t-27)$

$4t+27=0$ or $4t-27=0$

$t=-\dfrac{27}{4}$ $\quad t=\dfrac{27}{4}$

Since time cannot be negative, the object reaches the ground in $\dfrac{27}{4}=6.75$ seconds.

98. Area of large figure – Area of circle
$$\begin{aligned}&=[(6x)(5x)-2x^2]-\pi x^2\\&=30x^2-2x^2-\pi x^2\\&=28x^2-\pi x^2\\&=x^2(28-\pi)\end{aligned}$$

Chapter 6 Test

1. $x^2+11x+28=(x+7)(x+4)$

2. $49-m^2=(7^2-m^2)=(7-m)(7+m)$

3. $y^2+22y+121=y^2+2\cdot y\cdot 11+11^2=(y+11)^2$

4. $4(a+3)-y(a+3)=(a+3)(4-y)$

5. x^2+4 is the sum of two perfect squares (not the difference). The polynomial is prime.

6. $y^2-8y-48=(y-12)(y+4)$

7. x^2+x-10 is a prime polynomial.

8. $$\begin{aligned}9x^3+39x^2+12x&=3x(3x^2+13x+4)\\&=3x(3x+1)(x+4)\end{aligned}$$

9. $$\begin{aligned}3a^2+3ab-7a-7b&=3a(a+b)-7(a+b)\\&=(a+b)(3a-7)\end{aligned}$$

10. $3x^2-5x+2=(3x-2)(x-1)$

11. $x^2+14xy+24y^2=(x+12y)(x+2y)$

12. $$\begin{aligned}180-5x^2&=5(36-x^2)\\&=5(6^2-x^2)\\&=5(6+x)(6-x)\end{aligned}$$

13. $6t^2-t-5=(6t+5)(t-1)$

14. $$\begin{aligned}xy^2-7y^2-4x+28&=y^2(x-7)-4(x-7)\\&=(x-7)(y^2-4)\\&=(x-7)(y^2-2^2)\\&=(x-7)(y+2)(y-2)\end{aligned}$$

15. $$\begin{aligned}x-x^5&=x(1-x^4)\\&=x[1-(x^2)^2]\\&=x(1+x^2)(1-x^2)\\&=x(1+x^2)(1^2-x^2)\\&=x(1+x^2)(1+x)(1-x)\end{aligned}$$

16. $-xy^3-x^3y=xy(y^2+x^2)$

17. $$\begin{aligned}64x^3-1&=(4x)^3-1^3\\&=(4x-1)[(4x)^2+4x\cdot 1+1^2]\\&=(4x-1)(16x^2+4x+1)\end{aligned}$$

18. $$\begin{aligned}8y^3-64&=8(y^3-8)\\&=8(y^3-2^3)\\&=8(y-2)(y^2+y\cdot 2+2^2)\\&=8(y-2)(y^2+2y+4)\end{aligned}$$

19. $(x-3)(x+9)=0$

$x-3=0$ or $x+9=0$

$x=3$ $\quad x=-9$

The solutions are 3 and –9.

20. $$\begin{aligned}x^2+5x&=14\\x^2+5x-14&=0\\(x+7)(x-2)&=0\end{aligned}$$

$x+7=0$ or $x-2=0$

$x=-7$ $\quad x=2$

The solutions are –7 and 2.

21. $$\begin{aligned}x(x+6)&=7\\x^2+6x&=7\\x^2+6x-7&=0\\(x+7)(x-1)&=0\end{aligned}$$

$x+7=0$ or $x-1=0$

$x=-7$ $\quad x=1$

The solutions are –7 and 1.

22. $3x(2x-3)(3x+4)=0$

$3x=0$ or $2x-3=0$ or $3x+4=0$

$x=0$ $\quad 2x=3$ $\quad 3x=-4$

$x=\frac{3}{2}$ $\quad x=-\frac{4}{3}$

The solutions are 0, $\frac{3}{2}$, and $-\frac{4}{3}$.

23. $5t^3-45t=0$

$5t(t^2-9)=0$

$5t(t+3)(t-3)=0$

$5t=0$ or $t+3=0$ or $t-3=0$

$t=0$ $\quad t=-3$ $\quad t=3$

The solutions are 0, −3, and 3.

24. $t^2-2t-15=0$

$(t-5)(t+3)=0$

$t-5=0$ or $t+3=0$

$t=5$ $\quad t=-3$

The solutions are 5 and −3.

25. $6x^2=15x$

$6x^2-15x=0$

$3x(2x-5)=0$

$3x=0$ or $2x-5=0$

$x=0$ $\quad 2x=5$

$x=\frac{5}{2}$

The solutions are 0 and $\frac{5}{2}$.

26. Let x = the altitude of the triangle, then $x+9$ = the base.

$A=\frac{1}{2}bh$

$68=\frac{1}{2}(x+9)(x)$

$136=x^2+9x$

$0=x^2+9x-136$

$0=(x+17)(x-8)$

$x+17=0$ or $x-8=0$

$x=-17$ $\quad x=8$

Since the length of the base cannot be negative, the base is 8 + 9 = 17 feet.

27. Let x = the first number, then $17-x$ = the other number.

$x^2+(17-x)^2=145$

$x^2+289-34x+x^2=145$

$2x^2-34x+144=0$

$2(x^2-17x+72)=0$

$2(x-9)(x-8)=0$

$x-9=0$ or $x-8=0$

$x=9$ $\quad x=8$

The numbers are 8 and 9.

28. Find t when $h=0$.

$h=-16t^2+784$

$0=-16t^2+784$

$0=-16(t^2-49)$

$0=-16(t+7)(t-7)$

$t+7=0$ or $t-7=0$

$t=-7$ $\quad t=7$

Since the time cannot be negative, the object reaches the ground after 7 seconds.

29. Let x = length of the shorter leg, then $x+10$ = length of hypotenuse, and $x+5$ = length of longer leg.

$x^2+(x+5)^2=(x+10)^2$

$x^2+x^2+10x+25=x^2+20x+100$

$x^2-10x-75=0$

$(x+5)(x-15)=0$

$x+5=0$ or $x-15=0$

$x=-5$ $\quad x=15$

Since length cannot be negative, the lengths of the triangle sides are:
shorter leg = 15 cm, longer leg = 20 cm, hypotenuse = 25 cm.

Cumulative Review Chapters 1–6

1. a. $9\le 11$

b. $8>1$

c. $3\ne 4$

2. a. $|-5|>|-3|$

b. $|0|<|-2|$

3. a. $\frac{42}{49}=\frac{6\cdot 7}{7\cdot 7}=\frac{6}{7}$

b. $\frac{11}{27} = \frac{11}{3 \cdot 3 \cdot 3} = \frac{11}{27}$

c. $\frac{88}{20} = \frac{4 \cdot 22}{4 \cdot 5} = \frac{22}{5}$

4. Let $x = 20$ and $y = 10$.
$\frac{x}{y} + 5x = \frac{20}{10} + 5(20) = 2 + 100 = 102$

5. $\frac{8+2 \cdot 3}{2^2 - 1} = \frac{8+6}{4-1} = \frac{14}{3}$

6. Let $x = -20$ and $y = 10$.
$\frac{x}{y} + 5x = \frac{-20}{10} + 5(-20) = -2 - 100 = -102$

7. a. $3 + (-7) + (-8) = 3 + (-15) = -12$

b. $$\begin{aligned}[7+(-10)]+[-2+|-4|] &= -3+(-2+4)\\ &= -3+2\\ &= -1\end{aligned}$$

8. Let $x = -20$ and $y = -10$.
$\frac{x}{y} + 5x = \frac{-20}{-10} + 5(-20) = 2 - 100 = -98$

9. a. $(-6)(4) = -24$

b. $2(-1) = -2$

c. $(-5)(-10) = 50$

10. $5 - 2(3x - 7) = 5 - 6x + 14 = -6x + 19$

11. a. $7x - 3x = (7 - 3)x = 4x$

b. $10y^2 + y^2 = (10+1)y^2 = 11y^2$

c. $8x^2 + 2x - 3x = 8x^2 + (2-3)x = 8x^2 - x$

12. $$\begin{aligned}0.8y+0.2(y-1) &= 1.8\\ 0.8y+0.2y-0.2 &= 1.8\\ 1.0y-0.2 &= 1.8\\ y &= 2.0\end{aligned}$$

13. $$\begin{aligned}3-x &= 7\\ 3-3-x &= 7-3\\ -x &= 4\\ x &= -4\end{aligned}$$

14. $$\begin{aligned}\frac{x}{-7} &= -4\\ -7\left(\frac{x}{-7}\right) &= -7(-4)\\ x &= 28\end{aligned}$$

15. $$\begin{aligned}-3x &= 33\\ \frac{-3x}{-3} &= \frac{33}{-3}\\ x &= -11\end{aligned}$$

16. $$\begin{aligned}-\frac{2}{3}x &= -22\\ \left(-\frac{3}{2}\right)\left(-\frac{2}{3}\right)x &= \left(-\frac{3}{2}\right)(-22)\\ x &= 33\end{aligned}$$

17. $$\begin{aligned}8(2-t) &= -5t\\ 16-8t &= -5t\\ 16-8t+5t &= -5t+5t\\ 16-3t &= 0\\ 16-16-3t &= -16\\ -3t &= -16\\ \frac{-3t}{-3} &= \frac{-16}{-3}\\ t &= \frac{16}{3}\end{aligned}$$

18. $$\begin{aligned}-z &= \frac{7z+3}{5}\\ 5(-z) &= 5\left(\frac{7z+3}{5}\right)\\ -5z &= 7z+3\\ -5z-7z &= 7z-7z+3\\ -12z &= 3\\ \frac{-12z}{-12} &= \frac{3}{-12}\\ z &= -\frac{1}{4}\end{aligned}$$

19. Let x = the length of the shorter piece and
$3x$ = the length of the longer piece.
$$\begin{aligned}x+3x &= 48\\ 4x &= 48\\ x &= 12\end{aligned}$$
$3x = 3(12) = 36$
The pieces are 12 inches and 36 inches in length.

20. $3x + 9 \le 5(x - 1)$
$3x + 9 \le 5x - 5$
$-2x + 9 \le -5$
$2x \le -14$
$\frac{-2x}{-2} \ge \frac{-14}{-2}$
$x \ge 7$, $[7, \infty)$

21. $y = -\frac{1}{3}x + 2$

x	y
0	2
–3	3
3	1

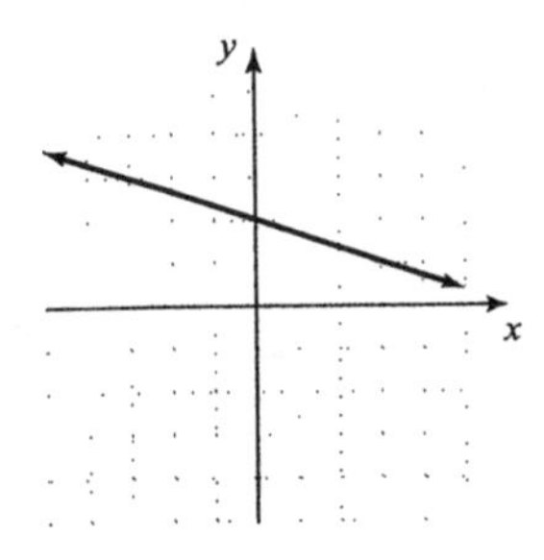

22. $-7x - 8y = -9$
$(-1, 2)$: $-7(-1) - 8(2) \stackrel{?}{=} -9$
$7 - 16 \stackrel{?}{=} -9$
$-9 = -9$ True
$(-1, 2)$ is a solution of the equation.

23. $3x - 4y = 4$
$-4y = -3x + 4$
$y = \frac{3}{4}x - 1$
$y = mx + b$
slope $= \frac{3}{4}$; y-intercept $= (0, -1)$

24. $(5, -6)$ and $(5, 2)$
$m = \frac{y_2 - y_1}{x_2 - x_1} = \frac{2 - (-6)}{5 - 5} = \frac{8}{0}$
The slope is undefined.

25. a. If $x = 5$, $2x^3 = 2(5)^3 = 2(125) = 250$.

b. If $x = -3$, $\frac{9}{x^2} = \frac{9}{(-3)^2} = \frac{9}{9} = 1$.

26. $7x - 3y = 2$
$-3y = -7x + 2$
$y = \frac{-7x}{-3} + \frac{2}{-3}$
$y = \frac{7}{3}x - \frac{2}{3}$
$y = mx + b$
slope $= \frac{7}{3}$; y-intercept $= \left(0, -\frac{2}{3}\right)$

27. a. $-3x^2$ has degree 2.

b. $5x^3yz$ has degree 3 + 1 + 1 = 5.

c. 2 has degree 0.

28. Vertical line has equation $x = c$.
Point (0, 7)
$x = 0$

29. $(2x^3 + 8x^2 - 6x) - (2x^3 - x^2 + 1)$
$= 2x^3 + 8x^2 - 6x - 2x^3 + x^2 - 1$
$= 9x^2 - 6x - 1$

30. $m = 4$, $b = \frac{1}{2}$
$y = mx + b$
$y = 4x + \frac{1}{2}$
$2y = 8x + 1$
$8x - 2y = -1$

31. $(3x + 2)(2x - 5)$
$= 3x(2x) + 3x(-5) + 2(2x) + 2(-5)$
$= 6x^2 - 15x + 4x - 10$
$= 6x^2 - 11x - 10$

32. $(-4, 0)$ and $(6, -1)$
$m = \frac{y_2 - y_1}{x_2 - x_1} = \frac{-1 - 0}{6 - (-4)} = -\frac{1}{10}$
$m = -\frac{1}{10}$, point $(-4, 0)$
$y - y_1 = m(x - x_1)$
$y - 0 = -\frac{1}{10}[x - (-4)]$
$y = -\frac{1}{10}x - \frac{4}{10}$
$10y = -x - 4$
$x + 10y = -4$

33. $(3y+1)^2=(3y)^2+2(3y)(1)+1^2=9y^2+6y+1$

34. $\begin{cases} -x+3y=18 \\ -3x+2y=19 \end{cases}$

Multiply the first equation by −3.

$$\begin{aligned} 3x-9y&=-54 \\ -3x+2y&=19 \\ \hline -7y&=-35 \\ y&=5 \end{aligned}$$

Substitute 5 for *y* in the first equation.

$$\begin{aligned} -x+3(5)&=18 \\ -x+15&=18 \\ -x&=3 \\ x&=-3 \end{aligned}$$

The solution to the system is (−3, 5).

35. a. $3^{-2}=\frac{1}{3^2}=\frac{1}{9}$

b. $2x^{-3}=\frac{2}{x^3}$

c. $2^{-1}+4^{-1}=\frac{1}{2}+\frac{1}{4}$

$$\begin{aligned} &=\frac{1\cdot 2}{2\cdot 2}+\frac{1}{4} \\ &=\frac{2}{4}+\frac{1}{4} \\ &=\frac{2+1}{4} \\ &=\frac{3}{4} \end{aligned}$$

d. $(-2)^{-4}=\frac{1}{(-2)^4}=\frac{1}{16}$

e. $\frac{1}{y^{-4}}=y^4$

f. $\frac{1}{7^{-2}}=7^2=49$

36. $\frac{(5a^7)^2}{a^5}=\frac{5^2a^{14}}{a^5}=25a^{14-5}=25a^9$

37. a. $367{,}000{,}000=3.67\times 10^8$

b. $0.000003=3.0\times 10^{-6}$

c. $20{,}520{,}000{,}000=2.052\times 10^{10}$

d. $0.00085=8.5\times 10^{-4}$

38. $(3x-7y)^2=(3x)^2-2(3x)(7y)+(7y)^2$

$$=9x^2-42xy+49y^2$$

39.

$$\begin{array}{r} x+4 \\ x+3\overline{)x^2+7x+12} \\ \underline{x^2+3x} \\ 4x+12 \\ \underline{4x+12} \\ 0 \end{array}$$

$$\frac{x^2+7x+12}{x+3}=x+4$$

40. $\frac{(xy)^{-3}}{(x^5y^6)^3}=\frac{x^{-3}y^{-3}}{x^{15}y^{18}}$

$$\begin{aligned} &=x^{-3-15}y^{-3-18} \\ &=x^{-18}y^{-21} \\ &=\frac{1}{x^{18}y^{21}} \end{aligned}$$

41. a. $x^3,\ x^7,\ x^5$: GCF $=x^3$

b. $y,\ y^4,\ y^7$: GCF $=y$

42. $z^3+7z+z^2+7=z(z^2+7)+1(z^2+7)$

$$=(z^2+7)(z+1)$$

43. $x^2+7x+12=(x+4)(x+3)$

44. $2x^3+2x^2-84x=2x(x^2+x-42)$

$$=2x(x+7)(x-6)$$

45. $8x^2-22x+5=8x^2-20x-2x+5$

$$\begin{aligned} &=4x(2x-5)-1(2x-5) \\ &=(2x-5)(4x-1) \end{aligned}$$

46. $-4x^2-23x+6=-1(4x^2+23x-6)$

$$\begin{aligned} &=-(4x^2-x+24x-6) \\ &=-[x(4x-1)+6(4x-1)] \\ &=-(4x-1)(x+6) \end{aligned}$$

47. $25a^2-9b^2=(5a)^2-(3b)^2=(5a+3b)(5a-3b)$

48. $9xy^2 - 16x = x(9y^2 - 16)$
$= x[(3y)^2 - 4^2]$
$= x(3y+4)(3y-4)$

49. $(x - 3)(x + 1) = 0$
$x - 3 = 0$ or $x + 1 = 0$
$x = 3$ $\quad x = -1$
The solutions are 3 and −1.

50. $x^2 - 13x = -36$
$x^2 - 13x + 36 = 0$
$(x - 9)(x - 4) = 0$
$x - 9 = 0$ or $x - 4 = 0$
$x = 9$ $\quad x = 4$
The solutions are 9 and 4.

Chapter 7

Section 7.1

Practice Exercises

1. a. Replace each x in the expression with 3 and then simplify.
$\frac{x+6}{3x-2}=\frac{3+6}{3(3)-2}=\frac{9}{9-2}=\frac{9}{7}$

   b. Replace each x in the expression with -3 and then simplify.
$\frac{x+6}{3x-2}=\frac{-3+6}{3(-3)-2}=\frac{3}{-9-2}=\frac{3}{-11}$ or $-\frac{3}{11}$

2. a. The denominator of $\frac{x}{x+6}$ is 0 when $x+6=0$ or when $x=-6$. Thus, when $x=-6$, the expression $\frac{x}{x+6}$ is undefined.

   b. The denominator of $\frac{x^4-3x^2+7x}{7}$ is never 0, so there are no values of x for which this expression is undefined.

   c. Set the denominator equal to zero.
$x^2+6x+8=0$
$(x+2)(x+4)=0$
$x+2=0$ or $x+4=0$
$x=-2$ $\quad x=-4$
Thus, when $x=-2$ or $x=-4$, the denominator x^2+6x+8 is 0. So the rational expression $\frac{x^2-5}{x^2+6x+8}$ is undefined when $x=-2$ or when $x=-4$.

   d. No matter which real number x is replaced by, the denominator x^4+5 does not equal 0, so there are no real numbers for which this expression is undefined.

3. Factor the numerator and denominator, then look for common factors.
$\frac{x^6-x^5}{6x-6}=\frac{x^5(x-1)}{6(x-1)}=\frac{x^5}{6}$

4. Factor the numerator and denominator, then look for common factors.
$\frac{x^2+5x+4}{x^2+2x-8}=\frac{(x+1)(x+4)}{(x-2)(x+4)}=\frac{x+1}{x-2}$

5. Factor the numerator and denominator, then look for common factors.
$\frac{x^3+9x^2}{x^2+18x+81}=\frac{x^2(x+9)}{(x+9)(x+9)}=\frac{x^2}{x+9}$

6. Factor the numerator and denominator, then look for common factors.
$\frac{x-7}{x^2-49}=\frac{x-7}{(x+7)(x-7)}=\frac{1}{x+7}$

7. a. Note that $s-t$ and $t-s$ are opposites. In other words, $t-s=-1(s-t)$.
$\frac{s-t}{t-s}=\frac{1\cdot(s-t)}{-1\cdot(s-t)}=\frac{1}{-1}=-1$

   b. By the commutative property of addition, $d+2c=2c+d$.
$\frac{2c+d}{d+2c}=\frac{2c+d}{2c+d}=1$

8. $$\frac{2x^2-5x-12}{16-x^2}=\frac{(x-4)(2x+3)}{(4-x)(4+x)}=\frac{(x-4)(2x+3)}{(-1)(x-4)(4+x)}=\frac{2x+3}{(-1)(4+x)}=-\frac{2x+3}{x+4}\text{ or }\frac{-2x-3}{x+4}$$

9. $-\frac{x+3}{6x-11}=\frac{-(x+3)}{6x-11}=\frac{-x-3}{6x-11}$
Also,
$-\frac{x+3}{6x-11}=\frac{x+3}{-(6x-11)}=\frac{x+3}{-6x+11}$ or $\frac{x+3}{11-6x}$
Thus, some equivalent forms of $-\frac{x+3}{6x-11}$ are $\frac{-(x+3)}{6x-11}$, $\frac{-x-3}{6x-11}$, $\frac{x+3}{-(6x-11)}$, $\frac{x+3}{-6x+11}$, and $\frac{x+3}{11-6x}$.

Vocabulary and Readiness Check

1. A rational expression is an expression that can be written in the form $\frac{P}{Q}$ where P and Q are polynomials and $Q \neq 0$.

2. The expression $\frac{x+3}{3+x}$ simplifies to 1.

3. The expression $\frac{x-3}{3-x}$ simplifies to −1.

4. A rational expression is undefined for values that make the denominator 0.

5. The expression $\frac{7x}{x-2}$ is undefined for $x = 2$.

6. The process of writing a rational expression in lowest terms is called simplifying.

7. For a rational expression, $-\frac{a}{b} = \frac{-a}{b} = \frac{a}{-b}$.

8. No, $\frac{x}{x+7}$ cannot be simplified.

9. Yes, $\frac{3+x}{x+3}$ can be simplified because $3 + x = x + 3$.

10. Yes, $\frac{5-x}{x-5}$ can be simplified because $5 - x = -1(x - 5)$.

11. No, $\frac{x+2}{x+8}$ cannot be simplified.

Exercise Set 7.1

1. $\frac{x+5}{x+2} = \frac{2+5}{2+2} = \frac{7}{4}$

3. $\frac{4z-1}{z-2} = \frac{4(-5)-1}{-5-2} = \frac{-20-1}{-7} = \frac{-21}{-7} = 3$

5. $\frac{y^3}{y^2-1} = \frac{(-2)^3}{(-2)^2-1} = \frac{-8}{4-1} = \frac{-8}{3} = -\frac{8}{3}$

7. $$\begin{aligned}\frac{x^2+8x+2}{x^2-x-6} &= \frac{2^2+8(2)+2}{2^2-2-6} \\ &= \frac{4+16+2}{4-8} \\ &= \frac{22}{-4} \\ &= \frac{11\cdot 2}{-2\cdot 2} \\ &= -\frac{11}{2}\end{aligned}$$

9. $2x = 0$
$x = 0$
The expression is undefined when $x = 0$.

11. $x + 2 = 0$
$x = -2$
The expression is undefined when $x = -2$.

13. $2x - 5 = 0$
$2x = 5$
$x = \frac{5}{2}$
The expression is undefined when $x = \frac{5}{2}$.

15. The denominator is never zero, so there are no values for which $\frac{x^2-5x-2}{4}$ is undefined.

17. $x^2 - 5x - 6 = 0$
$(x+1)(x-6) = 0$
$x + 1 = 0$ or $x - 6 = 0$
$x = -1$ $\quad$ $x = 6$
The expression is undefined when $x = -1, 6$.

19. The denominator is never zero, so there are no values for which $\frac{9x^3+4}{x^2+36}$ is undefined.

21. $3x^2 + 13x + 14 = 0$
$(x+2)(3x+7) = 0$
$x + 2 = 0$ or $3x + 7 = 0$
$x = -2$ $\quad$ $3x = -7$
$x = -\frac{7}{3}$
The expression is undefined when $x = -2, -\frac{7}{3}$.

23. $-\frac{x-10}{x+8}=\frac{-(x-10)}{x+8}=\frac{-x+10}{x+8}$ or $\frac{10-x}{x+8}$

$-\frac{x-10}{x+8}=\frac{x-10}{-(x+8)}=\frac{x-10}{-x-8}$

25. $-\frac{5y-3}{y-12}=\frac{-(5y-3)}{y-12}=\frac{-5y+3}{y-12}$ or $\frac{3-5y}{y-12}$

$-\frac{5y-3}{y-12}=\frac{5y-3}{-(y-12)}=\frac{5y-3}{-y+12}$ or $\frac{5y-3}{12-y}$

27. $\frac{x+7}{7+x}=\frac{x+7}{x+7}=1$

29. $\frac{x-7}{7-x}=\frac{x-7}{-1(x-7)}=\frac{1}{-1}=-1$

31. $\frac{2}{8x+16}=\frac{2}{8(x+2)}=\frac{2(1)}{2(4)(x+2)}=\frac{1}{4(x+2)}$

33. $\frac{x-2}{x^2-4}=\frac{x-2}{(x+2)(x-2)}=\frac{1}{x+2}$

35. $\frac{2x-10}{3x-30}=\frac{2(x-5)}{3(x-10)}$

The numerator and denominator have no common factors, so the expression cannot be simplified.

37. $\frac{-5a-5b}{a+b}=\frac{-5(a+b)}{a+b}=-5$

39. $\frac{7x+35}{x^2+5x}=\frac{7(x+5)}{x(x+5)}=\frac{7}{x}$

41. $\frac{x+5}{x^2-4x-45}=\frac{x+5}{(x+5)(x-9)}=\frac{1}{x-9}$

43. $\frac{5x^2+11x+2}{x+2}=\frac{(5x+1)(x+2)}{x+2}=5x+1$

45. $\frac{x^3+7x^2}{x^2+5x-14}=\frac{x^2(x+7)}{(x+7)(x-2)}=\frac{x^2}{x-2}$

47. $\frac{14x^2-21x}{2x-3}=\frac{7x(2x-3)}{2x-3}=7x$

49. $\frac{x^2+7x+10}{x^2-3x-10}=\frac{(x+2)(x+5)}{(x+2)(x-5)}=\frac{x+5}{x-5}$

51. $\frac{3x^2+7x+2}{3x^2+13x+4}=\frac{(3x+1)(x+2)}{(3x+1)(x+4)}=\frac{x+2}{x+4}$

53. $\frac{2x^2-8}{4x-8}=\frac{2(x^2-4)}{4(x-2)}=\frac{2(x+2)(x-2)}{2\cdot 2(x-2)}=\frac{x+2}{2}$

55. $\frac{4-x^2}{x-2}=\frac{(-1)(x^2-4)}{x-2}$

$=-\frac{(x+2)(x-2)}{x-2}$

$=-(x+2)$ or $-x-2$

57. $\frac{x^2-1}{x^2-2x+1}=\frac{(x+1)(x-1)}{(x-1)(x-1)}=\frac{x+1}{x-1}$

59. $\frac{m^2-6m+9}{m^2-m-6}=\frac{(m-3)(m-3)}{(m+2)(m-3)}=\frac{m-3}{m+2}$

61. $\frac{11x^2-22x^3}{6x-12x^2}=\frac{11x^2(1-2x)}{6x(1-2x)}=\frac{11x}{6}$

63. $\frac{x^2+xy+2x+2y}{x+2}=\frac{x(x+y)+2(x+y)}{x+2}$

$=\frac{(x+y)(x+2)}{x+2}$

$=x+y$

65. $\frac{5x+15-xy-3y}{2x+6}=\frac{5(x+3)-y(x+3)}{2(x+3)}$

$=\frac{(x+3)(5-y)}{2(x+3)}$

$=\frac{5-y}{2}$

67. $\frac{x^3+8}{x+2}=\frac{(x+2)(x^2-2x+4)}{x+2}=x^2-2x+4$

69. $\frac{x^3-1}{1-x}=\frac{(x-1)(x^2+x+1)}{-1(x-1)}$

$=-1(x^2+x+1)$

$=-x^2-x-1$

71. $\frac{2xy+5x-2y-5}{3xy+4x-3y-4} = \frac{x(2y+5)-1(2y+5)}{x(3y+4)-1(3y+4)}$
$= \frac{(2y+5)(x-1)}{(3y+4)(x-1)}$
$= \frac{2y+5}{3y+4}$

73. $\frac{9-x^2}{x-3} = \frac{-1(x^2-9)}{x-3}$
$= -\frac{(x+3)(x-3)}{x-3}$
$= -(x+3)$ or $-x-3$
The given answer is correct.

75. $\frac{7-34x-5x^2}{25x^2-1} = \frac{-1(5x^2+34x-7)}{(5x+1)(5x-1)}$
$= -\frac{(x+7)(5x-1)}{(5x+1)(5x-1)}$
$= -\frac{x+7}{5x+1}$
$= \frac{x+7}{-(5x+1)}$ or $\frac{x+7}{-5x-1}$
The given answer is correct.

77. $\frac{1}{3} \cdot \frac{9}{11} = \frac{1 \cdot 9}{3 \cdot 11} = \frac{3 \cdot 3}{3 \cdot 11} = \frac{3}{11}$

79. $\frac{1}{3} \div \frac{1}{4} = \frac{1}{3} \cdot \frac{4}{1} = \frac{4}{3}$

81. $\frac{13}{20} \div \frac{2}{9} = \frac{13}{20} \cdot \frac{9}{2} = \frac{13 \cdot 9}{20 \cdot 2} = \frac{117}{40}$

83. $\frac{5a-15}{5} = \frac{5(a-3)}{5} = a-3$
The statement is correct.

85. $\frac{1+2}{1+3} = \frac{3}{4}$
The statement is incorrect.

87. Answers may vary

89. Answers may vary

91. a. $R = \frac{150x^2}{x^2+3}$
$= \frac{150(1)^2}{1^2+3}$
$= \frac{150}{4}$
$= \$37.5$ million

b. $R = \frac{150x^2}{x^2+3}$
$= \frac{150(2)^2}{2^2+3}$
$= \frac{600}{7}$
$\approx \$85.7$ million

c. $85.7 - 37.5 = \$48.2$ million

93. Let $D = 1000$ and $A = 8$.
$C = \frac{DA}{A+12} = \frac{1000(8)}{8+12} = \frac{8000}{20} = 400$
The child should receive 400 mg.

95. $C = \frac{100W}{L}$; $W = 5, L = 6.4$
$C = \frac{100(5)}{6.4} = \frac{500}{6.4} = 78.125$
The skull is medium.

97. $\left[\dfrac{20C+0.5A+Y+80T-100I}{A}\right]\left(\dfrac{25}{6}\right)$; $C=362, A=557, Y=4397, T=31, I=9$

$$\left[\frac{20(362)+0.5(557)+4397+80(31)-100(9)}{557}\right]\left(\frac{25}{6}\right)=\left(\frac{13,495.5}{557}\right)\left(\frac{25}{6}\right)\approx 101.0$$

For the 2006 season, Peyton Manning's rating was 101.0.

99. $P=\dfrac{R-C}{R}$; $R=30.8, C=23.1$

$$P=\frac{30.8-23.1}{30.8}=\frac{7.7}{30.8}=0.25=25\%$$

In 2004, Best Buy's gross profit margin was 25%.

101. $y=\dfrac{x^2-16}{x-4}=\dfrac{(x+4)(x-4)}{x-4}=x+4,\ x\neq 4$

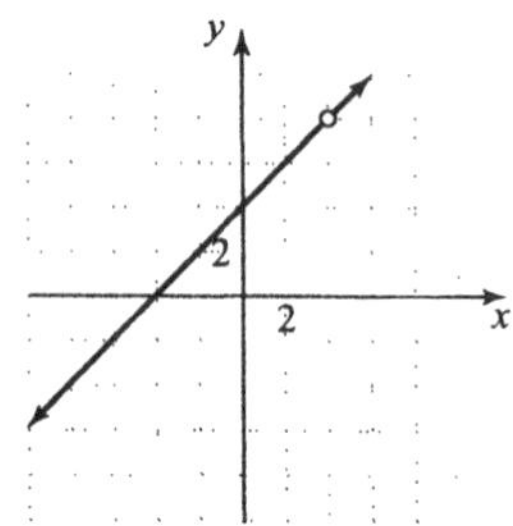

103. $y=\dfrac{x^2-6x+8}{x-2}=\dfrac{(x-2)(x-4)}{x-2}=x-4,\ x\neq 2$

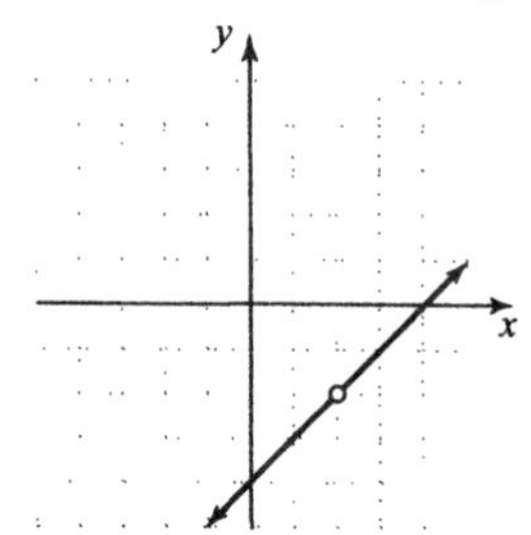

Section 7.2

Practice Exercises

1. a. $\dfrac{4a}{5}\cdot\dfrac{3}{b^2}=\dfrac{4a\cdot 3}{5\cdot b^2}=\dfrac{12a}{5b^2}$

b. $$\frac{-3p^4}{q^2}\cdot\frac{2q^3}{9p^4}=\frac{-3p^4\cdot 2q^3}{q^2\cdot 9p^4}$$
$$=\frac{-1\cdot 3\cdot p^4\cdot 2\cdot q\cdot q^2}{q^2\cdot 3\cdot 3\cdot p^4}$$
$$=-\frac{2q}{3}$$

2. $\frac{x^2-x}{5x}\cdot\frac{15}{x^2-1}=\frac{x(x-1)}{5x}\cdot\frac{3\cdot 5}{(x+1)(x-1)}$
$=\frac{x(x-1)\cdot 3\cdot 5}{5x\cdot(x+1)(x-1)}$
$=\frac{3}{x+1}$

3. $\frac{6-3x}{6x+6x^2}\cdot\frac{3x^2-2x-5}{x^2-4}$
$=\frac{3(2-x)}{2\cdot 3\cdot x(1+x)}\cdot\frac{(x+1)(3x-5)}{(x+2)(x-2)}$
$=\frac{3(2-x)(x+1)(3x-5)}{2\cdot 3x(1+x)(x+2)(x-2)}$
$=\frac{-1(x-2)(x+1)(3x-5)}{2x(x+1)(x+2)(x-2)}$
$=-\frac{3x-5}{2x(x+2)}$

4. $\frac{5a^3b^2}{24}\div\frac{10a^5}{6}=\frac{5a^3b^2}{24}\cdot\frac{6}{10a^5}$
$=\frac{5a^3b^2\cdot 6}{4\cdot 6\cdot 2\cdot 5\cdot a^2\cdot a^3}$
$=\frac{b^2}{8a^2}$

5. $\frac{(3x+1)(x-5)}{3}\div\frac{4x-20}{9}$
$=\frac{(3x+1)(x-5)}{3}\cdot\frac{9}{4x-20}$
$=\frac{(3x+1)(x-5)\cdot 3\cdot 3}{3\cdot 4(x-5)}$
$=\frac{3(3x+1)}{4}$

6. $\frac{10x-2}{x^2-9}\div\frac{5x^2-x}{x+3}=\frac{10x-2}{x^2-9}\cdot\frac{x+3}{5x^2-x}$
$=\frac{2(5x-1)(x+3)}{(x+3)(x-3)\cdot x(5x-1)}$
$=\frac{2}{x(x-3)}$

7. $\frac{3x^2-11x-4}{2x-8}\div\frac{9x+3}{6}=\frac{3x^2-11x-4}{2x-8}\cdot\frac{6}{9x+3}$
$=\frac{(3x+1)(x-4)\cdot 2\cdot 3}{2(x-4)\cdot 3(3x+1)}$
$=\frac{1}{1}$ or 1

8. a. $\frac{y+9}{8x}\cdot\frac{y+9}{2x}=\frac{(y+9)\cdot(y+9)}{8x\cdot 2x}=\frac{(y+9)^2}{16x^2}$

b. $\frac{y+9}{8x}\div\frac{y+9}{2}=\frac{y+9}{8x}\cdot\frac{2}{y+9}$
$=\frac{(y+9)\cdot 2}{2\cdot 4\cdot x\cdot(y+9)}$
$=\frac{1}{4x}$

c. $\frac{35x-7x^2}{x^2-25}\cdot\frac{x^2+3x-10}{x^2+4x}$
$=\frac{7x(5-x)}{(x+5)(x-5)}\cdot\frac{(x-2)(x+5)}{x(x+4)}$
$=\frac{7x\cdot(-1)(x-5)\cdot(x-2)(x+5)}{(x+5)(x-5)\cdot x(x+4)}$
$=-\frac{7(x-2)}{x+4}$

Vocabulary and Readiness Check

1. The expressions $\frac{x}{2y}$ and $\frac{2y}{x}$ are called <u>reciprocals</u>.

2. $\frac{a}{b}\cdot\frac{c}{d}=\underline{\frac{a\cdot c}{b\cdot d}}$ or $\underline{\frac{ac}{bd}}$

3. $\frac{a}{b}\div\frac{c}{d}=\underline{\frac{a\cdot d}{b\cdot c}}$ or $\underline{\frac{ad}{bc}}$

4. $\frac{x}{7}\cdot\frac{x}{6}=\underline{\frac{x^2}{42}}$

5. $\frac{x}{7}\div\frac{x}{6}=\underline{\frac{6}{7}}$

Exercise Set 7.2

1. $\frac{3x}{y^2}\cdot\frac{7y}{4x}=\frac{3\cdot x\cdot 7\cdot y}{y\cdot y\cdot 4\cdot x}=\frac{21}{4y}$

3. $\frac{8x}{2}\cdot\frac{x^5}{4x^2}=\frac{8x\cdot x^5}{2\cdot 4x^2}=\frac{2\cdot 4\cdot x\cdot x\cdot x^4}{2\cdot 4\cdot x\cdot x}=x^4$

5. $-\frac{5a^2b}{30a^2b^2}\cdot b^3=\frac{5a^2b\cdot b^3}{30a^2b^2}$
$=-\frac{5\cdot a^2\cdot b\cdot b\cdot b^2}{5\cdot 6\cdot a^2\cdot b^2}$
$=-\frac{b\cdot b}{6}$
$=-\frac{b^2}{6}$

7. $\frac{x}{2x-14}\cdot\frac{x^2-7x}{5}=\frac{x\cdot(x^2-7x)}{(2x-14)\cdot 5}$
$=\frac{x\cdot x(x-7)}{2(x-7)\cdot 5}$
$=\frac{x\cdot x}{2\cdot 5}$
$=\frac{x^2}{10}$

9. $\frac{6x+6}{5}\cdot\frac{10}{36x+36}=\frac{(6x+6)\cdot 10}{5\cdot(36x+36)}$
$=\frac{6(x+1)\cdot 2\cdot 5}{5\cdot 36(x+1)}$
$=\frac{6\cdot 5\cdot 2\cdot(x+1)}{6\cdot 5\cdot 2\cdot 3\cdot(x+1)}$
$=\frac{1}{3}$

11. $\frac{(m+n)^2}{m-n}\cdot\frac{m}{m^2+mn}=\frac{(m+n)(m+n)\cdot m}{(m-n)\cdot m(m+n)}=\frac{m+n}{m-n}$

13. $\frac{x^2-25}{x^2-3x-10}\cdot\frac{x+2}{x}=\frac{(x^2-25)\cdot(x+2)}{(x^2-3x-10)\cdot x}$
$=\frac{(x-5)(x+5)\cdot(x+2)}{(x-5)(x+2)\cdot x}$
$=\frac{x+5}{x}$

15. $\frac{x^2+6x+8}{x^2+x-20}\cdot\frac{x^2+2x-15}{x^2+8x+16}$
$=\frac{(x+2)(x+4)}{(x+5)(x-4)}\cdot\frac{(x+5)(x-3)}{(x+4)(x+4)}$
$=\frac{(x+2)(x+4)\cdot(x+5)(x-3)}{(x+5)(x-4)\cdot(x+4)(x+4)}$
$=\frac{(x+2)(x-3)}{(x-4)(x+4)}$

17. $\frac{5x^7}{2x^5}\div\frac{15x}{4x^3}=\frac{5x^7}{2x^5}\cdot\frac{4x^3}{15x}$
$=\frac{5\cdot x^2\cdot x^5\cdot 2\cdot 2\cdot x\cdot x^2}{2\cdot x^5\cdot 3\cdot 5\cdot x}$
$=\frac{2x^4}{3}$

19. $\frac{8x^2}{y^3}\div\frac{4x^2y^3}{6}=\frac{8x^2}{y^3}\cdot\frac{6}{4x^2y^3}=\frac{2\cdot 4\cdot x^2\cdot 6}{y^3\cdot 4x^2y^3}=\frac{12}{y^6}$

21. $\frac{(x-6)(x+4)}{4x}\div\frac{2x-12}{8x^2}$
$=\frac{(x-6)(x+4)}{4x}\cdot\frac{8x^2}{2x-12}$
$=\frac{(x-6)(x+4)\cdot 2\cdot 4\cdot x\cdot x}{4x\cdot 2(x-6)}$
$=x(x+4)$

23. $\frac{3x^2}{x^2-1}\div\frac{x^5}{(x+1)^2}=\frac{3x^2}{x^2-1}\cdot\frac{(x+1)^2}{x^5}$
$=\frac{3x^2\cdot(x+1)(x+1)}{(x-1)(x+1)\cdot x^2\cdot x^3}$
$=\frac{3(x+1)}{x^3(x-1)}$

25. $\frac{m^2-n^2}{m+n}\div\frac{m}{m^2+nm}=\frac{m^2-n^2}{m+n}\cdot\frac{m^2+nm}{m}$
$=\frac{(m-n)(m+n)\cdot m(m+n)}{(m+n)\cdot m}$
$=(m-n)(m+n)$
$=m^2-n^2$

27. $$\frac{x+2}{7-x} \div \frac{x^2-5x+6}{x^2-9x+14} = \frac{x+2}{7-x} \cdot \frac{x^2-9x+14}{x^2-5x+6}$$
$$= \frac{(x+2)\cdot(x-7)(x-2)}{-1(x-7)\cdot(x-3)(x-2)}$$
$$= -\frac{x+2}{x-3}$$

29. $$\frac{x^2+7x+10}{x-1} \div \frac{x^2+2x-15}{x-1}$$
$$= \frac{x^2+7x+10}{x-1} \cdot \frac{x-1}{x^2+2x-15}$$
$$= \frac{(x+5)(x+2)\cdot(x-1)}{(x-1)\cdot(x+5)(x-3)}$$
$$= \frac{x+2}{x-3}$$

31. $$\frac{5x-10}{12} \div \frac{4x-8}{8} = \frac{5x-10}{12} \cdot \frac{8}{4x-8}$$
$$= \frac{5(x-2)\cdot 2\cdot 4}{6\cdot 2\cdot 4(x-2)}$$
$$= \frac{5}{6}$$

33. $$\frac{x^2+5x}{8} \cdot \frac{9}{3x+15} = \frac{x(x+5)\cdot 3\cdot 3}{8\cdot 3(x+5)} = \frac{3x}{8}$$

35. $$\frac{7}{6p^2+q} \div \frac{14}{18p^2+3q} = \frac{7}{6p^2+q} \cdot \frac{18p^2+3q}{14}$$
$$= \frac{7\cdot 3(6p^2+q)}{(6p^2+q)\cdot 7\cdot 2}$$
$$= \frac{3}{2}$$

37. $$\frac{3x+4y}{x^2+4xy+4y^2} \cdot \frac{x+2y}{2} = \frac{(3x+4y)\cdot(x+2y)}{(x+2y)(x+2y)\cdot 2}$$
$$= \frac{3x+4y}{2(x+2y)}$$

39. $$\frac{(x+2)^2}{x-2} \div \frac{x^2-4}{2x-4} = \frac{(x+2)^2}{x-2} \cdot \frac{2x-4}{x^2-4}$$
$$= \frac{(x+2)(x+2)\cdot 2(x-2)}{(x-2)\cdot(x+2)(x-2)}$$
$$= \frac{2(x+2)}{x-2}$$

41. $$\frac{x^2-4}{24x} \div \frac{2-x}{6xy} = \frac{x^2-4}{24x} \cdot \frac{6xy}{2-x}$$
$$= \frac{(x+2)(x-2)\cdot 6x\cdot y}{4\cdot 6x\cdot(-1)(x-2)}$$
$$= -\frac{y(x+2)}{4}$$

43. $$\frac{a^2+7a+12}{a^2+5a+6} \cdot \frac{a^2+8a+15}{a^2+5a+4}$$
$$= \frac{(a+3)(a+4)\cdot(a+5)(a+3)}{(a+3)(a+2)\cdot(a+4)(a+1)}$$
$$= \frac{(a+5)(a+3)}{(a+2)(a+1)}$$

45. $$\frac{5x-20}{3x^2+x} \cdot \frac{3x^2+13x+4}{x^2-16}$$
$$= \frac{5(x-4)}{x(3x+1)} \cdot \frac{(3x+1)(x+4)}{(x+4)(x-4)}$$
$$= \frac{5(x-4)\cdot(3x+1)(x+4)}{x(3x+1)\cdot(x+4)(x-4)}$$
$$= \frac{5}{x}$$

47. $$\frac{8n^2-18}{2n^2-5n+3} \div \frac{6n^2+7n-3}{n^2-9n+8}$$
$$= \frac{8n^2-18}{2n^2-5n+3} \cdot \frac{n^2-9n+8}{6n^2+7n-3}$$
$$= \frac{2(2n+3)(2n-3)\cdot(n-8)(n-1)}{(n-1)(2n-3)\cdot(2n+3)(3n-1)}$$
$$= \frac{2(n-8)}{3n-1}$$

49. $$\frac{x^2-9}{2x} \div \frac{x+3}{8x^4} = \frac{x^2-9}{2x} \cdot \frac{8x^4}{x+3}$$
$$= \frac{(x+3)(x-3)}{2x} \cdot \frac{8x^4}{x+3}$$
$$= \frac{2x\cdot 4x^3\cdot(x+3)(x-3)}{2x\cdot(x+3)}$$
$$= 4x^3(x-3)$$

51. $\frac{a^2+ac+ba+bc}{a-b} \div \frac{a+c}{a+b}$
$= \frac{a(a+c)+b(a+c)}{a-b} \cdot \frac{a+b}{a+c}$
$= \frac{(a+c)(a+b)}{a-b} \cdot \frac{a+b}{a+c}$
$= \frac{(a+c)\cdot(a+b)\cdot(a+b)}{(a-b)\cdot(a+c)}$
$= \frac{(a+b)^2}{a-b}$

53. $\frac{3x^2+8x+5}{x^2+8x+7} \cdot \frac{x+7}{x^2+4} = \frac{(3x+5)(x+1)}{(x+7)(x+1)} \cdot \frac{x+7}{x^2+4}$
$= \frac{(3x+5)\cdot(x+1)\cdot(x+7)}{(x+7)\cdot(x+1)\cdot(x^2+4)}$
$= \frac{3x+5}{x^2+4}$

55. $\frac{x^3+8}{x^2-2x+4} \cdot \frac{4}{x^2-4}$
$= \frac{(x+2)(x^2-2x+4)}{x^2-2x+4} \cdot \frac{4}{(x+2)(x-2)}$
$= \frac{4\cdot(x+2)\cdot(x^2-2x+4)}{(x+2)(x-2)(x^2-2x+4)}$
$= \frac{4}{x-2}$

57. $\frac{a^2-ab}{6a^2+6ab} \div \frac{a^3-b^3}{a^2-b^2}$
$= \frac{a^2-ab}{6a^2+6ab} \cdot \frac{a^2-b^2}{a^3-b^3}$
$= \frac{a(a-b)}{6a(a+b)} \cdot \frac{(a-b)(a+b)}{(a-b)(a^2+ab+b^2)}$
$= \frac{a\cdot(a-b)\cdot(a-b)\cdot(a+b)}{6\cdot a\cdot(a+b)\cdot(a-b)\cdot(a^2+ab+b^2)}$
$= \frac{a-b}{6(a^2+ab+b^2)}$

59. $\frac{1}{5}+\frac{4}{5}=\frac{5}{5}=1$

61. $\frac{9}{9}-\frac{19}{9}=-\frac{10}{9}$

63. $\frac{6}{5}+\left(\frac{1}{5}-\frac{8}{5}\right)=\frac{6}{5}+\left(-\frac{7}{5}\right)=-\frac{1}{5}$

65. $x-2y=6$

x	y
0	−3
6	0

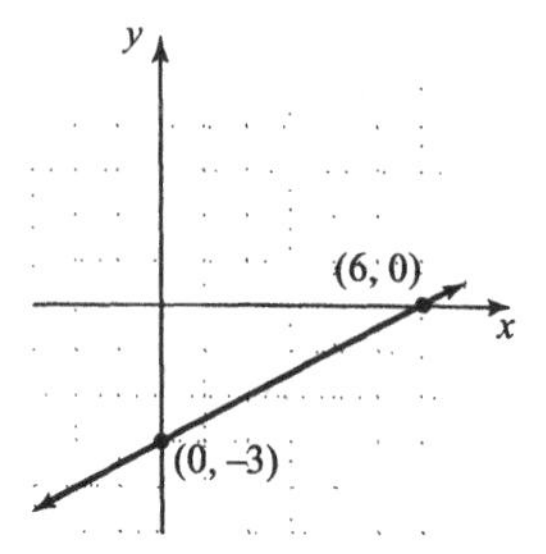

67. $\frac{4}{a}\cdot\frac{1}{b}=\frac{4\cdot 1}{a\cdot b}=\frac{4}{ab}$
The statement is true.

69. $\frac{x}{5}\cdot\frac{x+3}{4}=\frac{x\cdot(x+3)}{5\cdot 4}=\frac{x^2+3x}{20}$
The statement is false.

71. Area = length · width
$\frac{x+5}{9x}\cdot\frac{2x}{x^2-25}=\frac{(x+5)\cdot 2\cdot x}{9\cdot x\cdot(x+5)(x-5)}=\frac{2}{9(x-5)}$
The area of the rectangle is $\frac{2}{9(x-5)}$ square feet.

73. $\left(\frac{x^2-y^2}{x^2+y^2} \div \frac{x^2-y^2}{3x}\right)\cdot\frac{x^2+y^2}{6}$
$= \frac{x^2-y^2}{x^2+y^2}\cdot\frac{3x}{x^2-y^2}\cdot\frac{x^2+y^2}{6}$
$= \frac{(x^2-y^2)\cdot 3x\cdot(x^2+y^2)}{(x^2+y^2)\cdot(x^2-y^2)\cdot 2\cdot 3}$
$= \frac{x}{2}$

Aguirre

Aguirre

75. $\left(\dfrac{2a+b}{b^2}\cdot\dfrac{3a^2-2ab}{ab+2b^2}\right)\div\dfrac{a^2-3ab+2b^2}{5ab-10b^2}$

$=\dfrac{2a+b}{b^2}\cdot\dfrac{3a^2-2ab}{ab+2b^2}\cdot\dfrac{5ab-10b^2}{a^2-3ab+2b^2}$

$=\dfrac{(2a+b)\cdot(3a^2-2ab)\cdot(5ab-10b^2)}{b^2\cdot(ab+2b^2)\cdot(a^2-3ab+2b^2)}$

$=\dfrac{(2a+b)\cdot a(3a-2b)\cdot 5b(a-2b)}{b^2\cdot b(a+2b)\cdot(a-2b)(a-b)}$

$=\dfrac{5a(2a+b)(3a-2b)}{b^2(a+2b)(a-b)}$

77. Answers may vary

Section 7.3

Practice Exercises

1. $\dfrac{7a}{4b}+\dfrac{a}{4b}=\dfrac{7a+a}{4b}=\dfrac{8a}{4b}=\dfrac{2a}{b}$

2. $\dfrac{3x}{3x-2}-\dfrac{2}{3x-2}=\dfrac{3x-2}{3x-2}=\dfrac{1}{1}$ or 1

3. $\dfrac{4x^2+15x}{x+3}-\dfrac{8x+15}{x+3}=\dfrac{(4x^2+15x)-(8x+15)}{x+3}$

$=\dfrac{4x^2+15x-8x-15}{x+3}$

$=\dfrac{4x^2+7x-15}{x+3}$

$=\dfrac{(x+3)(4x-5)}{x+3}$

$=4x-5$

4. a. Find the prime factorization of each denominator.

$14=2\cdot 7$

$21=3\cdot 7$

The greatest number of times that the factor 2 appears is 1. The greatest number of times that the factor 3 appears is 1. The greatest number of times that the factor 7 appears is 1.

$\text{LCD}=2^1\cdot 3^1\cdot 7^1=42$

b. Factor each denominator.

$9y=3\cdot 3\cdot y=3^2\cdot y$

$15y^3=3\cdot 5\cdot y^3$

The greatest number of times that the factor 3 appears is 2. The greatest number of times that the factor 5 appears is 1. The greatest number of times that the factor y appears is 3.

$\text{LCD}=3^2\cdot 5^1\cdot y^3=9\cdot 5\cdot y^3=45y^3$

5. a. The denominators $y-5$ and $y-4$ are completely factored already. The factor $y-5$ appears once and the factor $y-4$ appears once.

$\text{LCD}=(y-5)(y-4)$

b. The denominators a and $a+2$ cannot be factored further. The factor a appears once and the factor $a+2$ appears once.

$\text{LCD}=a(a+2)$

6. Factor each denominator.

$(2x-1)^2=(2x-1)^2$

$6x-3=3(2x-1)$

The greatest number that the factor $2x-1$ appears in any one denominator is 2.

The greatest number of times that the factor 3 appears is 1.

$\text{LCD}=3(2x-1)^2$

7. Factor each denominator.

$x^2+5x+4=(x+1)(x+4)$

$x^2-16=(x-4)(x+4)$

$\text{LCD}=(x+1)(x+4)(x-4)$

8. The denominators $3-x$ and $x-3$ are opposites. That is, $3-x=-1(x-3)$. Use $x-3$ or $3-x$ as the LCD.

$\text{LCD}=x-3$ or $\text{LCD}=3-x$

9. a. Since $5y(7xy)=35xy^2$, multiply by 1 in the form of $\dfrac{7xy}{7xy}$.

$\dfrac{3x}{5y}=\dfrac{3x}{5y}\cdot 1=\dfrac{3x}{5y}\cdot\dfrac{7xy}{7xy}=\dfrac{3x(7xy)}{5y(7xy)}=\dfrac{21x^2y}{35xy^2}$

b. First, factor the denominator on the right.

$\dfrac{9x}{4x+7}=\dfrac{}{2(4x+7)}$

To obtain the denominator on the right from the denominator on the left, multiply by 1 in the form of $\dfrac{2}{2}$.

$$\frac{9x}{4x+7} = \frac{9x}{4x+7}\cdot\frac{2}{2}$$
$$= \frac{9x\cdot 2}{(4x+7)\cdot 2}$$
$$= \frac{18x}{2(4x+7)} \text{ or } \frac{18x}{8x+14}$$

10. First, factor the denominator $x^2 - 2x - 15$ as $(x + 3)(x - 5)$. If we multiply the original denominator $(x + 3)(x - 5)$ by $x - 2$, the result is the new denominator $(x - 2)(x + 3)(x - 5)$. Thus, we multiply by 1 in the form $\frac{x-2}{x-2}$.

$$\frac{3}{x^2-2x-15} = \frac{3}{(x+3)(x-5)}$$
$$= \frac{3}{(x+3)(x-5)}\cdot\frac{x-2}{x-2}$$
$$= \frac{3(x-2)}{(x+3)(x-5)(x-2)}$$
$$= \frac{3x-6}{(x-2)(x+30(x-5)}$$

Vocabulary and Readiness Check

1. $\frac{7}{11}+\frac{2}{11}=\underline{\frac{9}{11}}$

2. $\frac{7}{11}-\frac{2}{11}=\underline{\frac{5}{11}}$

3. $\frac{a}{b}+\frac{c}{b}=\underline{\frac{a+c}{b}}$

4. $\frac{a}{b}-\frac{c}{b}=\underline{\frac{a-c}{b}}$

5. $\frac{5}{x}-\frac{6+x}{x}=\underline{\frac{5-(6+x)}{x}}$

Exercise Set 7.3

1. $\frac{a+1}{13}+\frac{8}{13}=\frac{a+1+8}{13}=\frac{a+9}{13}$

3. $\frac{4m}{3n}+\frac{5m}{3n}=\frac{4m+5m}{3n}=\frac{9m}{3n}=\frac{3m}{n}$

5. $\frac{4m}{m-6}-\frac{24}{m-6}=\frac{4m-24}{m-6}=\frac{4(m-6)}{m-6}=4$

7. $\frac{9}{3+y}+\frac{y+1}{3+y}=\frac{9+y+1}{3+y}=\frac{10+y}{3+y}$

9.
$$\frac{5x^2+4x}{x-1}-\frac{6x+3}{x-1}=\frac{5x^2+4x-(6x+3)}{x-1}$$
$$=\frac{5x^2+4x-6x-3}{x-1}$$
$$=\frac{5x^2-2x-3}{x-1}$$
$$=\frac{(5x+3)(x-1)}{x-1}$$
$$=5x+3$$

11.
$$\frac{4a}{a^2+2a-15}-\frac{12}{a^2+2a-15}=\frac{4a-12}{a^2+2a-15}$$
$$=\frac{4(a-3)}{(a+5)(a-3)}$$
$$=\frac{4}{a+5}$$

13.
$$\frac{2x+3}{x^2-x-30}-\frac{x-2}{x^2-x-30}=\frac{2x+3-(x-2)}{x^2-x-30}$$
$$=\frac{2x+3-x+2}{x^2-x-30}$$
$$=\frac{x+5}{x^2-x-30}$$
$$=\frac{x+5}{(x-6)(x+5)}$$
$$=\frac{1}{x-6}$$

15. $\frac{2x+1}{x-3}+\frac{3x+6}{x-3}=\frac{2x+1+3x+6}{x-3}=\frac{5x+7}{x-3}$

17.
$$\frac{2x^2}{x-5}-\frac{25+x^2}{x-5}=\frac{2x^2-(25+x^2)}{x-5}$$
$$=\frac{2x^2-25-x^2}{x-5}$$
$$=\frac{x^2-25}{x-5}$$
$$=\frac{(x+5)(x-5)}{x-5}$$
$$=x+5$$

19. $\frac{5x+4}{x-1} - \frac{2x+7}{x-1} = \frac{5x+4-(2x+7)}{x-1}$
$= \frac{5x+4-2x-7}{x-1}$
$= \frac{3x-3}{x-1}$
$= \frac{3(x-1)}{x-1}$
$= 3$

21. $2x = 2 \cdot x$
$4x^3 = 2^2 \cdot x^3$
$\text{LCD} = 2^2 \cdot x^3 = 4x^3$

23. $8x = 2^3 \cdot x$
$2x + 4 = 2(x + 2)$
$\text{LCD} = 2^3 \cdot x \cdot (x+2) = 8x(x+2)$

25. $x + 3 = x + 3$
$x - 2 = x - 2$
$\text{LCD} = (x + 3)(x - 2)$

27. $x + 6 = x + 6$
$3x + 18 = 3(x + 6)$
$\text{LCD} = 3(x + 6)$

29. $(x-6)^2 = (x-6)^2$
$5x - 30 = 5(x - 6)$
$\text{LCD} = 5(x-6)^2$

31. $3x + 3 = 3 \cdot (x + 1)$
$2x^2 + 4x + 2 = 2(x^2 + 2x + 1) = 2 \cdot (x+1)^2$
$\text{LCD} = 2 \cdot 3(x+1)^2 = 6(x+1)^2$

33. $x - 8 = x - 8$
$8 - x = -(x - 8)$
$\text{LCD} = x - 8 \text{ or } 8 - x$

35. $x^2 + 3x - 4 = (x-1)(x+4)$
$x^2 + 2x - 3 = (x-1)(x+3)$
$\text{LCD} = (x - 1)(x + 4)(x + 3)$

37. $3x^2 + 4x + 1 = (3x+1)(x+1)$
$2x^2 - x - 1 = (x-1)(2x+1)$
$\text{LCD} = (3x + 1)(x + 1)(x - 1)(2x + 1)$

39. $x^2 - 16 = (x+4)(x-4)$
$2x^3 - 8x^2 = 2x^2(x-4)$
$\text{LCD} = 2x^2(x+4)(x-4)$

41. $\frac{3}{2x} = \frac{3(2x)}{2x(2x)} = \frac{6x}{4x^2}$

43. $\frac{6}{3a} = \frac{6(4b^2)}{3a(4b^2)} = \frac{24b^2}{12ab^2}$

45. $\frac{9}{2x+6} = \frac{9}{2(x+3)} = \frac{9(y)}{2(x+3)(y)} = \frac{9y}{2y(x+3)}$

47. $\frac{9a+2}{5a+10} = \frac{9a+2}{5(a+2)} = \frac{(9a+2)(b)}{5(a+2)(b)} = \frac{9ab+2b}{5b(a+2)}$

49. $\frac{x}{x^3+6x^2+8x} = \frac{x}{x(x+4)(x+2)}$
$= \frac{x(x+1)}{x(x+4)(x+2)(x+1)}$
$= \frac{x^2+x}{x(x+4)(x+2)(x+1)}$

51. $\frac{9y-1}{15x^2-30} = \frac{(9y-1)(2)}{(15x^2-30)2} = \frac{18y-2}{30x^2-60}$

53. $\frac{5x}{7} + \frac{9x}{7} = \frac{5x+9x}{7} = \frac{14x}{7} = \frac{2x}{1} = 2x$

55. $\frac{x+3}{4} \div \frac{2x-1}{4} = \frac{x+3}{4} \cdot \frac{4}{2x-1}$
$= \frac{(x+3) \cdot 4}{4 \cdot (2x-1)}$
$= \frac{x+3}{2x-1}$

57. $\frac{x^2}{x-6} - \frac{5x+6}{x-6} = \frac{x^2-(5x+6)}{x-6}$
$= \frac{x^2-5x-6}{x-6}$
$= \frac{(x+1)(x-6)}{x-6}$
$= x+1$

59. $\frac{-2x}{x^3-8x}+\frac{3x}{x^3-8x}=\frac{-2x+3x}{x^3-8x}$
$=\frac{x}{x(x^2-8)}$
$=\frac{1}{x^2-8}$

61. $\frac{12x-6}{x^2+3x}\cdot\frac{4x^2+13x+3}{4x^2-1}$
$=\frac{6(2x-1)\cdot(x+3)(4x+1)}{x(x+3)\cdot(2x+1)(2x-1)}$
$=\frac{6(4x+1)}{x(2x+1)}$

63. LCD = 21
$\frac{2}{3}+\frac{5}{7}=\frac{2(7)}{3(7)}+\frac{5(3)}{7(3)}=\frac{14}{21}+\frac{15}{21}=\frac{29}{21}$

65. $6=2\cdot 3$
$4=2^2$
$\text{LCD}=2^2\cdot 3=12$
$\frac{2}{6}-\frac{3}{4}=\frac{2(2)}{6(2)}-\frac{3(3)}{4(3)}=\frac{4}{12}-\frac{9}{12}=\frac{4-9}{12}=-\frac{5}{12}$

67. $12=2\cdot 2\cdot 3=2^2\cdot 3$
$20=2\cdot 2\cdot 5=2^2\cdot 5$
$\text{LCD}=2^2\cdot 3\cdot 5=60$
$\frac{1}{12}+\frac{3}{20}=\frac{1(5)}{12(5)}+\frac{3(3)}{20(3)}=\frac{5}{60}+\frac{9}{60}=\frac{14}{60}=\frac{7}{30}$

69. $4a-20=4(a-5)$
$(a-5)^2=(a-5)^2$
$\text{LCD}=4(a-5)^2$
The correct choice is d.

71. $\frac{3}{x}+\frac{y}{x}=\frac{3+y}{x}$
The correct choice is c.

73. $\frac{3}{x}\cdot\frac{y}{x}=\frac{3\cdot y}{x\cdot x}=\frac{3y}{x^2}$
The correct choice is b.

75. $\frac{5}{2-x}=\frac{5(-1)}{(2-x)(-1)}=-\frac{5}{x-2}$

77. $-\frac{7+x}{2-x}=\frac{7+x}{(-1)(2-x)}=\frac{7+x}{x-2}$

79. $P=\frac{5}{x-2}+\frac{5}{x-2}+\frac{5}{x-2}+\frac{5}{x-2}$
$=\frac{5+5+5+5}{x-2}$
$=\frac{20}{x-2}$
The perimeter is $\frac{20}{x-2}$ meters.

81. Answers may vary

83. $88=2^3\cdot 11$
$4332=2^3\cdot 3\cdot 19^2$
$\text{LCM}=2^3\cdot 3\cdot 11\cdot 19^2=95{,}304$
They will align again in 95,304 Earth days.

85. Answers may vary

87. Answers may vary

Section 7.4

Practice Exercises

1. a. Since 5 = 5 and 15 = 3 · 5, the LCD = 3 · 5 = 15.
$\frac{2x}{5}-\frac{6x}{15}=\frac{2x(3)}{5(3)}-\frac{6x}{15}$
$=\frac{6x}{15}-\frac{6x}{15}$
$=\frac{6x-6x}{15}$
$=\frac{0}{15}$
$=0$

b. Since $8a=2^3\cdot a$ and $12a^2=2^2\cdot 3\cdot a^2$, the $\text{LCD}=2^3\cdot 3\cdot a^2=24a^2$.
$\frac{7}{8a}+\frac{5}{12a^2}=\frac{7(3a)}{8a(3a)}+\frac{5(2)}{12a^2(2)}$
$=\frac{21a}{24a^2}+\frac{10}{24a^2}$
$=\frac{21a+10}{24a^2}$

2. Since $x^2-25=(x+5)(x-5)$, the LCD = $(x + 5)(x - 5)$.

$$\begin{aligned}\frac{12x}{x^2-25}-\frac{6}{x+5}&=\frac{12x}{(x+5)(x-5)}-\frac{6(x-5)}{(x+5)(x-5)}\\&=\frac{12x-6(x-5)}{(x+5)(x-5)}\\&=\frac{12x-6x+30}{(x+5)(x-5)}\\&=\frac{6x+30}{(x+5)(x-5)}\\&=\frac{6(x+5)}{(x+5)(x-5)}\\&=\frac{6}{x-5}\end{aligned}$$

3. The LCD is $5y(y+1)$.

$$\begin{aligned}\frac{3}{5y}+\frac{2}{y+1}&=\frac{3(y+1)}{5y(y+1)}+\frac{2(5y)}{(y+1)(5y)}\\&=\frac{3(y+1)+2(5y)}{5y(y+1)}\\&=\frac{3y+3+10y}{5y(y+1)}\\&=\frac{13y+3}{5y(y+1)}\end{aligned}$$

4. $x-5$ and $5-x$ are opposites. Write the denominator $5-x$ as $-(x-5)$ and simplify.

$$\begin{aligned}\frac{6}{x-5}-\frac{7}{5-x}&=\frac{6}{x-5}-\frac{7}{-(x-5)}\\&=\frac{6}{x-5}-\frac{-7}{x-5}\\&=\frac{6-(-7)}{x-5}\\&=\frac{13}{x-5}\end{aligned}$$

5. Note that 2 is the same as $\frac{2}{1}$. The LCD of $\frac{2}{1}$ and $\frac{b}{b+3}$ is $b+3$.

$$\begin{aligned}2+\frac{b}{b+3}&=\frac{2}{1}+\frac{b}{b+3}\\&=\frac{2(b+3)}{1(b+3)}+\frac{b}{b+3}\\&=\frac{2(b+3)+b}{b+3}\\&=\frac{2b+6+b}{b+3}\\&=\frac{3b+6}{b+3}\text{ or }\frac{3(b+2)}{b+3}\end{aligned}$$

6. First, factor the denominators.

$$\frac{5}{2x^2+3x}-\frac{3x}{4x+6}=\frac{5}{x(2x+3)}-\frac{3x}{2(2x+3)}$$

The LCD is $2x(2x+3)$.

$$\begin{aligned}\frac{5}{2x^2+3x}-\frac{3x}{4x+6}&=\frac{5(2)}{x(2x+3)(2)}-\frac{3x(x)}{2(2x+3)(x)}\\&=\frac{10-3x^2}{2x(2x+3)}\end{aligned}$$

7. First, factor the denominators.

$x^2+7x+12=(x+4)(x+3)$

$x^2-9=(x+3)(x-3)$

LCD = $(x+4)(x+3)(x-3)$

$$\begin{aligned}&\frac{2x}{x^2+7x+12}+\frac{3x}{x^2-9}\\&=\frac{2x}{(x+4)(x+3)}+\frac{3x}{(x+3)(x-3)}\\&=\frac{2x(x-3)}{(x+4)(x+3)(x-3)}+\frac{3x(x+4)}{(x+3)(x-3)(x+4)}\\&=\frac{2x(x-3)+3x(x+4)}{(x+4)(x+3)(x-3)}\\&=\frac{2x^2-6x+3x^2+12x}{(x+4)(x+3)(x-3)}\\&=\frac{5x^2+6x}{(x+4)(x+3)(x-3)}\text{ or }\frac{x(5x+6)}{(x+4)(x+3)(x-3)}\end{aligned}$$

Vocabulary and Readiness Check

1. The first step to perform on $\frac{3}{4}-\frac{y}{4}$ is to subtract the numerators and place the difference over the common denominator; choice d.

2. The first step to perform on $\frac{2}{a}\cdot\frac{3}{a+6}$ is to multiply the numerators and multiply the denominators; choice c.

3. The first step to perform on $\frac{x+1}{x}\div\frac{x-1}{x}$ is to multiply the first rational expression by the reciprocal of the second rational expression; choice a.

4. The first step to perform on $\frac{9}{x-2}-\frac{x}{x+2}$ is to find the LCD and write each expression as an equivalent expression with the LCD as denominator; choice b.

Exercise Set 7.4

1. LCD $= 2\cdot 3\cdot x = 6x$

$$\frac{4}{2x}+\frac{9}{3x}=\frac{4(3)}{2x(3)}+\frac{9(2)}{3x(2)}=\frac{12}{6x}+\frac{18}{6x}=\frac{30}{6x}=\frac{5(6)}{6x}=\frac{5}{x}$$

3. LCD $= 5b$

$$\frac{15a}{b}-\frac{6b}{5}=\frac{15a(5)}{b(5)}-\frac{6b(b)}{5(b)}=\frac{75a}{5b}-\frac{6b^2}{5b}=\frac{75a-6b^2}{5b}$$

5. LCD $= 2x^2$

$$\frac{3}{x}+\frac{5}{2x^2}=\frac{3(2x)}{x(2x)}+\frac{5}{2x^2}=\frac{6x}{2x^2}+\frac{5}{2x^2}=\frac{6x+5}{2x^2}$$

7. $2x+2=2(x+1)$
LCD $= 2(x+1)$

$$\frac{6}{x+1}+\frac{10}{2x+2}=\frac{6}{x+1}+\frac{10}{2(x+1)}=\frac{6(2)}{(x+1)2}+\frac{10}{2(x+1)}=\frac{12}{2(x+1)}+\frac{10}{2(x+1)}=\frac{12+10}{2(x+1)}=\frac{22}{2(x+1)}=\frac{2(11)}{2(x+1)}=\frac{11}{x+1}$$

9. $x^2-4=(x+2)(x-2)$
LCD $= (x+2)(x-2)$

$$\frac{3}{x+2}-\frac{2x}{x^2-4}=\frac{3(x-2)}{(x+2)(x-2)}-\frac{2x}{(x+2)(x-2)}=\frac{3(x-2)-2x}{(x+2)(x-2)}=\frac{3x-6-2x}{(x+2)(x-2)}=\frac{x-6}{(x+2)(x-2)}$$

11. LCD $= 4x(x-2)$

$$\frac{3}{4x}+\frac{8}{x-2}=\frac{3(x-2)}{4x(x-2)}+\frac{8(4x)}{(x-2)(4x)}=\frac{3x-6}{4x(x-2)}+\frac{32x}{4x(x-2)}=\frac{3x-6+32x}{4x(x-2)}=\frac{35x-6}{4x(x-2)}$$

13. $3-x=-(x-3)$

$$\frac{6}{x-3}+\frac{8}{3-x}=\frac{6}{x-3}+\frac{8}{-(x-3)}=\frac{6}{x-3}+\frac{-8}{x-3}=\frac{6+(-8)}{x-3}=-\frac{2}{x-3}$$

15. $3 - x = -(x - 3)$

$$\begin{aligned}\frac{9}{x-3}+\frac{9}{3-x} &= \frac{9}{x-3}+\frac{9}{-(x-3)}\\ &= \frac{9}{x-3}+\frac{-9}{x-3}\\ &= \frac{9+(-9)}{x-3}\\ &= \frac{0}{x-3}\\ &= 0\end{aligned}$$

17. $1-x^2 = -(x^2-1)$

$$\begin{aligned}\frac{-8}{x^2-1}-\frac{7}{1-x^2} &= \frac{8}{-(x^2-1)}-\frac{7}{1-x^2}\\ &= \frac{8}{1-x^2}-\frac{7}{1-x^2}\\ &= \frac{8-7}{1-x^2}\\ &= \frac{1}{1-x^2} \text{ or } -\frac{1}{x^2-1}\end{aligned}$$

19. LCD $= x$

$$\frac{5}{x}+2=\frac{5}{x}+\frac{2}{1}=\frac{5}{x}+\frac{2(x)}{1(x)}=\frac{5+2x}{x}$$

21. LCD $= x - 2$

$$\begin{aligned}\frac{5}{x-2}+6 &= \frac{5}{x-2}+\frac{6}{1}\\ &= \frac{5}{x-2}+\frac{6(x-2)}{1(x-2)}\\ &= \frac{5}{x-2}+\frac{6x-12}{x-2}\\ &= \frac{5+6x-12}{x-2}\\ &= \frac{6x-7}{x-2}\end{aligned}$$

23. LCD $= y + 3$

$$\begin{aligned}\frac{y+2}{y+3}-2 &= \frac{y+2}{y+3}-\frac{2}{1}\\ &= \frac{y+2}{y+3}-\frac{2(y+3)}{y+3}\\ &= \frac{y+2}{y+3}-\frac{2y+6}{y+3}\\ &= \frac{y+2-(2y+6)}{y+3}\\ &= \frac{y+2-2y-6}{y+3}\\ &= \frac{-y-4}{y+3}\\ &= \frac{-(y+4)}{y+3}\\ &= -\frac{y+4}{y+3}\end{aligned}$$

25. LCD $= 4x$

$$\begin{aligned}\frac{-x+2}{x}-\frac{x-6}{4x} &= \frac{(-x+2)(4)}{x(4)}-\frac{x-6}{4x}\\ &= \frac{4(-x+2)-(x-6)}{4x}\\ &= \frac{-4x+8-x+6}{4x}\\ &= \frac{-5x+14}{4x} \text{ or } -\frac{5x-14}{4x}\end{aligned}$$

27.
$$\begin{aligned}\frac{5x}{x+2}-\frac{3x-4}{x+2} &= \frac{5x-(3x-4)}{x+2}\\ &= \frac{5x-3x+4}{x+2}\\ &= \frac{2x+4}{x+2}\\ &= \frac{2(x+2)}{x+2}\\ &= 2\end{aligned}$$

29. LCD $= 21$

$$\begin{aligned}\frac{3x^4}{7}-\frac{4x^2}{21} &= \frac{3x^4(3)}{7(3)}-\frac{4x^2}{21}\\ &= \frac{3(3x^4)-4x^2}{21}\\ &= \frac{9x^4-4x^2}{21}\end{aligned}$$

31. LCD = $(x+3)^2$

$$\frac{1}{x+3}-\frac{1}{(x+3)^2}=\frac{1(x+3)}{(x+3)(x+3)}-\frac{1}{(x+3)^2}$$
$$=\frac{x+3}{(x+3)^2}-\frac{1}{(x+3)^2}$$
$$=\frac{x+3-1}{(x+3)^2}$$
$$=\frac{x+2}{(x+3)^2}$$

33. LCD = $5b(b-1)$

$$\frac{4}{5b}+\frac{1}{b-1}=\frac{4(b-1)}{5b(b-1)}+\frac{1(5b)}{(b-1)(5b)}$$
$$=\frac{4b-4}{5b(b-1)}+\frac{5b}{5b(b-1)}$$
$$=\frac{4b-4+5b}{5b(b-1)}$$
$$=\frac{9b-4}{5b(b-1)}$$

35. LCD = m

$$\frac{2}{m}+1=\frac{2}{m}+\frac{1}{1}=\frac{2}{m}+\frac{1(m)}{1(m)}=\frac{2+m}{m}$$

37. LCD = $(x-7)(x-2)$

$$\frac{2x}{x-7}-\frac{x}{x-2}=\frac{2x(x-2)}{(x-7)(x-2)}-\frac{x(x-7)}{(x-2)(x-7)}$$
$$=\frac{2x(x-2)-x(x-7)}{(x-7)(x-2)}$$
$$=\frac{2x^2-4x-x^2+7x}{(x-7)(x-2)}$$
$$=\frac{x^2+3x}{(x-7)(x-2)} \text{ or } \frac{x(x+3)}{(x-7)(x-2)}$$

39. $2x-1=-(1-2x)$

$$\frac{6}{1-2x}-\frac{4}{2x-1}=\frac{6}{1-2x}-\frac{4}{-(1-2x)}$$
$$=\frac{6}{1-2x}-\frac{-4}{1-2x}$$
$$=\frac{6-(-4)}{1-2x}$$
$$=\frac{10}{1-2x}$$

41. LCD = $(x-1)(x+1)^2$

$$\frac{7}{(x+1)(x-1)}+\frac{8}{(x+1)^2}$$
$$=\frac{7(x+1)}{(x+1)(x-1)(x+1)}+\frac{8(x-1)}{(x+1)^2(x-1)}$$
$$=\frac{7x+7}{(x+1)^2(x-1)}+\frac{8x-8}{(x+1)^2(x-1)}$$
$$=\frac{7x+7+8x-8}{(x+1)^2(x-1)}$$
$$=\frac{15x-1}{(x+1)^2(x-1)}$$

43. $x^2-1=(x+1)(x-1)$

$x^2-2x+1=(x-1)^2$

LCD = $(x+1)(x-1)^2$

$$\frac{x}{x^2-1}-\frac{2}{x^2-2x+1}$$
$$=\frac{x(x-1)}{(x-1)(x+1)(x-1)}-\frac{2(x+1)}{(x-1)^2(x+1)}$$
$$=\frac{x^2-x}{(x-1)^2(x+1)}-\frac{2x+2}{(x-1)^2(x+1)}$$
$$=\frac{x^2-x-(2x+2)}{(x-1)^2(x+1)}$$
$$=\frac{x^2-x-2x-2}{(x-1)^2(x+1)}$$
$$=\frac{x^2-3x-2}{(x-1)^2(x+1)}$$

45. $2a+6=2(a+3)$

LCD = $2(a+3)$

$$\frac{3a}{2a+6}-\frac{a-1}{a+3}=\frac{3a}{2(a+3)}-\frac{(a-1)(2)}{(a+3)(2)}$$
$$=\frac{3a}{2(a+3)}-\frac{2a-2}{2(a+3)}$$
$$=\frac{3a-(2a-2)}{2(a+3)}$$
$$=\frac{3a-2a+2}{2(a+3)}$$
$$=\frac{a+2}{2(a+3)}$$

47. $\text{LCD} = (2y+3)^2$

$$\frac{y-1}{2y+3}+\frac{3}{(2y+3)^2} = \frac{(y-1)(2y+3)}{(2y+3)(2y+3)}+\frac{3}{(2y+3)^2}$$
$$= \frac{(y-1)(2y+3)+3}{(2y+3)^2}$$
$$= \frac{2y^2+y-3+3}{(2y+3)^2}$$
$$= \frac{2y^2+y}{(2y+3)^2} \text{ or } \frac{y(2y+1)}{(2y+3)^2}$$

49. $2-x = -(x-2)$
$2x-4 = 2(x-2)$
$\text{LCD} = 2(x-2)$

$$\frac{5}{2-x}+\frac{x}{2x-4} = \frac{5}{-(x-2)}+\frac{x}{2(x-2)}$$
$$= \frac{-5}{x-2}+\frac{x}{2(x-2)}$$
$$= \frac{-5(2)}{(x-2)(2)}+\frac{x}{2(x-2)}$$
$$= \frac{-10}{2(x-2)}+\frac{x}{2(x-2)}$$
$$= \frac{x-10}{2(x-2)}$$

51. $x^2+6x+9 = (x+3)^2$
$\text{LCD} = (x+3)^2$

$$\frac{15}{x^2+6x+9}+\frac{2}{x+3} = \frac{15}{(x+3)^2}+\frac{2(x+3)}{(x+3)(x+3)}$$
$$= \frac{15+2(x+3)}{(x+3)^2}$$
$$= \frac{15+2x+6}{(x+3)^2}$$
$$= \frac{2x+21}{(x+3)^2}$$

53. $x^2-5x-6 = (x-3)(x-2)$
$\text{LCD} = (x-3)(x-2)$

$$\frac{13}{x^2-5x+6}-\frac{5}{x-3}$$
$$= \frac{13}{(x-3)(x-2)}-\frac{5(x-2)}{(x-3)(x-2)}$$
$$= \frac{13-(5x-10)}{(x-3)(x-2)}$$
$$= \frac{13-5x+10}{(x-3)(x-2)}$$
$$= \frac{-5x+23}{(x-3)(x-2)}$$

55. $m^2-100 = (m+10)(m-10)$
$\text{LCD} = 2(m+10)(m-10)$

$$\frac{70}{m^2-100}+\frac{7}{2(m+10)}$$
$$= \frac{70(2)}{(m+10)(m-10)(2)}+\frac{7(m-10)}{2(m+10)(m-10)}$$
$$= \frac{70(2)+7(m-10)}{2(m+10)(m-10)}$$
$$= \frac{140+7m-70}{2(m+10)(m-10)}$$
$$= \frac{7m+70}{2(m+10)(m-10)}$$
$$= \frac{7(m+10)}{2(m+10)(m-10)}$$
$$= \frac{7}{2(m-10)}$$

57. $x^2-5x-6 = (x-6)(x+1)$
$x^2-4x-5 = (x-5)(x+1)$
$\text{LCD} = (x-6)(x+1)(x-5)$

$$\frac{x+8}{x^2-5x-6}+\frac{x+1}{x^2-4x-5}$$
$$= \frac{(x+8)(x-5)}{(x-6)(x+1)(x-5)}+\frac{(x+1)(x-6)}{(x-5)(x+1)(x-6)}$$
$$= \frac{x^2+3x-40+x^2-5x-6}{(x-6)(x+1)(x-5)}$$
$$= \frac{2x^2-2x-46}{(x-6)(x+1)(x-5)}$$
$$\text{or } \frac{2(x^2-x-23)}{(x-6)(x+1)(x-5)}$$

59. $4n^2 - 12n + 8 = 4(n-1)(n-2)$

$3n^2 - 6n = 3n(n-2)$

LCD $= 4 \cdot 3n(n-1)(n-2) = 12n(n-1)(n-2)$

$$\begin{aligned}
&\frac{5}{4n^2-12n+8} - \frac{3}{3n^2-6n} \\
&= \frac{5(3n)}{4(n-1)(n-2)(3n)} - \frac{3(4)(n-1)}{3n(n-2)(4)(n-1)} \\
&= \frac{5(3n)-3(4)(n-1)}{12n(n-1)(n-2)} \\
&= \frac{15n-12n+12}{12n(n-1)(n-2)} \\
&= \frac{3n+12}{12n(n-1)(n-2)} \\
&= \frac{3(n+4)}{12n(n-1)(n-2)} \\
&= \frac{n+4}{4n(n-1)(n-2)}
\end{aligned}$$

61. $$\begin{aligned}
\frac{15x}{x+8} \cdot \frac{2x+16}{3x} &= \frac{15x}{x+8} \cdot \frac{2(x+8)}{3x} \\
&= \frac{2 \cdot 5 \cdot 3x \cdot (x+8)}{3x \cdot (x+8)} \\
&= 10
\end{aligned}$$

63. $$\begin{aligned}
\frac{8x+7}{3x+5} - \frac{2x-3}{3x+5} &= \frac{8x+7-(2x-3)}{3x+5} \\
&= \frac{8x+7-2x+3}{3x+5} \\
&= \frac{6x+10}{3x+5} \\
&= \frac{2(3x+5)}{3x+5} \\
&= 2
\end{aligned}$$

65. $$\begin{aligned}
\frac{5a+10}{18} \div \frac{a^2-4}{10a} &= \frac{5a+10}{18} \cdot \frac{10a}{a^2-4} \\
&= \frac{5(a+2) \cdot 2 \cdot 5a}{2 \cdot 9 \cdot (a-2)(a+2)} \\
&= \frac{25a}{9(a-2)}
\end{aligned}$$

67. $x^2 - 3x + 2 = (x-2)(x-1)$

LCD $= (x-2)(x-1)$

$$\begin{aligned}
&\frac{5}{x^2-3x+2} + \frac{1}{x-2} \\
&= \frac{5}{(x-2)(x-1)} + \frac{1}{(x-2)} \cdot \frac{(x-1)}{(x-1)} \\
&= \frac{5}{(x-2)(x-1)} + \frac{x-1}{(x-2)(x-1)} \\
&= \frac{5+x-1}{(x-2)(x-1)} \\
&= \frac{x+4}{(x-2)(x-1)}
\end{aligned}$$

69. $$\begin{aligned}
3x+5 &= 7 \\
3x+5-5 &= 7-5 \\
3x &= 2 \\
\frac{3x}{3} &= \frac{2}{3} \\
x &= \frac{2}{3}
\end{aligned}$$

71. $$\begin{aligned}
2x^2 - x - 1 &= 0 \\
(2x+1)(x-1) &= 0
\end{aligned}$$

$2x+1 = 0$ or $x-1=0$

$2x = -1$ $\quad x = 1$

$x = -\frac{1}{2}$

The solutions are $x = -\frac{1}{2}$ and $x = 1$.

73. $$\begin{aligned}
4(x+6)+3 &= -3 \\
4x+24+3 &= -3 \\
4x+27 &= -3 \\
4x &= -30 \\
x &= \frac{-30}{4} = -\frac{15}{2}
\end{aligned}$$

75. $x^2 - 1 = (x+1)(x-1)$

LCD $= x(x+1)(x-1)$

$$\frac{3}{x} - \frac{2x}{x^2-1} + \frac{5}{x+1} = \frac{3(x+1)(x-1)}{x(x+1)(x-1)} - \frac{2x(x)}{(x+1)(x-1)(x)} + \frac{5(x)(x-1)}{(x+1)(x)(x-1)}$$

$$= \frac{3(x+1)(x-1) - 2x(x) + 5x(x-1)}{x(x+1)(x-1)}$$

$$= \frac{3x^2 - 3 - 2x^2 + 5x^2 - 5x}{x(x+1)(x-1)}$$

$$= \frac{6x^2 - 5x - 3}{x(x+1)(x-1)}$$

77. $x^2 - 4 = (x+2)(x-2)$

$x^2 - 4x + 4 = (x-2)^2$

$x^2 - x - 6 = (x-3)(x+2)$

LCD $= (x+2)(x-2)^2(x-3)$

$$\frac{5}{x^2-4} + \frac{2}{x^2-4x+4} - \frac{3}{x^2-x-6} = \frac{5(x-2)(x-3)}{(x-2)(x+2)(x-2)(x-3)} + \frac{2(x+2)(x-3)}{(x-2)^2(x+2)(x-3)} - \frac{3(x-2)^2}{(x-3)(x+2)(x-2)^2}$$

$$= \frac{5(x^2-5x+6)}{(x-2)^2(x+2)(x-3)} + \frac{2(x^2-x-6)}{(x-2)^2(x+2)(x-3)} - \frac{3(x^2-4x+4)}{(x-2)^2(x+2)(x-3)}$$

$$= \frac{5x^2-25x+30}{(x-2)^2(x+2)(x-3)} + \frac{2x^2-2x-12}{(x-2)^2(x+2)(x-3)} - \frac{3x^2-12x+12}{(x-2)^2(x+2)(x-3)}$$

$$= \frac{5x^2-25x+30+2x^2-2x-12-3x^2+12x-12}{(x-2)^2(x+2)(x-3)}$$

$$= \frac{4x^2-15x+6}{(x-2)^2(x+2)(x-3)}$$

79. $x^2 + 9x + 14 = (x+2)(x+7)$

$x^2 + 10x + 21 = (x+3)(x+7)$

$x^2 + 5x + 6 = (x+2)(x+3)$

LCD $= (x+2)(x+7)(x+3)$

$$\frac{9}{x^2+9x+14} - \frac{3x}{x^2+10x+21} + \frac{x+4}{x^2+5x+6} = \frac{9(x+3)}{(x+2)(x+7)(x+3)} - \frac{3x(x+2)}{(x+3)(x+7)(x+2)} + \frac{(x+4)(x+7)}{(x+2)(x+3)(x+7)}$$

$$= \frac{9(x+3) - 3x(x+2) + (x+4)(x+7)}{(x+2)(x+7)(x+3)}$$

$$= \frac{9x+27-3x^2-6x+x^2+11x+28}{(x+2)(x+7)(x+3)}$$

$$= \frac{-2x^2+14x+55}{(x+2)(x+7)(x+3)}$$

81. The length of the other board is $\left(\frac{3}{x+4}-\frac{1}{x-4}\right)$ inches.

LCD = $(x+4)(x-4)$

$$\begin{aligned}\frac{3}{x+4}-\frac{1}{x-4} &= \frac{3(x-4)}{(x+4)(x-4)}-\frac{1(x+4)}{(x-4)(x+4)}\\ &= \frac{3(x-4)-(x+4)}{(x+4)(x-4)}\\ &= \frac{3x-12-x-4}{(x+4)(x-4)}\\ &= \frac{2x-16}{(x+4)(x-4)}\end{aligned}$$

The length of the other board is $\frac{2x-16}{(x+4)(x-4)}$ inches.

83. $1-\frac{G}{P}=\frac{1}{1}-\frac{G}{P}=\frac{1(P)}{1(P)}-\frac{G}{P}=\frac{P-G}{P}$

85. Answers may vary

87. $90°-\left(\frac{40}{x}\right)° = \left(90-\frac{40}{x}\right)°$

LCD = x

$\left(90\cdot\frac{x}{x}-\frac{40}{x}\right)° = \left(\frac{90x}{x}-\frac{40}{x}\right)° = \left(\frac{90x-40}{x}\right)°$

89. Answers may vary

The Bigger Picture

1. $-8.6+(-9.1)=-17.7$

2. $(-8.6)(-9.1)=78.26$

3. $14-(-14)=14+14=28$

4. $$\begin{aligned}3x^4-7+x^4-x^2-10 &= 3x^4+x^4-x^2-7-10\\ &= 4x^4-x-17\end{aligned}$$

5. $\frac{5x^2-5}{25x+25}=\frac{5(x+1)(x-1)}{5\cdot 5(x+1)}=\frac{x-1}{5}$

6. $$\begin{aligned}\frac{7x}{x^2+4x+3}\div\frac{x}{2x+6} &= \frac{7x}{x^2+4x+3}\cdot\frac{2x+6}{x}\\ &= \frac{7\cdot x\cdot 2\cdot(x+3)}{(x+3)(x+1)\cdot x}\\ &= \frac{14}{x+1}\end{aligned}$$

7. $9=3\cdot 3=3^2$

$6=2\cdot 3$

LCD = $2\cdot 3^2$

$\frac{2}{9}-\frac{5}{6}=\frac{2(2)}{9(2)}-\frac{5(3)}{6(3)}=\frac{4}{18}-\frac{15}{18}=-\frac{11}{18}$

8. $9=3\cdot 3=3^2$

LCD = $3^2\cdot 5$

$$\begin{aligned}\frac{x}{9}-\frac{x+3}{5} &= \frac{x(5)}{9(5)}-\frac{(x+3)(9)}{5(9)}\\ &= \frac{5x-9(x+3)}{45}\\ &= \frac{5x-9x-27}{45}\\ &= \frac{-4x-27}{45} \text{ or } -\frac{4x+27}{45}\end{aligned}$$

9. $$\begin{aligned}9x^3-2x^2-11x &= x(9x^2-2x-11)\\ &= x(9x-11)(x+1)\end{aligned}$$

10. $$\begin{aligned}12xy-21x+4y-7 &= 3x(4y-7)+1(4y-7)\\ &= (4y-7)(3x+1)\end{aligned}$$

11. $$\begin{aligned}7x-14 &= 5x+10\\ 2x-14 &= 10\\ 2x &= 24\\ x &= 12\end{aligned}$$

12. $$\begin{aligned}\frac{-x+2}{5} &< \frac{3}{10}\\ 10\left(\frac{-x+2}{5}\right) &< 10\left(\frac{3}{10}\right)\\ 2(-x+2) &< 3\\ -2x+4 &< 3\\ -2x &< -1\\ \frac{-2x}{-2} &> \frac{-1}{-2}\\ x &> \frac{1}{2}\end{aligned}$$

$\left(\frac{1}{2},\infty\right)$

13. $1+4(x+4)=3^2+x$
$1+4x+16=9+x$
$4x+17=9+x$
$3x+17=9$
$3x=-8$
$x=-\frac{8}{3}$

14. $x(x-2)=24$
$x^2-2x=24$
$x^2-2x-24=0$
$(x+4)(x-6)=0$
$x+4=0$ or $x-6=0$
$x=-4$ $x=6$
The solutions are $x=-4, 6$.

Section 7.5

Practice Exercises

1. The LCD of 3, 5, and 15 is 15.
$\frac{x}{3}+\frac{4}{5}=\frac{12}{5}$
$15\left(\frac{x}{3}+\frac{4}{5}\right)=15\left(\frac{2}{15}\right)$
$15\left(\frac{x}{3}\right)+15\left(\frac{4}{5}\right)=15\left(\frac{2}{15}\right)$
$5\cdot x+12=2$
$5x=-10$
$x=-2$

Check: $\frac{x}{3}+\frac{4}{5}=\frac{2}{15}$
$\frac{-2}{3}+\frac{4}{5}\stackrel{?}{=}\frac{2}{15}$
$\frac{2}{15}=\frac{2}{15}$ True

This number checks, so the solution is −2.

2. The LCD of 4, 3, and 12 is 12.
$\frac{x+4}{4}-\frac{x-3}{3}=\frac{11}{12}$
$12\left(\frac{x+4}{4}-\frac{x-3}{3}\right)=12\left(\frac{11}{12}\right)$
$12\left(\frac{x+4}{4}\right)-12\left(\frac{x-3}{3}\right)=12\left(\frac{11}{12}\right)$
$3(x+4)-4(x-3)=11$
$3x+12-4x+12=11$
$-x+24=11$
$-x=-13$
$x=13$

Check: $\frac{x+4}{4}-\frac{x-3}{3}=\frac{11}{12}$
$\frac{13+4}{4}-\frac{13-3}{3}\stackrel{?}{=}\frac{11}{12}$
$\frac{17}{4}-\frac{10}{3}\stackrel{?}{=}\frac{11}{12}$
$\frac{11}{12}=\frac{11}{12}$ True

The solution is 13.

3. In this equation, 0 cannot be a solution. The LCD is x.
$8+\frac{7}{x}=x+2$
$x\left(8+\frac{7}{x}\right)=x(x+2)$
$x(8)+x\left(\frac{7}{x}\right)=x\cdot x+x\cdot 2$
$8x+7=x^2+2x$
$0=x^2-6x-7$
$0=(x+1)(x-7)$
$x+1=0$ or $x-7=0$
$x=-1$ $x=7$
Neither −1 nor 7 makes the denominator in the original equation equal to 0.
Check:
$x=-1$
$8+\frac{7}{x}=x+2$
$8+\frac{7}{-1}\stackrel{?}{=}-1+2$
$8+(-7)\stackrel{?}{=}1$
$1=1$ True
$x=7$

$$8+\frac{7}{7} \stackrel{?}{=} 7+2$$
$$8+1 \stackrel{?}{=} 9$$
$$9=9 \quad \text{True}$$

Both -1 and 7 are solutions.

4. $x^2-5x-14=(x+2)(x-7)$

The LCD is $(x+2)(x-7)$.

$$\frac{6x}{x^2-5x-14}-\frac{3}{x+2}=\frac{1}{x-7}$$
$$(x+2)(x-7)\left(\frac{6x}{x^2-5x-14}-\frac{3}{x+2}\right)=(x+2)(x-7)\left(\frac{1}{x-7}\right)$$
$$(x+2)(x-7)\cdot\frac{6x}{x^2-5x-14}-(x+2)(x-7)\cdot\frac{3}{x+2}=(x+2)(x-7)\cdot\frac{1}{x-7}$$
$$6x-3(x-7)=x+2$$
$$6x-3x+21=x+2$$
$$3x+21=x+2$$
$$2x=-19$$
$$x=-\frac{19}{2}$$

Check by replacing x with $-\frac{19}{2}$ in the original equation. The solution is $-\frac{19}{2}$.

5. The LCD is $x-2$.

$$\frac{7}{x-2}=\frac{3}{x-2}+4$$
$$(x-2)\left(\frac{7}{x-2}\right)=(x-2)\left(\frac{3}{x-2}+4\right)$$
$$(x-2)\cdot\frac{7}{x-2}=(x-2)\cdot\frac{3}{x-2}+(x-2)\cdot 4$$
$$7=3+4x-8$$
$$7=4x-5$$
$$12=4x$$
$$3=x$$

Check by replacing x with 3 in the original equation. The solution is 3.

6. From the denominators in the equation, 5 can't be a solution. The LCD is $x-5$.

$$x+\frac{x}{x-5}=\frac{5}{x-5}-7$$
$$(x-5)\left(x+\frac{x}{x-5}\right)=(x-5)\left(\frac{5}{x-5}-7\right)$$
$$(x-5)(x)+(x-5)\left(\frac{x}{x-5}\right)=(x-5)\left(\frac{5}{x-5}\right)-(x-5)(7)$$
$$x^2-5x+x=5-7x+35$$
$$x^2-4x=40-7x$$
$$x^2+3x-40=0$$
$$(x+8)(x-5)=0$$

$$x+8=0 \quad \text{or} \quad x-5=0$$
$$x=-8 \qquad\qquad x=5$$

Since 5 can't be a solution, check by replacing x with -8 in the original equation. The only solution is -8.

7. The LCD is abx.

$$\frac{1}{a}+\frac{1}{b}=\frac{1}{x}$$
$$abx\left(\frac{1}{a}+\frac{1}{b}\right)=abx\left(\frac{1}{x}\right)$$
$$abx\left(\frac{1}{a}\right)+abx\left(\frac{1}{b}\right)=abx\cdot\frac{1}{x}$$
$$bx+ax=ab$$
$$ax=ab-bx$$
$$ax=b(a-x)$$
$$\frac{ax}{a-x}=b$$

Calculator Explorations

1. $y_1=\frac{x-4}{2}-\frac{x-3}{9},\ y_2=\frac{5}{18}$

Use INTERSECT

10

−10 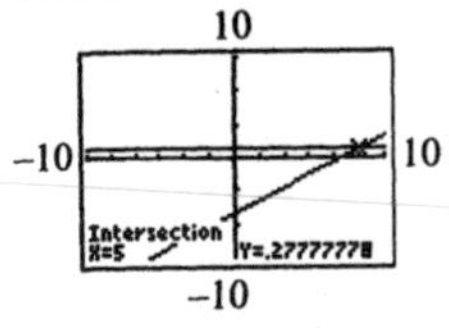 10

−10

The solution of the equation is 5.

2. $y_1=3-\frac{6}{x},\ y_2=x+8$

Use INTERSECT

10

−10 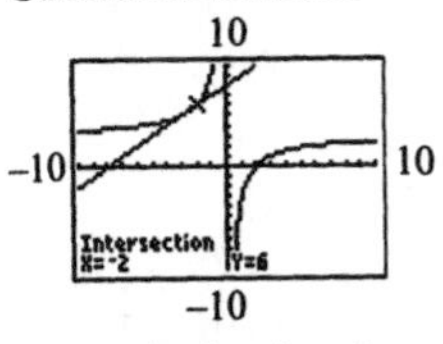 10

−10

One solution is -3.

10

−10 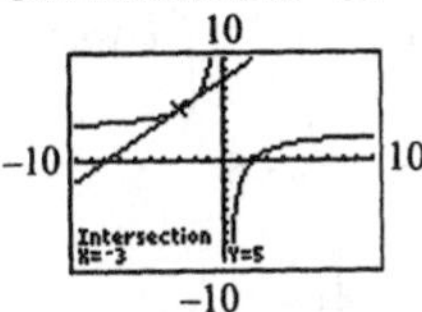 10

−10

The other solution is -2.

3. $y_1=\frac{2x}{x-4},\ y_2=\frac{8}{x-4}+1$

10

−10 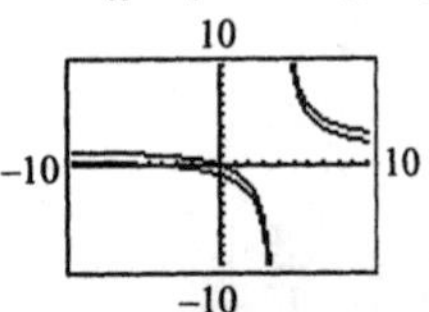 10

−10

Using TRACE and ZOOM, it is clear that the curves never intersect. The equation has no solution.

4. $y_1=x+\frac{14}{x-2},\ y_2=\frac{7x}{x-2}+1$

Use INTERSECT

15

−5 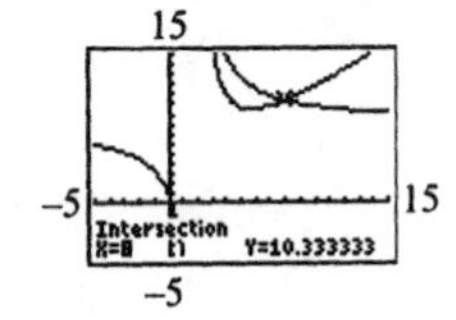 15

−5

The solution is 8.

Exercise Set 7.5

1. The LCD is 5.

$$\frac{x}{5}+3=9$$
$$5\left(\frac{x}{5}+3\right)=5(9)$$
$$5\left(\frac{x}{5}\right)+5(3)=5(9)$$
$$x+15=45$$
$$x=30$$

Check: $\frac{x}{5}+3=9$
$$\frac{30}{5}+3 \stackrel{?}{=} 9$$
$$6+3 \stackrel{?}{=} 9$$
$$9=9 \quad \text{True}$$

The solution is 30.

3. The LCD is 12.

$$\frac{x}{2}+\frac{5x}{4}=\frac{x}{12}$$
$$12\left(\frac{x}{2}+\frac{5x}{4}\right)=12\left(\frac{x}{12}\right)$$
$$12\left(\frac{x}{2}\right)+12\left(\frac{5x}{4}\right)=12\left(\frac{x}{12}\right)$$
$$6x+15x=x$$
$$21x=x$$
$$20x=0$$
$$x=0$$

Check: $\frac{x}{2}+\frac{5x}{4}=\frac{x}{12}$

$$\frac{0}{2}+\frac{5\cdot 0}{4}\stackrel{?}{=}\frac{0}{12}$$
$$0+\frac{0}{4}\stackrel{?}{=}0$$
$$0=0 \quad \text{True}$$

The solution is 0.

5. The LCD is x.

$$2-\frac{8}{x}=6$$
$$x\left(2-\frac{8}{x}\right)=x(6)$$
$$x\cdot 2-x\cdot\frac{8}{x}=x\cdot 6$$
$$2x-8=6x$$
$$-8=4x$$
$$-2=x$$

Check: $2-\frac{8}{x}=6$

$$2-\frac{8}{-2}\stackrel{?}{=}6$$
$$2-(-4)\stackrel{?}{=}6$$
$$6=6 \quad \text{True}$$

The solution is −2.

7. The LCD is x.

$$2+\frac{10}{x}=x+5$$
$$x\left(2+\frac{10}{x}\right)=x(x+5)$$
$$x(2)+x\left(\frac{10}{x}\right)=x(x+5)$$
$$2x+10=x^2+5x$$
$$0=x^2+3x-10$$
$$0=(x+5)(x-2)$$

$x+5=0$ or $x-2=0$

$x=-5$ $\quad$ $x=2$

Check:

$x=-5$: $2+\frac{10}{x}=x+5$

$$2+\frac{10}{-5}\stackrel{?}{=}-5+5$$
$$2+(-2)\stackrel{?}{=}-5+5$$
$$0=0 \quad \text{True}$$

$x=2$: $2+\frac{10}{x}=x+5$

$$2+\frac{10}{2}\stackrel{?}{=}2+5$$
$$2+5\stackrel{?}{=}2+5$$
$$7=7 \quad \text{True}$$

Both −5 and 2 are solutions.

9. The LCD is 10.

$$\frac{a}{5}=\frac{a-3}{2}$$
$$10\left(\frac{a}{5}\right)=10\left(\frac{a-3}{2}\right)$$
$$2a=5(a-3)$$
$$2a=5a-15$$
$$-3a=-15$$
$$a=5$$

Check: $\frac{a}{5}=\frac{a-3}{2}$

$$\frac{5}{5}\stackrel{?}{=}\frac{5-3}{2}$$
$$\frac{5}{5}\stackrel{?}{=}\frac{2}{2}$$
$$1=1 \quad \text{True}$$

The solution is 5.

11. The LCD is 10.

$$\frac{x-3}{5}+\frac{x-2}{2}=\frac{1}{2}$$
$$10\left(\frac{x-3}{5}+\frac{x-2}{2}\right)=10\left(\frac{1}{2}\right)$$
$$10\left(\frac{x-3}{5}\right)+10\left(\frac{x-2}{2}\right)=10\left(\frac{1}{2}\right)$$
$$2(x-3)+5(x-2)=5$$
$$2x-6+5x-10=5$$
$$7x-16=5$$
$$7x=21$$
$$x=3$$

Check: $\frac{x-3}{5}+\frac{x-2}{2}=\frac{1}{2}$

$$\frac{3-3}{5}+\frac{3-2}{2}\stackrel{?}{=}\frac{1}{2}$$
$$\frac{0}{5}+\frac{1}{2}\stackrel{?}{=}\frac{1}{2}$$
$$0+\frac{1}{2}\stackrel{?}{=}\frac{1}{2}$$
$$\frac{1}{2}=\frac{1}{2} \quad \text{True}$$

The solution is 3.

13. The LCD is $2a-5$.

$$\frac{3}{2a-5}=-1$$
$$(2a-5)\left(\frac{3}{2a-5}\right)=(2a-5)(-1)$$
$$3=-2a+5$$
$$-2=-2a$$
$$1=a$$

Check: $\frac{3}{2a-5}=-1$

$$\frac{3}{2(1)-5}\stackrel{?}{=}-1$$
$$\frac{3}{-3}\stackrel{?}{=}-1$$
$$-1=-1 \quad \text{True}$$

The solution is 1.

15. The LCD is $y-4$.

$$\frac{4y}{y-4}+5=\frac{5y}{y-4}$$
$$(y-4)\left(\frac{4y}{y-4}+5\right)=(y-4)\left(\frac{5y}{y-4}\right)$$
$$(y-4)\left(\frac{4y}{y-4}\right)+(y-4)(5)=(y-4)\left(\frac{5y}{y-4}\right)$$
$$4y+5y-20=5y$$
$$9y-20=5y$$
$$4y-20=0$$
$$4y=20$$
$$y=5$$

Check: $\frac{4y}{y-4}+5=\frac{5y}{y-4}$

$$\frac{4(5)}{5-4}+5\stackrel{?}{=}\frac{5(5)}{5-4}$$
$$\frac{20}{1}+5\stackrel{?}{=}\frac{25}{1}$$
$$25=25 \quad \text{True}$$

The solution is 5.

17. The LCD is $a - 3$.

$$\begin{aligned}
2+\frac{3}{a-3} &= \frac{a}{a-3}\\
(a-3)\left(2+\frac{3}{a-3}\right) &= (a-3)\left(\frac{a}{a-3}\right)\\
(a-3)(2)+(a-3)\left(\frac{3}{a-3}\right) &= a\\
2a-6+3 &= a\\
2a-3 &= a\\
-3 &= a-2a\\
-3 &= -a\\
\frac{-3}{-1} &= a\\
3 &= a
\end{aligned}$$

When a is 3, a denominator equals zero. The equation has no solution.

19. $x^2-9=(x+3)(x-3)$

The LCD is $(x + 3)(x - 3)$.

$$\begin{aligned}
\frac{1}{x+3}+\frac{6}{x^2-9} &= 1\\
(x+3)(x-3)\left(\frac{1}{x+3}+\frac{6}{(x+3)(x-3)}\right) &= (x+3)(x-3)(1)\\
(x+3)(x-3)\cdot\frac{1}{x+3}+(x+3)(x-3)\cdot\frac{6}{(x+3)(x-3)} &= (x+3)(x-3)\cdot 1\\
x-3+6 &= x^2-9\\
x+3 &= x^2-9\\
0 &= x^2-x-12\\
0 &= (x+3)(x-4)
\end{aligned}$$

$$\begin{aligned}
x+3=0 &\quad \text{or} \quad x-4=0\\
x=-3 &\quad\quad\quad\;\; x=4
\end{aligned}$$

When x is -3, a denominator equals zero. Check $x = 4$.

Check: $\frac{1}{x+3}+\frac{6}{x^2-9}=1$

$$\begin{aligned}
\frac{1}{4+3}+\frac{6}{4^2-9} &\stackrel{?}{=} 1\\
\frac{1}{7}+\frac{6}{7} &\stackrel{?}{=} 1\\
1 &= 1 \quad \text{True}
\end{aligned}$$

The solution is 4.

21. The LCD is $y + 4$.

$$\frac{2y}{y+4}+\frac{4}{y+4}=3$$
$$(y+4)\left(\frac{2y}{y+4}+\frac{4}{y+4}\right)=(y+4)(3)$$
$$(y+4)\cdot\frac{2y}{y+4}+(y+4)\cdot\frac{4}{y+4}=(y+4)\cdot 3$$
$$2y+4=3y+12$$
$$4=y+12$$
$$-8=y$$

Check: $\frac{2y}{y+4}+\frac{4}{y+4}=3$

$$\frac{2(-8)}{-8+4}+\frac{4}{-8+4}\stackrel{?}{=}3$$
$$\frac{-16}{-4}+\frac{4}{-4}\stackrel{?}{=}3$$
$$4-1\stackrel{?}{=}3$$
$$3=3 \quad \text{True}$$

The solution is -8.

23. The LCD is $(x + 2)(x - 2)$.

$$\frac{2x}{x+2}-2=\frac{x-8}{x-2}$$
$$(x+2)(x-2)\left(\frac{2x}{x+2}-2\right)=(x+2)(x-2)\left(\frac{x-8}{x-2}\right)$$
$$(x+2)(x-2)\cdot\frac{2x}{x+2}-(x+2)(x-2)(2)=(x+2)(x-2)\cdot\frac{x-8}{x-2}$$
$$2x(x-2)-2(x^2-4)=(x+2)(x-8)$$
$$2x^2-4x-2x^2+8=x^2-6x-16$$
$$-4x+8=x^2-6x-16$$
$$0=x^2-2x-24$$
$$0=(x+4)(x-6)$$

$x+4=0$ or $x-6=0$

$x=-4$ $\qquad$ $x=6$

Check $x = -4$: $\frac{2x}{x+2}-2=\frac{x-8}{x-2}$

$$\frac{2(-4)}{-4+2}-2\stackrel{?}{=}\frac{-4-8}{-4-2}$$
$$\frac{-8}{-2}-2\stackrel{?}{=}\frac{-12}{-6}$$
$$4-2\stackrel{?}{=}2$$
$$2=2 \quad \text{True}$$

Check $x = 6$: $\frac{2x}{x+2} - 2 = \frac{x-8}{x-2}$

$$\frac{2(6)}{6+2} - 2 \stackrel{?}{=} \frac{6-8}{6-2}$$
$$\frac{12}{8} - 2 \stackrel{?}{=} \frac{-2}{4}$$
$$\frac{3}{2} - 2 \stackrel{?}{=} -\frac{1}{2}$$
$$-\frac{1}{2} = -\frac{1}{2} \quad \text{True}$$

The solutions are –4 an 6.

25. The LCD is $2y$.

$$\frac{2}{y} + \frac{1}{2} = \frac{5}{2y}$$
$$2y\left(\frac{2}{y} + \frac{1}{2}\right) = 2y\left(\frac{5}{2y}\right)$$
$$2y\left(\frac{2}{y}\right) + 2y\left(\frac{1}{2}\right) = 2y\left(\frac{5}{2y}\right)$$
$$4 + y = 5$$
$$y = 1$$

The solution is 1.

27. The LCD is $(a - 6)(a - 1)$.

$$\frac{a}{a-6} = \frac{-2}{a-1}$$
$$(a-6)(a-1)\left(\frac{a}{a-6}\right) = (a-6)(a-1)\left(\frac{-2}{a-1}\right)$$
$$a(a-1) = -2(a-6)$$
$$a^2 - a = -2a + 12$$
$$a^2 + a - 12 = 0$$
$$(a+4)(a-3) = 0$$
$$a + 4 = 0 \quad \text{or} \quad a - 3 = 0$$
$$a = -4 \qquad\qquad a = 3$$

The solutions are –4 and 3.

29. The LCD is $6x$.

$$\frac{11}{2x} + \frac{2}{3} = \frac{7}{2x}$$
$$6x\left(\frac{11}{2x} + \frac{2}{3}\right) = 6x\left(\frac{7}{2x}\right)$$
$$6x \cdot \frac{11}{2x} + 6x \cdot \frac{2}{3} = 6x \cdot \frac{7}{2x}$$
$$33 + 4x = 21$$
$$4x = -12$$
$$x = -3$$

The solution is –3.

31. The LCD is $(x + 2)(x - 2)$.

$$\frac{2}{x-2}+1=\frac{x}{x+2}$$
$$(x+2)(x-2)\left(\frac{2}{x-2}+1\right)=(x+2)(x-2)\left(\frac{x}{x+2}\right)$$
$$(x+2)(x-2)\cdot\frac{2}{x-2}+(x+2)(x-2)\cdot 1=(x+2)(x-2)\cdot\frac{x}{x+2}$$
$$2(x+2)+(x+2)(x-2)=x(x-2)$$
$$2x+4+x^2-4=x^2-2x$$
$$x^2+2x=x^2-2x$$
$$2x=-2x$$
$$4x=0$$
$$x=0$$

The solution is 0.

33. The LCD is 6.

$$\frac{x+1}{3}-\frac{x-1}{6}=\frac{1}{6}$$
$$6\left(\frac{x+1}{3}-\frac{x-1}{6}\right)=6\left(\frac{1}{6}\right)$$
$$6\left(\frac{x+1}{3}\right)-6\left(\frac{x-1}{6}\right)=6\left(\frac{1}{6}\right)$$
$$2(x+1)-(x-1)=1$$
$$2x+2-x+1=1$$
$$x+3=1$$
$$x=-2$$

The solution is -2.

35. The LCD is $6(t - 4)$.

$$\frac{t}{t-4}=\frac{t+4}{6}$$
$$6(t-4)\left(\frac{t}{t-4}\right)=6(t-4)\left(\frac{t+4}{6}\right)$$
$$6t=(t-4)(t+4)$$
$$6t=t^2-16$$
$$0=t^2-6t-16$$
$$0=(t-8)(t+2)$$
$$t+2=0 \quad\text{or}\quad t-8=0$$
$$t=-2 \qquad\qquad t=8$$

The solutions are -2 and 8.

37. $2y + 2 = 2(y + 1)$
$4y + 4 = 2 \cdot 2(y + 1)$
The LCD is $4(y + 1)$.

$$\frac{y}{2y+2}+\frac{2y-16}{4y+4}=\frac{2y-3}{y+1}$$
$$4(y+1)\left(\frac{y}{2(y+1)}+\frac{2y-16}{4(y+1)}\right)=4(y+1)\left(\frac{2y-3}{y+1}\right)$$
$$4(y+1)\left(\frac{y}{2(y+1)}\right)+4(y+1)\left(\frac{2y-16}{4(y+1)}\right)=4(y+1)\left(\frac{2y-3}{y+1}\right)$$
$$2y+2y-16=4(2y-3)$$
$$4y-16=8y-12$$
$$-4y=4$$
$$y=-1$$

In the original equation, -1 makes a denominator 0. This equation has no solution.

39. $r^2+5r-14=(r+7)(r-2)$
The LCD is $(r+7)(r-2)$.

$$\frac{4r-4}{r^2+5r-14}+\frac{2}{r+7}=\frac{1}{r-2}$$
$$(r+7)(r-2)\left(\frac{4r-4}{(r+7)(r-2)}+\frac{2}{r+7}\right)=(r+7)(r-2)\left(\frac{1}{r-2}\right)$$
$$(r+7)(r-2)\left(\frac{4r-4}{(r+7)(r-2)}\right)+(r+7)(r-2)\left(\frac{2}{r+7}\right)=(r+7)(r-2)\left(\frac{1}{r-2}\right)$$
$$4r-4+2(r-2)=(r+7)(1)$$
$$4r-4+2r-4=r+7$$
$$6r-8=r+7$$
$$5r=15$$
$$r=3$$

The solution is 3.

41. $x^2+x-6=(x+3)(x-2)$
The LCD is $(x+3)(x-2)$.

$$\frac{x+1}{x+3}=\frac{x^2-11x}{x^2+x-6}-\frac{x-3}{x-2}$$
$$(x+3)(x-2)\left(\frac{x+1}{x+3}\right)=(x+3)(x-2)\left(\frac{x^2-11x}{(x+3)(x-2)}-\frac{x-3}{x-2}\right)$$
$$(x+3)(x-2)\cdot\frac{x+1}{x+3}=(x+3)(x-2)\cdot\frac{x^2-11x}{(x+3)(x-2)}-(x+3)(x-2)\cdot\frac{x-3}{x-2}$$
$$(x-2)(x+1)=x^2-11x-(x+3)(x-3)$$
$$x^2-x-2=x^2-11x-(x^2-9)$$
$$x^2-x-2=x^2-11x-x^2+9$$
$$x^2-x-2=-11x+9$$
$$x^2+10x-11=0$$
$$(x+11)(x-1)=0$$
$$x+11=0 \quad \text{or} \quad x-1=0$$
$$x=-11 \qquad\qquad x=1$$

The solutions are -11 and 1.

43. $$R = \frac{E}{I}$$
$$I(R) = I\left(\frac{E}{I}\right)$$
$$IR = E$$
$$I = \frac{E}{R}$$

45. $$T = \frac{2U}{B+E}$$
$$(B+E)(T) = (B+E)\left(\frac{2U}{B+E}\right)$$
$$BT + ET = 2U$$
$$BT = 2U - ET$$
$$B = \frac{2U - ET}{T}$$

47. $$B = \frac{705w}{h^2}$$
$$h^2(B) = h^2\left(\frac{705w}{h^2}\right)$$
$$Bh^2 = 705w$$
$$\frac{Bh^2}{705} = w$$

49. $$N = R + \frac{V}{G}$$
$$G(N) = G\left(R + \frac{V}{G}\right)$$
$$GN = GR + V$$
$$GN - GR = V$$
$$G(N - R) = V$$
$$G = \frac{V}{N - R}$$

51. $$\frac{C}{\pi r} = 2$$
$$\pi r\left(\frac{C}{\pi r}\right) = \pi r(2)$$
$$C = 2\pi r$$
$$\frac{C}{2\pi} = \frac{2\pi r}{2\pi}$$
$$\frac{C}{2\pi} = r$$

53. $$\frac{1}{y} + \frac{1}{3} = \frac{1}{x}$$
$$3xy\left(\frac{1}{y} + \frac{1}{3}\right) = 3xy\left(\frac{1}{x}\right)$$
$$3xy \cdot \frac{1}{y} + 3xy \cdot \frac{1}{3} = 3xy \cdot \frac{1}{x}$$
$$3x + xy = 3y$$
$$x(3 + y) = 3y$$
$$x = \frac{3y}{3 + y}$$

55. The reciprocal of x is $\frac{1}{x}$.

57. The reciprocal of x, added to the reciprocal of 2 is $\frac{1}{x} + \frac{1}{2}$.

59. If a tank is filled in 3 hours, then $\frac{1}{3}$ of the tank is filled in one hour.

61. The graph crosses the x-axis at $x = 2$. It crosses the y-axis at $y = -2$. The x-intercept is (2, 0) and the y-intercept is (0, −2).

63. The graph crosses the x-axis at $x = -4$, $x = -2$ and $x = 3$. It crosses the y-axis at $y = 4$. The x-intercepts are (−4, 0), (−2, 0) and (3, 0), and the y-intercept is (0, 4).

65. Answers may vary

67. expression
$$\frac{1}{x} + \frac{5}{9} = \frac{1(9)}{x(9)} + \frac{5x}{9x} = \frac{5x + 9}{9x}$$

69. equation
$$\frac{5}{x-1} - \frac{2}{x} = \frac{5}{x(x-1)}$$
$$x(x-1)\left(\frac{5}{x-1}\right) - x(x-1)\left(\frac{2}{x}\right) = x(x-1)\left(\frac{5}{x(x-1)}\right)$$
$$5x - 2(x-1) = 5$$
$$5x - 2x + 2 = 5$$
$$3x = 3$$
$$x = 1$$
1 makes a denominator zero. There is no solution.

71.
$$\frac{20x}{3}+\frac{32x}{6}=180$$
$$6\left(\frac{20x}{3}+\frac{32x}{6}\right)=6(180)$$
$$6\left(\frac{20x}{3}\right)+6\left(\frac{32x}{6}\right)=6(180)$$
$$40x+32x=1080$$
$$72x=1080$$
$$\frac{72x}{72}=\frac{1080}{72}$$
$$x=15$$
$$\frac{20x}{3}=\frac{20(15)}{3}=100$$
$$\frac{32x}{6}=\frac{32(15)}{6}=80$$
The angles are 100° and 80°.

73.
$$\frac{150}{x}+\frac{450}{x}=90$$
$$x\left(\frac{150}{x}+\frac{450}{x}\right)=x(90)$$
$$x\left(\frac{150}{x}\right)+x\left(\frac{450}{x}\right)=x(90)$$
$$150+450=90x$$
$$600=90x$$
$$\frac{600}{90}=\frac{90x}{90}$$
$$\frac{20}{3}=x$$
$$\frac{150}{x}=\frac{150}{\frac{20}{3}}=150\left(\frac{3}{20}\right)=\frac{45}{2}=22.5$$
$$\frac{450}{x}=\frac{450}{\frac{20}{3}}=450\left(\frac{3}{20}\right)=\frac{135}{2}=67.5$$
The angles are 22.5° and 67.5°.

75.

$$\frac{5}{a^2+4a+3}+\frac{2}{a^2+a-6}-\frac{3}{a^2-a-2}=0$$

$$\frac{5}{(a+3)(a+1)}+\frac{2}{(a+3)(a-2)}-\frac{3}{(a-2)(a+1)}=0$$

$$(a+3)(a+1)(a-2)\left(\frac{5}{(a+3)(a+1)}+\frac{2}{(a+3)(a-2)}-\frac{3}{(a-2)(a+1)}\right)=(a+3)(a+1)(a-2)(0)$$

$$(a+3)(a+1)(a-2)\left(\frac{5}{(a+3)(a+1)}\right)+(a+3)(a+1)(a-2)\left(\frac{2}{(a+3)(a-2)}\right)$$

$$-(a+3)(a+1)(a-2)\left(\frac{3}{(a-2)(a+1)}\right)=0$$

$$5(a-2)+2(a+1)-3(a+3)=0$$

$$5a-10+2a+2-3a-9=0$$

$$4a-17=0$$

$$4a=17$$

$$a=\frac{17}{4}$$

The solution is $\frac{17}{4}$.

Integrated Review

1. expression

$$\frac{1}{x}+\frac{2}{3}=\frac{1(3)}{x(3)}+\frac{2(x)}{3(x)}=\frac{3}{3x}+\frac{2x}{3x}=\frac{3+2x}{3x}$$

2. expression

$$\frac{3}{a}+\frac{5}{6}=\frac{3(6)}{a(6)}+\frac{5(a)}{6(a)}=\frac{18}{6a}+\frac{5a}{6a}=\frac{18+5a}{6a}$$

3. equation

$$\frac{1}{x}+\frac{2}{3}=\frac{3}{x}$$

$$3x\left(\frac{1}{x}+\frac{2}{3}\right)=3x\left(\frac{3}{x}\right)$$

$$3x\left(\frac{1}{x}\right)+3x\left(\frac{2}{3}\right)=3x\left(\frac{3}{x}\right)$$

$$3+2x=9$$

$$2x=6$$

$$x=3$$

The solution is 3.

4. equation

$$\frac{3}{a}+\frac{5}{6}=1$$
$$6a\left(\frac{3}{a}+\frac{5}{6}\right)=6a(1)$$
$$6a\left(\frac{3}{a}\right)+6a\left(\frac{5}{6}\right)=6a$$
$$18+5a=6a$$
$$18=a$$

The solution is 18.

5. expression

$$\frac{2}{x-1}-\frac{1}{x}=\frac{2(x)}{(x-1)(x)}-\frac{1(x-1)}{x(x-1)}$$
$$=\frac{2x-(x-1)}{x(x-1)}$$
$$=\frac{x+1}{x(x-1)}$$

6. expression

$$\frac{4}{x-3}-\frac{1}{x}=\frac{4(x)}{(x-3)(x)}-\frac{1(x-3)}{x(x-3)}$$
$$=\frac{4x-(x-3)}{x(x-3)}$$
$$=\frac{4x-x+3}{x(x-3)}$$
$$=\frac{3x+3}{x(x-3)}$$
$$=\frac{3(x+1)}{x(x-3)}$$

7. equation

$$\frac{2}{x+1}-\frac{1}{x}=1$$
$$x(x+1)\left(\frac{2}{x+1}-\frac{1}{x}\right)=x(x+1)(1)$$
$$x(x+1)\left(\frac{2}{x+1}\right)-x(x+1)\left(\frac{1}{x}\right)=x(x+1)$$
$$2x-(x+1)=x(x+1)$$
$$2x-x-1=x^2+x$$
$$x-1=x^2+x$$
$$-1=x^2$$

There is no real number solution.

8. equation

$$\frac{4}{x-3}-\frac{1}{x}=\frac{6}{x(x-3)}$$
$$x(x-3)\left(\frac{4}{x-3}-\frac{1}{x}\right)=x(x-3)\left(\frac{6}{x(x-3)}\right)$$
$$x(x-3)\left(\frac{4}{x-3}\right)-x(x-3)\left(\frac{1}{x}\right)=6$$
$$4x-(x-3)=6$$
$$4x-x+3=6$$
$$3x+3=6$$
$$3x=3$$
$$x=1$$

The solution is 1.

9. expression

$$\frac{15x}{x+8}\cdot\frac{2x+16}{3x}=\frac{15x\cdot(2x+16)}{(x+8)\cdot 3x}$$
$$=\frac{3\cdot 5\cdot x\cdot 2\cdot(x+8)}{(x+8)\cdot 3\cdot x}$$
$$=5\cdot 2$$
$$=10$$

10. expression

$$\frac{9z+5}{15}\cdot\frac{5z}{81z^2-25}=\frac{(9z+5)\cdot 5z}{15\cdot(81z^2-25)}$$
$$=\frac{(9z+5)\cdot 5\cdot z}{5\cdot 3\cdot(9z+5)(9z-5)}$$
$$=\frac{z}{3(9z-5)}$$

11. expression

$$\frac{2x+1}{x-3}+\frac{3x+6}{x-3}=\frac{2x+1+3x+6}{x-3}=\frac{5x+7}{x-3}$$

12. expression

$$\frac{4p-3}{2p+7}+\frac{3p+8}{2p+7}=\frac{4p-3+3p+8}{2p+7}=\frac{7p+5}{2p+7}$$

13. equation

$$\frac{x+5}{7}=\frac{8}{2}$$
$$14\left(\frac{x+5}{7}\right)=14\left(\frac{8}{2}\right)$$
$$2(x+5)=56$$
$$2x+10=56$$
$$2x=46$$
$$x=23$$

The solution is 23.

14. equation

$$\frac{1}{2}=\frac{x-1}{8}$$
$$8\left(\frac{1}{2}\right)=8\left(\frac{x-1}{8}\right)$$
$$4=x-1$$
$$5=x$$

The solution is 5.

15. expression

$$\frac{5a+10}{18}\div\frac{a^2-4}{10a}=\frac{5a+10}{18}\cdot\frac{10a}{a^2-4}$$
$$=\frac{5(a+2)\cdot 2\cdot 5\cdot a}{2\cdot 9(a+2)(a-2)}$$
$$=\frac{5\cdot 5\cdot a}{9(a-2)}$$
$$=\frac{25a}{9(a-2)}$$

16. expression

$$\frac{9}{x^2-1}+\frac{12}{3x+3}$$
$$=\frac{9(3)}{(x+1)(x-1)(3)}+\frac{12(x-1)}{3(x+1)(x-1)}$$
$$=\frac{27+12x-12}{3(x-1)(x+1)}$$
$$=\frac{15+12x}{3(x+1)(x-1)}$$
$$=\frac{3(5+4x)}{3(x+1)(x-1)}$$
$$=\frac{4x+5}{(x+1)(x-1)}$$

17. expression

$$\frac{x+2}{3x-1}+\frac{5}{(3x-1)^2}=\frac{(x+2)(3x-1)}{(3x-1)(3x-1)}+\frac{5}{(3x-1)^2}$$
$$=\frac{3x^2+5x-2+5}{(3x-1)^2}$$
$$=\frac{3x^2+5x+3}{(3x-1)^2}$$

18. expression

$$\frac{4}{(2x-5)^2}+\frac{x+1}{2x-5}=\frac{4}{(2x-5)^2}+\frac{(x+1)(2x-5)}{(2x-5)(2x-5)}$$
$$=\frac{4+2x^2-3x-5}{(2x-5)^2}$$
$$=\frac{2x^2-3x-1}{(2x-5)^2}$$

19. expression

$$\frac{x-7}{x}-\frac{x+2}{5x}=\frac{(x-7)(5)}{x(5)}-\frac{x+2}{5x}$$
$$=\frac{5x-35-x-2}{5x}$$
$$=\frac{4x-37}{5x}$$

20. equation

$$\frac{9}{x^2-4}+\frac{2}{x+2}=\frac{-1}{x-2}$$
$$(x^2-4)\left(\frac{9}{x^2-4}\right)+(x^2-4)\left(\frac{2}{x+2}\right)=(x^2-4)\left(\frac{-1}{x-2}\right)$$
$$9+(x-2)(2)=(x+2)(-1)$$
$$9+2x-4=-x-2$$
$$2x+5=-x-2$$
$$3x+5=-2$$
$$3x=-7$$
$$x=-\frac{7}{3}$$

The solution is $-\frac{7}{3}$.

21. equation

$$\frac{3}{x+3}=\frac{5}{x^2-9}-\frac{2}{x-3}$$
$$(x^2-9)\left(\frac{3}{x+3}\right)=(x^2-9)\left(\frac{5}{x^2-9}\right)-(x^2-9)\left(\frac{2}{x-3}\right)$$
$$(x-3)(3)=5-(x+3)(2)$$
$$3x-9=5-2x-6$$
$$3x-9=-2x-1$$
$$5x-9=-1$$
$$5x=8$$
$$x=\frac{8}{5}$$

The solution is $\frac{8}{5}$.

22. expression

$$\frac{10x-9}{x}-\frac{x-4}{3x}=\frac{(10x-9)(3)}{x(3)}-\frac{x-4}{3x}$$
$$=\frac{30x-27-x+4}{3x}$$
$$=\frac{29x-23}{3x}$$

Section 7.6

Practice Exercises

1. Solve the equation as a rational equation.

$$\frac{36}{x}=\frac{4}{11}$$
$$11x\cdot\frac{36}{x}=11x\cdot\frac{4}{11}$$
$$11\cdot 36=x\cdot 4$$
$$396=4x$$
$$\frac{396}{4}=\frac{4x}{4}$$
$$99=x$$

Solve the proportion using cross products.

$$\frac{36}{x}=\frac{4}{11}$$
$$36\cdot 11=x\cdot 4$$
$$396=4x$$
$$\frac{396}{4}=\frac{4x}{4}$$
$$99=x$$

Check: Both methods give a solution of 99. To check, substitute 99 for x in the original proportion. The solution is 99.

2.
$$\frac{3x+2}{9}=\frac{x-1}{2}$$
$$2(3x+2)=9(x-1)$$
$$6x+4=9x-9$$
$$6x=9x-13$$
$$-3x=-13$$
$$\frac{-3x}{-3}=\frac{-13}{-3}$$
$$x=\frac{13}{3}$$

Check: Verify that $\frac{13}{3}$ is the solution.

3. Let x = price of seven 2-liter bottles of Diet Pepsi.

$$\frac{\text{4 bottles}}{\text{7 bottles}} = \frac{\text{price of 4 bottles}}{\text{price of 7 bottles}}$$

$$\frac{4}{7} = \frac{5.16}{x}$$
$$4x = 7(5.16)$$
$$4x = 36.12$$
$$x = 9.03$$

Check: Verify that 4 bottles is to 7 bottles as \$5.16 is to \$9.03.
Seven 2-liter bottles of Diet Pepsi cost \$9.03.

4. Since the triangles are similar, their corresponding sides are in proportion.

$$\frac{20}{8} = \frac{15}{x}$$
$$20x = 8 \cdot 15$$
$$20x = 120$$
$$x = 6$$

Check: To check, replace x with 6 in the original proportion and see that a true statement results.
The missing length is 6 meters.

5. Let x = the unknown number.

In words	the quotient of x and 5	minus	$\frac{3}{2}$	is	the quotient of x and 10
	↓	↓	↓	↓	↓
Translate:	$\frac{x}{5}$	$-$	$\frac{3}{2}$	$=$	$\frac{x}{10}$

The LCD is 10.

$$10\left(\frac{x}{5} - \frac{3}{2}\right) = 10\left(\frac{x}{10}\right)$$
$$10\left(\frac{x}{5}\right) - 10\left(\frac{3}{2}\right) = 10\left(\frac{x}{10}\right)$$
$$2x - 15 = x$$
$$x - 15 = 0$$
$$x = 15$$

Check: To check, verify that "the quotient of 15 and 5 minus $\frac{3}{2}$ is the quotient of 15 and 10," or $\frac{15}{5} - \frac{3}{2} = \frac{15}{10}$.

6. Let x = the time in hours it takes Cindy and Mary to complete the job together. Then $\frac{1}{x}$ = the part of the job they complete in 1 hour.

	Hours to Complete Total Job	Part of Job Completed in 1 Hour
Cindy	3	$\frac{1}{3}$
Mary	4	$\frac{1}{4}$
Together	x	$\frac{1}{x}$

The part of the job Cindy completes in 1 hour, added to the part of the job Mary completes in 1 hour is equal to the part of the job they complete together in 1 hour.

$$\frac{1}{3}+\frac{1}{4}=\frac{1}{x}$$
$$12x\left(\frac{1}{3}\right)+12x\left(\frac{1}{4}\right)=12x\left(\frac{1}{x}\right)$$
$$4x+3x=12$$
$$7x=12$$
$$x=\frac{12}{7} \text{ or } 1\frac{5}{7}$$

Check: The proposed solution is reasonable since $1\frac{5}{7}$ hours is more than half of Cindy's time and less than half of Mary's time. Check $1\frac{5}{7}$ hours in the originally stated problem.

Cindy and Mary can complete the garden planting in $1\frac{5}{7}$ hours.

7. Let x = the speed of the bus. Then since the car's speed is 15 mph faster than that of the bus, the speed of the car is $x + 15$.

Since distance = rate · time, or $d = r \cdot t$, then $t=\frac{d}{r}$.

The bus travels 180 miles in the same time that the car travels 240 miles.

	Distance =	Rate ·	Time
Bus	180	x	$\frac{180}{x}$
Car	240	$x + 15$	$\frac{240}{x+15}$

Since the car and the bus traveled the same amount of time, $\frac{180}{x}=\frac{240}{x+15}$.

$$\frac{180}{x}=\frac{240}{x+15}$$
$$180(x+15)=240x$$
$$180x+2700=240x$$
$$2700=60x$$
$$45=x$$

The speed of the bus is 45 miles per hour. The speed of the car must then be $x + 15$ or 60 miles per hour.

Check: Find the time it takes the car to travel 240 miles and the time it takes the bus to travel 180 miles.

Car: $t=\frac{d}{r}=\frac{240}{60}=4$ hours

Bus: $t=\frac{d}{r}=\frac{180}{45}=4$ hours

Since the times are the same, the proposed solution is correct. The speed of the bus is 45 miles per hour and the speed of the car is 60 miles per hour.

Vocabulary and Readiness Check

1. If both people work together, they can complete the job in less time than either person working alone. That is, in less than 5 hours; choice c.

2. If both inlet pipes are on, they can fill the pond in less time than either pipe alone. That is, in less than 25 hours; choice a.

Exercise Set 7.6

1.
$$\frac{2}{3}=\frac{x}{6}$$
$$12=3x$$
$$4=x$$

3. $\frac{x}{10} = \frac{5}{9}$
$9x = 50$
$x = \frac{50}{9}$

5. $\frac{x+1}{2x+3} = \frac{2}{3}$
$3(x+1) = 2(2x+3)$
$3x+3 = 4x+6$
$3 = x+6$
$-3 = x$

7. $\frac{9}{5} = \frac{12}{3x+2}$
$9(3x+2) = 5(12)$
$27x+18 = 60$
$27x = 42$
$x = \frac{42}{27} = \frac{14}{9}$

9. Let x = the elephant's weight on Pluto.
$\frac{100}{3} = \frac{4100}{x}$
$100x = 3(4100)$
$100x = 12{,}300$
$x = 123$
The elephant's weight is 123 pounds.

11. Let x = the number of calories in 43.2 grams.
$\frac{110}{28.8} = \frac{x}{43.2}$
$110(43.2) = 28.8x$
$4752 = 28.8x$
$165 = x$
There are 165 calories in 43.2 grams.

13. $\frac{16}{10} = \frac{34}{y}$
$16y = 340$
$y = 21.25$

15. $\frac{28}{20} = \frac{8}{y}$
$28y = 160$
$y = \frac{160}{28} = \frac{40}{7}$
$y = 5\frac{5}{7}$ feet

17. $3 \cdot \frac{1}{x} = 9 \cdot \frac{1}{6}$
$\frac{3}{x} = \frac{9}{6}$
$6x\left(\frac{3}{x}\right) = 6x\left(\frac{9}{6}\right)$
$18 = 9x$
$x = 2$
The unknown number is 2.

19. $\frac{3+2x}{x+1} = \frac{3}{2}$
$2(x+1)\left(\frac{3+2x}{x+1}\right) = 2(x+1)\left(\frac{3}{2}\right)$
$2(3+2x) = 3(x+1)$
$6+4x = 3x+3$
$x = -3$
The unknown number is −3.

21. Let x be the number of hours for the two surveyors to survey the roadbed together.

	Hours to Complete Total Job	Part of Job Completed in 1 Hour
Experienced	4	$\frac{1}{4}$
Apprentice	5	$\frac{1}{5}$
Together	x	$\frac{1}{x}$

$\frac{1}{4} + \frac{1}{5} = \frac{1}{x}$
$20x\left(\frac{1}{4}\right) + 20x\left(\frac{1}{5}\right) = 20x\left(\frac{1}{x}\right)$
$5x + 4x = 20$
$9x = 20$
$x = \frac{20}{9}$ or $2\frac{2}{9}$

The experienced surveyor and apprentice surveyor, working together, can survey the road in $2\frac{2}{9}$ hours.

23. Let x be the number of minutes it takes the belts working together.

	Minutes to Complete Total Job	Part of Job Completed in 1 Minute
Larger belt	2	$\frac{1}{2}$
Smaller belt	6	$\frac{1}{6}$
Both belts	x	$\frac{1}{x}$

$$\frac{1}{2}+\frac{1}{6}=\frac{1}{x}$$
$$6x\left(\frac{1}{2}\right)+6x\left(\frac{1}{6}\right)=6x\left(\frac{1}{x}\right)$$
$$3x+x=6$$
$$4x=6$$
$$x=\frac{6}{4}=\frac{3}{2}=1\frac{1}{2}$$

Both belts together can move the cans to the storage area in $1\frac{1}{2}$ minutes.

25. Let r be the jogger's rate. Then, since distance = rate · time, or $d = r \cdot t$, then $t=\frac{d}{r}$.

	Distance =	Rate ·	Time
Trip to Park	12	r	$\frac{12}{r}$
Return Trip	18	r	$\frac{18}{r}$

Since her time on the return trip is 1 hour longer than on the trip to the park, $\frac{18}{r}=\frac{12}{r}+1$.

$$r\left(\frac{18}{r}\right)=r\left(\frac{12}{r}\right)+r(1)$$
$$18=12+r$$
$$6=r$$

She jogs at 6 miles per hour.

27. Let r be his speed on the first portion. Then his speed on the cooldown portion is $r - 2$.

	Distance =	Rate ·	Time
1st portion	20	r	$\frac{20}{r}$
Cooldown portion	16	$r-2$	$\frac{16}{r-2}$

$$\frac{20}{r}=\frac{16}{r-2}$$
$$20(r-2)=16r$$
$$20r-40=16r$$
$$-40=-4r$$
$$r=10$$

and $r - 2 = 10 - 2 = 8$

His speed was 10 miles per hour during the first portion and 8 miles per hour during the cooldown portion.

29. Let x = the minimum floor space needed by 40 students.

$$\frac{1}{9}=\frac{40}{x}$$
$$1x=9(40)$$
$$x=360$$

40 students need 360 square feet.

31.
$$\frac{1}{4}=\frac{x}{8}$$
$$8\left(\frac{1}{4}\right)=8\left(\frac{x}{8}\right)$$
$$2=x$$

The unknown number is 2.

33. Let x be the amount of time it takes Marcus and Tony working together.

	Hours to Complete Total Job	Part of Job Completed in 1 Hour
Marcus	6	$\frac{1}{6}$
Tony	4	$\frac{1}{4}$
Together	x	$\frac{1}{x}$

$$\frac{1}{6}+\frac{1}{4}=\frac{1}{x}$$
$$12x\left(\frac{1}{6}\right)+12x\left(\frac{1}{4}\right)=12x\left(\frac{1}{x}\right)$$
$$2x+3x=12$$
$$5x=12$$
$$x=\frac{12}{5}=2\frac{2}{5}$$
$$45\left(\frac{12}{5}\right)=108$$

Together Marcus and Tony work for $2\frac{2}{5}$ hours at $45 per hour. The labor estimate should be $108.

35. Let w be the speed of the wind.

	Distance =	Rate	· Time
With wind	400	$230+w$	$\frac{400}{230+w}$
Against wind	336	$230-w$	$\frac{336}{230-w}$

Since the time with the wind is the same as the time against the wind, $\frac{336}{230-w}=\frac{400}{230+w}$.

$$\frac{336}{230-w}=\frac{400}{230+w}$$
$$336(230+w)=400(230-w)$$
$$77,280+336w=92,000-400w$$
$$736w=14,720$$
$$w=20$$

The speed of the wind is 20 miles per hour.

37. $\frac{y}{25}=\frac{3}{2}$

$$y\cdot 2=25\cdot 3$$
$$y\cdot 2=75$$
$$y=\frac{75}{2}$$
$$y=37\frac{1}{2}$$

The unknown length is $37\frac{1}{2}$ feet.

39. Let x = the number of rushing yards in one game.

$$\frac{x}{1}=\frac{4045}{12}$$
$$12x=1(4045)$$
$$12x=4045$$
$$x\approx 337$$

Ken averaged 337 yards per game.

41.
$$\frac{2}{x-3}-\frac{4}{x+3}=8\cdot\frac{1}{x^2-9}$$
$$(x-3)(x+3)\left(\frac{2}{x-3}-\frac{4}{x+3}\right)=(x-3)(x+3)\left(\frac{8}{x^2-9}\right)$$
$$(x-3)(x+3)\left(\frac{2}{x-3}\right)-(x-3)(x+3)\left(\frac{4}{x+3}\right)=8$$
$$2(x+3)-4(x-3)=8$$
$$2x+6-4x+12=8$$
$$-2x=-10$$
$$x=5$$
The unknown number is 5.

43. Let r be the rate of the plane in still air.

	Distance =	Rate	· Time
With wind	630	$r+35$	$\frac{630}{r+35}$
Against wind	455	$r-35$	$\frac{455}{r-35}$

$$\frac{630}{r+35}=\frac{455}{r-35}$$
$$630(r-35)=455(r+35)$$
$$630r-22{,}050=455r+15{,}925$$
$$175r=37{,}975$$
$$r=217$$
The speed in still air is 217 mph.

45. Let x = the number of gallons of water needed.
$$\frac{8}{2}=\frac{36}{x}$$
$$8x=2(36)$$
$$8x=72$$
$$x=9$$
Nine gallons of water are needed for the entire box.

47.

	r ×	t =	d
With wind	$16+x$	$\frac{48}{16+x}$	48
Into Wind	$16-x$	$\frac{16}{16-x}$	16

Since the times are the same, $\frac{48}{16+x}=\frac{16}{16-x}$.

$$\frac{48}{16+x}=\frac{16}{16-x}$$
$$48(16-x)=16(16+x)$$
$$768-48x=256+16x$$
$$512=64x$$
$$8=x$$

The rate of the wind is 8 miles per hour.

49. Let x be the slower speed. Then $x + 40$ is the faster speed.

	r	× t	= d
Slower	x	$\frac{70}{x}$	70
Faster	$x+40$	$\frac{300}{x+40}$	300

Since the time spent at the faster speed was twice that spent at the slower speed, $\frac{300}{x+4}=2\left(\frac{70}{x}\right)$.

$$\frac{300}{x+40}=\frac{140}{x}$$
$$300x=140(x+40)$$
$$300x=140x+5600$$
$$160x=5600$$
$$x=35$$
$$x+40=35+40=75$$

The slower speed was 35 miles per hour and the faster speed was 75 miles per hour.

51. Let x be the amount of time it takes the second worker to do the job alone.

	Hours to Complete Total Job	Part of Job Completed in 1 Hour
Custodian	3	$\frac{1}{3}$
2nd Worker	x	$\frac{1}{x}$
Together	$1\frac{1}{2}$ or $\frac{3}{2}$	$\frac{2}{3}$

$$\frac{1}{3}+\frac{1}{x}=\frac{2}{3}$$
$$3x\left(\frac{1}{3}\right)+3x\left(\frac{1}{x}\right)=3x\left(\frac{2}{3}\right)$$
$$x+3=2x$$
$$3=x$$

It takes the second worker 3 hours to do the job alone.

53. Let x be the missing dimension.

$$\frac{x}{8}=\frac{20}{6}$$
$$6x=8\cdot 20$$
$$x=\frac{160}{6}$$
$$x=\frac{80}{3}=26\frac{2}{3}$$

The side is $26\frac{2}{3}$ feet long.

55.
$$\frac{3}{2}=\frac{324}{x}$$
$$3\cdot x=2\cdot 324$$
$$3x=648$$
$$x=\frac{648}{3}=216$$

There should be 216 other nuts in the can.

57. Let x be the speed of the plane in still air.

	r	× t	= d
With wind	$x+30$	$\frac{2160}{x+30}$	2160
Against Wind	$x-30$	$\frac{1920}{x-30}$	1920

Since the times are the same, $\frac{2160}{x+30}=\frac{1920}{x-30}$.

$$\frac{2160}{x+30}=\frac{1920}{x-30}$$
$$2160(x-30)=1920(x+30)$$
$$2160x-64,800=1920x+57,600$$
$$240x=122,400$$
$$x=510$$

The speed of the plane in still air is 510 miles per hour.

59. Let x be the number of hours it would take the third pipe alone to fill the pool.

	Hours to Complete Total Job	Part of Job Completed in 1 Hour
1st Pipe	20	$\frac{1}{20}$
2nd Pipe	15	$\frac{1}{15}$
3rd Pipe	x	$\frac{1}{x}$
3 Pipes Together	6	$\frac{1}{6}$

$$\frac{1}{20}+\frac{1}{15}+\frac{1}{x}=\frac{1}{6}$$
$$60x\left(\frac{1}{20}\right)+60x\left(\frac{1}{15}\right)+60x\left(\frac{1}{x}\right)=60x\left(\frac{1}{6}\right)$$
$$3x+4x+60=10x$$
$$7x+60=10x$$
$$60=3x$$
$$20=x$$

It takes the third pipe 20 hours to fill the pool.

61. Let m be the speed of the motorcycle. Then the speed of the car is $m + 10$.

	r	$\times$ t	$=$ d
Motorcycle	m	$\frac{240}{m}$	240
Car	$m+10$	$\frac{280}{m+10}$	280

Since the times are the same, $\frac{240}{m}=\frac{280}{m+10}$.

$$\frac{240}{m}=\frac{280}{m+10}$$
$$240(m+10)=280m$$
$$240m+2400=280m$$
$$2400=40m$$
$$60=m$$

$m + 10 = 60 + 10 = 70$

The speed of the motorcycle is 60 miles per hour and the speed of the car is 70 miles per hour.

63. Let x be the amount of time it takes the third cook alone to prepare the pies.

	Time	In one hour
First cook	6	$\frac{1}{6}$
Second cook	7	$\frac{1}{7}$
Third cook	x	$\frac{1}{x}$
Together	2	$\frac{1}{2}$

$$\frac{1}{6}+\frac{1}{7}+\frac{1}{x}=\frac{1}{2}$$
$$42x\left(\frac{1}{6}+\frac{1}{7}+\frac{1}{x}\right)=42x\left(\frac{1}{2}\right)$$
$$42x\left(\frac{1}{6}\right)+42x\left(\frac{1}{7}\right)+42x\left(\frac{1}{x}\right)=21x$$
$$7x+6x+42=21x$$
$$13x+42=21x$$
$$42=21x-13x$$
$$42=8x$$
$$\frac{42}{8}=x$$
$$\frac{21}{4}=x$$
$$5\frac{1}{4}=x$$

The third cook can prepare the pies in $5\frac{1}{4}$ hours.

65. Let x be the number of minutes it takes the second pump to fill the tank. Then it takes $3x$ minutes for the first pump to fill the tank.

	Minutes to Complete Total Job	Part of Job Completed in 1 Minute
1st Pump	$3x$	$\frac{1}{3x}$
2nd Pump	x	$\frac{1}{x}$
Together	21	$\frac{1}{21}$

$$\frac{1}{3x}+\frac{1}{x}=\frac{1}{21}$$
$$21x\left(\frac{1}{3x}\right)+21x\left(\frac{1}{x}\right)=21x\left(\frac{1}{21}\right)$$
$$7+21=x$$
$$28=x,\ 3x=3(28)=84$$

The 1st pump takes 28 minutes and the 2nd takes 84 minutes.

67. $\frac{9}{12}=\frac{3.75}{x}$
$9x=45$
$x=5$
The missing length is 5.

69. $\frac{16}{24}=\frac{9}{x}$
$16x=216$
$x=13.5$
The missing length is 13.5.

71. $(-2, 5), (4, -3)$
$m=\frac{-3-5}{4-(-2)}=\frac{-8}{6}=-\frac{4}{3}$
Since the slope is negative, the line moves downward.

73. $(-3, -6), (1, 5)$
$m=\frac{5-(-6)}{1-(-3)}=\frac{11}{4}$
Since the slope is positive, the line moves upward.

75. $(3, 7), (3, -2)$
$m=\frac{-2-7}{3-3}=\frac{-9}{0}$
The slope is undefined. Since the slope is undefined, the line is vertical.

77. The capacity in 2001 was approximately 4400 megawatts. The capacity in 2003 was approximately 6400 megawatts. The increase in capacity was approximately $6400-4400=2000$ megawatts.

79. The capacity in 2007 was approximately 14,630 megawatts, or 14.63(1000 megawatts).
$14.63(560{,}000)\approx 8{,}190{,}000$
In 2007, the number of megawatts generate4d from wind would serve the electricity needs of 8,190,000 people.

81. Answers may vary

83. None; answers may vary

85. $\frac{1}{6}x+\frac{1}{12}x+\frac{1}{7}x+5+\frac{1}{2}x+4=x$
$$\frac{1}{6}x+\frac{1}{12}x+\frac{1}{7}x+\frac{1}{2}x+9=x$$
$$84\left(\frac{1}{6}x+\frac{1}{12}x+\frac{1}{7}x+\frac{1}{2}x+9\right)=84x$$
$$14x+7x+12x+42x+756=84x$$
$$75x+756=84x$$
$$756=9x$$
$$\frac{756}{9}=\frac{9x}{9}$$
$$84=x$$
He died when he was 84 years old.

87.
$$4+\frac{1}{2}x+\frac{1}{6}x+3+\frac{1}{10}x=x$$
$$30\left(7+\frac{1}{2}x+\frac{1}{6}x+\frac{1}{10}x\right)=(30)(x)$$
$$30\cdot 7+30\left(\frac{1}{2}x\right)+30\left(\frac{1}{6}x\right)+30\left(\frac{1}{10}x\right)=30x$$
$$210+15x+5x+3x=30x$$
$$210+23x=30x$$
$$210=30x-23x$$
$$210=7x$$
$$30=x$$
You are 30 years old.

89. Let d be the distance that the giraffe runs before the hyena catches it. Then the hyena runs $d+0.5$ miles.

	Distance =	Rate	· Time
Hyena	$d+0.5$	40	$\frac{d+0.5}{40}$
Giraffe	d	32	$\frac{d}{32}$

$$\frac{d+0.5}{40}=\frac{d}{32}$$
$$32(d+0.5)=40d$$
$$32d+16=40d$$
$$16=8d$$
$$2=d,\ \frac{d}{32}=\frac{2}{32}=\frac{1}{16}$$

It will take the hyena $\frac{1}{16}$ hour or 3.75 minutes to overtake the giraffe.

The Bigger Picture

1. $(3x-2)(4x^2-x-5)$
$=3x(4x^2-x-5)-2(4x^2-x-5)$
$=12x^3-3x^2-15x-8x^2+2x+10$
$=12x^3-11x^2-13x+10$

2. $(2x-y)^2=(2x)^2-2(2x)(y)+y^2$
$=4x^2-4xy+y^2$

3. $8y^3-20y^5=4y^3(2-5y^2)$

4. $9m^2-11mn+2n^2=9m^2-2mn-9mn+2n^2$
$=m(9m-2n)-n(9m-2n)$
$=(9m-2n)(m-n)$

5. $\frac{7}{x}=\frac{9}{x-10}$
$7(x-10)=9x$
$7x-70=9x$
$-70=2x$
$-35=x$

6. $\frac{7}{x}+\frac{9}{x-10}=\frac{7(x-10)}{x(x-10)}+\frac{9(x)}{(x-10)x}$
$=\frac{7(x-10)+9x}{x(x-10)}$
$=\frac{7x-70+9x}{x(x-10)}$
$=\frac{16x-70}{x(x-10)}$ or $\frac{2(8x-35)}{x(x-10)}$

7. $(-3x^5)\left(\frac{1}{2}x^7\right)(8x)=\left(-3\cdot\frac{1}{2}\cdot 8\right)(x^5\cdot x^7\cdot x^1)$
$=-12x^{5+7+1}$
$=-12x^{13}$

8. $5x-1=|-4|+|-5|$
$5x-1=4+5$
$5x-1=9$
$5x=10$
$x=2$

9. $\frac{8-12}{12\div 3\cdot 2}=\frac{-4}{4\cdot 2}=\frac{-4}{8}=-\frac{1}{2}$

10. $-2(3y-4)\le 5y-7-7y-1$
$-6y+8\le -2y-8$
$8\le 4y-8$
$16\le 4y$
$4\le y$
$[4, \infty)$

11.
$$\frac{7}{x}+\frac{5}{2x+3}=\frac{-2}{x}$$
$$x(2x+3)\left(\frac{7}{x}+\frac{5}{2x+3}\right)=x(2x+3)\left(\frac{-2}{x}\right)$$
$$x(2x+3)\cdot\frac{7}{x}+x(2x+3)\cdot\frac{5}{2x+3}=x(2x+3)\cdot\frac{-2}{x}$$
$$7(2x+3)+5x=-2(2x+3)$$
$$14x+21+5x=-4x-6$$
$$19x+21=-4x-6$$
$$23x+21=-6$$
$$23x=-27$$
$$x=-\frac{27}{23}$$

12. $\frac{(a^{-3}b^2)^{-5}}{ab^4}=\frac{a^{(-3)(-5)}b^{2(-5)}}{ab^4}=\frac{a^{15}b^{-10}}{ab^4}=\frac{a^{14}}{b^{14}}$

Section 7.7

Practice Exercises

1. Use the ordered pair (2, 10) in the direct variation equation $y = kx$.
$y=kx$
$10=k\cdot 2$
$\frac{10}{2}=\frac{k\cdot 2}{2}$
$5=k$
Since $k = 5$, the equation is $y = 5x$. To check, see that each given y is 5 times the given x.

2. Since y varies directly as x, the relationship is of the form $y = kx$. Let $y = 12$ and $x = 48$.
$12=k\cdot 48$
$\frac{12}{48}=\frac{k\cdot 48}{48}$
$\frac{1}{4}=k$

The constant of variation is $\frac{1}{4}$ and the equation is $y=\frac{1}{4}x$. Now replace x with 20.

$y = \frac{1}{4}x$
$y = \frac{1}{4} \cdot 20$
$y = 5$
Thus, when x is 20, y is 5.

3. The constant of variation is the same as the slope of the line.
$\text{slope} = \frac{6-0}{8-0} = \frac{6}{8} = \frac{3}{4}$
Thus, $k = \frac{3}{4}$ and the variation equation is
$y = \frac{3}{4}x.$

4. Use the ordered pair (2, 4) in the inverse variation equation $y = \frac{k}{x}$.
$y = \frac{k}{x}$
$4 = \frac{k}{2}$
$2 \cdot 4 = 2 \cdot \frac{k}{2}$
$8 = k$
Since $k = 8$, the equation is $y = \frac{8}{x}$.

5. Since y varies inversely as x, the constant of variation is the product of the given x and y.
$k = xy = (42)(0.05) = 2.1$
The constant of variation is 2.1 and the equation is $y = \frac{2.1}{x}$. Now replace x with 70.
$y = \frac{2.1}{x}$
$y = \frac{2.1}{70}$
$y = 0.03$
Thus, when x is 70, y is 0.03.

6. Since the area A varies directly as the square of one of its legs x, the equation is $A = kx^2$.
Let $A = 32$ and $x = 8$.
$A = k \cdot x^2$
$32 = k \cdot 8^2$
$32 = k \cdot 64$
$\frac{1}{2} = k$
The formula for the area of an isosceles right triangle is then $A = \frac{1}{2}x^2$ where x is the length of one leg. Substitute 3.6 for x.
$A = \frac{1}{2}x^2$
$A = \frac{1}{2}(3.6)^2$
$A = 6.48$
The area of an isosceles right triangle whose legs measure 3.6 units is 6.48 square units.

7. Since the volume of gas varies inversely with pressure, the equation is $V = \frac{k}{P}$.
To find k, let $V = 50$ and $P = 20$.
$V = \frac{k}{P}$
$50 = \frac{k}{20}$
$1000 = k$
The equation of variation is $V = \frac{1000}{P}$.
Now let $P = 40$.
$V = \frac{1000}{P}$
$V = \frac{1000}{40}$
$V = 25$
At a pressure of 40 atmospheres, the volume of the oxygen is 25 ml.

Vocabulary and Readiness Check

1. $y = \frac{k}{x}$, where k is a constant is inverse variation.

2. $y = kx$, where k is a constant is direct variation.

3. $y = 5x$ is direct variation.

4. $y = \frac{5}{x}$ is inverse variation.

5. $y = \frac{7}{x^2}$ is inverse variation.

6. $y = 6.5x^4$ is direct variation.

7. $y = \frac{11}{x}$ is inverse variation.

8. $y = 18x$ is direct variation.

9. $y = 12x^2$ is direct variation.

10. $y = \frac{20}{x^3}$ is inverse variation.

Exercise Set 7.7

1. $y = kx$
$3 = k(6)$
$\frac{3}{6} = k$
$\frac{1}{2} = k$
$y = \frac{1}{2}x$

3. $y = kx$
$-12 = k(-2)$
$6 = k$
$y = 6x$

5. $k = \text{slope} = \frac{3-0}{1-0} = \frac{3}{1} = 3$
$y = 3x$

7. $k = \text{slope} = \frac{2-0}{3-0} = \frac{2}{3}$
$y = \frac{2}{3}x$

9. $y = \frac{k}{x}$
$7 = \frac{k}{1}$
$7 = k$
$y = \frac{7}{x}$

11. $y = \frac{k}{x}$
$0.05 = \frac{k}{10}$
$0.5 = k$
$y = \frac{0.5}{x}$

13. $y = kx$

15. $h = \frac{k}{t}$

17. $z = kx^2$

19. $y = \frac{k}{z^3}$

21. $x = \frac{k}{\sqrt{y}}$

23. $y = kx$
$20 = k(5)$
$4 = k$

$y = 4x$
$y = 4(10)$
$y = 40$

25. $y = \frac{k}{x}$
$5 = \frac{k}{60}$
$300 = k$

$y = \frac{300}{x}$
$y = \frac{300}{100}$
$y = 3$

27. $z = kx^2$
$96 = k(4)^2$
$96 = 16k$
$6 = k$

$z = 6x^2$
$z = 6(3)^2$
$z = 6(9)$
$z = 54$

29. $a = \frac{k}{b^3}$
$\frac{3}{2} = \frac{k}{2^3}$
$\frac{3}{2} = \frac{k}{8}$
$2k = 24$
$k = 12$

$a = \frac{12}{b^3}$
$a = \frac{12}{3^3}$
$a = \frac{12}{27}$
$a = \frac{4}{9}$

31. $p = kh$
$112.50 = k(18)$
$6.25 = k$

$p = 6.25h$
$p = 6.25(10)$
$p = 62.5$
Your pay is \$62.50 for 10 hours.

33. $x = \frac{k}{n}$
$9.00 = \frac{k}{5000}$
$45,000 = k$

$c = \frac{45,000}{n}$
$c = \frac{45,000}{7500}$
$c = 6$
The cost is \$6.00 to manufacture 7500 headphones.

35. $d = kw$
$4 = k(60)$
$\frac{4}{60} = k$
$\frac{1}{15} = k$

$d = \frac{1}{15}w$
$d = \frac{1}{15}(80)$
$d = \frac{80}{15}$
$d = 5\frac{1}{3}$

The spring stretches $5\frac{1}{3}$ inches with an 80-pound weight.

37. $w = \frac{k}{d^2}$
$180 = \frac{k}{4000^2}$
$180 = \frac{k}{16,000,000}$
$2,880,000,000 = k$

$w = \frac{2,880,000,000}{d^2}$
$w = \frac{2,880,000,000}{4010^2}$
$w = \frac{2,880,000,000}{16,080,100}$
$w \approx 179.1$
His weight 10 miles above the earth's surface is 179.1 pounds.

39. $d = kt^2$
$64 = k(2)^2$
$64 = 4k$
$16 = k$

$d = 16t^2$
$d = 16(10)^2$
$d = 16(100)$
$d = 1600$
He falls 1600 feet in 10 seconds.

41. $\dfrac{\frac{3}{4}+\frac{1}{4}}{\frac{3}{8}+\frac{13}{8}}=\dfrac{\frac{3+1}{4}}{\frac{3+13}{8}}=\dfrac{\frac{4}{4}}{\frac{16}{8}}=\dfrac{1}{2}$

43. $\dfrac{\frac{2}{5}+\frac{1}{5}}{\frac{7}{10}+\frac{7}{10}}=\dfrac{\frac{2+1}{5}}{\frac{7+7}{10}}$
$=\dfrac{\frac{3}{5}}{\frac{14}{10}}$
$=\dfrac{3}{5}\div\dfrac{14}{10}$
$=\dfrac{3}{5}\cdot\dfrac{10}{14}$
$=\dfrac{3\cdot 2\cdot 5}{5\cdot 2\cdot 7}$
$=\dfrac{3}{7}$

45. $y=kx$
If x is tripled, y is also tripled.

47. $P=k\sqrt{l}$
The result of quadrupling l is $4l$.
$\sqrt{4l}=\sqrt{4}\cdot\sqrt{l}=2\sqrt{l}$
Thus, when the length of the pendulum is quadrupled, the period is doubled.

Section 7.8

Practice Exercises

1. $\dfrac{\frac{3}{4}}{\frac{6}{11}}=\dfrac{3}{4}\div\dfrac{6}{11}=\dfrac{3}{4}\cdot\dfrac{11}{6}=\dfrac{3\cdot 11}{4\cdot 2\cdot 3}=\dfrac{11}{8}$

2. $\dfrac{\frac{3}{4}+\frac{2}{3}}{\frac{3}{4}-\frac{1}{5}}=\dfrac{\frac{3(3)}{4(3)}+\frac{2(4)}{3(4)}}{\frac{3(5)}{4(5)}-\frac{1(4)}{5(4)}}$
$=\dfrac{\frac{9}{12}+\frac{8}{12}}{\frac{15}{20}-\frac{4}{20}}$
$=\dfrac{\frac{17}{12}}{\frac{11}{20}}$
$=\dfrac{17}{12}\cdot\dfrac{20}{11}$
$=\dfrac{17\cdot 4\cdot 5}{3\cdot 4\cdot 11}$
$=\dfrac{85}{33}$

3. $\dfrac{\frac{4}{x}-\frac{1}{2}}{\frac{1}{5}-\frac{x}{10}}=\dfrac{\frac{8}{2x}-\frac{x}{2x}}{\frac{2}{10}-\frac{x}{10}}$
$=\dfrac{\frac{8-x}{2x}}{\frac{2-x}{10}}$
$=\dfrac{8-x}{2x}\cdot\dfrac{10}{2-x}$
$=\dfrac{2\cdot 5(8-x)}{2\cdot x(2-x)}$
$=\dfrac{5(8-x)}{x(2-x)}$

4. The LCD of $\frac{3}{4}$, $\frac{2}{3}$, $\frac{3}{4}$, and $\frac{1}{5}$ is 60.
$\dfrac{\frac{3}{4}+\frac{2}{3}}{\frac{3}{4}-\frac{1}{5}}=\dfrac{60\left(\frac{3}{4}+\frac{2}{3}\right)}{60\left(\frac{3}{4}-\frac{1}{5}\right)}$
$=\dfrac{60\left(\frac{3}{4}\right)+60\left(\frac{2}{3}\right)}{60\left(\frac{3}{4}\right)-60\left(\frac{1}{5}\right)}$
$=\dfrac{45+40}{45-12}$
$=\dfrac{85}{33}$

5. The LCD of $\frac{a-b}{b}$, $\frac{a}{b}$, and $\frac{4}{1}$ is b.
$\dfrac{\frac{a-b}{b}}{\frac{a}{b}+4}=\dfrac{b\left(\frac{a-b}{b}\right)}{b\left(\frac{a}{b}+4\right)}=\dfrac{b\left(\frac{a-b}{b}\right)}{b\left(\frac{a}{b}\right)+b(4)}=\dfrac{a-b}{a+4b}$

6. The LCD of $\frac{4}{3b}$, $\frac{b}{a}$, $\frac{a}{3}$, and $\frac{b}{1}$ is $3ab$.
$\dfrac{\frac{4}{3b}+\frac{b}{a}}{\frac{a}{3}-b}=\dfrac{3ab\left(\frac{4}{3b}+\frac{b}{a}\right)}{3ab\left(\frac{a}{3}-b\right)}$
$=\dfrac{3ab\left(\frac{4}{3b}\right)+3ab\left(\frac{b}{a}\right)}{3ab\left(\frac{a}{3}\right)-3ab(b)}$
$=\dfrac{4a+3b^2}{a^2b-3ab^2}$ or $\dfrac{4a+3b^2}{ab(a-3b)}$

Vocabulary and Readiness Check

1. $\dfrac{\frac{y}{2}}{\frac{5x}{2}}=\dfrac{2\left(\frac{y}{2}\right)}{2\left(\frac{5x}{2}\right)}=\dfrac{y}{5x}$

2. $\dfrac{\frac{10}{x}}{\frac{z}{x}}=\dfrac{x\left(\frac{10}{x}\right)}{x\left(\frac{z}{x}\right)}=\dfrac{10}{z}$

3. $\dfrac{\frac{3}{x}}{\frac{5}{x^2}}=\dfrac{x^2\left(\frac{3}{x}\right)}{x^2\left(\frac{5}{x^2}\right)}=\dfrac{3x}{5}$

4. $\dfrac{\frac{a}{10}}{\frac{b}{20}}=\dfrac{20\left(\frac{a}{10}\right)}{20\left(\frac{b}{20}\right)}=\dfrac{2a}{b}$

Exercise Set 7.8

1. $\dfrac{\frac{1}{2}}{\frac{3}{4}}=\dfrac{1}{2}\cdot\dfrac{4}{3}=\dfrac{1\cdot2\cdot2}{2\cdot3}=\dfrac{2}{3}$

3. $\dfrac{-\frac{4x}{9}}{-\frac{2x}{3}}=-\dfrac{4x}{9}\cdot-\dfrac{3}{2x}=\dfrac{2\cdot2\cdot3\cdot x}{3\cdot3\cdot2\cdot x}=\dfrac{2}{3}$

5. $\dfrac{\frac{1+x}{6}}{\frac{1+x}{3}}=\dfrac{1+x}{6}\cdot\dfrac{3}{1+x}=\dfrac{3\cdot(1+x)}{2\cdot3\cdot(1+x)}=\dfrac{1}{2}$

7.
$$\dfrac{\frac{1}{2}+\frac{2}{3}}{\frac{5}{9}-\frac{5}{6}}=\dfrac{\frac{1}{2}\cdot\frac{3}{3}+\frac{2}{3}\cdot\frac{2}{2}}{\frac{5}{9}\cdot\frac{2}{2}-\frac{5}{6}\cdot\frac{3}{3}}$$
$$=\dfrac{\frac{3}{6}+\frac{4}{6}}{\frac{10}{18}-\frac{15}{18}}$$
$$=\dfrac{\frac{7}{6}}{-\frac{5}{18}}$$
$$=\dfrac{7}{6}\cdot-\dfrac{18}{5}$$
$$=-\dfrac{7\cdot3\cdot6}{6\cdot5}$$
$$=-\dfrac{21}{5}$$

9.
$$\dfrac{2+\frac{7}{10}}{1+\frac{3}{5}}=\dfrac{10\left(2+\frac{7}{10}\right)}{10\left(1+\frac{3}{5}\right)}$$
$$=\dfrac{10(2)+10\left(\frac{7}{10}\right)}{10(1)+10\left(\frac{3}{5}\right)}$$
$$=\dfrac{20+7}{10+6}$$
$$=\dfrac{27}{16}$$

11. $\dfrac{\frac{1}{3}}{\frac{1}{2}-\frac{1}{4}}=\dfrac{12\left(\frac{1}{3}\right)}{12\left(\frac{1}{2}-\frac{1}{4}\right)}=\dfrac{12\left(\frac{1}{3}\right)}{12\left(\frac{1}{2}\right)-12\left(\frac{1}{4}\right)}=\dfrac{4}{6-3}=\dfrac{4}{3}$

13. $\dfrac{-\frac{2}{9}}{-\frac{14}{3}}=-\dfrac{2}{9}\cdot-\dfrac{3}{14}=\dfrac{2\cdot3}{3\cdot3\cdot2\cdot7}=\dfrac{1}{21}$

15. $\dfrac{-\frac{5}{12x^2}}{\frac{25}{16x^3}}=-\dfrac{5}{12x^2}\cdot\dfrac{16x^3}{25}=-\dfrac{5\cdot4\cdot4\cdot x^2\cdot x}{4\cdot3\cdot x^2\cdot5\cdot5}=-\dfrac{4x}{15}$

17. $\dfrac{\frac{m}{n}-1}{\frac{m}{n}+1}=\dfrac{n\left(\frac{m}{n}-1\right)}{n\left(\frac{m}{n}+1\right)}=\dfrac{n\left(\frac{m}{n}\right)-n(1)}{n\left(\frac{m}{n}\right)+n(1)}=\dfrac{m-n}{m+n}$

19.
$$\dfrac{\frac{1}{5}-\frac{1}{x}}{\frac{7}{10}+\frac{1}{x^2}}=\dfrac{10x^2\left(\frac{1}{5}-\frac{1}{x}\right)}{10x^2\left(\frac{7}{10}+\frac{1}{x^2}\right)}$$
$$=\dfrac{10x^2\left(\frac{1}{5}\right)-10x^2\left(\frac{1}{x}\right)}{10x^2\left(\frac{7}{10}\right)+10x^2\left(\frac{1}{x^2}\right)}$$
$$=\dfrac{2x^2-10x}{7x^2+10}$$
$$=\dfrac{2x(x-5)}{7x^2+10}$$

21. $\dfrac{1+\frac{1}{y-2}}{y+\frac{1}{y-2}} = \dfrac{(y-2)\left(1+\frac{1}{y-2}\right)}{(y-2)\left(y+\frac{1}{y-2}\right)}$

$= \dfrac{(y-2)(1)+(y-2)\left(\frac{1}{y-2}\right)}{(y-2)(y)+(y-2)\left(\frac{1}{y-2}\right)}$

$= \dfrac{y-2+1}{y^2-2y+1}$

$= \dfrac{y-1}{(y-1)^2}$

$= \dfrac{1}{y-1}$

23. $\dfrac{\frac{4y-8}{16}}{\frac{6y-12}{4}} = \dfrac{4y-8}{16} \cdot \dfrac{4}{6y-12} = \dfrac{4(y-2)\cdot 4}{4\cdot 4\cdot 6(y-2)} = \dfrac{1}{6}$

25. $\dfrac{\frac{x}{y}+1}{\frac{x}{y}-1} = \dfrac{y\left(\frac{x}{y}+1\right)}{y\left(\frac{x}{y}-1\right)} = \dfrac{y\left(\frac{x}{y}\right)+y(1)}{y\left(\frac{x}{y}\right)-y(1)} = \dfrac{x+y}{x-y}$

27. $\dfrac{1}{2+\frac{1}{3}} = \dfrac{3(1)}{3\left(2+\frac{1}{3}\right)} = \dfrac{3(1)}{3(2)+3\left(\frac{1}{3}\right)} = \dfrac{3}{6+1} = \dfrac{3}{7}$

29. $\dfrac{\frac{ax+ab}{x^2-b^2}}{\frac{x+b}{x-b}} = \dfrac{ax+ab}{x^2-b^2} \cdot \dfrac{x-b}{x+b}$

$= \dfrac{a(x+b)\cdot(x-b)}{(x+b)(x-b)\cdot(x+b)}$

$= \dfrac{a}{x+b}$

31. $\dfrac{\frac{-3+y}{4}}{\frac{8+y}{28}} = \dfrac{-3+y}{4} \cdot \dfrac{28}{8+y} = \dfrac{4\cdot 7\cdot(-3+y)}{4\cdot(8+y)} = \dfrac{7(y-3)}{8+y}$

33. $\dfrac{3+\frac{12}{x}}{1-\frac{16}{x^2}} = \dfrac{x^2\left(3+\frac{12}{x}\right)}{x^2\left(1-\frac{16}{x^2}\right)}$

$= \dfrac{x^2(3)+x^2\left(\frac{12}{x}\right)}{x^2(1)-x^2\left(\frac{16}{x^2}\right)}$

$= \dfrac{3x^2+12x}{x^2-16}$

$= \dfrac{3x(x+4)}{(x-4)(x+4)}$

$= \dfrac{3x}{x-4}$

35. $\dfrac{\frac{8}{x+4}+2}{\frac{12}{x+4}-2} = \dfrac{(x+4)\left(\frac{8}{x+4}+2\right)}{(x+4)\left(\frac{12}{x+4}-2\right)}$

$= \dfrac{(x+4)\left(\frac{8}{x+4}\right)+(x+4)(2)}{(x+4)\left(\frac{12}{x+4}\right)-(x+4)(2)}$

$= \dfrac{8+2x+8}{12-2x-8}$

$= \dfrac{16+2x}{4-2x}$

$= \dfrac{2(8+x)}{2(2-x)}$

$= \dfrac{8+x}{2-x}$

$= -\dfrac{x+8}{x-2}$

37. $\dfrac{\frac{s}{r}+\frac{r}{s}}{\frac{s}{r}-\frac{r}{s}} = \dfrac{rs\left(\frac{s}{r}+\frac{r}{s}\right)}{rs\left(\frac{s}{r}-\frac{r}{s}\right)} = \dfrac{rs\left(\frac{s}{r}\right)+rs\left(\frac{r}{s}\right)}{rs\left(\frac{s}{r}\right)-rs\left(\frac{r}{s}\right)} = \dfrac{s^2+r^2}{s^2-r^2}$

39. $$\frac{\frac{6}{x-5}+\frac{x}{x-2}}{\frac{3}{x-6}-\frac{2}{x-5}} = \frac{(x-5)(x-2)(x-6)\left(\frac{6}{x-5}+\frac{x}{x-2}\right)}{(x-5)(x-2)(x-6)\left(\frac{3}{x-6}-\frac{2}{x-5}\right)}$$
$$= \frac{(x-5)(x-2)(x-6\left(\frac{6}{x-5}\right)+(x-5)(x-2)(x-6)\left(\frac{x}{x-2}\right)}{(x-5)(x-2)(x-6)\left(\frac{3}{x-6}\right)-(x-5)(x-2)(x-6)\left(\frac{2}{x-5}\right)}$$
$$= \frac{6(x-2)(x-6)+x(x-5)(x-6)}{3(x-5)(x-2)-2(x-2)(x-6)}$$
$$= \frac{6x^2-48x+72+x^3-11x^2+30x}{3x^2-21x+30-2x^2+16x-24}$$
$$= \frac{x^3-5x^2-18x+72}{x^2-5x+6}$$
$$= \frac{(x-6)(x-3)(x+4)}{(x-2)(x-3)}$$
$$= \frac{(x-6)(x+4)}{x-2}$$

41. $\sqrt{81}=\sqrt{9^2}=9$

43. $\sqrt{1}=\sqrt{1^2}=1$

45. $\sqrt{\frac{1}{25}}=\sqrt{\left(\frac{1}{5}\right)^2}=\frac{1}{5}$

47. $\sqrt{\frac{4}{9}}=\sqrt{\left(\frac{2}{3}\right)^2}=\frac{2}{3}$

49. Answers may vary

51. $$\frac{\frac{1}{3}+\frac{3}{4}}{2}=\frac{12\left(\frac{1}{3}+\frac{3}{4}\right)}{12(2)}=\frac{12\left(\frac{1}{3}\right)+12\left(\frac{3}{4}\right)}{12(2)}=\frac{4+9}{24}=\frac{13}{24}$$

53. $$\frac{1}{\frac{1}{R_1}+\frac{1}{R_2}}=\frac{R_1R_2(1)}{R_1R_2\left(\frac{1}{R_1}+\frac{1}{R_2}\right)}$$
$$=\frac{R_1R_2}{R_1R_2\left(\frac{1}{R_1}\right)+R_1R_2\left(\frac{1}{R_2}\right)}$$
$$=\frac{R_1R_2}{R_2+R_1}$$

55. $\dfrac{x^{-1}+2^{-1}}{x^{-2}-4^{-1}} = \dfrac{\frac{1}{x}+\frac{1}{2}}{\frac{1}{x^2}-\frac{1}{4}}$

$= \dfrac{4x^2\left(\frac{1}{x}+\frac{1}{2}\right)}{4x^2\left(\frac{1}{x^2}-\frac{1}{4}\right)}$

$= \dfrac{4x+2x^2}{4-x^2}$

$= \dfrac{2x^2+4x}{-(x^2-4)}$

$= -\dfrac{2x(x+2)}{(x+2)(x-2)}$

$= -\dfrac{2x}{x-2}$ or $\dfrac{2x}{2-x}$

57. $\dfrac{y^{-2}}{1-y^{-2}} = \dfrac{\frac{1}{y^2}}{1-\frac{1}{y^2}}$

$= \dfrac{y^2\left(\frac{1}{y^2}\right)}{y^2\left(1-\frac{1}{y^2}\right)}$

$= \dfrac{y^2\left(\frac{1}{y^2}\right)}{y^2(1)-y^2\left(\frac{1}{y^2}\right)}$

$= \dfrac{1}{y^2-1}$

59. $t = \dfrac{d}{r}$

$t = \dfrac{\frac{20x}{3}}{\frac{5x}{9}} = \dfrac{20x}{3}\cdot\dfrac{9}{5x} = \dfrac{4\cdot 5x\cdot 3\cdot 3}{3\cdot 5x} = 12$ hours

Chapter 7 Vocabulary Check

1. A ratio is the quotient of two numbers.

2. $\dfrac{x}{2} = \dfrac{7}{16}$ is an example of a proportion.

3. If $\dfrac{a}{b} = \dfrac{c}{d}$, then ad and bc are called cross products.

4. A rational expression is an expression that can be written in the form $\dfrac{P}{Q}$, where P and Q are polynomials and Q is not 0.

5. In a complex fraction, the numerator or denominator or both may contain fractions.

6. In equation $y = \dfrac{k}{x}$ is an example of inverse variation.

7. The equation $y = kx$ is an example of direct variation.

Chapter 7 Review

1. The rational expression is undefined when

$x^2 - 4 = 0$
$(x-2)(x+2) = 0$
$x-2 = 0$ or $x+2 = 0$
$x = 2$ $\quad x = -2$

2. The rational expression is undefined when

$4x^2 - 4x - 15 = 0$
$(2x+3)(2x-5) = 0$
$2x+3 = 0$ or $2x-5 = 0$
$2x = -3$ $\quad 2x = 5$
$x = -\dfrac{3}{2}$ $\quad x = \dfrac{5}{2}$

3. $\dfrac{2-z}{z+5} = \dfrac{2-(-2)}{-2+5} = \dfrac{2+2}{3} = \dfrac{4}{3}$

4. $\dfrac{x^2+xy-y^2}{x+y} = \dfrac{5^2+5\cdot 7-7^2}{5+7}$

$= \dfrac{25+35-49}{12}$

$= \dfrac{11}{12}$

5. $\dfrac{2x+6}{x^2+3x} = \dfrac{2(x+3)}{x(x+3)} = \dfrac{2}{x}$

6. $\dfrac{3x-12}{x^2-4x} = \dfrac{3(x-4)}{x(x-4)} = \dfrac{3}{x}$

7. $\dfrac{x+2}{x^2-3x-10} = \dfrac{x+2}{(x-5)(x+2)} = \dfrac{1}{x-5}$

8. $\frac{x+4}{x^2+5x+4} = \frac{x+4}{(x+1)(x+4)} = \frac{1}{x+1}$

9. $\frac{x^3-4x}{x^2+3x+2} = \frac{x(x^2-4)}{(x+2)(x+1)}$
$= \frac{x(x-2)(x+2)}{(x+2)(x+1)}$
$= \frac{x(x-2)}{x+1}$

10. $\frac{5x^2-125}{x^2+2x-15} = \frac{5(x^2-25)}{(x-3)(x+5)}$
$= \frac{5(x-5)(x+5)}{(x-3)(x+5)}$
$= \frac{5(x-5)}{x-3}$

11. $\frac{x^2-x-6}{x^2-3x-10} = \frac{(x-3)(x+2)}{(x-5)(x+2)} = \frac{x-3}{x-5}$

12. $\frac{x^2-2x}{x^2+2x-8} = \frac{x(x-2)}{(x+4)(x-2)} = \frac{x}{x+4}$

13. $\frac{x^2+xa+xb+ab}{x^2-xc+bx-bc} = \frac{x(x+a)+b(x+a)}{x(x-c)+b(x-c)}$
$= \frac{(x+a)(x+b)}{(x-c)(x+b)}$
$= \frac{x+a}{x-c}$

14. $\frac{x^2+5x-2x-10}{x^2-3x-2x+6} = \frac{x(x+5)-2(x+5)}{x(x-3)-2(x-3)}$
$= \frac{(x+5)(x-2)}{(x-3)(x-2)}$
$= \frac{x+5}{x-3}$

15. $\frac{4-x}{x^3-64} = -\frac{x-4}{x^3-64}$
$= -\frac{x-4}{(x-4)(x^2+4x+16)}$
$= -\frac{1}{x^2+4x+16}$

16. $\frac{x^2-4}{x^3+8} = \frac{(x+2)(x-2)}{(x+2)(x^2-2x+4)} = \frac{x-2}{x^2-2x+4}$

17. $\frac{15x^3y^2}{z} \cdot \frac{z}{5xy^3} = \frac{15x^3y^2 \cdot z}{z \cdot 5xy^3}$
$= \frac{3 \cdot 5 \cdot x^2 \cdot x \cdot y^2 \cdot z}{z \cdot 5 \cdot x \cdot y^2 \cdot y}$
$= \frac{3x^2}{y}$

18. $\frac{-y^3}{8} \cdot \frac{9x^2}{y^3} = -\frac{y^3 \cdot 9x^2}{8 \cdot y^3} = -\frac{9x^2}{8}$

19. $\frac{x^2-9}{x^2-4} \cdot \frac{x-2}{x+3} = \frac{(x^2-9) \cdot (x-2)}{(x^2-4) \cdot (x+3)}$
$= \frac{(x-3)(x+3)(x-2)}{(x+2)(x-2)(x+3)}$
$= \frac{x-3}{x+2}$

20. $\frac{2x+5}{x-6} \cdot \frac{2x}{-x+6} = \frac{2x+5}{x-6} \cdot \frac{2x}{-(x-6)}$
$= \frac{2x+5}{x-6} \cdot \frac{-2x}{x-6}$
$= \frac{(2x+5) \cdot (-2x)}{(x-6) \cdot (x-6)}$
$= \frac{-2x(2x+5)}{(x-6)^2}$

21. $\frac{x^2-5x-24}{x^2-x-12} \div \frac{x^2-10x+16}{x^2+x-6}$
$= \frac{x^2-5x-24}{x^2-x-12} \cdot \frac{x^2+x-6}{x^2-10x+16}$
$= \frac{(x-8)(x+3) \cdot (x+3)(x-2)}{(x-4)(x+3) \cdot (x-8)(x-2)}$
$= \frac{x+3}{x-4}$

22. $\frac{4x+4y}{xy^2} \div \frac{3x+3y}{x^2y} = \frac{4x+4y}{xy^2} \cdot \frac{x^2y}{3x+3y}$
$= \frac{4(x+y) \cdot x \cdot x \cdot y}{x \cdot y \cdot y \cdot 3(x+y)}$
$= \frac{4x}{3y}$

23. $\frac{x^2+x-42}{x-3}\cdot\frac{(x-3)^2}{x+7}$
$=\frac{(x+7)(x-6)\cdot(x-3)(x-3)}{(x-3)\cdot(x+7)}$
$=(x-6)(x-3)$

24. $\frac{2a+2b}{3}\cdot\frac{a-b}{a^2-b^2}=\frac{2(a+b)\cdot(a-b)}{3\cdot(a+b)(a-b)}=\frac{2}{3}$

25. $\frac{2x^2-9x+9}{8x-12}\div\frac{x^2-3x}{2x}=\frac{2x^2-9x+9}{8x-12}\cdot\frac{2x}{x^2-3x}$
$=\frac{(2x-3)(x-3)\cdot 2x}{4(2x-3)\cdot x(x-3)}$
$=\frac{2}{4}$
$=\frac{1}{2}$

26. $\frac{x^2-y^2}{x^2+xy}\div\frac{3x^2-2xy-y^2}{3x^2+6x}$
$=\frac{x^2-y^2}{x^2+xy}\cdot\frac{3x^2+6x}{3x^2-2xy-y^2}$
$=\frac{(x-y)(x+y)\cdot 3x(x+2)}{x(x+y)\cdot(3x+y)(x-y)}$
$=\frac{3(x+2)}{3x+y}$

27. $\frac{x-y}{4}\div\frac{y^2-2y-xy+2x}{16x+24}$
$=\frac{x-y}{4}\cdot\frac{16x+24}{y^2-2y-xy+2x}$
$=\frac{x-y}{4}\cdot\frac{8(2x+3)}{y(y-2)-x(y-2)}$
$=\frac{x-y}{4}\cdot\frac{8(2x+3)}{(y-2)(y-x)}$
$=-\frac{y-x}{4}\cdot\frac{8(2x+3)}{(y-2)(y-x)}$
$=-\frac{2\cdot 4(y-x)(2x+3)}{4(y-2)(y-x)}$
$=-\frac{2(2x+3)}{y-2}$

28. $\frac{5+x}{7}\div\frac{xy+5y-3x-15}{7y-35}$
$=\frac{5+x}{7}\cdot\frac{7y-35}{xy+5y-3x-15}$
$=\frac{(5+x)\cdot 7(y-5)}{7\cdot(x+5)(y-3)}$
$=\frac{y-5}{y-3}$

29. $\frac{x}{x^2+9x+14}+\frac{7}{x^2+9x+14}=\frac{x+7}{x^2+9x+14}$
$=\frac{x+7}{(x+7)(x+2)}$
$=\frac{1}{x+2}$

30. $\frac{x}{x^2+2x-15}+\frac{5}{x^2+2x-15}=\frac{x+5}{x^2+2x-15}$
$=\frac{x+5}{(x+5)(x-3)}$
$=\frac{1}{x-3}$

31. $\frac{4x-5}{3x^2}-\frac{2x+5}{3x^2}=\frac{4x-5-(2x+5)}{3x^2}$
$=\frac{4x-5-2x-5}{3x^2}$
$=\frac{2x-10}{3x^2}$

32. $\frac{9x+7}{6x^2}-\frac{3x+4}{6x^2}=\frac{9x+7-(3x+4)}{6x^2}$
$=\frac{9x+7-3x-4}{6x^2}$
$=\frac{6x+3}{6x^2}$
$=\frac{3(2x+1)}{3\cdot 2x^2}$
$=\frac{2x+1}{2x^2}$

33. $2x=2\cdot x$
$7x=7\cdot x$
$\text{LCD}=2\cdot 7\cdot x=14x$

34. $x^2 - 5x - 24 = (x-8)(x+3)$
$x^2 + 11x + 24 = (x+8)(x+3)$
$\text{LCD} = (x-8)(x+3)(x+8)$

35. $\dfrac{5}{7x} = \dfrac{5}{7x} \cdot \dfrac{2x^2y}{2x^2y} = \dfrac{5 \cdot 2x^2y}{7x \cdot 2x^2y} = \dfrac{10x^2y}{14x^3y}$

36. $\dfrac{9}{4y} = \dfrac{9}{4y} \cdot \dfrac{4y^2x}{4y^2x} = \dfrac{9 \cdot 4y^2x}{4y \cdot 4y^2x} = \dfrac{36y^2x}{16y^3x}$

37. $\dfrac{x+2}{x^2+11x+18} = \dfrac{x+2}{(x+9)(x+2)}$
$= \dfrac{(x+2)(x-5)}{(x+9)(x+2)(x-5)}$
$= \dfrac{x^2-3x-10}{(x+2)(x-5)(x+9)}$

38. $\dfrac{3x-5}{x^2+4x+4} = \dfrac{3x-5}{(x+2)^2}$
$= \dfrac{(3x-5)(x+3)}{(x+2)^2(x+3)}$
$= \dfrac{3x^2+4x-15}{(x+2)^2(x+3)}$

39. $\dfrac{4}{5x^2} - \dfrac{6}{y} = \dfrac{4(y)}{5x^2(y)} - \dfrac{6(5x^2)}{y(5x^2)} = \dfrac{4y-30x^2}{5x^2y}$

40. $\dfrac{2}{x-3} - \dfrac{4}{x-1} = \dfrac{2(x-1)}{(x-3)(x-1)} - \dfrac{4(x-3)}{(x-1)(x-3)}$
$= \dfrac{2(x-1)-4(x-3)}{(x-3)(x-1)}$
$= \dfrac{2x-2-4x+12}{(x-3)(x-1)}$
$= \dfrac{-2x+10}{(x-3)(x-1)}$

41. $\dfrac{4}{x+3} - 2 = \dfrac{4}{x+3} - \dfrac{2(x+3)}{x+3}$
$= \dfrac{4-2(x+3)}{x+3}$
$= \dfrac{4-2x-6}{x+3}$
$= \dfrac{-2x-2}{x+3}$

42. $\dfrac{3}{x^2+2x-8} + \dfrac{2}{x^2-3x+2}$
$= \dfrac{3}{(x+4)(x-2)} + \dfrac{2}{(x-1)(x-2)}$
$= \dfrac{3(x-1)}{(x+4)(x-2)(x-1)} + \dfrac{2(x+4)}{(x-1)(x-2)(x+4)}$
$= \dfrac{3(x-1)+2(x+4)}{(x+4)(x-2)(x-1)}$
$= \dfrac{3x-3+2x+8}{(x+4)(x-2)(x-1)}$
$= \dfrac{5x+5}{(x+4)(x-2)(x-1)}$

43. $\dfrac{2x-5}{6x+9} - \dfrac{4}{2x^2+3x} = \dfrac{2x-5}{3(2x+3)} - \dfrac{4}{x(2x+3)}$
$= \dfrac{(2x-5)(x)}{3(2x+3)(x)} - \dfrac{4(3)}{x(2x+3)(3)}$
$= \dfrac{2x^2-5x-12}{3x(2x+3)}$
$= \dfrac{(2x+3)(x-4)}{3x(2x+3)}$
$= \dfrac{x-4}{3x}$

3X
-8X
-5

44. $\dfrac{x-1}{x^2-2x+1} - \dfrac{x+1}{x-1} = \dfrac{x-1}{(x-1)^2} - \dfrac{x+1}{x-1}$
$= \dfrac{1}{x-1} - \dfrac{x+1}{x-1}$
$= \dfrac{1-(x+1)}{x-1}$
$= \dfrac{1-x-1}{x-1}$
$= \dfrac{-x}{x-1}$
$= -\dfrac{x}{x-1}$

45. $P = 2l + 2w$
$P = 2\left(\dfrac{x}{8}\right) + 2\left(\dfrac{x+2}{4x}\right)$
$= \dfrac{x}{4} + \dfrac{2(x+2)}{4x}$
$= \dfrac{x \cdot x}{4 \cdot x} + \dfrac{2x+4}{4x}$
$= \dfrac{x^2+2x+4}{4x}$

$A = l \cdot w$

$A = \frac{x}{8} \cdot \frac{x+2}{4x} = \frac{x \cdot (x+2)}{8 \cdot 4x} = \frac{x+2}{32}$

The perimeter is $\frac{x^2+2x+4}{4x}$ units and the area is $\frac{x+2}{32}$ square units.

46. $P = \frac{3x}{4x-4} + \frac{2x}{3x-3} + \frac{x}{x-1}$

$= \frac{3x}{4(x-1)} + \frac{2x}{3(x-1)} + \frac{x}{x-1}$

$= \frac{3x(3)}{4(x-1)(3)} + \frac{2x(4)}{3(x-1)(4)} + \frac{x(12)}{(x-1)(12)}$

$= \frac{9x+8x+12x}{12(x-1)}$

$= \frac{29x}{12(x-1)}$

$A = \frac{1}{2} \cdot b \cdot h$

$A = \frac{1}{2} \cdot \frac{x}{x-1} \cdot \frac{6y}{5} = \frac{1 \cdot x \cdot 2 \cdot 3y}{2 \cdot (x-1) \cdot 5} = \frac{3xy}{5(x-1)}$

The perimeter is $\frac{29x}{12(x-1)}$ units and the area is $\frac{3xy}{5(x-1)}$ square units.

47. $\frac{n}{10} = 9 - \frac{n}{5}$

$10\left(\frac{n}{10}\right) = 10\left(9 - \frac{n}{5}\right)$

$10\left(\frac{n}{10}\right) = 10(9) - 10\left(\frac{n}{5}\right)$

$n = 90 - 2n$

$3n = 90$

$n = 30$

48. $\frac{2}{x+1} - \frac{1}{x-2} = -\frac{1}{2}$

$2(x+1)(x-2)\left(\frac{2}{x+1} - \frac{1}{x-2}\right) = 2(x+1)(x-2)\left(-\frac{1}{2}\right)$

$2(x+1)(x-2)\left(\frac{2}{x+1}\right) - 2(x+1)(x-2)\left(\frac{1}{x-2}\right) = 2(x+1)(x-2)\left(-\frac{1}{2}\right)$

$4(x-2) - 2(x+1) = -(x+1)(x-2)$

$4x - 8 - 2x - 2 = -(x^2 - x - 2)$

$2x - 10 = -x^2 + x + 2$

$x^2 + x - 12 = 0$

$(x+4)(x-3) = 0$

$x + 4 = 0$ or $x - 3 = 0$

$x = -4$ $\quad$ $x = 3$

49.

$$\frac{y}{2y+2}+\frac{2y-16}{4y+4}=\frac{y-3}{y+1}$$
$$\frac{y}{2(y+1)}+\frac{2y-16}{4(y+1)}=\frac{y-3}{y+1}$$
$$4(y+1)\left(\frac{y}{2(y+1)}+\frac{2y-16}{4(y+1)}\right)=4(y+1)\left(\frac{y-3}{y+1}\right)$$
$$4(y+1)\left(\frac{y}{2(y+1)}\right)+4(y+1)\left(\frac{2y-16}{4(y+1)}\right)=4(y+1)\left(\frac{y-3}{y+1}\right)$$
$$2y+2y-16=4(y-3)$$
$$4y-16=4y-12$$
$$-16=-12 \quad \text{False}$$

This equation has no solution.

50.

$$\frac{2}{x-3}-\frac{4}{x+3}=\frac{8}{x^2-9}$$
$$(x-3)(x+3)\left(\frac{2}{x-3}-\frac{4}{x+3}\right)=(x-3)(x+3)\left(\frac{8}{(x-3)(x+3)}\right)$$
$$(x-3)(x+3)\left(\frac{2}{x-3}\right)-(x-3)(x+3)\left(\frac{4}{x+3}\right)=8$$
$$2(x+3)-4(x-3)=8$$
$$2x+6-4x+12=8$$
$$-2x+18=8$$
$$-2x=-10$$
$$x=5$$

51.

$$\frac{x-3}{x+1}-\frac{x-6}{x+5}=0$$
$$(x+1)(x+5)\left(\frac{x-3}{x+1}-\frac{x-6}{x+5}\right)=(x+1)(x+5)(0)$$
$$(x+1)(x+5)\left(\frac{x-3}{x+1}\right)-(x+1)(x+5)\left(\frac{x-6}{x+5}\right)=0$$
$$(x+5)(x-3)-(x+1)(x-6)=0$$
$$x^2+2x-15-(x^2-5x-6)=0$$
$$x^2+2x-15-x^2+5x+6=0$$
$$7x-9=0$$
$$7x=9$$
$$x=\frac{9}{7}$$

52. $x+5=\frac{6}{x}$
$x(x+5)=x\left(\frac{6}{x}\right)$
$x^2+5x=6$
$x^2+5x-6=0$
$(x+6)(x-1)=0$
$x+6=0$ or $x-1=0$
$x=-6$ $\quad x=1$

53. $\frac{4A}{5b}=x^2$
$4A=5bx^2$
$\frac{4A}{5x^2}=\frac{5bx^2}{5x^2}$
$\frac{4A}{5x^2}=b$

54. $\frac{x}{7}+\frac{y}{8}=10$
$56\left(\frac{x}{7}\right)+56\left(\frac{y}{8}\right)=56(10)$
$8x+7y=560$
$7y=560-8x$
$y=\frac{560-8x}{7}$

55. $\frac{x}{2}=\frac{12}{4}$
$4x=24$
$x=6$

56. $\frac{20}{1}=\frac{x}{25}$
$500=x$

57. $\frac{2}{x-1}=\frac{3}{x+3}$
$2(x+3)=3(x-1)$
$2x+6=3x-3$
$6=x-3$
$9=x$

58. $\frac{4}{y-3}=\frac{2}{y-3}$
$4(y-3)=2(y-3)$
$4y-12=2y-6$
$2y-12=-6$
$2y=6$
$y=3$
$y=3$ doesn't check, so this equation has no solution.

59. Let x = the number of parts processed in 45 minutes.
$\frac{300}{20}=\frac{x}{45}$
$13,500=20x$
$675=x$
675 parts can be processed in 45 minutes.

60. Let x = the charge for 3 hours.
$\frac{90.00}{8}=\frac{x}{3}$
$270.00=8x$
$33.75=x$
He charges \$33.75 for 3 hours.

61. $5\cdot\frac{1}{x}=\frac{3}{2}\cdot\frac{1}{x}+\frac{7}{6}$
$\frac{5}{x}=\frac{3}{2x}+\frac{7}{6}$
$6x\left(\frac{5}{x}\right)=6x\left(\frac{3}{2x}\right)+6x\left(\frac{7}{6}\right)$
$30=9+7x$
$21=7x$
$x=3$
The unknown number is 3.

62. $\frac{1}{x}=\frac{1}{4-x}$
$4-x=x$
$4=2x$
$2=x$
The unknown number is 2.

63. Let r be the rate of the faster car. Then the rate of the slower car is $r-10$.

	Distance =	Rate ·	Time
Fast car	90	r	$\frac{90}{r}$
Slow car	60	$r-10$	$\frac{60}{r-10}$

$$\frac{90}{r} = \frac{60}{r-10}$$
$$90(r-10) = 60r$$
$$90r - 900 = 60r$$
$$-900 = -30r$$
$$30 = r$$
$$r - 10 = 30 - 10 = 20$$

The rate of the fast car is 30 miles per hour and the rate of the slower car is 20 miles per hour.

64. Let r be the speed of the boat in still water.

	Distance	= Rate	· Time
Upstream	48	$r-4$	$\frac{48}{r-4}$
Downstream	72	$r+4$	$\frac{72}{r+4}$

$$\frac{48}{r-4} = \frac{72}{r+4}$$
$$48(r+4) = 72(r-4)$$
$$48r + 192 = 72r - 288$$
$$480 = 24r$$
$$r = 20$$

The speed of the boat in still water is 20 miles per hour.

65. Let x be the time it takes Maria working alone.

	Hours to Complete Total Job	Part of Job Completed in 1 Hour
Mark	7	$\frac{1}{7}$
Maria	x	$\frac{1}{x}$
Together	5	$\frac{1}{5}$

$$\frac{1}{7} + \frac{1}{x} = \frac{1}{5}$$
$$35x\left(\frac{1}{7}\right) + 35x\left(\frac{1}{x}\right) = 35x\left(\frac{1}{5}\right)$$
$$5x + 35 = 7x$$
$$35 = 2x$$
$$x = \frac{35}{2} \text{ or } 17\frac{1}{2}$$

It takes Maria $17\frac{1}{2}$ hours to complete the job alone.

66. Let x be the number of days it takes the pipes to fill the pond together.

	Days to Complete Total Job	Part of Job Completed in 1 Day
Pipe A	20	$\frac{1}{20}$
Pipe B	15	$\frac{1}{15}$
Together	x	$\frac{1}{x}$

$$\frac{1}{20} + \frac{1}{25} = \frac{1}{x}$$
$$60x\left(\frac{1}{20}\right) + 60x\left(\frac{1}{15}\right) = 60x\left(\frac{1}{x}\right)$$
$$3x + 4x = 60$$
$$7x = 60$$
$$x = \frac{60}{7} = 8\frac{4}{7}$$

Both pipes fill the pond in $8\frac{4}{7}$ days.

67. $\frac{2}{3} = \frac{10}{x}$
$$2x = 30$$
$$x = 15$$

The missing length is 15.

68. $\frac{12}{4} = \frac{18}{x}$
$$12x = 72$$
$$x = 6$$

The missing length is 6.

69. $y = kx$
$$40 = k(4)$$
$$10 = k$$

$$y = 10x$$
$$y = 10(11)$$
$$y = 110$$

70.
$$y=\frac{k}{x}$$
$$4=\frac{k}{6}$$
$$24=k$$

$$y=\frac{24}{x}$$
$$y=\frac{24}{48}$$
$$y=\frac{1}{2}$$

71.
$$y=\frac{k}{x^3}$$
$$12.5=\frac{k}{2^3}$$
$$12.5=\frac{k}{8}$$
$$100=k$$

$$y=\frac{100}{x^3}$$
$$y=\frac{100}{3^3}$$
$$y=\frac{100}{27}$$

72.
$$y=kx^2$$
$$175=k(5)^2$$
$$175=25k$$
$$7=k$$

$$y=7x^2$$
$$y=7(10)^2$$
$$y=7(100)$$
$$y=700$$

73.
$$c=\frac{k}{a}$$
$$6600=\frac{k}{3000}$$
$$19,800,000=k$$

$$c=\frac{19,800,000}{a}$$
$$c=\frac{19,800,000}{5000}$$
$$c=3960$$

It costs \$3960 to manufacture 5000 ml of medicine.

74.
$$d=kw$$
$$8=k(150)$$
$$\frac{8}{150}=k$$
$$\frac{4}{75}=k$$

$$d=\frac{4}{75}w$$
$$d=\frac{4}{75}(90)$$
$$d=\frac{360}{75}$$
$$d=4\frac{4}{5}$$

A 90-pound weight would stretch the spring $4\frac{4}{5}$ inches.

75. $$\frac{\frac{5x}{27}}{-\frac{10xy}{21}}=\frac{5x}{27}\cdot-\frac{21}{10xy}=-\frac{5x\cdot3\cdot7}{3\cdot9\cdot5\cdot2\cdot x\cdot y}=-\frac{7}{18y}$$

76. $$\frac{\frac{3}{5}+\frac{2}{7}}{\frac{1}{5}+\frac{5}{6}}=\frac{\frac{21}{35}+\frac{10}{35}}{\frac{6}{30}+\frac{25}{30}}=\frac{\frac{31}{35}}{\frac{31}{30}}=\frac{31}{35}\cdot\frac{30}{31}=\frac{31\cdot5\cdot6}{5\cdot7\cdot31}=\frac{6}{7}$$

77. $$\frac{3-\frac{1}{y}}{2-\frac{1}{y}}=\frac{y\left(3-\frac{1}{y}\right)}{y\left(2-\frac{1}{y}\right)}=\frac{y(3)-y\left(\frac{1}{y}\right)}{y(2)-y\left(\frac{1}{y}\right)}=\frac{3y-1}{2y-1}$$

78. $\dfrac{\frac{6}{x+2}+4}{\frac{8}{x+2}-4}=\dfrac{(x+2)\left(\frac{6}{x+2}+4\right)}{(x+2)\left(\frac{8}{x+2}-4\right)}$
$=\dfrac{(x+2)\left(\frac{6}{x+2}\right)+(x+2)(4)}{(x+2)\left(\frac{8}{x+2}\right)-(x+2)(4)}$
$=\dfrac{6+4x+8}{8-4x-8}$
$=\dfrac{4x+14}{-4x}$
$=-\dfrac{2(2x+7)}{2\cdot 2x}$
$=-\dfrac{2x+7}{2x}$

79. $\dfrac{4x+12}{8x^2+24x}=\dfrac{4(x+3)}{2\cdot 4\cdot x(x+3)}=\dfrac{1}{2x}$

80. $\dfrac{x^3-6x^2+9x}{x^2+4x-21}=\dfrac{x(x-3)^2}{(x+7)(x-3)}=\dfrac{x(x-3)}{x+7}$

81. $\dfrac{x^2+9x+20}{x^2-25}\cdot\dfrac{x^2-9x+20}{x^2+8x+16}$
$=\dfrac{(x+4)(x+5)\cdot(x-4)(x-5)}{(x+5)(x-5)\cdot(x+4)(x+4)}$
$=\dfrac{x-4}{x+4}$

82. $\dfrac{x^2-x-72}{x^2-x-30}\div\dfrac{x^2+6x-27}{x^2-9x+18}$
$=\dfrac{x^2-x-72}{x^2-x-30}\cdot\dfrac{x^2-9x+18}{x^2+6x-27}$
$=\dfrac{(x-9)(x+8)\cdot(x-3)(x-6)}{(x+5)(x-6)\cdot(x+9)(x-3)}$
$=\dfrac{(x-9)(x+8)}{(x+5)(x+9)}$

83. $\dfrac{x}{x^2-36}+\dfrac{6}{x^2-36}=\dfrac{x+6}{x^2-36}$
$=\dfrac{x+6}{(x+6)(x-6)}$
$=\dfrac{1}{x-6}$

84. $\dfrac{5x-1}{4x}-\dfrac{3x-2}{4x}=\dfrac{5x-1-(3x-2)}{4x}$
$=\dfrac{5x-1-3x+2}{4x}$
$=\dfrac{2x+1}{4x}$

85. $\dfrac{4}{3x^2+8x-3}+\dfrac{2}{3x^2-7x+2}$
$=\dfrac{4}{(x+3)(3x-1)}+\dfrac{2}{(x-2)(3x-1)}$
$=\dfrac{4(x-2)}{(x+3)(3x-1)(x-2)}+\dfrac{2(x+3)}{(x-2)(3x-1)(x+3)}$
$=\dfrac{4(x-2)+2(x+3)}{(x+3)(3x-1)(x-2)}$
$=\dfrac{4x-8+2x+6}{(x+3)(3x-1)(x-2)}$
$=\dfrac{6x-2}{(x+3)(3x-1)(x-2)}$
$=\dfrac{2(3x-1)}{(x+3)(3x-1)(x-2)}$
$=\dfrac{2}{(x+3)(x-2)}$

86. $\dfrac{3x}{x^2+9x+14}-\dfrac{6x}{x^2+4x-21}$
$=\dfrac{3x}{(x+7)(x+2)}-\dfrac{6x}{(x+7)(x-3)}$
$=\dfrac{3x(x-3)}{(x+7)(x+2)(x-3)}-\dfrac{6x(x+2)}{(x+7)(x-3)(x+2)}$
$=\dfrac{3x(x-3)-6x(x+2)}{(x+7)(x+2)(x-3)}$
$=\dfrac{3x^2-9x-6x^2-12x}{(x+7)(x+2)(x-3)}$
$=\dfrac{-3x^2-21x}{(x+7)(x+2)(x-3)}$
$=\dfrac{-3x(x+7)}{(x+7)(x+2)(x-3)}$
$=-\dfrac{3x}{(x+2)(x-3)}$

87.
$$\frac{4}{a-1}+2=\frac{3}{a-1}$$
$$(a-1)\left(\frac{4}{a-1}\right)+(a-1)(2)=(a-1)\left(\frac{3}{a-1}\right)$$
$$4+2(a-1)=3$$
$$4+2a-2=3$$
$$2+2a=3$$
$$2a=1$$
$$a=\frac{1}{2}$$

88.
$$\frac{x}{x+3}+4=\frac{x}{x+3}$$
$$(x+3)\left(\frac{x}{x+3}\right)+(x+3)(4)=(x+3)\left(\frac{x}{x+3}\right)$$
$$x+4(x+3)=x$$
$$x+4x+12=x$$
$$5x+12=x$$
$$12=-4x$$
$$-3=x$$
Since $x=-3$ makes a denominator 0, the solution does not check. This equation has no solution.

89.
$$\frac{2x}{3}-\frac{1}{6}=\frac{x}{2}$$
$$6\left(\frac{2x}{3}\right)-6\left(\frac{1}{6}\right)=6\left(\frac{x}{2}\right)$$
$$4x-1=3x$$
$$-1=-x$$
$$1=x$$
The unknown number is 1.

90. Let x be the number of days it takes them to paint the house working together.

	Days to Complete Total Job	Part of Job Completed in 1 Day
Mr. Crocker	3	$\frac{1}{3}$
Son	4	$\frac{1}{4}$
Together	x	$\frac{1}{x}$

$$\frac{1}{3}+\frac{1}{4}=\frac{1}{x}$$
$$12x\left(\frac{1}{3}\right)+12x\left(\frac{1}{4}\right)=12x\left(\frac{1}{x}\right)$$
$$4x+3x=12$$
$$7x=12$$
$$x=\frac{12}{7}\text{ or }1\frac{5}{7}$$
Working together, Mr. Crocker and his son can paint the house in $1\frac{5}{7}$ days.

91. $\frac{5}{3}=\frac{10}{x}$
$5x=30$
$x=6$
The missing length is 6.

92. $\frac{6}{18}=\frac{4}{x}$
$6x=72$
$x=12$
The missing length is 12.

93. $$\frac{\frac{1}{4}}{\frac{1}{3}+\frac{1}{2}}=\frac{12\left(\frac{1}{4}\right)}{12\left(\frac{1}{3}+\frac{1}{2}\right)}=\frac{12\left(\frac{1}{4}\right)}{12\left(\frac{1}{3}\right)+12\left(\frac{1}{2}\right)}=\frac{3}{4+6}=\frac{3}{10}$$

94. $$\frac{4+\frac{2}{x}}{6+\frac{3}{x}}=\frac{x\left(4+\frac{2}{x}\right)}{x\left(6+\frac{3}{x}\right)}$$
$$=\frac{x(4)+x\left(\frac{2}{x}\right)}{x(6)+x\left(\frac{3}{x}\right)}$$
$$=\frac{4x+2}{6x+3}$$
$$=\frac{2(2x+1)}{3(2x+1)}$$
$$=\frac{2}{3}$$

Chapter 7 Test

1. The rational expression is undefined when
$$x^2+4x+3=0$$
$$(x+3)(x+1)=0$$
$$x+3=0\quad\text{or}\quad x+1=0$$
$$x=-3\qquad\qquad x=-1$$

2. a. $C = \frac{100x + 3000}{x}$
$= \frac{100(200) + 3000}{200}$
$= \frac{20{,}000 + 3000}{200}$
$= \frac{23{,}000}{200}$
$= 115$

The average cost per desk is $115.

b. $C = \frac{100x + 3000}{x}$
$= \frac{100(1000) + 3000}{1000}$
$= \frac{100{,}000 + 3000}{1000}$
$= \frac{103{,}000}{1000}$
$= 103$

The average cost per desk is $103.

3. $\frac{3x-6}{5x-10} = \frac{3(x-2)}{5(x-2)} = \frac{3}{5}$

4. $\frac{x+6}{x^2+12x+36} = \frac{x+6}{(x+6)^2} = \frac{1}{x+6}$

5. $\frac{x+3}{x^3+27} = \frac{x+3}{(x+3)(x^2-3x+9)} = \frac{1}{x^2-3x+9}$

6. $\frac{2m^3-2m^2-12m}{m^2-5m+6} = \frac{2m(m^2-m-6)}{(m-3)(m-2)}$
$= \frac{2m(m-3)(m+2)}{(m-3)(m-2)}$
$= \frac{2m(m+2)}{m-2}$

7. $\frac{ay+3a+2y+6}{ay+3a+5y+15} = \frac{(y+3)(a+2)}{(y+3)(a+5)} = \frac{a+2}{a+5}$

8. $\frac{y-x}{x^2-y^2} = \frac{-(x-y)}{(x-y)(x+y)} = -\frac{1}{x+y}$

9. $\frac{3}{x-1} \cdot (5x-5) = \frac{3}{x-1} \cdot 5(x-1) = \frac{3 \cdot 5(x-1)}{x-1} = 15$

10. $\frac{y^2-5y+6}{2y+4}$..

11. $\frac{15x}{2x+5} - \frac{6-4x}{2x+5} = \frac{15x-(6-}{2x+5}$
$= \frac{15x-6+4x}{2x+5}$
$= \frac{19x-6}{2x+5}$

12. $\frac{5a}{a^2-a-6} - \frac{2}{a-3}$
$= \frac{5a}{(a-3)(a+2)} - \frac{2(a+2)}{(a-3)(a+2)}$
$= \frac{5a-2(a+2)}{(a-3)(a+2)}$
$= \frac{5a-2a-4}{(a-3)(a+2)}$
$= \frac{3a-4}{(a-3)(a+2)}$

13. $\frac{6}{x^2-1} + \frac{3}{x+1} = \frac{6}{(x+1)(x-1)} + \frac{3(x-1)}{(x+1)(x-1)}$
$= \frac{6+3x-3}{(x+1)(x-1)}$
$= \frac{3x+3}{(x+1)(x-1)}$
$= \frac{3(x+1)}{(x+1)(x-1)}$
$= \frac{3}{x-1}$

14. $\frac{x^2-9}{x^2-3x} \div \frac{xy+5x+3y+15}{2x+10}$
$= \frac{x^2-9}{x^2-3x} \cdot \frac{2x+10}{xy+5x+3y+15}$
$= \frac{(x-3)(x+3) \cdot 2(x+5)}{x(x-3) \cdot (x+3)(y+5)}$
$= \frac{2(x+5)}{x(y+5)}$

15. $$\frac{x+2}{x^2+11x+18}+\frac{5}{x^2-3x-10}=\frac{x+2}{(x+9)(x+2)}+\frac{5}{(x-5)(x+2)}$$
$$=\frac{(x+2)(x-5)}{(x+9)(x+2)(x-5)}+\frac{5(x+9)}{(x-5)(x+2)(x+9)}$$
$$=\frac{(x+2)(x-5)+5(x+9)}{(x+9)(x+2)(x-5)}$$
$$=\frac{x^2-3x-10+5x+45}{(x+9)(x+2)(x-5)}$$
$$=\frac{x^2+2x+35}{(x+9)(x+2)(x-5)}$$

16. $$\frac{4}{y}-\frac{5}{3}=-\frac{1}{5}$$
$$15y\left(\frac{4}{y}-\frac{5}{3}\right)=15y\left(-\frac{1}{5}\right)$$
$$15y\left(\frac{4}{y}\right)-15y\left(\frac{5}{3}\right)=15y\left(-\frac{1}{5}\right)$$
$$60-25y=-3y$$
$$60=22y$$
$$\frac{60}{22}=y$$
$$y=\frac{30}{11}$$

17. $$\frac{5}{y+1}=\frac{4}{y+2}$$
$$5(y+2)=4(y+1)$$
$$5y+10=4y+4$$
$$y=-6$$

18. $$\frac{a}{a-3}=\frac{3}{a-3}-\frac{3}{2}$$
$$2(a-3)\left(\frac{a}{a-3}\right)=2(a-3)\left(\frac{3}{a-3}-\frac{3}{2}\right)$$
$$2a=2(a-3)\left(\frac{3}{a-3}\right)-2(a-3)\left(\frac{3}{2}\right)$$
$$2a=6-3(a-3)$$
$$2a=6-3a+9$$
$$2a=15-3a$$
$$5a=15$$
$$a=3$$
In the original equation, 3 makes a denominator 0. This equation has no solution.

19.
$$x-\frac{14}{x-1}=4-\frac{2x}{x-1}$$
$$(x-1)\left(x-\frac{14}{x-1}\right)=(x-1)\left(4-\frac{2x}{x-1}\right)$$
$$x(x-1)-14=4(x-1)-2x$$
$$x^2-x-14=4x-4-2x$$
$$x^2-x-14=2x-4$$
$$x^2-3x-10=0$$
$$(x-5)(x+2)=0$$
$$x-5=0 \quad \text{or} \quad x+2=0$$
$$x=5 \qquad\qquad x=-2$$

20.
$$\frac{10}{x^2-25}=\frac{3}{x+5}+\frac{1}{x-5}$$
$$\frac{10}{(x+5)(x-5)}=\frac{3}{x+5}+\frac{1}{x-5}$$
$$(x+5)(x-5)\left(\frac{10}{(x+5)(x-5)}\right)=(x+5)(x-5)\left(\frac{3}{x+5}\right)+(x+5)(x-5)\left(\frac{1}{x-5}\right)$$
$$10=3(x-5)+1(x+5)$$
$$10=3x-15+x+5$$
$$10=4x-10$$
$$20=4x$$
$$5=x$$
In the original equation 5 makes a denominator 0. This equation has no solution.

21. $\dfrac{\frac{5x^2}{yz^2}}{\frac{10x}{z^3}}=\dfrac{5x^2}{yz^2}\cdot\dfrac{z^3}{10x}=-\dfrac{5\cdot x\cdot x\cdot z\cdot z^2}{y\cdot z^2\cdot 2\cdot 5\cdot x}=\dfrac{xz}{2y}$

22.
$$\frac{5-\frac{1}{y^2}}{\frac{1}{y}+\frac{2}{y^2}}=\frac{y^2\left(5-\frac{1}{y^2}\right)}{y^2\left(\frac{1}{y}+\frac{2}{y^2}\right)}$$
$$=\frac{y^2(5)-y^2\left(\frac{1}{y^2}\right)}{y^2\left(\frac{1}{y}\right)+y^2\left(\frac{2}{y^2}\right)}$$
$$=\frac{5y^2-1}{y+2}$$

23. $y = kx$
$10 = k(15)$
$\frac{10}{15} = k$
$\frac{2}{3} = k$

$y = \frac{2}{3}x$
$y = \frac{2}{3}(42)$
$y = \frac{84}{3}$
$y = 28$

24. $y = \frac{k}{x^2}$
$8 = \frac{k}{5^2}$
$8 = \frac{k}{25}$
$200 = k$

$y = \frac{200}{x^2}$
$y = \frac{200}{15^2}$
$y = \frac{200}{225}$
$y = \frac{8}{9}$

25. Let x = the number of defective bulbs.
$\frac{85}{3} = \frac{510}{x}$
$85x = 1530$
$x = 18$
Expect to find 18 defective bulbs.

26. $x + 5 \cdot \frac{1}{x} = 6$
$x + \frac{5}{x} = 6$
$x\left(x + \frac{5}{x}\right) = x(6)$
$x(x) + x\left(\frac{5}{x}\right) = x(6)$
$x^2 + 5 = 6x$
$x^2 - 6x + 5 = 0$
$(x-5)(x-1) = 0$
$x - 5 = 0$ or $x - 1 = 0$
$x = 5$ $\quad$ $x = 1$
The unknown number is 5 or 1.

27. Let r be the speed of the boat in still water.

	Distance =	Rate ·	Time
Upstream	14	$r - 2$	$\frac{14}{r-2}$
Downstream	16	$r + 2$	$\frac{16}{r+2}$

$\frac{14}{r-2} = \frac{16}{r+2}$
$14(r+2) = 16(r-2)$
$14r + 28 = 16r - 32$
$60 = 2r$
$r = 30$
The speed of the boat in still water is 30 miles per hour.

28. Let x be the number of hours it takes to fill the tank using both pipes.

	Hours to Complete Total Job	Part of Job Completed in 1 Hour
1st Pipe	12	$\frac{1}{12}$
2nd Pipe	15	$\frac{1}{15}$
Together	x	$\frac{1}{x}$

$$\frac{1}{12}+\frac{1}{15}=\frac{1}{x}$$
$$60x\left(\frac{1}{12}\right)+60x\left(\frac{1}{15}\right)=60x\left(\frac{1}{x}\right)$$
$$5x+4x=60$$
$$9x=60$$
$$x=\frac{60}{9}=\frac{20}{3}=6\frac{2}{3}$$

Together, the pipes can fill the tank in $6\frac{2}{3}$ hours.

29. $\frac{8}{x}=\frac{10}{15}$
$$8(15)=10x$$
$$120=10x$$
$$12=x$$
The missing length is 12.

Cumulative Review Chapters 1–7

1. a. $\frac{15}{x}=4$

b. $12-3=x$

c. $4x+17\neq 21$

d. $3x<48$

2. a. $12-x=-45$

b. $12x=-45$

c. $x-10=2x$

3. Let x = the amount invested at 9% for one year.

	Principal	· Rate =	Interest
9%	x	0.09	$0.09x$
7%	$20,000-x$	0.07	$0.07(20,000-x)$
Total	20,000		1550

$$0.09x+0.07(20,000-x)=1550$$
$$0.09x+1400-0.07x=1550$$
$$0.02x+1400=1550$$
$$0.02x=150$$
$$x=7500$$
$20,000-x=20,000-7500=12,500$
He invested \$7500 at 9% and \$12,500 at 7%.

4. Let x be the number of bankruptcies in 1994 then $2x-80,000$ is the number in 2002.
$$x+2x-80,000=2,290,000$$
$$3x-80,000=2,290,000$$
$$3x=2,370,000$$
$$x=790,000$$
$2x-80,000=2(790,000)-80,000=1,500,000$
There were 790,000 bankruptcies in 1994 and 1,500,000 in 2002.

5. $x-3y=6$

x	y
0	−2
6	0

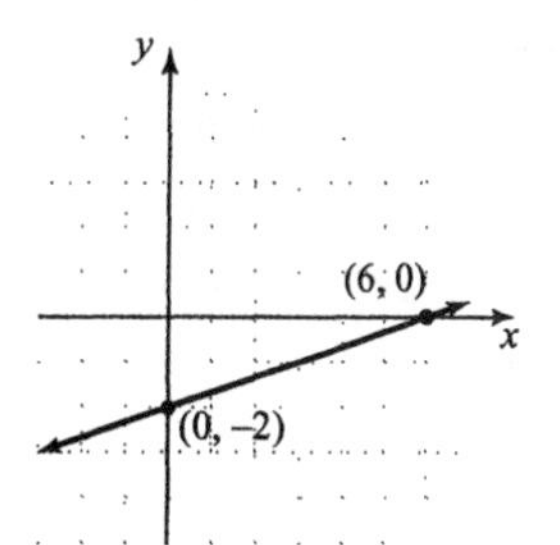

6. $7x+2y=9$
$$2y=-7x+9$$
$$y=-\frac{7}{2}x+\frac{9}{2}$$
$$y=mx+b$$
$$m=-\frac{7}{2}$$

7. a. $4^2\cdot 4^5=4^{2+5}=4^7$

b. $x^4\cdot x^6=x^{4+6}=x^{10}$

c. $y^3\cdot y=y^{3+1}=y^4$

d. $y^3\cdot y^2\cdot y^7=y^{3+2+7}=y^{12}$

e. $(-5)^7\cdot(-5)^8=(-5)^{7+8}=(-5)^{15}$

f. $a^2\cdot b^2=a^2b^2$

8. a. $\frac{x^9}{x^7}=x^{9-7}=x^2$

b. $\frac{x^{19}y^5}{xy} = x^{19-1} \cdot y^{5-1} = x^{18}y^4$

c. $(x^5y^2)^3 = x^{5\cdot3}y^{2\cdot3} = x^{15}y^6$

d. $(-3a^2b)(5a^3b) = -15a^{2+3}b^{1+1} = -15a^5b^2$

9. $[(8z+11)+(9z-2)]-(5z-7)$
$= 8z+11+9z-2-5z+7$
$= 12z+16$

10. $(x+1)-(9x^2-6x+2) = x+1-9x^2+6x-2$
$= -9x^2+7x-1$

11. $(3a+b)^3$
$= (3a+b)(3a+b)^2$
$= (3a+b)[(3a)^2+2(3a)(b)+(b)^2]$
$= (3a+b)(9a^2+6ab+b^2)$
$= 27a^3+18a^2b+3ab^2+9a^2b+6ab^2+b^3$
$= 27a^3+27a^2b+9ab^2+b^3$

12. $(2x+1)(5x^2-x+2)$
$= 2x(5x^2-x+2)+1(5x^2-x+2)$
$= 10x^3-2x^2+4x+5x^2-x+2$
$= 10x^3+3x^2+3x+2$

13. a. $(t+2)^2 = (t)^2+2(t)(2)+(2)^2 = t^2+4t+4$

b. $(p-q)^2 = (p)^2-2(p)(q)+(q)^2$
$= p^2-2pq+q^2$

c. $(2x+5)^2 = (2x)^2+2(2x)(5)+(5)^2$
$= 4x^2+20x+25$

d. $(x^2-7y)^2 = (x^2)^2-2(x^2)(7y)+(7y)^2$
$= x^4-14x^2y+49y^2$

14. a. $(x+9)^2 = (x)^2+2(x)(9)+(9)^2$
$= x^2+18x+81$

b. $(2x+1)(2x-1) = (2x)^2-(1)^2 = 4x^2-1$

c. $8x(x^2+1)(x^2-1) = 8x[(x^2)^2-(1)^2]$
$= 8x[x^4-1]$
$= 8x^5-8x$

15. a. $\frac{1}{x^{-3}} = x^3$

b. $\frac{1}{3^{-4}} = 3^4 = 81$

c. $\frac{p^{-4}}{q^{-9}} = \frac{q^9}{p^4}$

d. $\frac{5^{-3}}{2^{-5}} = \frac{2^5}{5^3} = \frac{32}{125}$

16. a. $5^{-3} = \frac{1}{5^3} = \frac{1}{125}$

b. $\frac{9}{x^{-7}} = 9x^7$

c. $\frac{11^{-1}}{7^{-2}} = \frac{7^2}{11^1} = \frac{49}{11}$

17.

$$\begin{array}{r|l} & 4x^2-4x+6 \\ 2x+3 & 8x^3+4x^2+0x+7 \\ & \underline{8x^3+12x^2} \\ & -8x^2+0x \\ & \underline{-8x^2-12x} \\ & 12x+7 \\ & \underline{12x+18} \\ & -11 \end{array}$$

$\frac{4x^2+7+8x^3}{2x+3} = 4x^2-4x+6-\frac{11}{2x+3}$

18.

$$\begin{array}{r|l} & 4x^2+16x+55 \\ x-4 & 4x^3+0x^2-9x+2 \\ & \underline{4x^3-16x^2} \\ & 16x^2-9x \\ & \underline{16x^2-64x} \\ & 55x+2 \\ & \underline{55x-220} \\ & 222 \end{array}$$

$\frac{4x^3-9x+2}{x-4} = 4x^2+16x+55+\frac{222}{x-4}$

19. a. $28 = 2 \cdot 2 \cdot 7$
$40 = 2 \cdot 2 \cdot 2 \cdot 5$
$\text{GCF} = 2^2 = 4$

b. $55 = 5 \cdot 11$
$21 = 3 \cdot 7$
$\text{GCF} = 1$

c. $15 = 3 \cdot 5$
$18 = 2 \cdot 3 \cdot 3$
$66 = 2 \cdot 3 \cdot 11$
$\text{GCF} = 3$

20. $9x^2 = 3 \cdot 3 \cdot x^2$
$6x^3 = 2 \cdot 3 \cdot x^3$
$21x^5 = 3 \cdot 7 \cdot x^5$
$\text{GCF} = 3x^2$

21. $-9a^5 + 18a^2 - 3a = -3a(3a^4 - 6a + 1)$

22. $7x^6 - 7x^5 + 7x^4 = 7x^4(x^2 - x + 1)$

23.
$$\begin{aligned} 3m^2 - 24m - 60 &= 3(m^2 - 8m - 20) \\ &= 3(m^2 - 10m + 2m - 20) \\ &= 3[m(m-10) + 2(m-10)] \\ &= 3(m-10)(m+2) \end{aligned}$$

24.
$$\begin{aligned} -2a^2 + 10a + 12 &= -2(a^2 - 5a - 6) \\ &= -2(a+1)(a-6) \end{aligned}$$

25.
$$\begin{aligned} 3x^2 + 11x + 6 &= 3x^2 + 2x + 9x + 6 \\ &= x(3x+2) + 3(3x+2) \\ &= (3x+2)(x+3) \end{aligned}$$

26.
$$\begin{aligned} 10m^2 - 7m + 1 &= 10m^2 - 2m - 5m + 1 \\ &= 2m(5m-1) - 1(5m-1) \\ &= (2m-1)(5m-1) \end{aligned}$$

27. $x^2 + 12x + 36 = x^2 + 2 \cdot x \cdot 6 + 6^2 = (x+6)^2$

28.
$$\begin{aligned} 4x^2 + 12x + 9 &= (2x)^2 + 2(2x)(3) + (3)^2 \\ &= (2x+3)^2 \end{aligned}$$

29. $x^2 + 4$ is a prime polynomial.

30. $x^2 - 4 = (x)^2 - (2)^2 = (x+2)(x-2)$

31.
$$\begin{aligned} x^3 + 8 &= x^3 + 2^3 \\ &= (x+2)(x^2 - x \cdot 2 + 2^2) \\ &= (x+2)(x^2 - 2x + 4) \end{aligned}$$

32.
$$\begin{aligned} 27y^3 - 1 &= (3y)^3 - (1)^3 \\ &= (3y-1)[(3y)^2 + 3y(1) + (1)^2] \\ &= (3y-1)(9y^2 + 3y + 1) \end{aligned}$$

33.
$$\begin{aligned} 2x^3 + 3x^2 - 2x - 3 &= x^2(2x+3) - 1(2x+3) \\ &= (2x+3)(x^2 - 1) \\ &= (2x+3)(x^2 - 1^2) \\ &= (2x+3)(x+1)(x-1) \end{aligned}$$

34.
$$\begin{aligned} 3x^3 + 5x^2 - 12x - 20 &= x^2(3x+5) - 4(3x+5) \\ &= (3x+5)(x^2 - 4) \\ &= (3x+5)(x^2 - 2^2) \\ &= (3x+5)(x+2)(x-2) \end{aligned}$$

35.
$$\begin{aligned} 12m^2 - 3n^2 &= 3(4m^2 - n^2) \\ &= 3[(2m)^2 - (n)^2] \\ &= 3(2m+n)(2m-n) \end{aligned}$$

36.
$$\begin{aligned} x^5 - x &= x(x^4 - 1) \\ &= x[(x^2)^2 - 1^2] \\ &= x(x^2+1)(x^2-1) \\ &= x(x^2+1)(x+1)(x-1) \end{aligned}$$

37.
$$\begin{aligned} x(2x-7) &= 4 \\ 2x^2 - 7x &= 4 \\ 2x^2 - 7x - 4 &= 0 \\ 2x^2 - 8x + x - 4 &= 0 \\ 2x(x-4) + 1(x-4) &= 0 \\ (x-4)(2x+1) &= 0 \end{aligned}$$
$2x + 1 = 0$ or $x - 4 = 0$
$2x = -1$ $\quad x = 4$
$x = -\frac{1}{2}$

38.
$$\begin{aligned} 3x^2 + 5x &= 2 \\ 3x^2 + 5x - 2 &= 0 \\ 3x^2 + 6x - x - 2 &= 0 \\ 3x(x+2) - 1(x+2) &= 0 \\ (x+2)(3x-1) &= 0 \end{aligned}$$

$3x-1=0$ or $x+2=0$
$3x=1$ $\quad x=-2$
$x=\frac{1}{3}$

39. $y=x^2-5x+4$
$0=x^2-5x+4$
$0=(x-4)(x-1)$
$x-1=0$ or $x-4=0$
$x=1$ $\quad x=4$
The x-intercepts are (1, 0) and (4, 0).

40. $y=x^2-x-6$
$0=x^2-x-6$
$0=(x-3)(x+2)$
$x+2=0$ or $x-3=0$
$x=-2$ $\quad x=3$
The x-intercepts are (–2, 0) and (3, 0).

41. Let x = the base and $2x - 2$ = the height.
$A=\frac{1}{2}bh$
$30=\frac{1}{2}x(2x-2)$
$30=\frac{1}{2}(2x)(x-1)$
$30=x(x-1)$
$30=x^2-x$
$0=x^2-x-30$
$0=(x+5)(x-6)$
$x-6=0$ or $x+5=0$
$x=6$ $\quad x=-5$
Length cannot be negative, so $x = 6$.
$2x - 2 = 2(6) - 2 = 10$
The base is 6 meters and the height is 10 meters.

42. Let x = the base and $3x + 5$ = the height.
$A=bh$
$182=x(3x+5)$
$182=3x^2+5x$
$0=3x^2+5x-182$
$0=3x^2+26x-21x-182$
$0=x(3x+26)-7(3x+26)$
$0=(x-7)(3x+26)$
$x-7=0$ or $3x+26=0$
$x=7$ $\quad x=-\frac{26}{3}$
Length cannot be negative so $x = 7$.
$3x + 5 = 3(7) + 5 = 26$
The base is 7 ft and the height is 26 ft.

43. $\frac{5x-5}{x^3-x^2}=\frac{5(x-1)}{x^2(x-1)}=\frac{5}{x^2}$

44. $\frac{2x^2-50}{4x^4-20x^3}=\frac{2(x^2-25)}{4x^3(x-5)}$
$=\frac{2(x+5)(x-5)}{4x^3(x-5)}$
$=\frac{x+5}{2x^3}$

45. $\frac{6x+2}{x^2-1}\div\frac{3x^2+x}{x-1}=\frac{6x+2}{x^2-1}\cdot\frac{x-1}{3x^2+x}$
$=\frac{2(3x+1)}{(x+1)(x-1)}\cdot\frac{x-1}{x(3x+1)}$
$=\frac{2}{x(x+1)}$

46. $\frac{6x^2-18x}{3x^2-2x}\cdot\frac{15x-10}{x^2-10}=\frac{6x(x-3)\cdot 5(3x-2)}{x(3x-2)\cdot(x+3)(x-3)}$
$=\frac{30}{x+3}$

47. $\frac{\frac{x+1}{y}}{\frac{x}{y}+2}=\frac{y\left(\frac{x+1}{y}\right)}{y\left(\frac{x}{y}+2\right)}=\frac{x+1}{y\left(\frac{x}{y}\right)+2y}=\frac{x+1}{x+2y}$

48. $\frac{\frac{m}{3}+\frac{n}{6}}{\frac{m+n}{12}}=\frac{12}{12}\cdot\frac{\frac{m}{3}+\frac{n}{6}}{\frac{m+n}{12}}$
$=\frac{12\left(\frac{m}{3}\right)+12\left(\frac{n}{6}\right)}{12\left(\frac{m+n}{12}\right)}$
$=\frac{4m+2n}{m+n}$ or $\frac{2(2m+n)}{m+n}$

Chapter 8

Section 8.1

Practice Exercises

1. a. $\sqrt{\frac{4}{81}}=\frac{2}{9}$ because $\left(\frac{2}{9}\right)^2=\frac{4}{81}$ and $\frac{2}{9}$ is positive.

   b. $-\sqrt{25}=-5$
   The negative sign in front of the radical indicates the negative square root of 25.

   c. $\sqrt{144}=12$ because $12^2=144$ and 12 is positive.

   d. $\sqrt{0.49}=0.7$ because $(0.7)^2=0.49$ and 0.7 is positive.

   e. $-\sqrt{1}=-1$
   The negative sign in front of the radical indicates the negative square root of 1.

2. a. $\sqrt[3]{0}=0$ because $0^3=0$.

   b. $\sqrt[3]{-64}=-4$ because $(-4)^3=-64$.

   c. $\sqrt[3]{\frac{1}{8}}=\frac{1}{2}$ because $\left(\frac{1}{2}\right)^3=\frac{1}{8}$.

3. a. $\sqrt[4]{81}=3$ because $3^4=81$ and 3 is positive.

   b. $\sqrt[5]{100,000}=10$ because $10^5=100,000$.

   c. $\sqrt[6]{-64}$ is not a real number since the index 6 is even and the radicand -64 is negative.

   d. $\sqrt[3]{-125}=-5$ because $(-5)^3=-125$.

4. $\sqrt{17}\approx 4.123105626$
   To three decimal places, $\sqrt{17}=4.123$.

5. a. $\sqrt{x^{10}}=x^5$ because $(x^5)^2=x^{10}$.

   b. $\sqrt{y^{14}}=y^7$ because $(y^7)^2=y^{14}$.

   c. $\sqrt[3]{125z^9}=5z^3$ because $(5z^3)^3=125z^9$.

   d. $\sqrt{49x^2}=7x$ because $(7x)^2=49x^2$.

   e. $\sqrt{\frac{z^4}{36}}=\frac{z^2}{6}$ because $\left(\frac{z^2}{6}\right)^2=\frac{z^4}{36}$.

   f. $\sqrt[3]{-8a^6b^{12}}=-2a^2b^4$ because $(-2a^2b^4)^3=-8a^6b^{12}$.

Calculator Explorations

1. $\sqrt{7}\approx 2.646$
   This is reasonable; 7 is between 4 and 9 so $\sqrt{7}$ is between $\sqrt{4}=2$ and $\sqrt{9}=3$.

2. $\sqrt{14}\approx 3.742$
   This is reasonable; 14 is between 9 and 16 so $\sqrt{14}$ is between $\sqrt{9}=3$ and $\sqrt{16}=4$.

3. $\sqrt{11}\approx 3.317$
   This is reasonable; 11 is between 9 and 16 so $\sqrt{11}$ is between $\sqrt{9}=3$ and $\sqrt{16}=4$.

4. $\sqrt{200}\approx 14.142$
   This is reasonable; 200 is between 196 and 225 so $\sqrt{200}$ is between $\sqrt{196}=14$ and $\sqrt{225}=15$.

5. $\sqrt{82}\approx 9.055$
   This is reasonable; 82 is between 81 and 100 so $\sqrt{82}$ is between $\sqrt{81}=9$ and $\sqrt{100}=10$.

6. $\sqrt{46}\approx 6.782$
   This is reasonable; 46 is between 36 and 49 so $\sqrt{46}$ is between $\sqrt{36}=6$ and $\sqrt{49}=7$.

7. $\sqrt[3]{40}\approx 3.420$
   This is reasonable; 40 is between 27 and 64 so $\sqrt[3]{40}$ is between $\sqrt[3]{27}=3$ and $\sqrt[3]{64}=4$.

8. $\sqrt[3]{71}\approx 4.141$
   This is reasonable; 71 is between 64 and 125 so $\sqrt[3]{71}$ is between $\sqrt[3]{64}=4$ and $\sqrt[3]{125}=5$.

9. $\sqrt[4]{20} \approx 2.115$
This is reasonable; 20 is between 16 and 81 so $\sqrt[4]{20}$ is between $\sqrt[4]{16} = 2$ and $\sqrt[4]{81} = 3$.

10. $\sqrt[4]{15} \approx 1.968$
This is reasonable; 15 is between 1 and 16 so $\sqrt[4]{15}$ is between $\sqrt[4]{1} = 1$ and $\sqrt[4]{16} = 2$.

11. $\sqrt[5]{18} \approx 1.783$
This is reasonable; 18 is between 1 and 32 so $\sqrt[5]{18}$ is between $\sqrt[5]{1} = 1$ and $\sqrt[5]{32} = 2$.

12. $\sqrt[6]{2} \approx 1.122$
This is reasonable; 2 is between 1 and 64 so $\sqrt[6]{2}$ is between $\sqrt[6]{1} = 1$ and $\sqrt[6]{64} = 2$.

Vocabulary and Readiness Check

1. In the expression $\sqrt[4]{16}$, the number 4 is called the index, the number 16 is called the radicand, and $\sqrt{\ }$ is called the radical sign.

2. The symbol $\sqrt{\ }$ is used to denote the positive, or principal, square root.

3. False; $\sqrt{-16}$ is not a real number since the index is even and the radicand is negative.

4. True

5. True

6. True

7. True

8. False; $\sqrt{x^{16}} = x^8$ because $(x^8)^2 = x^{16}$.

Exercise Set 8.1

1. $\sqrt{16} = 4$, because $4^2 = 16$ and 4 is positive.

3. $\sqrt{\frac{1}{25}} = \frac{1}{5}$, because $\left(\frac{1}{5}\right)^2 = \frac{1}{25}$ and $\frac{1}{5}$ is positive.

5. $-\sqrt{100} = -10$, because $10^2 = 100$ and the negative sign indicates the negative square root.

7. $\sqrt{-4}$ is not a real number.

9. $-\sqrt{121} = -11$, because $11^2 = 121$ and the negative sign indicates the negative square root.

11. $\sqrt{\frac{9}{25}} = \frac{3}{5}$, because $\left(\frac{3}{5}\right)^2 = \frac{9}{25}$ and $\frac{3}{5}$ is positive.

13. $\sqrt{900} = 30$ because $30^2 = 900$ and 30 is positive.

15. $\sqrt{144} = 12$, because $12^2 = 144$ and 12 is positive.

17. $\sqrt{\frac{1}{100}} = \frac{1}{10}$ because $\left(\frac{1}{10}\right)^2 = \frac{1}{100}$ and $\frac{1}{10}$ is positive.

19. $\sqrt{0.25} = 0.5$ because $(0.5)^2 = 0.25$ and 0.5 is positive.

21. $\sqrt[3]{125} = 5$, because $(5)^3 = 125$.

23. $\sqrt[3]{-64} = -4$, because $(-4)^3 = -64$.

25. $-\sqrt[3]{8} = -2$ because $2^3 = 8$.

27. $\sqrt[3]{\frac{1}{8}} = \frac{1}{2}$, because $\left(\frac{1}{2}\right)^3 = \frac{1}{8}$.

29. $\sqrt[3]{-125} = -5$, because $(-5)^3 = -125$.

31. $\sqrt[5]{32} = 2$, because $(2)^5 = 32$.

33. $\sqrt{81} = 9$, because $9^2 = 81$ and 9 is positive.

35. $\sqrt[4]{-16}$ is not a real number.

37. $\sqrt[3]{-\frac{27}{64}} = -\frac{3}{4}$ because $\left(-\frac{3}{4}\right)^3 = -\frac{27}{64}$.

39. $-\sqrt[4]{625} = -5$, because $\sqrt[4]{625} = 5$.

41. $\sqrt[6]{1} = 1$, because $(1)^6 = 1$.

43. $\sqrt{7} \approx 2.646$

45. $\sqrt{37} \approx 6.083$

47. $\sqrt{136} \approx 11.662$

49. $\sqrt{2} \approx 1.41$
$90\sqrt{2} \approx 90 \cdot 1.41 = 126.90$ feet

51. $\sqrt{x^4} = x^2$, because $(x^2)^2 = x^4$.

53. $\sqrt{9x^8} = 3x^4$, because $(3x^4)^2 = 9x^8$.

55. $\sqrt{81x^2} = 9x$, because $(9x)^2 = 81x^2$.

57. $\sqrt{\frac{x^6}{36}} = \frac{x^3}{6}$ because $\left(\frac{x^3}{6}\right)^2 = \frac{x^6}{36}$.

59. $\sqrt{\frac{25y^2}{9}} = \frac{5y}{3}$ because $\left(\frac{5y}{3}\right)^2 = \frac{25y^2}{9}$.

61. $\sqrt{16a^6b^4} = 4a^3b^2$ because $(4a^3b^2)^2 = 16a^6b^4$.

63. $\sqrt[3]{a^6b^{18}} = a^2b^6$ because $(a^2b^6)^3 = a^6b^{18}$.

65. $\sqrt[3]{-8x^3y^{27}} = -2xy^9$ because
$(-2xy^9)^3 = -8x^3y^{27}$.

67. $50 = 25 \cdot 2$; 25 is a perfect square

69. $32 = 16 \cdot 2$; 16 is a perfect square or $32 = 4 \cdot 8$; 4 is a perfect square

71. $28 = 4 \cdot 7$; 4 is a perfect square

73. $27 = 9 \cdot 3$; 9 is a perfect square

75. a and b are real numbers since both have odd indices.

77. Let $A = 49$.
The length of a side $= \sqrt{A}$
$\sqrt{A} = \sqrt{49} = 7$
The length of a side is 7 miles.

79. Let $A = 9.0601$.
The length of a side is $\sqrt{A}$.
$\sqrt{A} = \sqrt{9.0601} = 3.01$
The length of a side is 3.01 inches.

81. $\sqrt{\sqrt{81}} = \sqrt{9} = 3$

83. $\sqrt{\sqrt{10,000}} = \sqrt{100} = 10$

85. Let $L = 30$ and $g = 32$. Use 3.14 for π.
$T = 2\pi\sqrt{\frac{L}{g}}$
$T = 2(3.14)\sqrt{\frac{30}{32}} \approx 6.1$
The period is 6.1 seconds.

87. Answers may vary

89.

x	$y = \sqrt{x}$
0	$\sqrt{0} = 0$
1	$\sqrt{1} = 1$
3	$\sqrt{3} = 1.7$
4	$\sqrt{4} = 2$
9	$\sqrt{9} = 3$

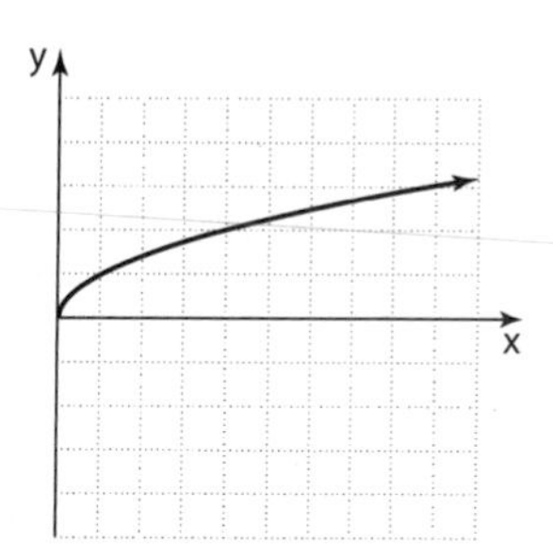

91. $\sqrt{x^2} = \sqrt{(x)^2} = |x|$

93. $\sqrt{(x+2)^2} = |x+2|$

95. $y = \sqrt{x-2}$

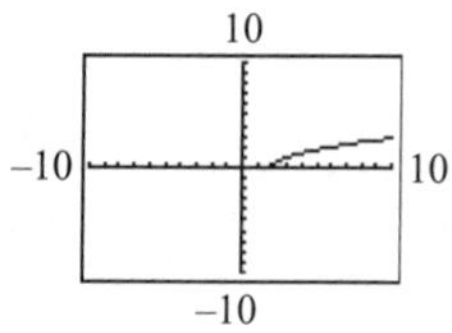

The graph starts at (2, 0) because $x - 2 \geq 0$ or $x \geq 2$.

97. $y = \sqrt{x+4}$

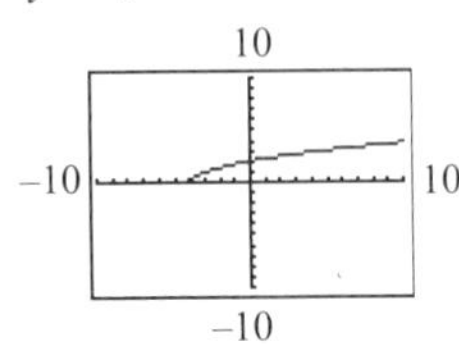

The graph starts at (−4, 0) because $x + 4 \geq 0$ or $x \geq -4$.

Section 8.2

Practice Exercises

1. a. $\sqrt{24} = \sqrt{4 \cdot 6} = \sqrt{4} \cdot \sqrt{6} = 2\sqrt{6}$

b. $\sqrt{60} = \sqrt{4 \cdot 15} = \sqrt{4} \cdot \sqrt{15} = 2\sqrt{15}$

c. $\sqrt{42}$ is in simplest form since the radicand 42 contains no perfect square factors other than 1.

d. $\sqrt{300} = \sqrt{100 \cdot 3} = \sqrt{100} \cdot \sqrt{3} = 10\sqrt{3}$

2. $$\begin{aligned} 5\sqrt{40} &= 5 \cdot \sqrt{40} \\ &= 5 \cdot \sqrt{4 \cdot 10} \\ &= 5 \cdot \sqrt{4} \cdot \sqrt{10} \\ &= 5 \cdot 2 \cdot \sqrt{10} \\ &= 10\sqrt{10} \end{aligned}$$

3. a. $\sqrt{\dfrac{5}{49}} = \dfrac{\sqrt{5}}{\sqrt{49}} = \dfrac{\sqrt{5}}{7}$

b. $\sqrt{\dfrac{9}{100}} = \dfrac{\sqrt{9}}{\sqrt{100}} = \dfrac{3}{10}$

c. $\sqrt{\dfrac{18}{25}} = \dfrac{\sqrt{18}}{\sqrt{25}} = \dfrac{\sqrt{9} \cdot \sqrt{2}}{5} = \dfrac{3\sqrt{2}}{5}$

4. a. $\sqrt{x^7} = \sqrt{x^6 \cdot x} = \sqrt{x^6} \cdot \sqrt{x} = x^3\sqrt{x}$

b. $$\begin{aligned} \sqrt{12a^4} &= \sqrt{4 \cdot 3 \cdot a^4} \\ &= \sqrt{4a^4 \cdot 3} \\ &= \sqrt{4a^4} \cdot \sqrt{3} \\ &= 2a^2\sqrt{3} \end{aligned}$$

c. $\sqrt{\dfrac{98}{z^8}} = \dfrac{\sqrt{98}}{\sqrt{z^8}} = \dfrac{\sqrt{49 \cdot 2}}{z^4} = \dfrac{\sqrt{49} \cdot \sqrt{2}}{z^4} = \dfrac{7\sqrt{2}}{z^4}$

d. $$\begin{aligned} \sqrt{\frac{11y^9}{49}} &= \frac{\sqrt{11y^9}}{\sqrt{49}} \\ &= \frac{\sqrt{y^8 \cdot 11y}}{7} \\ &= \frac{\sqrt{y^8} \cdot \sqrt{11y}}{7} \\ &= \frac{y^4\sqrt{11y}}{7} \end{aligned}$$

5. a. $\sqrt[3]{24} = \sqrt[3]{8 \cdot 3} = \sqrt[3]{8} \cdot \sqrt[3]{3} = 2\sqrt[3]{3}$

b. The number 38 contains no perfect cube factors, so $\sqrt[3]{38}$ cannot be simplified further.

c. $\sqrt[3]{\dfrac{5}{27}} = \dfrac{\sqrt[3]{5}}{\sqrt[3]{27}} = \dfrac{\sqrt[3]{5}}{3}$

d. $\sqrt[3]{\dfrac{15}{64}} = \dfrac{\sqrt[3]{15}}{\sqrt[3]{64}} = \dfrac{\sqrt[3]{15}}{4}$

6. a. $\sqrt[4]{32} = \sqrt[4]{16 \cdot 2} = \sqrt[4]{16} \cdot \sqrt[4]{2} = 2\sqrt[4]{2}$

b. $\sqrt[4]{\dfrac{5}{81}} = \dfrac{\sqrt[4]{5}}{\sqrt[4]{81}} = \dfrac{\sqrt[4]{5}}{3}$

c. $\sqrt[5]{96} = \sqrt[5]{32 \cdot 3} = \sqrt[5]{32} \cdot \sqrt[5]{3} = 2\sqrt[5]{3}$

Vocabulary and Readiness Check

1. If $\sqrt{a}$ and $\sqrt{b}$ are real numbers, then $\sqrt{a \cdot b} = \underline{\sqrt{a} \cdot \sqrt{b}}.$

2. If $\sqrt{a}$ and $\sqrt{b}$ are real numbers, then $\sqrt{\dfrac{a}{b}} = \underline{\dfrac{\sqrt{a}}{\sqrt{b}}}.$

3. $\sqrt{16 \cdot 25} = \sqrt{\underline{16}} \cdot \sqrt{\underline{25}} = \underline{4} \cdot \underline{5} = \underline{20}$

4. $\sqrt{36 \cdot 3} = \sqrt{\underline{36}} \cdot \sqrt{\underline{3}} = \underline{6} \cdot \sqrt{\underline{3}} = \underline{6\sqrt{3}}$

Exercise Set 8.2

1. $\sqrt{20} = \sqrt{4 \cdot 5} = \sqrt{4} \cdot \sqrt{5} = 2\sqrt{5}$

3. $\sqrt{50} = \sqrt{25 \cdot 2} = \sqrt{25} \cdot \sqrt{2} = 5\sqrt{2}$

5. $\sqrt{33}$ can't be simplified.

7. $\sqrt{98} = \sqrt{49 \cdot 2} = \sqrt{49} \cdot \sqrt{2} = 7\sqrt{2}$

9. $\sqrt{60} = \sqrt{4 \cdot 15} = \sqrt{4} \cdot \sqrt{15} = 2\sqrt{15}$

11. $\sqrt{180} = \sqrt{36 \cdot 5} = \sqrt{36} \cdot \sqrt{5} = 6\sqrt{5}$

13. $\sqrt{52} = \sqrt{4 \cdot 13} = \sqrt{4} \cdot \sqrt{13} = 2\sqrt{13}$

15. $3\sqrt{25} = 3 \cdot \sqrt{25} = 3 \cdot 5 = 15$

17. $$\begin{aligned} 7\sqrt{63} &= 7 \cdot \sqrt{63} \\ &= 7 \cdot \sqrt{9 \cdot 7} \\ &= 7 \cdot \sqrt{9} \cdot \sqrt{7} \\ &= 7 \cdot 3 \cdot \sqrt{7} \\ &= 21\sqrt{7} \end{aligned}$$

19. $$\begin{aligned} -5\sqrt{27} &= -5 \cdot \sqrt{27} \\ &= -5 \cdot \sqrt{9 \cdot 3} \\ &= -5 \cdot \sqrt{9} \cdot \sqrt{3} \\ &= -5 \cdot 3 \cdot \sqrt{3} \\ &= -15\sqrt{3} \end{aligned}$$

21. $\sqrt{\frac{8}{25}} = \frac{\sqrt{8}}{\sqrt{25}} = \frac{\sqrt{4 \cdot 2}}{5} = \frac{\sqrt{4} \cdot \sqrt{2}}{5} = \frac{2\sqrt{2}}{5}$

23. $\sqrt{\frac{27}{121}} = \frac{\sqrt{27}}{\sqrt{121}} = \frac{\sqrt{9 \cdot 3}}{11} = \frac{\sqrt{9} \cdot \sqrt{3}}{11} = \frac{3\sqrt{3}}{11}$

25. $\sqrt{\frac{9}{4}} = \frac{\sqrt{9}}{\sqrt{4}} = \frac{3}{2}$

27. $\sqrt{\frac{125}{9}} = \frac{\sqrt{125}}{\sqrt{9}} = \frac{\sqrt{25 \cdot 5}}{3} = \frac{\sqrt{25} \cdot \sqrt{5}}{3} = \frac{5\sqrt{5}}{3}$

29. $\sqrt{\frac{11}{36}} = \frac{\sqrt{11}}{\sqrt{36}} = \frac{\sqrt{11}}{6}$

31. $$\begin{aligned} -\sqrt{\frac{27}{144}} &= -\frac{\sqrt{27}}{\sqrt{144}} \\ &= -\frac{\sqrt{9 \cdot 3}}{12} \\ &= -\frac{\sqrt{9} \cdot \sqrt{3}}{12} \\ &= -\frac{3\sqrt{3}}{12} \\ &= -\frac{\sqrt{3}}{4} \end{aligned}$$

33. $\sqrt{x^7} = \sqrt{x^6 \cdot x} = \sqrt{x^6} \cdot \sqrt{x} = x^3\sqrt{x}$

35. $\sqrt{x^{13}} = \sqrt{x^{12} \cdot x} = \sqrt{x^{12}} \cdot \sqrt{x} = x^6\sqrt{x}$

37. $\sqrt{36a^3} = \sqrt{36a^2 \cdot a} = \sqrt{36a^2} \cdot \sqrt{a} = 6a\sqrt{a}$

39. $\sqrt{96x^4} = \sqrt{16x^4 \cdot 6} = \sqrt{16x^4} \cdot \sqrt{6} = 4x^2\sqrt{6}$

41. $\sqrt{\frac{12}{m^2}} = \frac{\sqrt{12}}{\sqrt{m^2}} = \frac{\sqrt{4 \cdot 3}}{m} = \frac{\sqrt{4} \cdot \sqrt{3}}{m} = \frac{2\sqrt{3}}{m}$

43. $\sqrt{\frac{9x}{y^{10}}} = \frac{\sqrt{9x}}{\sqrt{y^{10}}} = \frac{\sqrt{9 \cdot x}}{y^5} = \frac{\sqrt{9} \cdot \sqrt{x}}{y^5} = \frac{3\sqrt{x}}{y^5}$

45. $\sqrt{\frac{88}{x^{12}}} = \frac{\sqrt{88}}{\sqrt{x^{12}}} = \frac{\sqrt{4 \cdot 22}}{x^6} = \frac{\sqrt{4} \cdot \sqrt{22}}{x^6} = \frac{2\sqrt{22}}{x^6}$

47. $8\sqrt{4} = 8 \cdot \sqrt{4} = 8 \cdot 2 = 16$

49. $\sqrt{\frac{36}{121}} = \frac{\sqrt{36}}{\sqrt{121}} = \frac{6}{11}$

51. $\sqrt{175} = \sqrt{25 \cdot 7} = \sqrt{25} \cdot \sqrt{7} = 5\sqrt{7}$

53. $\sqrt{\frac{20}{9}} = \frac{\sqrt{20}}{\sqrt{9}} = \frac{\sqrt{4 \cdot 5}}{3} = \frac{\sqrt{4} \cdot \sqrt{5}}{3} = \frac{2\sqrt{5}}{3}$

55. $\sqrt{24m^7} = \sqrt{4m^6 \cdot 6m} = \sqrt{4m^6} \cdot \sqrt{6m} = 2m^3\sqrt{6m}$

57. $\sqrt{\frac{23y^3}{4x^6}} = \frac{\sqrt{23y^3}}{\sqrt{4x^6}}$
$= \frac{\sqrt{y^2 \cdot 23y}}{2x^3}$
$= \frac{\sqrt{y^2} \cdot \sqrt{23y}}{2x^3}$
$= \frac{y\sqrt{23y}}{2x^3}$

59. $\sqrt[3]{24} = \sqrt[3]{8 \cdot 3} = \sqrt[3]{8} \cdot \sqrt[3]{3} = 2\sqrt[3]{3}$

61. $\sqrt[3]{250} = \sqrt[3]{125 \cdot 2} = \sqrt[3]{125} \cdot \sqrt[3]{2} = 5\sqrt[3]{2}$

63. $\sqrt[3]{\frac{5}{64}} = \frac{\sqrt[3]{5}}{\sqrt[3]{64}} = \frac{\sqrt[3]{5}}{4}$

65. $\sqrt[3]{\frac{23}{8}} = \frac{\sqrt[3]{23}}{\sqrt[3]{8}} = \frac{\sqrt[3]{23}}{2}$

67. $\sqrt[3]{\frac{15}{64}} = \frac{\sqrt[3]{15}}{\sqrt[3]{64}} = \frac{\sqrt[3]{15}}{4}$

69. $\sqrt[3]{80} = \sqrt[3]{8 \cdot 10} = \sqrt[3]{8} \cdot \sqrt[3]{10} = 2\sqrt[3]{10}$

71. $\sqrt[4]{48} = \sqrt[4]{16 \cdot 3} = \sqrt[4]{16} \cdot \sqrt[4]{3} = 2\sqrt[4]{3}$

73. $\sqrt[4]{\frac{8}{81}} = \frac{\sqrt[4]{8}}{\sqrt[4]{81}} = \frac{\sqrt[4]{8}}{3}$

75. $\sqrt[5]{96} = \sqrt[5]{32 \cdot 3} = \sqrt[5]{32} \cdot \sqrt[5]{3} = 2\sqrt[5]{3}$

77. $\sqrt[5]{\frac{5}{32}} = \frac{\sqrt[5]{5}}{\sqrt[5]{32}} = \frac{\sqrt[5]{5}}{2}$

79. $\sqrt[3]{80} = \sqrt[3]{8 \cdot 10} = \sqrt[3]{8} \cdot \sqrt[3]{10} = 2\sqrt[3]{10}$
The length of each side is $2\sqrt[3]{10}$ inches.

81. $6x + 8x = (6 + 8)x = 14x$

83. $(2x+3)(x-5) = 2x^2 - 10x + 3x - 15$
$= 2x^2 - 7x - 15$

85. $9y^2 - 9y^2 = 0$

87. $\sqrt{x^6y^3} = \sqrt{x^6y^2y} = \sqrt{x^6y^2} \cdot \sqrt{y} = x^3y\sqrt{y}$

89. $\sqrt{98x^5y^4} = \sqrt{49x^4y^4 \cdot 2x}$
$= \sqrt{49x^4y^4} \cdot \sqrt{2x}$
$= 7x^2y^2\sqrt{2x}$

91. $\sqrt[3]{-8x^6} = -2x^2$ because $(-2x^2)^3 = -8x^6$.

93. Answers may vary

95. Let $A = 120$. The length of a side is $\sqrt{\frac{A}{6}}$.
$\sqrt{\frac{A}{6}} = \sqrt{\frac{120}{6}} = \sqrt{20} = \sqrt{4} \cdot \sqrt{5} = 2\sqrt{5}$
The length of a side is $2\sqrt{5}$ inches.

97. Let $A = 30.375$. The length of a side is $\sqrt{\frac{A}{6}}$.
$\sqrt{\frac{A}{6}} = \sqrt{\frac{30.375}{6}} = \sqrt{5.0625} = 2.25$
The length of one side of a Rubik's cube is 2.25 inches.

99. Let $n = 1000$
$C = 100\sqrt[3]{n} + 700$
$C = 100\sqrt[3]{1000} + 700$
$= 100(10) + 700$
$= 1700$
The cost is $1700.

101. Let $h = 169$ and $w = 64$.
$B = \sqrt{\frac{hw}{3600}}$
$B = \sqrt{\frac{(169)(64)}{3600}}$
$= \sqrt{\frac{10,816}{3600}}$
$= \sqrt{\frac{676}{225}}$
$= \frac{26}{15}$
≈ 1.7
The surface area is about 1.7 sq m.

Section 8.3

Practice Exercises

1. a. $3\sqrt{2}+5\sqrt{2}=(3+5)\sqrt{2}=8\sqrt{2}$

b. $\sqrt{6}-8\sqrt{6}=1\sqrt{6}-8\sqrt{6}=(1-8)\sqrt{6}=-7\sqrt{6}$

c. $6\sqrt[4]{5}-2\sqrt[4]{5}+11\sqrt[4]{7}=(6-2)\sqrt[4]{5}+11\sqrt[4]{7}$
$=4\sqrt[4]{5}+11\sqrt[4]{7}$
This expression cannot be simplified further since the radicals are not the same.

d. $4\sqrt{13}-5\sqrt[3]{13}$ cannot be simplified further since the indices are not the same.

2. a.
$$\begin{aligned}\sqrt{45}+\sqrt{20}&=\sqrt{9\cdot 5}+\sqrt{4\cdot 5}\\&=\sqrt{9}\cdot\sqrt{5}+\sqrt{4}\cdot\sqrt{5}\\&=3\sqrt{5}+2\sqrt{5}\\&=5\sqrt{5}\end{aligned}$$

b.
$$\begin{aligned}&\sqrt{36}+3\sqrt{24}-\sqrt{40}-\sqrt{150}\\&=6+3\sqrt{4\cdot 6}-\sqrt{4\cdot 10}-\sqrt{25\cdot 6}\\&=6+3\sqrt{4}\cdot\sqrt{6}-\sqrt{4}\cdot\sqrt{10}-\sqrt{25}\cdot\sqrt{6}\\&=6+3\cdot 2\sqrt{6}-2\sqrt{10}-5\sqrt{6}\\&=6+6\sqrt{6}-2\sqrt{10}-5\sqrt{6}\\&=6-2\sqrt{10}+\sqrt{6}\end{aligned}$$

c.
$$\begin{aligned}\sqrt{98}-5\sqrt{8}&=\sqrt{49\cdot 2}-5\sqrt{4\cdot 2}\\&=\sqrt{49}\cdot\sqrt{2}-5\sqrt{4}\cdot\sqrt{2}\\&=7\sqrt{2}-5\cdot 2\sqrt{2}\\&=7\sqrt{2}-10\sqrt{2}\\&=-3\sqrt{2}\end{aligned}$$

3.
$$\begin{aligned}\sqrt{x^3}-8x\sqrt{x}+3\sqrt{x^2}&=\sqrt{x^2\cdot x}-8x\sqrt{x}+3x\\&=\sqrt{x^2}\cdot\sqrt{x}-8x\sqrt{x}+3x\\&=x\sqrt{x}-8x\sqrt{x}+3x\\&=-7x\sqrt{x}+3x\end{aligned}$$

4.
$$\begin{aligned}4\sqrt[3]{81x^6}-\sqrt[3]{24x^6}&=4\sqrt[3]{27x^6\cdot 3}-\sqrt[3]{8x^6\cdot 3}\\&=4\cdot\sqrt[3]{27x^6}\cdot\sqrt[3]{3}-\sqrt[3]{8x^6}\cdot\sqrt[3]{3}\\&=4\cdot 3x^2\cdot\sqrt[3]{3}-2x^2\cdot\sqrt[3]{3}\\&=12x^2\sqrt[3]{3}-2x^2\sqrt[3]{3}\\&=10x^2\sqrt[3]{3}\end{aligned}$$

Vocabulary and Readiness Check

1. Radicals that have the same index and same radicand are called like radicals.

2. The expressions $7\sqrt[3]{2x}$ and $-\sqrt[3]{2x}$ are called like radicals.

3. $11\sqrt{2}+6\sqrt{2}=17\sqrt{2}$

4. $\sqrt{5}$ is the same as $1\sqrt{5}$.

5. $\sqrt{5}+\sqrt{5}=2\sqrt{5}$

6. $9\sqrt{7}-\sqrt{7}=8\sqrt{7}$

Exercise Set 8.3

1. $4\sqrt{3}-8\sqrt{3}=(4-8)\sqrt{3}=-4\sqrt{3}$

3.
$$\begin{aligned}3\sqrt{6}+8\sqrt{6}-2\sqrt{6}-5&=(3+8-2)\sqrt{6}-5\\&=9\sqrt{6}-5\end{aligned}$$

5.
$$\begin{aligned}\sqrt{11}+\sqrt{11}+11&=1\sqrt{11}+1\sqrt{11}+11\\&=(1+1)\sqrt{11}+11\\&=2\sqrt{11}+11\end{aligned}$$

7. $6\sqrt{5}-5\sqrt{5}+\sqrt{2}=(6-5)\sqrt{5}+\sqrt{2}=\sqrt{5}+\sqrt{2}$

9.
$$\begin{aligned}\sqrt[3]{16}+\sqrt[3]{16}-4\sqrt[3]{16}&=1\sqrt[3]{16}+1\sqrt[3]{16}-4\sqrt[3]{16}\\&=(1+1-4)\sqrt[3]{16}\\&=-2\sqrt[3]{16}\end{aligned}$$

11. $2\sqrt[3]{3}+5\sqrt[3]{3}-\sqrt{3}=(2+5)\sqrt[3]{3}-\sqrt{3}=7\sqrt[3]{3}-\sqrt{3}$

13. $2\sqrt[3]{2}-7\sqrt[3]{2}-6=(2-7)\sqrt[3]{2}-6=-5\sqrt[3]{2}-6$

15.
$$\begin{aligned}\sqrt{12}+\sqrt{27}&=\sqrt{4\cdot 3}+\sqrt{9\cdot 3}\\&=\sqrt{4}\cdot\sqrt{3}+\sqrt{9}\cdot\sqrt{3}\\&=2\sqrt{3}+3\sqrt{3}\\&=5\sqrt{3}\end{aligned}$$

17.
$$\begin{aligned}\sqrt{45}+3\sqrt{20}&=\sqrt{9\cdot 5}+3\sqrt{4\cdot 5}\\&=\sqrt{9}\cdot\sqrt{5}+3\sqrt{4}\cdot\sqrt{5}\\&=3\sqrt{5}+3(2)\sqrt{5}\\&=3\sqrt{5}+6\sqrt{5}\\&=9\sqrt{5}\end{aligned}$$

19. $2\sqrt{54}-\sqrt{20}+\sqrt{45}-\sqrt{24}$
$=2\sqrt{9\cdot 6}-\sqrt{4\cdot 5}+\sqrt{9\cdot 5}-\sqrt{4\cdot 6}$
$=2\sqrt{9}\cdot\sqrt{6}-\sqrt{4}\cdot\sqrt{5}+\sqrt{9}\cdot\sqrt{5}-\sqrt{4}\cdot\sqrt{6}$
$=2(3)\sqrt{6}-2\sqrt{5}+3\sqrt{5}-2\sqrt{6}$
$=6\sqrt{6}-2\sqrt{5}+3\sqrt{5}-2\sqrt{6}$
$=4\sqrt{6}+\sqrt{5}$

21. $4x-3\sqrt{x^2}+\sqrt{x}=4x-3x+\sqrt{x}=x+\sqrt{x}$

23. $\sqrt{25x}+\sqrt{36x}-11\sqrt{x}$
$=\sqrt{25}\cdot\sqrt{x}+\sqrt{36}\cdot\sqrt{x}-11\sqrt{x}$
$=5\sqrt{x}+6\sqrt{x}-11\sqrt{x}$
$=0$

25. $\sqrt{16x}-\sqrt{x^3}=\sqrt{16x}-\sqrt{x^2\cdot x}$
$=\sqrt{16}\cdot\sqrt{x}-\sqrt{x^2}\cdot\sqrt{x}$
$=4\sqrt{x}-x\sqrt{x}$
$=(4-x)\sqrt{x}$

27. $12\sqrt{5}-\sqrt{5}-4\sqrt{5}=(12-1-4)\sqrt{5}=7\sqrt{5}$

29. $\sqrt{5}+\sqrt[3]{5}$ cannot be simplified.

31. $4+8\sqrt{2}-9=8\sqrt{2}+4-9=8\sqrt{2}-5$

33. $8-\sqrt{2}-5\sqrt{2}=8+(-1-5)\sqrt{2}=8-6\sqrt{2}$

35. $5\sqrt{32}-\sqrt{72}=5\sqrt{16\cdot 2}-\sqrt{36\cdot 2}$
$=5\sqrt{16}\sqrt{2}-\sqrt{36}\sqrt{2}$
$=5(4)\sqrt{2}-6\sqrt{2}$
$=20\sqrt{2}-6\sqrt{2}$
$=14\sqrt{2}$

37. $\sqrt{8}+\sqrt{9}+\sqrt{18}+\sqrt{81}$
$=\sqrt{4\cdot 2}+\sqrt{9}+\sqrt{9\cdot 2}+\sqrt{81}$
$=\sqrt{4}\cdot\sqrt{2}+3+\sqrt{9}\cdot\sqrt{2}+9$
$=2\sqrt{2}+3+3\sqrt{2}+9$
$=5\sqrt{2}+12$

39. $\sqrt{\frac{5}{9}}+\sqrt{\frac{5}{81}}=\frac{\sqrt{5}}{\sqrt{9}}+\frac{\sqrt{5}}{\sqrt{81}}$
$=\frac{\sqrt{5}}{3}+\frac{\sqrt{5}}{9}$
$=\frac{3\sqrt{5}}{9}+\frac{\sqrt{5}}{9}$
$=\frac{3\sqrt{5}+\sqrt{5}}{9}$
$=\frac{4\sqrt{5}}{9}$

41. $\sqrt{\frac{3}{4}}-\sqrt{\frac{3}{64}}=\frac{\sqrt{3}}{\sqrt{4}}-\frac{\sqrt{3}}{\sqrt{64}}$
$=\frac{\sqrt{3}}{2}-\frac{\sqrt{3}}{8}$
$=\frac{4\sqrt{3}}{8}-\frac{\sqrt{3}}{8}$
$=\frac{4\sqrt{3}-\sqrt{3}}{8}$
$=\frac{3\sqrt{3}}{8}$

43. $2\sqrt{45}-2\sqrt{20}=2\sqrt{9\cdot 5}-2\sqrt{4\cdot 5}$
$=2\sqrt{9}\cdot\sqrt{5}-2\sqrt{4}\cdot\sqrt{5}$
$=2(3)\sqrt{5}-2(2)\sqrt{5}$
$=6\sqrt{5}-4\sqrt{5}$
$=2\sqrt{5}$

45. $\sqrt{35}-\sqrt{140}=\sqrt{35}-\sqrt{4\cdot 35}$
$=\sqrt{35}-\sqrt{4}\cdot\sqrt{35}$
$=\sqrt{35}-2\sqrt{35}$
$=-\sqrt{35}$

47. $5\sqrt{2x}+\sqrt{98x}=5\sqrt{2x}+\sqrt{49\cdot 2x}$
$=5\sqrt{2x}+\sqrt{49}\cdot\sqrt{2x}$
$=5\sqrt{2x}+7\sqrt{2x}$
$=12\sqrt{2x}$

49. $5\sqrt{x}+4\sqrt{4x}-13\sqrt{x}=5\sqrt{x}+4\sqrt{4}\cdot\sqrt{x}-13\sqrt{x}$
$=5\sqrt{x}+4(2)\sqrt{x}-13\sqrt{x}$
$=5\sqrt{x}+8\sqrt{x}-13\sqrt{x}$
$=13\sqrt{x}-13\sqrt{x}$
$=0$

51. $\sqrt{3x^3}+3x\sqrt{x}=\sqrt{x^2\cdot 3x}+3x\sqrt{x}$
$=\sqrt{x^2}\cdot\sqrt{3x}+3x\sqrt{x}$
$=x\sqrt{3x}+3x\sqrt{x}$

53. $\sqrt[3]{81}+\sqrt[3]{24}=\sqrt[3]{27\cdot 3}+\sqrt[3]{8\cdot 3}$
$=\sqrt[3]{27}\sqrt[3]{3}+\sqrt[3]{8}\sqrt[3]{3}$
$=3\sqrt[3]{3}+2\sqrt[3]{3}$
$=5\sqrt[3]{3}$

55. $4\sqrt[3]{9}-\sqrt[3]{243}=4\sqrt[3]{9}-\sqrt[3]{27\cdot 9}$
$=4\sqrt[3]{9}-\sqrt[3]{27}\sqrt[3]{9}$
$=4\sqrt[3]{9}-3\sqrt[3]{9}$
$=\sqrt[3]{9}$

57. $2\sqrt[3]{8}+2\sqrt[3]{16}=2(2)+2\sqrt[3]{8\cdot 2}$
$=4+2\cdot\sqrt[3]{8}\cdot\sqrt[3]{2}$
$=4+2\cdot 2\cdot\sqrt[3]{2}$
$=4+4\sqrt[3]{2}$

59. $\sqrt[3]{8}+\sqrt[3]{54}-5=2+\sqrt[3]{27\cdot 2}-5$
$=2+\sqrt[3]{27}\cdot\sqrt[3]{2}-5$
$=-3+3\sqrt[3]{2}$

61. $\sqrt{32x^2}+\sqrt[3]{32}+\sqrt{4x^2}$
$=\sqrt{16x^2\cdot 2}+\sqrt[3]{8\cdot 4}+\sqrt{4x^2}$
$=\sqrt{16x^2}\cdot\sqrt{2}+\sqrt[3]{8}\cdot\sqrt[3]{4}+\sqrt{4x^2}$
$=4x\sqrt{2}+2\sqrt[3]{4}+2x$

63. $\sqrt{40x}+\sqrt[3]{40}-2\sqrt{10x}-\sqrt[3]{5}$
$=\sqrt{4\cdot 10x}+\sqrt[3]{8\cdot 5}-2\sqrt{10x}-\sqrt[3]{5}$
$=\sqrt{4}\cdot\sqrt{10x}+\sqrt[3]{8}\cdot\sqrt[3]{5}-2\sqrt{10x}-\sqrt[3]{5}$
$=2\sqrt{10x}+2\sqrt[3]{5}-2\sqrt{10x}-\sqrt[3]{5}$
$=\sqrt[3]{5}$

65. $(x+6)^2=x^2+2(6)x+6^2=x^2+12x+36$

67. $(2x-1)^2=(2x)^2+2(-1)(2x)+(-1)^2$
$=4x^2-4x+1$

69. $\begin{cases}x=2y\\x+5y=14\end{cases}$

Substitute $2y$ for x in the second equation.

$2y+5y=14$
$7y=14$
$y=2$

Let $y=2$ in the first equation.
$x=2(2)=4$
The solution is (4, 2).

71. Answers may vary

73. Let $l=3\sqrt{5}$ and $w=\sqrt{5}$.
Perimeter $=2l+2w$
$=2(3\sqrt{5})+2(\sqrt{5})$
$=6\sqrt{5}+2\sqrt{5}$
$=8\sqrt{5}$ inches

75. Let $l=8$ and $w=3$.
Area = area of 2 triangles + area of 2 rectangles
$=2\left(\frac{3\sqrt{27}}{4}\right)+2lw$
$=\frac{3\sqrt{9}\cdot\sqrt{3}}{2}+2(8)(3)$
$=\left(\frac{9\sqrt{3}}{2}+48\right)$ square feet

77. $\sqrt{\frac{x^3}{16}}-x\sqrt{\frac{9x}{25}}+\frac{\sqrt{81x^3}}{2}$
$=\frac{\sqrt{x^2\cdot x}}{\sqrt{16}}-x\frac{\sqrt{9x}}{\sqrt{25}}+\frac{\sqrt{81x^2\cdot x}}{2}$
$=\frac{x\sqrt{x}}{4}-x\frac{3\sqrt{x}}{5}+\frac{9x\sqrt{x}}{2}$
$=\frac{5x\sqrt{x}}{4\cdot 5}-4x\frac{3\sqrt{x}}{4\cdot 5}+\frac{10\cdot 9x\sqrt{x}}{2\cdot 10}$
$=\frac{5x\sqrt{x}-12x\sqrt{x}+90x\sqrt{x}}{20}$
$=\frac{83x\sqrt{x}}{20}$

Section 8.4

Practice Exercises

1. a. $\sqrt{11}\cdot\sqrt{7}=\sqrt{11\cdot 7}=\sqrt{77}$

b. $9\sqrt{10}\cdot 8\sqrt{3}=9\cdot 8\sqrt{10\cdot 3}=72\sqrt{30}$

c. $\sqrt{5}\cdot\sqrt{10}=\sqrt{5\cdot 10}$
$=\sqrt{50}$
$=\sqrt{25\cdot 2}$
$=\sqrt{25}\cdot\sqrt{2}$
$=5\sqrt{2}$

d. $\sqrt{17}\cdot\sqrt{17}=\sqrt{17\cdot 17}=\sqrt{289}=17$

e. $\sqrt{15y}\cdot\sqrt{5y^3}=\sqrt{15y\cdot 5y^3}$
$=\sqrt{75y^4}$
$=\sqrt{25y^4\cdot 3}$
$=\sqrt{25y^4}\cdot\sqrt{3}$
$=5y^2\sqrt{3}$

2. $\left(2\sqrt{7}\right)^2=2^2\cdot\left(\sqrt{7}\right)^2=4\cdot 7=28$

3. $\sqrt[3]{10}\cdot\sqrt[3]{50}=\sqrt[3]{10\cdot 50}$
$=\sqrt[3]{500}$
$=\sqrt[3]{125\cdot 4}$
$=\sqrt[3]{125}\cdot\sqrt[3]{4}$
$=5\sqrt[3]{4}$

4. a. $\sqrt{3}\left(\sqrt{3}-\sqrt{5}\right)=\sqrt{3}\cdot\sqrt{3}-\sqrt{3}\cdot\sqrt{5}=3-\sqrt{15}$

b. $\sqrt{2z}\left(\sqrt{z}+7\sqrt{2}\right)=\sqrt{2z}\cdot\sqrt{z}+\sqrt{2z}\cdot 7\sqrt{2}$
$=\sqrt{2z\cdot z}+7\sqrt{2z\cdot 2}$
$=\sqrt{2\cdot z^2}+7\sqrt{4\cdot z}$
$=\sqrt{2}\cdot\sqrt{z^2}+7\sqrt{4}\cdot\sqrt{z}$
$=z\sqrt{2}+7\cdot 2\cdot\sqrt{z}$
$=z\sqrt{2}+14\sqrt{z}$

c. $\left(\sqrt{x}-\sqrt{7}\right)\left(\sqrt{x}+\sqrt{2}\right)$
$=\sqrt{x}\cdot\sqrt{x}+\sqrt{x}\cdot\sqrt{2}-\sqrt{7}\cdot\sqrt{x}-\sqrt{7}\cdot\sqrt{2}$
$=\sqrt{x^2}+\sqrt{2x}-\sqrt{7x}-\sqrt{14}$
$=x+\sqrt{2x}-\sqrt{7x}-\sqrt{14}$

5. a. $\left(\sqrt{7}+4\right)\left(\sqrt{7}-4\right)=\left(\sqrt{7}\right)^2-4^2$
$=7-16$
$=-9$

b. $\left(\sqrt{3x}-5\right)^2=\left(\sqrt{3x}\right)^2-2\left(\sqrt{3x}\right)(5)+(5)^2$
$=3x-10\sqrt{3x}+25$

6. a. $\dfrac{\sqrt{21}}{\sqrt{7}}=\sqrt{\dfrac{21}{7}}=\sqrt{3}$

b. $\dfrac{\sqrt{48}}{\sqrt{6}}=\sqrt{\dfrac{48}{6}}=\sqrt{8}=\sqrt{4\cdot 2}=\sqrt{4}\cdot\sqrt{2}=2\sqrt{2}$

c. $\dfrac{\sqrt{45y^5}}{\sqrt{5y}}=\sqrt{\dfrac{45y^5}{5y}}=\sqrt{9y^4}=3y^2$

7. $\dfrac{\sqrt[3]{625}}{\sqrt[3]{5}}=\sqrt[3]{\dfrac{625}{5}}=\sqrt[3]{125}=5$

8. a. $\dfrac{4}{\sqrt{5}}=\dfrac{4}{\sqrt{5}}\cdot\dfrac{\sqrt{5}}{\sqrt{5}}=\dfrac{4\cdot\sqrt{5}}{\sqrt{5}\cdot\sqrt{5}}=\dfrac{4\sqrt{5}}{5}$

b. $\dfrac{\sqrt{3}}{\sqrt{18}}=\dfrac{\sqrt{3}}{\sqrt{9\cdot 2}}$
$=\dfrac{\sqrt{3}}{3\sqrt{2}}$
$=\dfrac{\sqrt{3}}{3\sqrt{2}}\cdot\dfrac{\sqrt{2}}{\sqrt{2}}$
$=\dfrac{\sqrt{3}\cdot\sqrt{2}}{3\sqrt{2}\cdot\sqrt{2}}$
$=\dfrac{\sqrt{6}}{3\cdot 2}$
$=\dfrac{\sqrt{6}}{6}$

c. $\sqrt{\dfrac{3}{14x}}=\dfrac{\sqrt{3}}{\sqrt{14x}}$
$=\dfrac{\sqrt{3}}{\sqrt{14x}}\cdot\dfrac{\sqrt{14x}}{\sqrt{14x}}$
$=\dfrac{\sqrt{3}\cdot\sqrt{14x}}{\sqrt{14x}\cdot\sqrt{14x}}$
$=\dfrac{\sqrt{42x}}{14x}$

9. a. $\dfrac{3}{\sqrt[3]{25}}=\dfrac{3\cdot\sqrt[3]{5}}{\sqrt[3]{25}\cdot\sqrt[3]{5}}=\dfrac{3\sqrt[3]{5}}{\sqrt[3]{125}}=\dfrac{3\sqrt[3]{5}}{5}$

b. $\dfrac{\sqrt[3]{6}}{\sqrt[3]{5}} = \dfrac{\sqrt[3]{6}\cdot\sqrt[3]{25}}{\sqrt[3]{5}\cdot\sqrt[3]{25}} = \dfrac{\sqrt[3]{150}}{\sqrt[3]{125}} = \dfrac{\sqrt[3]{150}}{5}$

10. a.
$$\begin{aligned}\frac{4}{1+\sqrt{5}} &= \frac{4(1-\sqrt{5})}{(1+\sqrt{5})(1-\sqrt{5})}\\ &= \frac{4(1-\sqrt{5})}{1^2-(\sqrt{5})^2}\\ &= \frac{4(1-\sqrt{5})}{1-5}\\ &= \frac{4(1-\sqrt{5})}{-4}\\ &= -\frac{4(1-\sqrt{5})}{4}\\ &= -1(1-\sqrt{5})\\ &= -1+\sqrt{5}\end{aligned}$$

b.
$$\begin{aligned}\frac{\sqrt{3}+2}{\sqrt{3}-1} &= \frac{(\sqrt{3}+2)(\sqrt{3}+1)}{(\sqrt{3}-1)(\sqrt{3}+1)}\\ &= \frac{3+\sqrt{3}+2\sqrt{3}+2}{3-1}\\ &= \frac{5+3\sqrt{3}}{2}\end{aligned}$$

c. $\dfrac{8}{5-\sqrt{x}} = \dfrac{8(5+\sqrt{x})}{(5-\sqrt{x})(5+\sqrt{x})} = \dfrac{8(5+\sqrt{x})}{25-x}$

11.
$$\begin{aligned}\frac{14-\sqrt{28}}{6} &= \frac{14-\sqrt{4\cdot 7}}{6}\\ &= \frac{14-2\sqrt{7}}{6}\\ &= \frac{2(7-\sqrt{7})}{2\cdot 3}\\ &= \frac{7-\sqrt{7}}{3}\end{aligned}$$

Vocabulary and Readiness Check

1. $\sqrt{7}\cdot\sqrt{3} = \underline{\sqrt{21}}$

2. $\sqrt{10}\cdot\sqrt{10} = \underline{\sqrt{100}\text{ or }10}$

3. $\dfrac{\sqrt{15}}{\sqrt{3}} = \underline{\sqrt{\dfrac{15}{3}}\text{ or }\sqrt{5}}$

4. The process of eliminating the radical in the denominator of a radical expression is called rationalizing the denominator.

5. The conjugate of $2+\sqrt{3}$ is $\underline{2-\sqrt{3}}$.

Exercise Set 8.4

1. $\sqrt{8}\cdot\sqrt{2} = \sqrt{8\cdot 2} = \sqrt{16} = 4$

3.
$$\begin{aligned}\sqrt{10}\cdot\sqrt{5} &= \sqrt{10\cdot 5}\\ &= \sqrt{50}\\ &= \sqrt{25\cdot 2}\\ &= \sqrt{25}\cdot\sqrt{2}\\ &= 5\sqrt{2}\end{aligned}$$

5. $(\sqrt{6})^2 = \sqrt{6}\cdot\sqrt{6} = 6$

7. $\sqrt{2x}\cdot\sqrt{2x} = (\sqrt{2x})^2 = 2x$

9. $(2\sqrt{5})^2 = 2^2(\sqrt{5})^2 = 4(5) = 20$

11. $(6\sqrt{x})^2 = 6^2(\sqrt{x})^2 = 36x$

13.
$$\begin{aligned}\sqrt{3x^5}\cdot\sqrt{6x} &= \sqrt{3x^5\cdot 6x}\\ &= \sqrt{18x^6}\\ &= \sqrt{9x^6\cdot 2}\\ &= \sqrt{9x^6}\cdot\sqrt{2}\\ &= 3x^3\sqrt{2}\end{aligned}$$

15.
$$\begin{aligned}\sqrt{2xy^2}\cdot\sqrt{8xy} &= \sqrt{2xy^2\cdot 8xy}\\ &= \sqrt{16x^2y^3}\\ &= \sqrt{16x^2y^2\cdot y}\\ &= \sqrt{16x^2y^2}\cdot\sqrt{y}\\ &= 4xy\sqrt{y}\end{aligned}$$

17. $\sqrt{6}(\sqrt{5}+\sqrt{7}) = \sqrt{6}\cdot\sqrt{5}+\sqrt{6}\cdot\sqrt{7} = \sqrt{30}+\sqrt{42}$

19. $\sqrt{10}(\sqrt{2}+\sqrt{5}) = \sqrt{10}\cdot\sqrt{2}+\sqrt{10}\cdot\sqrt{5}$
$= \sqrt{20}+\sqrt{50}$
$= \sqrt{4\cdot 5}+\sqrt{25\cdot 2}$
$= \sqrt{4}\cdot\sqrt{5}+\sqrt{25}\cdot\sqrt{2}$
$= 2\sqrt{5}+5\sqrt{2}$

21. $\sqrt{7y}(\sqrt{y}-2\sqrt{7}) = \sqrt{7y}\cdot\sqrt{y}-\sqrt{7y}\cdot 2\sqrt{7}$
$= \sqrt{7y^2}-2\sqrt{49y}$
$= \sqrt{y^2}\cdot\sqrt{7}-2\cdot\sqrt{49}\cdot\sqrt{y}$
$= y\sqrt{7}-2\cdot 7\cdot\sqrt{y}$
$= y\sqrt{7}-14\sqrt{y}$

23. $(\sqrt{3}+6)(\sqrt{3}-6) = (\sqrt{3})^2-6^2 = 3-36 = -33$

25. $(\sqrt{3}+\sqrt{5})(\sqrt{2}-\sqrt{5})$
$= \sqrt{3}\cdot\sqrt{2}-\sqrt{3}\cdot\sqrt{5}+\sqrt{5}\cdot\sqrt{2}-(\sqrt{5})^2$
$= \sqrt{6}-\sqrt{15}+\sqrt{10}-5$

27. $(2\sqrt{11}+1)(\sqrt{11}-6)$
$= 2\sqrt{11}\cdot\sqrt{11}-2\sqrt{11}\cdot 6+1\cdot\sqrt{11}-1\cdot 6$
$= 2\cdot 11-12\sqrt{11}+\sqrt{11}-6$
$= 22-11\sqrt{11}-6$
$= 16-11\sqrt{11}$

29. $(\sqrt{x}+6)(\sqrt{x}-6) = (\sqrt{x})^2-(6)^2 = x-36$

31. $(\sqrt{x}-7)^2 = (\sqrt{x})^2-2\cdot\sqrt{x}\cdot 7+7^2$
$= x-14\sqrt{x}+49$

33. $(\sqrt{6y}+1)^2 = (\sqrt{6y})^2+2\cdot\sqrt{6y}\cdot 1+1^2$
$= 6y+2\sqrt{6y}+1$

35. $\frac{\sqrt{32}}{\sqrt{2}} = \sqrt{\frac{32}{2}} = \sqrt{16} = 4$

37. $\frac{\sqrt{21}}{\sqrt{3}} = \sqrt{\frac{21}{3}} = \sqrt{7}$

39. $\frac{\sqrt{90}}{\sqrt{5}} = \sqrt{\frac{90}{5}} = \sqrt{18} = \sqrt{9\cdot 2} = \sqrt{9}\cdot\sqrt{2} = 3\sqrt{2}$

41. $\frac{\sqrt{75y^5}}{\sqrt{3y}} = \sqrt{\frac{75y^5}{3y}} = \sqrt{25y^4} = 5y^2$

43. $\frac{\sqrt{150}}{\sqrt{2}} = \sqrt{\frac{150}{2}} = \sqrt{75} = \sqrt{25\cdot 3} = \sqrt{25}\cdot\sqrt{3} = 5\sqrt{3}$

45. $\frac{\sqrt{72y^5}}{\sqrt{3y^3}} = \sqrt{\frac{72y^5}{3y^3}}$
$= \sqrt{24y^2}$
$= \sqrt{4y^2\cdot 6}$
$= \sqrt{4y^2}\cdot\sqrt{6}$
$= 2y\sqrt{6}$

47. $\frac{\sqrt{24x^3y^4}}{\sqrt{2xy}} = \sqrt{\frac{24x^3y^4}{2xy}}$
$= \sqrt{12x^2y^3}$
$= \sqrt{4x^2y^2\cdot 3y}$
$= \sqrt{4x^2y^2}\cdot\sqrt{3y}$
$= 2xy\sqrt{3y}$

49. $\frac{\sqrt{3}}{\sqrt{5}} = \frac{\sqrt{3}\cdot\sqrt{5}}{\sqrt{5}\cdot\sqrt{5}} = \frac{\sqrt{15}}{\sqrt{25}} = \frac{\sqrt{15}}{5}$

51. $\frac{7}{\sqrt{2}} = \frac{7\cdot\sqrt{2}}{\sqrt{2}\cdot\sqrt{2}} = \frac{7\sqrt{2}}{2}$

53. $\frac{1}{\sqrt{6y}} = \frac{1\cdot\sqrt{6y}}{\sqrt{6y}\cdot\sqrt{6y}} = \frac{\sqrt{6y}}{6y}$

55. $\sqrt{\frac{3}{x}} = \frac{\sqrt{3}}{\sqrt{x}} = \frac{\sqrt{3}\cdot\sqrt{x}}{\sqrt{x}\cdot\sqrt{x}} = \frac{\sqrt{3x}}{x}$

57. $\sqrt{\frac{1}{8}} = \frac{\sqrt{1}}{\sqrt{8}} = \frac{1}{\sqrt{4}\cdot\sqrt{2}}$
$= \frac{1}{2\sqrt{2}}$
$= \frac{1\cdot\sqrt{2}}{2\cdot\sqrt{2}\cdot\sqrt{2}}$
$= \frac{\sqrt{2}}{2\cdot 2}$
$= \frac{\sqrt{2}}{4}$

59. $\sqrt{\frac{2}{15}} = \frac{\sqrt{2}}{\sqrt{15}} = \frac{\sqrt{2}\cdot\sqrt{15}}{\sqrt{15}\cdot\sqrt{15}} = \frac{\sqrt{30}}{15}$

61. $\frac{8y}{\sqrt{5}} = \frac{8y\cdot\sqrt{5}}{\sqrt{5}\cdot\sqrt{5}} = \frac{8y\sqrt{5}}{5}$

63.
$$\begin{aligned}\sqrt{\frac{y}{12x}} &= \frac{\sqrt{y}}{\sqrt{12x}}\\ &= \frac{\sqrt{y}}{\sqrt{4}\cdot\sqrt{3x}}\\ &= \frac{\sqrt{y}}{2\sqrt{3x}}\\ &= \frac{\sqrt{y}\cdot\sqrt{3x}}{2\cdot\sqrt{3x}\cdot\sqrt{3x}}\\ &= \frac{\sqrt{3xy}}{2\cdot 3x}\\ &= \frac{\sqrt{3xy}}{6x}\end{aligned}$$

65.
$$\begin{aligned}\frac{3}{\sqrt{2}+1} &= \frac{3\cdot\left(\sqrt{2}-1\right)}{\left(\sqrt{2}+1\right)\left(\sqrt{2}-1\right)}\\ &= \frac{3\left(\sqrt{2}-1\right)}{\left(\sqrt{2}\right)^2-1^2}\\ &= \frac{3\left(\sqrt{2}-1\right)}{2-1}\\ &= \frac{3\left(\sqrt{2}-1\right)}{1}\\ &= 3\sqrt{2}-3\end{aligned}$$

67.
$$\begin{aligned}\frac{\sqrt{5}+1}{\sqrt{6}-\sqrt{5}} &= \frac{\left(\sqrt{5}+1\right)\left(\sqrt{6}+\sqrt{5}\right)}{\left(\sqrt{6}-\sqrt{5}\right)\left(\sqrt{6}+\sqrt{5}\right)}\\ &= \frac{\sqrt{30}+5+\sqrt{6}+\sqrt{5}}{\left(\sqrt{6}\right)^2-\left(\sqrt{5}\right)^2}\\ &= \frac{\sqrt{30}+5+\sqrt{6}+\sqrt{5}}{6-5}\\ &= \frac{\sqrt{30}+5+\sqrt{6}+\sqrt{5}}{1}\\ &= \sqrt{30}+5+\sqrt{6}+\sqrt{5}\end{aligned}$$

69.
$$\begin{aligned}\frac{3}{\sqrt{x}-4} &= \frac{3\left(\sqrt{x}+4\right)}{\left(\sqrt{x}-4\right)\left(\sqrt{x}+4\right)}\\ &= \frac{3\cdot\sqrt{x}+3\cdot 4}{\left(\sqrt{x}\right)^2-4^2}\\ &= \frac{3\sqrt{x}-12}{x-16}\end{aligned}$$

71.
$$\begin{aligned}\sqrt{\frac{3}{20}} &= \frac{\sqrt{3}}{\sqrt{20}}\\ &= \frac{\sqrt{3}}{\sqrt{4}\cdot\sqrt{5}}\\ &= \frac{\sqrt{3}}{2\sqrt{5}}\\ &= \frac{\sqrt{3}\cdot\sqrt{5}}{2\sqrt{5}\cdot\sqrt{5}}\\ &= \frac{\sqrt{15}}{2(5)}\\ &= \frac{\sqrt{15}}{10}\end{aligned}$$

73.
$$\begin{aligned}\frac{4}{2-\sqrt{5}} &= \frac{4\cdot\left(2+\sqrt{5}\right)}{\left(2-\sqrt{5}\right)\left(2+\sqrt{5}\right)}\\ &= \frac{4\left(2+\sqrt{5}\right)}{2^2-\left(\sqrt{5}\right)^2}\\ &= \frac{4\left(2+\sqrt{5}\right)}{4-5}\\ &= \frac{4\left(2+\sqrt{5}\right)}{-1}\\ &= -4\left(2+\sqrt{5}\right)\\ &= -8-4\sqrt{5}\end{aligned}$$

75. $\frac{3x}{\sqrt{2x}} = \frac{3x\cdot\sqrt{2x}}{\sqrt{2x}\cdot\sqrt{2x}} = \frac{3x\sqrt{2x}}{2x} = \frac{3\sqrt{2x}}{2}$

77. $\frac{5}{2+\sqrt{x}} = \frac{5(2-\sqrt{x})}{(2+\sqrt{x})(2-\sqrt{x})}$
$= \frac{10-5\sqrt{x}}{2^2-(\sqrt{x})^2}$
$= \frac{10-5\sqrt{x}}{4-x}$

79. $\frac{6+2\sqrt{3}}{2} = \frac{2(3+\sqrt{3})}{2} = 3+\sqrt{3}$

81. $\frac{18-12\sqrt{5}}{6} = \frac{6(3-2\sqrt{5})}{6} = 3-2\sqrt{5}$

83. $\frac{15\sqrt{3}+5}{5} = \frac{5(3\sqrt{3}+1)}{5} = 3\sqrt{3}+1$

85. $\sqrt[3]{12}\cdot\sqrt[3]{4} = \sqrt[3]{12\cdot 4}$
$= \sqrt[3]{48}$
$= \sqrt[3]{8\cdot 6}$
$= \sqrt[3]{8}\cdot\sqrt[3]{6}$
$= 2\sqrt[3]{6}$

87. $2\sqrt[3]{5}\cdot 6\sqrt[3]{2} = 2\cdot 6\cdot\sqrt[3]{5\cdot 2} = 12\sqrt[3]{10}$

89. $\sqrt[3]{15}\cdot\sqrt[3]{25} = \sqrt[3]{375} = \sqrt[3]{125\cdot 3} = \sqrt[3]{125}\cdot\sqrt[3]{3} = 5\sqrt[3]{3}$

91. $\frac{\sqrt[3]{54}}{\sqrt[3]{2}} = \sqrt[3]{\frac{54}{2}} = \sqrt[3]{27} = 3$

93. $\frac{\sqrt[3]{120}}{\sqrt[3]{5}} = \sqrt[3]{\frac{120}{5}} = \sqrt[3]{24} = \sqrt[3]{8\cdot 3} = \sqrt[3]{8}\cdot\sqrt[3]{3} = 2\sqrt[3]{3}$

95. $\sqrt[3]{\frac{5}{4}} = \frac{\sqrt[3]{5}}{\sqrt[3]{4}} = \frac{\sqrt[3]{5}\cdot\sqrt[3]{2}}{\sqrt[3]{4}\cdot\sqrt[3]{2}} = \frac{\sqrt[3]{10}}{\sqrt[3]{8}} = \frac{\sqrt[3]{10}}{2}$

97. $\frac{6}{\sqrt[3]{2}} = \frac{6\cdot\sqrt[3]{4}}{\sqrt[3]{2}\cdot\sqrt[3]{4}} = \frac{6\sqrt[3]{4}}{\sqrt[3]{8}} = \frac{6\sqrt[3]{4}}{2} = 3\sqrt[3]{4}$

99. $\sqrt[3]{\frac{1}{9}} = \frac{\sqrt[3]{1}}{\sqrt[3]{9}} = \frac{1\cdot\sqrt[3]{3}}{\sqrt[3]{9}\cdot\sqrt[3]{3}} = \frac{\sqrt[3]{3}}{\sqrt[3]{27}} = \frac{\sqrt[3]{3}}{3}$

101. $\sqrt[3]{\frac{2}{9}} = \frac{\sqrt[3]{2}}{\sqrt[3]{9}} = \frac{\sqrt[3]{2}\cdot\sqrt[3]{3}}{\sqrt[3]{9}\cdot\sqrt[3]{3}} = \frac{\sqrt[3]{6}}{\sqrt[3]{27}} = \frac{\sqrt[3]{6}}{3}$

103. $x+5 = 7^2$
$x+5 = 49$
$x = 44$

105. $4z^2+6z-12 = (2z)^2$
$4z^2+6z-12 = 4z^2$
$6z-12 = 0$
$6z = 12$
$z = 2$

107. $9x^2+5x+4 = (3x+1)^2$
$9x^2+5x+4 = 9x^2+6x+1$
$5x+4 = 6x+1$
$4 = x+1$
$3 = x$

109. Let $l = 13\sqrt{2}$ and $w = 5\sqrt{6}$.
$A = lw$
$= 13\sqrt{2}\cdot 5\sqrt{6}$
$= 13\cdot 5\cdot\sqrt{2\cdot 6}$
$= 65\sqrt{12}$
$= 65\sqrt{4\cdot 3}$
$= 65\sqrt{4}\cdot\sqrt{3}$
$= 65(2)\sqrt{3}$
$= 130\sqrt{3}$ square meters

111. $\sqrt{\frac{A}{\pi}} = \frac{\sqrt{A}}{\sqrt{\pi}} = \frac{\sqrt{A}\cdot\sqrt{\pi}}{\sqrt{\pi}\cdot\sqrt{\pi}} = \frac{\sqrt{A\pi}}{\pi}$

113. True

115. False; $\sqrt{3x}\cdot\sqrt{3x} = (\sqrt{3x})^2 = 3x$.

117. False; $\sqrt{11}+\sqrt{2}$ cannot be simplified further because the radicands are different.

119. Answers may vary

121. Answers may vary

123. $\frac{\sqrt{3}+1}{\sqrt{2}-1}=\frac{(\sqrt{3}+1)(\sqrt{3}-1)}{(\sqrt{2}-1)(\sqrt{3}-1)}$
$=\frac{(\sqrt{3})^2-1^2}{\sqrt{2}\cdot\sqrt{3}-\sqrt{2}-\sqrt{3}+1^2}$
$=\frac{3-1}{\sqrt{6}-\sqrt{2}-\sqrt{3}+1}$
$=\frac{2}{\sqrt{6}-\sqrt{2}-\sqrt{3}+1}$

Integrated Review

1. $\sqrt{36}=6$, because $6^2=36$ and 6 is positive.
2. $\sqrt{48}=\sqrt{16\cdot 3}=\sqrt{16}\cdot\sqrt{3}=4\sqrt{3}$
3. $\sqrt{x^4}=x^2$, because $(x^2)^2=x^4$.
4. $\sqrt{y^7}=\sqrt{y^6\cdot y}=\sqrt{y^6}\sqrt{y}=y^3\sqrt{y}$
5. $\sqrt{16x^2}=4x$, because $(4x)^2=16x^2$.
6. $\sqrt{18x^{11}}=\sqrt{9x^{10}\cdot 2x}=\sqrt{9x^{10}}\sqrt{2x}=3x^5\sqrt{2x}$
7. $\sqrt[3]{8}=2$, because $(2)^3=8$.
8. $\sqrt[4]{81}=3$, because $(3)^4=81$.
9. $\sqrt[3]{-27}=-3$, because $(-3)^3=-27$.
10. $\sqrt{-4}$ is not a real number.
11. $\sqrt{\frac{11}{9}}=\frac{\sqrt{11}}{\sqrt{9}}=\frac{\sqrt{11}}{3}$
12. $\sqrt[3]{\frac{7}{64}}=\frac{\sqrt[3]{7}}{\sqrt[3]{64}}=\frac{\sqrt[3]{7}}{4}$
13. $-\sqrt{16}=-4$ because $4^2=16$.
14. $-\sqrt{25}=-5$ because $5^2=25$.
15. $\sqrt{\frac{9}{49}}=\frac{\sqrt{9}}{\sqrt{49}}=\frac{3}{7}$
16. $\sqrt{\frac{1}{64}}=\frac{\sqrt{1}}{\sqrt{64}}=\frac{1}{8}$
17. $\sqrt{a^8b^2}=a^4b$
18. $\sqrt{x^{10}y^{20}}=x^5y^{10}$
19. $\sqrt{25m^6}=5m^3$
20. $\sqrt{9n^{16}}=3n^8$
21. $5\sqrt{7}+\sqrt{7}=(5+1)\sqrt{7}=6\sqrt{7}$
22. $\sqrt{50}-\sqrt{8}=\sqrt{25\cdot 2}-\sqrt{4\cdot 2}$
$=\sqrt{25}\cdot\sqrt{2}-\sqrt{4}\cdot\sqrt{2}$
$=5\sqrt{2}-2\sqrt{2}$
$=(5-2)\sqrt{2}$
$=3\sqrt{2}$
23. $5\sqrt{2}-5\sqrt{3}$ cannot be simplified.
24. $2\sqrt{x}+\sqrt{25x}-\sqrt{36x}+3x$
$=2\sqrt{x}+\sqrt{25\cdot x}-\sqrt{36\cdot x}+3x$
$=2\sqrt{x}+\sqrt{25}\cdot\sqrt{x}-\sqrt{36}\cdot\sqrt{x}+3x$
$=2\sqrt{x}+5\sqrt{x}-6\sqrt{x}+3x$
$=(2+5-6)\sqrt{x}+3x$
$=\sqrt{x}+3x$
25. $\sqrt{2}\cdot\sqrt{15}=\sqrt{2\cdot 15}=\sqrt{30}$
26. $\sqrt{3}\cdot\sqrt{3}=\sqrt{3\cdot 3}=\sqrt{9}=3$
27. $(2\sqrt{7})^2=2^2(\sqrt{7})^2=4(7)=28$
28. $(3\sqrt{5})^2=3^2(\sqrt{5})^2=9(5)=45$
29. $\sqrt{3}(\sqrt{11}+1)=\sqrt{3}\cdot\sqrt{11}+\sqrt{3}\cdot 1=\sqrt{33}+\sqrt{3}$
30. $\sqrt{6}(\sqrt{3}-2)=\sqrt{6}\cdot\sqrt{3}-\sqrt{6}\cdot 2$
$=\sqrt{18}-2\sqrt{6}$
$=\sqrt{9\cdot 2}-2\sqrt{6}$
$=\sqrt{9}\cdot\sqrt{2}-2\sqrt{6}$
$=3\sqrt{2}-2\sqrt{6}$

31. $\sqrt{8y}\cdot\sqrt{2y}=\sqrt{8y\cdot 2y}=\sqrt{16y^2}=4y$

32. $$\begin{aligned}\sqrt{15x^2}\cdot\sqrt{3x^2}&=\sqrt{15x^2\cdot 3x^2}\\&=\sqrt{45x^4}\\&=\sqrt{9x^4}\cdot\sqrt{5}\\&=3x^2\sqrt{5}\end{aligned}$$

33. $$\begin{aligned}\left(\sqrt{x}-5\right)\left(\sqrt{x}+2\right)&=\sqrt{x^2}+2\sqrt{x}-5\sqrt{x}-10\\&=x-3\sqrt{x}-10\end{aligned}$$

34. $$\begin{aligned}\left(3+\sqrt{2}\right)^2&=3^2+2(3)\sqrt{2}+\left(\sqrt{2}\right)^2\\&=9+6\sqrt{2}+2\\&=11+6\sqrt{2}\end{aligned}$$

35. $\dfrac{\sqrt{8}}{\sqrt{2}}=\sqrt{\dfrac{8}{2}}=\sqrt{4}=2$

36. $\dfrac{\sqrt{45}}{\sqrt{15}}=\sqrt{\dfrac{45}{15}}=\sqrt{3}$

37. $$\begin{aligned}\frac{\sqrt{24x^5}}{\sqrt{2x}}&=\sqrt{\frac{24x^5}{2x}}\\&=\sqrt{12x^4}\\&=\sqrt{4x^4\cdot 3}\\&=\sqrt{4x^4}\cdot\sqrt{3}\\&=2x^2\sqrt{3}\end{aligned}$$

38. $$\begin{aligned}\frac{\sqrt{75a^4b^5}}{\sqrt{5ab}}&=\sqrt{\frac{75a^4b^5}{5ab}}\\&=\sqrt{15a^3b^4}\\&=\sqrt{a^2b^4\cdot 15a}\\&=\sqrt{a^2b^4}\cdot\sqrt{15a}\\&=ab^2\sqrt{15a}\end{aligned}$$

39. $\sqrt{\dfrac{1}{6}}=\dfrac{\sqrt{1}}{\sqrt{6}}=\dfrac{1}{\sqrt{6}}\cdot\dfrac{\sqrt{6}}{\sqrt{6}}=\dfrac{\sqrt{6}}{6}$

40. $$\begin{aligned}\frac{x}{\sqrt{20}}&=\frac{x}{\sqrt{4}\cdot\sqrt{5}}\\&=\frac{x}{2\sqrt{5}}\\&=\frac{x}{2\sqrt{5}}\cdot\frac{\sqrt{5}}{\sqrt{5}}\\&=\frac{x\sqrt{5}}{2(5)}\\&=\frac{x\sqrt{5}}{10}\end{aligned}$$

41. $$\begin{aligned}\frac{4}{\sqrt{6}+1}&=\frac{4}{\sqrt{6}+1}\cdot\frac{\sqrt{6}-1}{\sqrt{6}-1}\\&=\frac{4\left(\sqrt{6}-1\right)}{6-1}\\&=\frac{4\sqrt{6}-4}{5}\end{aligned}$$

42. $\dfrac{\sqrt{2}+1}{\sqrt{x}-5}=\dfrac{\sqrt{2}+1}{\sqrt{x}-5}\cdot\dfrac{\sqrt{x}+5}{\sqrt{x}+5}=\dfrac{\sqrt{2x}+5\sqrt{2}+\sqrt{x}+5}{x-25}$

Section 8.5

Practice Exercises

1. $$\begin{aligned}\sqrt{x-5}&=2\\\left(\sqrt{x-5}\right)^2&=2^2\\x-5&=4\\x&=9\end{aligned}$$

Check: $$\begin{aligned}\sqrt{x-5}&=2\\\sqrt{9-5}&\stackrel{?}{=}2\\\sqrt{4}&\stackrel{?}{=}2\\2&=2 \quad \text{True}\end{aligned}$$

The solution is 9.

2. $$\begin{aligned}\sqrt{x}+5&=3\\\sqrt{x}&=-2\end{aligned}$$

Since $\sqrt{x}$ is the principal or nonnegative square root of x, $\sqrt{x}$ cannot equal -2. The equation has no solution.

3. $\sqrt{7x-4}=\sqrt{x}$

$$\left(\sqrt{7x-4}\right)=\left(\sqrt{x}\right)^2$$
$$7x-4=x$$
$$-4=-6x$$
$$\frac{-4}{-6}=x$$
$$\frac{2}{3}=x$$

Check: $\sqrt{7x-4}=\sqrt{x}$

$$\sqrt{7\cdot\frac{2}{3}-4}\stackrel{?}{=}\sqrt{\frac{2}{3}}$$
$$\sqrt{\frac{14}{3}-4}\stackrel{?}{=}\sqrt{\frac{2}{3}}$$
$$\sqrt{\frac{14}{3}-\frac{12}{3}}\stackrel{?}{=}\sqrt{\frac{2}{3}}$$
$$\sqrt{\frac{2}{3}}=\sqrt{\frac{2}{3}} \quad \text{True}$$

The solution is $\frac{2}{3}$.

4. $\sqrt{16y^2+4y-28}=4y$

$$\left(\sqrt{16y^2+4y-28}\right)^2=(4y)^2$$
$$16y^2+4y-28=16y^2$$
$$4y-28=0$$
$$4y=28$$
$$y=7$$

Check: $\sqrt{16y^2+4y-28}=4y$

$$\sqrt{16\cdot 7^2+4\cdot 7-28}\stackrel{?}{=}4\cdot 7$$
$$\sqrt{16\cdot 49+28-28}\stackrel{?}{=}28$$
$$\sqrt{784}\stackrel{?}{=}28$$
$$28=28 \quad \text{True}$$

The solution is 7.

5. $\sqrt{x+15}-x=-5$

$$\sqrt{x+15}=x-5$$
$$\left(\sqrt{x+15}\right)^2=(x-5)^2$$
$$x+15=x^2-10x+25$$
$$0=x^2-11x+10$$
$$0=(x-1)(x-10)$$

$0=x-1$ or $0=x-10$

$1=x$ $\qquad$ $10=x$

Check:

Let $x=1$.

$$\sqrt{x+15}-x=-5$$
$$\sqrt{1+15}-1\stackrel{?}{=}-5$$
$$\sqrt{16}-1\stackrel{?}{=}-5$$
$$4-1\stackrel{?}{=}-5$$
$$3\stackrel{?}{=}-5 \quad \text{False}$$

Let $x=10$.

$$\sqrt{x+15}-x=-5$$
$$\sqrt{10+15}-10\stackrel{?}{=}-5$$
$$\sqrt{25}-10\stackrel{?}{=}-5$$
$$5-10\stackrel{?}{=}-5$$
$$-5=-5 \quad \text{True}$$

Since replacing x with 1 resulted in a false statement, 1 is an extraneous solution. The only solution is 10.

6. $\sqrt{x}-4=\sqrt{x-16}$

$$\left(\sqrt{x}-4\right)^2=\left(\sqrt{x-16}\right)^2$$
$$x-8\sqrt{x}+16=x-16$$
$$-8\sqrt{x}=-32$$
$$\sqrt{x}=4$$
$$x=16$$

Check the proposed solution in the original equation. The solution is 16.

Exercise Set 8.5

1. $\sqrt{x}=9$

$$\left(\sqrt{x}\right)^2=9^2$$
$$x=81$$

3. $\sqrt{x+5}=2$

$$\left(\sqrt{x+5}\right)^2=2^2$$
$$x+5=4$$
$$x=-1$$

5. $\sqrt{x}-2=5$

$$\sqrt{x}=7$$
$$\left(\sqrt{x}\right)^2=7^2$$
$$x=49$$

7. $3\sqrt{x}+5=2$

$$3\sqrt{x}=-3$$

The square root cannot be negative, therefore there is no solution.

9. $\sqrt{x} = \sqrt{3x-8}$

$$\left(\sqrt{x}\right)^2 = \left(\sqrt{3x-8}\right)^2$$
$$x = 3x - 8$$
$$-2x = -8$$
$$x = 4$$

11. $\sqrt{4x-3} = \sqrt{x+3}$

$$\left(\sqrt{4x-3}\right)^2 = \left(\sqrt{x+3}\right)^2$$
$$4x - 3 = x + 3$$
$$3x = 6$$
$$x = 2$$

13. $\sqrt{9x^2+2x-4} = 3x$

$$\left(\sqrt{9x^2+2x-4}\right)^2 = (3x)^2$$
$$9x^2 + 2x - 4 = 9x^2$$
$$2x - 4 = 0$$
$$2x = 4$$
$$x = 2$$

15. $\sqrt{x} = x - 6$

$$\left(\sqrt{x}\right)^2 = (x-6)^2$$
$$x = x^2 - 12x + 36$$
$$0 = x^2 - 13x + 36$$
$$0 = (x-9)(x-4)$$
$$x - 9 = 0 \quad \text{or} \quad x - 4 = 0$$
$$x = 9 \qquad x = 4 \text{ (extraneous)}$$

17. $\sqrt{x+7} = x + 5$

$$\left(\sqrt{x+7}\right)^2 = (x+5)^2$$
$$x + 7 = x^2 + 10x + 25$$
$$0 = x^2 + 9x + 18$$
$$0 = (x+3)(x+6)$$
$$x + 3 = 0 \quad \text{or} \quad x + 6 = 0$$
$$x = -3 \qquad x = -6 \text{ (extraneous)}$$

19. $\sqrt{3x+7} - x = 3$

$$\sqrt{3x+7} = x + 3$$
$$\left(\sqrt{3x+7}\right)^2 = (x+3)^2$$
$$3x + 7 = x^2 + 6x + 9$$
$$0 = x^2 + 3x + 2$$
$$0 = (x+1)(x+2)$$
$$0 = x + 1 \quad \text{or} \quad 0 = x + 2$$
$$-1 = x \qquad -2 = x$$

Both solutions check.

21. $\sqrt{16x^2+2x+2} = 4x$

$$\left(\sqrt{16x^2+2x+2}\right)^2 = (4x)^2$$
$$16x^2 + 2x + 2 = 16x^2$$
$$2x + 2 = 0$$
$$2x = -2$$
$$x = -1$$

A check shows that $x = -1$ is an extraneous solution. Therefore, there is no solution.

23. $\sqrt{2x^2+6x+9} = 3$

$$\left(\sqrt{2x^2+6x+9}\right)^2 = (3)^2$$
$$2x^2 + 6x + 9 = 9$$
$$2x^2 + 6x = 0$$
$$2x(x+3) = 0$$
$$2x = 0 \quad \text{or} \quad x + 3 = 0$$
$$x = 0 \qquad x = -3$$

25. $\sqrt{x-7} = \sqrt{x} - 1$

$$\left(\sqrt{x-7}\right)^2 = \left(\sqrt{x}-1\right)^2$$
$$x - 7 = x - 2\sqrt{x} + 1$$
$$2\sqrt{x} = 8$$
$$\sqrt{x} = 4$$
$$\left(\sqrt{x}\right)^2 = (4)^2$$
$$x = 16$$

27. $\sqrt{x} + 2 = \sqrt{x+24}$

$$\left(\sqrt{x}+2\right)^2 = \left(\sqrt{x+24}\right)^2$$
$$x + 4\sqrt{x} + 4 = x + 24$$
$$4\sqrt{x} = 20$$
$$\sqrt{x} = 5$$
$$x = 25$$

29. $\sqrt{x+8} = \sqrt{x}+2$

$$\left(\sqrt{x+8}\right)^2 = \left(\sqrt{x}+2\right)^2$$
$$x+8 = x+4\sqrt{x}+4$$
$$4 = 4\sqrt{x}$$
$$1 = \sqrt{x}$$
$$1^2 = \left(\sqrt{x}\right)^2$$
$$1 = x$$

31. $\sqrt{2x+6} = 4$

$$\left(\sqrt{2x+6}\right)^2 = 4^2$$
$$2x+6 = 16$$
$$2x = 10$$
$$x = 5$$

33. $\sqrt{x+6}+1 = 3$

$$\sqrt{x+6} = 2$$
$$\left(\sqrt{x+6}\right)^2 = 2^2$$
$$x+6 = 4$$
$$x = -2$$

35. $\sqrt{x+6}+5 = 3$

$$\sqrt{x+6} = -2$$

The square root cannot be negative, therefore there is no solution.

37. $\sqrt{16x^2-3x+6} = 4x$

$$\left(\sqrt{16x^2-3x+6}\right)^2 = (4x)^2$$
$$16x^2-3x+6 = 16x^2$$
$$-3x+6 = 0$$
$$-3x = -6$$
$$x = 2$$

39. $-\sqrt{x} = -6$

$$\sqrt{x} = 6$$
$$\left(\sqrt{x}\right)^2 = 6^2$$
$$x = 36$$

41. $\sqrt{x+9} = \sqrt{x}-3$

$$\left(\sqrt{x+9}\right)^2 = \left(\sqrt{x}-3\right)^2$$
$$x+9 = x-6\sqrt{x}+9$$
$$0 = -6\sqrt{x}$$
$$0 = \sqrt{x}$$
$$0 = x$$

$x = 0$ does not check. The equation has no solution.

43. $\sqrt{2x+1}+3 = 5$

$$\sqrt{2x+1} = 2$$
$$\left(\sqrt{2x+1}\right)^2 = 2^2$$
$$2x+1 = 4$$
$$2x = 3$$
$$x = \frac{3}{2}$$

45. $\sqrt{x}+3 = 7$

$$\sqrt{x} = 4$$
$$\left(\sqrt{x}\right)^2 = 4^2$$
$$x = 16$$

47. $\sqrt{4x} = \sqrt{2x+6}$

$$\left(\sqrt{4x}\right)^2 = \left(\sqrt{2x+6}\right)^2$$
$$4x = 2x+6$$
$$2x = 6$$
$$x = 3$$

49. $\sqrt{2x+1} = x-7$

$$\left(\sqrt{2x+1}\right)^2 = (x-7)^2$$
$$2x+1 = x^2-14x+49$$
$$0 = x^2-16x+48$$
$$0 = (x-12)(x-4)$$

$x-12 = 0$ or $x-4 = 0$

$x = 12$ $\quad$ $x = 4$ (extraneous)

51. $x = \sqrt{2x-2}+1$

$x-1 = \sqrt{2x-2}$

$(x-1)^2 = \left(\sqrt{2x-2}\right)^2$

$x^2-2x+1 = 2x-2$

$x^2-4x+3 = 0$

$(x-1)(x-3) = 0$

$x-1=0$ or $x-3=0$

$x=1$ $\quad x=3$

53. $\sqrt{1-8x}-x=4$

$\sqrt{1-8x} = x+4$

$\left(\sqrt{1-8x}\right)^2 = (x+4)^2$

$1-8x = x^2+8x+16$

$0 = x^2+16x+15$

$0 = (x+1)(x+15)$

$x+1=0$ or $x+15=0$

$x=-1$ $\quad x=-15$ (extraneous)

55. $3x-8=19$

$3x=27$

$x=9$

57. Let x = width and $2x$ = length.

$2(2x+x)=24$

$2(3x)=24$

$6x=24$

$x=4$

$2x=2(4)=8$

The length is 8 inches.

59. $\sqrt{x-3}+3 = \sqrt{3x+4}$

$\left(\sqrt{x-3}+3\right)^2 = \left(\sqrt{3x+4}\right)^2$

$(x-3)+6\sqrt{x-3}+9 = 3x+4$

$x+6\sqrt{x-3}+6 = 3x+4$

$6\sqrt{x-3} = 2x-2$

$3\sqrt{x-3} = x-1$

$\left(3\sqrt{x-3}\right)^2 = (x-1)^2$

$9(x-3) = x^2-2x+1$

$9x-27 = x^2-2x+1$

$0 = x^2-11x+28$

$0 = (x-4)(x-7)$

$0=x-4$ or $0=x-7$

$4=x$ $\quad 7=x$

Both solutions check.

61. Answers may vary

63. $b = \sqrt{\frac{V}{2}}$

a. $b = \sqrt{\frac{20}{2}} \approx 3.2$

$b = \sqrt{\frac{200}{2}} = 10$

$b = \sqrt{\frac{2000}{2}} \approx 31.6$

V	20	200	2000
b	3.2	10	31.6

b. No; it increases by a factor of $\sqrt{10}$.

65. $y_1 = \sqrt{x-2}$, $y_2 = x-5$

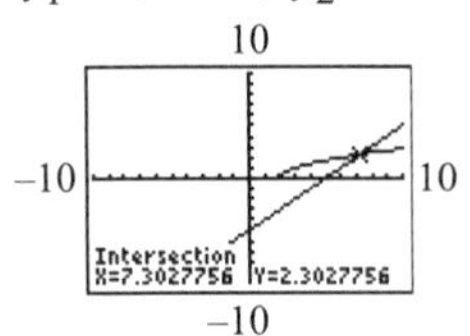

The solution is 7.30.

67. $y_1 = -\sqrt{x+4}$, $y_2 = 5x-6$

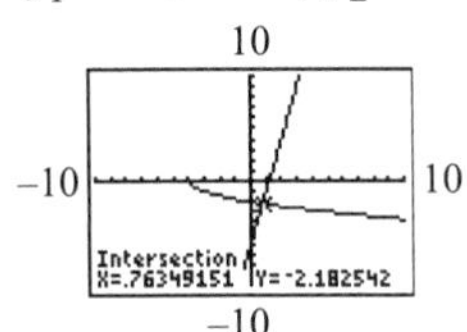

The solution is 0.76.

The Bigger Picture

1. $\sqrt{56} = \sqrt{4\cdot 14} = \sqrt{4}\cdot\sqrt{14} = 2\sqrt{14}$

2. $\sqrt{\frac{20x^5}{49}} = \frac{\sqrt{20x^5}}{\sqrt{49}} = \frac{\sqrt{4x^4\cdot 5x}}{7} = \frac{\sqrt{4x^4}\cdot\sqrt{5x}}{7} = \frac{2x^2\sqrt{5x}}{7}$

3. $(-5x^{12}y^{-3})(3x^{-7}y^{14}) = -5 \cdot 3x^{12-7}y^{-3+14}$
$= -15x^5y^{11}$

4. $\sqrt{\frac{10}{11}} = \frac{\sqrt{10}}{\sqrt{11}} = \frac{\sqrt{10} \cdot \sqrt{11}}{\sqrt{11} \cdot \sqrt{11}} = \frac{\sqrt{110}}{11}$

5. $\frac{8}{\sqrt{5}-1} = \frac{8(\sqrt{5}+1)}{(\sqrt{5}-1)(\sqrt{5}+1)}$
$= \frac{8(\sqrt{5}+1)}{5-1}$
$= \frac{8(\sqrt{5}+1)}{4}$
$= 2(\sqrt{5}+1)$ or $2\sqrt{5}+2$

6. $\frac{1}{2}(6x^2-4)+\frac{1}{3}(6x^2-9)-14$
$= 3x^2-2+2x^2-3-14$
$= 5x^2-19$

7. $9x-7 = 7x-9$
$2x-7 = -9$
$2x = -2$
$x = -1$

8. $\frac{x}{5} = \frac{x-3}{11}$
$55\left(\frac{x}{5}\right) = 55\left(\frac{x-3}{11}\right)$
$11x = 5(x-3)$
$11x = 5x-15$
$6x = -15$
$x = \frac{-15}{6}$
$x = -\frac{5}{2}$

9. $-5(2y+1) \le 3y-2-2y+1$
$-10y-5 \le y-1$
$-5 \le 11y-1$
$-4 \le 11y$
$-\frac{4}{11} \le y$
$\left[-\frac{4}{11}, \infty\right)$

10. $x(x+1) = 42$
$x^2+x = 42$
$x^2+x-42 = 0$
$(x+7)(x-6) = 0$
$x+7=0$ or $x-6=0$
$x=-7$ $\quad x=6$

11. $-\frac{6}{x-7}+\frac{8}{x} = \frac{-4}{x-7}$
$x(x-7)\left(-\frac{6}{x-7}+\frac{8}{x}\right) = x(x-7)\left(\frac{-4}{x-7}\right)$
$-6x+8(x-7) = -4x$
$-6x+8x-56 = -4x$
$2x-56 = -4x$
$-56 = -6x$
$\frac{-56}{-6} = x$
$\frac{28}{3} = x$

12. $1+4(x-2) = x(x-6)-x^2+13$
$1+4x-8 = x^2-6x-x^2+13$
$4x-7 = -6x+13$
$10x-7 = 13$
$10x = 20$
$x = 2$

Section 8.6

Practice Exercises

1. Use the Pythagorean theorem.
Let $a = 5$ and $b = 12$.
$a^2+b^2 = c^2$
$5^2+12^2 = c^2$
$25+144 = c^2$
$169 = c^2$
Since c is a length, it is the principal square root of 169.
$\sqrt{169} = c$
$13 = c$
The hypotenuse has a length of 13 inches.

2. Let $a = 3$ and $c = 7$ with b the unknown leg.

$$a^2 + b^2 = c^2$$
$$3^2 + b^2 = 7^2$$
$$9 + b^2 = 49$$
$$b^2 = 40$$
$$b = \sqrt{40} = 2\sqrt{10} \approx 6.32$$

The exact length of the leg is $2\sqrt{10}$ meters which is approximately 6.32 meters.

3. The points A, B, and C form a right triangle. The hypotenuse, $\overline{BC}$, has a length of 95 feet and the leg $\overline{AC}$ has a length of 60 feet. In the Pythagorean theorem, let $a = 60$ and $c = 95$ with b the unknown length.

$$a^2 + b^2 = c^2$$
$$60^2 + b^2 = 95^2$$
$$3600 + b^2 = 9025$$
$$b^2 = 5425$$
$$b = \sqrt{5425} \approx 74$$

The length of the bridge is exactly $\sqrt{5425}$ feet which is approximately 74 feet.

4. Use the distance formula with $(x_1, y_1) = (-2, 5)$ and $(x_2, y_2) = (-4, -7)$.

$$d = \sqrt{(x_2 - x_1)^2 + (y_2 - y_1)^2}$$
$$= \sqrt{[-4-(-2)]^2 + (-7-5)^2}$$
$$= \sqrt{(-2)^2 + (-12)^2}$$
$$= \sqrt{4+144}$$
$$= \sqrt{148}$$
$$= 2\sqrt{37}$$

The distance is $2\sqrt{37}$ units.

5. Use the formula $v = \sqrt{2gh}$ with $g = 32$ and $h = 12$.

$$v = \sqrt{2gh} = \sqrt{2 \cdot 32 \cdot 12} = \sqrt{768} = 16\sqrt{3}$$

The velocity of the object is exactly $16\sqrt{3}$ feet per second, or approximately 27.71 feet per second.

Exercise Set 8.6

1.
$$a^2 + b^2 = c^2$$
$$2^2 + 3^2 = c^2$$
$$4 + 9 = c^2$$
$$13 = c^2$$
$$\sqrt{13} = c$$

The length is $\sqrt{13} \approx 3.61$.

3.
$$a^2 + b^2 = c^2$$
$$3^2 + b^2 = 6^2$$
$$9 + b^2 = 36$$
$$b^2 = 27$$
$$b = \sqrt{27}$$
$$b = 3\sqrt{3}$$

The length is $3\sqrt{3} \approx 5.20$.

5.
$$a^2 + b^2 = c^2$$
$$7^2 + 24^2 = c^2$$
$$49 + 576 = c^2$$
$$625 = c^2$$
$$\sqrt{625} = c$$
$$25 = c$$

The length is 25.

7.
$$a^2 + b^2 = c^2$$
$$a^2 + \left(\sqrt{3}\right)^2 = 5^2$$
$$a^2 + 3 = 25$$
$$a^2 = 22$$
$$a = \sqrt{22}$$

The length is $\sqrt{22} \approx 4.69$.

9.
$$a^2 + b^2 = c^2$$
$$4^2 + b^2 = 13^2$$
$$16 + b^2 = 169$$
$$b^2 = 153$$
$$b = \sqrt{153}$$
$$b = 3\sqrt{17}$$

The length is $3\sqrt{17} \approx 12.37$.

11.
$$\begin{aligned} a^2+b^2 &= c^2 \\ 4^2+5^2 &= c^2 \\ 16+25 &= c^2 \\ 41 &= c^2 \\ \sqrt{41} &= c \end{aligned}$$
The length is $\sqrt{41} \approx 6.40$.

13.
$$\begin{aligned} a^2+b^2 &= c^2 \\ a^2+2^2 &= 6^2 \\ a^2+4 &= 36 \\ a^2 &= 32 \\ a &= \sqrt{32} \\ a &= 4\sqrt{2} \end{aligned}$$
The length is $4\sqrt{2} \approx 5.66$.

15.
$$\begin{aligned} a^2+b^2 &= c^2 \\ \left(\sqrt{10}\right)^2+b^2 &= 10^2 \\ 10+b^2 &= 100 \\ b^2 &= 90 \\ b &= \sqrt{90} \\ b &= 3\sqrt{10} \end{aligned}$$
The length is $3\sqrt{10} \approx 9.49$.

17.
$$\begin{aligned} a^2+b^2 &= c^2 \\ 40^2+b^2 &= 65^2 \\ 1600+b^2 &= 4225 \\ b^2 &= 4225-1600 \\ b^2 &= 2625 \\ \sqrt{b^2} &= \sqrt{2625} \\ b &\approx 51.2 \text{ ft} \end{aligned}$$

19.
$$\begin{aligned} a^2+b^2 &= c^2 \\ 5^2+20^2 &= c^2 \\ 25+400 &= c^2 \\ 425 &= c^2 \\ \sqrt{425} &= c \end{aligned}$$
The length is $\sqrt{425} \approx 20.6$ feet.

21.
$$\begin{aligned} a^2+b^2 &= c^2 \\ 6^2+10^2 &= c^2 \\ 36+100 &= c^2 \\ 136 &= c^2 \\ \sqrt{136} &= c \end{aligned}$$
The length is $\sqrt{136} \approx 11.7$ feet.

23. (3, 6) and (5, 11)
$$\begin{aligned} d &= \sqrt{(x_2-x_1)^2+(y_2-y_1)^2} \\ &= \sqrt{(5-3)^2+(11-6)^2} \\ &= \sqrt{2^2+5^2} \\ &= \sqrt{4+25} \\ &= \sqrt{29} \end{aligned}$$

25. (−3, 1) and (5, −2)
$$\begin{aligned} d &= \sqrt{(x_2-x_1)^2+(y_2-y_1)^2} \\ &= \sqrt{[5-(-3)]^2+(-2-1)^2} \\ &= \sqrt{(8)^2+(-3)^2} \\ &= \sqrt{64+9} \\ &= \sqrt{73} \end{aligned}$$

27. (3, −2) and (1, −8)
$$\begin{aligned} d &= \sqrt{(x_2-x_1)^2+(y_2-y_1)^2} \\ &= \sqrt{(1-3)^2+[-8-(-2)]^2} \\ &= \sqrt{(-2)^2+(-6)^2} \\ &= \sqrt{4+36} \\ &= \sqrt{40} \\ &= \sqrt{4\cdot 10} \\ &= 2\sqrt{10} \end{aligned}$$

29. $\left(\frac{1}{2}, 2\right)$ and $(2, -1)$

$$\begin{aligned} d &= \sqrt{(x_2 - x_1)^2 + (y_2 - y_1)^2} \\ &= \sqrt{\left(2 - \frac{1}{2}\right)^2 + (-1-2)^2} \\ &= \sqrt{\left(\frac{3}{2}\right)^2 + (-3)^2} \\ &= \sqrt{\frac{9}{4} + 9} \\ &= \sqrt{\frac{45}{4}} \\ &= \frac{\sqrt{45}}{\sqrt{4}} \\ &= \frac{3\sqrt{5}}{2} \end{aligned}$$

31. $(3, -2)$ and $(5, 7)$

$$\begin{aligned} d &= \sqrt{(x_2 - x_1)^2 + (y_2 - y_1)^2} \\ &= \sqrt{(5-3)^2 + [7-(-2)]^2} \\ &= \sqrt{2^2 + 9^2} \\ &= \sqrt{4 + 81} \\ &= \sqrt{85} \end{aligned}$$

33.

$$\begin{aligned} b &= \sqrt{\frac{3V}{h}} \\ 6 &= \sqrt{\frac{3V}{2}} \\ 6^2 &= \left(\sqrt{\frac{3V}{2}}\right)^2 \\ 36 &= \frac{3V}{2} \\ 24 &= V \end{aligned}$$

The volume is 24 cubic feet.

35.

$$\begin{aligned} s &= \sqrt{30fd} \\ s &= \sqrt{30(0.35)(280)} \\ &= \sqrt{2940} \\ &\approx 54 \end{aligned}$$

It was moving at 54 mph.

37.

$$\begin{aligned} v &= \sqrt{2.5r} \\ v &= \sqrt{2.5(300)} \\ &= \sqrt{750} \\ &\approx 27 \end{aligned}$$

It can travel at 27 mph.

39.

$$\begin{aligned} d &= 3.5\sqrt{h} \\ d &= 3.5\sqrt{305.4} \approx 61.2 \end{aligned}$$

You can see 61.2 km.

41. $2^5 = 2 \cdot 2 \cdot 2 \cdot 2 \cdot 2 = 32$

43. $\left(-\frac{1}{5}\right)^2 = \left(-\frac{1}{5}\right)\left(-\frac{1}{5}\right) = \frac{1}{25}$

45. $x^2 \cdot x^3 = x^{2+3} = x^5$

47. $y^3 \cdot y = y^{3+1} = y^4$

49. Let y = length of whole base and z = length of unlabeled section of base.

Find y:

$$\begin{aligned} y^2 + 3^2 &= 7^2 \\ y^2 + 9 &= 49 \\ y^2 &= 40 \\ y &= \sqrt{40} = 2\sqrt{10} \end{aligned}$$

Find z.

$$\begin{aligned} z^2 + 3^2 &= 5^2 \\ z^2 + 9 &= 25 \\ z^2 &= 16 \\ z &= \sqrt{16} = 4 \end{aligned}$$

Find x.

$$x = y - z = 2\sqrt{10} - 4$$

51.

$$\begin{aligned} a^2 + b^2 &= c^2 \\ [60(3)]^2 + [30(3)]^2 &= c^2 \\ 180^2 + 90^2 &= c^2 \\ 32,400 + 8100 &= c^2 \\ 40,500 &= c^2 \\ \sqrt{40,500} &= c \\ 201 &\approx c \end{aligned}$$

They are about 201 miles apart.

53. Answers may vary

Section 8.7

Practice Exercises

1. a. $36^{1/2} = \sqrt{36} = 6$

b. $125^{1/3} = \sqrt[3]{125} = 5$

c. $-\left(\frac{1}{81}\right)^{1/4} = -\sqrt[4]{\frac{1}{81}} = -\frac{1}{3}$

d. $(-1000)^{1/3} = \sqrt[3]{-1000} = -10$

e. $32^{1/5} = \sqrt[5]{32} = 2$

2. a. $9^{3/2} = (9^{1/2})^3 = \left(\sqrt{9}\right)^3 = 3^3 = 27$

b. $8^{5/3} = (8^{1/3})^5 = \left(\sqrt[3]{8}\right)^5 = 2^5 = 32$

c. $-625^{1/4} = -\sqrt[4]{625} = -5$

3. a. $25^{-1/2} = \frac{1}{25^{1/2}} = \frac{1}{\sqrt{25}} = \frac{1}{5}$

b.
$$1000^{-2/3} = \frac{1}{1000^{2/3}} = \frac{1}{\left(\sqrt[3]{1000}\right)^2} = \frac{1}{10^2} = \frac{1}{100}$$

c. $-49^{1/2} = -\sqrt{49} = -7$

d.
$$1024^{-2/5} = \frac{1}{1024^{2/5}} = \frac{1}{\left(\sqrt[5]{1024}\right)^2} = \frac{1}{4^2} = \frac{1}{16}$$

4. a. $6^{3/5} \cdot 6^{7/5} = 6^{(3/5)+(7/5)} = 6^{10/5} = 6^2 = 36$

b. $\frac{7^{1/6}}{7^{3/6}} = 7^{(1/6)-(3/6)} = 7^{-2/6} = 7^{-1/3} = \frac{1}{7^{1/3}}$

c. $(z^{3/8})^{16} = z^{(3/8)16} = z^6$

d. $\frac{a^{3/7}}{a^{-4/7}} = a^{(3/7)-(-4/7)} = a^{7/7} = a^1$ or a

e. $\left(\frac{x^{5/8}}{y^{2/3}}\right)^{12} = \frac{x^{(5/8)12}}{y^{(2/3)12}} = \frac{x^{15/2}}{y^8}$

Exercise Set 8.7

1. $8^{1/3} = \sqrt[3]{8} = 2$

3. $9^{1/2} = \sqrt{9} = 3$

5. $16^{3/4} = \left(\sqrt[4]{16}\right)^3 = 2^3 = 8$

7. $32^{2/5} = \left(\sqrt[5]{32}\right)^2 = 2^2 = 4$

9. $-16^{-1/4} = -\frac{1}{16^{1/4}} = -\frac{1}{\sqrt[4]{16}} = -\frac{1}{2}$

11. $16^{-3/2} = \frac{1}{16^{3/2}} = \frac{1}{\left(\sqrt{16}\right)^3} = \frac{1}{4^3} = \frac{1}{64}$

13. $81^{-3/2} = \frac{1}{81^{3/2}} = \frac{1}{\left(\sqrt{81}\right)^3} = \frac{1}{9^3} = \frac{1}{729}$

15. $\left(\frac{4}{25}\right)^{-1/2} = \frac{1}{\left(\frac{4}{25}\right)^{1/2}} = \frac{1}{\sqrt{\frac{4}{25}}} = \frac{1}{\frac{2}{5}} = \frac{5}{2}$

17. Answers may vary

19. $2^{1/3} \cdot 2^{2/3} = 2^{3/3} = 2^1 = 2$

21. $\frac{4^{3/4}}{4^{1/4}} = 4^{\frac{3}{4}-\frac{1}{4}} = 4^{2/4} = 4^{1/2} = \sqrt{4} = 2$

23. $\frac{x^{1/6}}{x^{5/6}} = x^{\frac{1}{6}-\frac{5}{6}} = x^{-4/6} = x^{-2/3} = \frac{1}{x^{2/3}}$

25. $(x^{1/2})^6 = x^{6/2} = x^3$

27. Answers may vary

29. $81^{1/2} = \sqrt{81} = 9$

31. $(-8)^{1/3} = \sqrt[3]{-8} = -2$

33. $-81^{1/4} = -\left(\sqrt[4]{81}\right) = -(3) = -3$

35. $\left(\frac{1}{81}\right)^{1/2} = \sqrt{\frac{1}{81}} = \frac{1}{9}$

37. $\left(\frac{27}{64}\right)^{1/3} = \frac{27^{1/3}}{64^{1/3}} = \frac{\sqrt[3]{27}}{\sqrt[3]{64}} = \frac{3}{4}$

39. $9^{3/2} = \left(\sqrt{9}\right)^3 = (3)^3 = 27$

41. $64^{3/2} = \left(\sqrt{64}\right)^3 = (8)^3 = 512$

43. $-8^{2/3} = -(8^{2/3}) = -\left(\sqrt[3]{8}\right)^2 = -(2)^2 = -(4) = -4$

45. $4^{5/2} = \left(\sqrt{4}\right)^5 = (2)^5 = 32$

47. $\left(\frac{4}{9}\right)^{3/2} = \frac{4^{3/2}}{9^{3/2}} = \frac{\left(\sqrt{4}\right)^3}{\left(\sqrt{9}\right)^3} = \frac{2^3}{3^3} = \frac{8}{27}$

49. $\left(\frac{1}{81}\right)^{3/4} = \frac{1^{3/4}}{81^{3/4}} = \frac{\left(\sqrt[4]{1}\right)^3}{\left(\sqrt[4]{81}\right)^3} = \frac{1^3}{3^3} = \frac{1}{27}$

51. $4^{-1/2} = \frac{1}{4^{1/2}} = \frac{1}{\sqrt{4}} = \frac{1}{2}$

53. $215^{-1/3} = \frac{1}{215^{1/3}} = \frac{1}{\sqrt[3]{215}} = \frac{1}{5}$

55. $625^{-3/4} = \frac{1}{625^{3/4}} = \frac{1}{\left(\sqrt[4]{625}\right)^3} = \frac{1}{5^3} = \frac{1}{125}$

57. $3^{4/3} \cdot 3^{2/3} = 3^{\frac{4}{3}+\frac{2}{3}} = 3^{6/3} = 3^2 = 9$

59. $\frac{6^{2/3}}{6^{1/3}} = 6^{\frac{2}{3}-\frac{1}{3}} = 6^{1/3}$

61. $(x^{2/3})^9 = x^{\frac{2}{3}\cdot 9} = x^6$

63. $\frac{6^{1/3}}{6^{-5/3}} = 6^{\frac{1}{3}-\left(-\frac{5}{3}\right)} = 6^{6/3} = 6^2 = 36$

65. $\frac{3^{-3/5}}{3^{2/5}} = 3^{-\frac{3}{5}-\frac{2}{5}} = 3^{-5/5} = 3^{-1} = \frac{1}{3}$

67. $\left(\frac{x^{1/3}}{y^{3/4}}\right)^2 = \frac{(x^{1/3})^2}{(y^{3/4})^2} = \frac{x^{2/3}}{y^{3/2}}$

69. $\left(\frac{x^{2/5}}{y^{3/4}}\right)^8 = \frac{(x^{2/5})^8}{(y^{3/4})^8} = \frac{x^{16/5}}{y^6}$

71. $\begin{cases} x + y < 6 \\ y \ge 2x \end{cases}$

$x + y < 6$
Test (0, 0)
$0 + 0 \overset{?}{<} 6$
$0 < 6$ True
Shade below.

$y \ge 2x$
Test (0, 1)
$1 \overset{?}{\ge} 2(0)$
$1 \ge 0$ True
Shade above.

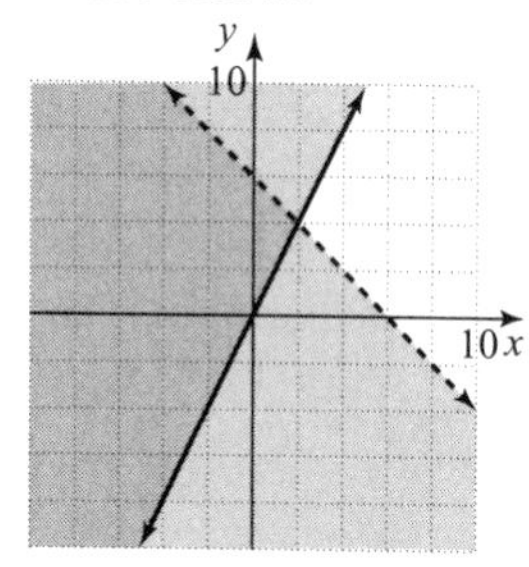

73.
$x^2 - 4 = 3x$
$x^2 - 3x - 4 = 0$
$(x-4)(x+1) = 0$
$x - 4 = 0$ or $x + 1 = 0$
$x = 4$ $\quad x = -1$
The solutions are −1 and 4.

75. $$\begin{aligned} 2x^2-5x-3&=0\\ 2x^2-6x+x-3&=0\\ 2x(x-3)+1(x-3)&=0\\ (2x+1)(x-3)&=0 \end{aligned}$$
$2x+1=0$ or $x-3=0$
$x=-\frac{1}{2}$ $\quad x=3$

The solutions are $-\frac{1}{2}$ and 3.

77. Let $N=1.5$ and $P_O=10{,}000$.
$P=P_O(1.08)^N=10{,}000(1.08)^{1.5}=11{,}224$
The population will be 11,224 people.

79. $5^{3/4}\approx 3.344$

81. $18^{3/5}\approx 5.665$

Chapter 8 Vocabulary Check

1. The expressions $5\sqrt{x}$ and $7\sqrt{x}$ are examples of like radicals.
2. In the expression $\sqrt[3]{45}$ the number 3 is the index, the number 45 is the radicand, and $\sqrt{}$ is called the radical sign.
3. The conjugate of $(a+b)$ is $(a-b)$.
4. The principal square root of 25 is 5.
5. The process of eliminating the radical in the denominator of a radical expression is called rationalizing the denominator.

Chapter 8 Review

1. $\sqrt{81}=9$, because $9^2=81$ and 9 is positive.
2. $-\sqrt{49}=-7$, because $\sqrt{49}=7$.
3. $\sqrt[3]{27}=3$, because $(3)^3=27$.
4. $\sqrt[4]{16}=2$, because $2^4=16$.
5. $-\sqrt{\frac{9}{64}}=-\frac{3}{8}$, because $\sqrt{\frac{9}{64}}=\frac{3}{8}$.
6. $\sqrt{\frac{36}{81}}=\frac{6}{9}=\frac{2}{3}$, because $\left(\frac{6}{9}\right)^2=\frac{36}{81}$.
7. $\sqrt[4]{\frac{16}{81}}=\frac{2}{3}$ because $\left(\frac{2}{3}\right)^4=\frac{16}{81}$.
8. $\sqrt[3]{-\frac{27}{64}}=-\frac{3}{4}$ because $\left(-\frac{3}{4}\right)^3=-\frac{27}{64}$.
9. **c** is not a real number since the index is even and the radicand is negative.
10. **a** and **c** are not real numbers since the indices are even and the radicands are odd.
11. $\sqrt{x^{12}}=x^6$, because $(x^6)^2=x^{12}$.
12. $\sqrt{x^8}=x^4$, because $(x^4)^2=x^8$.
13. $\sqrt{9x^6}=3x^3$, because $(3x^3)^2=9x^6$.
14. $\sqrt{25x^4}=5x^2$, because $(5x^2)^2=25x^4$.
15. $\sqrt{\frac{16}{y^{10}}}=\frac{4}{y^5}$ because $\left(\frac{4}{y^5}\right)^2=\frac{16}{y^{10}}$.
16. $\sqrt{\frac{y^{12}}{49}}=\frac{y^6}{7}$ because $\left(\frac{y^6}{7}\right)^2=\frac{y^{12}}{49}$.
17. $\sqrt{54}=\sqrt{9\cdot 6}=\sqrt{9}\cdot\sqrt{6}=3\sqrt{6}$
18. $\sqrt{88}=\sqrt{4\cdot 22}=\sqrt{4}\cdot\sqrt{22}=2\sqrt{22}$
19. $\sqrt{150x^3}=\sqrt{25x^2\cdot 6x}=\sqrt{25x^2}\sqrt{6x}=5x\sqrt{6x}$
20. $$\begin{aligned}\sqrt{92y^5}&=\sqrt{4\cdot 23\cdot y^4\cdot y}\\&=\sqrt{4y^4}\cdot\sqrt{23y}\\&=2y^2\sqrt{23y}\end{aligned}$$
21. $\sqrt[3]{54}=\sqrt[3]{27\cdot 2}=\sqrt[3]{27}\sqrt[3]{2}=3\sqrt[3]{2}$
22. $\sqrt[3]{88}=\sqrt[3]{8\cdot 11}=\sqrt[3]{8}\cdot\sqrt[3]{11}=2\sqrt[3]{11}$
23. $\sqrt[4]{48}=\sqrt[4]{16\cdot 3}=\sqrt[4]{16}\sqrt[4]{3}=2\sqrt[4]{3}$

24. $\sqrt[4]{162} = \sqrt[4]{81 \cdot 2} = \sqrt[4]{81} \cdot \sqrt[4]{2} = 3\sqrt[4]{2}$

25. $\sqrt{\frac{18}{25}} = \frac{\sqrt{18}}{\sqrt{25}} = \frac{\sqrt{9 \cdot 2}}{5} = \frac{\sqrt{9} \cdot \sqrt{2}}{5} = \frac{3\sqrt{2}}{5}$

26. $\sqrt{\frac{75}{64}} = \frac{\sqrt{75}}{\sqrt{64}} = \frac{\sqrt{25 \cdot 3}}{8} = \frac{\sqrt{25} \cdot \sqrt{3}}{8} = \frac{5\sqrt{3}}{8}$

27. $\sqrt{\frac{45y^2}{4x^4}} = \frac{\sqrt{45y^2}}{\sqrt{4x^4}}$
$= \frac{\sqrt{9y^2 \cdot 5}}{2x^2}$
$= \frac{\sqrt{9y^2} \cdot \sqrt{5}}{2x^2}$
$= \frac{3y\sqrt{5}}{2x^2}$

28. $\sqrt{\frac{20x^5}{9x^2}} = \sqrt{\frac{20x^3}{9}}$
$= \frac{\sqrt{4 \cdot 5 \cdot x^2 \cdot x}}{\sqrt{9}}$
$= \frac{\sqrt{4x^2} \cdot \sqrt{5x}}{3}$
$= \frac{2x\sqrt{5x}}{3}$

29. $\sqrt[4]{\frac{9}{16}} = \frac{\sqrt[4]{9}}{\sqrt[4]{16}} = \frac{\sqrt[4]{9}}{2}$

30. $\sqrt[3]{\frac{40}{27}} = \frac{\sqrt[3]{8 \cdot 5}}{\sqrt[3]{27}} = \frac{\sqrt[3]{8} \cdot \sqrt[3]{5}}{3} = \frac{2\sqrt[3]{5}}{3}$

31. $\sqrt[3]{\frac{3}{8}} = \frac{\sqrt[3]{3}}{\sqrt[3]{8}} = \frac{\sqrt[3]{3}}{2}$

32. $\sqrt[4]{\frac{5}{81}} = \frac{\sqrt[4]{5}}{\sqrt[4]{81}} = \frac{\sqrt[4]{5}}{3}$

33. $3\sqrt[3]{2} + 2\sqrt[3]{3} - 4\sqrt[3]{2} = (3-4)\sqrt[3]{2} + 2\sqrt[3]{3}$
$= -\sqrt[3]{2} + 2\sqrt[3]{2}$

34. $5\sqrt{2} + 2\sqrt[3]{2} - 8\sqrt{2} = (5-8)\sqrt{2} + 2\sqrt[3]{2}$
$= -3\sqrt{2} + 2\sqrt[3]{2}$

35. $\sqrt{6} + 2\sqrt[3]{6} - 4\sqrt[3]{6} + 5\sqrt{6} = (1+5)\sqrt{6} + (2-4)\sqrt[3]{6}$
$= 6\sqrt{6} - 2\sqrt[3]{6}$

36. $3\sqrt{5} - \sqrt[3]{5} - 2\sqrt{5} + 3\sqrt[3]{5} = (3-2)\sqrt{5} + (-1+3)\sqrt[3]{5}$
$= \sqrt{5} + 2\sqrt[3]{5}$

37. $\sqrt{28x} + \sqrt{63x} + \sqrt[3]{56}$
$= \sqrt{4 \cdot 7x} + \sqrt{9 \cdot 7x} + \sqrt[3]{8 \cdot 7}$
$= \sqrt{4} \cdot \sqrt{7x} + \sqrt{9} \cdot \sqrt{7x} + \sqrt[3]{8} \cdot \sqrt[3]{7}$
$= 2\sqrt{7x} + 3\sqrt{7x} + 2\sqrt[3]{7}$
$= 5\sqrt{7x} + 2\sqrt[3]{7}$

38. $\sqrt{75y} + \sqrt{48y} - \sqrt[4]{16} = \sqrt{25 \cdot 3y} + \sqrt{16 \cdot 3y} - 2$
$= \sqrt{25} \cdot \sqrt{3y} + \sqrt{16} \cdot \sqrt{3y} - 2$
$= 5\sqrt{3y} + 4\sqrt{3y} - 2$
$= 9\sqrt{3y} - 2$

39. $\sqrt{\frac{5}{9}} - \sqrt{\frac{5}{36}} = \frac{\sqrt{5}}{\sqrt{9}} - \frac{\sqrt{5}}{\sqrt{36}}$
$= \frac{\sqrt{5}}{3} - \frac{\sqrt{5}}{6}$
$= \frac{2\sqrt{5}}{6} - \frac{\sqrt{5}}{6}$
$= \frac{2\sqrt{5} - \sqrt{5}}{6}$
$= \frac{\sqrt{5}}{6}$

40. $\sqrt{\frac{11}{25}} + \sqrt{\frac{11}{16}} = \frac{\sqrt{11}}{\sqrt{25}} + \frac{\sqrt{11}}{\sqrt{16}}$
$= \frac{\sqrt{11}}{5} + \frac{\sqrt{11}}{4}$
$= \frac{4\sqrt{11}}{20} + \frac{5\sqrt{11}}{20}$
$= \frac{4\sqrt{11} + 5\sqrt{11}}{20}$
$= \frac{9\sqrt{11}}{20}$

41. $2\sqrt[3]{125} - 5\sqrt[3]{8} = 2(5) - 5(2) = 10 - 10 = 0$

42. $3\sqrt[3]{16}-2\sqrt[3]{2}=3\sqrt[3]{8\cdot 2}-2\sqrt[3]{2}$
$=3\sqrt[3]{8}\cdot\sqrt[3]{2}-2\sqrt[3]{2}$
$=3\cdot 2\sqrt[3]{2}-2\sqrt[3]{2}$
$=6\sqrt[3]{2}-2\sqrt[3]{2}$
$=4\sqrt[3]{2}$

43. $3\sqrt{10}\cdot 2\sqrt{5}=3\cdot 2\sqrt{10\cdot 5}$
$=6\sqrt{50}$
$=6\sqrt{25\cdot 2}$
$=6\sqrt{25}\sqrt{2}$
$=6(5)\sqrt{2}$
$=30\sqrt{2}$

44. $2\sqrt[3]{4}\cdot 5\sqrt[3]{6}=2\cdot 5\sqrt[3]{4\cdot 6}$
$=10\sqrt[3]{24}$
$=10\sqrt[3]{8\cdot 3}$
$=10\cdot 2\sqrt[3]{3}$
$=20\sqrt[3]{3}$

45. $\sqrt{3}\left(2\sqrt{6}-3\sqrt{12}\right)=2\sqrt{18}-3\sqrt{36}$
$=2\sqrt{9\cdot 2}-3\sqrt{36}$
$=2\sqrt{9}\sqrt{2}-3(6)$
$=2(3)\sqrt{2}-18$
$=6\sqrt{2}-18$

46. $4\sqrt{5}\left(2\sqrt{10}-5\sqrt{5}\right)=8\sqrt{50}-20\sqrt{25}$
$=8\sqrt{25\cdot 2}-20(5)$
$=8\cdot\sqrt{25}\cdot\sqrt{2}-100$
$=8\cdot 5\cdot\sqrt{2}-100$
$=40\sqrt{2}-100$

47. $\left(\sqrt{3}+2\right)\left(\sqrt{6}-5\right)=\sqrt{18}-5\sqrt{3}+2\sqrt{6}-10$
$=\sqrt{9\cdot 2}-5\sqrt{3}+2\sqrt{6}-10$
$=\sqrt{9}\cdot\sqrt{2}-5\sqrt{3}+2\sqrt{6}-10$
$=3\sqrt{2}-5\sqrt{3}+2\sqrt{6}-10$

48. $\left(2\sqrt{5}+1\right)\left(4\sqrt{5}-3\right)=8\sqrt{25}-6\sqrt{5}+4\sqrt{5}-3$
$=8\cdot 5-2\sqrt{5}-3$
$=40-3-2\sqrt{5}$
$=37-2\sqrt{5}$

49. $\left(\sqrt{x}-2\right)^2=\left(\sqrt{x}\right)^2-2\cdot\sqrt{x}\cdot 2+2^2$
$=x-4\sqrt{x}+4$

50. $\left(\sqrt{y}+4\right)^2=\left(\sqrt{y}\right)^2+2\cdot\sqrt{y}\cdot 4+4^2$
$=y+8\sqrt{y}+16$

51. $\frac{\sqrt{27}}{\sqrt{3}}=\sqrt{\frac{27}{3}}=\sqrt{9}=3$

52. $\frac{\sqrt{20}}{\sqrt{5}}=\sqrt{\frac{20}{5}}=\sqrt{4}=2$

53. $\frac{\sqrt{160}}{\sqrt{8}}=\sqrt{\frac{160}{8}}=\sqrt{20}=\sqrt{4\cdot 5}=\sqrt{4}\cdot\sqrt{5}=2\sqrt{5}$

54. $\frac{\sqrt{96}}{\sqrt{3}}=\sqrt{\frac{96}{3}}=\sqrt{32}=\sqrt{16\cdot 2}=\sqrt{16}\cdot\sqrt{2}=4\sqrt{2}$

55. $\frac{\sqrt{30x^6}}{\sqrt{2x^3}}=\sqrt{\frac{30x^6}{2x^3}}=\sqrt{15x^3}=\sqrt{x^2\cdot 15x}=x\sqrt{15x}$

56. $\frac{\sqrt{54x^5y^2}}{\sqrt{3xy^2}}=\sqrt{\frac{54x^5y^2}{3xy^2}}$
$=\sqrt{18x^4}$
$=\sqrt{9x^4\cdot 2}$
$=\sqrt{9x^4}\cdot\sqrt{2}$
$=3x^2\sqrt{2}$

57. $\frac{\sqrt{2}}{\sqrt{11}}=\frac{\sqrt{2}\cdot\sqrt{11}}{\sqrt{11}\cdot\sqrt{11}}=\frac{\sqrt{22}}{11}$

58. $\frac{\sqrt{3}}{\sqrt{13}}=\frac{\sqrt{3}\cdot\sqrt{13}}{\sqrt{13}\cdot\sqrt{13}}=\frac{\sqrt{39}}{13}$

59. $\sqrt{\frac{5}{6}}=\frac{\sqrt{5}}{\sqrt{6}}=\frac{\sqrt{5}\cdot\sqrt{6}}{\sqrt{6}\cdot\sqrt{6}}=\frac{\sqrt{30}}{\sqrt{36}}=\frac{\sqrt{30}}{6}$

60. $\sqrt{\frac{7}{10}}=\frac{\sqrt{7}}{\sqrt{10}}=\frac{\sqrt{7}\cdot\sqrt{10}}{\sqrt{10}\cdot\sqrt{10}}=\frac{\sqrt{70}}{\sqrt{100}}=\frac{\sqrt{70}}{10}$

61. $\frac{1}{\sqrt{5x}}=\frac{1\cdot\sqrt{5x}}{\sqrt{5x}\cdot\sqrt{5x}}=\frac{\sqrt{5x}}{5x}$

62. $\frac{5}{\sqrt{3y}}=\frac{5\cdot\sqrt{3y}}{\sqrt{3y}\cdot\sqrt{3y}}=\frac{5\sqrt{3y}}{3y}$

63. $\sqrt{\frac{3}{x}} = \frac{\sqrt{3}}{\sqrt{x}} = \frac{\sqrt{3}\cdot\sqrt{x}}{\sqrt{x}\cdot\sqrt{x}} = \frac{\sqrt{3x}}{x}$

64. $\sqrt{\frac{6}{y}} = \frac{\sqrt{6}}{\sqrt{y}} = \frac{\sqrt{6}\cdot\sqrt{y}}{\sqrt{y}\cdot\sqrt{y}} = \frac{\sqrt{6y}}{y}$

65.
$$\frac{3}{\sqrt{5}-2} = \frac{3\cdot(\sqrt{5}+2)}{(\sqrt{5}-2)(\sqrt{5}+2)}$$
$$= \frac{3(\sqrt{5}+2)}{(\sqrt{5})^2-2^2}$$
$$= \frac{3(\sqrt{5}+2)}{5-4}$$
$$= \frac{3(\sqrt{5}+2)}{1}$$
$$= 3\sqrt{5}+6$$

66.
$$\frac{8}{\sqrt{10}-3} = \frac{8\cdot(\sqrt{10}+3)}{(\sqrt{10}-3)(\sqrt{10}+3)}$$
$$= \frac{8(\sqrt{10}+3)}{(\sqrt{10})^2-3^2}$$
$$= \frac{8(\sqrt{10}+3)}{10-9}$$
$$= \frac{8(\sqrt{10}+3)}{1}$$
$$= 8\sqrt{10}+24$$

67.
$$\frac{\sqrt{2}+1}{\sqrt{3}-1} = \frac{(\sqrt{2}+1)(\sqrt{3}+1)}{(\sqrt{3}-1)(\sqrt{3}+1)}$$
$$= \frac{\sqrt{2}\cdot\sqrt{3}+\sqrt{2}+\sqrt{3}+1^2}{3-1}$$
$$= \frac{\sqrt{6}+\sqrt{2}+\sqrt{3}+1}{2}$$

68.
$$\frac{\sqrt{3}-2}{\sqrt{5}+2} = \frac{(\sqrt{3}-2)(\sqrt{5}-2)}{(\sqrt{5}+2)(\sqrt{5}-2)}$$
$$= \frac{\sqrt{3}\cdot\sqrt{5}-2\sqrt{3}-2\sqrt{5}+2^2}{5-4}$$
$$= \sqrt{15}-2\sqrt{3}-2\sqrt{5}+4$$

69. $\frac{10}{\sqrt{x}+5} = \frac{10(\sqrt{x}-5)}{(\sqrt{x}+5)(\sqrt{x}-5)} = \frac{10\sqrt{x}-50}{x-25}$

70. $\frac{8}{\sqrt{x}-1} = \frac{8(\sqrt{x}+1)}{(\sqrt{x}-1)(\sqrt{x}+1)} = \frac{8\sqrt{x}+8}{x-1}$

71. $\sqrt[3]{\frac{7}{9}} = \frac{\sqrt[3]{7}}{\sqrt[3]{9}} = \frac{\sqrt[3]{7}\cdot\sqrt[3]{3}}{\sqrt[3]{9}\cdot\sqrt[3]{3}} = \frac{\sqrt[3]{21}}{\sqrt[3]{27}} = \frac{\sqrt[3]{21}}{3}$

72. $\sqrt[3]{\frac{3}{4}} = \frac{\sqrt[3]{3}}{\sqrt[3]{4}} = \frac{\sqrt[3]{3}\cdot\sqrt[3]{2}}{\sqrt[3]{4}\cdot\sqrt[3]{2}} = \frac{\sqrt[3]{6}}{\sqrt[3]{8}} = \frac{\sqrt[3]{6}}{2}$

73. $\sqrt[3]{\frac{3}{2}} = \frac{\sqrt[3]{3}}{\sqrt[3]{2}} = \frac{\sqrt[3]{3}\cdot\sqrt[3]{4}}{\sqrt[3]{2}\cdot\sqrt[3]{4}} = \frac{\sqrt[3]{12}}{\sqrt[3]{8}} = \frac{\sqrt[3]{12}}{2}$

74. $\sqrt[3]{\frac{5}{4}} = \frac{\sqrt[3]{5}}{\sqrt[3]{4}} = \frac{\sqrt[3]{5}\cdot\sqrt[3]{2}}{\sqrt[3]{4}\cdot\sqrt[3]{2}} = \frac{\sqrt[3]{10}}{\sqrt[3]{8}} = \frac{\sqrt[3]{10}}{2}$

75.
$$\sqrt{2x} = 6$$
$$(\sqrt{2x})^2 = 6^2$$
$$2x = 36$$
$$x = 18$$

76.
$$\sqrt{x+3} = 4$$
$$(\sqrt{x+3})^2 = 4^2$$
$$x+3 = 16$$
$$x = 13$$

77.
$$\sqrt{x}+3 = 8$$
$$\sqrt{x} = 5$$
$$(\sqrt{x})^2 = 5^2$$
$$x = 25$$

78.
$$\sqrt{x}+8 = 3$$
$$\sqrt{x} = -5$$
The square root cannot be negative, therefore there is no solution.

79. $\sqrt{2x+1} = x-7$

$\left(\sqrt{2x+1}\right)^2 = (x-7)^2$

$2x+1 = x^2 - 14x + 49$

$0 = x^2 - 16x + 48$

$0 = (x-12)(x-4)$

$x-12 = 0$ or $x-4 = 0$

$x = 12$ $x = 4$ (extraneous)

80. $\sqrt{3x+1} = x-1$

$\left(\sqrt{3x+1}\right)^2 = (x-1)^2$

$3x+1 = x^2 - 2x + 1$

$0 = x^2 - 5x$

$0 = x(x-5)$

$x = 0$ or $x-5 = 0$

$x = 0$ (extraneous) $x = 5$

81. $\sqrt{x}+3 = \sqrt{x+15}$

$\left(\sqrt{x}+3\right)^2 = \left(\sqrt{x+15}\right)^2$

$x + 6\sqrt{x} + 9 = x + 15$

$6\sqrt{x} = 6$

$\sqrt{x} = 1$

$x = 1$

82. $\sqrt{x-5} = \sqrt{x}-1$

$\left(\sqrt{x-5}\right)^2 = \left(\sqrt{x}-1\right)^2$

$x-5 = x - 2\sqrt{x} + 1$

$-6 = -2\sqrt{x}$

$3 = \sqrt{x}$

$9 = x$

83. $a^2 + b^2 = c^2$

$6^2 + 9^2 = c^2$

$36 + 81 = c^2$

$117 = c^2$

$\sqrt{117} = c$

$\sqrt{9 \cdot 13} = c$

$3\sqrt{13} = c$

The length is $3\sqrt{13} \approx 10.82$.

84. $a^2 + b^2 = c^2$

$5^2 + b^2 = 9^2$

$25 + b^2 = 81$

$b^2 = 56$

$b = \sqrt{56}$

$b = \sqrt{4 \cdot 14}$

$b = 2\sqrt{14}$

The length is $2\sqrt{14} \approx 7.48$.

85. $a^2 + b^2 = c^2$

$20^2 + 12^2 = c^2$

$400 + 144 = c^2$

$544 = c^2$

$\sqrt{544} = c$

$\sqrt{16 \cdot 34} = c$

$4\sqrt{34} = c$

They are $4\sqrt{34}$ feet apart, approximately 23.32 feet.

86. $a^2 + b^2 = c^2$

$a^2 + 5^2 = 10^2$

$a^2 + 25 = 100$

$a^2 = 75$

$a = \sqrt{75}$

$a = \sqrt{25 \cdot 3}$

$a = 5\sqrt{3}$

The length is $5\sqrt{3}$ inches, approximately 8.66 inches.

87. (6, −2) and (−3, 5)

$d = \sqrt{(x_2 - x_1)^2 + (y_2 - y_1)^2}$

$d = \sqrt{(-3-6)^2 + [5-(-2)]^2}$

$d = \sqrt{(-9)^2 + (5+2)^2}$

$d = \sqrt{81 + 7^2}$

$d = \sqrt{81 + 49}$

$d = \sqrt{130}$

88. (2, 8) and (−6, 10)

$$d = \sqrt{(x_2 - x_1)^2 + (y_2 - y_1)^2}$$

$$\begin{aligned}\sqrt{(-6-2)^2 + (10-8)^2} &= \sqrt{(-8)^2 + 2^2}\\ &= \sqrt{64+4}\\ &= \sqrt{68}\\ &= \sqrt{4 \cdot 17}\\ &= 2\sqrt{17}\end{aligned}$$

89. $r = \sqrt{\dfrac{S}{4\pi}}$

$r = \sqrt{\dfrac{72}{4\pi}} \approx 2.4$

The radius is about 2.4 inches.

90.

$$\begin{aligned} r &= \sqrt{\frac{S}{4\pi}}\\ 6 &= \sqrt{\frac{S}{4\pi}}\\ 6^2 &= \left(\sqrt{\frac{S}{4\pi}}\right)^2\\ 36 &= \frac{S}{4\pi}\\ 144\pi &= S\end{aligned}$$

The surface area is 144π square inches.

91. $\sqrt{a^5} = a^{5/2}$

92. $\sqrt[5]{a^3} = a^{3/5}$

93. $\sqrt[6]{x^{15}} = x^{15/6} = x^{5/2}$

94. $\sqrt[4]{x^{12}} = x^{12/4} = x^3$

95. $16^{1/2} = \sqrt{16} = 4$

96. $36^{1/2} = \sqrt{36} = 6$

97. $(-8)^{1/3} = \sqrt[3]{-8} = -2$

98. $(-32)^{1/5} = \sqrt[5]{-32} = -2$

99. $-64^{3/2} = -(64^{3/2}) = -\left(\sqrt{64}\right)^3 = -(8)^3 = -512$

100. $-8^{2/3} = -\sqrt[3]{8^2} = -\sqrt[3]{64} = -4$

101. $\left(\dfrac{16}{81}\right)^{3/4} = \dfrac{16^{3/4}}{81^{3/4}} = \dfrac{\left(\sqrt[4]{16}\right)^3}{\left(\sqrt[4]{81}\right)^3} = \dfrac{2^3}{3^3} = \dfrac{8}{27}$

102. $\left(\dfrac{9}{25}\right)^{3/2} = \left(\sqrt{\dfrac{9}{25}}\right)^3 = \left(\dfrac{3}{5}\right)^3 = \dfrac{27}{125}$

103. $8^{1/3} \cdot 8^{4/3} = 8^{5/3} = \left(\sqrt[3]{8}\right)^5 = 2^5 = 32$

104. $4^{3/2} \cdot 4^{1/2} = 4^{\frac{3}{2}+\frac{1}{2}} = 4^{4/2} = 4^2 = 16$

105. $\dfrac{3^{1/6}}{3^{5/6}} = 3^{\frac{1}{6}-\frac{5}{6}} = 3^{-4/6} = 3^{-2/3} = \dfrac{1}{3^{2/3}}$

106. $\dfrac{2^{1/4}}{2^{-3/5}} = 2^{\frac{1}{4}-\left(-\frac{3}{5}\right)} = 2^{\frac{1}{4}+\frac{3}{5}} = 2^{\frac{5+12}{20}} = 2^{17/20}$

107. $(x^{-1/3})^6 = x^{-\frac{1}{3}\cdot 6} = x^{-2} = \dfrac{1}{x^2}$

108. $\left(\dfrac{x^{1/2}}{y^{1/3}}\right)^2 = \dfrac{x^{2/2}}{y^{2/3}} = \dfrac{x}{y^{2/3}}$

109. $\sqrt{144} = 12$ because $12^2 = 144$ and 12 is positive.

110. $-\sqrt[3]{64} = -\left(\sqrt[3]{64}\right) = -4$ because $4^3 = 64$.

111. $\sqrt{16x^{16}} = 4x^8$ because $(4x^8)^2 = 16x^{16}$ and $4x^8$ is positive.

112. $\sqrt{4x^{24}} = 2x^{12}$ because $(2x^{12})^2 = 4x^{24}$ and $2x^{12}$ is positive.

113. $\sqrt{18x^7} = \sqrt{9x^6 \cdot 2x} = \sqrt{9x^6} \cdot \sqrt{2x} = 3x^3\sqrt{2x}$

114. $\sqrt{48y^6} = \sqrt{16y^6 \cdot 3} = \sqrt{16y^6} \cdot \sqrt{3} = 4y^3\sqrt{3}$

115. $25^{-1/2} = \dfrac{1}{25^{1/2}} = \dfrac{1}{\sqrt{25}} = \dfrac{1}{5}$

116. $64^{-2/3} = \frac{1}{64^{2/3}} = \frac{1}{\left(\sqrt[3]{64}\right)^2} = \frac{1}{4^2} = \frac{1}{16}$

117. $\sqrt{\frac{y^4}{81}} = \frac{\sqrt{y^4}}{\sqrt{81}} = \frac{y^2}{9}$

118. $\sqrt{\frac{x^9}{9}} = \frac{\sqrt{x^9}}{\sqrt{9}} = \frac{\sqrt{x^8 \cdot x}}{3} = \frac{x^4\sqrt{x}}{3}$

119. $\sqrt{12} + \sqrt{75} = \sqrt{4 \cdot 3} + \sqrt{25 \cdot 3} = 2\sqrt{3} + 5\sqrt{3} = 7\sqrt{3}$

120. $$\begin{aligned}\sqrt{63} + \sqrt{28} - \sqrt[3]{27} &= \sqrt{9 \cdot 7} + \sqrt{4 \cdot 7} - 3 \\ &= 3\sqrt{7} + 2\sqrt{7} - 3 \\ &= 5\sqrt{7} - 3\end{aligned}$$

121. $$\begin{aligned}\sqrt{\frac{3}{16}} - \sqrt{\frac{3}{4}} &= \frac{\sqrt{3}}{\sqrt{16}} - \frac{\sqrt{3}}{\sqrt{4}} \\ &= \frac{\sqrt{3}}{4} - \frac{\sqrt{3}}{2} \\ &= \frac{\sqrt{3}}{4} - \frac{2\sqrt{3}}{4} \\ &= \frac{\sqrt{3} - 2\sqrt{3}}{4} \\ &= -\frac{\sqrt{3}}{4}\end{aligned}$$

122. $$\begin{aligned}&\sqrt{45x^3} + x\sqrt{20x} - \sqrt{5x^3} \\ &= \sqrt{9x^2 \cdot 5x} + x\sqrt{4 \cdot 5x} - \sqrt{x^2 \cdot 5x} \\ &= 3x\sqrt{5x} + x \cdot 2\sqrt{5x} - x\sqrt{5x} \\ &= 3x\sqrt{5x} + 2x\sqrt{5x} - x\sqrt{5x} \\ &= 4x\sqrt{5x}\end{aligned}$$

123. $$\begin{aligned}\sqrt{7} \cdot \sqrt{14} &= \sqrt{7 \cdot 14} \\ &= \sqrt{98} \\ &= \sqrt{49 \cdot 2} \\ &= \sqrt{49} \cdot \sqrt{2} \\ &= 7\sqrt{2}\end{aligned}$$

124. $$\begin{aligned}\sqrt{3}\left(\sqrt{9} - \sqrt{2}\right) &= \sqrt{3}\left(3 - \sqrt{2}\right) \\ &= \sqrt{3} \cdot 3 - \sqrt{3} \cdot \sqrt{2} \\ &= 3\sqrt{3} - \sqrt{6}\end{aligned}$$

125. $$\begin{aligned}\left(\sqrt{2} + 4\right)\left(\sqrt{5} - 1\right) &= \sqrt{2} \cdot \sqrt{5} - \sqrt{2} + 4\sqrt{5} - 4 \\ &= \sqrt{10} - \sqrt{2} + 4\sqrt{5} - 4\end{aligned}$$

126. $\left(\sqrt{x} + 3\right)^2 = \left(\sqrt{x}\right)^2 + 2 \cdot \sqrt{x} \cdot 3 + 3^2 = x + 6\sqrt{x} + 9$

127. $\frac{\sqrt{120}}{\sqrt{5}} = \sqrt{\frac{120}{5}} = \sqrt{24} = \sqrt{4 \cdot 6} = 2\sqrt{6}$

128. $\frac{\sqrt{60x^9}}{\sqrt{15x^7}} = \sqrt{\frac{60x^9}{15x^7}} = \sqrt{4x^2} = 2x$

129. $\sqrt{\frac{2}{7}} = \frac{\sqrt{2}}{\sqrt{7}} = \frac{\sqrt{2} \cdot \sqrt{7}}{\sqrt{7} \cdot \sqrt{7}} = \frac{\sqrt{14}}{7}$

130. $\frac{3}{\sqrt{2x}} = \frac{3 \cdot \sqrt{2x}}{\sqrt{2x} \cdot \sqrt{2x}} = \frac{3\sqrt{2x}}{2x}$

131. $\frac{3}{\sqrt{x} - 6} = \frac{3\left(\sqrt{x} + 6\right)}{\left(\sqrt{x} - 6\right)\left(\sqrt{x} + 6\right)} = \frac{3\sqrt{x} + 18}{x - 36}$

132. $$\begin{aligned}\frac{\sqrt{7} - 5}{\sqrt{5} + 3} &= \frac{\left(\sqrt{7} - 5\right)\left(\sqrt{5} - 3\right)}{\left(\sqrt{5} + 3\right)\left(\sqrt{5} - 3\right)} \\ &= \frac{\sqrt{7} \cdot \sqrt{5} - 3\sqrt{7} - 5\sqrt{5} + 5 \cdot 3}{5 - 9} \\ &= \frac{\sqrt{35} - 3\sqrt{7} - 5\sqrt{5} + 15}{-4}\end{aligned}$$

133. $$\begin{aligned}\sqrt{4x} &= 2 \\ \left(\sqrt{4x}\right)^2 &= 2^2 \\ 4x &= 4 \\ x &= 1\end{aligned}$$

134. $$\begin{aligned}\sqrt{x - 4} &= 3 \\ \left(\sqrt{x - 4}\right)^2 &= 3^2 \\ x - 4 &= 9 \\ x &= 13\end{aligned}$$

135. $\sqrt{4x+8}+6=x$
$\sqrt{4x+8}=x-6$
$\left(\sqrt{4x+8}\right)^2=(x-6)^2$
$4x+8=x^2-12x+36$
$0=x^2-16x+28$
$0=(x-2)(x-14)$
$0=x-2$ or $0=x-14$
$2=x$ (extraneous) $14=x$

136. $\sqrt{x-8}=\sqrt{x}-2$
$\left(\sqrt{x-8}\right)^2=\left(\sqrt{x}-2\right)^2$
$x-8=x-4\sqrt{x}+4$
$-12=-4\sqrt{x}$
$3=\sqrt{x}$
$9=x$

137. The unknown side is the hypotenuse. Let $a = 3$ and $b = 7$.
$a^2+b^2=c^2$
$3^2+7^2=c^2$
$9+49=c^2$
$58=c^2$
$\sqrt{58}=c$
The unknown length is $\sqrt{58}$ units or approximately 7.62 units.

138. Use the Pythagorean theorem with $a = 2$ and $c = 6$.
$a^2+b^2=c^2$
$2^2+b^2=6^2$
$4+b^2=36$
$b^2=32$
$b=\sqrt{32}=4\sqrt{2}$
The length of the rectangle is $4\sqrt{2}$ inches or approximately 5.66 inches.

139. Let x be the length of a side of the cube. Since the cube includes only the four walls and roof, there are 5 sides, each of which has area x^2 square feet, for a total surface area of $5x^2$ square feet.
$5x^2=5120$
$x^2=1024$
$x=\sqrt{1024}$
$x=32$
The length of the sides of the cube is 32 feet.

Chapter 8 Test

1. $\sqrt{16}=4$, because $4^2=16$ and 4 is positive.

2. $\sqrt[3]{-125}=-5$, because $(-5)^3=-125$.

3. $16^{3/4}=\left(\sqrt[4]{16}\right)^3=2^3=8$

4. $\left(\frac{9}{16}\right)^{1/2}=\frac{9^{1/2}}{16^{1/2}}=\frac{\sqrt{9}}{\sqrt{16}}=\frac{3}{4}$

5. $\sqrt[4]{-81}$ is not a real number.

6. $27^{-2/3}=\frac{1}{27^{2/3}}=\frac{1}{\left(\sqrt[3]{27}\right)^2}=\frac{1}{3^2}=\frac{1}{9}$

7. $\sqrt{54}=\sqrt{9\cdot 6}=\sqrt{9}\cdot\sqrt{6}=3\sqrt{6}$

8. $\sqrt{92}=\sqrt{4\cdot 23}=\sqrt{4}\cdot\sqrt{23}=2\sqrt{23}$

9. $\sqrt{3x^6}=\sqrt{x^6}\cdot\sqrt{3}=x^3\sqrt{3}$

10. $\sqrt{8x^4y^7}=\sqrt{4x^4y^6\cdot 2y}$
$=\sqrt{4x^4y^6}\sqrt{2y}$
$=2x^2y^3\sqrt{2y}$

11. $\sqrt{9x^9}=\sqrt{9x^8}\sqrt{x}=3x^4\sqrt{x}$

12. $\sqrt[3]{8}=2$ because $2^3=8$.

13. $\sqrt[3]{40}=\sqrt[3]{8\cdot 5}=\sqrt[3]{8}\sqrt[3]{5}=2\sqrt[3]{5}$

14. $\sqrt{x^{10}}=x^5$ because $(x^5)^2=10$.

15. $\sqrt{y^7}=\sqrt{y^6\cdot y}=\sqrt{y^6}\cdot\sqrt{y}=y^3\sqrt{y}$

16. $\sqrt{\frac{5}{16}}=\frac{\sqrt{5}}{\sqrt{16}}=\frac{\sqrt{5}}{4}$

17. $\sqrt{\frac{y^3}{25}}=\frac{\sqrt{y^3}}{\sqrt{25}}=\frac{\sqrt{y^2\cdot y}}{5}=\frac{y\sqrt{y}}{5}$

18. $\sqrt[3]{\frac{2}{27}}=\frac{\sqrt[3]{2}}{\sqrt[3]{27}}=\frac{\sqrt[3]{2}}{3}$

19. $3\sqrt{8x}=3\sqrt{4\cdot 2x}=3\sqrt{4}\sqrt{2x}=3(2)\sqrt{2x}=6\sqrt{2x}$

20. $\sqrt{13}+\sqrt{13}-4\sqrt{13}=(1+1-4)\sqrt{13}=-2\sqrt{13}$

21.
$$\begin{aligned}\sqrt{12}-2\sqrt{75}&=\sqrt{4\cdot 3}-2\sqrt{25\cdot 3}\\&=\sqrt{4}\sqrt{3}-2\sqrt{25}\sqrt{3}\\&=2\sqrt{3}-2(5)\sqrt{3}\\&=2\sqrt{3}-10\sqrt{3}\\&=-8\sqrt{3}\end{aligned}$$

22.
$$\begin{aligned}&\sqrt{2x^2}+\sqrt[3]{54}-x\sqrt{18}\\&=\sqrt{x^2\cdot 2}+\sqrt[3]{27\cdot 2}-x\sqrt{9\cdot 2}\\&=\sqrt{x^2}\sqrt{2}+\sqrt[3]{27}\sqrt[3]{2}-x\sqrt{9}\sqrt{2}\\&=x\sqrt{2}+3\sqrt[3]{2}-x(3)\sqrt{2}\\&=x\sqrt{2}+3\sqrt[3]{2}-3x\sqrt{2}\\&=-2x\sqrt{2}+3\sqrt[3]{2}\end{aligned}$$

23.
$$\begin{aligned}\sqrt{\frac{3}{4}}+\sqrt{\frac{3}{25}}&=\frac{\sqrt{3}}{\sqrt{4}}+\frac{\sqrt{3}}{\sqrt{25}}\\&=\frac{\sqrt{3}}{2}+\frac{\sqrt{3}}{5}\\&=\frac{5\sqrt{3}}{10}+\frac{2\sqrt{3}}{10}\\&=\frac{7\sqrt{3}}{10}\end{aligned}$$

24.
$$\begin{aligned}\sqrt{7}\cdot\sqrt{14}&=\sqrt{7\cdot 14}\\&=\sqrt{98}\\&=\sqrt{49\cdot 2}\\&=\sqrt{49}\cdot\sqrt{2}\\&=7\sqrt{2}\end{aligned}$$

25.
$$\begin{aligned}\sqrt{2}\left(\sqrt{6}-\sqrt{5}\right)&=\sqrt{2}\cdot\sqrt{6}-\sqrt{2}\cdot\sqrt{5}\\&=\sqrt{12}-\sqrt{10}\\&=\sqrt{4\cdot 3}-\sqrt{10}\\&=\sqrt{4}\cdot\sqrt{3}-\sqrt{10}\\&=2\sqrt{3}-\sqrt{10}\end{aligned}$$

26.
$$\begin{aligned}\left(\sqrt{x}+2\right)\left(\sqrt{x}-3\right)&=\sqrt{x}\cdot\sqrt{x}-3\sqrt{x}+2\sqrt{x}-6\\&=x-\sqrt{x}-6\end{aligned}$$

27. $\frac{\sqrt{50}}{\sqrt{10}}=\sqrt{\frac{50}{10}}=\sqrt{5}$

28.
$$\begin{aligned}\frac{\sqrt{40x^4}}{\sqrt{2x}}&=\sqrt{\frac{40x^4}{2x}}\\&=\sqrt{20x^3}\\&=\sqrt{4x^2\cdot 5x}\\&=\sqrt{4x^2}\cdot\sqrt{5x}\\&=2x\sqrt{5x}\end{aligned}$$

29. $\sqrt{\frac{2}{3}}=\frac{\sqrt{2}}{\sqrt{3}}=\frac{\sqrt{2}\cdot\sqrt{3}}{\sqrt{3}\cdot\sqrt{3}}=\frac{\sqrt{6}}{\sqrt{9}}=\frac{\sqrt{6}}{3}$

30. $\sqrt[3]{\frac{5}{9}}=\frac{\sqrt[3]{5}}{\sqrt[3]{9}}=\frac{\sqrt[3]{5}\cdot\sqrt[3]{3}}{\sqrt[3]{9}\cdot\sqrt[3]{3}}=\frac{\sqrt[3]{15}}{\sqrt[3]{27}}=\frac{\sqrt[3]{15}}{3}$

31.
$$\begin{aligned}\sqrt{\frac{5}{12x^2}}&=\frac{\sqrt{5}}{\sqrt{12x^2}}\\&=\frac{\sqrt{5}}{\sqrt{4x^2\cdot 3}}\\&=\frac{\sqrt{5}}{2x\sqrt{3}}\\&=\frac{\sqrt{5}\cdot\sqrt{3}}{2x\sqrt{3}\cdot\sqrt{3}}\\&=\frac{\sqrt{15}}{2x(3)}\\&=\frac{\sqrt{15}}{6x}\end{aligned}$$

32. $\frac{2\sqrt{3}}{\sqrt{3}-3}=\frac{2\sqrt{3}(\sqrt{3}+3)}{(\sqrt{3}-3)(\sqrt{3}+3)}$
$=\frac{2\sqrt{3}(\sqrt{3}+3)}{3-9}$
$=\frac{2(3)+6\sqrt{3}}{-6}$
$=\frac{6+6\sqrt{3}}{-6}$
$=\frac{6(1+\sqrt{3})}{-6}$
$=-1(1+\sqrt{3})$
$=-1-\sqrt{3}$

33. $\sqrt{x}+8=11$
$\sqrt{x}=3$
$(\sqrt{x})^2=3^2$
$x=9$

34. $\sqrt{3x-6}=\sqrt{x+4}$
$(\sqrt{3x-6})^2=(\sqrt{x+4})^2$
$3x-6=x+4$
$2x=10$
$x=5$

35. $\sqrt{2x-2}=x-5$
$(\sqrt{2x-2})^2=(x-5)^2$
$2x-2=x^2-10x+25$
$0=x^2-12x+27$
$0=(x-9)(x-3)$
$x-9=0$ or $x-3=0$
$x=9$ $\quad x=3$ (extraneous)

36. $a^2+b^2=c^2$
$8^2+b^2=12^2$
$64+b^2=144$
$b^2=80$
$b=\sqrt{80}$
$b=\sqrt{16\cdot 5}$
$b=4\sqrt{5}$
The length is $4\sqrt{5}$ inches.

37. (−3, 6) and (−2, 8)
$d=\sqrt{(x_2-x_1)^2+(y_2-y_1)^2}$
$d=\sqrt{[-2-(-3)]^2+(8-6)^2}$
$d=\sqrt{(-2+3)^2+(2)^2}$
$d=\sqrt{1^2+4}$
$d=\sqrt{1+4}$
$d=\sqrt{5}$

38. $16^{-3/4}\cdot 16^{-1/4}=16^{-4/4}=16^{-1}=\frac{1}{16}$

39. $\left(\frac{x^{2/3}}{y^{2/5}}\right)^5=\frac{x^{10/3}}{y^{10/5}}=\frac{x^{10/3}}{y^2}$

Cumulative Review Chapters 1–8

1. a. $\frac{(-12)(-3)+3}{-7-(-2)}=\frac{36+3}{-7+2}=\frac{39}{-5}=-\frac{39}{5}$

b. $\frac{2(-3)^2-20}{-5+4}=\frac{2(9)-20}{-1}$
$=-(18-20)$
$=-(-2)$
$=2$

2. a. $\frac{4(-3)-(-6)}{-8+4}=\frac{-12+6}{-4}=\frac{-6}{-4}=\frac{3}{2}$

b. $\frac{3+(-3)(-2)^3}{-1-(-4)}=\frac{3+(-3)(-8)}{-1+4}$
$=\frac{3+24}{3}$
$=\frac{27}{3}$
$=9$

3. $2x+3x-5+7=10x+3-6x-4$
$5x+2=4x-1$
$x+2=-1$
$x=-3$

4. $6y-11+4+2y=8+15y-8y$
$8y-7=8+7y$
$y-7=8$
$y=15$

5. $y = 3x$
$x = -1: y = 3(-1) = -3$
$y = 0: 0 = 3x \Rightarrow x = 0$
$y = -9: -9 = 3x \Rightarrow x = -3$

x	y
−1	−3
0	0
−3	−9

6. $2x + y = 6$
$x = 0: 2(0) + y = 6 \Rightarrow y = 6$
$y = -2: 2x + (-2) = 6 \Rightarrow 2x = 8 \Rightarrow x = 4$
$x = 3: 2(3) + y = 6 \Rightarrow 6 + y = 6 \Rightarrow y = 0$

x	y
0	6
4	−2
3	0

7. $m = \frac{1}{4},\ b = -3$
$y = mx + b$
$y = \frac{1}{4}x - 3$

8. $m = -2, b = 4$
$y = mx + b$
$y = -2x + 4$
$2x + y = 4$

9. $y = 5$ is horizontal so a parallel line is also horizontal.
$y = c$
Point $(-2, -3)$
$y = -3$

10. $y = m_1 x + b_1$
$y = 2x + 4 \Rightarrow m_1 = 2$

Perpendicular line: $m_2 = -\frac{1}{m_1} = -\frac{1}{2}$

Point on line 2: (1, 5)

$y = mx + b$
$5 = -\frac{1}{2}(1) + b$
$\frac{10}{2} + \frac{1}{2} = b$
$\frac{11}{2} = b$

The equation is $y = -\frac{1}{2}x + \frac{11}{2}$.

11. a. b. c.

12. a. c. d.

13. $\begin{cases} 2x - 3y = 6 \\ x = 2y \end{cases}$

(12, 6)

$$\begin{aligned} 2x - 3y &= 6 \\ 2(12) - 3(6) &\stackrel{?}{=} 6 \\ 24 - 18 &\stackrel{?}{=} 6 \\ 6 &= 6 \quad \text{True} \end{aligned} \qquad \begin{aligned} x &= 2y \\ 12 &\stackrel{?}{=} 2(6) \\ 12 &\stackrel{?}{=} 12 \\ 12 &= 12 \quad \text{True} \end{aligned}$$

(12, 6) is a solution.

14. $\begin{cases} 2x + y = 4 \\ x + y = 2 \end{cases}$

a. (1, 1)

$$\begin{aligned} 2x + y &= 4 \\ 2(1) + (1) &\stackrel{?}{=} 4 \\ 2 + 1 &\stackrel{?}{=} 4 \\ 3 &= 4 \quad \text{False} \end{aligned} \qquad \begin{aligned} x + y &= 2 \\ 1 + 1 &\stackrel{?}{=} 2 \\ 2 &\stackrel{?}{=} 2 \\ 2 &= 2 \quad \text{True} \end{aligned}$$

(1, 1) is not a solution.

b. (2, 0)

$$\begin{aligned} 2x + y &= 4 \\ 2(2) + (0) &\stackrel{?}{=} 4 \\ 4 + 0 &\stackrel{?}{=} 4 \\ 4 &= 4 \quad \text{True} \end{aligned} \qquad \begin{aligned} x + y &= 2 \\ 2 + 0 &\stackrel{?}{=} 2 \\ 2 &\stackrel{?}{=} 2 \\ 2 &= 2 \quad \text{True} \end{aligned}$$

(2, 0) is a solution.

15. $\begin{cases} 2x + y = 10 \\ x = y + 2 \end{cases}$

Substitute $y + 2$ for x in the first equation.
$2(y + 2) + y = 10$
$2y + 4 + y = 10$
$3y = 6$
$y = 2$
Let $y = 2$ in the second equation.
$x = (2) + 2$
$x = 4$
The solution of the system is (4, 2).

16. $\begin{cases} 3y = x+10 \\ 2x+5y = 24 \end{cases}$

Solve the first equation for x.

$3y = x+10$
$3y-10 = x$

Substitute $3y - 10$ for x in the second equation.

$2(3y-10)+5y = 24$
$6y-20+5y = 24$
$11y = 44$
$y = 4$

Let $y = 4$ in the first equation.

$x = 3(4) - 10$
$x = 2$

The solution of the system is (2, 4).

17. $\begin{cases} -x-\frac{y}{2}=\frac{5}{2} \\ -\frac{x}{2}+\frac{y}{4}=0 \end{cases}$

Multiply the first equation by 2 and the second equation by 4.

$-2x - y = 5$
$-2x + y = 0$
$-4x \quad = 5$

$x = -\frac{5}{4}$

Let $x = -\frac{5}{4}$ in the second equation.

$-\frac{-\frac{5}{4}}{2}+\frac{y}{4}=0$

$\frac{5}{8}+\frac{y}{4}=0$

$5+2y=0$
$2y=-5$

$y=-\frac{5}{2}$

The solution of the system is $\left(-\frac{5}{4}, -\frac{5}{2}\right)$.

18. $\begin{cases} \frac{x}{2}+y=\frac{5}{6} \\ 2x-y=\frac{5}{6} \end{cases}$

Multiply both equations by 6.

$3x+6y=5$
$12x-6y=5$
$15x \quad =10$

$x=\frac{2}{3}$

Let $x=\frac{2}{3}$ in the second equation.

$2\left(\frac{2}{3}\right)-y=\frac{5}{6}$

$\frac{4}{3}-y=\frac{5}{6}$

$8-6y=5$
$-6y=-3$

$y=\frac{1}{2}$

The solution of the system is $\left(\frac{2}{3}, \frac{1}{2}\right)$.

19. Let x = the amount of 25% saline.

	No. of liters	· Strength	= Amt of Saline
25%	x	0.25	$0.25x$
5%	$10 - x$	0.05	$0.05(10 - x)$
20%	20	0.2	0.2(10)

$0.25x+0.05(10-x)=0.2(10)$
$0.25x+0.5-0.05x=2$
$0.2x+0.5=2$
$0.2x=1.5$
$x=7.5$

$10 - x = 10 - 7.5 = 2.5$

Mix 7.5 liters of 25% saline with 2.5 liters of 5% saline.

20. Let x = slower speed.

	r	· t	= d
Slower	x	0.2	$0.2x$
Faster	$x + 15$	0.2	$0.2(x + 15)$

$0.2x+0.2(x+15)=11$
$0.2x+0.2x+3=11$
$0.4x=8$
$x=20$

$x + 15 = 20 + 15 = 35$

The slower streetcar travels at 20 mph.
The faster streetcar travels at 35 mph.

21. $\begin{cases} 3x \ge y \\ x+2y \le 8 \end{cases}$

$3x \ge y$
Test (0, 1)
$3(0) \overset{?}{\ge} 1$
False
Shade below

$x + 2y \le 8$
Test (0, 0)
$0+2(0) \overset{?}{\le} 8$
True
Shade below

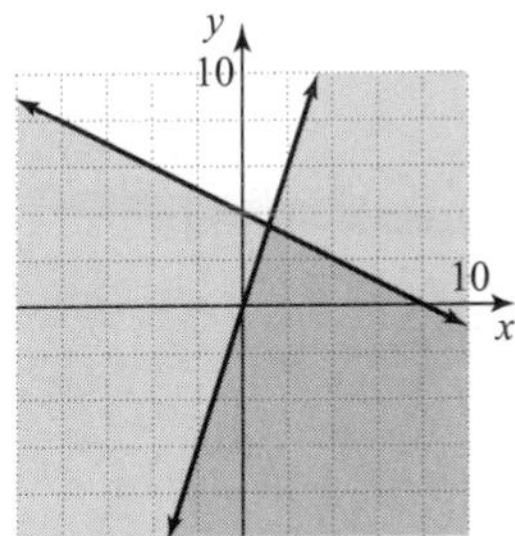

22. $\begin{cases} x+y \le 1 \\ 2x-y \ge 2 \end{cases}$

$x + y \le 1$
Test (0, 0)
$0+0 \overset{?}{\le} 1$
True
Shade below

$2x - y \ge 2$
Test (0, 0)
$2(0)-0 \overset{?}{\ge} 2$
False
Shade below

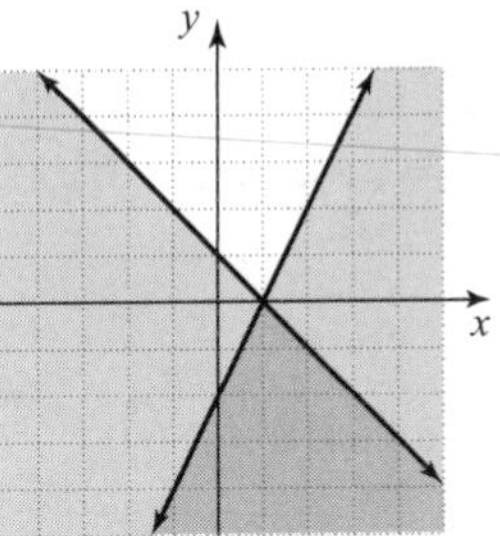

23. $-9x^2+3xy-5y^2+7xy = -9x^2-5y^2+10xy$

24. $4a^2+3a-2a^2+7a-5 = 2a^2+10a-5$

25. $x^2+7xy+6y^2 = (x+6y)(x+y)$

26. $3x^2+15x+18 = 3(x^2+5x+6) = 3(x+2)(x+3)$

27. $\dfrac{4-x^2}{3x^2-5x-2} = \dfrac{-(x^2-4)}{3x^2-5x-2}$
$= \dfrac{-(x+2)(x-2)}{(3x+1)(x-2)}$
$= -\dfrac{x+2}{3x+1}$ or $\dfrac{-x-2}{3x+1}$

28. $\dfrac{2x^2+7x+3}{x^2-9} = \dfrac{(2x+1)(x+3)}{(x+3)(x-3)} = \dfrac{2x+1}{x-3}$

29. $\dfrac{3x^3y^7}{40} \div \dfrac{4x^3}{y^2} = \dfrac{3x^3y^7}{40} \cdot \dfrac{y^2}{4x^3} = \dfrac{3x^3y^7 \cdot y^2}{40 \cdot 4x^3} = \dfrac{3y^9}{160}$

30. $\dfrac{12x^2y^3}{5} \div \dfrac{3y^3}{x} = \dfrac{12x^2y^3}{5} \cdot \dfrac{x}{3y^3}$
$= \dfrac{12x^2y^3 \cdot x}{5 \cdot 3y^3}$
$= \dfrac{4x^3}{5}$

31. $\dfrac{2y}{2y-7} - \dfrac{7}{2y-7} = \dfrac{2y-7}{2y-7} = 1$

32. $\dfrac{-4x^2}{x+1} - \dfrac{4x}{x+1} = \dfrac{-4x^2-4x}{x+1} = -\dfrac{4x(x+1)}{x+1} = -4x$

33. $\dfrac{2x}{x^2+2x+1} + \dfrac{x}{x^2-1}$
$= \dfrac{2x}{(x+1)(x+1)} + \dfrac{x}{(x+1)(x-1)}$
$= \dfrac{2x}{(x+1)(x+1)} \cdot \dfrac{x-1}{x-1} + \dfrac{x}{(x+1)(x-1)} \cdot \dfrac{x+1}{x+1}$
$= \dfrac{2x(x-1)+x(x+1)}{(x+1)(x+1)(x-1)}$
$= \dfrac{2x^2-2x+x^2+x}{(x+1)(x+1)(x-1)}$
$= \dfrac{3x^2-x}{(x+1)(x+1)(x-1)}$
$= \dfrac{x(3x-1)}{(x+1)^2(x-1)}$

34. $\frac{3x}{x^2+5x+6}+\frac{1}{x^2+2x-3}$
$=\frac{3x}{(x+2)(x+3)}+\frac{1}{(x+3)(x-1)}$
$=\frac{3x}{(x+2)(x+3)}\cdot\frac{x-1}{x-1}+\frac{1}{(x+3)(x-1)}\cdot\frac{x+2}{x+2}$
$=\frac{3x(x-1)+1(x+2)}{(x+2)(x+3)(x-1)}$
$=\frac{3x^2-3x+x+2}{(x+2)(x+3)(x-1)}$
$=\frac{3x^2-2x+2}{(x+2)(x+3)(x-1)}$

35.
$$\frac{x}{2}+\frac{8}{3}=\frac{1}{6}$$
$$6\left(\frac{x}{2}+\frac{8}{3}\right)=6\left(\frac{1}{6}\right)$$
$$6\left(\frac{x}{2}\right)+6\left(\frac{8}{3}\right)=1$$
$$3x+16=1$$
$$3x=-15$$
$$x=-5$$

36.
$$\frac{1}{21}+\frac{x}{7}=\frac{5}{3}$$
$$21\left(\frac{1}{21}+\frac{x}{7}\right)=21\left(\frac{5}{3}\right)$$
$$21\left(\frac{1}{21}\right)+21\left(\frac{x}{7}\right)=35$$
$$1+3x=35$$
$$3x=34$$
$$x=\frac{34}{3}$$

37. $\frac{2}{3}=\frac{10}{x}$
$2x=30$
$x=15$
The missing length is 15 yards.

38. Let x be the missing length.
$\frac{2}{5}=\frac{5}{x}$
$2x=25$
$x=\frac{25}{2}$

The missing length is $\frac{25}{2}$ units.

39. $\frac{\frac{1}{z}-\frac{1}{2}}{\frac{1}{3}-\frac{z}{6}}=\frac{6z\left(\frac{1}{z}-\frac{1}{2}\right)}{6z\left(\frac{1}{3}-\frac{z}{6}\right)}$
$=\frac{6z\left(\frac{1}{z}\right)-6z\left(\frac{1}{2}\right)}{6z\left(\frac{1}{3}\right)-6z\left(\frac{z}{6}\right)}$
$=\frac{6-3z}{2z-z^2}$
$=\frac{3(2-z)}{z(2-z)}$
$=\frac{3}{z}$

40. $\frac{x+3}{\frac{1}{x}+\frac{1}{3}}=\frac{3x(x+3)}{3x\left(\frac{1}{x}+\frac{1}{3}\right)}$
$=\frac{3x(x+3)}{3x\left(\frac{1}{x}\right)+3x\left(\frac{1}{3}\right)}$
$=\frac{3x(x+3)}{3+x}$
$=3x$

41. a. $\sqrt{54}=\sqrt{9\cdot 6}=\sqrt{9}\cdot\sqrt{6}=3\sqrt{6}$

b. $\sqrt{12}=\sqrt{4\cdot 3}=\sqrt{4}\cdot\sqrt{3}=2\sqrt{3}$

c. $\sqrt{200}=\sqrt{100\cdot 2}=\sqrt{100}\cdot\sqrt{2}=10\sqrt{2}$

d. $\sqrt{35}=\sqrt{35}$

42. a. $\sqrt{40}=\sqrt{4\cdot 10}=\sqrt{4}\cdot\sqrt{10}=2\sqrt{10}$

b. $\sqrt{500}=\sqrt{100\cdot 5}=\sqrt{100}\cdot\sqrt{5}=10\sqrt{5}$

c. $\sqrt{63}=\sqrt{9\cdot 7}=\sqrt{9}\cdot\sqrt{7}=3\sqrt{7}$

d. $\sqrt{169}=13$

43. a. $\left(\sqrt{5}-7\right)\left(\sqrt{5}+7\right)=\left(\sqrt{5}\right)^2-7^2$
$=5-49$
$=-44$

b. $\left(\sqrt{7x}+2\right)^2=\left(\sqrt{7x}\right)^2+2\left(\sqrt{7x}\right)(2)+2^2$
$=7x+4\sqrt{7x}+4$

44. a. $\left(\sqrt{6}+2\right)^2 = \left(\sqrt{6}\right)^2 + 2\left(\sqrt{6}\right)(2) + 2^2$
$= 6 + 4\sqrt{6} + 4$
$= 10 + 4\sqrt{6}$

b. $\left(\sqrt{x}+5\right)\left(\sqrt{x}-5\right) = \left(\sqrt{x}\right)^2 - 5^2 = x - 25$

45. $\sqrt{x} + 6 = 4$
$\sqrt{x} = -2$
The square root of a real number cannot be negative. There is no solution.

46. $\sqrt{x+4} = \sqrt{3x-1}$
$\left(\sqrt{x+4}\right)^2 = \left(\sqrt{3x-1}\right)^2$
$x + 4 = 3x - 1$
$-2x + 4 = -1$
$-2x = -5$
$x = \frac{5}{2}$

47. $a = 6, b = 8$
$c^2 = a^2 + b^2$
$c^2 = 6^2 + 8^2$
$c^2 = 36 + 64$
$c^2 = 100$
$\sqrt{c^2} = \sqrt{100}$
$c = 10$
The hypotenuse is 10 inches long.

48. $c = 13, b = 9$
$a^2 + b^2 = c^2$
$a^2 + 9^2 = 13^2$
$a^2 + 81 = 169$
$a^2 = 88$
$\sqrt{c^2} = \sqrt{88}$
$c = \sqrt{4 \cdot 22}$
$c = \sqrt{4} \cdot \sqrt{22}$
$c = 2\sqrt{22}$
The other leg is $2\sqrt{22}$ inches long.

49. a. $4^{3/2} = \left(\sqrt{4}\right)^3 = 2^3 = 8$

b. $27^{2/3} = \left(\sqrt[3]{27}\right)^2 = 3^2 = 9$

c. $-16^{3/4} = -\left(\sqrt[4]{16}\right)^3 = -2^3 = -8$

50. a. $9^{5/2} = \left(\sqrt{9}\right)^5 = 3^5 = 243$

b. $-81^{1/4} = -\left(\sqrt[4]{81}\right) = -3$

c. $(-64)^{2/3} = \left(\sqrt[3]{-64}\right)^2 = (-4)^2 = 16$

Chapter 9

Section 9.1

Practice Exercises

1. $x^2 - 16 = 0$

$x^2 = 16$

$x = \sqrt{16}$ or $x = -\sqrt{16}$

$x = 4$ $\quad$ $x = -4$

Check:

$x^2 - 16 = 0$ $\quad$ $x^2 - 16 = 0$

$4^2 - 16 \stackrel{?}{=} 0$ $\quad$ $(-4)^2 - 16 \stackrel{?}{=} 0$

$0 = 0$ True $\quad$ $0 = 0$ True

The solutions are 4 and –4.

2. $5x^2 = 13$

$x^2 = \frac{13}{5}$

$x = \sqrt{\frac{13}{5}}$ or $x = -\sqrt{\frac{13}{5}}$

$x = \frac{\sqrt{5}\cdot\sqrt{13}}{\sqrt{5}\cdot\sqrt{5}}$ $\quad$ $x = -\frac{\sqrt{5}\cdot\sqrt{13}}{\sqrt{5}\cdot\sqrt{5}}$

$x = \frac{\sqrt{65}}{5}$ $\quad$ $x = -\frac{\sqrt{65}}{5}$

The solutions are $-\frac{\sqrt{65}}{5}$ and $\frac{\sqrt{65}}{5}$.

3. $(x-5)^2 = 36$

$x - 5 = \sqrt{36}$ or $x - 5 = -\sqrt{36}$

$x - 5 = 6$ $\quad$ $x - 5 = -6$

$x = 11$ $\quad$ $x = -1$

Check:

$(x-5)^2 \stackrel{?}{=} 36$ $\quad$ $(x-5)^2 = 36$

$(11-5)^2 \stackrel{?}{=} 36$ $\quad$ $(-1-5)^2 \stackrel{?}{=} 36$

$6^2 \stackrel{?}{=} 36$ $\quad$ $(-6)^2 \stackrel{?}{=} 36$

$36 = 36$ True $\quad$ $36 = 36$ True

The solutions are –1 and 11.

4. $(x+2)^2 = 12$

$x + 2 = \sqrt{12}$ or $x + 2 = -\sqrt{12}$

$x + 2 = 2\sqrt{3}$ $\quad$ $x + 2 = -2\sqrt{3}$

$x = -2 + 2\sqrt{3}$ $\quad$ $x = -2 - 2\sqrt{3}$

The solutions are $-2 \pm 2\sqrt{3}$.

5. $(x-8)^2 = -5$

This equation has no real solution because the square root of –5 is not a real number.

6. $(3x-5)^2 = 17$

$3x - 5 = \sqrt{17}$ or $3x - 5 = -\sqrt{17}$

$3x = 5 + \sqrt{17}$ $\quad$ $3x = 5 - \sqrt{17}$

$x = \frac{5+\sqrt{17}}{3}$ $\quad$ $x = \frac{5-\sqrt{17}}{3}$

The solutions are $\frac{5 \pm \sqrt{17}}{3}$.

7. Use $h = 16t^2$, where t is the time in seconds and h is the height in feet.

Let h = 84,700 feet; $84,700 = 16t^2$

$84,700 = 16t^2$

$5293.75 = t^2$

$\sqrt{5293.75} = t$ or $-\sqrt{5293.75} = t$

$72.8 \approx t$ $\quad$ $-72.8 \approx t$

Reject –72.8 since the time of the fall is not a negative number.

The free fall lasted approximately 72.8 seconds.

Exercise Set 9.1

1. $x^2 = 64$

$x = \sqrt{64} = 8$ or $x = -\sqrt{64} = -8$

The solutions are ±8.

3. $x^2 = 21$

$x = \sqrt{21}$ or $x = -\sqrt{21}$

The solutions are $\pm\sqrt{21}$.

5. $x^2 = \frac{1}{25}$

$x = \sqrt{\frac{1}{25}} = \frac{1}{5}$ or $x = -\sqrt{\frac{1}{25}} = -\frac{1}{5}$

The solutions are $\pm\frac{1}{5}$.

7. $x^2 = -4$

This equation has no real solution because $\sqrt{-4}$ is not a real number.

9. $3x^2 = 13$

$x^2 = \frac{13}{3}$

$x = \sqrt{\frac{13}{3}}$ or $x = -\sqrt{\frac{13}{3}}$

$x = \sqrt{\frac{13}{3}} \cdot \frac{\sqrt{3}}{\sqrt{3}}$ $\quad x = -\sqrt{\frac{13}{3}} \cdot \frac{\sqrt{3}}{\sqrt{3}}$

$x = \frac{\sqrt{39}}{3}$ $\quad x = -\frac{\sqrt{39}}{3}$

The solutions are $\pm\frac{\sqrt{39}}{3}$.

11. $7x^2 = 4$

$x^2 = \frac{4}{7}$

$x = \sqrt{\frac{4}{7}}$ or $x = -\sqrt{\frac{4}{7}}$

$x = \frac{2}{\sqrt{7}} \cdot \frac{\sqrt{7}}{\sqrt{7}}$ $\quad x = -\frac{2}{\sqrt{7}} \cdot \frac{\sqrt{7}}{\sqrt{7}}$

$x = \frac{2\sqrt{7}}{7}$ $\quad x = -\frac{2\sqrt{7}}{7}$

The solutions are $\pm\frac{2\sqrt{7}}{7}$.

13. $x^2 - 2 = 0$

$x^2 = 2$

$x = \sqrt{2}$ or $x = -\sqrt{2}$

The solutions are $\pm\sqrt{2}$.

15. $2x^2 - 10 = 0$

$2x^2 = 10$

$x^2 = 5$

$x = \sqrt{5}$ or $x = -\sqrt{5}$

The solutions are $\pm\sqrt{5}$.

17. Answers may vary

19. $(x-5)^2 = 49$

$x - 5 = \sqrt{49}$ or $x - 5 = -\sqrt{49}$

$x - 5 = 7$ $\quad x - 5 = -7$

$x = 5 + 7 = 12$ $\quad x = 5 - 7 = -2$

The solutions are –2 and 12.

21. $(x+2)^2 = 7$

$x + 2 = \sqrt{7}$ or $x + 2 = -\sqrt{7}$

$x = -2 + \sqrt{7}$ $\quad x = -2 - \sqrt{7}$

The solutions are $-2 \pm \sqrt{7}$.

23. $\left(m - \frac{1}{2}\right)^2 = \frac{1}{4}$

$m - \frac{1}{2} = \sqrt{\frac{1}{4}}$ or $m - \frac{1}{2} = -\sqrt{\frac{1}{4}}$

$m - \frac{1}{2} = \frac{1}{2}$ $\quad m - \frac{1}{2} = -\frac{1}{2}$

$m = \frac{1}{2} + \frac{1}{2} = 1$ $\quad m = \frac{1}{2} - \frac{1}{2} = 0$

The solutions are 0 and 1.

25. $(p+2)^2 = 10$

$p + 2 = \sqrt{10}$ or $p + 2 = -\sqrt{10}$

$p = -2 + \sqrt{10}$ $\quad p = -2 - \sqrt{10}$

The solutions are $-2 \pm \sqrt{10}$.

27. $(3y+2)^2 = 100$

$3y + 2 = \sqrt{100}$ or $3y + 2 = -\sqrt{100}$

$3y + 2 = 10$ $\quad 3y + 2 = -10$

$3y = -2 + 10$ $\quad 3y = -2 - 10$

$3y = 8$ $\quad 3y = -12$

$y = \frac{8}{3}$ $\quad y = -4$

The solutions are –4 and $\frac{8}{3}$.

29. $(z-4)^2 = -9$

This equation has no real solution because $\sqrt{-9}$ is not a real number.

31. $(2x-11)^2 = 50$

$2x - 11 = \sqrt{50}$ or $2x - 11 = -\sqrt{50}$

$2x - 11 = 5\sqrt{2}$ $\quad 2x - 11 = -5\sqrt{2}$

$2x = 11 + 5\sqrt{2}$ $\quad 2x = 11 - 5\sqrt{2}$

$x = \frac{11 + 5\sqrt{2}}{2}$ $\quad x = \frac{11 - 5\sqrt{2}}{2}$

The solutions are $\frac{11 \pm 5\sqrt{2}}{2}$.

33. $(3x-7)^2 = 32$

$3x-7 = \sqrt{32}$ or $3x-7 = -\sqrt{32}$

$3x-7 = 4\sqrt{2}$ $\quad 3x-7 = -4\sqrt{2}$

$3x = 7+4\sqrt{2}$ $\quad 3x = 7-4\sqrt{2}$

$x = \frac{7+4\sqrt{2}}{3}$ $\quad x = \frac{7-4\sqrt{2}}{3}$

The solutions are $\frac{7\pm 4\sqrt{2}}{3}$.

35. $(2p-5)^2 = 121$

$2p-5 = \sqrt{121}$ or $2p-5 = -\sqrt{121}$

$2p-5 = 11$ $\quad 2p-5 = -11$

$2p = 16$ $\quad 2p = -6$

$p = 8$ $\quad p = -3$

The solutions are 8 and −3.

37. $x^2 - 2 = 0$

$x^2 = 2$

$x = \pm\sqrt{2}$

The solutions are $\pm\sqrt{2}$.

39. $(x+6)^2 = 24$

$x+6 = \sqrt{24}$ or $x+6 = -\sqrt{24}$

$x+6 = 2\sqrt{6}$ $\quad x+6 = -2\sqrt{6}$

$x = -6+2\sqrt{6}$ $\quad x = -6-2\sqrt{6}$

The solutions are $-6\pm 2\sqrt{6}$.

41. $\frac{1}{2}n^2 = 5$

$n^2 = 10$

$n = \pm\sqrt{10}$

The solutions are $\pm\sqrt{10}$.

43. $(4x-1)^2 = 5$

$4x-1 = \sqrt{5}$ or $4x-1 = -\sqrt{5}$

$4x = 1+\sqrt{5}$ $\quad 4x = 1-\sqrt{5}$

$x = \frac{1+\sqrt{5}}{4}$ $\quad x = \frac{1-\sqrt{5}}{4}$

The solutions are $\frac{1\pm\sqrt{5}}{4}$.

45. $3z^2 = 36$

$z^2 = 12$

$z = \pm\sqrt{12} = \pm 2\sqrt{3}$

The solutions are $\pm 2\sqrt{3}$.

47. $(8-3x)^2 - 45 = 0$

$(8-3x)^2 = 45$

$8-3x = \sqrt{45}$ or $8-3x = -\sqrt{45}$

$8-3x = 3\sqrt{5}$ $\quad 8-3x = -3\sqrt{5}$

$-3x = -8+3\sqrt{5}$ $\quad -3x = -8-3\sqrt{5}$

$x = \frac{8-3\sqrt{5}}{3}$ $\quad x = \frac{8+3\sqrt{5}}{3}$

The solutions are $\frac{8\pm 3\sqrt{5}}{3}$.

49. Let $A = 20$.

$A = s^2$

$20 = s^2$

$\sqrt{20} = 2\sqrt{5} = s$ or $-\sqrt{20} = -2\sqrt{5} = s$

$4.47 \approx s$ $\quad -4.47 \approx s$

The length of a side is not a negative number so the length is $2\sqrt{5}$ inches or approximately 4.47 inches.

51. Let $A = 31,329$.

$A = s^2$

$31,329 = s^2$

$\sqrt{31,329} = s$ or $-\sqrt{31,329} = s$

$177 = s$ $\quad -177 = s$

The length of a side is not a negative number, so the length is 177 meters.

53. Let $d = 400$.

$d = 16t^2$

$400 = 16t^2$

$\frac{400}{16} = t^2$

$25 = t^2$

$\sqrt{25} = t$ or $-\sqrt{25} = t$

$5 = t$ $\quad -5 = t$

The length of time is not a negative number so the fall lasted 5 seconds.

55. Let $h = 115$.

$$h = 16t^2$$
$$115 = 16t^2$$
$$\frac{115}{16} = t^2$$
$$7.1875 = t^2$$
$$\sqrt{7.1875} = t \quad \text{or} \quad -\sqrt{7.1875} = t$$
$$2.7 \approx t \qquad\qquad -2.7 \approx t$$

The length of the dive is not a negative number so a dive lasts approximately 2.7 seconds.

57. Let $h = 4000$.

$$h = 16t^2$$
$$4000 = 16t^2$$
$$\frac{4000}{16} = t^2$$
$$250 = t^2$$
$$\sqrt{250} = t \quad \text{or} \quad -\sqrt{250} = t$$
$$15.8 \approx t \qquad\qquad -15.8 \approx t$$

The length of the fall is not a negative number so the fall would last approximately 15.8 seconds.

59. Let $A = 36\pi$.

$$A = \pi r^2$$
$$36\pi = \pi r^2$$
$$36 = r^2$$
$$\sqrt{36} = r \quad \text{or} \quad -\sqrt{36} = r$$
$$6 = r \qquad\qquad -6 = r$$

The radius of the circle is not a negative number so the radius is 6 inches.

61. $x^2 + 6x + 9 = x^2 + 2 \cdot x \cdot 3 + 3^2 = (x+3)^2$

63. $x^2 - 4x + 4 = x^2 - 2 \cdot x \cdot 2 + 2^2 = (x-2)^2$

65. $x^2 + 4x + 4 = 16$

$$(x+2)^2 = 16$$
$$x + 2 = \sqrt{16} \quad \text{or} \quad x + 2 = -\sqrt{16}$$
$$x + 2 = 4 \qquad\qquad x + 2 = -4$$
$$x = 2 \qquad\qquad x = -6$$

The solutions are –6 and 2.

67. $x^2 + 14x + 49 = 31$

$$(x+7)^2 = 31$$
$$x + 7 = \sqrt{31} \quad \text{or} \quad x + 7 = -\sqrt{31}$$
$$x = -7 + \sqrt{31} \qquad\qquad x = -7 - \sqrt{31}$$

The solutions are $-7 \pm \sqrt{31}$.

69. $x^2 = 1.78$

$$x = \sqrt{1.78} \quad \text{or} \quad x = -\sqrt{1.78}$$
$$x \approx 1.33 \qquad\qquad x \approx -1.33$$

The solutions are ±1.33.

71. $y = -120(x-4)^2 + 6200$

Let $y = 6080$.

$$6080 = -120(x-4)^2 + 6200$$
$$-120 = -120(x-4)^2$$
$$1 = (x-4)^2$$
$$\sqrt{1} = x - 4 \quad \text{or} \quad -\sqrt{1} = x - 4$$
$$1 = x - 4 \qquad\qquad -1 = x - 4$$
$$5 = x \qquad\qquad 3 = x$$

Since 2003 + 3 = 2006 and we want a year after 2006, the answer is 2003 + 5 = 2008.

73. $y = 75x^2 + 6400$

Let $y = 7600$.

$$7600 = 75x^2 + 6400$$
$$1200 = 75x^2$$
$$16 = x^2$$
$$\sqrt{16} = x \quad \text{or} \quad -\sqrt{16} = x$$
$$4 = x \qquad\qquad -4 = x$$

Since we are looking for a year in the future, $x = 4$, and the year is 2003 + 4 = 2007.

Section 9.2

Practice Exercises

1. a. $z^2 + 8z$

$$\frac{8}{2} = 4,\ 4^2 = 16$$
$$z^2 + 8z + 16 = (z+4)^2$$

b. $x^2 - 12x$

$$\frac{-12}{2} = -6,\ (-6)^2 = 36$$
$$x^2 - 12x + 36 = (x-6)^2$$

c. $b^2 + 5b$

$$\left(\frac{5}{2}\right)^2 = \frac{25}{4}$$
$$b^2 + 5b + \frac{25}{4} = \left(b + \frac{5}{2}\right)^2$$

2. $x^2+2x-5=0$

$x^2+2x=5$

$x^2+2x+\left(\frac{2}{2}\right)^2=5+\left(\frac{2}{2}\right)^2$

$x^2+2x+1=5+1$

$(x+1)^2=6$

$x+1=\sqrt{6}$ or $x+1=-\sqrt{6}$

$x=-1+\sqrt{6}$ $\quad$ $x=-1-\sqrt{6}$

The solutions are $-1\pm\sqrt{6}$.

3. $x^2-8x=-8$

$x^2-8x+\left(\frac{-8}{2}\right)^2=-8+\left(-\frac{8}{2}\right)^2$

$x^2-8x+16=-8+16$

$(x-4)^2=8$

$x-4=\sqrt{8}$ or $x-4=-\sqrt{8}$

$x-4=2\sqrt{2}$ $\quad$ $x-4=-2\sqrt{2}$

$x=4+2\sqrt{2}$ $\quad$ $x=4-2\sqrt{2}$

The solutions are $4\pm2\sqrt{2}$.

4. $9x^2-36x-13=0$

$x^2-4x-\frac{13}{9}=0$

$x^2-4x=\frac{13}{9}$

$x^2-4x+\left(\frac{-4}{2}\right)^2=\frac{13}{9}+\left(\frac{-4}{2}\right)^2$

$x^2-4x+4=\frac{13}{9}+4$

$(x-2)^2=\frac{13}{9}+\frac{36}{9}=\frac{49}{9}$

$x-2=\sqrt{\frac{49}{9}}$ or $x-2=-\sqrt{\frac{49}{9}}$

$x-2=\frac{7}{3}$ $\quad$ $x-2=-\frac{7}{3}$

$x=2+\frac{7}{3}$ $\quad$ $x=2-\frac{7}{3}$

$x=\frac{13}{3}$ $\quad$ $x=-\frac{1}{3}$

The solutions are $-\frac{1}{3}$ and $\frac{13}{3}$.

5. $2x^2+12x=-20$

$x^2+6x=-10$

$x^2+6x+\left(\frac{6}{2}\right)^2=-10+\left(\frac{6}{2}\right)^2$

$x^2+6x+9=-10+9$

$(x+3)^2=-1$

There is no real solution since the square root of -1 is not a real number.

6. $2x^2=6x-3$

$x^2=3x-\frac{3}{2}$

$x^2-3x=-\frac{3}{2}$

$x^2-3x+\left(\frac{-3}{2}\right)^2=-\frac{3}{2}+\left(\frac{-3}{2}\right)^2$

$x^2-3x+\frac{9}{4}=-\frac{6}{4}+\frac{9}{4}$

$\left(x-\frac{3}{2}\right)^2=\frac{3}{4}$

$x-\frac{3}{2}=\sqrt{\frac{3}{4}}$ or $x-\frac{3}{2}=-\sqrt{\frac{3}{4}}$

$x-\frac{3}{2}=\frac{\sqrt{3}}{2}$ $\quad$ $x-\frac{3}{2}=-\frac{\sqrt{3}}{2}$

$x=\frac{3+\sqrt{3}}{2}$ $\quad$ $x=\frac{3-\sqrt{3}}{2}$

The solutions are $\frac{3\pm\sqrt{3}}{2}$.

Vocabulary and Readiness Check

1. By the zero factor property, if the product of two numbers is zero, then at least one of these two numbers must be zero.

2. If a is a positive number, and if $x^2=a$ then $x=\pm\sqrt{a}$.

3. An equation that can be written in the form $ax^2+bx+c=0$ where a, b, and c are real numbers and a is not zero is called a quadratic equation.

4. The process of solving a quadratic equation by writing it in the form $(x+a)^2=c$ is called completing the square.

5. To complete the square on $x^2 + 6x$, add $\underline{9}$.

6. To complete the square on $x^2 + bx$, add $\underline{\left(\frac{b}{2}\right)^2}$.

7. $p^2 + 8p$

$\left(\frac{8}{2}\right)^2 = 4^2 = 16$

8. $p^2 + 6p$

$\left(\frac{6}{2}\right)^2 = 3^2 = 9$

9. $x^2 + 20x$

$\left(\frac{20}{2}\right)^2 = 10^2 = 100$

10. $x^2 + 18x$

$\left(\frac{18}{2}\right)^2 = 9^2 = 81$

11. $y^2 + 14y$

$\left(\frac{14}{2}\right)^2 = 7^2 = 49$

12. $y^2 + 2y$

$\left(\frac{2}{2}\right)^2 = 1^2 = 1$

Exercise Set 9.2

1. $x^2 + 4x \Rightarrow \left(\frac{4}{2}\right)^2 = 2^2 = 4$

$x^2 + 4x + 4 = (x+2)^2$

3. $k^2 - 12k \Rightarrow \left(\frac{-12}{2}\right)^2 = (-6)^2 = 36$

$k^2 - 12k + 36 = (k-6)^2$

5. $x^2 - 3x \Rightarrow \left(\frac{-3}{2}\right)^2 = \frac{9}{4}$

$x^2 - 3x + \frac{9}{4} = \left(x - \frac{3}{2}\right)^2$

7. $m^2 - m \Rightarrow \left(\frac{-1}{2}\right)^2 = \frac{1}{4}$

$m^2 - m + \frac{1}{4} = \left(m - \frac{1}{2}\right)^2$

9. $x^2 + 8x = -12$

$x^2 + 8x + 16 = -12 + 16$

$(x+4)^2 = 4$

$x + 4 = \sqrt{4}$ or $x + 4 = -\sqrt{4}$

$x = -4 + 2$ $\quad$ $x = -4 - 2$

$x = -2$ $\quad$ $x = -6$

The solutions are −6 and −2.

11. $x^2 + 2x - 7 = 0$

$x^2 + 2x = 7$

$x^2 + 2x + 1 = 7 + 1$

$(x+1)^2 = 8$

$x + 1 = \sqrt{8}$ or $x + 1 = -\sqrt{8}$

$x + 1 = 2\sqrt{2}$ $\quad$ $x + 1 = -2\sqrt{2}$

$x = -1 + 2\sqrt{2}$ $\quad$ $x = -1 - 2\sqrt{2}$

The solutions are $-1 \pm 2\sqrt{2}$.

13. $x^2 - 6x = 0$

$x^2 - 6x + 9 = 0 + 9$

$(x-3)^2 = 9$

$x - 3 = \sqrt{9}$ or $x - 3 = -\sqrt{9}$

$x = 3 + 3$ $\quad$ $x = 3 - 3$

$x = 6$ $\quad$ $x = 0$

The solutions are 0 and 6.

15.
$$z^2+5z=7$$
$$z^2+5z+\frac{25}{4}=7+\frac{25}{4}$$
$$\left(z+\frac{5}{2}\right)^2=\frac{53}{4}$$
$$z+\frac{5}{2}=\sqrt{\frac{53}{4}} \quad \text{or} \quad z+\frac{5}{2}=-\sqrt{\frac{53}{4}}$$
$$z+\frac{5}{2}=\frac{\sqrt{53}}{2} \quad\quad z+\frac{5}{2}=-\frac{\sqrt{53}}{2}$$
$$z=-\frac{5}{2}+\frac{\sqrt{53}}{2} \quad\quad z=-\frac{5}{2}-\frac{\sqrt{53}}{2}$$
The solutions are $\frac{-5\pm\sqrt{53}}{2}$.

17.
$$x^2-2x-1=0$$
$$x^2-2x=1$$
$$x^2-2x+1=1+1$$
$$(x-1)^2=2$$
$$x-1=\sqrt{2} \quad \text{or} \quad x-1=-\sqrt{2}$$
$$x=1+\sqrt{2} \quad\quad x=1-\sqrt{2}$$
The solutions are $1\pm\sqrt{2}$.

19.
$$y^2+5y+4=0$$
$$y^2+5y=-4$$
$$y^2+5y+\frac{25}{4}=-4+\frac{25}{4}$$
$$\left(y+\frac{5}{2}\right)^2=\frac{9}{4}$$
$$y+\frac{5}{2}=\sqrt{\frac{9}{4}} \quad \text{or} \quad y+\frac{5}{2}=-\sqrt{\frac{9}{4}}$$
$$y=-\frac{5}{2}+\frac{3}{2} \quad\quad y=-\frac{5}{2}-\frac{3}{2}$$
$$y=-1 \quad\quad y=-4$$
The solutions are −4 and −1.

21.
$$3x^2-6x=24$$
$$x^2-2x=8$$
$$x^2-2x+1=8+1$$
$$(x-1)^2=9$$
$$x-1=\sqrt{9} \quad \text{or} \quad x-1=-\sqrt{9}$$
$$x=1+3 \quad\quad x=1-3$$
$$x=4 \quad\quad x=-2$$
The solutions are −2 and 4.

23.
$$5x^2+10x+6=0$$
$$5x^2+10x=-6$$
$$x^2+2x=-\frac{6}{5}$$
$$x^2+2x+1=-\frac{6}{5}+1$$
$$(x+1)^2=-\frac{1}{5}$$
This equation has no real solution because $\sqrt{-\frac{1}{5}}$ is not a real number.

25.
$$2x^2=6x+5$$
$$2x^2-6x=5$$
$$x^2-3x=\frac{5}{2}$$
$$x^2-3x+\frac{9}{4}=\frac{5}{2}+\frac{9}{4}$$
$$\left(x-\frac{3}{2}\right)^2=\frac{19}{4}$$
$$x-\frac{3}{2}=\sqrt{\frac{19}{4}} \quad \text{or} \quad x-\frac{3}{2}=-\sqrt{\frac{19}{4}}$$
$$x-\frac{3}{2}=\frac{\sqrt{19}}{2} \quad\quad x-\frac{3}{2}=-\frac{\sqrt{19}}{2}$$
$$x=\frac{3}{2}+\frac{\sqrt{19}}{2} \quad\quad x=\frac{3}{2}-\frac{\sqrt{19}}{2}$$
The solutions are $\frac{3\pm\sqrt{19}}{2}$.

27.
$$2y^2+8y+5=0$$
$$2y^2+8y=-5$$
$$y^2+4y=-\frac{5}{2}$$
$$y^2+4y+4=-\frac{5}{2}+4$$
$$(y+2)^2=\frac{3}{2}$$
$$y+2=\sqrt{\frac{3}{2}} \quad \text{or} \quad y+2=-\sqrt{\frac{3}{2}}$$
$$y+2=\frac{\sqrt{3}}{\sqrt{2}}\cdot\frac{\sqrt{2}}{\sqrt{2}} \quad\quad y+2=-\frac{\sqrt{3}}{\sqrt{2}}\cdot\frac{\sqrt{2}}{\sqrt{2}}$$
$$y=-2+\frac{\sqrt{6}}{2} \quad\quad y=-2-\frac{\sqrt{6}}{2}$$
The solutions are $-2\pm\frac{\sqrt{6}}{2}$.

29. $x^2 + 6x - 25 = 0$

$x^2 + 6x = 25$

$x^2 + 6x + 9 = 25 + 9$

$(x+3)^2 = 34$

$x + 3 = \sqrt{34}$ or $x + 3 = -\sqrt{34}$

$x = -3 + \sqrt{34}$ $\quad$ $x = -3 - \sqrt{34}$

The solutions are $-3 \pm \sqrt{34}$.

31. $x^2 - 3x - 3 = 0$

$x^2 - 3x = 3$

$x^2 - 3x + \frac{9}{4} = 3 + \frac{9}{4}$

$\left(x - \frac{3}{2}\right)^2 = \frac{21}{4}$

$x - \frac{3}{2} = \sqrt{\frac{21}{4}}$ or $x - \frac{3}{2} = -\sqrt{\frac{21}{4}}$

$x - \frac{3}{2} = \frac{\sqrt{21}}{2}$ $\quad$ $x - \frac{3}{2} = -\frac{\sqrt{21}}{2}$

$x = \frac{3}{2} + \frac{\sqrt{21}}{2}$ $\quad$ $x = \frac{3}{2} - \frac{\sqrt{21}}{2}$

The solutions are $\frac{3 \pm \sqrt{21}}{2}$.

33. $2y^2 - 3y + 1 = 0$

$2y^2 - 3y = -1$

$y^2 - \frac{3}{2}y = -\frac{1}{2}$

$y^2 - \frac{3}{2}y + \frac{9}{16} = -\frac{1}{2} + \frac{9}{16}$

$\left(y - \frac{3}{4}\right)^2 = \frac{1}{16}$

$y - \frac{3}{4} = \sqrt{\frac{1}{16}}$ or $y - \frac{3}{4} = -\sqrt{\frac{1}{16}}$

$y = \frac{3}{4} + \frac{1}{4}$ $\quad$ $y = \frac{3}{4} - \frac{1}{4}$

$y = 1$ $\quad$ $y = \frac{1}{2}$

The solutions are $\frac{1}{2}$ and 1.

35. $x(x+3) = 18$

$x^2 + 3x = 18$

$x^2 + 3x + \frac{9}{4} = 18 + \frac{9}{4}$

$\left(x + \frac{3}{2}\right)^2 = \frac{81}{4}$

$x + \frac{3}{2} = \sqrt{\frac{81}{4}}$ or $x + \frac{3}{2} = -\sqrt{\frac{81}{4}}$

$x = -\frac{3}{2} + \frac{9}{2}$ $\quad$ $x = -\frac{3}{2} - \frac{9}{2}$

$x = 3$ $\quad$ $x = -6$

The solutions are –6 and 3.

37. $3z^2 + 6z + 4 = 0$

$3z^2 + 6z = -4$

$z^2 + 2z = -\frac{4}{3}$

$z^2 + 2z + 1 = -\frac{4}{3} + 1$

$(z+1)^2 = -\frac{1}{3}$

This equation has no real solution because $\sqrt{-\frac{1}{3}}$ is not a real number.

39. $4x^2 + 16x = 48$

$x^2 + 4x = 12$

$x^2 + 4x + 4 = 12 + 4$

$(x+2)^2 = 16$

$x + 2 = \sqrt{16}$ or $x + 2 = -\sqrt{16}$

$x = 4 - 2$ $\quad$ $x = -4 - 2$

$x = 2$ $\quad$ $x = -6$

The solutions are 2 and –6.

41. $\frac{3}{4} - \sqrt{\frac{25}{16}} = \frac{3}{4} - \frac{5}{4} = -\frac{2}{4} = -\frac{1}{2}$

43. $\frac{1}{2} - \sqrt{\frac{9}{4}} = \frac{1}{2} - \frac{3}{2} = -\frac{2}{2} = -1$

45. $\frac{6 + 4\sqrt{5}}{2} = \frac{2\left(3 + 2\sqrt{5}\right)}{2} = 3 + 2\sqrt{5}$

47. $\frac{3 - 9\sqrt{2}}{6} = \frac{3\left(1 - 3\sqrt{2}\right)}{3 \cdot 2} = \frac{1 - 3\sqrt{2}}{2}$

49. Answers may vary

51. a. $x^2+6x+9=11$

$(x+3)^2=11$

$x+3=\sqrt{11}$ or $x+3=-\sqrt{11}$

$x=-3+\sqrt{11}$ $x=-3-\sqrt{11}$

The solutions are $-3\pm\sqrt{11}$.

b. Answers may vary

53. $x^2+kx+16$

$\left(\frac{k}{2}\right)^2=16$

$\frac{k^2}{4}=16$

$k^2=64$

$k=\pm\sqrt{64}$

$k=\pm 8$

55. $y=250x^2-750x+7800$

Let $y = 8800$.

$8800=250x^2-750x+7800$

$0=250x^2-750x-1000$

$0=x^2-3x-4$

$0=(x-4)(x+1)$

$x-4=0$ or $x+1=0$

$x=4$ $x=-1$

Since we want a year after 2002, $x = 4$, and the year is 2002 + 4 = 2006.

57. $x^2+8x=-12$

$y_1=x^2+8x$

$y_2=-12$

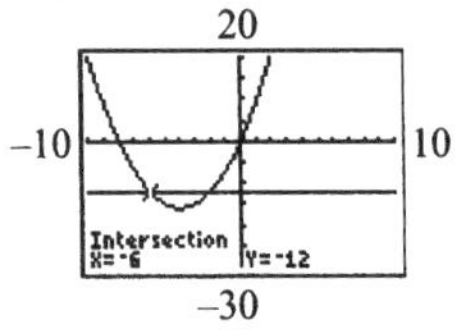

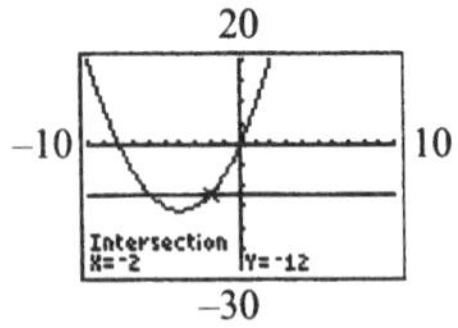

20

−10 10

Intersection X=-6 Y=-12

−30

20

−10 10

Intersection X=-2 Y=-12

−30

The x-coordinates of the intersections, −6 and −2, are the solutions.

59. $2x^2=6x+5$

$y_1=2x^2$

$y_2=6x+5$

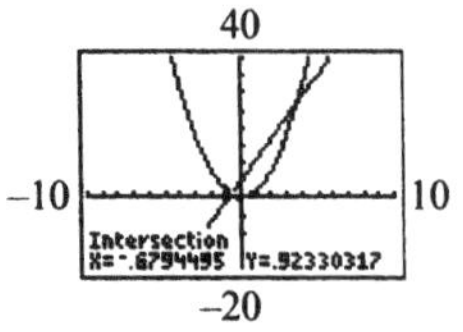

40

−10 10

Intersection X=-.6794495 Y=.92330317

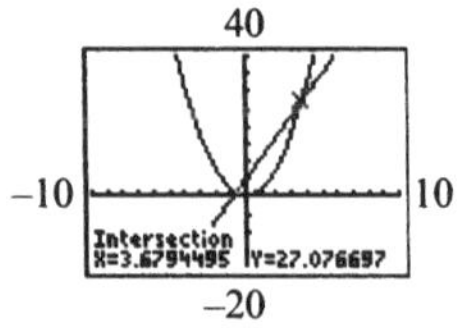

−20

40

−10 10

Intersection X=3.6794495 Y=27.076697

−20

The x-coordinates of the intersections, −0.68 and 3.68, are the approximate solutions.

Section 9.3

Practice Exercises

1. $5x^2+x-2=0$

$a=5,\ b=1,\ c=-2$

$x=\frac{-b\pm\sqrt{b^2-4ac}}{2a}$

$x=\frac{-1\pm\sqrt{1^2-4(5)(-2)}}{2(5)}$

$=\frac{-1\pm\sqrt{1+40}}{10}$

$=\frac{-1\pm\sqrt{41}}{10}$

The solutions are $\frac{-1\pm\sqrt{41}}{10}$.

2. $3x^2+2x=8$

$3x^2+2x-8=0$

$a=3,\ b=2,\ c=-8$

$x=\frac{-b\pm\sqrt{b^2-4ac}}{2a}$

$x=\frac{-2\pm\sqrt{2^2-4(3)(-8)}}{2(3)}$

$=\frac{-2\pm\sqrt{4+96}}{6}$

$=\frac{-2\pm\sqrt{100}}{6}$

$=\frac{-2\pm 10}{6}$

$$x = \frac{-2+10}{6} = \frac{8}{6} = \frac{4}{3} \text{ or } x = \frac{-2-10}{6} = \frac{-12}{6} = -2$$

The solutions are $\frac{4}{3}$ and -2.

3. $3x^2 = 5$

$3x^2 - 5 = 0$

$a = 3, b = 0, c = -5$

$$x = \frac{0 \pm \sqrt{0^2 - 4(3)(-5)}}{2(3)} = \frac{\pm\sqrt{60}}{6} = \frac{\pm 2\sqrt{15}}{6} = \frac{\pm\sqrt{15}}{3}$$

The solutions are $\pm\frac{\sqrt{15}}{3}$.

4. $x^2 = 3x - 4$

$x^2 - 3x + 4 = 0$

$a = 1, b = -3, c = 4$

$$x = \frac{-(-3) \pm \sqrt{(-3)^2 - 4(1)(4)}}{2(1)} = \frac{3 \pm \sqrt{9-16}}{2} = \frac{3 \pm \sqrt{-7}}{2}$$

There is no real number solution because $\sqrt{-7}$ is not a real number.

5. $\frac{1}{5}x^2 - x = 1$

$\frac{1}{5}x^2 - x - 1 = 0$

$x^2 - 5x - 5 = 0$

$a = 1, b = -5, c = -5$

$$x = \frac{-(-5) \pm \sqrt{(-5)^2 - 4(1)(-5)}}{2(1)} = \frac{5 \pm \sqrt{25+20}}{2} = \frac{5 \pm \sqrt{45}}{2} = \frac{5 \pm 3\sqrt{5}}{2}$$

The solutions are $\frac{5 \pm 3\sqrt{5}}{2}$.

6. The exact solutions are $\frac{-1 \pm \sqrt{41}}{10}$.

$\frac{-1+\sqrt{41}}{10} \approx 0.5403124237 \approx 0.5$

$\frac{-1-\sqrt{41}}{10} \approx -0.7403124237 \approx -0.7$

7. $5x^2 + x - 2 = 0$

$a = 5, b = 1, c = -2$

$b^2 - 4ac = 1^2 - 4(5)(-2) = 1 + 40 = 41$

Since $41 > 0$, there are two distinct real solutions.

8. a. $x^2 - 10x + 35 = 0$

$a = 1, b = -10, c = 35$

$$b^2 - 4ac = (-10)^2 - 4(1)(35) = 100 - 140 = -40$$

Since $-40 < 0$, there is no real solution.

b. $5x^2 + 3x = 0$

$a = 5, b = 3, c = 0$

$b^2 - 4ac = 3^2 - 4(5)(0) = 9$

Since $9 > 0$, there are two distinct real solutions.

Vocabulary and Readiness Check

1. The quadratic formula is $x = \frac{-b \pm \sqrt{b^2 - 4ac}}{2a}$.

2. In $5x^2 - 7x + 1 = 0$, $a = \underline{5}$, $b = \underline{-7}$, $c = \underline{1}$.

3. In $x^2 + 3x - 7 = 0$, $a = \underline{1}$, $b = \underline{3}$, $c = \underline{-7}$.

4. In $x^2 - 6 = 0$, $a = \underline{1}$, $b = \underline{0}$, $c = \underline{-6}$.

5. In $x^2 + x - 1 = 0$, $a = \underline{1}$, $b = \underline{1}$, $c = \underline{-1}$.

6. In $9x^2 - 4 = 0$, $a = \underline{9}$, $b = \underline{0}$, $c = \underline{-4}$.

Exercise Set 9.3

1. $$\frac{-1 \pm \sqrt{1^2 - 4(1)(-2)}}{2(1)} = \frac{-1 \pm \sqrt{1+8}}{2} = \frac{-1 \pm \sqrt{9}}{2} = \frac{-1 \pm 3}{2}$$

$$\frac{-1+3}{2} = \frac{2}{2} = 1; \frac{-1-3}{2} = \frac{-4}{2} = -2$$

3. $$\frac{-5 \pm \sqrt{5^2 - 4(1)(2)}}{2(1)} = \frac{-5 \pm \sqrt{25-8}}{2} = \frac{-5 \pm \sqrt{17}}{2}$$

5. $$\frac{-(-4) \pm \sqrt{(-4)^2 - 4(2)(1)}}{2(2)} = \frac{4 \pm \sqrt{16-8}}{4} = \frac{4 \pm \sqrt{8}}{4} = \frac{4 \pm \sqrt{4}\sqrt{2}}{4} = \frac{4 \pm 2\sqrt{2}}{4} = \frac{2\left(2 \pm \sqrt{2}\right)}{2 \cdot 2} = \frac{2 \pm \sqrt{2}}{2}$$

7. $x^2 - 3x + 2 = 0$

$a = 1$, $b = -3$, and $c = 2$

$$x = \frac{-(-3) \pm \sqrt{(-3)^2 - 4(1)(2)}}{2(1)} = \frac{3 \pm \sqrt{9-8}}{2} = \frac{3 \pm \sqrt{1}}{2} = \frac{3 \pm 1}{2}$$

$$x = \frac{3+1}{2} = 2 \text{ or } x = \frac{3-1}{2} = 1$$

The solutions are 1 and 2.

9. $3k^2 + 7k + 1 = 0$

$a = 3$, $b = 7$, and $c = 1$

$$k = \frac{-7 \pm \sqrt{7^2 - 4(3)(1)}}{2(3)} = \frac{-7 \pm \sqrt{49-12}}{6} = \frac{-7 \pm \sqrt{37}}{6}$$

The solutions are $\frac{-7 \pm \sqrt{37}}{6}$.

11. $49x^2 - 4 = 0$

$a = 49$, $b = 0$, and $c = -4$

$$x = \frac{-0 \pm \sqrt{0^2 - 4(49)(-4)}}{2(49)} = \frac{\pm\sqrt{784}}{98} = \frac{\pm 28}{98} = \pm\frac{2}{7}$$

The solutions are $\pm\frac{2}{7}$.

13. $5z^2 - 4z + 3 = 0$

$a = 5$, $b = -4$, and $c = 3$

$$z = \frac{-(-4) \pm \sqrt{(-4)^2 - 4(5)(3)}}{2(5)} = \frac{4 \pm \sqrt{16-60}}{10} = \frac{4 \pm \sqrt{-44}}{10}$$

There is no real solution because $\sqrt{-44}$ is not a real number.

15. $y^2 = 7y + 30$

$y^2 - 7y - 30 = 0$

$a = 1$, $b = -7$, and $c = -30$

$$y = \frac{-(-7) \pm \sqrt{(-7)^2 - 4(1)(-30)}}{2(1)} = \frac{7 \pm \sqrt{49 + 120}}{2} = \frac{7 \pm \sqrt{169}}{2} = \frac{7 \pm 13}{2}$$

$$y = \frac{7+13}{2} = 10 \quad \text{or} \quad y = \frac{7-13}{2} = -3$$

The solutions are -3 and 10.

17. $2x^2 = 10$

$2x^2 - 10 = 0$

$a = 2$, $b = 0$, and $c = -10$

$$x = \frac{-0 \pm \sqrt{0^2 - 4(2)(-10)}}{2(2)} = \frac{\pm\sqrt{80}}{4} = \frac{\pm 4\sqrt{5}}{4} = \pm\sqrt{5}$$

The solutions are $\pm\sqrt{5}$.

19. $m^2 - 12 = m$

$m^2 - m - 12 = 0$

$a = 1$, $b = -1$, and $c = -12$

$$m = \frac{-(-1) \pm \sqrt{(-1)^2 - 4(1)(-12)}}{2(1)} = \frac{1 \pm \sqrt{1 + 48}}{2} = \frac{1 \pm \sqrt{49}}{2} = \frac{1 \pm 7}{2}$$

$$m = \frac{1+7}{2} = 4 \quad \text{or} \quad m = \frac{1-7}{2} = -3$$

The solutions are -3 and 4.

21. $3 - x^2 = 4x$

$-x^2 - 4x + 3 = 0$

$a = -1$, $b = -4$, and $c = 3$

$$x = \frac{-(-4) \pm \sqrt{(-4)^2 - 4(-1)(3)}}{2(-1)} = \frac{4 \pm \sqrt{16 + 12}}{-2} = \frac{4 \pm \sqrt{28}}{-2} = \frac{4 \pm 2\sqrt{7}}{-2} = \frac{2\left(2 \pm \sqrt{7}\right)}{-2} = -2 \pm \sqrt{7}$$

The solutions are $-2 \pm \sqrt{7}$.

23. $2a^2 - 7a + 3 = 0$

$a = 2$, $b = -7$, $c = 3$

$$a = \frac{-(-7) \pm \sqrt{(-7)^2 - 4(2)(3)}}{2(2)} = \frac{7 \pm \sqrt{49 - 24}}{4} = \frac{7 \pm \sqrt{25}}{4} = \frac{7 \pm 5}{4}$$

$$a = \frac{7+5}{4} = 3 \quad \text{or} \quad a = \frac{7-5}{4} = \frac{1}{2}$$

The solutions are 3 and $\frac{1}{2}$.

25. $x^2 - 5x - 2 = 0$

$a = 1$, $b = -5$, $c = -2$

$$x = \frac{-(-5) \pm \sqrt{(-5)^2 - 4(1)(-2)}}{2(1)} = \frac{5 \pm \sqrt{25 + 8}}{2} = \frac{5 \pm \sqrt{33}}{2}$$

The solutions are $\frac{5 \pm \sqrt{33}}{2}$.

27. $3x^2 - x - 14 = 0$

$a = 3, b = -1, c = -14$

$$x = \frac{-(-1) \pm \sqrt{(-1)^2 - 4(3)(-14)}}{2(3)} = \frac{1 \pm \sqrt{1+168}}{6} = \frac{1 \pm \sqrt{169}}{6} = \frac{1 \pm 13}{6}$$

$$x = \frac{1+13}{6} = \frac{7}{3} \quad \text{or} \quad x = \frac{1-13}{6} = -2$$

The solutions are $\frac{7}{3}$ and -2.

29. $6x^2 + 9x = 2$

$6x^2 + 9x - 2 = 0$

$a = 6$, $b = 9$, and $c = -2$

$$x = \frac{-9 \pm \sqrt{9^2 - 4(6)(-2)}}{2(6)} = \frac{-9 \pm \sqrt{81+48}}{12} = \frac{-9 \pm \sqrt{129}}{12}$$

The solutions are $\frac{-9 \pm \sqrt{129}}{12}$.

31. $7p^2 + 2 = 8p$

$7p^2 - 8p + 2 = 0$

$a = 7$, $b = -8$, and $c = 2$

$$p = \frac{-(-8) \pm \sqrt{(-8)^2 - 4(7)(2)}}{2(7)} = \frac{8 \pm \sqrt{64-56}}{14} = \frac{8 \pm \sqrt{8}}{14} = \frac{8 \pm 2\sqrt{2}}{14} = \frac{2\left(4 \pm \sqrt{2}\right)}{2 \cdot 7} = \frac{4 \pm \sqrt{2}}{7}$$

The solutions are $\frac{4 \pm \sqrt{2}}{7}$.

33. $a^2 - 6a + 2 = 0$

$a = 1$, $b = -6$, and $c = 2$

$$a = \frac{-(-6) \pm \sqrt{(-6)^2 - 4(1)(2)}}{2(1)} = \frac{6 \pm \sqrt{36-8}}{2} = \frac{6 \pm \sqrt{28}}{2} = \frac{6 \pm 2\sqrt{7}}{2} = \frac{2\left(3 \pm \sqrt{7}\right)}{2} = 3 \pm \sqrt{7}$$

The solutions are $3 \pm \sqrt{7}$.

35. $2x^2 - 6x + 3 = 0$

$a = 2$, $b = -6$, and $c = 3$

$$x = \frac{-(-6) \pm \sqrt{(-6)^2 - 4(2)(3)}}{2(2)} = \frac{6 \pm \sqrt{36-24}}{4} = \frac{6 \pm \sqrt{12}}{4} = \frac{6 \pm 2\sqrt{3}}{4} = \frac{2\left(3 \pm \sqrt{3}\right)}{2 \cdot 2} = \frac{3 \pm \sqrt{3}}{2}$$

The solutions are $\frac{3 \pm \sqrt{3}}{2}$.

37. $3x^2 = 1 - 2x$

$3x^2 + 2x - 1 = 0$

$a = 3$, $b = 2$, and $c = -1$

$$x = \frac{-2 \pm \sqrt{2^2 - 4(3)(-1)}}{2(3)} = \frac{-2 \pm \sqrt{4+12}}{6} = \frac{-2 \pm \sqrt{16}}{6} = \frac{-2 \pm 4}{6}$$

$$x = \frac{-2+4}{6} = \frac{1}{3} \quad \text{or} \quad x = \frac{-2-4}{6} = -1$$

The solutions are -1 and $\frac{1}{3}$.

39. $20y^2 = 3 - 11y$

$20y^2 + 11y - 3 = 0$

$a = 20$, $b = 11$, and $c = -3$

$$y = \frac{-11 \pm \sqrt{11^2 - 4(20)(-3)}}{2(20)} = \frac{-11 \pm \sqrt{121+240}}{40} = \frac{-11 \pm \sqrt{361}}{40} = \frac{-11 \pm 19}{40}$$

$$y = \frac{-11+19}{40} = \frac{1}{5} \quad \text{or} \quad y = \frac{-11-19}{40} = -\frac{3}{4}$$

The solutions are $-\frac{3}{4}$ and $\frac{1}{5}$.

41. $x^2 + x + 1 = 0$

$a = 1$, $b = 1$, and $c = 1$

$$x = \frac{-1 \pm \sqrt{1^2 - 4(1)(1)}}{2(1)} = \frac{-1 \pm \sqrt{1-4}}{2} = \frac{-1 \pm \sqrt{-3}}{2}$$

There is no real solution because $\sqrt{-3}$ is not a real number.

43. $4y^2 = 6y + 1$

$4y^2 - 6y - 1 = 0$

$a = 4$, $b = -6$, and $c = -1$

$$y = \frac{-(-6) \pm \sqrt{(-6)^2 - 4(4)(-1)}}{2(4)} = \frac{6 \pm \sqrt{36+16}}{8} = \frac{6 \pm \sqrt{52}}{8} = \frac{6 \pm 2\sqrt{13}}{8} = \frac{2\left(3 \pm \sqrt{13}\right)}{2 \cdot 4} = \frac{3 \pm \sqrt{13}}{4}$$

The solutions are $\frac{3 \pm \sqrt{13}}{4}$.

45. $3p^2 - \frac{2}{3}p + 1 = 0$

$9p^2 - 2p + 3 = 0$

$a = 9$, $b = -2$, and $c = 3$

$$p = \frac{-(-2) \pm \sqrt{(-2)^2 - 4(9)(3)}}{2(9)} = \frac{2 \pm \sqrt{4-108}}{18} = \frac{2 \pm \sqrt{-104}}{18}$$

There is no real solution because $\sqrt{-104}$ is not a real number.

47. $\frac{m^2}{2} = m + \frac{1}{2}$

$m^2 = 2m + 1$

$m^2 - 2m - 1 = 0$

$a = 1$, $b = -2$ and $c = -1$

$$m = \frac{-(-2) \pm \sqrt{(-2)^2 - 4(1)(-1)}}{2(1)}$$
$$= \frac{2 \pm \sqrt{4+4}}{2}$$
$$= \frac{2 \pm \sqrt{8}}{2}$$
$$= \frac{2 \pm 2\sqrt{2}}{2}$$
$$= \frac{2\left(1 \pm \sqrt{2}\right)}{2}$$
$$= 1 \pm \sqrt{2}$$

The solutions are $1 \pm \sqrt{2}$.

49. $4p^2 + \frac{3}{2} = -5p$

$8p^2 + 3 = -10p$

$8p^2 + 10p + 3 = 0$

$a = 8$, $b = 10$, and $c = 3$

$$p = \frac{-10 \pm \sqrt{10^2 - 4(8)(3)}}{2(8)}$$
$$= \frac{-10 \pm \sqrt{100-96}}{16}$$
$$= \frac{-10 \pm \sqrt{4}}{16}$$
$$= \frac{-10 \pm 2}{16}$$

$p = \frac{-10+2}{16} = -\frac{1}{2}$ or $p = \frac{-10-2}{16} = -\frac{3}{4}$

The solutions are $-\frac{3}{4}$ and $-\frac{1}{2}$.

51. $5x^2 = \frac{7}{2}x + 1$

$10x^2 = 7x + 2$

$10x^2 - 7x - 2 = 0$

$a = 10$, $b = -7$, and $c = -2$

$$x = \frac{-(-7) \pm \sqrt{(-7)^2 - 4(10)(-2)}}{2(10)}$$
$$= \frac{7 \pm \sqrt{49+80}}{20}$$
$$= \frac{7 \pm \sqrt{129}}{20}$$

The solutions are $\frac{7 \pm \sqrt{129}}{20}$.

53. $28x^2 + 5x + \frac{11}{4} = 0$

$112x^2 + 20x + 11 = 0$

$a = 112$, $b = 20$, and $c = 11$

$$p = \frac{-20 \pm \sqrt{20^2 - 4(112)(11)}}{2(112)}$$
$$= \frac{-20 \pm \sqrt{400 - 4928}}{224}$$
$$= \frac{-20 \pm \sqrt{-4528}}{224}$$

There is no real solution because $\sqrt{-4528}$ is not a real number.

55. $5z^2 - 2z = \frac{1}{5}$

$25z^2 - 10z = 1$

$25z^2 - 10z - 1 = 0$

$a = 25$, $b = -10$, and $c = -1$

$$x = \frac{-(-10) \pm \sqrt{(-10)^2 - 4(25)(-1)}}{2(25)}$$
$$= \frac{10 \pm \sqrt{100+100}}{50}$$
$$= \frac{10 \pm \sqrt{200}}{50}$$
$$= \frac{10 \pm 10\sqrt{2}}{50}$$
$$= \frac{10\left(1+\sqrt{2}\right)}{10 \cdot 5}$$
$$= \frac{1 \pm \sqrt{2}}{5}$$

The solutions are $\frac{1 \pm \sqrt{2}}{5}$.

57. $x^2 + 3\sqrt{2}x - 5 = 0$

$a = 1$, $b = 3\sqrt{2}$, and $c = -5$

$$x = \frac{-3\sqrt{2} \pm \sqrt{\left(3\sqrt{2}\right)^2 - 4(1)(-5)}}{2(1)}$$
$$= \frac{-3\sqrt{2} \pm \sqrt{18+20}}{2}$$
$$= \frac{-3\sqrt{2} \pm \sqrt{38}}{2}$$

The solutions are $\frac{-3\sqrt{2} \pm \sqrt{38}}{2}$.

59. $3x^2 = 21$

$3x^2 - 21 = 0$

$a = 3, b = 0, c = -21$

$$x = \frac{-0 \pm \sqrt{0^2 - 4(3)(-21)}}{2(3)} = \frac{\pm\sqrt{252}}{6} = \frac{\pm 6\sqrt{7}}{6} = \pm\sqrt{7} \approx \pm 2.6$$

The solutions are $\pm\sqrt{7} \approx \pm 2.6$.

61. $x^2 + 6x + 1 = 0$

$a = 1, b = 6, c = 1$

$$x = \frac{-6 \pm \sqrt{6^2 - 4(1)(1)}}{2(1)} = \frac{-6 \pm \sqrt{36 - 4}}{2} = \frac{-6 \pm \sqrt{32}}{2} = \frac{-6 \pm 4\sqrt{2}}{2} = -3 \pm 2\sqrt{2}$$

$x = -3 + 2\sqrt{2} \approx 0.2$ or $x = -3 - 2\sqrt{2} \approx -5.8$

The solutions are $-3 \pm 2\sqrt{2}$, approximately -5.8 and -0.2.

63. $x^2 = 9x + 4$

$x^2 - 9x - 4 = 0$

$a = 1, b = -9, c = -4$

$$x = \frac{-(-9) \pm \sqrt{(-9)^2 - 4(1)(-4)}}{2(1)} = \frac{9 \pm \sqrt{81 + 16}}{2} = \frac{9 \pm \sqrt{97}}{2}$$

$x = \frac{9 + \sqrt{97}}{2} \approx 9.4$ or $x = \frac{9 - \sqrt{97}}{2} \approx -0.4$

The solutions are $\frac{9 \pm \sqrt{97}}{2}$, approximately 9.4 and -0.4.

65. $3x^2 - 2x - 2 = 0$

$a = 3, b = -2, c = -2$

$$x = \frac{-(-2) \pm \sqrt{(-2)^2 - 4(3)(-2)}}{2(3)} = \frac{2 \pm \sqrt{4 + 24}}{6} = \frac{2 \pm \sqrt{28}}{6} = \frac{2 \pm 2\sqrt{7}}{6} = \frac{1 \pm \sqrt{7}}{3}$$

$x = \frac{1 + \sqrt{7}}{3} \approx 1.2$ or $x = \frac{1 - \sqrt{7}}{3} \approx -0.5$

The solutions are $\frac{1 \pm \sqrt{7}}{3}$, approximately 1.2 and -0.5.

67. $x^2 + 3x - 1 = 0$

$a = 1, b = 3, c = -1$

$b^2 - 4ac = 3^2 - 4(1)(-1) = 9 + 4 = 13$

Since the discriminant is a positive number, this equation has two distinct real solutions.

69. $3x^2 + x + 5 = 0$

$a = 3, b = 1, c = 5$

$b^2 - 4ac = 1^2 - 4(3)(5) = 1 - 60 = -59$

Since the discriminant is a negative number, this equation has no real solution.

71. $4x^2 + 4x = -1$

$4x^2 + 4x + 1 = 0$

$a = 4, b = 4, c = 1$

$b^2 - 4ac = 4^2 - 4(4)(1) = 16 - 16 = 0$

Since the discriminant is 0, this equation has one real solution.

73. $9x^2 + 2x = 0$

$a = 9, b = 2, c = 0$

$b^2 - 4ac = 2^2 - 4(9)(0) = 4 - 0 = 4$

Since the discriminant is a positive number this equation has two distinct real solutions.

75. $5x^2 + 1 = 0$

$a = 5, b = 0, c = 1$

$b^2 - 4ac = 0^2 - 4(5)(1) = 0 - 20 = -20$

Since the discriminant is a negative number, this equation has no real solution.

77. $x^2 + 36 = -12x$
$x^2 + 12x + 36 = 0$
$a = 1, b = 12, c = 36$
$b^2 - 4ac = 12^2 - 4(1)(36) = 144 - 144 = 0$
Since the discriminant is 0, this equation has one real solution.

79. $\sqrt{48} = \sqrt{16 \cdot 3} = \sqrt{16} \cdot \sqrt{3} = 4\sqrt{3}$

81. $\sqrt{50} = \sqrt{25 \cdot 2} = \sqrt{25} \cdot \sqrt{2} = 5\sqrt{2}$

83. Let x = the base and $4x$ = the height.
$\frac{1}{2}bh = A$
$\frac{1}{2}(x)(4x) = 18$
$\frac{1}{2}(4x^2) = 18$
$2x^2 = 18$
$x^2 = 9$
$x = \pm\sqrt{9}$
$x = \pm 3$
Since the length can't be negative, base = 3 feet and height = 4(3) = 12 feet.

85. $5x^2 + 2 = x;\ a = 5$
$5x^2 - x + 2 = 0$
$a = 5, b = -1, c = 2$
The value of b is -1, choice **c**.

87. $7y^2 = 3y;\ b = 3$
$0 = -7y^2 + 3y$
$a = -7, b = 3, c = 0$
The value of a is -7, choice **b**.

89. Let x = the width; then $2x + 0.5$ = the length.
$A = lw$
$50.8 = (2x + 0.5)x$
$50.8 = 2x^2 + 0.5x$
$508 = 20x^2 + 5x$
$0 = 20x^2 + 5x - 508$
$a = 20, b = 5, c = -508$

$x = \frac{-5 \pm \sqrt{5^2 - 4(20)(-508)}}{2(20)}$
$= \frac{-5 \pm \sqrt{25 + 40,640}}{40}$
$= \frac{-5 \pm \sqrt{40,665}}{40}$
≈ -5.2 or 4.9
Lengths are positive, so $x \approx 4.9$.
$2x + 0.5 \approx 2(4.9) + 0.5 = 9.8 + 0.5 = 10.3$
The length is 10.3 feet and the width is 4.9 feet.

91. $1.2x^2 - 5.2x - 3.9 = 0$
$a = 1.2, b = -5.2, c = -3.9$
$x = \frac{-(-5.2) \pm \sqrt{(-5.2)^2 - 4(1.2)(-3.9)}}{2(1.2)}$
$x = \frac{5.2 \pm \sqrt{45.76}}{2.4}$
$x \approx -0.7, 5.0$

93. Let $h = 30$.
$y = -16t^2 + 120t + 80$
$30 = -16t^2 + 120t + 80$
$0 = -16t^2 + 120t + 50$
$a = -16, b = 120$, and $c = 50$
$t = \frac{-120 \pm \sqrt{120^2 - 4(-16)(50)}}{2(-16)}$
$= \frac{-120 \pm \sqrt{14,400 + 3200}}{-32}$
$= \frac{-120 \pm \sqrt{17,600}}{-32}$
Since the time cannot be negative,
$t = \frac{-120 - \sqrt{17,600}}{-32} \approx 7.9.$
The rocket will be 30 feet from the ground after 7.9 seconds.

95. Answers may vary

97. $y = 3.6x^2 + 578x + 13,538$
Let $y = 23,018$.
$23,018 = 3.6x^2 + 578x + 13,538$
$0 = 3.6x^2 + 578x - 9480$
$a = 3.6, b = 578, c = -9480$

$$x = \frac{-578 \pm \sqrt{578^2 - 4(3.6)(-9480)}}{2(3.6)}$$
$$= \frac{-578 \pm \sqrt{470,596}}{7.2}$$
$$= \frac{-578 \pm 686}{7.2}$$
$$x = \frac{-578 + 686}{7.2} = 15 \quad \text{or}$$
$$x = \frac{-578 - 686}{7.2} \approx -176$$

The year is after 2000, so $x = 15$, and the year is $2000 + 15 = 2015$.

99. $y = -1100x^2 + 11,800x + 46,769$

Let $y = 78,269$.

$$78,269 = -1100x^2 + 11,800x + 46,769$$
$$0 = -1100x^2 + 11,800x - 31,500$$

$a = -1100$, $b = 11,800$, $c = -31,500$

$$x = \frac{-11,800 \pm \sqrt{11,800^2 - 4(-1100)(-31,500)}}{2(-1100)}$$
$$= \frac{-11,800 \pm \sqrt{640,000}}{-2200}$$
$$= \frac{-11,800 \pm 800}{-2200}$$
$$x = \frac{-11,800 + 800}{-2200} = 5 \quad \text{or}$$
$$x = \frac{-11,800 - 800}{-2200} \approx 5.7$$

$2003 + 5 = 2008$ and $2003 + 5.7 = 2008.7$.

The year is 2008.

The Bigger Picture

1. $7.9 - 9.7 = 7.9 + (-9.7) = -1.8$

2. $5 + (-3) + (-7) = 2 + (-7) = -5$

3. $(-4)^2 - 5^2 = 16 - 25 = -9$

4. $$7x - 2 + \frac{1}{3}(9x - 3) + 5 = 7x - 2 + 3x - 1 + 5$$
$$= 7x + 3x - 2 - 1 + 5$$
$$= 10x + 2$$

5. $$\left(\frac{1}{2}x + 5\right)\left(\frac{1}{2}x - 5\right) = \left(\frac{1}{2}x\right)^2 - 5^2 = \frac{1}{4}x^2 - 25$$

6. $$\frac{9x^2y + 3xy - 12y}{3xy} = \frac{9x^2y}{3xy} + \frac{3xy}{3xy} - \frac{12y}{3xy}$$
$$= 3x + 1 - \frac{4}{x}$$

7. $$\frac{x^2}{(x-5)(x-4)} - \frac{3x+10}{(x-5)(x-4)} = \frac{x^2 - (3x+10)}{(x-5)(x-4)}$$
$$= \frac{x^2 - 3x - 10}{(x-5)(x-4)}$$
$$= \frac{(x-5)(x+2)}{(x-5)(x-4)}$$
$$= \frac{x+2}{x-4}$$

8. $$\frac{x}{x-10} + \frac{5}{x+3} = \frac{x(x+3)}{(x-10)(x+3)} + \frac{5(x-10)}{(x+3)(x-10)}$$
$$= \frac{x^2 + 3x + 5x - 50}{(x-10)(x+3)}$$
$$= \frac{x^2 + 8x - 50}{(x-10)(x+3)}$$

9. $\sqrt{50} = \sqrt{25 \cdot 2} = \sqrt{25}\sqrt{2} = 5\sqrt{2}$

10. $$\frac{\sqrt{30a^2b^3}}{\sqrt{3ab}} = \sqrt{\frac{30a^2b^3}{30ab}}$$
$$= \sqrt{10ab^2}$$
$$= \sqrt{b^2 \cdot 10a}$$
$$= \sqrt{b^2}\sqrt{10a}$$
$$= b\sqrt{10a}$$

11. $$\sqrt{\frac{2}{3}} = \frac{\sqrt{2}}{\sqrt{3}} = \frac{\sqrt{2} \cdot \sqrt{3}}{\sqrt{3} \cdot \sqrt{3}} = \frac{\sqrt{6}}{3}$$

12. $$\frac{7x-14}{x^2-4} \cdot \frac{x^2+5x+6}{49} = \frac{7(x-2) \cdot (x+2)(x+3)}{(x+2)(x-2) \cdot 7 \cdot 7}$$
$$= \frac{x+3}{7}$$

13. $x^2+3x-5=0$

$a=1, b=3, c=-5$

$$x=\frac{-b\pm\sqrt{b^2-4ac}}{2a}$$

$$x=\frac{-3\pm\sqrt{3^2-4(1)(-5)}}{2(1)}$$

$$=\frac{-3\pm\sqrt{9+20}}{2}$$

$$=\frac{-3\pm\sqrt{29}}{2}$$

14. $x^2+x=x^2+6$

$x^2+x-x^2=x^2+6-x^2$

$x=6$

15. $-2x\le 5.6$

$\frac{-2x}{-2}\ge\frac{5.6}{-2}$

$x\ge -2.8$

$\{x|x\ge -2.8\}$ or $[-2.8, \infty)$

16. $2x^2+15x=8$

$2x^2+15x-8=0$

$(2x-1)(x+8)=0$

$2x-1=0$ or $x+8=0$

$2x=1$ $\quad$ $x=-8$

$x=\frac{1}{2}$

17. $\sqrt{x+2}+4=x$

$\sqrt{x+2}=x-4$

$\left(\sqrt{x+2}\right)^2=(x-4)^2$

$x+2=x^2-8x+16$

$0=x^2-9x+14$

$0=(x-7)(x-2)$

$x-7=0$ or $x-2=0$

$x=7$ $\quad$ $x=2$

The value $x=2$ does not check, so $x=7$ is the only solution.

18. $\frac{5}{x}-\frac{3}{x-4}=\frac{7+x}{x(x-4)}$

$x(x-4)\left(\frac{5}{x}-\frac{3}{x-4}\right)=x(x-4)\frac{7+x}{x(x-4)}$

$5(x-4)-3x=7+x$

$5x-20-3x=7+x$

$2x-20=7+x$

$x-20=7$

$x=27$

Integrated Review

Practice Exercises

1. $y^2-3y-4=0$

$(y-4)(y+1)=0$

$y-4=0$ or $y+1=0$

$y=4$ $\quad$ $y=-1$

The solutions are 4 and -1.

2. $(2x+5)^2=45$

$2x+5=\pm\sqrt{45}$

$2x+5=\pm 3\sqrt{5}$

$2x=-5\pm 3\sqrt{5}$

$x=\frac{-5\pm 3\sqrt{5}}{2}$

The solutions are $\frac{-5\pm 3\sqrt{5}}{2}$.

3. $x^2-\frac{5}{2}x=-\frac{3}{2}$

$x^2-\frac{5}{2}x+\frac{3}{2}=0$

$2x^2-5x+3=0$

$(2x-3)(x-1)=0$

$2x-3=0$ or $x-1=0$

$2x=3$ $\quad$ $x=1$

$x=\frac{3}{2}$

The solutions are $\frac{3}{2}$ and 1.

Integrated Review

1. $5x^2 - 11x + 2 = 0$
$(5x-1)(x-2) = 0$
$5x - 1 = 0$ or $x - 2 = 0$
$5x = 1$ $\quad x = 2$
$x = \frac{1}{5}$

The solutions are $\frac{1}{5}$ and 2.

2. $5x^2 + 13x - 6 = 0$
$(5x-2)(x+3) = 0$
$5x - 2 = 0$ or $x + 3 = 0$
$5x = 2$ $\quad x = -3$
$x = \frac{2}{5}$

The solutions are $\frac{2}{5}$ and -3.

3. $x^2 - 1 = 2x$
$x^2 - 2x = 1$
$x^2 - 2x + 1 = 1 + 1$
$(x-1)^2 = 2$
$x - 1 = \pm\sqrt{2}$
$x = 1 \pm \sqrt{2}$
The solutions are $1 \pm \sqrt{2}$.

4. $x^2 + 7 = 6x$
$x^2 - 6x = -7$
$x^2 - 6x + 9 = -7 + 9$
$(x-3)^2 = 2$
$x - 3 = \pm\sqrt{2}$
$x = 3 \pm \sqrt{2}$
The solutions are $3 \pm \sqrt{2}$.

5. $a^2 = 20$
$a = \pm\sqrt{20}$
$= \pm 2\sqrt{5}$
The solutions are $\pm 2\sqrt{5}$.

6. $a^2 = 72$
$a = \pm\sqrt{72}$
$= \pm 6\sqrt{2}$
The solutions are $\pm 6\sqrt{2}$.

7. $x^2 - x + 4 = 0$
$x^2 - x = -4$
$x^2 - x + \frac{1}{4} = -4 + \frac{1}{4}$
$\left(x - \frac{1}{2}\right)^2 = -\frac{15}{4}$
There is no real solution.

8. $x^2 - 2x + 7 = 0$
$x^2 - 2x = -7$
$x^2 - 2x + 1 = -7 + 1$
$(x-1)^2 = -6$
There is no real solution.

9. $3x^2 - 12x + 12 = 0$
$x^2 - 4x + 4 = 0$
$(x-2)^2 = 0$
$x - 2 = 0$
$x = 2$
The solution is 2.

10. $5x^2 - 30x + 45 = 0$
$x^2 - 6x + 9 = 0$
$(x-3)^2 = 0$
$x - 3 = 0$
$x = 3$
The solution is 3.

11. $9 - 6p + p^2 = 0$
$(p-3)^2 = 0$
$p - 3 = 0$
$p = 3$
The solution is 3.

12. $49 - 28p + 4p^2 = 0$
$(2p-7)^2 = 0$
$2p - 7 = 0$
$2p = 7$
$p = \frac{7}{2}$

The solution is $\frac{7}{2}$.

13. $4y^2 - 16 = 0$
$$4y^2 = 16$$
$$y^2 = 4$$
$$y = \pm\sqrt{4}$$
$$y = \pm 2$$
The solutions are ±2.

14. $3y^2 - 27 = 0$
$$3y^2 = 27$$
$$y^2 = 9$$
$$y = \pm\sqrt{9}$$
$$y = \pm 3$$
The solutions are ±3.

15. $x^4 - 3x^3 + 2x^2 = 0$
$$x^2(x^2 - 3x + 2) = 0$$
$$x^2(x-1)(x-2) = 0$$
$$x^2 = 0 \quad \text{or} \quad x - 1 = 0 \quad \text{or} \quad x - 2 = 0$$
$$x = 0 \qquad x = 1 \qquad x = 2$$
The solutions are 0, 1, and 2.

16. $x^3 + 7x^2 + 12x = 0$
$$x(x^2 + 7x + 12) = 0$$
$$x(x+4)(x+3) = 0$$
$$x = 0 \quad \text{or} \quad x + 4 = 0 \quad \text{or} \quad x + 3 = 0$$
$$x = -4 \qquad x = -3$$
The solutions are –4, –3, and 0.

17. $(2z+5)^2 = 25$
$$2z + 5 = \pm\sqrt{25}$$
$$2z = -5 \pm 5$$
$$z = \frac{-5 \pm 5}{2}$$
$$z = \frac{-5-5}{2} = -5 \quad \text{or} \quad z = \frac{-5+5}{2} = 0$$
The solutions are 0 and –5.

18. $(3z-4)^2 = 16$
$$3z - 4 = \pm\sqrt{16}$$
$$3z = 4 \pm 4$$
$$z = \frac{4 \pm 4}{3}$$
$$z = \frac{4-4}{3} = 0 \quad \text{or} \quad z = \frac{4+4}{3} = \frac{8}{3}$$
The solutions are 0 and $\frac{8}{3}$.

19. $30x = 25x^2 + 2$
$$0 = 25x^2 - 30x + 2 = 0$$
$a = 25$, $b = -30$, and $c = 2$
$$x = \frac{-(-30) \pm \sqrt{(-30)^2 - 4(25)(2)}}{2(25)}$$
$$= \frac{30 \pm \sqrt{900 - 200}}{50}$$
$$= \frac{30 \pm \sqrt{700}}{50}$$
$$= \frac{30 \pm 10\sqrt{7}}{50}$$
$$= \frac{3 \pm \sqrt{7}}{5}$$
The solutions are $\frac{3 \pm \sqrt{7}}{5}$.

20. $12x = 4x^2 + 4$
$$0 = 4x^2 - 12x + 4$$
$$0 = x^2 - 3x + 1$$
$a = 1$, $b = -3$, and $c = 1$
$$x = \frac{-(-3) \pm \sqrt{(-3)^2 - 4(1)(1)}}{2(1)}$$
$$= \frac{3 \pm \sqrt{9 - 4}}{2}$$
$$= \frac{3 \pm \sqrt{5}}{2}$$
The solutions are $\frac{3 \pm \sqrt{5}}{2}$.

21. $\frac{2}{3}m^2 - \frac{1}{3}m - 1 = 0$
$$2m^2 - m - 3 = 0$$
$$(2m-3)(m+1) = 0$$
$$2m - 3 = 0 \quad \text{or} \quad m + 1 = 0$$
$$2m = 3 \qquad m = -1$$
$$m = \frac{3}{2}$$
The solutions are –1 and $\frac{3}{2}$.

22. $\frac{5}{8}m^2 + m - \frac{1}{2} = 0$

$5m^2 + 8m - 4 = 0$

$(5m - 2)(m + 2) = 0$

$5m - 2 = 0$ or $m + 2 = 0$

$5m = 2$ $\quad m = -2$

$m = \frac{2}{5}$

The solutions are -2 and $\frac{2}{5}$.

23. $x^2 - \frac{1}{2}x - \frac{1}{5} = 0$

$10x^2 - 5x - 2 = 0$

$a = 10, b = -5$, and $c = -2$

$$x = \frac{-(-5) \pm \sqrt{(-5)^2 - 4(10)(-2)}}{2(10)}$$

$$= \frac{5 \pm \sqrt{25 + 80}}{20}$$

$$= \frac{5 \pm \sqrt{105}}{20}$$

The solutions are $\frac{5 \pm \sqrt{105}}{20}$.

24. $x^2 + \frac{1}{2}x - \frac{1}{8} = 0$

$8x^2 + 4x - 1 = 0$

$a = 8, b = 4$, and $c = -1$

$$x = \frac{-4 \pm \sqrt{4^2 - 4(8)(-1)}}{2(8)}$$

$$= \frac{-4 \pm \sqrt{16 + 32}}{16}$$

$$= \frac{-4 \pm \sqrt{48}}{16}$$

$$= \frac{-4 \pm 4\sqrt{3}}{16}$$

$$= \frac{-1 \pm \sqrt{3}}{4}$$

The solutions are $\frac{-1 \pm \sqrt{3}}{4}$.

25. $4x^2 - 27x + 35 = 0$

$(4x - 7)(x - 5) = 0$

$4x - 7 = 0$ or $x - 5 = 0$

$4x = 7$ $\quad x = 5$

$x = \frac{7}{4}$

The solutions are $\frac{7}{4}$ and 5.

26. $9x^2 - 16x + 7 = 0$

$(9x - 7)(x - 1) = 0$

$9x - 7 = 0$ or $x - 1 = 0$

$9x = 7$ $\quad x = 1$

$x = \frac{7}{9}$

The solutions are $\frac{7}{9}$ and 1.

27. $(7 - 5x)^2 = 18$

$7 - 5x = \pm\sqrt{18}$

$7 - 5x = \pm 3\sqrt{2}$

$-5x = -7 \pm 3\sqrt{2}$

$$\frac{-5x}{-5} = \frac{-7 \pm 3\sqrt{2}}{-5}$$

$$x = \frac{7 \pm 3\sqrt{2}}{5}$$

The solutions are $\frac{7 \pm 3\sqrt{2}}{5}$.

28. $(5 - 4x)^2 = 75$

$5 - 4x = \pm\sqrt{75}$

$5 - 4x = \pm 5\sqrt{3}$

$-4x = -5 \pm 5\sqrt{3}$

$$\frac{-4x}{-4} = \frac{-5 \pm 5\sqrt{3}}{-4}$$

$$x = \frac{5 \pm 5\sqrt{3}}{4}$$

The solutions are $\frac{5 \pm 5\sqrt{3}}{4}$.

29. $3z^2 - 7z = 12$

$3z^2 - 7z - 12 = 0$

$a = 3, b = -7$, and $c = -12$

$$z = \frac{-(-7) \pm \sqrt{(-7)^2 - 4(3)(-12)}}{2(3)}$$
$$= \frac{7 \pm \sqrt{49+144}}{6}$$
$$= \frac{7 \pm \sqrt{193}}{6}$$

The solutions are $\frac{7 \pm \sqrt{193}}{6}$.

30. $6z^2 + 7z = 6$

$6z^2 + 7z - 6 = 0$

$a = 6$, $b = 7$, and $c = -6$

$$z = \frac{-7 \pm \sqrt{7^2 - 4(6)(-6)}}{2(6)}$$
$$= \frac{-7 \pm \sqrt{49+144}}{12}$$
$$= \frac{-7 \pm \sqrt{193}}{12}$$

The solutions are $\frac{-7 \pm \sqrt{193}}{12}$.

31. $x = x^2 - 110$

$0 = x^2 - x - 110$

$0 = (x+10)(x-11)$

$x + 10 = 0$ or $x - 11 = 0$

$x = -10$ $\quad x = 11$

The solutions are -10 and 11.

32. $x = 56 - x^2$

$x^2 + x - 56 = 0$

$(x+8)(x-7) = 0$

$x + 8 = 0$ or $x - 7 = 0$

$x = -8$ $\quad x = 7$

The solutions are -8 and 7.

33. $\frac{3}{4}x^2 - \frac{5}{2}x - 2 = 0$

$3x^2 - 10x - 8 = 0$

$(3x+2)(x-4) = 0$

$3x + 2 = 0$ or $x - 4 = 0$

$3x = -2$ $\quad x = 4$

$x = -\frac{2}{3}$

The solutions are $-\frac{2}{3}$ and 4.

34. $x^2 - \frac{6}{5}x - \frac{8}{5} = 0$

$5x^2 - 6x - 8 = 0$

$(5x+4)(x-2) = 0$

$5x + 4 = 0$ or $x - 2 = 0$

$5x = -4$ $\quad x = 2$

$x = -\frac{4}{5}$

The solutions are $-\frac{4}{5}$ and 2.

35. $x^2 - 0.6x + 0.05 = 0$

$100x^2 - 60x + 5 = 0$

$20x^2 - 12x + 1 = 0$

$(10x-1)(2x-1) = 0$

$10x - 1 = 0$ or $2x - 1 = 0$

$10x = 1$ $\quad 2x = 1$

$x = \frac{1}{10} = 0.1$ $\quad x = \frac{1}{2} = 0.5$

The solutions are 0.1 and 0.5.

36. $x^2 - 0.1x + 0.06 = 0$

$100x^2 - 10x + 6 = 0$

$50x^2 - 5x + 3 = 0$

$(5x+1)(10x-3) = 0$

$5x + 1 = 0$ or $10x - 3 = 0$

$5x = -1$ $\quad 10x = 3$

$x = -\frac{1}{5} = -0.2$ $\quad x = \frac{3}{10} = 0.3$

The solutions are -0.2 and 0.3.

37. $10x^2 - 11x + 2 = 0$

$a = 10$, $b = -11$, and $c = 2$

$$x = \frac{-(-11) \pm \sqrt{(-11)^2 - 4(10)(2)}}{2(10)}$$
$$= \frac{11 \pm \sqrt{121-80}}{20}$$
$$= \frac{11 \pm \sqrt{41}}{20}$$

The solutions are $\frac{11 \pm \sqrt{41}}{20}$.

38. $20x^2 - 11x + 1 = 0$

$a = 20$, $b = -11$, and $c = 1$

$$x = \frac{-(-11) \pm \sqrt{(-11)^2 - 4(20)(1)}}{2(20)} = \frac{11 \pm \sqrt{121 - 80}}{40} = \frac{11 \pm \sqrt{41}}{40}$$

The solutions are $\frac{11 \pm \sqrt{41}}{40}$.

39. $\frac{1}{2}z^2 - 2z + \frac{3}{4} = 0$

$2z^2 - 8z + 3 = 0$

$a = 2$, $b = -8$, and $c = 3$

$$z = \frac{-(-8) \pm \sqrt{(-8)^2 - 4(2)(3)}}{2(2)} = \frac{8 \pm \sqrt{64 - 24}}{4} = \frac{8 \pm \sqrt{40}}{4} = \frac{8 \pm 2\sqrt{10}}{4} = \frac{4 \pm \sqrt{10}}{2}$$

The solutions are $\frac{4 \pm \sqrt{10}}{2}$.

40. $\frac{1}{5}z^2 - \frac{1}{2}z - 2 = 0$

$2z^2 - 5z - 20 = 0$

$a = 2$, $b = -5$, and $c = -20$

$$z = \frac{-(-5) \pm \sqrt{(-5)^2 - 4(2)(-20)}}{2(2)} = \frac{5 \pm \sqrt{25 + 160}}{4} = \frac{5 \pm \sqrt{185}}{4}$$

The solutions are $\frac{5 \pm \sqrt{185}}{4}$.

41. Answers may vary

Section 9.4

Practice Exercises

1. a. $\sqrt{-36} = \sqrt{-1 \cdot 36} = \sqrt{-1} \cdot \sqrt{36} = i \cdot 6 = 6i$

b. $\sqrt{-15} = \sqrt{-1 \cdot 15} = \sqrt{-1} \cdot \sqrt{15} = i\sqrt{15}$

c. $\sqrt{-48} = \sqrt{-1 \cdot 48} = \sqrt{-1} \cdot \sqrt{48} = i \cdot 4\sqrt{3} = 4i\sqrt{3}$

2. a. $6 = 6 + 0i$

b. $0 = 0 + 0i$

c. $\sqrt{24} = 2\sqrt{6} = 2\sqrt{6} + 0i$

d. $\sqrt{-1} = i = 0 + i$

e. $5 + \sqrt{-9} = 5 + \sqrt{-1 \cdot 9} = 5 + \sqrt{-1} \cdot \sqrt{9} = 5 + i \cdot 3 = 5 + 3i$

3. a. $(4 + 3i) + (-8 - 2i) = [4 + (-8)] + (3i - 2i) = -4 + i$

b. $(5i) + (6 - 9i) = 6 + (5i - 9i) = 6 - 4i$

c. $(3 - 2i) - 4 = (3 - 4) - 2i = -1 - 2i$

4. $3i - (13 - 5i) = 3i - 13 + 5i = -13 + 8i$

5. a. $2i(3 - 4i) = 2i(3) + 2i(-4i) = 6i - 8i^2 = 6i - 8(-1) = 6i + 8 = 8 + 6i$

b. $(3 + i)(2 - 3i) = 6 - 9i + 2i - 3i^2 = 6 - 7i - 3(-1) = 6 - 7i + 3 = 9 - 7i$

c. $(5 - 2i)(5 + 2i) = 25 + 10i - 10i - 4i^2 = 25 - 4(-1) = 25 + 4 = 29 + 0i$

6. $\frac{3-i}{2+5i}=\frac{(3-i)}{(2+5i)}\cdot\frac{(2-5i)}{(2-5i)}$
$=\frac{6-15i-2i+5i^2}{4-25i^2}$
$=\frac{6-17i+5(-1)}{4-25(-1)}$
$=\frac{6-17i-5}{4+25}$
$=\frac{1-17i}{29}$
$=\frac{1}{29}-\frac{17}{29}i$

7. $(x-3)^2=-16$
$x-3=\pm\sqrt{-16}$
$x-3=\pm 4i$
$x=3\pm 4i$
The solutions are $3 \pm 4i$.

8. $y^2=3y-5$
$y^2-3y+5=0$
$a=1, b=-3, c=5$
$y=\frac{-(-3)\pm\sqrt{(-3)^2-4(1)(5)}}{2(1)}$
$=\frac{3\pm\sqrt{9-20}}{2}$
$=\frac{3\pm\sqrt{-11}}{2}$
$=\frac{3\pm i\sqrt{11}}{2}$
The solutions are $\frac{3\pm i\sqrt{11}}{2}$.

9. $x^2+x=-4$
$x^2+x+4=0$
$a=1, b=1, c=4$
$x=\frac{-1\pm\sqrt{1^2-4(1)(4)}}{2(1)}$
$=\frac{-1\pm\sqrt{1-16}}{2}$
$=\frac{-1\pm\sqrt{-15}}{2}$
$=\frac{-1\pm i\sqrt{15}}{2}$
The solutions are $\frac{-1\pm i\sqrt{15}}{2}$.

Vocabulary and Readiness Check

1. A number that can be written in the form $a + bi$ is called a complex number.
2. A complex number that can be written in the form $0 + bi$ is also called an imaginary number.
3. A complex number that can be written in the form $a + 0i$ is also called a real number.
4. The conjugate of $a + bi$ is $a - bi$.
5. The form $a + bi$ is called standard form.

Exercise Set 9.4

1. $\sqrt{-9}=\sqrt{-1\cdot 9}=\sqrt{-1}\sqrt{9}=i\cdot 3=3i$

3. $\sqrt{-100}=\sqrt{-1\cdot 100}=\sqrt{-1}\sqrt{100}=i\cdot 10=10i$

5. $\sqrt{-50}=\sqrt{-1\cdot 25\cdot 2}$
$=\sqrt{-1}\sqrt{25}\sqrt{2}$
$=i\cdot 5\sqrt{2}$
$=5i\sqrt{2}$

7. $\sqrt{-63}=\sqrt{-1\cdot 9\cdot 7}=\sqrt{-1}\sqrt{9}\sqrt{7}=i\cdot 3\sqrt{7}=3i\sqrt{7}$

9. $(2-i)+(-5+10i)=2-5+(-i+10i)=-3+9i$

11. $(-11+3i)-(1-3i)=-11+3i-1+3i$
$=-11-1+(3i+3i)$
$=-12+6i$

13. $(3-4i)-(2-i)=3-4i-2+i$
$=3-2+(-4i+i)$
$=1-3i$

15. $(16+2i)+(-7-6i)=16+2i-7-6i$
$=16-7+(2i-6i)$
$=9-4i$

17. $4i(3-2i)=12i-8i^2$
$=12i-8(-1)$
$=12i+8$
$=8+12i$

19. $(6-2i)(4+i)=24+6i-8i-2i^2$
$=24-2i-2(-1)$
$=24-2i+2$
$=26-2i$

21. $(3+8i)(3-8i) = 3^2-(8i)^2$
$= 9-64i^2$
$= 9-64(-1)$
$= 9+64$
$= 73$

23. Answers may vary

25. $\frac{8-12i}{4} = \frac{4(2-3i)}{4} = 2-3i$

27. $\frac{7-i}{4-3i} = \frac{(7-i)}{(4-3i)} \cdot \frac{(4+3i)}{(4+3i)}$
$= \frac{28+21i-4i-3i^2}{16-9i^2}$
$= \frac{28+17i-3(-1)}{16-9(-1)}$
$= \frac{28+17i+3}{16+9}$
$= \frac{31+17i}{25}$
$= \frac{31}{25}+\frac{17}{25}i$

29. $(x+1)^2 = -9$
$x+1 = \pm\sqrt{-9}$
$x+1 = \pm 3i$
$x = -1\pm 3i$
The solutions are $-1 \pm 3i$.

31. $(2z-3)^2 = -12$
$2z-3 = \pm\sqrt{-12}$
$2z-3 = \pm\sqrt{-1}\sqrt{4}\sqrt{3}$
$2z-3 = \pm 2i\sqrt{3}$
$2z = 3\pm 2i\sqrt{3}$
$z = \frac{3\pm 2i\sqrt{3}}{2}$
The solutions are $\frac{3\pm 2i\sqrt{3}}{2}$.

33. $y^2+6y+13=0$
$a=1, b=6, c=13$
$y = \frac{-6\pm\sqrt{6^2-4(1)(13)}}{2(1)}$
$= \frac{-6\pm\sqrt{36-52}}{2}$
$= \frac{-6\pm\sqrt{-16}}{2}$
$= \frac{-6\pm 4i}{2}$
$= \frac{2(-3\pm 2i)}{2}$
$= -3\pm 2i$
The solutions are $-3 \pm 2i$.

35. $4x^2+7x+4=0$
$a=4, b=7, c=4$
$x = \frac{-7\pm\sqrt{7^2-4(4)(4)}}{2(4)}$
$= \frac{-7\pm\sqrt{49-64}}{8}$
$= \frac{-7\pm\sqrt{-15}}{8}$
$= \frac{-7\pm i\sqrt{15}}{8}$
The solutions are $\frac{-7\pm i\sqrt{15}}{8}$.

37. $2m^2-4m+5=0$
$a=2, b=-4, c=5$
$m = \frac{-(-4)\pm\sqrt{(-4)^2-4(2)(5)}}{2(2)}$
$= \frac{4\pm\sqrt{16-40}}{4}$
$= \frac{4\pm\sqrt{-24}}{4}$
$= \frac{4\pm\sqrt{-1\cdot 4\cdot 6}}{4}$
$= \frac{4\pm 2i\sqrt{6}}{4}$
$= \frac{2\left(2\pm i\sqrt{6}\right)}{4}$
$= \frac{2\pm i\sqrt{6}}{2}$
The solutions are $\frac{2\pm i\sqrt{6}}{2}$.

39. $3 + (12 - 7i) = 3 + 12 - 7i = 15 - 7i$

41. $$\begin{aligned}-9i(5i-7) &= -45i^2 + 63i \\ &= -45(-1) + 63i \\ &= 45 + 63i\end{aligned}$$

43. $$\begin{aligned}(2-i)-(3-4i) &= 2-i-3+4i \\ &= 2-3+(-i+4i) \\ &= -1+3i\end{aligned}$$

45. $$\begin{aligned}\frac{15+10i}{5i} &= \frac{(15+10i)}{5i}\cdot\frac{(-i)}{(-i)} \\ &= \frac{-15i-10i^2}{-5i^2} \\ &= \frac{-15i-10(-1)}{-5(-1)} \\ &= \frac{-15i+10}{5} \\ &= \frac{5(-3i+2)}{5} \\ &= -3i+2 \\ &= 2-3i\end{aligned}$$

47. $$\begin{aligned}-5+i-(2+3i) &= -5+i-2-3i \\ &= -5-2+(i-3i) \\ &= -7-2i\end{aligned}$$

49. $$\begin{aligned}(4-3i)(4+3i) &= (4)^2-(3i)^2 \\ &= 16-9i^2 \\ &= 16-9(-1) \\ &= 16+9 \\ &= 25\end{aligned}$$

51. $$\begin{aligned}\frac{4-i}{1+2i} &= \frac{(4-i)(1-2i)}{(1+2i)(1-2i)} \\ &= \frac{4-8i-i+2i^2}{(1)^2-(2i)^2} \\ &= \frac{4-9i+2(-1)}{1-4i^2} \\ &= \frac{4-9i-2}{1-4(-1)} \\ &= \frac{2-9i}{5} \\ &= \frac{2}{5}-\frac{9}{5}i\end{aligned}$$

53. $$\begin{aligned}(5+2i)^2 &= 5^2+2(5)(2i)+(2i)^2 \\ &= 25+20i+4i^2 \\ &= 25+20i+4(-1) \\ &= 25+20i-4 \\ &= 21+20i\end{aligned}$$

55. $$\begin{aligned}(y-4)^2 &= -64 \\ y-4 &= \pm\sqrt{-64} \\ y-4 &= \pm 8i \\ y &= 4\pm 8i\end{aligned}$$
The solutions are $4 \pm 8i$.

57. $$\begin{aligned}4x^2 &= -100 \\ x^2 &= -25 \\ x &= \pm\sqrt{-25} \\ x &= \pm 5i\end{aligned}$$
The solutions are $\pm 5i$.

59. $z^2+6z+10=0$
$a = 1, b = 6, c = 10$
$$\begin{aligned}z &= \frac{-6\pm\sqrt{6^2-4(1)(10)}}{2(1)} \\ &= \frac{-6\pm\sqrt{36-40}}{2} \\ &= \frac{-6\pm\sqrt{-4}}{2} \\ &= \frac{-6\pm 2i}{2} \\ &= \frac{2(-3\pm i)}{2} \\ &= -3\pm i\end{aligned}$$
The solutions are $-3 \pm i$.

61. $2a^2-5a+9=0$
$a = 2, b = -5, c = 9$
$$\begin{aligned}a &= \frac{-(-5)\pm\sqrt{(-5)^2-4(2)(9)}}{2(2)} \\ &= \frac{5\pm\sqrt{25-72}}{4} \\ &= \frac{5\pm\sqrt{-47}}{4} \\ &= \frac{5\pm i\sqrt{47}}{4}\end{aligned}$$
The solutions are $\frac{5\pm i\sqrt{47}}{4}$.

63. $(2x+8)^2 = -20$
$$2x+8 = \pm\sqrt{-20}$$
$$2x+8 = \pm 2i\sqrt{5}$$
$$2x = -8 \pm 2i\sqrt{5}$$
$$x = \frac{-8 \pm 2i\sqrt{5}}{2}$$
$$x = \frac{2\left(-4 \pm i\sqrt{5}\right)}{2}$$
$$x = -4 \pm i\sqrt{5}$$
The solutions are $-4 \pm i\sqrt{5}$.

65. $3m^2 + 108 = 0$
$$3m^2 = -108$$
$$m^2 = -36$$
$$m = \pm\sqrt{-36}$$
$$m = \pm 6i$$
The solutions are $\pm 6i$.

67. $x^2 + 14x + 50 = 0$
$a = 1$, $b = 14$, $c = 50$
$$x = \frac{-14 \pm \sqrt{14^2 - 4(1)(50)}}{2(1)}$$
$$= \frac{-14 \pm \sqrt{196 - 200}}{2}$$
$$= \frac{-14 \pm \sqrt{-4}}{2}$$
$$= \frac{-14 \pm 2i}{2}$$
$$= \frac{2(-7 \pm i)}{2}$$
$$= -7 \pm i$$
The solutions are $-7 \pm i$.

69. $y = -3$
$y = -3$ for all values of x.

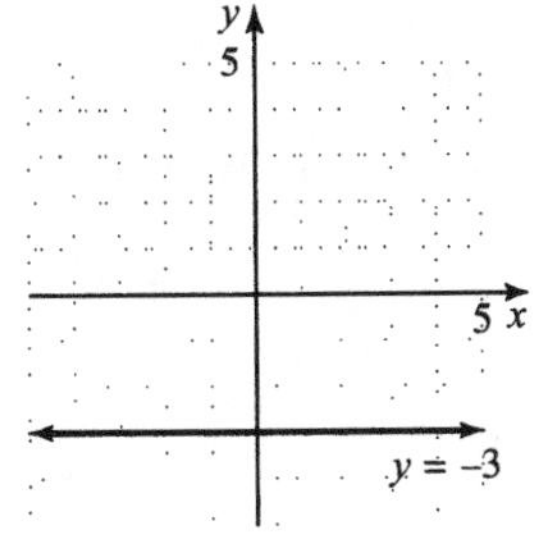

71. $y = 3x - 2$

x	y
0	–2
1	1

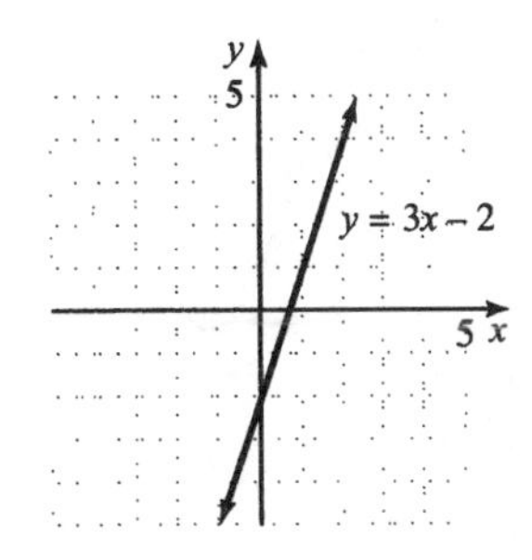

73. Approximately 51% of households had computers in 2000.

75. Let $x = 10$.
$y = 4.05x + 42$
$y = 4.05(10) + 42 = 40.5 + 42 = 82.5$
Expect 82.5% of the households to have computers in 2008.

77. False, i is not a real number.

79. False, no purely imaginary numbers are real numbers.

Section 9.5

Practice Exercises

1. $y = -\frac{1}{2}x^2$

x	y
0	0
1	$-\frac{1}{2}$
2	–2
4	–8
–1	$-\frac{1}{2}$
–2	–2
–4	–8

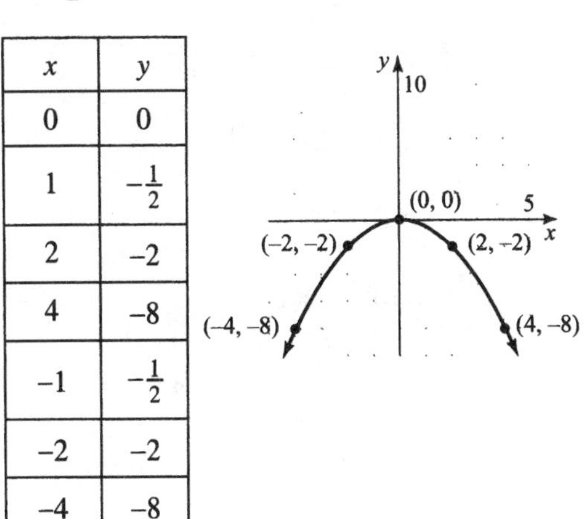

2. $y = x^2 + 1$

y-intercept: $x = 0,\ y = 0^2 + 1 = 1$
x-intercept: $y = 0$,
$0 = x^2 + 1$
$-1 = x^2$
No x-intercept

x	y
0	1
1	2
2	5
3	10
−1	2
−2	5
−3	10

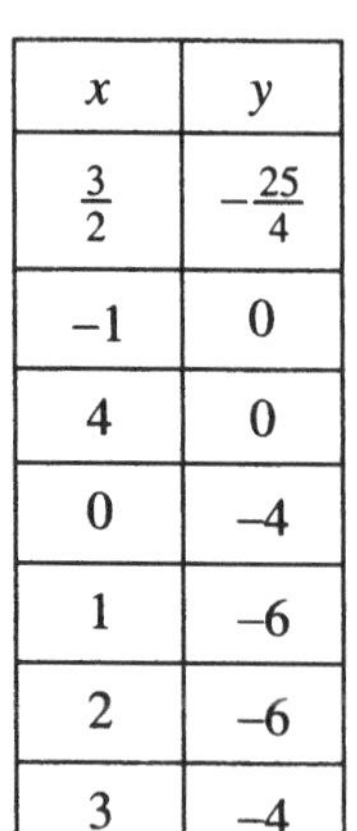

3. $y = x^2 - 3x - 4$
$a = 1, b = -3, c = -4$
Vertex:
$$x = \frac{-b}{2a} = \frac{-(-3)}{2(1)} = \frac{3}{2}$$
$$y = x^2 - 3x - 4$$
$$= \left(\frac{3}{2}\right)^2 - 3\left(\frac{3}{2}\right) - 4$$
$$= \frac{9}{4} - \frac{9}{2} - 4$$
$$= \frac{9}{4} - \frac{18}{4} - \frac{16}{4}$$
$$= -\frac{25}{4}$$
The vertex is $\left(\frac{3}{2}, -\frac{25}{4}\right)$.
x-intercepts: $y = 0$,
$0 = x^2 - 3x - 4$
$0 = (x + 1)(x - 4)$
$x + 1 = 0$ or $x - 4 = 0$
$x = -1$ $\quad x = 4$
y-intercept: $x = 0$,
$y = x^2 - 3x - 4 = 0^2 - 3(0) - 4 = -4$

x	y
$\frac{3}{2}$	$-\frac{25}{4}$
−1	0
4	0
0	−4
1	−6
2	−6
3	−4

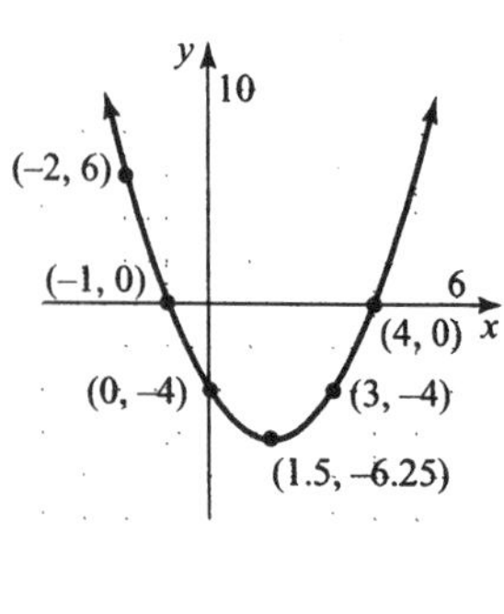

4. $y = x^2 + 4x - 7$
$a = 1, b = 4, c = -7$
vertex:
$$x = \frac{-b}{2a} = \frac{-4}{2(1)} = \frac{-4}{2} = -2$$
$$y = x^2 + 4x - 7$$
$$= (-2)^2 + 4(-2) - 7$$
$$= 4 - 8 - 7$$
$$= -11$$
The vertex is $(-2, -11)$.
x-intercepts: $y = 0$,
$0 = x^2 + 4x - 7$
$$x = \frac{-4 \pm \sqrt{4^2 - 4(1)(-7)}}{2(1)}$$
$$= \frac{-4 \pm \sqrt{16 + 28}}{2}$$
$$= \frac{-4 \pm \sqrt{44}}{2}$$
$$= \frac{-4 \pm 2\sqrt{11}}{2}$$
$$= -2 \pm \sqrt{11}$$
y-intercept: $x = 0$,
$y = x^2 + 4x - 7 = 0^2 + 4(0) - 7 = -7$

x	y
−2	−11
$-2+\sqrt{11}$	0
$-2-\sqrt{11}$	0
0	−7
1	−2
2	5
3	14

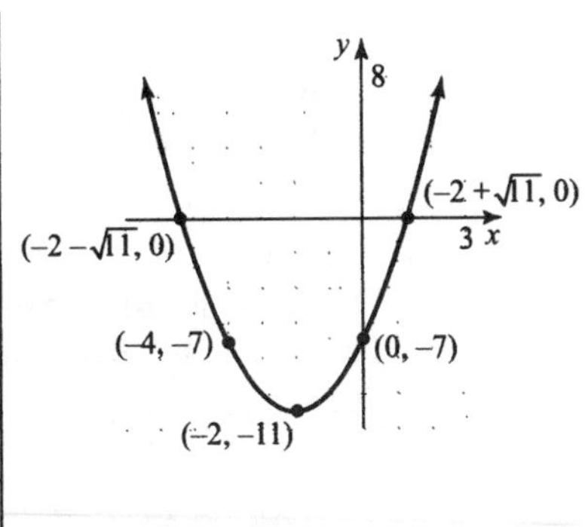

Calculator Explorations

1. $x^2 - 7x - 3 = 0$

$y_1 = x^2 - 7x - 3$

$y_2 = 0$

The x-coordinates of the intersections, −0.41 and 7.41, are the solutions.

2. $2x^2 - 11x - 1 = 0$

$y_1 = 2x^2 - 11x - 1$

$y_2 = 0$

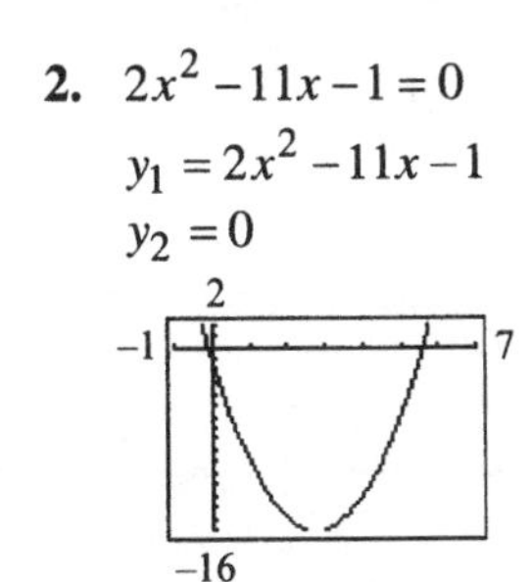

The x-coordinates of the intersections, −0.09 and 5.59 are the solutions.

3. $-1.7x^2 + 5.6x - 3.7 = 0$

$y_1 = -1.7x^2 + 5.6x - 3.7$

$y_2 = 0$

The x-coordinates of the intersections, 0.91 and 2.38, are the solutions.

4. $-5.8x^2 + 2.3x - 3.9 = 0$

$y_1 = -5.8x^2 + 2.3x - 3.9$

$y_2 = 0$

There are no x-intercepts so there are no real solutions.

5. $5.8x^2 - 2.6x - 1.9 = 0$

$y_1 = 5.8x^2 - 2.6x - 1.9$

$y_2 = 0$

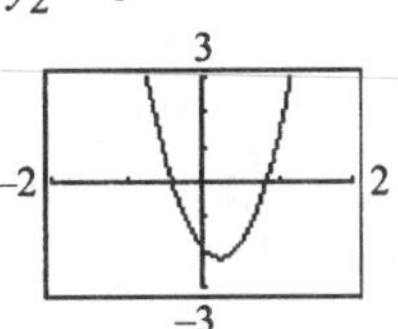

The x-coordinates of the intersections, −0.39 and 0.84, are the solutions.

6. $7.5x^2 - 3.7x - 1.1 = 0$

$y_1 = 7.5x^2 - 3.7x - 1.1$

$y_2 = 0$

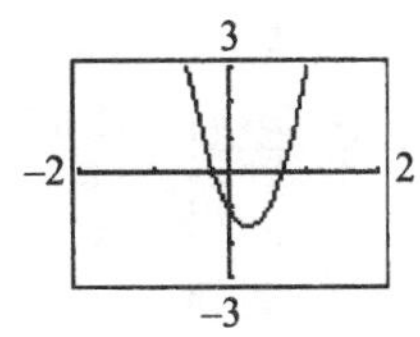

The x-coordinates of the intersections, −0.21 and 0.70 are the solutions.

Exercise Set 9.5

1. $y = 2x^2$

x	y
−2	8
−1	2
0	0
1	2
2	8

3. $y = -x^2$

x	y
−2	−4
−1	−1
0	0
1	−1
2	−4

5. $y = x^2 - 1$

y-intercept: $x = 0$, $y = 0^2 - 1 = -1$, $(0, -1)$

vertex: $(0, -1)$

x-intercepts: $y = 0$,

$0 = x^2 - 1 = (x+1)(x-1)$

$x + 1 = 0$ or $x - 1 = 0$

$x = -1$ $\quad$ $x = 1$

$(-1, 0)$ and $(1, 0)$

x	y
−2	3
−1	0
0	−1
1	0
2	3

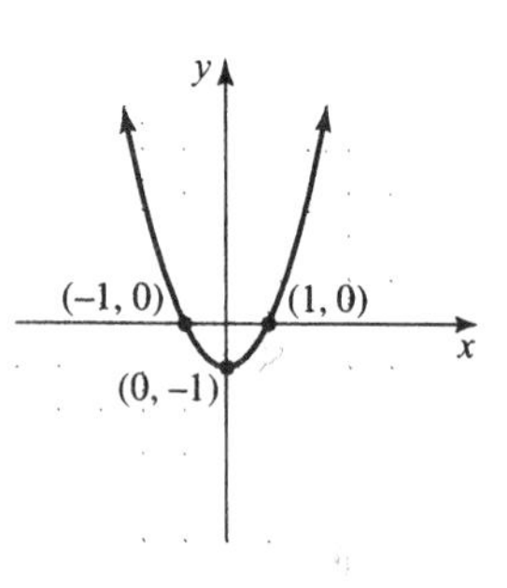

7. $y = x^2 + 4$

y-intercept: $x = 0$, $y = 0^2 + 4 = 4$, $(0, 4)$

vertex: $(0, 4)$

x-intercepts: $y = 0$,

$0 = x^2 + 4$

$-4 = x^2$

There are no x-intercepts because there is no real solution to this equation.

x	y
−2	8
−1	5
0	4
1	1
2	8

9. $y = -x^2 + 4x - 4$

$a = -1$, $b = 4$, $c = -4$

vertex: $x = \dfrac{-b}{2a} = \dfrac{-4}{2(-1)} = 2$

$y = -2^2 + 4(2) - 4 = -4 + 8 - 4 = 0$

$(2, 0)$

x-intercept: $y = 0$, $x = 2$, $(2, 0)$

y-intercept: $x = 0$, $y = -0^2 + 4(0) - 4 = -4$, $(0, -4)$

x	y
2	0
0	−4
1	−1
3	−1
4	−4

11. $y = x^2 + 5x + 4$

$a = 1, b = 5, c = 4$

vertex: $x = \frac{-b}{2a} = \frac{-5}{2(1)} = -\frac{5}{2}$

$y = \left(-\frac{5}{2}\right)^2 + 5\left(-\frac{5}{2}\right) + 4 = -\frac{9}{4}$

$\left(-\frac{5}{2}, -\frac{9}{4}\right)$

y-intercept: $x = 0,\ y = 0^2 + 5(0) + 4 = 4,\ (0, 4)$

x-intercepts: $y = 0$,

$0 = x^2 + 5x + 4$

$0 = (x + 4)(x + 1)$

$x = -4$ or $x = -1$

$(-4, 0)$ and $(-1, 0)$

x	y
-5	4
-4	0
$-\frac{5}{2}$	$-\frac{9}{4}$
-1	0
0	4

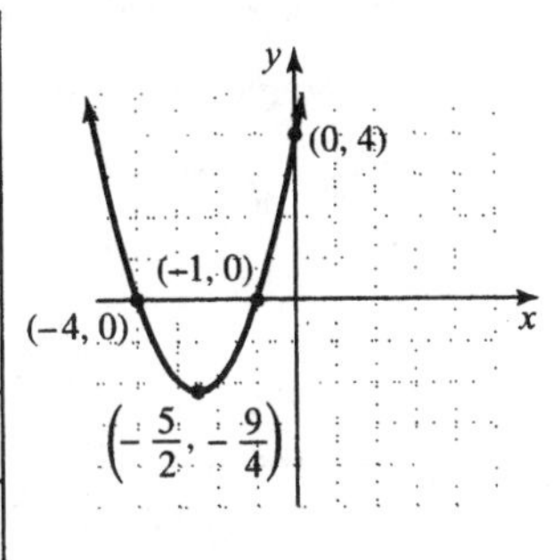

13. $y = x^2 - 4x + 5$

$a = 1, b = -4, c = 5$

vertex: $x = \frac{-b}{2a} = \frac{-(-4)}{2(1)} = \frac{4}{2} = 2$

$y = 2^2 - 4(2) + 5 = 4 - 8 + 5 = 1$

$(2, 1)$

x-intercept: $y = 0,\ 0 = x^2 - 4x + 5$

$$x = \frac{-(-4) \pm \sqrt{(-4)^2 - 4(1)(5)}}{2(1)}$$

$$= \frac{4 \pm \sqrt{16 - 20}}{2}$$

$$= \frac{4 \pm \sqrt{-4}}{2}$$

There are no x-intercepts.

y-intercept: $x = 0,\ y = 0^2 - 4(0) + 5 = 5,\ (0, 5)$

x	y
2	1
0	5
1	2
-1	10
3	2

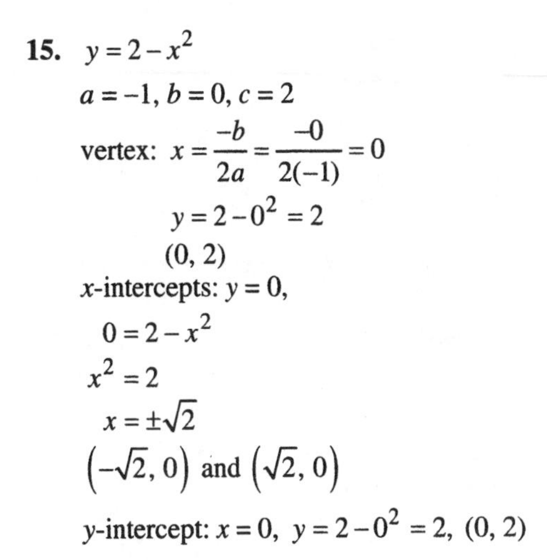

15. $y = 2 - x^2$

$a = -1, b = 0, c = 2$

vertex: $x = \frac{-b}{2a} = \frac{-0}{2(-1)} = 0$

$y = 2 - 0^2 = 2$

$(0, 2)$

x-intercepts: $y = 0$,

$0 = 2 - x^2$

$x^2 = 2$

$x = \pm\sqrt{2}$

$\left(-\sqrt{2}, 0\right)$ and $\left(\sqrt{2}, 0\right)$

y-intercept: $x = 0,\ y = 2 - 0^2 = 2,\ (0, 2)$

x	y
0	2
$\sqrt{2}$	0
$-\sqrt{2}$	0
0	2
1	1

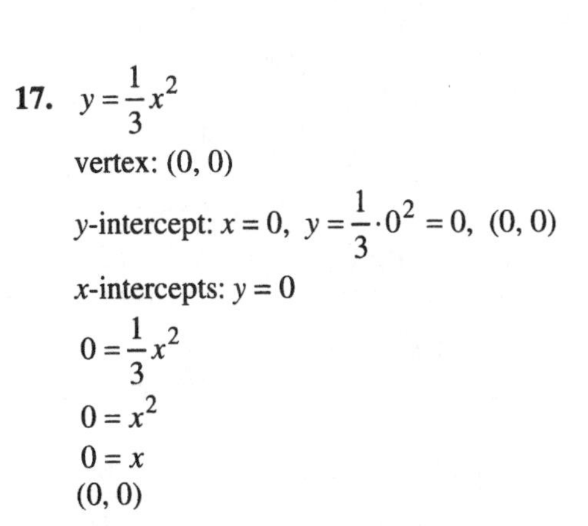

17. $y = \frac{1}{3}x^2$

vertex: $(0, 0)$

y-intercept: $x = 0,\ y = \frac{1}{3} \cdot 0^2 = 0,\ (0, 0)$

x-intercepts: $y = 0$

$0 = \frac{1}{3}x^2$

$0 = x^2$

$0 = x$

$(0, 0)$

x	y
−6	12
−3	3
0	0
3	3
6	12

19. $y = x^2 + 6x$

$a = 1, b = 6, c = 0$

vertex: $x = \frac{-b}{2a} = \frac{-6}{2(1)} = -3$

$y = (-3)^2 + 6(-3) = -9$

$(-3, -9)$

y-intercept: $x = 0,\ y = 0^2 + 6(0) = 0,\ (0, 0)$

x-intercepts: $y = 0$,

$0 = x^2 + 6x$

$0 = x(x + 6)$

$x = -6$ or $x = 0$

$(-6, 0)$ and $(0, 0)$

x	y
−7	7
−6	0
−3	−9
0	0
1	7

21. $y = x^2 + 2x - 8$

$a = 1, b = 2, c = -8$

vertex: $x = \frac{-b}{2a} = \frac{-2}{2(1)} = -1$

$y = (-1)^2 + 2(-1) - 8 = -9$

$(-1, -9)$

y-intercept: $x = 0,\ y = 0^2 + 2(0) - 8 = -8, (0, -8)$

x-intercepts: $y = 0$,

$0 = x^2 + 2x - 8$

$0 = (x + 4)(x - 2)$

$x = -4$ or $x = 2$

$(-4, 0)$ and $(2, 0)$

x	y
−4	0
−2	−8
−1	−9
0	−8
2	0

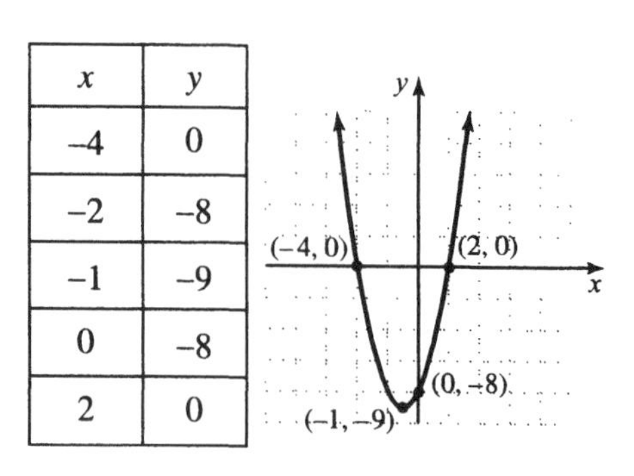

23. $y = -\frac{1}{2}x^2$

vertex: $(0, 0)$

y-intercept: $x = 0,\ y = -\frac{1}{2} \cdot 0^2 = 0,\ (0, 0)$

x-intercepts: $y = 0$

$0 = -\frac{1}{2}x^2$

$0 = x^2$

$0 = x$

$(0, 0)$

x	y
−4	−8
−2	−2
0	0
2	−2
4	−8

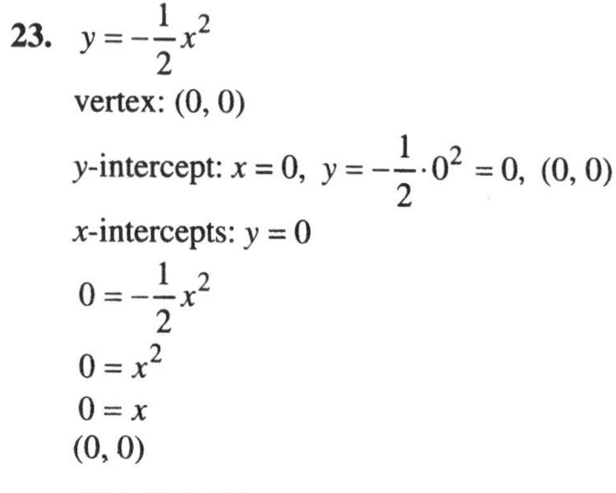

25. $y = 2x^2 - 11x + 5$

$a = 2, b = -11, c = 5$

vertex: $x = \frac{-b}{2a} = \frac{-(-11)}{2(2)} = \frac{11}{4}$

$$y = 2\left(\frac{11}{4}\right)^2 - 11\left(\frac{11}{4}\right) + 5 = -\frac{81}{8}$$

$\left(\frac{11}{4}, -\frac{81}{8}\right)$

y-intercept: $x = 0,\ y = 2 \cdot 0^2 - 11 \cdot 0 + 5 = 5,\ (0, 5)$

x-intercepts: $y = 0$

$0 = 2x^2 - 11x + 5$

$0 = (2x-1)(x-5)$

$x = \frac{1}{2}$ or $x = 5$

$\left(\frac{1}{2}, 0\right)$ and $(5, 0)$

x	y
0	5
$\frac{1}{2}$	0
1	−4
$\frac{11}{4}$	$-\frac{81}{8}$
5	0

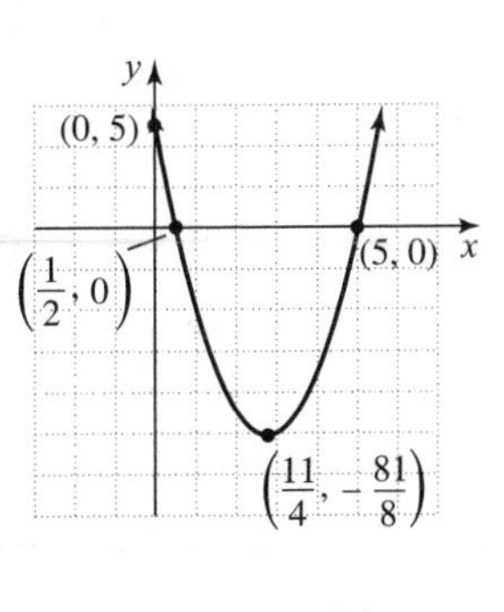

27. $y = -x^2 + 4x - 3$

$a = -1, b = 4, c = -3$

vertex: $x = \frac{-b}{2a} = \frac{-4}{2(-1)} = 2$

$y = -(2)^2 + 4(2) - 3 = 1$

$(2, 1)$

y-intercept: $x = 0,\ y = -0^2 + 4(0) - 3 = -3,$ $(0, -3)$

x-intercepts: $y = 0,$

$0 = -x^2 + 4x - 3$

$0 = x^2 - 4x + 3$

$0 = (x-1)(x-3)$

$x = 1$ or $x = 3$

$(1, 0)$ and $(3, 0)$

x	y
0	−3
1	0
2	1
3	0
4	−3

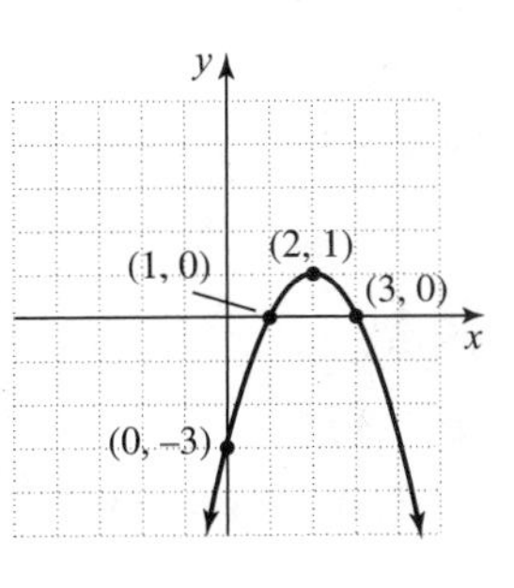

29. $y = x^2 + 2x - 2$

$a = 1, b = 2, c = -2$

vertex: $x = \frac{-b}{2a} = \frac{-2}{2(1)} = -1$

$y = (-1)^2 + 2(-1) - 2 = -3$

$(-1, -3)$

y-intercept: $x = 0,\ y = 0^2 + 2(0) - 2 = -2,\ (0, -2)$

x-intercepts: $y = 0,$

$0 = x^2 + 2x - 2$

$$x = \frac{-2 \pm \sqrt{2^2 - 4(1)(-2)}}{2(1)} = \frac{-2 \pm \sqrt{12}}{2} = \frac{-2 \pm 2\sqrt{3}}{2} = -1 \pm \sqrt{3}$$

$\left(-1-\sqrt{3}, 0\right)$ and $\left(-1+\sqrt{3}, 0\right)$

x	y
$-1-\sqrt{3}$	0
−2	−2
−1	−3
0	−2
$-1+\sqrt{3}$	0

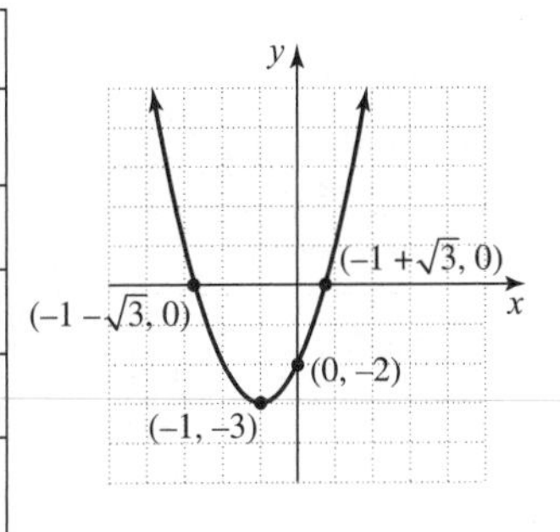

31. $y = x^2 - 3x + 1$

$a = 1, b = -3, c = 1$

vertex: $x = \frac{-b}{2a} = \frac{-(-3)}{2(1)} = \frac{3}{2} = 1\frac{1}{2}$

$y = \left(\frac{3}{2}\right)^2 - 3\left(\frac{3}{2}\right) + 1 = -\frac{5}{4} = -1\frac{1}{4}$

$\left(1\frac{1}{2}, -1\frac{1}{4}\right)$

y-intercept: $x = 0,\ y = 0^2 - 3(0) + 1 = 1,\ (0, 1)$

x-intercepts: $y = 0,$

$0 = x^2 - 3x + 1$

$$x = \frac{-(-3) \pm \sqrt{(-3)^2 - 4(1)(1)}}{2(1)} = \frac{3 \pm \sqrt{5}}{2}$$

$\left(\frac{3-\sqrt{5}}{2}, 0\right)$ and $\left(\frac{3+\sqrt{5}}{2}, 0\right)$

x	y
-1	5
0	1
$\frac{3-\sqrt{5}}{2}$	0
$1\frac{1}{2}$	$-1\frac{1}{4}$
$\frac{3+\sqrt{5}}{2}$	0

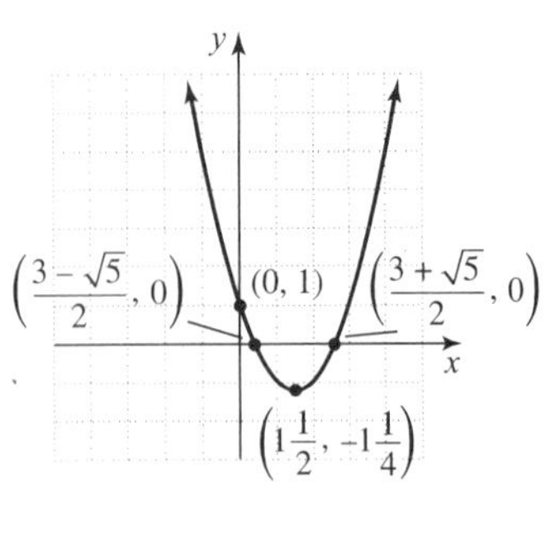

33. $\frac{\frac{1}{7}}{\frac{2}{5}} = \frac{1}{7} \div \frac{2}{5} = \frac{1}{7} \cdot \frac{5}{2} = \frac{5}{14}$

35. $\frac{\frac{1}{x}}{\frac{2}{x^2}} = \frac{1}{x} \div \frac{2}{x^2} = \frac{1}{x} \cdot \frac{x^2}{2} = \frac{x}{2}$

37. $\frac{2x}{1-\frac{1}{x}} = \frac{2x}{\frac{x-1}{x}} = 2x \div \frac{x-1}{x} = \frac{2x}{1} \cdot \frac{x}{x-1} = \frac{2x^2}{x-1}$

39. $\frac{\frac{a-b}{2b}}{\frac{b-a}{8b^2}} = \frac{a-b}{2b} \div \frac{b-a}{8b^2}$

$= \frac{a-b}{2b} \cdot \frac{8b^2}{b-a}$

$= \frac{a-b}{2b} \cdot \frac{8b^2}{-1(a-b)}$

$= -4b$

41. domain: $(-\infty, \infty)$; range: $(-\infty, 3]$

43. domain: $(-\infty, \infty)$; range: $(-\infty, 1]$

45. a. The maximum height is about 256 feet.

b. The fireball is at its maximum height after about 4 seconds.

c. The fireball returns to the ground after about 8 seconds.

47. C

49. A

Chapter 9 Vocabulary Check

1. If $x^2 = a$, then $x = \sqrt{a}$ or $x = -\sqrt{a}$. This property is called the square root property.

2. A number that can be written in the form $a + bi$ is called a complex number.

3. The formula $\frac{-b}{2a}$ where $y = ax^2 + bx + c$ is called the vertex formula.

4. A complex number that can be written in the form $0 + bi$ is also called an imaginary number.

5. The conjugate of $2 + 3i$ is $2 - 3i$.

6. $\sqrt{-1} = i$.

7. The process of solving a quadratic equation by writing it in the form $(x+a)^2 = c$ is called completing the square.

8. The formula $x = \frac{-b \pm \sqrt{b^2 - 4ac}}{2a}$ is called the quadratic formula.

Chapter 9 Review

1. $x^2 = 36$
$x = \pm\sqrt{36}$
$x = \pm 6$
The solutions are ± 6.

2. $x^2 = 81$
$x = \pm\sqrt{81}$
$x = \pm 9$
The solutions are ± 9.

3. $k^2 = 50$
$k = \pm\sqrt{50}$
$k = \pm 5\sqrt{2}$
The solutions are $\pm 5\sqrt{2}$.

4. $k^2 = 45$
$k = \pm\sqrt{45}$
$k = \pm 3\sqrt{5}$
The solutions are $\pm 3\sqrt{5}$.

5. $(x-11)^2 = 49$
$x-11 = \pm\sqrt{49}$
$x-11 = \pm 7$
$x = 11 \pm 7$
$x = 11 - 7 = 4$ or $x = 11 + 7 = 18$
The solutions are 4 and 18.

6. $(x+3)^2 = 100$
$x+3 = \pm\sqrt{100}$
$x+3 = \pm 10$
$x = -3 \pm 10$
$x = -3 - 10 = -13$ or $x = -3 + 10 = 7$
The solutions are −13 and 7.

7. $(4p+5)^2 = 41$
$4p+5 = \pm\sqrt{41}$
$4p = -5 \pm \sqrt{41}$
$p = \dfrac{-5 \pm \sqrt{41}}{4}$
The solutions are $\dfrac{-5 \pm \sqrt{41}}{4}$.

8. $(3p+7)^2 = 37$
$3p+7 = \pm\sqrt{37}$
$3p = -7 \pm \sqrt{37}$
$p = \dfrac{-7 \pm \sqrt{37}}{3}$
The solutions are $-\dfrac{7 \pm \sqrt{37}}{3}$.

9. Let $h = 100$.
$16t^2 = h$
$16t^2 = 100$
$t^2 = \dfrac{100}{16}$
$t = \pm\sqrt{\dfrac{100}{16}} = \pm\dfrac{10}{4} = \pm 2.5$
The length of time is not a negative number so the dive lasted 2.5 seconds.

10. Let $h = 5 \cdot 5280 = 26{,}400$
$16t^2 = h$
$16t^2 = 26{,}400$
$t^2 = \dfrac{26{,}400}{16} = 1650$
$t = \pm\sqrt{1650} \approx \pm 40.6$
The length of time is not a negative number so the fall lasted 40.6 seconds.

11. $a^2 + 4a \Rightarrow \left(\dfrac{4}{2}\right)^2 = 2^2 = 4$
$a^2 + 4a + 4 = (a+2)^2$

12. $a^2 - 12a \Rightarrow \left(\dfrac{-12}{2}\right)^2 = (-6)^2 = 36$
$a^2 - 12a + 36 = (a-6)^2$

13. $m^2 - 3m \Rightarrow \left(\dfrac{-3}{2}\right)^2 = \dfrac{9}{4}$
$m^2 - 3m + \dfrac{9}{4} = \left(m - \dfrac{3}{2}\right)^2$

14. $m^2 + 5m \Rightarrow \left(\dfrac{5}{2}\right)^2 = \dfrac{25}{4}$
$m^2 + 5m + \dfrac{25}{4} = \left(m + \dfrac{5}{2}\right)^2$

15. $x^2 - 9x = -8$
$x^2 - 9x + \left(-\dfrac{9}{2}\right)^2 = -8 + \left(-\dfrac{9}{2}\right)^2$
$x^2 - 9x + \dfrac{81}{4} = -\dfrac{32}{4} + \dfrac{81}{4}$
$\left(x - \dfrac{9}{2}\right)^2 = \dfrac{49}{4}$
$x - \dfrac{9}{2} = \sqrt{\dfrac{49}{4}}$ or $x - \dfrac{9}{2} = -\sqrt{\dfrac{49}{4}}$
$x = \dfrac{9}{2} + \dfrac{7}{2}$ $\quad$ $x = \dfrac{9}{2} - \dfrac{7}{2}$
$x = \dfrac{16}{2} = 8$ $\quad$ $x = \dfrac{2}{2} = 1$
The solutions are 8 and 1.

16.
$$x^2+8x=20$$
$$x^2+8x+\left(\frac{8}{2}\right)^2=20+\left(\frac{8}{2}\right)^2$$
$$x^2+8x+16=20+16$$
$$(x+4)^2=36$$
$$x+4=\sqrt{36} \quad \text{or} \quad x+4=-\sqrt{36}$$
$$x=6-4 \qquad x=-6-4$$
$$x=2 \qquad x=-10$$
The solutions are 2 and −10.

17.
$$x^2+4x=1$$
$$x^2+4x+\left(\frac{4}{2}\right)^2=1+\left(\frac{4}{2}\right)^2$$
$$x^2+4x+4=1+4$$
$$(x+2)^2=5$$
$$x+2=\sqrt{5} \quad \text{or} \quad x+2=-\sqrt{5}$$
$$x=-2+\sqrt{5} \qquad x=-2-\sqrt{5}$$
The solutions are $-2\pm\sqrt{5}$.

18.
$$x^2-8x=3$$
$$x^2-8x+\left(\frac{-8}{2}\right)^2=3+\left(\frac{-8}{2}\right)^2$$
$$x^2-8x+16=3+16$$
$$(x-4)^2=19$$
$$x-4=\sqrt{19} \quad \text{or} \quad x-4=-\sqrt{19}$$
$$x=4+\sqrt{19} \qquad x=4-\sqrt{19}$$
The solutions are $4\pm\sqrt{19}$.

19.
$$x^2-6x+7=0$$
$$x^2-6x=-7$$
$$x^2-6x+\left(\frac{-6}{2}\right)^2=-7+\left(\frac{-6}{2}\right)^2$$
$$x^2-6x+9=-7+9$$
$$(x-3)^2=2$$
$$x-3=\sqrt{2} \quad \text{or} \quad x-3=-\sqrt{2}$$
$$x=3+\sqrt{2} \qquad x=3-\sqrt{2}$$
The solutions are $3\pm\sqrt{2}$.

20.
$$x^2+6x+7=0$$
$$x^2+6x=-7$$
$$x^2+6x+\left(\frac{6}{2}\right)^2=-7+\left(\frac{6}{2}\right)^2$$
$$x^2+6x+9=-7+9$$
$$(x+3)^2=2$$
$$x+3=\sqrt{2} \quad \text{or} \quad x+3=-\sqrt{2}$$
$$x=-3+\sqrt{2} \qquad x=-3-\sqrt{2}$$
The solutions are $-3\pm\sqrt{2}$.

21.
$$2y^2+y-1=0$$
$$y^2+\frac{1}{2}y-\frac{1}{2}=0$$
$$y^2+\frac{1}{2}y=\frac{1}{2}$$
$$y^2+\frac{1}{2}y+\left(\frac{1}{4}\right)^2=\frac{1}{2}+\left(\frac{1}{4}\right)^2$$
$$y^2+\frac{1}{2}y+\frac{1}{16}=\frac{1}{2}+\frac{1}{16}$$
$$\left(y+\frac{1}{4}\right)^2=\frac{9}{16}$$
$$y+\frac{1}{4}=\sqrt{\frac{9}{16}} \quad \text{or} \quad y+\frac{1}{4}=-\sqrt{\frac{9}{16}}$$
$$y=-\frac{1}{4}+\frac{3}{4} \qquad y=-\frac{1}{4}-\frac{3}{4}$$
$$y=\frac{1}{2} \qquad y=-1$$
The solutions are $\frac{1}{2}$ and −1.

22.
$$y^2+3y-1=0$$
$$y^2+3y=1$$
$$y^2+3y+\left(\frac{3}{2}\right)^2=1+\left(\frac{3}{2}\right)^2$$
$$y^2+3y+\frac{9}{4}=1+\frac{9}{4}$$
$$\left(y+\frac{3}{2}\right)^2=\frac{13}{4}$$

$$y+\frac{3}{2}=\sqrt{\frac{13}{4}} \quad \text{or} \quad y+\frac{3}{2}=-\sqrt{\frac{13}{4}}$$

$$y+\frac{3}{2}=\frac{\sqrt{13}}{2} \qquad y+\frac{3}{2}=-\frac{\sqrt{13}}{2}$$

$$y=-\frac{3}{2}+\frac{\sqrt{13}}{2} \qquad y=-\frac{3}{2}-\frac{\sqrt{13}}{2}$$

The solutions are $\frac{-3\pm\sqrt{13}}{2}$.

23. $9x^2+30x+25=0$

$a = 9$, $b = 30$, and $c = 25$

$$x=\frac{-30\pm\sqrt{30^2-4(9)(25)}}{2(9)}$$
$$=\frac{-30\pm\sqrt{900-900}}{18}$$
$$=\frac{-30\pm\sqrt{0}}{18}$$
$$=-\frac{5}{3}$$

The solution is $-\frac{5}{3}$.

24. $16x^2-72x+81=0$

$a = 16$, $b = -72$, and $c = 81$

$$x=\frac{-(-72)\pm\sqrt{(-72)^2-4(16)(81)}}{2(16)}$$
$$=\frac{72\pm\sqrt{5184-5184}}{32}$$
$$=\frac{72\pm\sqrt{0}}{32}$$
$$=\frac{9}{4}$$

The solution is $\frac{9}{4}$.

25. $7x^2=35$

$$x^2=5$$
$$x^2-5=0$$

$a = 1$, $b = 0$, $c = -5$

$$x=\frac{-0\pm\sqrt{0^2-4(1)(-5)}}{2(1)}$$
$$=\pm\frac{\sqrt{20}}{2}$$
$$=\pm\frac{2\sqrt{5}}{2}$$
$$=\pm\sqrt{5}$$

The solutions are $\pm\sqrt{5}$.

26. $11x^2=33$

$$x^2=3$$
$$x^2-3=0$$

$a = 1$, $b = 0$, $c = -3$

$$x=\frac{-0\pm\sqrt{0^2-4(1)(-3)}}{2(1)}$$
$$=\pm\frac{\sqrt{12}}{2}$$
$$=\pm\frac{2\sqrt{3}}{2}$$
$$=\pm\sqrt{3}$$

The solutions are $\pm\sqrt{3}$.

27. $x^2-10x+7=0$

$a = 1$, $b = -10$, and $c = 7$

$$x=\frac{-(-10)\pm\sqrt{(-10)^2-4(1)(7)}}{2(1)}$$
$$=\frac{10\pm\sqrt{100-28}}{2}$$
$$=\frac{10\pm\sqrt{72}}{2}$$
$$=\frac{10\pm6\sqrt{2}}{2}$$
$$=5\pm3\sqrt{2}$$

The solutions are $5\pm3\sqrt{2}$.

28. $x^2+4x-7=0$
$a=1$, $b=4$, and $c=-7$
$$x=\frac{-4\pm\sqrt{4^2-4(1)(-7)}}{2(1)}=\frac{-4\pm\sqrt{16+28}}{2}=\frac{-4\pm\sqrt{44}}{2}=\frac{-4\pm2\sqrt{11}}{2}=-2\pm\sqrt{11}$$
The solutions are $-2\pm\sqrt{11}$.

29. $3x^2+x-1=0$
$a=3$, $b=1$, $c=-1$
$$x=\frac{-1\pm\sqrt{1^2-4(3)(-1)}}{2(3)}=\frac{-1\pm\sqrt{1+12}}{6}=\frac{-1\pm\sqrt{13}}{6}$$
The solutions are $\frac{-1\pm\sqrt{13}}{6}$.

30. $x^2+3x-1=0$
$a=1$, $b=3$, and $c=-1$
$$x=\frac{-3\pm\sqrt{3^2-4(1)(-1)}}{2(1)}=\frac{-3\pm\sqrt{9+4}}{2}=\frac{-3\pm\sqrt{13}}{2}$$
The solutions are $\frac{-3\pm\sqrt{13}}{2}$.

31. $2x^2+x+5=0$
$a=2$, $b=1$, and $c=5$
$$x=\frac{-1\pm\sqrt{1^2-4(2)(5)}}{2(2)}=\frac{-1\pm\sqrt{1-40}}{4}=\frac{-1\pm\sqrt{-39}}{4}$$
There is no real solution because $\sqrt{-39}$ is not a real number.

32. $7x^2-3x+1=0$
$a=7$, $b=-3$ and $c=1$
$$x=\frac{-(-3)\pm\sqrt{(-3)^2-4(7)(1)}}{2(7)}=\frac{3\pm\sqrt{9-28}}{14}=\frac{3\pm\sqrt{-19}}{14}$$
There is no real solution because $\sqrt{-19}$ is not a real number.

33. From Exercise 29, the exact solutions are $\frac{-1\pm\sqrt{13}}{6}$.
$$\frac{-1+\sqrt{13}}{6}\approx 0.4$$
$$\frac{-1-\sqrt{13}}{6}\approx -0.8$$

34. From Exercise 30, the exact solutions are $\frac{-3\pm\sqrt{13}}{2}$.
$$\frac{-3+\sqrt{13}}{2}\approx 0.3$$
$$\frac{-3-\sqrt{13}}{2}\approx -3.3$$

35. $x^2-7x-1=0$
$a=1$, $b=-7$, $c=-1$
$b^2-4ac=(-7)^2-4(1)(-1)=49+4=53$
Since the discriminant is a positive number, this equation has two distinct real solutions.

36. $x^2+x+5=0$
$a=1$, $b=1$, $c=5$
$b^2-4ac=1^2-4(1)(5)=1-20=-19$
Since the discriminant is a negative number, this equation has no real solution.

37. $9x^2+1=6x$
$9x^2-6x+1=0$
$a=9$, $b=-6$, $c=1$
$b^2-4ac=(-6)^2-4(9)(1)=36-36=0$
Since the discriminant is 0, this equation has one real solution.

38. $x^2 + 6x = 5$
$x^2 + 6x - 5 = 0$
$a = 1, b = 6, c = -5$
$b^2 - 4ac = 6^2 - 4(1)(-5) = 36 + 20 = 56$
Since the discriminant is a positive number, this equation has two distinct real solutions.

39. $5x^2 + 4 = 0$
$a = 5, b = 0, c = 4$
$b^2 - 4ac = 0^2 - 4(5)(4) = 0 - 80 = -80$
Since the discriminant is a negative number, this equation has no real solution.

40. $x^2 + 25 = 10x$
$x^2 - 10x + 25 = 0$
$a = 1, b = -10, c = 25$
$b^2 - 4ac = (-10)^2 - 4(1)(25) = 100 - 100 = 0$
Since the discriminant is 0, this equation has one real solution.

41. $y = -x^2 + 51x + 218$
Let $y = 658$.
$658 = -x^2 + 51x + 218$
$0 = -x^2 + 51x - 440$
$0 = x^2 - 51x + 440$
$0 = (x - 11)(x - 40)$
$x - 11 = 0$ or $x - 40 = 0$
$x = 11$ $\quad x = 40$
Since we're looking for the first year that occurs, $x = 11$, and the year is 2000 + 11 = 2011.

42. $y = 2.1x^2 + 31x + 125$
Let $y = 645$.
$645 = 2.1x^2 + 31x + 125$
$0 = 2.1x^2 + 31x - 520$
$0 = 21x^2 + 310x - 5200$
$0 = (x - 10)(21x + 520)$
$x = 10$ or $x = -\frac{520}{21}$
Since we're looking for a future time, $x = 10$, and the year is 2000 + 10 = 2010.

43. $\sqrt{-144} = \sqrt{-1 \cdot 144} = \sqrt{-1} \cdot \sqrt{144} = i \cdot 12 = 12i$

44. $\sqrt{-36} = \sqrt{36 \cdot i} = \sqrt{36} \cdot \sqrt{-1} = 6i$

45. $\sqrt{-108} = \sqrt{-1 \cdot 36 \cdot 3}$
$= \sqrt{-1} \cdot \sqrt{36} \cdot \sqrt{3}$
$= i \cdot 6\sqrt{3}$
$= 6i\sqrt{3}$

46. $\sqrt{-500} = \sqrt{100 \cdot 5 \cdot -1}$
$= \sqrt{100} \cdot \sqrt{-1} \cdot \sqrt{5}$
$= 10i\sqrt{5}$

47. $2i(3 - 5i) = (2i)(3) - (2i)(5i)$
$= 6i - 10i^2$
$= 6i - 10(-1)$
$= 10 + 6i$

48. $i(-7 - i) = (i)(-7) - (i)(i)$
$= -7i - i^2$
$= -7i - (-1)$
$= 1 - 7i$

49. $(7 - i) + (14 - 9i) = 7 - i + 14 - 9i$
$= 7 + 14 + (-i - 9i)$
$= 21 - 10i$

50. $(10 - 4i) + (9 - 21i) = 10 - 4i + 9 - 21i$
$= 10 + 9 + (-4i - 21i)$
$= 19 - 25i$

51. $3 - (11 + 2i) = 3 - 11 - 2i = -8 - 2i$

52. $(-4 - 3i) + 5i = -4 - 3i + 5i = -4 + 2i$

53. $(2 - 3i)(3 - 2i) = 6 - 4i - 9i + 6i^2$
$= 6 - 13i + 6(-1)$
$= 6 - 13i - 6$
$= -13i$

54. $(2 + 5i)(5 - i) = 10 - 2i + 25i - 5i^2$
$= 10 - 2i + 25i - 5(-1)$
$= 10 + 23i + 5$
$= 15 + 23i$

55. $(3 - 4i)(3 + 4i) = (3)^2 - (4i)^2$
$= 9 - 16i^2$
$= 9 - 16(-1)$
$= 9 + 16$
$= 25$

56. $(7-2i)(7-2i) = 49-14i-14i+4i^2$
$= 49-28i+4(-1)$
$= 49-28i-4$
$= 49-4-28i$
$= 45-28i$

57. $\frac{2-6i}{4i} = \frac{(2-6i)(-i)}{4i(-i)}$
$= \frac{-2i+6i^2}{-4i^2}$
$= \frac{-2i+6(-1)}{-4(-1)}$
$= \frac{-2i-6}{4}$
$= \frac{2(-i-3)}{4}$
$= -\frac{i}{2}-\frac{3}{2}$
$= -\frac{3}{2}-\frac{1}{2}i$

58. $\frac{5-i}{2i} = \frac{(5-i)(-i)}{2i(-i)}$
$= \frac{-5i+i^2}{-2i^2}$
$= \frac{-5i+(-1)}{-2(-1)}$
$= \frac{-1-5i}{2}$
$= -\frac{1}{2}-\frac{5}{2}i$

59. $\frac{4-i}{1+2i} = \frac{(4-i)(1-2i)}{(1+2i)(1-2i)}$
$= \frac{4(1)+4(-2i)-i(1)-i(-2i)}{(1)^2-(2i)^2}$
$= \frac{4-8i-i+2i^2}{1-4i^2}$
$= \frac{4-9i+2(-1)}{1-4(-1)}$
$= \frac{4-9i-2}{1+4}$
$= \frac{2-9i}{5}$
$= \frac{2}{5}-\frac{9}{5}i$

60. $\frac{1+3i}{2-7i} = \frac{1+3i}{2-7i}\cdot\frac{2+7i}{2+7i}$
$= \frac{1(2)+1(7i)+3i(2)+3i(7i)}{2^2-(7i)^2}$
$= \frac{2+7i+6i+21i^2}{4-49i^2}$
$= \frac{2+13i+21(-1)}{4-49(-1)}$
$= \frac{2+13i-21}{4+49}$
$= \frac{-19+13i}{53}$
$= -\frac{19}{53}+\frac{13}{53}i$

61. $3x^2 = -48$
$x^2 = -16$
$x = \pm\sqrt{-16}$
$x = \pm 4i$
The solutions are $\pm 4i$.

62. $5x^2 = -125$
$x^2 = -25$
$x = \pm\sqrt{-25}$
$x = \pm 5i$
The solutions are $\pm 5i$.

63. $x^2-4x+13=0$
$a = 1, b = -4, c = 13$
$x = \frac{-(-4)\pm\sqrt{(-4)^2-4(1)(13)}}{2(1)}$
$= \frac{4\pm\sqrt{16-52}}{2}$
$= \frac{4\pm\sqrt{-36}}{2}$
$= \frac{4\pm 6i}{2}$
$= \frac{2(2\pm 3i)}{2}$
$= 2\pm 3i$
The solutions are $2 \pm 3i$.

64. $x^2 + 4x + 11 = 0$

$a = 1, b = 4, c = 11$

$$x = \frac{-4 \pm \sqrt{4^2 - 4(1)(11)}}{2(1)}$$
$$= \frac{-4 \pm \sqrt{16 - 44}}{2}$$
$$= \frac{-4 \pm \sqrt{-28}}{2}$$
$$= \frac{-4 \pm \sqrt{4 \cdot 7 \cdot -1}}{2}$$
$$= \frac{-4 \pm 2i\sqrt{7}}{2}$$
$$= \frac{2\left(-2 \pm i\sqrt{7}\right)}{2}$$
$$= -2 \pm i\sqrt{7}$$

The solutions are $-2 \pm i\sqrt{7}$.

65. 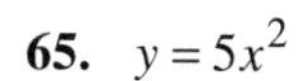 $y = 5x^2$

The vertex, x-intercept, and y-intercept are the same, (0, 0).

x	y
−2	20
−1	5
0	0
1	5
2	20

66. $y = -\frac{1}{2}x^2$

The vertex, x-intercept, and y-intercept are the same, (0, 0).

x	y
−2	−2
−1	$-\frac{1}{2}$
0	0
1	$-\frac{1}{2}$
2	−2

67. $y = x^2 - 25$

$a = 1, b = 0, c = -25$

vertex: $x = \frac{-b}{2a} = \frac{-0}{2(1)} = 0,\ \ y = 0^2 - 25 = -25,$

(0, −25)

y-intercept: $x = 0,\ y = 0^2 - 25 = -25,$ (0, −25)

x-intercept: $y = 0,$

$0 = x^2 - 25$

$25 = x^2$

$\pm 5 = x$

(−5, 0), (5, 0)

x	y
0	−25
−5	0
5	0
−1	−24
1	−24

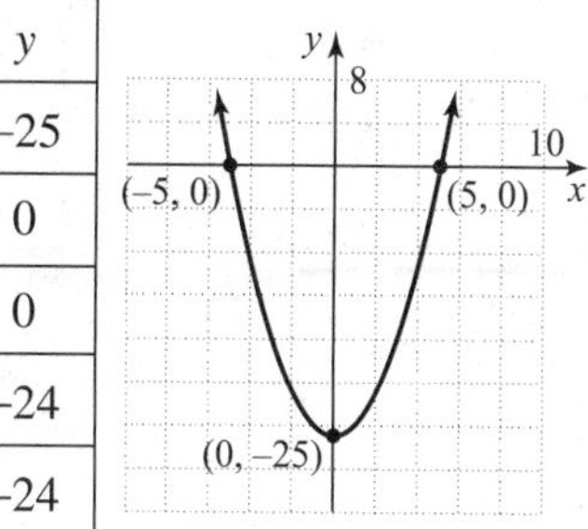

68. $y = x^2 - 36$

$a = 1, b = 0, c = -36$

vertex: $x = \frac{-b}{2a} = \frac{-0}{2(1)} = 0,\ \ y = 0^2 - 36 = -36,$

(0, −36)

y-intercept: (0, −36)

x-intercept: $y = 0,$

$0 = x^2 - 36$

$36 = x^2$

$\pm 6 = x$

(−6, 0), (6, 0)

x	y
0	−36
−6	0
6	0
−3	−27
3	−27

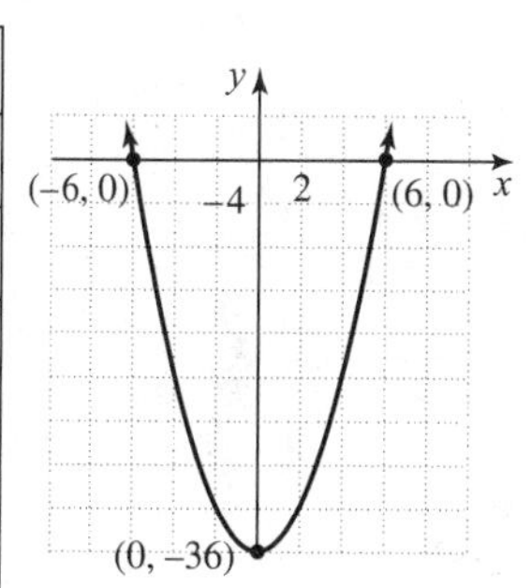

69. $y = x^2 + 3$

$a = 1, b = 0, c = 3$

vertex: $x = \frac{-b}{2a} = \frac{-0}{2(1)} = 0, \quad y = 0^2 + 3 = 0, \ (0, 3)$

y-intercept: (0, 3)

x-intercept: $y = 0$,

$0 = x^2 + 3$

$-3 = x^2$

There are no x-intercepts.

x	y
−3	12
−1	4
0	3
1	4
3	12

(0, 3)

70. $y = x^2 + 8$

$a = 1, b = 0, c = 8$

vertex: $x = \frac{-b}{2a} = \frac{-0}{2(1)} = 0, \ y = 0^2 + 8 = 8, \ (0, 8)$

y-intercept: (0, 8)

x-intercept: $y = 0$,

$0 = x^2 + 8$

$-8 = x^2$

There are no x-intercepts.

x	y
−2	12
−1	9
0	8
1	9
2	12

(0, 8)

71. $y = -4x^2 + 8$

$a = -4, b = 0, c = 8$

vertex: $x = \frac{-b}{2a} = \frac{-0}{2(-4)} = 0, \quad y = -4(0)^2 + 8 = 8,$

(0, 8)

y-intercept: (0, 8)

x-intercept: $y = 0$,

$0 = -4x^2 + 8$

$4x^2 = 8$

$x^2 = 2$

$x = \pm\sqrt{2}$

$\left(-\sqrt{2}, 0\right), \left(\sqrt{2}, 0\right)$

x	y
−2	−8
−1	4
0	8
1	4
2	−8

(0, 8)
$(-\sqrt{2}, 0)$
$(\sqrt{2}, 0)$

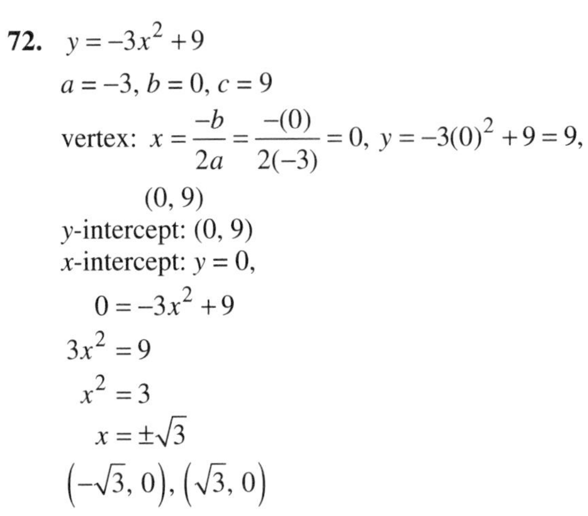

72. $y = -3x^2 + 9$

$a = -3, b = 0, c = 9$

vertex: $x = \frac{-b}{2a} = \frac{-(0)}{2(-3)} = 0, \ y = -3(0)^2 + 9 = 9,$

(0, 9)

y-intercept: (0, 9)

x-intercept: $y = 0$,

$0 = -3x^2 + 9$

$3x^2 = 9$

$x^2 = 3$

$x = \pm\sqrt{3}$

$\left(-\sqrt{3}, 0\right), \left(\sqrt{3}, 0\right)$

x	y
$-\sqrt{3}$	0
−1	6
0	9
1	6
$\sqrt{3}$	0

(0, 9)
$(-\sqrt{3}, 0)$
$(\sqrt{3}, 0)$

73. $y = x^2 + 3x - 10$

$a = 1, b = 3, c = -10$

vertex: $x = \frac{-b}{2a} = \frac{-3}{2(1)} = -\frac{3}{2},$

$y = \left(-\frac{3}{2}\right)^2 + 3\left(-\frac{3}{2}\right) - 10 = -\frac{49}{4},$

$\left(-\frac{3}{2}, -\frac{49}{4}\right)$

y-intercept: $x = 0,\ y = 0^2 + 3(0) - 10 = -10,$

$(0, -10)$

x-intercept: $y = 0,$

$0 = x^2 + 3x - 10$

$0 = (x+5)(x-2)$

$x + 5 = 0$ or $x - 2 = 0$

$x = -5$ $\quad x = 2$

$(-5, 0), (2, 0)$

x	y
-5	0
2	0
$-\frac{3}{2}$	$-\frac{49}{4}$
0	-10
1	-6

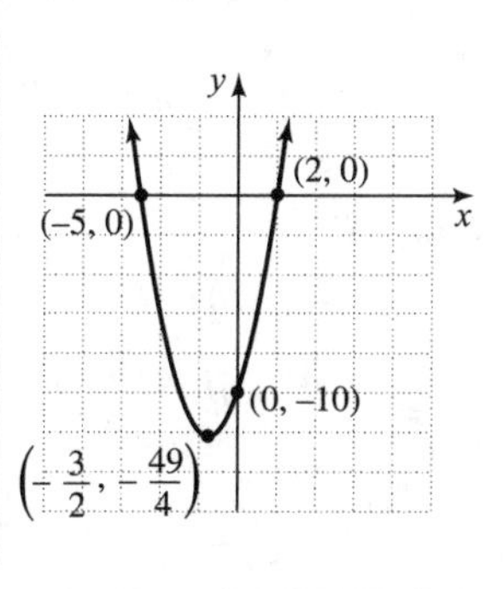

74. $y = x^2 + 3x - 4$

$a = 1, b = 3, c = -4$

vertex: $x = \frac{-b}{2a} = \frac{-3}{2(1)} = -\frac{3}{2},$

$y = \left(-\frac{3}{2}\right)^2 + 3\left(-\frac{3}{2}\right) - 4 = -\frac{25}{4},$

$\left(-\frac{3}{2}, -\frac{25}{4}\right)$

y-intercept: $x = 0,\ y = 0^2 + 3(0) - 4 = -4,\ (0, -4)$

x-intercept: $y = 0,$

$0 = x^2 + 3x - 4$

$0 = (x+4)(x-1)$

$x + 4 = 0$ or $x - 1 = 0$

$x = -4$ $\quad x = 1$

$(-4, 0), (1, 0)$

x	y
$-\frac{3}{2}$	$-\frac{25}{4}$
-4	0
1	0
0	-4
-1	-6

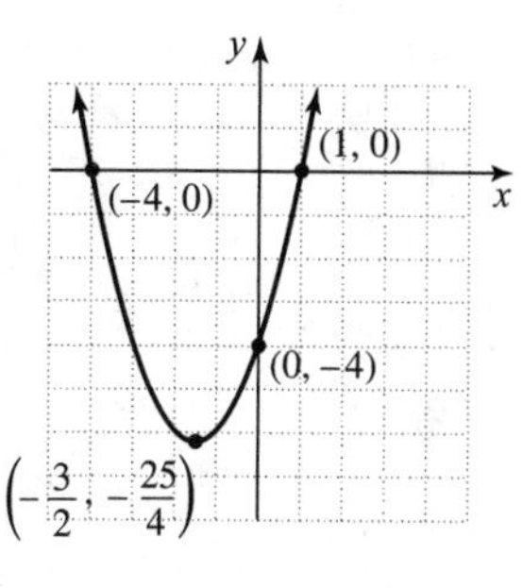

75. $y = -x^2 - 5x - 6$

$a = -1, b = -5, c = -6$

vertex: $x = \frac{-b}{2a} = \frac{-(-5)}{2(-1)} = -\frac{5}{2},$

$y = -\left(-\frac{5}{2}\right)^2 - 5\left(-\frac{5}{2}\right) - 6 = \frac{1}{4},$

$\left(-\frac{5}{2}, \frac{1}{4}\right)$

y-intercept: $x = 0,\ y = -0^2 - 5(0) - 6 = -6,$

$(0, -6)$

x-intercept: $y = 0,$

$0 = -x^2 - 5x - 6$

$0 = x^2 + 5x + 6$

$0 = (x+3)(x+2)$

$x + 3 = 0$ or $x + 2 = 0$

$x = -3$ $\quad x = -2$

$(-3, 0), (-2, 0)$

x	y
$-\frac{5}{2}$	$\frac{1}{4}$
0	-6
-2	0
-3	0
-1	-2

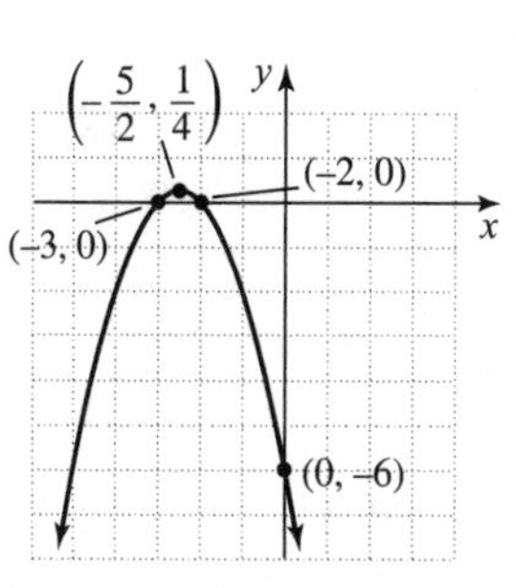

76. $y = 3x^2 - x - 2$

$a = 3, b = -1, c = -2$

vertex: $x = \frac{-b}{2a} = \frac{-(-1)}{2(3)} = \frac{1}{6}$

$y = 3\left(\frac{1}{6}\right)^2 - \left(\frac{1}{6}\right) - 2 = -\frac{25}{12}$

$\left(\frac{1}{6}, -\frac{25}{12}\right)$

y-intercept: $x = 0,\ y = 3 \cdot 0^2 - 0 - 2 = -2\ (0, -2)$

x-intercepts: $y = 0,$

$0 = 3x^2 - x - 2$

$0 = (3x + 2)(x - 1)$

$x = -\frac{2}{3}$ or $x = 1$

$\left(-\frac{2}{3}, 0\right), (1, 0)$

x	y
$-\frac{2}{3}$	0
0	-2
$\frac{1}{6}$	$-\frac{25}{12}$
1	0

77. $y = 2x^2 - 11x - 6$

$a = 2, b = -11, c = -6$

vertex: $x = \frac{-b}{2a} = \frac{-(-11)}{2(2)} = \frac{11}{4}$

$y = 2\left(\frac{11}{4}\right)^2 - 11\left(\frac{11}{4}\right) - 6 = -\frac{169}{8}$

$\left(\frac{11}{4}, -\frac{169}{8}\right)$

y-intercept: $x = 0,\ y = 2 \cdot 0^2 - 11 \cdot 0 - 6 = -6,$

$(0, -6)$

x-intercepts: $y = 0,$

$0 = 2x^2 - 11x - 6$

$0 = (2x + 1)(x - 6)$

$x = -\frac{1}{2}$ or $x = 6$

$\left(-\frac{1}{2}, 0\right), (6, 0)$

x	y
$-\frac{1}{2}$	0
0	-6
$\frac{11}{4}$	$-\frac{169}{8}$
$\frac{11}{2}$	-6
6	0

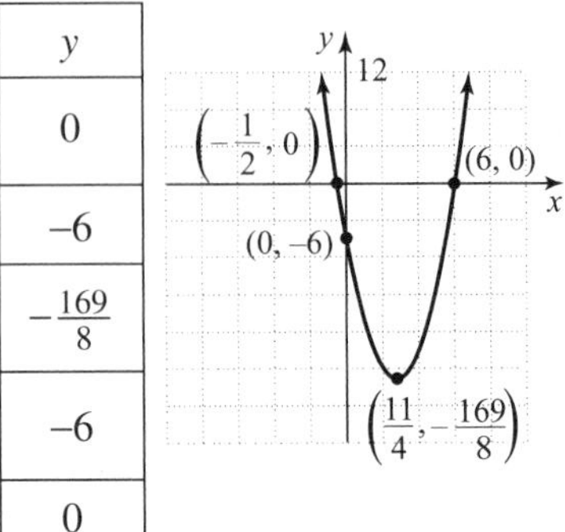

78. $y = -x^2 + 4x + 8$

$a = -1, b = 4, c = 8$

vertex: $x = \frac{-b}{2a} = \frac{-4}{2(-1)} = 2,$

$y = -2^2 + 4(2) + 8 = 12,\ (2, 12)$

y-intercept: $x = 0,\ y = -0^2 + 4(0) + 8 = 8,\ (0, 8)$

x-intercept: $y = 0,$

$0 = -x^2 + 4x + 8$

$0 = x^2 - 4x - 8$

$x = \frac{-(-4) \pm \sqrt{(-4)^2 - 4(1)(-8)}}{2(1)}$

$= \frac{4 \pm \sqrt{16 + 32}}{2}$

$= \frac{4 \pm \sqrt{48}}{2}$

$= \frac{4 \pm 4\sqrt{3}}{2}$

$= 2 \pm 2\sqrt{3}$

$\left(2 + 2\sqrt{3}, 0\right), \left(2 - 2\sqrt{3}, 0\right)$

x	y
2	12
0	8
$2 + 2\sqrt{3}$	0
$2 - 2\sqrt{3}$	0
1	11

79. A

80. D

81. B

82. C

83. The equation has one solution because the graph intersects the x-axis at one point.

84. The equation has two solutions because the graph intersects the x-axis at two points.

85. The equation has no real solution because the graph does not intersect the x-axis.

86. The equation has two solutions because the graph intersects the x-axis at two points.

87. $x^2 = 49$

$x = \pm\sqrt{49}$

$x = \pm 7$

The solutions are ± 7.

88. $y^2 = 75$

$y = \pm\sqrt{75}$

$y = \pm 5\sqrt{3}$

The solutions are $\pm 5\sqrt{3}$.

89. $(x-7)^2 = 64$

$x - 7 = \pm\sqrt{64}$

$x = 7 \pm 8$

$x = 7 + 8 = 15$ or $x = 7 - 8 = -1$

The solutions are 15 and -1.

90.

$$x^2 + 4x = 6$$

$$x^2 + 4x + \left(\frac{4}{2}\right)^2 = 6 + \left(\frac{4}{2}\right)^2$$

$$x^2 + 4x + 4 = 6 + 4$$

$$(x+2)^2 = 10$$

$x + 2 = \sqrt{10}$ or $x + 2 = -\sqrt{10}$

$x = -2 + \sqrt{10}$ $\quad$ $x = -2 - \sqrt{10}$

The solutions are $-2 \pm \sqrt{10}$.

91.

$$3x^2 + x = 2$$

$$x^2 + \frac{1}{3}x = \frac{2}{3}$$

$$x^2 + \frac{1}{3}x + \left(\frac{1}{6}\right)^2 = \frac{2}{3} + \left(\frac{1}{6}\right)^2$$

$$\left(x + \frac{1}{6}\right)^2 = \frac{25}{36}$$

$x + \frac{1}{6} = \sqrt{\frac{25}{36}}$ or $x + \frac{1}{6} = -\sqrt{\frac{25}{36}}$

$x = -\frac{1}{6} + \frac{5}{6}$ $\quad$ $x = -\frac{1}{6} - \frac{5}{6}$

$x = \frac{2}{3}$ $\quad$ $x = -1$

The solutions are -1 and $\frac{2}{3}$.

92.

$$4x^2 - x - 2 = 0$$

$$4x^2 - x = 2$$

$$x^2 - \frac{1}{4}x = \frac{1}{2}$$

$$x^2 - \frac{1}{4}x + \left(\frac{-1}{8}\right)^2 = \frac{1}{2} + \left(\frac{-1}{8}\right)^2$$

$$\left(x - \frac{1}{8}\right)^2 = \frac{33}{64}$$

$x - \frac{1}{8} = \sqrt{\frac{33}{64}}$ or $x - \frac{1}{8} = -\sqrt{\frac{33}{64}}$

$x - \frac{1}{8} = \frac{\sqrt{33}}{8}$ $\quad$ $x - \frac{1}{8} = -\frac{\sqrt{33}}{8}$

$x = \frac{1}{8} + \frac{\sqrt{33}}{8}$ $\quad$ $x = \frac{1}{8} - \frac{\sqrt{33}}{8}$

The solutions are $\frac{1 \pm \sqrt{33}}{8}$.

93. $4x^2 - 3x - 2 = 0$

$a = 4, b = -3, c = -2$

$$x = \frac{-(-3) \pm \sqrt{(-3)^2 - 4(4)(-2)}}{2(4)}$$

$$= \frac{3 \pm \sqrt{9 + 32}}{8}$$

$$= \frac{3 \pm \sqrt{41}}{8}$$

The solutions are $\frac{3 \pm \sqrt{41}}{8}$.

94. $5x^2 + x - 2 = 0$
$a = 5, b = 1, c = -2$

$$x = \frac{-1 \pm \sqrt{1^2 - 4(5)(-2)}}{2(5)} = \frac{-1 \pm \sqrt{1+40}}{10} = \frac{-1 \pm \sqrt{41}}{10}$$

The solutions are $\frac{-1 \pm \sqrt{41}}{10}$.

95. $4x^2 + 12x + 9 = 0$
$a = 4, b = 12, c = 9$

$$x = \frac{-12 \pm \sqrt{12^2 - 4(4)(9)}}{2(4)} = \frac{-12 \pm \sqrt{144-144}}{8} = -\frac{3}{2}$$

The solution is $-\frac{3}{2}$.

96. $2x^2 + x + 4 = 0$
$a = 2, b = 1, c = 4$

$$x = \frac{-1 \pm \sqrt{1^2 - 4(2)(4)}}{2(2)} = \frac{-1 \pm \sqrt{1-32}}{4} = \frac{-1 \pm \sqrt{-31}}{4}$$

Since $\sqrt{-31}$ is not a real number, there is no real solution.

97. $y = 4 - x^2$
$a = -1, b = 0, c = 4$

vertex: $x = \frac{-b}{2a} = \frac{-0}{2(-1)} = 0, \quad y = 4 - 0^2 = 4,$
(0, 4)

y-intercept: $x = 0, \; y = 4 - 0^2 = 4, \; (0, 4)$

x-intercept: $y = 0,$

$0 = 4 - x^2$

$x^2 = 4$

$x = \pm 2$

(−2, 0), (2, 0)

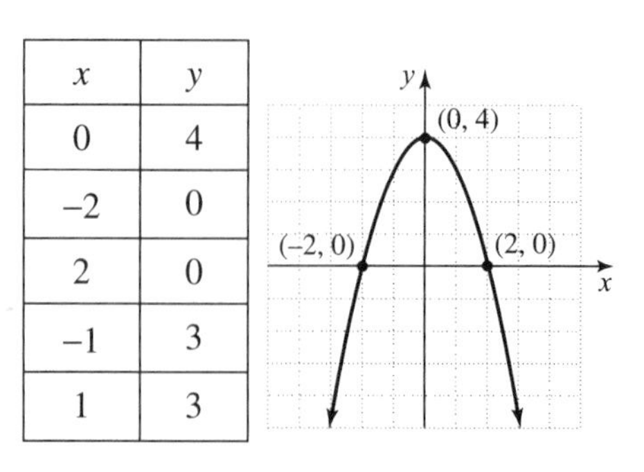

x	y
0	4
−2	0
2	0
−1	3
1	3

98. $y = x^2 + 4$
$a = 1, b = 0, c = 4$

vertex: $x = \frac{-b}{2a} = \frac{-0}{2(1)} = 0, \quad y = 0^2 + 4 = 4, \; (0, 4)$

y-intercept: $x = 0, \; y = 0^2 + 4 = 4, \; (0, 4)$

x-intercept: $y = 0,$

$0 = x^2 + 4$

$-4 = x^2$

There are no x-intercepts.

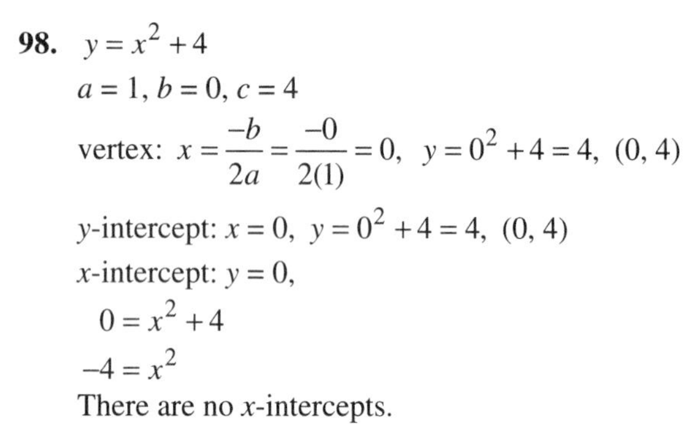

x	y
0	4
−1	5
1	5
−2	8
2	8

99. $y = x^2 + 6x + 8$
$a = 1, b = 6, c = 8$

vertex: $x = \frac{-b}{2a} = \frac{-6}{2(1)} = -3,$

$y = (-3)^2 + 6(-3) + 8 = -1, \; (-3, -1)$

y-intercept: $x = 0, \; y = 0^2 + 6(0) + 8 = 8, \; (0, 8)$

x-intercepts: $y = 0,$

$0 = x^2 + 6x + 8$

$0 = (x+2)(x+4)$

$x + 2 = 0$ or $x + 4 = 0$

$x = -2 \qquad x = -4$

(−2, 0), (−4, 0)

x	y
−3	−1
0	8
−2	0
−4	0
−1	3

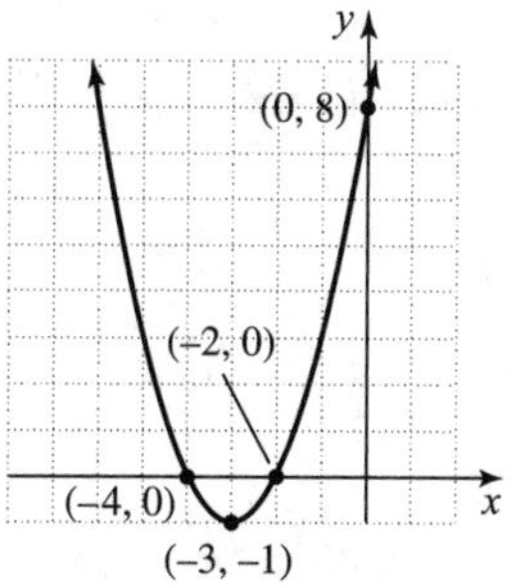

100. $y = x^2 - 2x - 4$

$a = 1, b = -2, c = -4$

vertex: $x = \frac{-b}{2a} = \frac{-(-2)}{2(1)} = 1,$

$y = 1^2 - 2(1) - 4 = -5,\ (1, -5)$

y-intercept: $x = 0,\ y = 0^2 - 2(0) - 4 = -4,\ (0, -4)$

x-intercepts $y = 0$,

$0 = x^2 - 2x - 4$

$$x = \frac{-(-2) \pm \sqrt{(-2)^2 - 4(1)(-4)}}{2(1)}$$
$$= \frac{2 \pm \sqrt{4+16}}{2}$$
$$= \frac{2 \pm \sqrt{20}}{2}$$
$$= \frac{2 \pm 2\sqrt{5}}{2}$$
$$= 1 \pm \sqrt{5}$$

$\left(1-\sqrt{5}, 0\right), \left(1+\sqrt{5}, 0\right)$

x	y
1	−5
0	−4
$1-\sqrt{5}$	0
$1+\sqrt{5}$	0
−1	−1

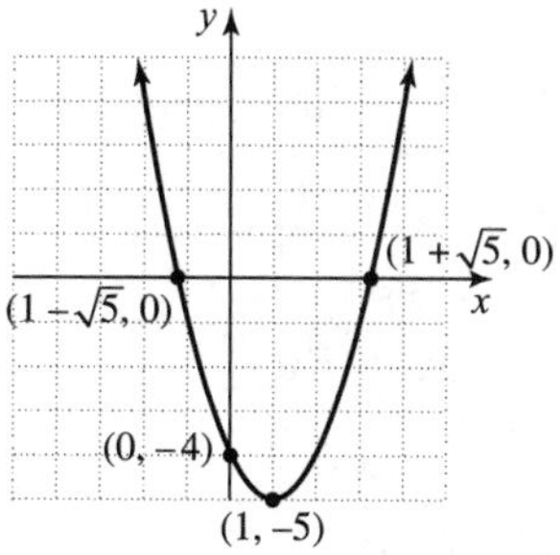

Chapter 9 Test

1. $5k^2 = 80$

$k^2 = 16$

$k = \pm\sqrt{16}$

$k = \pm 4$

The solutions are ±4.

2. $(3m-5)^2 = 8$

$3 - 5 = \pm\sqrt{8}$

$3m - 5 = \pm 2\sqrt{2}$

$3m = 5 \pm 2\sqrt{2}$

$m = \frac{5 \pm 2\sqrt{2}}{3}$

The solutions are $\frac{5 \pm 2\sqrt{2}}{3}$.

3. $x^2 - 26x + 160 = 0$

$x^2 - 26x = -160$

$x^2 - 26x + 169 = -160 + 169$

$(x-13)^2 = 9$

$x - 13 = \pm\sqrt{9}$

$x - 13 = \pm 3$

$x = 13 - 3$ or $x = 13 + 3$

$x = 10$ $\quad$ $x = 16$

The solutions are 10 and 16.

4. $3x^2 + 12x - 4 = 0$

$x^2 + 4x - \frac{4}{3} = 0$

$x^2 + 4x = \frac{4}{3}$

$x^2 + 4x + 4 = \frac{4}{3} + 4$

$(x+2)^2 = \frac{16}{3}$

$x + 2 = \pm\sqrt{\frac{16}{3}}$

$x = -2 \pm \frac{4}{\sqrt{3}}$

$x = -2 \pm \frac{4\sqrt{3}}{3}$

$x = \frac{-6 \pm 4\sqrt{3}}{3}$

The solutions are $\frac{-6 \pm 4\sqrt{3}}{3}$.

5. $x^2 - 3x - 10 = 0$

$a = 1$, $b = -3$, and $c = -10$

$$x = \frac{-(-3) \pm \sqrt{(-3)^2 - 4(1)(-10)}}{2(1)} = \frac{3 \pm \sqrt{9+40}}{2} = \frac{3 \pm \sqrt{49}}{2} = \frac{3 \pm 7}{2}$$

$$x = \frac{3-7}{2} = -2 \quad \text{or} \quad x = \frac{3+7}{2} = 5$$

The solutions are −2 and 5.

6. $p^2 - \frac{5}{3}p - \frac{1}{3} = 0$

$3p^2 - 5p - 1 = 0$

$a = 3$, $b = -5$, and $c = -1$

$$p = \frac{-(-5) \pm \sqrt{(-5)^2 - 4(3)(-1)}}{2(3)} = \frac{5 \pm \sqrt{25+12}}{6} = \frac{5 \pm \sqrt{37}}{6}$$

The solutions are $\frac{5 \pm \sqrt{37}}{6}$.

7. $(3x - 5)(x + 2) = -6$

$3x^2 + x - 10 = -6$

$3x^2 + x - 4 = 0$

$(3x + 4)(x - 1) = 0$

$3x + 4 = 0$ or $x - 1 = 0$

$x = -\frac{4}{3}$ $\quad$ $x = 1$

The solutions are $-\frac{4}{3}$ and 1.

8. $(3x - 1)^2 = 16$

$3x - 1 = \pm\sqrt{16}$

$3x - 1 = \pm 4$

$3x = 1 \pm 4$

$x = \frac{1 \pm 4}{3}$

$$x = \frac{1-4}{3} = -1 \quad \text{or} \quad x = \frac{1+4}{3} = \frac{5}{3}$$

The solutions are −1 and $\frac{5}{3}$.

9. $3x^2 - 7x - 2 = 0$

$a = 3$, $b = -7$, and $c = -2$

$$x = \frac{-(-7) \pm \sqrt{(-7)^2 - 4(3)(-2)}}{2(3)} = \frac{7 \pm \sqrt{49+24}}{6} = \frac{7 \pm \sqrt{73}}{6}$$

The solutions are $\frac{7 \pm \sqrt{73}}{6}$.

10. $x^2 - 4x + 5 = 0$

$a = 1$, $b = -4$, and $c = 5$

$$x = \frac{-(-4) \pm \sqrt{(-4)^2 - 4(1)(5)}}{2(1)} = \frac{4 \pm \sqrt{16-20}}{2} = \frac{4 \pm \sqrt{-4}}{2} = \frac{4 \pm 2i}{2} = 2 \pm i$$

The solutions are $2 \pm i$.

11. $3x^2 - 7x + 2 = 0$

$(3x - 1)(x - 2) = 0$

$3x - 1 = 0$ or $x - 2 = 0$

$x = \frac{1}{3}$ $\quad$ $x = 2$

The solutions are $\frac{1}{3}$ and 2.

12. $2x^2 - 6x + 1 = 0$

$a = 1$, $b = -6$, and $c = 1$

$$x = \frac{-(-6) \pm \sqrt{(-6)^2 - 4(2)(1)}}{2(2)}$$
$$= \frac{6 \pm \sqrt{36 - 8}}{4}$$
$$= \frac{6 \pm \sqrt{28}}{4}$$
$$= \frac{6 \pm 2\sqrt{7}}{4}$$
$$= \frac{3 \pm \sqrt{7}}{2}$$

The solutions are $\frac{3 \pm \sqrt{7}}{2}$.

13.
$$9x^3 = x$$
$$9x^3 - x = 0$$
$$x(9x^2 - 1) = 0$$
$$x(3x + 1)(3x - 1) = 0$$

$x = 0$ or $3x + 1 = 0$ or $3x - 1 = 0$

$x = 0$ $\quad x = -\frac{1}{3}$ $\quad x = \frac{1}{3}$

The solutions are 0 and $\pm\frac{1}{3}$.

14. $\sqrt{-25} = i\sqrt{25} = 5i$

15. $\sqrt{-200} = i\sqrt{100 \cdot 2} = i\sqrt{100} \cdot \sqrt{2} = 10i\sqrt{2}$

16. $(3 + 2i) + (5 - i) = 3 + 2i + 5 - i = 8 + i$

17. $(3 + 2i) - (3 - 2i) = 3 + 2i - 3 + 2i = 4i$

18.
$$(3 + 2i)(3 - 2i) = (3)^2 - (2i)^2$$
$$= 9 - 4i^2$$
$$= 9 - 4(-1)$$
$$= 9 + 4$$
$$= 13$$

19.
$$\frac{3 - i}{1 + 2i} = \frac{3 - i}{1 + 2i} \cdot \frac{1 - 2i}{1 - 2i}$$
$$= \frac{3 - 6i - i + 2i^2}{1^2 - (2i)^2}$$
$$= \frac{3 - 7i + 2(-1)}{1 - 4i^2}$$
$$= \frac{3 - 7i - 2}{1 - 4(-1)}$$
$$= \frac{1 - 7i}{1 + 4}$$
$$= \frac{1}{5} - \frac{7}{5}i$$

20. $y = -5x^2$

$a = -5$, $b = 0$, $c = 0$

vertex: $x = \frac{-b}{2a} = \frac{-0}{2(-5)} = 0$

$y = -5(0)^2 = 0$, $(0, 0)$

y-intercept: $x = 0$, $y = 0$, $(0, 0)$

x-intercept: $y = 0$, $x = 0$, $(0, 0)$

x	y
−2	−20
−1	−5
0	0
1	−5
2	−20

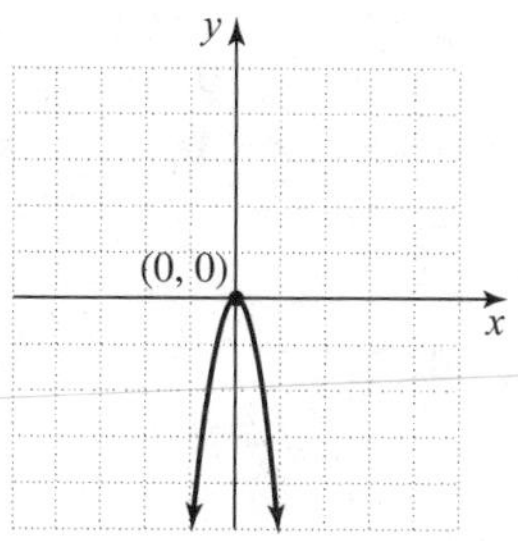

y

(0, 0)

x

21. $y = x^2 - 4$

$a = 1$, $b = 0$, $c = -4$

vertex: $x = \frac{-b}{2a} = \frac{-0}{2(1)} = 0$, $y = 0^2 - 4 = -4$,

$(0, -4)$

y-intercept: $x = 0$, $y = 0^2 - 4 = -4$, $(0, -4)$

x-intercept: $y = 0$,

$$0 = x^2 - 4$$
$$4 = x^2$$
$$\pm 2 = x$$

$(2, 0)$, $(-2, 0)$

x	y
−2	0
−1	−3
0	−4
1	−3
2	0

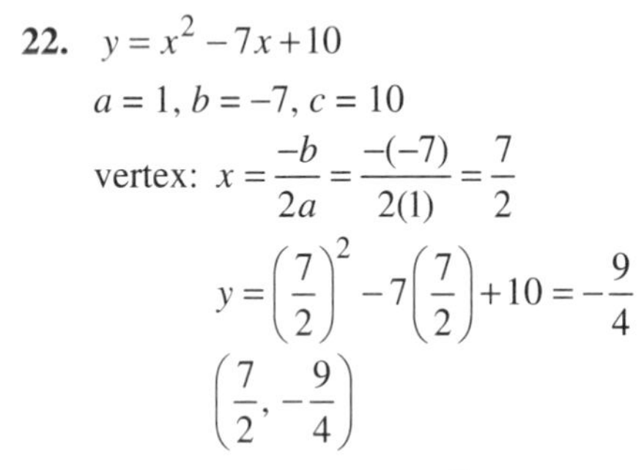

22. $y = x^2 - 7x + 10$

$a = 1, b = -7, c = 10$

vertex: $x = \frac{-b}{2a} = \frac{-(-7)}{2(1)} = \frac{7}{2}$

$y = \left(\frac{7}{2}\right)^2 - 7\left(\frac{7}{2}\right) + 10 = -\frac{9}{4}$

$\left(\frac{7}{2}, -\frac{9}{4}\right)$

y-intercept: $x = 0,\ y = 0^2 - 7 \cdot 0 + 10 = 10,\ (0, 10)$

x-intercepts: $y = 0$,

$0 = x^2 - 7x + 10$

$0 = (x-2)(x-5)$

$x = 2$ or $x = 5$

(2, 0), (5, 0)

x	y
0	10
2	0
$\frac{7}{2}$	$-\frac{9}{4}$
5	0
7	10

23. $y = 2x^2 + 4x - 1$

$a = 2, b = 4, c = -1$

vertex: $x = \frac{-b}{2a} = \frac{-4}{2(2)} = -1,$

$y = 2(-1)^2 + 4(-1) - 1 = -3,\ (-1, -3)$

x-intercept: $y = 0$,

$0 = 2x^2 + 4x - 1$

$$x = \frac{-4 \pm \sqrt{4^2 - 4(2)(-1)}}{2(2)} = \frac{-4 \pm \sqrt{16+8}}{4} = \frac{-4 \pm \sqrt{24}}{4} = \frac{-4 \pm 2\sqrt{6}}{4} = \frac{-2 \pm \sqrt{6}}{2}$$

$\left(\frac{-2-\sqrt{6}}{2}, 0\right), \left(\frac{-2+\sqrt{6}}{2}, 0\right)$

y-intercept: $x = 0, y = -1, (0, -1)$

x	y
−2	−1
−1	−3
0	−1
1	5
2	15

24. Let x = the length of the base;
then $4x$ = the height.

$A = \frac{1}{2}bh$

$18 = \frac{1}{2}x(4x)$

$18 = 2x^2$

$9 = x^2$

$\pm 3 = x$

The length cannot be negative, so the base is 3 feet and the height is 4(3) = 12 feet.

25. $d = \frac{n^2 - 3n}{2};\ d = 9$

$9 = \frac{n^2 - 3n}{2}$

$18 = n^2 - 3n$

$0 = n^2 - 3n - 18$

$0 = (n-6)(n+3)$

$n - 6 = 0$ or $n + 3 = 0$

$n = 6$ $\qquad n = -3$

The number of sides cannot be negative, so $n = 6$ sides.

26. Let $h = 120.75$.

$16t^2 = h$

$16t^2 = 120.75$

$t^2 = \frac{120.75}{16}$

$t = \pm\sqrt{\frac{120.75}{16}} \approx \pm 2.7$

The length of time is not a negative number so the dive lasted 2.7 seconds.

Cumulative Review Chapters 1–9

1. a. $\frac{x-y}{12+x} = \frac{2-(-5)}{12+2} = \frac{7}{14} = \frac{1}{2}$

b. $x^2 - 3y = (2)^2 - 3(-5) = 4 + 15 = 19$

2. a. $\frac{x-y}{7-x} = \frac{(-4)-7}{7-(-4)} = \frac{-11}{11} = -1$

b. $x^2 + 2y = (-4)^2 + 2(7) = 16 + 14 = 30$

3. a. $2x + 3x + 5 + 2 = 5x + 7$

b. $-5a - 3 + a + 2 = -4a - 1$

c. $4y - 3y^2 = 4y - 3y^2$

d. $2.3x + 5x - 6 = 7.3x - 6$

e. $-\frac{1}{2}b + b = \frac{1}{2}b$

4. a. $4x - 3 + 7 - 5x = -x + 4$

b. $-6y + 3y - 8 + 8y = 5y - 8$

c. $2 + 8.1a + a - 6 = 9.1a - 4$

d. $2x^2 - 2x = 2x^2 - 2x$

5. a. x-intercept: $(-3, 0)$
y-intercept: $(0, 2)$

b. x-intercepts: $(-4, 0)$, $(-1, 0)$
y-intercept: $(0, 1)$

c. x-intercept: $(0, 0)$
y-intercept: $(0, 0)$

d. x-intercept: $(2, 0)$
y-intercept: none

e. x-intercepts: $(-1, 0)$, $(3, 0)$
y-intercepts: $(0, 2)$, $(0, -1)$

6. a. x-intercept: $(4, 0)$
y-intercept: $(0, 1)$

b. x-intercepts: $(-2, 0)$, $(0, 0)$, $(3, 0)$
y-intercept: $(0, 0)$

c. x-intercept: none
y-intercept: $(0, -3)$

d. x-intercepts: $(-3, 0)$, $(3, 0)$
y-intercepts: $(0, -3)$, $(0, 3)$

7. $y = -\frac{1}{5}x + 1$: $m_1 = -\frac{1}{5}$

$2x + 10y = 30$

$y = -\frac{1}{5}x + 3$: $m_2 = -\frac{1}{5}$

$m_1 = m_2$

They are parallel.

8. $y = 3x + 7$: $m_1 = 3$

$x + 3y = -15$

$y = -\frac{1}{3}x - 5$: $m_2 = -\frac{1}{3}$

$m_1 \cdot m_2 = -1$

They are perpendicular.

9. $\begin{cases} 2x + y = 7 \\ 2y = -4x \end{cases}$

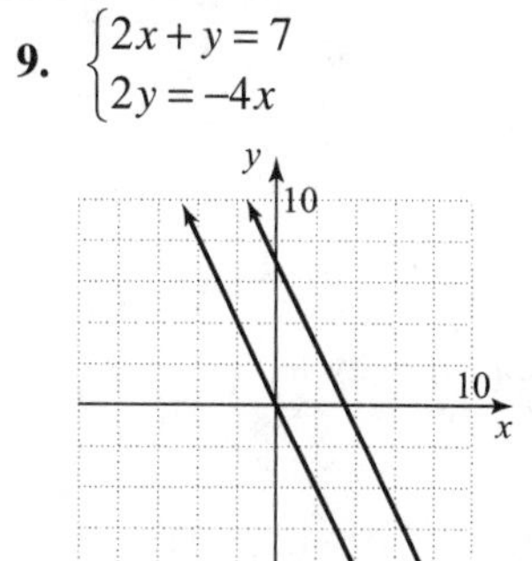

There is no solution, the lines are parallel.

10. $\begin{cases} y = x+2 \\ 2x+y=5 \end{cases}$

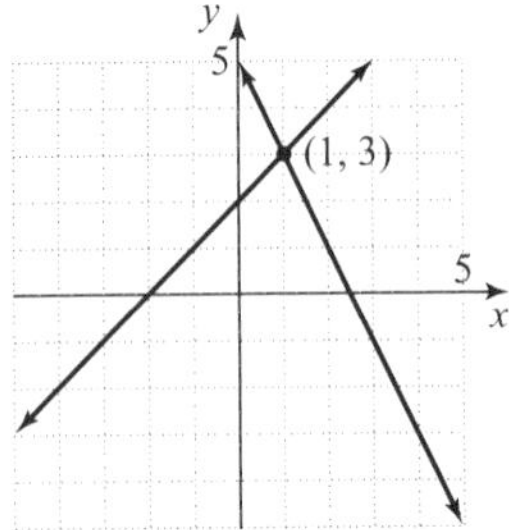

The solution is (1, 3).

11. $\begin{cases} 7x-3y=-14 \\ -3x+y=6 \end{cases}$

Multiply the second equation by 3.

$$\begin{aligned} 7x-3y&=-14 \\ -9x+3y&=18 \\ \hline -2x\quad&=4 \\ x&=-2 \end{aligned}$$

Let $x = -2$ in the second equation.

$$\begin{aligned} -3(-2)+y&=6 \\ 6+y&=6 \\ y&=0 \end{aligned}$$

The solution of the system is (−2, 0).

12. $\begin{cases} 5x+y=3 \\ y=-5x \end{cases}$

Substitute $-5x$ for y in the first equation.

$$\begin{aligned} 5x+(-5x)&=3 \\ 0&=3 \end{aligned}$$

There is no solution.

13. $\begin{cases} 3x-2y=2 \\ -9x+6y=-6 \end{cases}$

Multiply the first equation by 3.

$$\begin{aligned} 9x-6y&=6 \\ -9x+6y&=-6 \\ \hline 0&=0 \end{aligned}$$

The system has an infinite number of solutions.

14. $\begin{cases} -2x+y=7 \\ 6x-3y=-21 \end{cases}$

Multiply the first equation by 3.

$$\begin{aligned} -6x+3y&=21 \\ 6x-3y&=-21 \\ \hline 0&=0 \end{aligned}$$

The system has an infinite number of solutions.

15. Let x = the rate of Albert.

	Rate	· Time =	Distance
Albert	x	2	$2x$
Louis	$x + 1$	2	$2(x + 1)$
Total			15

$$\begin{aligned} 2x+2(x+1)&=15 \\ 2x+2x+2&=15 \\ 4x+2&=15 \\ 4x&=13 \\ x&=3.25 \end{aligned}$$

$x + 1 = 3.25 + 1 = 4.25$

Albert: 3.25 mph; Louis: 4.25 mph

16. Let x = the number of dimes, and $15 - x$ = the number of quarters.

	No. of Coins	· Value =	Amt. of Money
Dimes	x	0.1	$0.1x$
Quarters	$15 - x$	0.25	$0.25(15 - x)$
Total	15		2.85

$$\begin{aligned} 0.1x+0.25(15-x)&=2.85 \\ 0.1x+3.75-0.25x&=2.85 \\ -0.15x+3.75&=2.85 \\ -0.15x&=-0.9 \\ x&=6 \end{aligned}$$

$15 - x = 15 - 6 = 9$

There are 6 dimes and 9 quarters in the purse.

17. $\begin{cases} -3x+4y<12 \\ x \ge 2 \end{cases}$

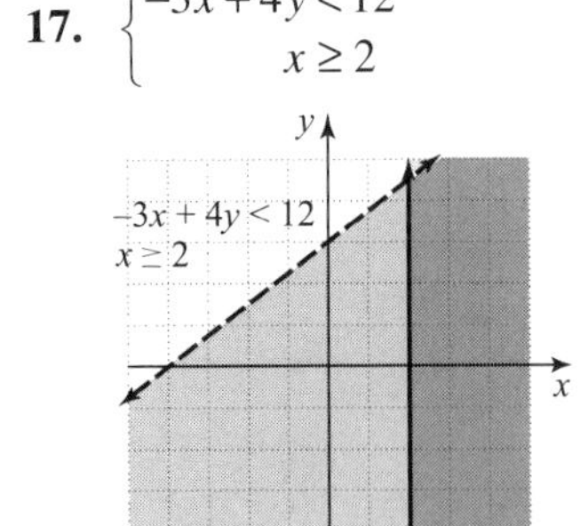

18. $\begin{cases} 2x - y \le 6 \\ y \ge 2 \end{cases}$

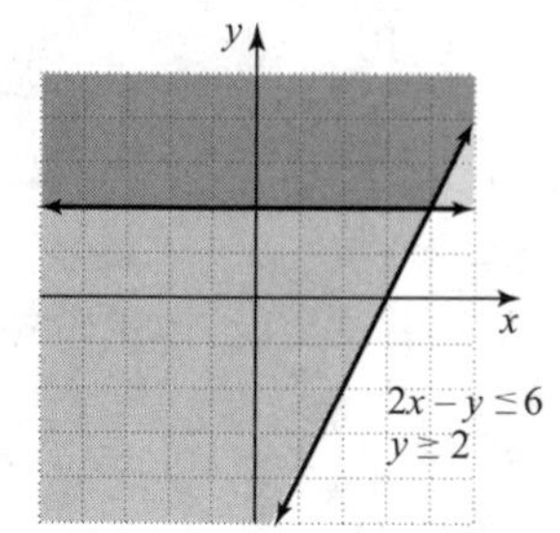

19. a. $\left(\frac{st}{2}\right)^4 = \frac{(st)^4}{2^4} = \frac{s^4t^4}{16}$

b. $(9y^5z^7)^2 = 9^2 y^{5\cdot 2} z^{7\cdot 2} = 81y^{10}z^{14}$

c. $\left(\frac{-5x^2}{y^3}\right)^2 = \frac{(-5)^2 x^4}{y^6} = \frac{25x^4}{y^6}$

20. a. $\left(\frac{-6x}{y^3}\right)^3 = \frac{(-6)^3 x^3}{y^9} = \frac{-216x^3}{y^9}$

b. $\frac{a^2b^7}{(2b^2)^5} = \frac{a^2b^7}{2^5b^{10}}$

$= \frac{a^2}{32}\cdot b^{7-10}$

$= \frac{a^2}{32}\cdot b^{-3}$

$= \frac{a^2}{32b^3}$

c. $\frac{(3y)^2}{y^2} = \frac{3^2y^2}{y^2} = 9y^{2-2} = 9$

d. $\frac{(x^2y^4)^2}{xy^3} = \frac{x^4y^8}{xy^3} = x^{4-1}y^{8-3} = x^3y^5$

21. $(5x-1)(2x^2+15x+18) = 0$

$(5x-1)(2x+3)(x+6) = 0$

$5x - 1 = 0$ or $2x + 3 = 0$ or $x + 6 = 0$

$x = \frac{1}{5}$ $\quad x = -\frac{3}{2}$ $\quad x = -6$

The solutions are $\frac{1}{5}$, $-\frac{3}{2}$, and -6.

22. $(x+1)(2x^2-3x-5) = 0$

$(x+1)(2x-5)(x+1) = 0$

$x + 1 = 0$ or $2x - 5 = 0$

$x = -1$ $\quad x = \frac{5}{2}$

The solutions are -1 and $\frac{5}{2}$.

23. $\frac{45}{x} = \frac{5}{7}$

$45\cdot 7 = 5x$

$315 = 5x$

$63 = x$

24. $\frac{2x+7}{3} = \frac{x-6}{2}$

$6\left(\frac{2x+7}{3}\right) = 6\left(\frac{x-6}{2}\right)$

$2(2x+7) = 3(x-6)$

$4x + 14 = 3x - 18$

$x + 14 = -18$

$x = -32$

25. a. $\sqrt[4]{16} = \sqrt[4]{2^4} = 2$

b. $\sqrt[5]{-32} = \sqrt[5]{(-2)^5} = -2$

c. $-\sqrt[3]{8} = -\sqrt[3]{(2)^3} = -2$

d. $\sqrt[4]{-81}$ is not a real number.

26. a. $\sqrt[3]{27} = \sqrt[3]{3^3} = 3$

b. $\sqrt[4]{256} = \sqrt[4]{4^4} = 4$

c. $\sqrt[3]{-125} = \sqrt[3]{(-5)^3} = -5$

d. $\sqrt[5]{1} = \sqrt[5]{1^5} = 1$

27. a. $\sqrt{\frac{25}{36}} = \frac{\sqrt{25}}{\sqrt{36}} = \frac{5}{6}$

b. $\sqrt{\frac{3}{64}} = \frac{\sqrt{3}}{\sqrt{64}} = \frac{\sqrt{3}}{8}$

c. $\sqrt{\frac{40}{81}} = \frac{\sqrt{40}}{\sqrt{81}} = \frac{\sqrt{4 \cdot 10}}{9} = \frac{2\sqrt{10}}{9}$

28. a. $\sqrt{\frac{4}{25}} = \frac{\sqrt{4}}{\sqrt{25}} = \frac{2}{5}$

b. $\sqrt{\frac{16}{121}} = \frac{\sqrt{16}}{\sqrt{121}} = \frac{4}{11}$

c. $\sqrt{\frac{2}{49}} = \frac{\sqrt{2}}{\sqrt{49}} = \frac{\sqrt{2}}{7}$

29. a.
$$\begin{aligned}\sqrt{50} + \sqrt{8} &= \sqrt{25 \cdot 2} + \sqrt{4 \cdot 2} \\ &= 5\sqrt{2} + 2\sqrt{2} \\ &= 7\sqrt{2}\end{aligned}$$

b.
$$\begin{aligned}7\sqrt{12} - \sqrt{75} &= 7\sqrt{4 \cdot 3} - \sqrt{25 \cdot 3} \\ &= 7 \cdot 2\sqrt{3} - 5\sqrt{3} \\ &= 14\sqrt{3} - 5\sqrt{3} \\ &= 9\sqrt{3}\end{aligned}$$

c.
$$\begin{aligned}&\sqrt{25} - \sqrt{27} - 2\sqrt{18} - \sqrt{16} \\ &= 5 - \sqrt{9 \cdot 3} - 2\sqrt{9 \cdot 2} - 4 \\ &= 1 - 3\sqrt{3} - 2 \cdot 3\sqrt{2} \\ &= 1 - 3\sqrt{3} - 6\sqrt{2}\end{aligned}$$

30. a.
$$\begin{aligned}\sqrt{80} + \sqrt{20} &= \sqrt{16 \cdot 5} + \sqrt{4 \cdot 5} \\ &= 4\sqrt{5} + 2\sqrt{5} \\ &= 6\sqrt{5}\end{aligned}$$

b.
$$\begin{aligned}2\sqrt{98} - 2\sqrt{18} &= 2\sqrt{49 \cdot 2} - 2\sqrt{9 \cdot 2} \\ &= 2 \cdot 7\sqrt{2} - 2 \cdot 3\sqrt{2} \\ &= 14\sqrt{2} - 6\sqrt{2} \\ &= 8\sqrt{2}\end{aligned}$$

c.
$$\begin{aligned}\sqrt{32} + \sqrt{121} - \sqrt{12} &= \sqrt{16 \cdot 2} + 11 - \sqrt{4 \cdot 3} \\ &= 11 + 4\sqrt{2} - 2\sqrt{3}\end{aligned}$$

31. a. $\sqrt{7} \cdot \sqrt{3} = \sqrt{7 \cdot 3} = \sqrt{21}$

b. $\sqrt{3} \cdot \sqrt{3} = 3$

c. $\sqrt{3} \cdot \sqrt{15} = \sqrt{3 \cdot 15} = \sqrt{9 \cdot 5} = 3\sqrt{5}$

d. $2\sqrt{3} \cdot 5\sqrt{2} = 10\sqrt{3 \cdot 2} = 10\sqrt{6}$

e. $\sqrt{2x^3} \cdot \sqrt{6x} = \sqrt{12x^4} = \sqrt{4x^4 \cdot 3} = 2x^2\sqrt{3}$

32. a. $\sqrt{2} \cdot \sqrt{5} = \sqrt{2 \cdot 5} = \sqrt{10}$

b.
$$\begin{aligned}\sqrt{56} \cdot \sqrt{7} &= \sqrt{56 \cdot 7} \\ &= \sqrt{392} \\ &= \sqrt{196 \cdot 2} \\ &= 14\sqrt{2}\end{aligned}$$

c. $\left(4\sqrt{3}\right)^2 = 4^2\left(\sqrt{3}\right)^2 = 16 \cdot 3 = 48$

d. $3\sqrt{8} \cdot 7\sqrt{2} = 21\sqrt{8 \cdot 2} = 21\sqrt{16} = 21 \cdot 4 = 84$

33.
$$\begin{aligned}\sqrt{x} &= \sqrt{5x-2} \\ \left(\sqrt{x}\right)^2 &= \left(\sqrt{5x-2}\right)^2 \\ x &= 5x - 2 \\ 0 &= 4x - 2 \\ 2 &= 4x \\ \frac{2}{4} &= x \\ \frac{1}{2} &= x\end{aligned}$$

34.
$$\begin{aligned}\sqrt{x-4} + 7 &= 2 \\ \sqrt{x-4} &= -5\end{aligned}$$

The square root of a real number cannot be negative. There is no solution.

35.
$$\begin{aligned}a^2 + b^2 &= c^2 \\ \left(\overline{PQ}\right)^2 + \left(\overline{QR}\right)^2 &= \left(\overline{PR}\right)^2 \\ \left(\overline{PQ}\right)^2 + (240)^2 &= (320)^2 \\ \left(\overline{PQ}\right)^2 + 57,600 &= 102,400 \\ \left(\overline{PQ}\right)^2 &= 44,800 \\ \overline{PQ} &= \sqrt{44,800} \approx 212\end{aligned}$$

The distance is approximately 212 feet.

36. (−7, 4) and (2, 5)
$$\begin{aligned}d &= \sqrt{(x_2 - x_1)^2 + (y_2 - y_1)^2} \\ d &= \sqrt{(2-(-7))^2 + (5-4)^2} \\ &= \sqrt{9^2 + 1^2} \\ &= \sqrt{81 + 1} \\ &= \sqrt{82}\end{aligned}$$

37. a. $25^{1/2} = \sqrt{25} = 5$

b. $8^{1/3} = \sqrt[3]{8} = 2$

c. $-16^{1/4} = -\sqrt[4]{16} = -2$

d. $(-27)^{1/3} = \sqrt[3]{-27} = -3$

e. $\left(\frac{1}{9}\right)^{1/2} = \sqrt{\frac{1}{9}} = \frac{1}{3}$

38. a. $-49^{1/2} = -\sqrt{49} = -7$

b. $256^{1/4} = \sqrt[4]{256} = \sqrt[4]{(4)^4} = 4$

c. $(-64)^{1/3} = \sqrt[3]{-64} = -4$

d. $\left(\frac{25}{36}\right)^{1/2} = \sqrt{\frac{25}{36}} = \frac{5}{6}$

e. $(32)^{1/5} = \sqrt[5]{32} = \sqrt[5]{(2)^5} = 2$

39.
$$2x^2 = 7$$
$$x^2 = \frac{7}{2}$$
$$x = \pm\sqrt{\frac{7}{2}}$$
$$x = \pm\frac{\sqrt{7}}{\sqrt{2}} \cdot \frac{\sqrt{2}}{\sqrt{2}} = \pm\frac{\sqrt{14}}{2}$$

40.
$$3(x-4)^2 = 9$$
$$(x-4)^2 = 3$$
$$x-4 = \pm\sqrt{3}$$
$$x = 4 \pm \sqrt{3}$$

41.
$$x^2 - 10x = -14$$
$$x^2 - 10x + 25 = -14 + 25$$
$$(x-5)^2 = 11$$
$$x-5 = \pm\sqrt{11}$$
$$x = 5 \pm \sqrt{11}$$

42.
$$x^2 + 4x = 8$$
$$x^2 + 4x + 4 = 8 + 4$$
$$(x+2)^2 = 12$$
$$x+2 = \pm\sqrt{12}$$
$$x = -2 \pm \sqrt{4 \cdot 3}$$
$$x = -2 \pm 2\sqrt{3}$$

43.
$$2x^2 - 9x = 5$$
$$2x^2 - 9x - 5 = 0$$
$a = 2$, $b = -9$, and $c = -5$
$$x = \frac{-(-9) \pm \sqrt{(-9)^2 - 4(2)(-5)}}{2(2)}$$
$$= \frac{9 \pm \sqrt{81+40}}{4}$$
$$= \frac{9 \pm \sqrt{121}}{4}$$
$$= \frac{9 \pm 11}{4}$$
$$x = \frac{9+11}{4} = 5 \quad \text{or} \quad x = \frac{9-11}{4} = -\frac{1}{2}$$
The solutions are 5 and $-\frac{1}{2}$.

44.
$$2x^2 + 5x = 7$$
$$2x^2 + 5x - 7 = 0$$
$a = 2$, $b = 5$, and $c = -7$
$$x = \frac{-5 \pm \sqrt{5^2 - 4(2)(-7)}}{2(2)}$$
$$= \frac{-5 \pm \sqrt{25+56}}{4}$$
$$= \frac{-5 \pm \sqrt{81}}{4}$$
$$= \frac{-5 \pm 9}{4}$$
$$x = \frac{-5+9}{4} = 1 \quad \text{or} \quad x = \frac{-5-9}{4} = -\frac{7}{2}$$
The solutions are 1 and $-\frac{7}{2}$.

45. a. $\sqrt{-4} = i\sqrt{4} = 2i$

b. $\sqrt{-11} = i\sqrt{11}$

c. $\sqrt{-20} = i\sqrt{20} = i\sqrt{4 \cdot 5} = 2i\sqrt{5}$

46. a. $\sqrt{-7} = i\sqrt{7}$

b. $\sqrt{-16} = i\sqrt{16} = 4i$

c. $\sqrt{-27} = i\sqrt{27} = i\sqrt{9 \cdot 3} = 3i\sqrt{3}$

47. $y = x^2 - 4$

y-intercept: $x = 0,\ y = 0^2 - 4 = -4,\ (0, -4)$

vertex: $(0, -4)$

x-intercepts: $y = 0$,

$0 = x^2 - 4$

$0 = (x+2)(x-2)$

$x + 2 = 0$ or $x - 2 = 0$

$x = -2$ $\quad$ $x = 2$

$(-2, 0)$ and $(2, 0)$

x	y
−2	0
−1	−3
0	−4
1	−3
2	0

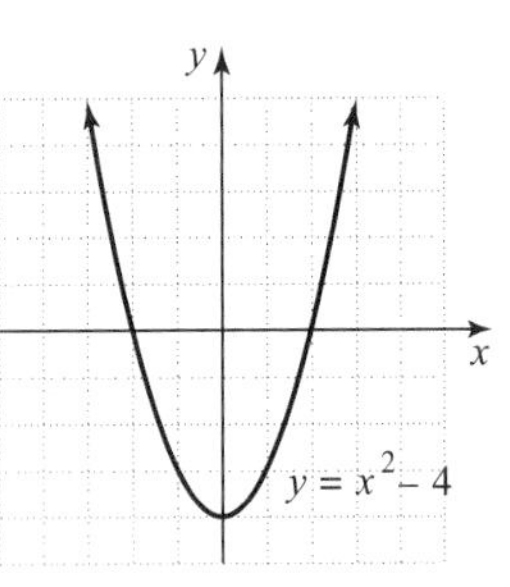

48. $y = x^2 + 2x + 3$

$a = 1, b = 2, c = 3$

vertex: $x = \frac{-b}{2a} = \frac{-2}{2(1)} = -1$

$y = (-1)^2 + 2(-1) + 3 = 2$

$(-1, 2)$

y-intercept: $x = 0,\ y = 0^2 + 2(0) + 3 = 3,\ (0, 3)$

x-intercepts: $y = 0,\ 0 = x^2 + 2x + 3$

$$x = \frac{-2 \pm \sqrt{2^2 - 4(1)(3)}}{2(1)} = \frac{-2 \pm \sqrt{4 - 12}}{2} = \frac{-2 \pm \sqrt{-8}}{2}$$

There are no real solutions to the equation, so there are no x-intercepts.

x	y
−3	6
−2	3
−1	2
0	3
1	6

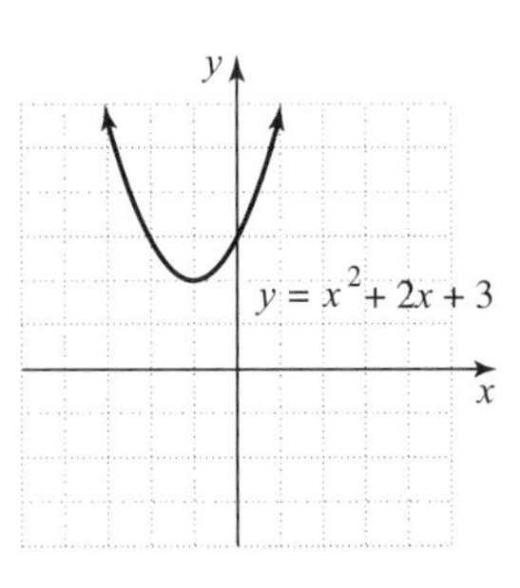

Appendix A

A.2 Practice Final Exam

1. $-3^4 = -(3^4) = -81$

2. $4^{-3} = \frac{1}{4^3} = \frac{1}{64}$

3.
$$\begin{aligned} 6[5+2(3-8)-3] &= 6[5+2(-5)-3] \\ &= 6[5+(-10)-3] \\ &= 6[-5-3] \\ &= 6[-8] \\ &= -48 \end{aligned}$$

4.
$$\begin{aligned} &(5x^3+x^2+5x-2)-(8x^3-4x^2+x-7) \\ &= 5x^3+x^2+5x-2-8x^3+4x^2-x+7 \\ &= 5x^3-8x^3+x^2+4x^2+5x-x-2+7 \\ &= -3x^3+5x^2+4x+5 \end{aligned}$$

5.
$$\begin{aligned} (4x-2)^2 &= (4x)^2-2(4x)(2)+2^2 \\ &= 16x^2-16x+4 \end{aligned}$$

6.
$$\begin{aligned} &(3x+7)(x^2+5x+2) \\ &= 3x(x^2+5x+2)+7(x^2+5x+2) \\ &= 3x^3+15x^2+6x+7x^2+35x+14 \\ &= 3x^3+22x^2+41x+14 \end{aligned}$$

7. $y^2-8y-48 = (y-12)(y+4)$

8.
$$\begin{aligned} 9x^3+39x^2+12x &= 3x(3x^2+13x+4) \\ &= 3x(3x+1)(x+4) \end{aligned}$$

9.
$$\begin{aligned} 180-5x^2 &= 5(36-x^2) \\ &= 5(6^2-x^2) \\ &= 5(6+x)(6-x) \end{aligned}$$

10.
$$\begin{aligned} 3a^2+3ab-7a-7b &= 3a(a+b)-7(a+b) \\ &= (a+b)(3a-7) \end{aligned}$$

11.
$$\begin{aligned} 8y^3-64 &= 8(y^3-8) \\ &= 8(y^3-2^3) \\ &= 8(y-2)(y^2+y\cdot 2+2^2) \\ &= 8(y-2)(y^2+2y+4) \end{aligned}$$

12. $\left(\frac{x^2y^3}{x^3y^{-4}}\right)^2 = \left(\frac{y^{3-(-4)}}{x^{3-2}}\right)^2 = \left(\frac{y^7}{x^1}\right)^2 = \frac{y^{7\cdot 2}}{x^{1\cdot 2}} = \frac{y^{14}}{x^2}$

13. $\dfrac{5-\frac{1}{y^2}}{\frac{1}{y}+\frac{2}{y^2}} = \dfrac{y^2\left(5-\frac{1}{y^2}\right)}{y^2\left(\frac{1}{y}+\frac{2}{y^2}\right)} = \dfrac{5y^2-1}{y+2}$

14.
$$\begin{aligned} &\frac{x^2-9}{x^2-3x} \div \frac{xy+5x+3y+15}{2x+10} \\ &= \frac{x^2-9}{x^2-3x} \cdot \frac{2x+10}{xy+5x+3y+15} \\ &= \frac{(x+3)(x-3)\cdot 2(x+5)}{x(x-3)\cdot(y+5)(x+3)} \\ &= \frac{2(x+5)}{x(y+5)} \end{aligned}$$

15. $a^2-a-6 = (a-3)(a+2)$

LCD = $(a-3)(a+2)$

$$\begin{aligned} &\frac{5a}{a^2-a-6} - \frac{2}{a-3} \\ &= \frac{5a}{(a-3)(a+2)} - \frac{2\cdot(a+2)}{(a-3)\cdot(a+2)} \\ &= \frac{5a-2(a+2)}{(a-3)(a+2)} \\ &= \frac{5a-2a-4}{(a-3)(a+2)} \\ &= \frac{3a-4}{(a-3)(a+2)} \end{aligned}$$

16.
$$\begin{aligned} 4(n-5) &= -(4-2n) \\ 4n-20 &= -4+2n \\ 2n-20 &= -4 \\ 2n &= 16 \\ n &= 8 \end{aligned}$$

17.
$$\begin{aligned} x(x+6) &= 7 \\ x^2+6x &= 7 \\ x^2+6x-7 &= 0 \\ (x+7)(x-1) &= 0 \end{aligned}$$
$x+7=0$ or $x-1=0$

$x=-7$ $\quad$ $x=1$

18. $3x-5 \ge 7x+3$
$-5 \ge 4x+3$
$-8 \ge 4x$
$-2 \ge x$
$x \le -2$
$(-\infty, -2]$

19. $2x^2-6x+1=0$
$a = 2, b = -6, c = 1$
$$x = \frac{-b \pm \sqrt{b^2-4ac}}{2a}$$
$$x = \frac{-(-6) \pm \sqrt{(-6)^2-4(2)(1)}}{2(2)}$$
$$= \frac{6 \pm \sqrt{36-8}}{4}$$
$$= \frac{6 \pm \sqrt{28}}{4}$$
$$= \frac{6 \pm 2\sqrt{7}}{4}$$
$$= \frac{3 \pm \sqrt{7}}{2}$$

20. $$\frac{4}{y} - \frac{5}{3} = -\frac{1}{5}$$
$$15y\left(\frac{4}{y}\right) - 15y\left(\frac{5}{3}\right) = 15y\left(-\frac{1}{5}\right)$$
$$60-25y = -3y$$
$$60 = 22y$$
$$\frac{60}{22} = y$$
$$\frac{30}{11} = y$$

21. $$\frac{5}{y+1} = \frac{4}{y+2}$$
$5(y+2) = 4(y+1)$
$5y+10 = 4y+4$
$y+10 = 4$
$y = -6$

22. $$\frac{a}{a-3} = \frac{3}{a-3} - \frac{3}{2}$$
$$2(a-3)\left(\frac{a}{a-3}\right) = 2(a-3)\left(\frac{3}{a-3}\right) - 2(a-3)\left(\frac{3}{2}\right)$$
$$2a = 6-3(a-3)$$
$$2a = 6-3a+9$$
$$2a = 15-3a$$
$$5a = 15$$
$$a = 3$$
In the original equation, $a = 3$ makes a denominator 0, so it is extraneous. The equation has no solution.

23. $\sqrt{2x-2} = x-5$
$\left(\sqrt{2x-2}\right)^2 = (x-5)^2$
$2x-2 = x^2-10x+25$
$0 = x^2-12x+27$
$0 = (x-9)(x-3)$
$x-9=0$ or $x-3=0$
$x=9$ $\quad x=3$
$x = 3$ is extraneous. The only solution is $x = 9$.

24. $5x - 7y = 10$

x	y
2	0
0	$-\frac{10}{7}$

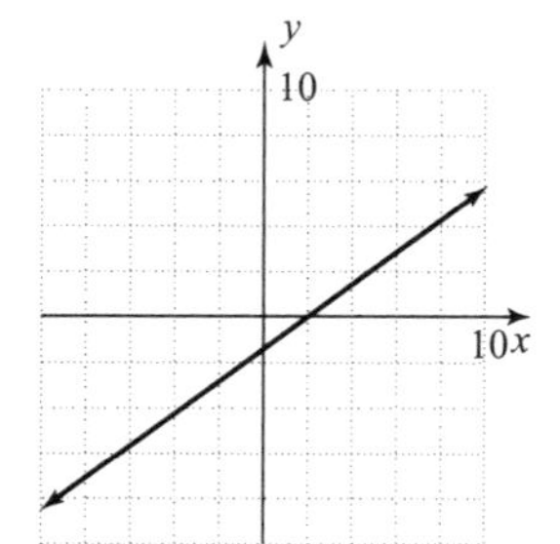

25. $x-3=0$
$x=3$

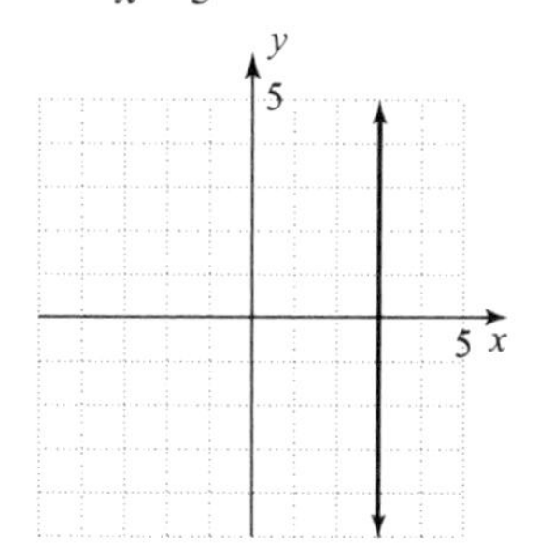

26. $y > -4x$

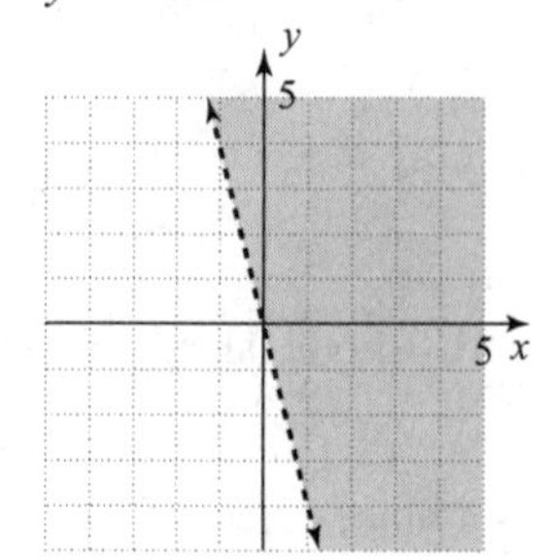

27. $m = \dfrac{y_2 - y_1}{x_2 - x_1}$

$m = \dfrac{2-(-5)}{-1-6} = \dfrac{2+5}{-7} = \dfrac{7}{-7} = -1$

28. $-3x + y = 5$

$y = 3x + 5$

$m = 3$

29. $(x_1, y_1) = (2, -5), (x_2, y_2) = (1, 3)$

$m = \dfrac{y_2 - y_1}{x_2 - x_1} = \dfrac{3-(-5)}{1-2} = \dfrac{3+5}{-1} = \dfrac{8}{-1} = -8$

$y - y_1 = m(x - x_1)$
$y - (-5) = -8(x-2)$
$y + 5 = -8x + 16$
$8x + y = 11$

30. A line parallel to $x = 7$ is a vertical line. The vertical line through $(-5, -1)$ has equation $x = -5$.

31. $\begin{cases} 3x - 2y = -14 \\ y = x + 5 \end{cases}$

Substitute $x + 5$ for y in the first equation.

$3x - 2y = -14$
$3x - 2(x+5) = -14$
$3x - 2x - 10 = -14$
$x = -4$

Substitute -4 for x in the second equation.

$y = x + 5$
$y = -4 + 5$
$y = 1$

The solution is $(-4, 1)$.

32. $\begin{cases} 4x - 6y = 7 \\ -2x + 3y = 0 \end{cases}$

Multiply the second equation by 2 and add the result to the first equation.

$$\begin{array}{r} 4x - 6y = 7 \\ -4x + 6y = 0 \\ \hline 0 = 7 \end{array}$$

The statement $0 = 7$ is false, so the system has no solution.

33. $h(x) = x^3 - x$

a. $h(-1) = (-1)^3 - (-1) = -1 + 1 = 0$

b. $h(0) = 0^3 - 0 = 0 - 0 = 0$

c. $h(4) = 4^3 - 4 = 64 - 4 = 60$

34. Domain: $(-\infty, \infty)$
Range: $(-\infty, 4]$

35. $\sqrt{16} = 4$ because $4^2 = 16$ and 4 is positive.

36. $27^{-\frac{2}{3}} = \dfrac{1}{27^{2/3}} = \dfrac{1}{\left(\sqrt[3]{27}\right)^2} = \dfrac{1}{3^2} = \dfrac{1}{9}$

37. $\left(\dfrac{9}{16}\right)^{\frac{1}{2}} = \dfrac{9^{1/2}}{16^{1/2}} = \dfrac{\sqrt{9}}{\sqrt{16}} = \dfrac{3}{4}$

38. $\sqrt{54} = \sqrt{9 \cdot 6} = \sqrt{9} \cdot \sqrt{6} = 3\sqrt{6}$

39. $\sqrt{9x^9} = \sqrt{9x^8 \cdot x} = \sqrt{9x^8} \cdot \sqrt{x} = 3x^4\sqrt{x}$

40. $\sqrt{12} - 2\sqrt{75} = \sqrt{4 \cdot 3} - 2\sqrt{25 \cdot 3}$
$= 2\sqrt{3} - 2 \cdot 5\sqrt{3}$
$= 2\sqrt{3} - 10\sqrt{3}$
$= -8\sqrt{3}$

41. $\dfrac{\sqrt{40x^4}}{\sqrt{2x}} = \sqrt{\dfrac{40x^4}{2x}}$
$= \sqrt{20x^3}$
$= \sqrt{4x^2 \cdot 5x}$
$= 2x\sqrt{5x}$

42. $\sqrt{2}\left(\sqrt{6}-\sqrt{5}\right) = \sqrt{2}\cdot\sqrt{6} - \sqrt{2}\cdot\sqrt{5}$
$$= \sqrt{12} - \sqrt{10}$$
$$= \sqrt{4\cdot 3} - \sqrt{10}$$
$$= 2\sqrt{3} - \sqrt{10}$$

43. $\sqrt{\dfrac{5}{12x^2}} = \dfrac{\sqrt{5}}{\sqrt{12x^2}}$
$$= \frac{\sqrt{5}}{\sqrt{4x^2\cdot 3}}$$
$$= \frac{\sqrt{5}}{2x\sqrt{3}}$$
$$= \frac{\sqrt{5}\cdot\sqrt{3}}{2x\sqrt{3}\cdot\sqrt{3}}$$
$$= \frac{\sqrt{15}}{6x}$$

44. $\dfrac{2\sqrt{3}}{\sqrt{3}-3} = \dfrac{2\sqrt{3}\left(\sqrt{3}+3\right)}{\left(\sqrt{3}-3\right)\left(\sqrt{3}+3\right)}$
$$= \frac{2\cdot 3+6\sqrt{3}}{3-9}$$
$$= \frac{6+6\sqrt{3}}{-6}$$
$$= \frac{6\left(1+\sqrt{3}\right)}{-6}$$
$$= -\left(1+\sqrt{3}\right)$$
$$= -1-\sqrt{3}$$

45. Let x be the number.
$$x+5\cdot\frac{1}{x}=6$$
$$x\left(x+\frac{5}{x}\right)=x\cdot 6$$
$$x^2+5=6x$$
$$x^2-6x+5=0$$
$$(x-1)(x-5)=0$$
$$x-1=0 \quad\text{or}\quad x-5=0$$
$$x=1 \qquad\qquad x=5$$
The number is 1 or 5.

46. Let x be the smaller area code. Then the other area code is $2x$.
$$x+2x=1203$$
$$3x=1203$$
$$x=401$$
$2x = 2(401) = 802$
The area codes are 401 (Rhode Island) and 802 (Vermont).

47. Let x be the number of hours since the trains left. One train will have traveled $50x$ miles and the other will have traveled $64x$ miles.
$$50x+64x=285$$
$$114x=285$$
$$x=\frac{285}{114}=\frac{5}{2} \text{ or } 2\frac{1}{2}$$
The trains are 285 miles apart after $2\frac{1}{2}$ hours.

48. Let x be the amount of 12% solution.

Amount	Percent	Total Saline
x	12% = 0.12	$0.12x$
80	22% = 0.22	$0.22\cdot 80 = 17.6$
$80+x$	16% = 0.16	$0.16(80+x)$ $= 12.8+0.16x$

$$0.12x+17.6=12.8+0.16x$$
$$4.8=0.04x$$
$$120=x$$
120 cc of 12% saline solution should be added.

Appendix B

Exercise Set B.3

1. $90° - 19° = 71°$

3. $90° - 70.8° = 19.2°$

5. $90° - 11\frac{1}{4}° = 78\frac{3}{4}°$

7. $180° - 150° = 30°$

9. $180° - 30.2° = 149.8°$

11. $180° - 79\frac{1}{2}° = 100\frac{1}{2}°$

13. $m\angle 1 = 110°$
$m\angle 2 = 180° - 110° = 70°$
$m\angle 3 = m\angle 2 = 70°$
$m\angle 4 = m\angle 2 = 70°$
$m\angle 5 = m\angle 1 = 110°$
$m\angle 6 = m\angle 4 = 70°$
$m\angle 7 = m\angle 5 = 110°$

15. $180° - 11° - 79° = 90°$

17. $180° - 25° - 65° = 90°$

19. $180° - 30° - 60° = 90°$

21. $90° - 45° = 45°$
$45°, 90°$

23. $90° - 17° = 73°$
$73°, 90°$

25. $90° - 39\frac{3}{4}° = 50\frac{1}{4}°$

$50\frac{1}{4}°, 90°$

27.
$$\frac{12}{4} = \frac{18}{x}$$
$$4x\left(\frac{12}{4}\right) = 4x\left(\frac{18}{x}\right)$$
$$12x = 72$$
$$x = 6$$

29.
$$\frac{6}{9} = \frac{3}{x}$$
$$9x\left(\frac{6}{9}\right) = 9x\left(\frac{3}{x}\right)$$
$$6x = 27$$
$$x = 4.5$$

31.
$$a^2 + b^2 = c^2$$
$$6^2 + 8^2 = c^2$$
$$36 + 64 = c^2$$
$$100 = c^2$$
$$10 = c$$

33.
$$a^2 + b^2 = c^2$$
$$5^2 + b^2 = 13^2$$
$$25 + b^2 = 169$$
$$b^2 = 144$$
$$b = 12$$

Exercise Set B.4

1. Volume $= lwh = 6(4)(3) = 72$ cu in.
Surface area $= 2lh + 2wh + 2lw$
$= 2(6)(3) + 2(4)(3) + 2(6)(4)$
$= 36 + 24 + 48$
$= 108$ sq in.

3. Volume $= s^3 = 8^3 = 512$ cu cm
Surface area $= 6s^2 = 6(8)^2 = 384$ sq cm

5. Volume $= \frac{1}{3}\pi r^2 h = \frac{1}{3}\pi(2)^2(3) = 4\pi$ cu yd
$\approx 4\left(\frac{22}{7}\right) = 12\frac{4}{7}$ cu yd

Surface area $= \pi r\sqrt{r^2 + h^2} + \pi r^2$
$= \pi(2)\sqrt{2^2 + 3^2} + \pi 2^2$
$= \left(2\sqrt{13}\pi + 4\pi\right)$ sq yd
$\approx 2\sqrt{13}(3.14) + 4(3.14)$
$= 35.20$ sq yd

7. Volume $= \frac{4}{3}\pi r^3 = \frac{4}{3}\pi(5)^3 = \frac{500}{3}\pi$ cu in.

$\approx \frac{500}{3}\left(\frac{22}{7}\right) = 523\frac{17}{21}$ cu in.

Surface area $= 4\pi r^2 = 4\pi(5)^2 = 100\pi$ sq in.

$\approx 100\left(\frac{22}{7}\right) = 314\frac{2}{7}$ sq in.

9. Volume $= \frac{1}{3}s^2h = \frac{1}{3}(6)^2(4) = 48$ cu cm

Surface area $= B + \frac{1}{2}pl$

$= 36 + \frac{1}{2}(24)(5)$

$= 96$ sq cm

11. Volume $= s^3 = \left(1\frac{1}{3}\right)^3 = 2\frac{10}{27}$ cu in.

13. Surface area $= 2lh + 2wh + 2lw$

$= 2(2)(1.4) + 2(3)(1.4) + 2(2)(3)$

$= 5.6 + 8.4 + 12$

$= 26$ sq ft

15. Volume $= \frac{1}{3}s^2h = \frac{1}{3}(5)^2(1.3) = 10\frac{5}{6}$ cu in.

17. Volume $= \frac{1}{3}s^2h = \frac{1}{3}(12)^2(20) = 960$ cu cm

19. Surface area $= 4\pi r^2 = 4\pi(7)^2 = 196\pi$ sq in.

21. Volume $= lwh = 2\left(2\frac{1}{2}\right)\left(1\frac{1}{2}\right) = 7\frac{1}{2}$ cu ft

23. Volume $= \frac{1}{3}\pi r^2 h$

$\approx \frac{1}{3}\left(\frac{22}{7}\right)(2)^2(3) = 12\frac{4}{7}$ cu cm

Appendix C

1. $$\frac{30}{10}=\frac{15}{y}$$
$$30y=150$$
$$y=5$$

3. $$\frac{8}{15}=\frac{z}{6}$$
$$48=15z$$
$$\frac{48}{15}=z$$
$$z=\frac{16}{5} \text{ or } 3.2$$

5. $$\frac{-3.5}{12.5}=\frac{-7}{n}$$
$$-3.5n=-87.5$$
$$n=25$$

7. $$\frac{n}{0.6}=\frac{0.05}{12}$$
$$12n=0.030$$
$$n=\frac{0.030}{12}$$
$$n=0.0025$$

9. $$\frac{8}{\frac{2}{3}}=\frac{24}{n}$$
$$8n=16$$
$$n=2$$

11. $$\frac{7}{9}=\frac{35}{3x}$$
$$21x=315$$
$$x=15$$

13. $$\frac{7x}{18}=\frac{5}{3}$$
$$21x=90$$
$$x=\frac{90}{21}$$
$$x=\frac{30}{7}$$

15. $$\frac{11}{7}=\frac{4}{x+1}$$
$$11(x+1)=28$$
$$11x+11=28$$
$$11x=17$$
$$x=\frac{17}{11}$$

17. $$\frac{x-3}{2x+1}=\frac{4}{9}$$
$$9(x-3)=4(2x+1)$$
$$9x-27=8x+4$$
$$x-27=4$$
$$x=31$$

19. $$\frac{2x+1}{4}=\frac{6x-1}{5}$$
$$5(2x+1)=4(6x-1)$$
$$10x+5=24x-4$$
$$5=14x-4$$
$$9=14x$$
$$\frac{9}{14}=x$$

21. Let x be the number of applications in the 14-oz bottle.
$$\frac{3}{4}=\frac{14}{x}$$
$$3x=56$$
$$x=\frac{56}{3}\approx 18.7$$
There should be 18 full applications in the 14-oz bottle.

23. Let x be the number of weeks that 8 reams last.
$$\frac{5}{3}=\frac{8}{x}$$
$$5x=24$$
$$x=\frac{24}{5}=4.8$$
A case of paper will last about 5 weeks.

25. Let x be the number of servings for 4 cups of milk.

$$\frac{1\frac{1}{2}}{4} = \frac{4}{x}$$
$$\frac{3}{2}x = 16$$
$$x = 16 \cdot \frac{2}{3} = \frac{32}{3} = 10\frac{2}{3}$$

Ming can make $10\frac{2}{3}$ servings of pancakes.

27. Let x be the number of calories in a 24-oz size.

$$\frac{16}{80} = \frac{24}{x}$$
$$16x = 1920$$
$$x = 120$$

There are 120 calories in the 24-oz size.

29. a. Let x be the number of teaspoons needed to treat 450 square feet.

$$\frac{25}{1} = \frac{450}{x}$$
$$25x = 450$$
$$x = 18$$

18 teaspoons of granules are needed to treat 450 square feet.

b. 18 tsp = 6(3 tsp) = 6(1 tbsp) = 6 tbsp
6 tablespoons of granules must be used.

31. Let x be the number of people that 3750 square feet of lawn provides oxygen for.

$$\frac{625}{1} = \frac{3750}{x}$$
$$625x = 3750$$
$$x = 6$$

3750 square feet of lawn provides oxygen for 6 people.

33. Let x be the approximate height of the Statue of Liberty.

$$\frac{2}{5\frac{1}{3}} = \frac{42}{x}$$
$$2x = 224$$
$$x = 112$$

The approximate height of the Statue of Liberty is 112 feet. This is a difference of 11 inches from the actual height of 111 feet, 1 inch.

35. Let x be the amount of cholesterol in 5 ounces of lobster.

$$\frac{3.5}{72} = \frac{5}{x}$$
$$3.5x = 360$$
$$x = \frac{360}{3.5} \approx 102.9$$

There are about 102.9 milligrams of cholesterol in 5 ounces of lobster.

37. Let x be the number of visits in which medication is prescribed.

$$\frac{10}{7} = \frac{620}{x}$$
$$10x = 4340$$
$$x = 434$$

Out of 620 emergency room visits for injury, medication would be prescribed in 434 of the visits.

39. Let x be the number of people expected to have worked in the restaurant industry.

$$\frac{3}{1} = \frac{84}{x}$$
$$3x = 84$$
$$x = 28$$

In an office of 84 people, 28 people are likely to have worked in the restaurant industry.

Appendix D

1.
$$\begin{array}{r} 9.076 \\ +\,8.004 \\ \hline 17.080 \end{array}$$

3.
$$\begin{array}{r} 27.004 \\ -\,14.200 \\ \hline 12.804 \end{array}$$

5.
$$\begin{array}{r} 107.92 \\ +\quad 3.04 \\ \hline 110.96 \end{array}$$

7.
$$\begin{array}{r} 10.0 \\ -\,7.6 \\ \hline 2.4 \end{array}$$

9.
$$\begin{array}{r} 126.32 \\ -\,97.89 \\ \hline 28.43 \end{array}$$

11.
$$\begin{array}{r} 3.25 \\ \times\quad 70 \\ \hline 227.50 \end{array}$$

13.
$$\begin{array}{r} 2.7 \\ 3\overline{)8.1} \\ \underline{6} \\ 2\,1 \\ \underline{2\,1} \\ 0 \end{array}$$

15.
$$\begin{array}{r} 55.4050 \\ -\quad 6.1711 \\ \hline 49.2339 \end{array}$$

17. $0.75\overline{)60}$ becomes
$$\begin{array}{r} 80 \\ 75\overline{)6000} \\ \underline{600} \\ 00 \\ \underline{00} \end{array}$$

19. $7.612 \div 100 = 0.07612$

21. $2.7\overline{)12.312}$ becomes
$$\begin{array}{r} 4.56 \\ 27\overline{)123.12} \\ \underline{108} \\ 15\,1 \\ \underline{13\,5} \\ 1\,62 \\ \underline{1\,62} \\ 0 \end{array}$$

23.
$$\begin{array}{r} 569.20 \\ 71.25 \\ +\quad 8.01 \\ \hline 648.46 \end{array}$$

25.
$$\begin{array}{r} 768.00 \\ -\quad 0.17 \\ \hline 767.83 \end{array}$$

27.
$$\begin{array}{r} 12.000 \\ +\,0.062 \\ \hline 12.062 \end{array}$$

29.
$$\begin{array}{r} 76.00 \\ -\,14.52 \\ \hline 61.48 \end{array}$$

31. $0.43\overline{)3.311}$ becomes
$$\begin{array}{r} 7.7 \\ 43\overline{)331.1} \\ \underline{301} \\ 30\,1 \\ \underline{30\,1} \\ 0 \end{array}$$

33.
$$\begin{array}{r} 762.12 \\ 89.70 \\ +\,11.55 \\ \hline 863.37 \end{array}$$

35.
$$\begin{array}{r} 23.400 \\ -\,0.821 \\ \hline 22.579 \end{array}$$

37.
$$\begin{array}{r} 476.12 \\ -\,112.97 \\ \hline 363.15 \end{array}$$

39.
$$\begin{array}{r} 0.007 \\ +\,7.000 \\ \hline 7.007 \end{array}$$

Appendix E

1. 21, 28, 16, 42, 38

$\bar{x} = \frac{21+28+16+42+38}{5} = \frac{145}{5} = 29$

16, 21, 28, 38, 42

median = 28

no mode

3. 7.6, 8.2, 8.2, 9.6, 5.7, 9.1

$\bar{x} = \frac{7.6+8.2+8.2+9.6+5.7+9.1}{6} = \frac{48.4}{6} = 8.1$

5.7, 7.6, 8.2, 8.2, 9.1, 9.6

median $= \frac{8.2+8.2}{2} = 8.2$

mode = 8.2

5. 0.2, 0.3, 0.5, 0.6, 0.6, 0.9, 0.2, 0.7, 1.1

$\bar{x} = \frac{0.2+0.3+0.5+0.6+0.6+0.9+0.2+0.7+1.1}{9}$

$= \frac{5.1}{9}$

$= 0.6$

median = 0.6

mode = 0.2 and 0.6

7. 231, 543, 601, 293, 588, 109, 334, 268

$\bar{x} = \frac{231+543+601+293+588+109+334+268}{8}$

$= \frac{2967}{8}$

$= 370.9$

109, 231, 268, 293, 334, 543, 588, 601

median $= \frac{293+334}{2} = 313.5$

no mode

9. 1454, 1250, 1136, 1127, 1107

$\bar{x} = \frac{1454+1250+1136+1127+1107}{5}$

$= \frac{6074}{5}$

$= 1214.8$ feet

11. 1454, 1250, 1136, 1127, 1107, 1046, 1023, 1002

median $= \frac{1127+1107}{2} = 1117$ feet

13. $\bar{x} = \frac{7.8+6.9+7.5+4.7+6.9+7.0}{6}$

$= \frac{40.8}{6}$

$= 6.8$ seconds

15. 4.7, 6.9, 6.9, 7.0, 7.5, 7.8

mode = 6.9

17. 74, 77, 85, 86, 91, 95

median $= \frac{85+86}{2} = 85.5$

19. Sum = 78 + 80 + 66 + 68 + 71 + 64 + 82 + 71 + 70 + 65 + 70 + 75 + 77 + 86 + 72

= 1095

$\bar{x} = \frac{1095}{15} = 73$

21. 64, 65, 66, 68, 70, 70, 71, 71, 72, 75, 77, 78, 80, 82, 86

mode = 70 and 71

23. 64, 65, 66, 68, 70, 70, 71, 71, 72, ↑ 75, 77, 78, 80, 82, 86

mean = 73

9 rates were lower than the mean.

25. __, __, 16, 18, __;

Since the mode is 21, at least two of the missing numbers must be 21. The mean is 20. Let the one unknown number be x.

$\bar{x} = \frac{21+21+16+18+x}{5} = 20$

$\frac{76+x}{5} = 20$

$76 + x = 100$

$x = 24$

The missing numbers are 21, 21, 24.